GUIDE TO PARTS

Part 1 — Materials
1-1 to 1-8

Part 2 — Product Information and Capability
2-1 to 2-56

Part 3 — Analysis and Design of Precast,
Prestressed Concrete Structures
3-1 to 3-75

Part 4 — Design of Precast and Prestressed
Concrete Components
4-1 to 4-75

Part 5 — Product Handling and Erection Bracing
5-1 to 5-36

Part 6 — Design of Connections
6-1 to 6-69

Part 7 — Special Topics for Architectural Precast Concrete
7-1 to 7-30

Part 8 — Tolerances for Precast and Prestressed Concrete
8-1 to 8-19

Part 9 — Thermal, Acoustical, Fire and
Other Considerations
9-1 to 9-76

Part 10 — Specifications and Standard Practices
10-1 to 10-36

Part 11 — General Design Information
11-1 to 11-30

Index

pci design handbook

Precast and Prestressed Concrete

THIRD EDITION

prestressed concrete institute

201 North Wells Street, Chicago, Illinois 60606

Library of Congress Catalog Card Number 85-60484

ISBN 0-937040-23-1

Printed in U.S.A

FOREWORD

The Prestressed Concrete Institute, a non-profit corporation, was founded in 1954 for the purpose of advancing the design, manufacture and use of prestressed and precast concrete and architectural precast concrete in the United States and Canada.

To meet this purpose, PCI continually disseminates information on the latest concepts, techniques and design data to the architectural and engineering professions through regional and national programs and technical publications.

The third edition of the *PCI Design Handbook* combines most of the material from the second edition with material from the *PCI Manual for Structural Design of Architectural Precast Concrete*, first edition. In addition, it reflects the changes and advancements in design of prestressed and architectural precast concrete including the latest provisions of the ACI Building Code (ACI 318-83). The two structural manuals were combined to provide the user with a single source of information on the design of both architectural precast concrete and structural precast and prestressed concrete.

The primary objective of this Handbook is to enable the designer to improve and shorten design procedures for precast concrete products and structures. It is intended to provide the professional designer with sufficient information to permit the safe design of precast and prestressed concrete, in accordance with commonly accepted industry practice.

Although precast concrete has been an established construction material for many years, it continues to advance at a rapid pace. As a result, the Handbook includes procedures and practices that may not be common to all areas. Some of the recommendations are under further review and study by PCI committees or are the subject of on-going research. The designer must recognize that no handbook or code can substitute for experienced engineering judgment.

Substantial effort has been made to ensure that all data and information in this Handbook are accurate. However, PCI cannot accept responsibility for any errors or oversights in the use of material or in the preparation of engineering plans. This publication is intended for use by professional personnel competent to evalute the significance and limitations of its contents and able to accept responsibility for the application of the material it contains.

Users of this Handbook are encouraged to offer comments to PCI on the contents of this publication and suggestions for changes in the next edition. Questions concerning the source and derivation of any material in the Handbook should be directed to PCI.

This edition of the Handbook was produced under the direction of the PCI Industry Handbook Committee, with Daniel P. Jenny, chairman, and Leslie D. Martin, editor-in-chief. Other members of the committee were Roger J. Becker, Robert D. Finfrock, Sepp Firnkas, Sidney Freedman, Jack D. Gillum, Edward C. Gloppen, Herman C. Himes, James K. Iverson, Edward S. Knowles, David J. Matlock, Ray A. McCann, John P. McGrew, Ben G. Olson, Walter J. Prebis, Rangaswamy Ramadev, Kurt L. Salm, Edward F. Schaack, Irwin J. Speyer, Kenneth Vick, and Helmuth Wilden. Consulting firms that provided valuable input to the committee were The Consulting Engineers Group, Inc.; H. Wilden and Associates, Inc.; GCE of Colorado, Inc.; Portland Cement Association; Shiner Associates, Inc.; and Computerized Structural Design, Inc.

PART 1
MATERIALS

		Page No.
1.1	Concrete	1–2
	1.1.1 Compressive Strength	1–2
	1.1.2 Tensile Strength	1–3
	1.1.3 Shear Strength	1–3
	1.1.4 Modulus of Elasticity	1–3
	1.1.5 Poisson's Ratio	1–3
	1.1.6 Volume Changes	1–3
	1.1.7 Freeze-Thaw and Chemical Resistance	1–4
1.2	Reinforcement	1–5
	1.2.1 Prestressing Tendons	1–5
	1.2.2 Deformed Reinforcing Bars	1–5
	1.2.3 Welded Wire Fabric	1–5
	1.2.4 Protection of Reinforcement	1–6
1.3	Grout, Mortar and Drypack	1–7
	1.3.1 Sand-Cement Mixtures	1–7
	1.3.2 Non-Shrink Grout	1–8
	1.3.3 Epoxy Grouts	1–8
1.4	References	1–8

MATERIALS

This Part of the Handbook is a brief review of the materials used in precast and prestressed concrete. For more complete information, see Refs. 1 through 4.

1.1 Concrete

The 28-day design strength of concrete used in precast and prestressed products is usually in the 5000 psi to 6000 psi range. The strength at which prestress force is transferred to concrete is usually about 3000 psi and may be more or less as required by design. However, a practical limit is that which can be attained in about 16 hours, in order to remove the product from the forms on a daily basis.

Aggregates for structural products are usually the same as those used for all other quality concrete in the local area. Lightweight structural concrete is also sometimes used for prestressed products and the characteristics of the mix should be determined from the local plant.

Aggregates commonly selected for exposed concrete facings are quartz, granite or marble which offer a wide variety of color and texture. Lower cost sand and gravel aggregates may also be used to produce architectural concrete. Special attention should be paid to sand and gravel aggregates to determine that they do not have rusting or staining problems when the aggregate is exposed to the environment. During production, architectural precast concrete panels generally do not receive accelerated heat curing, as do precast, prestressed concrete structural members. They are removed from forms, at an age of about 16 hours, after the concrete has reached a strength adequate to handle the pieces.

1.1.1 Compressive Strength

The compressive strength of concrete, made with aggregate of adequate strength, is governed by either the strength of the cement paste or the bond between the paste and the aggregate particles. At early ages the bond strength is lower than the paste strength; at later ages the reverse may be the case. For a given cement and acceptable aggregates the strength that may be developed by a workable, properly placed mixture of cement, aggregate, and water (under the same mixing, curing, and testing conditions) is influenced by (a) the ratio of water to cement, (b) the ratio of cement to aggregate, (c) grading, surface texture, shape, strength, and stiffness of aggregate particles, and (d) maximum size of the aggregate. Mix factors, partially or totally independent of water-cement ratio, which affect the strength are (a) type and brand of cement, (b) amount and type of admixture or pozzolan, and (c) mineral composition of the aggregate.

Compressive strength is measured by testing 6 × 12 in. cylinders in accordance with standard ASTM procedures. The precast concrete industry also uses 4 × 8 in. cylinders and 4 in. cube specimens. Correction factors need to be applied to these non-standard specimens to correlate with the standard 6 × 12 in. cylinders.

Because of the need for early strength gain, Type III cement is often used by precasters so that molds may be reused daily. Structural precast concrete and much architectural concrete is made with gray cement that meets ASTM C150. Type III and Type I white and buff portland cements are frequently used in architectural products. These are usually assumed to have the same characteristics (other than color) as gray cement. Pigments are also available to achieve colored concrete, and have little or no effect on strength at the recommended dosages. Cement types and experience with color should be coordinated with the local producers.

High strength concrete mixes (6000 psi or more) are available in some areas. Local suppliers should be contacted to furnish mix and design information.

Initial curing of precast concrete takes place in the form, usually by covering to prevent loss of moisture and, in some instances, especially structural products, by the application of radiant heat or live steam. Additional curing has been shown to rarely be necessary to attain the specified strength. Control techniques for the most effective and economical accelerated curing was the subject of recent research.[5]

Most concrete subjected to freezing and thawing should be air-entrained, although the dense mixes used in precast products have a high resistance to freezing and thawing without artificially entrained air. Many precast concrete mixtures have difficulty in entraining air contents normally specified for the leaner mixes most often used in cast-in-place flatwork. Thus, it is recommended that a "normal dosage" of the air-entraining agent be used instead of specifying a particular range of air content. For precast concrete elements constructed above grade in a vertical position, air contents as low as 2 to 3% will usually provide the required durability. The precast concrete industry does not use air-entraining portland cements. Instead, admixtures are added to the concrete during the mixing cycle to entrain the air. In some concrete mixes, reductions of strength may be anticipated with the use of air entrainment.

1.1.2 Tensile Strength

A critical measure of performance of architectural precast concrete is its resistance to cracking, which is a function of tensile strength. Reinforcement does not prevent cracking, but controls crack widths after cracking occurs. Tensile stresses which would theoretically result in cracking are permitted by ACI 318-83 in prestressed concrete.

The flexural strength in tension is measured by the modulus of rupture. It can be determined by test, but for structural design the modulus of rupture is generally assumed to be a function of compressive strength as given by:

$$f_r = K\lambda\sqrt{f'_c} \qquad \text{(Eq. 1.1.1)}$$

where:
f_r = modulus of rupture, psi
f'_c = compressive strength, psi
K = a constant, usually between 8 and 10 but implied to be equal to 7.5 by ACI 318-83
λ = 1.0 for normal weight, 0.85 for sand-lightweight and 0.75 for all-lightweight concrete.

1.1.3 Shear Strength

The shear (or diagonal tension) strength of concrete is also a function of compressive strength. The equations for shear strength given by ACI 318-83 are given in Part 4. The shear strength of lightweight concrete is a function of the splitting tensile strength, which is determined by test. However, in lieu of test, the ACI Building Code permits the use of the coefficient, λ, as described above.

1.1.4 Modulus of Elasticity

Modulus of elasticity (E) is the ratio of normal stress to corresponding strain for tensile or compressive stresses. It is the material property which determines the deformability under load. Thus it is used to calculate deflections, axial shortening and elongation, buckling and relative distribution of applied forces in composite and non-homogeneous structural members.

The modulus of elasticity of concrete and other masonry materials is not as well defined as, for example, steel. It is therefore defined by some approximation, such as the "secant modulus". Thus, calculations which involve its use have an inherent imprecision, but this is seldom bad enough to affect practical performance. While it may be desirable in some rare instances to determine modulus of elasticity by test, especially with some lightweight concretes, the equation given in ACI 318 is usually adequate for design:

$$E_c = w^{1.5}\, 33\, \sqrt{f'_c} \qquad \text{(Eq. 1.1.2)}$$

where:
E_c = modulus of elasticity of concrete, psi
w = unit weight of concrete, pcf

1.1.5 Poisson's Ratio

Poisson's ratio is the ratio of transverse strain to axial strain resulting from uniformly distributed axial load. Values generally range between 0.11 and 0.27, and is usually assumed to be 0.20 for both normal and lightweight concrete.

1.1.6 Volume Changes

Volume changes of precast concrete are caused by variations in temperature, shrinkage due to air-drying, and by creep caused by sustained stress. If precast concrete is free to deform, volume changes are of little consequence. If these members are restrained by foundations, connections, steel reinforcement, or connecting members, significant stresses may develop over time.

The volume changes due to temperature variations can be positive (expansion) or negative (contraction), while volume changes from shrinkage and creep are only negative.

Precast concrete members are usually 30 days or older when they are erected. Thus, much of the creep and shrinkage will have taken place during yard storage. However, connection details and joints must be designed to accommodate the changes which will occur after the precast member is erected and connected to the structure. In most cases, the shortening that takes place prior to making the final connections will reduce the shrinkage and creep strains to manageable proportions.

Typical creep, shrinkage, and temperature strains and design examples are given in Part 3.

Temperature effects

The coefficient of thermal expansion of concrete varies with the aggregate used as shown in Table 1.1.[1] Ranges for normal weight concrete are 5 to 7×10^{-6} in/in/deg F when made with siliceous aggregates and 3.5 to 5×10^{-6} when made with calcareous aggregates. The approximate values for structural lightweight concretes are 3.6 to 6×10^{-6} in/in/deg F, depending on the type of aggregate and amount of natural sand. Coefficients of 6×10^{-6} in/in/deg F for normal weight and 5×10^{-6} for sand-lightweight concrete are frequently used. If greater accuracy is needed, tests should be made on the specific concrete.

Fig. 1.1 Volume surface ratios for precast structural concrete members

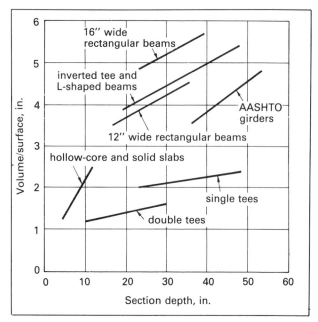

Since the thermal coefficient for steel is also about 6×10^{-6} in/in/deg F, the addition of steel reinforcement does not significantly affect the concrete coefficient.

Shrinkage and creep

Precast concrete members are subject to air-drying as soon as they are removed from the molds. During this exposure to the atmosphere, the concrete slowly loses some of its original water causing shrinkage volume change to occur.

When concrete is subjected to a sustained load, the deformation may be divided into two parts: (1) an elastic deformation which occurs immediately, and (2) a time-dependent deformation which begins immediately and continues over time. This long-term deformation is called creep.

Creep and shrinkage strains vary with relative humidity, volume-surface ratio (ratio of area to perimeter — see Fig. 1.1), level of sustained load including prestress, concrete strength at time of load application, amount and location of steel reinforcement, and other characteristics of the material and design.

Table 1.1 Average coefficients of linear thermal expansion of rock (aggregate) and concrete

Type of Rock (Aggregate)	Average Coefficient of Thermal Expansion $\times 10^{-6}$ in./in./deg F	
	Aggregate	Concrete*
Quartzite, Cherts	6.1 - 7.0	6.6 - 7.1
Sandstones	5.6 - 6.7	5.6 - 6.5
Quartz Sands & Gravels	5.5 - 7.1	6.0 - 8.7
Granites & Gneisses	3.2 - 5.3	3.8 - 5.3
Syenites, Diorites, Andesite, Gabbros, Diabas, Basalt	3.0 - 4.5	4.4 - 5.3
Limestones	2.0 - 3.6	3.4 - 5.1
Marbles	2.2 - 3.9	2.3
Dolomites	3.9 - 5.5	—
Expanded Shale, Clay & Slate	—	3.6 - 4.3
Expanded Slag	—	3.9 - 6.2
Blast-Furnace Slag	—	5.1 - 5.9
Pumice	—	5.2 - 6.0
Perlite	—	4.2 - 6.5
Vermiculite	—	4.6 - 7.9
Barite	—	10.0
Limonite, Magnetite	—	4.6 - 6.0
None (Neat Cement)		10.3
Cellular Concrete		5.0 - 7.0
1:1 (Cement: Sand)	—	7.5
1:3 **		6.2
1:6		5.6

*Coefficients for concretes made with aggregates from different sources vary from these values, especially those for gravels, granites, and limestones. Fine aggregates generally the same material as coarse aggregates.

**Tests made on 2-yr. old samples.

1.1.7 Freeze-Thaw and Chemical Resistance

Cycles of freezing and thawing can cause damage to concrete ranging from minor surface scale to severe cracking. Corrosion of reinforcing steel, prestressing strand or connection hardware can also result, which may affect the integrity of the structure.

The effects of freezing and thawing can be resisted by high quality concrete and air entrainment. Adequate concrete cover over steel and surface drainage is essential in structures exposed to weather.

Freeze-thaw damage is magnified when deicing chemicals are used. Deicers can be applied indirectly in various ways, such as by drippings from the undersides of vehicles.

Some proprietary surface treatments have been found to be effective deterrents to freeze-thaw, deicer, and other chemical damage. These are relatively expensive and should be selected and applied with care. Most commercial floor treatments are not effective.

Other foreign materials can damage concrete, such as sulfates in soils or groundwater and industrial acids. The former can be resisted by specifying cements with low C_3A contents. Presence of acids frequently require a topping of concrete or other material.

1.2 Reinforcement

Reinforcement used in structural and architectural precast concrete includes prestressing tendons, deformed steel bars, and welded wire fabric.

1.2.1 Prestressing Tendons

Tendons for prestressing concrete may be wires, strands, or bars. In precast, prestressed structural concrete, nearly all tendons are 7-wire strands conforming to ASTM A416. The strands are pretensioned, that is, they are tensioned prior to placement of the concrete. After the concrete has reached a predetermined strength, the strands are cut and the prestress force is transferred to the concrete through bond.

In the past, most strand has been ordinary stress-relieved, but low-relaxation strand is becoming more common.[10] Low-relaxation strand is specified in the supplement to ASTM A416, and differs from ordinary strand in only two respects: first, it must meet certain relaxation loss requirements, and second, the minimum yield strength, as measured by the 1% extension method is 90% of the minimum breaking strength, as opposed to 85% for normal stress-relieved strand. Most of the load tables in Part 2 are based on the use of low-relaxation strand.

Architectural precast concrete is also sometimes prestressed. Depending on the facilities available at the plant, the prestressing tendons may be either pretensioned or post-tensioned. In post-tensioning, which is more common in cast-in-place construction, the tendons are either placed in a conduit or are coated so they will not bond to the concrete. The tendons are then tensioned after the concrete has reached the predetermined strength. When the tendons are placed in a conduit, they are usually grouted after tensioning (bonded post-tensioning). When they are greased and wrapped, or coated, they usually are not grouted (unbonded post-tensioning). More information on post-tensioning can be found in the Post-Tensioning Manual.[11]

Prestressing wire or bars are rarely used as primary reinforcement in precast products, but bars meeting ASTM A722 are frequently used in connections.

The properties of prestressing strand, wire and bars are given in Part 11.

1.2.2 Deformed Reinforcing Bars

Reinforcing bars are hot-rolled from steels with varying carbon content. They are usually required to meet ASTM A615, A616, or A617. These specifications are of the performance type, and do not closely control the chemistry or manufacture of the bars. Bars are usually specified to have a minimum yield strength of 40,000 (Grade 40) or 60,000 (Grade 60) psi. Grade 40 bars will usually have a lower carbon content than Grade 60, but not necessarily.

ASTM A706 specifies a bar with controlled chemistry that is weldable. For bars that are to be welded, see Sect. 6.5.1.

In order for the reinforcing bar to develop its full strength in the concrete, a minimum length of embedment is required, or the bars may be hooked. Information on bar sizes, bend and hook dimensions and development length are given in Part 11 and Ref. 6.

1.2.3 Welded Wire Fabric

Welded wire fabric is a prefabricated reinforcement consisting of parallel cold-drawn wires welded together in square or rectangular grids. Each wire intersection is electrically resistance-welded by a continuous automatic welder. Pressure and heat fuse the intersecting wires into a homogeneous section and fix all wires in their proper position.

Smooth wires, deformed wires or a combination of both may be used in welded wire fabric. Smooth wire sizes are specified by the letter W followed by a number indicating the cross-sectional area of the wire in hundredths of a square inch. For example: W16 denotes a smooth wire with cross-sectional area of 0.16 sq in. Similarly, deformed wire sizes are specified by the letter D followed by a number which indicates hundredths of a square inch.

Smooth wire fabric bonds to concrete by the mechanical anchorage at each welded wire intersection. Deformed wire fabric utilizes wire deformations plus welded intersections for bond and anchorage.

Welded wire fabric for architectural precast concrete is normally supplied in flat sheets. Use of fabric from rolls, particularly in thin precast sections, is not recommended because the rolled fabric cannot be flattened to the required tolerance.

In addition to using fabric in flat sheets, many plants have equipment for bending it into various shapes, such as U-shaped stirrups, four-sided cages, etc.

Available wire sizes, common stock sizes and other information on welded wire fabric is given in Part 11 and Ref. 7.

1.2.4 Protection of Reinforcement

Reinforcing steel is protected from corrosion by embedment in concrete. A protective iron oxide film forms on the surface of the bar as a result of the high alkalinity of the cement paste. As long as the alkalinity is maintained, this film is effective in preventing corrosion.

The protective high alkalinity of the cement paste is usually lost only by leaching or carbonation. Therefore, concrete of low permeability and of sufficient thickness (cover) over the steel will usually provide adequate protection. Low permeability is obtained in well-consolidated concrete having a low water-cement ratio and high cement content, a characteristic of precast concrete.

Cracking may allow oxygen and moisture to reach the embedded steel, providing conditions where rusting of the steel and staining of the surface may occur. A sufficient amount of closely spaced reinforcement, however, limits the width of cracks and the intrusion of water, maintaining the protection of the steel. Prestressing may also be used to limit cracking.

Concrete cover is the minimum clear distance from the reinforcement to the face of the concrete. For exposed aggregate surfaces, the concrete cover to surface of steel is not measured from the original surface. Instead, the depth of mortar removed between the pieces of coarse aggregate should be subtracted. Attention must also be given to scoring, false joints, and drips, as these reduce the cover.

In determining cover, consideration should be given to the following:
1. Structural or nonstructural use of precast member.
2. Maximum aggregate size (cover should be greater than maximum aggregate size, particularly if a face mix is used).

Table 1.2 Minimum cover requirements for precast concrete

Condition	Minimum Cover		Prestressed Reinforcement[a,b]
	Nonprestressed Reinforcement		
Exposed to earth or weather			
Wall panels Other members	#11 and smaller #6 through #11 # 5, W31 or D31 wire, and smaller	¾ in. 1 ½ in. 1 ¼ in.	1 in. 1 ½ in.
Not exposed to earth or weather			

Wall panels, slabs and joists
  Beams and columns
    Main steel
	#11 and smaller
	
	Diameter of bar, but not less than ⅝ in., and need not exceed 1½ in.
		⅝ in.	¾ in.
			
			1 ½ in.

Ties, stirrups or spirals	All sizes	⅜ in.	1 in.

a. When the tensile stress in the concrete under service load exceeds $6\sqrt{f_c'}$ for members exposed to earth, weather, or corrosive environment, cover shall be increased by 50% over that specified.

b. Also applies to ducts and end fittings.

3. The means of securing the steel in a controlled position and maintaining this control during placement of concrete.
4. Accessibility for placement of concrete, and the proportioning of the concrete mix relative to the structural environment.
5. The type of finish treatment of the concrete surface.
6. The environment of the concrete surface: interior or exposed to weather, ocean atmosphere, or aggressive industrial fumes.
7. Fire code requirements.

Minimum cover requirements for precast concrete manufactured under plant control conditions are specified in Sect. 7.7 of ACI 318-83. These requirements are listed in Table 1.2.

Cover requirements over reinforcement should be increased to 1-1/2 in. for nongalvanized reinforcement, or 3/4 in. for galvanized reinforcement when the precast elements are acid treated, exposed to a corrosive environment, or subjected to severe exposure conditions. In addition, 3/4 in. cover is realistic only if the maximum aggregate size does not exceed 1/2 in.

Galvanized reinforcing bars or welded wire fabric are often used when minimum cover requirements cannot be achieved, or when the concrete is exposed to a severe environment. However, a detrimental chemical reaction can take place when the concrete is damp and chlorides are present. Therefore, galvanizing is not recommended for members subjected to deicing salts or similar treatment.

Galvanized welded wire fabric is usually available as a stock item in some sizes. Individual wires are galvanized before they are welded together to form the fabric. Zinc at each wire intersection is burned off during welding, but the resulting black spots have not caused appreciable problems. After welding, the fabric is normally shipped to the plant without further treatment. There is no ASTM specification for galvanized welded wire fabric.

When galvanized reinforcement is used in concrete it should not be coupled directly to ungalvanized steel reinforcement, copper, or other dissimilar metals. Polyethylene and similar tapes can be used to provide local insulation between any dissimilar metals. Galvanized reinforcement should be fastened with ties of soft stainless steel or other non-corrosive wire rather than black wire.

When galvanized reinforcement is placed close to nongalvanized metal forms, the concrete may have a tendency to stick to the forms. This may also happen if nongalvanized reinforcement is used close to galvanized forms or form liners. Sticking may be prevented by passivating the surface of the galvanized steel reinforcement using a chromate treatment. Chromate coating of galvanized surfaces can be readily done in most galvanizing plants. The addition of chromic oxide (CrO_3) to the concrete mix (150 parts per million based on weight of mixing water, or about 8 oz. of a 10% solution per cu. yd. of concrete) may also be effective in eliminating sticking and reflection of steel. Precautions in handling of chromic oxide shall be taken to avoid dermatitis.

Epoxy coated reinforcing bars, welded wire fabric and prestressing strand are available for use in products which require special corrosion protection. This adds significant cost to the products.

1.3 Grout, Mortar and Drypack

When water, sand and a cementitious material are mixed together without coarse aggregate, the result is called mortar, grout, or drypack, depending on consistency. These materials have numerous applications with precast concrete: sometimes only for fire or corrosion protection, or for cosmetic treatment, other times to transfer loads in horizontal and vertical joints.

Different cementitious materials are used:
1. Portland cement
2. Shrinkage-compensating portland cement
3. Expansive portland cement made with special additives
4. Gypsum or gypsum/portland cements
5. Epoxy resins.

1.3.1 Sand-Cement Mixtures

Most grout is a simple mixture of portland cement, sand, and water. Proportions are usually one part cement to 2.25 to 3 parts sand. The amount of water depends on the method of placement.

Flowable grouts are high-slump mixes used to fill voids that are either formed in the field or cast into the precast member. They are used at joints that are heavily congested but not confined, thus requiring some formwork. These grouts usually have a high water-cement ratio, resulting in low strength and high shrinkage. There is also a tendency for the solids to settle, leaving a layer of water on the top. Special ingredients or treatments can improve these characteristics.

For very small spaces in confined areas, grouts may be pumped or pressure injected. The confinement must be of sufficient strength to resist the pressure. Less water may be used than for flowable grouts, hence less shrinkage and higher strength.

A stiffer grout, or mortar, is used when the joint

is not totally confined, for example in vertical joints between wall panels. This material will usually develop strengths of 3000 to 6000 psi, and have much less shrinkage than flowable grouts.

Drypack is the common name used for very stiff sand-cement mixes. They are used if a relatively high strength is desired, for example, under column base plates. Compaction is attained by hand tamping, using a rod or stick.

When freeze-thaw durability is a factor, the grout should be air-entrained. Air content of plastic grout or mortar of 9 or 10% is required for adequate protection.

Typical portland cement mortars have very slow early strength gain when placed in cold weather. Heating the material is usually not effective, because the heat is rapidly dissipated to surrounding concrete. Thus, unless a heated enclosure can be provided, special proprietary mixes, usually containing gypsum, may be indicated. However, these mixes may deteriorate under prolonged exposure to water.

1.3.2 Non-Shrink Grout

Shrinkage of sand-cement grout can be reduced by using proprietary non-shrink mixes, or by adding aluminum powder to the mix. Non-shrink grouts can be classified by the method of expansion:
1. Gas-liberating
2. Metal-oxidizing
3. Gypsum-forming
4. Expansive cement.

Some expansive ingredients may cause undesirable effects in some applications, so manufacturers' recommendations should be followed.

Aluminum powder added to ordinary sand-cement grout forms a gas-liberating mixture. Extremely small amounts of powder are required — about a teaspoonful per bag of cement. Thus these mixes are very sensitive, and trial mixes should be tested.

1.3.3 Epoxy Grouts

Epoxy grouts are used when very high strength is desired, or positive bonding to the concrete is necessary. They are mixtures of epoxy resins and a filler material, usually sand.

The physical properties of epoxy compounds vary widely. The user should be familiar with the compound to be used, either through experience or test.[9] Of particular importance in some applications is the thermal expansion, which can be up to 7 times that of concrete.

Epoxy resins without fillers are sometimes pressure-injected into cracked concrete as a repair measure.

1.4 References

1. Fintel, M., Editor, *Handbook of Concrete Engineering*, Van Nostrand Reinhold Company, New York, NY, Second Edition, 1984.

2. *Design and Control of Concrete Mixtures*, 12th Edition, Portland Cement Association, Skokie, IL, 1979.

3. *ACI Manual of Concrete Practice* (5 Volumes), American Concrete Institute, Detroit, MI, 1984.

4. *Concrete Manual, A Water Resources Technical Publication*, Eighth Edition, U.S. Department of Interior, Bureau of Reclamation, Engineering and Research Center, Denver, CO, 1975.

5. Pfeifer, D. W., Marusin, Stella, and Landgren, J. R., "Energy-Efficient Accelerated Curing of Concrete", Technical Report No. 1, Prestressed Concrete Institute, Chicago, IL, 1981.

6. *Manual of Standard Practice*, Concrete Reinforcing Steel Institute, Schaumburg, IL, 1976.

7. *Welded Wire Fabric Manual of Standard Practice*, Wire Reinforcement Institute, Inc., McLean, VA, 1979.

8. *Corps of Engineers Specification for Non-shrink Grout*, CRD-C588-78A, U.S. Army Corps of Engineers, 1978.

9. ACI Committee 503, "Use of Epoxy Compounds with Concrete", *ACI Journal*, V. 70, No. 9, Sept., 1973.

10. Martin, L. D. and Pellow, D. L. "Low-Relaxation Strand—Practical Applications in Precast Prestressed Concrete", *PCI Journal*, V. 28, No. 4, July-August, 1983.

11. *Post-Tensioning Manual*, Third Edition, Post-Tensioning Institute, Phoenix, AZ, 1981.

PART 2
PRODUCT INFORMATION AND CAPABILITY

		Page No.
2.1	General	2—2
	2.1.1 Notation	2—2
	2.1.2 Introduction	2—3
2.2	Explanation of Load Tables	2—3
	2.2.1 Safe Superimposed Load	2—3
	2.2.2 Limiting Criteria	2—3
	2.2.3 Estimated Cambers	2—4
	2.2.4 Design Criteria	2—4
	2.2.5 Concrete Strengths and Unit Weights	2—4
	2.2.6 Prestressing Strands	2—4
	2.2.7 Losses	2—5
	2.2.8 Strand Placement	2—5
	2.2.9 Columns and Load Bearing Wall Panels	2—5
	2.2.10 Piles	2—5
2.3	Stemmed Deck Members	
	Double tee load tables	2—6
	Single tee load tables	2—25
2.4	Flat Deck Members	
	Hollow-core slab load tables	2—29
	Hollow-core slab section properties	2—35
	Solid slab load tables	2—38
2.5	Beam Load Tables	
	Rectangular beams	2—44
	L-shaped beams	2—45
	Inverted tee beams	2—46
2.6	Columns and Load Bearing Wall Panels	
	Precast, prestressed columns	2—47
	Precast, reinforced columns	2—49
	Double tee wall panels	2—51
	Hollow-core wall panels	2—52
	Solid wall panels	2—53
2.7	Piles	
	Piles	2—55
	Sheet piles	2—56

PRODUCT INFORMATION AND CAPABILITY

2.1 General

2.1.1 Notation

A = cross-sectional area

A_g = gross cross-sectional area

b_w = web width

D = unfactored dead loads

E_c = modulus of elasticity of concrete

e_c = eccentricity of prestress force from the centroid of the section at the center of the span

e_e = eccentricity of prestress force from the centroid of the section at the end of the span

f'_c = specified compressive strength of concrete

f'_{ci} = compressive strength of concrete at time of initial prestress

f_{pu} = ultimate strength of prestressing steel

f_{se} = effective stress in prestressing steel after losses

h = overall depth of member

I = moment of inertia

L = unfactored live loads

ℓ = span

M_n = nominal moment strength of a member

M_{nb} = nominal moment strength under balanced conditions

M_o = nominal moment strength of a compression member with zero axial load

M_u = applied factored moment at section

P_n = nominal axial load strength of a compression member at given eccentricity

P_{nb} = nominal axial load strength under balanced conditions

P_o = nominal axial load strength of a compression member with zero eccentricity

P_u = factored axial load

t = thickness

V_{ci} = nominal shear strength provided by concrete when diagonal cracking results from combined shear and moment

V_{cw} = nominal shear strength provided by concrete when diagonal cracking results from excessive principal tensile stress in the web

V_u = factored shear force

V/S = volume-surface ratio

y_b = distance from bottom fiber to center of gravity of section

y_t = distance from top fiber to center of gravity of section

Z = section modulus

Z_b = section modulus with respect to the bottom fiber of a cross section

Z_t = section modulus with respect to the top fiber of a cross section

δ = moment magnification factor

ϕ = strength reduction factor

2.1.2 Introduction

This part presents data on the shapes that are standard in the precast concrete industry. Other shapes and standard shapes with depth and width variations are also available in many areas of the country. Designers should contact the manufacturers in the geographic area of the proposed structure to determine the properties and dimensions of products available. Manufacturers will usually have their own load tables for the members they produce. This section, plus the design aids and techniques provided in other parts of this Handbook, should enable the designer to quickly and expeditiously complete his design.

2.2 Explanation of Load Tables

The load tables on the following pages show dimensions, section properties and engineering capabilities of the shapes most commonly used throughout the industry. These shapes include double and single tees, hollow-core and solid flat slabs, beams, girders, columns, piles and wall panels. The dimensions of the shapes shown in the tables may vary among manufacturers. Adjustment of these minor variations can be made by the designer. Hollow-core slabs of different thicknesses, core sizes and shapes are available in the market under various trade names. Cross-sections and section properties of proprietary hollow-core slabs are shown on pages 2–35 through 2–37. Load tables, on pages 2–29 through 2–34, are developed for non-proprietary hollow-core sections of thicknesses most commonly used in the industry.

Load tables for stemmed deck members, flat deck members and beams show the allowable superimposed service load, estimated camber at the time of erection and the estimated long term camber after the member has essentially stabilized, but before the application of superimposed live loads. For the deck members, the table at the top of the page gives the information for the member with no topping, and the table at the bottom of the page is for the same member with 2 in. of normal weight concrete topping acting compositely with the precast section. Values in the tables assume a uniform 2 in. topping over the full span length, and assume the member to be unshored at the time the topping is placed. Safe loads and cambers shown in the tables are based on the dimensions and section properties shown on the page, and will vary for members with different dimensions.

For beams, a single load table is used for several sizes of members. The values shown are based on sections containing the maximum practical number of prestressing strands, but in some cases, more strands could be used.

2.2.1 Safe Superimposed Load

The values for safe superimposed service load are based on the capacity of the member as governed by the ACI Building Code limitations on flexural strength, service load flexural stresses, or, in the case of flat deck members without shear reinforcement, shear strength. A portion of the safe load shown is assumed to be dead load for the purpose of applying load factors and determining cambers and deflections. For untopped deck members, 10 psf of the capacity shown is assumed as superimposed dead load, typical for roof members. For topped deck members, 15 psf of the capacity shown is assumed as superimposed dead load, typical for floor members. The capacity shown is in addition to the weight of the topping. For beams, 50 percent of the capacity shown is assumed as dead load, normally conservative for beams which support concrete decks.

Example: For an 8DT24/88-D1 untopped, (p. 2–16) with a 52 ft span, the capacity shown is 68 psf. The member can safely carry service loads of 10 psf dead and 58 psf live.

2.2.2 Limiting Criteria

The criteria used to determine the safe superimposed load and strand placement are based on "Building Code Requirements for Reinforced Concrete (ACI 318-83)." For design procedures, see Part 4 of this Handbook. A summary of the Code provisions used in the development of these load tables is as follows:

1. Capacity governed by design flexural strength:

   Load factors: 1.4D + 1.7L

   Strength reduction factor, $\phi = 0.90$

   Calculation of design moments assumes simple spans with roller supports. If the strands are fully developed (see Sect. 4.2.3), the critical moment is assumed to be at midspan in members with straight strands, and at 0.4ℓ (ℓ = span) in products with strands depressed at midspan. (Note: The actual critical point can be determined by analysis, but will seldom vary significantly from 0.4ℓ.) Flexural strength is calculated using strain compatibility as discussed in Part 4.

2. Capacity governed by service load stresses:

   Flexural stresses immediately after transfer of prestress, before long time losses:

   a) Compression: $0.6 f'_{ci}$
   b) End tension: $6 \sqrt{f'_{ci}}$
      Midspan tension: $3 \sqrt{f'_{ci}}$

Note 1: End stresses are calculated 50 strand diameters from the end of the member, the theoretical point of full transfer.

Note 2: Release tension is not used as a limiting criterion for beams. Supplemental top reinforcement must be provided, designed as described in Sect. 4.2.2.2.

Stresses at service loads, after all losses:

a) Compression: 0.45 f'_c

b) Tension:

Stemmed deck members and beams:

$12\sqrt{f'_c}$

Note that in final design, deflections must be determined based on bilinear moment-deflection relationships. See Sect. 4.6.3.

Flat deck members: $6\sqrt{f'_c}$

Critical point for service load moment is assumed at midspan for members with straight strands and at 0.4ℓ for members with strands depressed at midspan, as described above.

3. For *flat deck members*, the capacity may be limited by the design shear strength. In this case, the safe superimposed load is that which will yield a factored shear force V_u of no more than ϕV_{ci} or ϕV_{cw}, as permitted by ACI 318-83 for slabs without shear reinforcement. See Sect. 4.3 for the design procedures.

Stemmed deck members and beams do not have this limitation. It may be necessary to provide shear reinforcement, designed as described in Sect. 4.3. For the majority of such deck members, minimum or no reinforcement may be required as provided by Sect. 11.5.5 of ACI 318-83. For loads which are heavier than normal, special transverse reinforcement may be required to resist the moment in the overhanging flange.

4. *Flat deck members* show no values beyond a span/depth ratio of 50 for untopped members and 40 for topped members. These are suggested maximums for roof and floor members respectively, unless a detailed analysis is made.

2.2.3 Estimated Cambers

The estimated cambers shown are calculated using the multipliers shown in Sect. 4.6.5 of this Handbook. *These values are estimates, and should not be used as absolute values.* Non-structural components attached to members which could be affected by camber variations, such as partitions or folding doors, should be placed with adequate allowance for error. Calculation of topping quantities should also recognize that the values can vary.

2.2.4 Design Criteria

The design of prestressed concrete is dependent on many variables that must be defined prior to executing the design, e.g. concrete strength at release and 28 days—modulus of elasticity of concrete—weight of concrete—size, grade and type of strand—single or double depressed strand patterns—determination of losses—initial tension applied to strands—strand placement—etc. The tables presented show allowable superimposed loads based on one set of variables. Higher allowable loads or greater spans may be achieved by selecting a different set of conditions.

2.2.5 Concrete Strength and Unit Weights

Twenty-eight day cylinder strength for concrete in the precast units is assumed to be 5000 psi unless otherwise indicated. Tables for units with composite topping are based on the topping concrete being normal weight concrete with a cylinder strength of 3000 psi.

Concrete strength at time of strand tension release is 3500 psi unless the value falls below the heavy line shown in the load table, indicating that a cylinder strength greater than 3500 psi is required. No values are shown when the required release strength exceeds 4500 psi.

The concrete strengths used in the tables are not intended to be limitations or recommendations for actual use. Some precasting manufacturers may choose to use higher or lower strengths, resulting in slightly different table values. For low levels of prestress, the concrete release strength is usually governed by handling stresses. For products such as columns, piles, wall panels and lightly stressed flexural members a release strength of 2000 psi is realistic.

Unit weights of concrete are assumed to be 150 lb. per cu. ft. for normal weight and 115 lb. per cu. ft. for lightweight.

2.2.6 Prestressing Strands

Prestressing strands are available in diameters from 1/4 in. to 0.6 in., grades 250 ksi and 270 ksi, either stress-relieved or low-relaxation (see Table 11.2.3). The grade of strand indicates the ultimate strength and the type of strand, i.e., stress-relieved or low-relaxation, defines the manufacturing process.

It is the most common practice in the precast industry to apply an initial (jacking) tension of 70 percent of the nominal strength of the strand when using stress-relieved strand. Stress relaxation losses in stress-relieved strand have been shown to be proportionally higher when an initial stress higher than 70 percent is used, but with low-relaxation strand, the relaxation loss remains more or less proportional up to an initial stress of 75 percent of ultimate.[1] Thus, in the tables, the strands are assumed to have been stressed initially to 75 percent of ultimate when low-relaxation strands are indicated and 70 percent when stress-relieved is shown. Stress at transfer of prestress has been assumed to be reduced by 10 percent of the initial stress.

2.2.7 Losses

Losses were calculated in accordance with the recommendations in "Estimating Prestress Losses."[2] This procedure includes consideration of initial stress level ($0.7 f_{pu}$ or higher), type of strand, exposure conditions and type of construction. Lower limits of 35,000 psi for stress-relieved and 30,000 psi for low-relaxation strands were also used. This lower limit is arbitrary; other designers may choose not to impose this limit. Additional information on losses is given in Sect. 4.5.

2.2.8 Strand Placement

Quantity, size and profile of strands are shown in the load tables under the column headed "Strand Pattern", for example, 88-S. The first digit indicates the total number of strands in the unit, the second digit is the diameter of the strand in 16ths of an inch (8/16 equals 1/2 in. diameter), and the "S" indicates that the strands are straight. A "D1" indicates that the strands are depressed at midspan. Some precast producers choose to depress the strand at 2 points, which provides a somewhat higher capacity.

For *stemmed deck members* and *beams,* the eccentricities of strands at the ends and midspan are shown in the load tables. Strands have been placed so that the stress at 50 strand diameters from the end (theoretical transfer point) will not exceed those specified above.

For *flat deck members,* the load table values are based on strand centered 1-1/2 in. from the bottom of the slab. Strand placement can vary from as low as 7/8 in. to as high as 2-1/8 in. from the bottom, which will change the capacity and camber values shown. The higher strand placements give improved fire resistance ratings (see Sect. 9.3 of this Handbook for more information on fire resistance). The lower strand placement may require higher release strengths, or top tension reinforcement at the ends. The designer should contact the local supplier of flat deck members for available and recommended strand placement locations.

2.2.9 Columns and Load Bearing Wall Panels

Interaction curves for selected precast, pre-stressed columns, precast reinforced columns and various types of commonly used wall panels are provided on pp. 2—47 to 2—54.

These interaction curves are for strength design loads and moments and the appropriate load factors must be applied to the service loads and moments before entering the charts. Also, the curves are for *short* members. The effects of slenderness must be determined as described in Sect. 3.5 before the design is complete.

For prestressed columns and wall panels, curves are shown for both partially developed strands, usually appropriate for end moment capacity, and fully developed strands, usually appropriate for mid-span moment capacity. This is discussed more completely in Sect. 4.7. The columns and wall panels which use reinforcing bars assume full development of the bars.

The column curves are terminated at a value of $P_u = 0.80 \phi P_o$, the maximum allowable load for tied columns under ACI 318-83. Most of the wall panel curves show the lower portion of the curve only (flexure controlling). Actual design loads will rarely exceed the values shown.

The curves for double tee wall panels are for bending in the direction that causes tension in the stem. They are conservative, in the range shown, for bending in the opposite direction.

The curves for hollow-core wall panels are based on a generic section as shown. They can be used with small error for all sections commonly marketed for wall panel use.

2.2.10 Piles

Allowable concentric service loads on pre-stressed concrete piles, based on the structural capacity of the pile alone are shown in Table 2.7.1. The ability of the soil to carry these loads must be evaluated separately. Values for concrete strengths up to 8000 psi are shown. Available strengths should be checked with local manufacturers. The design of prestressed concrete piles is discussed in Sect. 4.7.6 of this Handbook.

Section properties and allowable service load bending moments for prestressed concrete sheet pile units are shown in Table 2.7.2. These units are available in some areas for use in earth retaining structures.

References:

1. Martin, L.D. and Pellow, D.L. "Low-Relaxation Strand—Practical Applications in Precast Prestressed Concrete," *PCI Journal,* V. 28, No. 4, July-August, 1983, pp. 84-101.
2. Zia, Paul, Preston, H.K., Scott, N.L. and Workman, E.B., "Estimating Prestress Losses," *Concrete International,* V. 1, No. 6, June 1979, pp. 32-38.
3. PCI Committee on Prestressed Concrete Piling, "Recommended Practice for Design, Manufacture and Installation of Prestressed Concrete Piling," *PCI Journal,* V. 22, No. 2, March-April, 1977, pp. 20-49.

DOUBLE TEE

8'-0" x 12"
Normal Weight Concrete

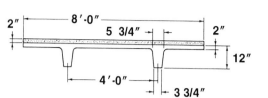

f'_c = 5000 psi
f_{pu} = 270,000 psi
Low-relaxation strand

Section Properties

		Untopped	Topped
A	=	287 in.²	—
I	=	2872 in.⁴	4389 in.⁴
Y_b	=	9.13 in.	10.45 in.
Y_t	=	2.87 in.	3.55 in.
Z_b	=	315 in.³	420 in.³
Z_t	=	1001 in.³	1236 in.³
wt	=	299 plf	499 plf
		37 psf	62 psf
V/S	=	1.22 in.	

8DT12
No Topping

Table of safe superimposed service load (psf) and cambers

Strand Pattern	e_e / e_c	12	14	16	18	20	22	24	26	28	30	32	34	36	38	40	42	44	46
28-S	7.13 7.13	178 0.2 0.2	137 0.2 0.3	108 0.3 0.3	81 0.3 0.4	60 0.3 0.4	45 0.3 0.4	33 0.3 0.3											
48-S	5.13 5.13			188 0.4 0.5	143 0.5 0.6	110 0.5 0.7	86 0.6 0.7	68 0.7 0.8	53 0.7 0.8	42 0.7 0.7	33 0.7 0.6								
68-S	3.13 3.13			159 0.4 0.5	123 0.5 0.6	97 0.5 0.6	77 0.6 0.6	61 0.6 0.6	49 0.6 0.5	39 0.5 0.4	31 0.5 0.1								
68-D1	3.13 6.63									92 1.4 1.6	76 1.5 1.7	64 1.6 1.7	53 1.6 1.6	44 1.7 1.5	37 1.7 1.3	30 1.6 1.0			
88-D1	1.13 6.38																37 1.8 1.1	31 1.7 0.6	

8DT12 + 2
2" Normal Weight Topping

Table of safe superimposed service load (psf) and cambers

Strand Pattern	e_e / e_c	12	14	16	18	20	22	24	26	28	30	32	34
28-S	7.13 7.13	200 0.2 0.2	150 0.2 0.2	116 0.2 0.2	83 0.3 0.2	58 0.3 0.1	39 0.3 0.0						
48-S	5.13 5.13			164 0.5 0.5	124 0.5 0.5	94 0.6 0.5	71 0.7 0.4	53 0.7 0.3	39 0.7 0.1				
68-S	3.13 3.13			198 0.4 0.4	151 0.5 0.4	117 0.5 0.4	90 0.6 0.3	70 0.6 0.2	50 0.6 -0.1				
68-D1	3.13 6.63									98 1.4 0.9	79 1.5 0.8	64 1.6 0.6	50 1.6 0.3

Strength based on strain compatibility; bottom tension limited to $12\sqrt{f'_c}$; see pages 2-3–2-5 for explanation

Values below heavy line require release strengths higher than 3500 psi.

DOUBLE TEE

8'-0" x 12"
Lightweight Concrete

No of strand (6)

S = straight D = depressed

68-D1

No. of depression points

Diameter of strand in 16ths

Safe loads shown include dead load of 10 psf for untopped members and 15 psf for topped members. Remainder is live load. Long-time cambers include superimposed dead load but do not include live load.

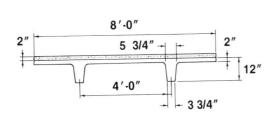

f'_c = 5000 psi
f_{pu} = 270,000 psi
Low-relaxation strand

Section Properties

	Untopped	Topped
A =	287 in.^{2}	—
I =	2872 in.^{4}	4819 in.^{4}
Y_b =	9.13 in.	10.82 in.
Y_t =	2.87 in.	3.18 in.
Z_b =	315 in.^{3}	445 in.^{3}
Z_t =	1001 in.^{3}	1515 in.^{3}
wt =	229 plf	429 plf
	29 psf	54 psf
V/S =	1.22 in.	

Key
186 — Safe superimposed service load, psf
0.2 — Estimated camber at erection, in.
0.3 — Estimated long-time camber, in.

8LDT12
No Topping

Table of safe superimposed service load (psf) and cambers

Strand Pattern	e_e e_c	Span, ft.															
		12	14	16	18	20	22	24	26	28	30	32	34	36	38	40	42
28-S	7.13 7.13	186 0.2 0.3	144 0.3 0.4	116 0.4 0.5	89 0.5 0.6	68 0.6 0.7	52 0.6 0.7	40 0.7 0.7	31 0.7 0.6								
48-S	5.13 5.13		195 0.6 0.8	150 0.7 0.9	118 0.9 1.1	93 1.0 1.2	75 1.1 1.3	61 1.2 1.4	49 1.3 1.4	40 1.3 1.3	33 1.4 1.2						
68-S	3.13 3.13			166 0.6 0.8	131 0.8 1.0	104 0.9 1.1	84 1.0 1.1	68 1.0 1.2	56 1.1 1.1	46 1.1 1.0	38 1.1 0.9	31 1.0 0.6					
68-D1	3.13 6.63											71 2.6 2.9	60 2.8 2.9	51 3.0 2.9	44 3.1 2.7	37 3.2 2.4	32 3.2 2.1

8LDT12 + 2
2" Normal Weight Topping

Table of safe superimposed service load (psf) and cambers

Strand Pattern	e_e e_c	Span, ft.												
		14	16	18	20	22	24	26	28	30	32	34	36	38
28-S	7.13 7.13	157 0.3 0.3	123 0.4 0.3	90 0.5 0.3	65 0.5 0.3	47 0.6 0.2	32 0.6 0.1							
48-S	5.13 5.13			172 0.7 0.7	131 0.9 0.8	101 1.0 0.8	78 1.1 0.8	61 1.2 0.7	47 1.3 0.5	35 1.3 0.2				
68-S	3.13 3.13			158 0.8 0.7	124 0.9 0.7	97 1.0 0.6	77 1.0 0.5	60 1.1 0.3	47 1.1 0.0					
68-D1	3.13 6.63										71 2.6 1.3	58 2.8 1.0	46 3.0 0.6	33 3.1 0.0

Strength based on strain compatibility; bottom tension limited to $12\sqrt{f'_c}$; see pages 2-3–2-5 for explanation

Values below heavy line require release strengths higher than 3500 psi.

DOUBLE TEE

8'-0" x 14"
Normal Weight Concrete

Strand Pattern Designation

No of strand (6)

S = straight   D = depressed

68-D1

No. of depression points

Diameter of strand in 16ths

Safe loads shown include dead load of 10 psf for untopped members and 15 psf for topped members. Remainder is live load. Long-time cambers include superimposed dead load but do not include live load.

f'_c = 5000 psi
f_{pu} = 270,000 psi
Low-relaxation strand

Section Properties

	Untopped	Topped
A =	306 in.²	—
I =	4508 in.⁴	6539 in.⁴
Y_b =	10.51 in.	11.97 in.
Y_t =	3.49 in.	4.03 in.
Z_b =	429 in.³	546 in.³
Z_t =	1292 in.³	1623 in.³
wt =	319 plf	519 plf
	40 psf	65 psf
V/S =	1.25 in.	

Key

168 — Safe superimposed service load, psf
0.2 — Estimated camber at erection, in.
0.2 — Estimated long-time camber, in.

8DT14
No Topping

Table of safe superimposed service load (psf) and cambers

Strand Pattern	e_e / e_c	14	16	18	20	22	24	26	28	30	32	34	36	38	40	42	44	46	48
28-S	8.51 / 8.51	168 / 0.2 / 0.2	134 / 0.2 / 0.2	102 / 0.2 / 0.3	77 / 0.3 / 0.3	58 / 0.3 / 0.3	44 / 0.3 / 0.3	33 / 0.3 / 0.2											
48-S	7.51 / 7.51				163 / 0.5 / 0.7	129 / 0.6 / 0.8	103 / 0.7 / 0.9	83 / 0.8 / 0.9	68 / 0.8 / 1.0	55 / 0.9 / 1.0	45 / 0.9 / 0.9	36 / 0.9 / 0.8							
68-S	4.51 / 4.51				175 / 0.5 / 0.6	139 / 0.5 / 0.7	112 / 0.6 / 0.8	91 / 0.6 / 0.8	74 / 0.7 / 0.8	61 / 0.7 / 0.8	50 / 0.7 / 0.7	40 / 0.7 / 0.6	33 / 0.6 / 0.4						
68-D1	4.51 / 8.01							142 / 1.1 / 1.4	118 / 1.2 / 1.5	99 / 1.3 / 1.6	83 / 1.4 / 1.6	70 / 1.5 / 1.7	59 / 1.6 / 1.6	50 / 1.6 / 1.6	42 / 1.6 / 1.4	35 / 1.5 / 1.2	30 / 1.4 / 0.9		
88-D1	2.51 / 7.76														60 / 2.0 / 2.0	52 / 2.0 / 1.8	44 / 2.0 / 1.6	38 / 2.0 / 1.3	32 / 1.9 / 1.0

8DT14+2
2" Normal Weight Topping

Table of safe superimposed service load (psf) and cambers

Strand Pattern	e_e / e_c	14	16	18	20	22	24	26	28	30	32	34	36
28-S	8.51 / 8.51	181 / 0.2 / 0.1	141 / 0.2 / 0.2	103 / 0.2 / 0.2	74 / 0.2 / 0.1	52 / 0.3 / 0.1	36 / 0.3 / 0.0						
48-S	7.51 / 7.51				176 / 0.5 / 0.5	136 / 0.6 / 0.6	106 / 0.7 / 0.6	83 / 0.8 / 0.6	65 / 0.8 / 0.5	50 / 0.9 / 0.4	38 / 0.9 / 0.2		
68-S	4.51 / 4.51				159 / 0.5 / 0.5	126 / 0.6 / 0.5	100 / 0.6 / 0.5	79 / 0.7 / 0.4	62 / 0.7 / 0.2	49 / 0.7 / 0.0			
68-D1	4.51 / 8.01							152 / 1.1 / 1.0	124 / 1.2 / 1.0	102 / 1.3 / 1.0	83 / 1.4 / 0.9	68 / 1.5 / 0.7	55 / 1.6 / 0.5 · 44 / 1.6 / 0.2

Strength based on strain compatibility; bottom tension limited to $12\sqrt{f'_c}$; see pages 2-3-2-5 for explanation

Values below heavy line require release strengths higher than 3500 psi.

DOUBLE TEE

8'-0" x 14"
Lightweight Concrete

Strand Pattern Designation

No of strand (6)
S = straight D = depressed

68-D1

No. of depression points
Diameter of strand in 16ths

Safe loads shown include dead load of 10 psf for untopped members and 15 psf for topped members. Remainder is live load. Long-time cambers include superimposed dead load but do not include live load.

Section Properties

		Untopped	Topped
A	=	306 in.²	—
I	=	4508 in.⁴	7173 in.⁴
Y_b	=	10.51 in.	12.40 in.
Y_t	=	3.49 in.	3.60 in.
Z_b	=	429 in.³	578 in.³
Z_t	=	1292 in.³	1992 in.³
wt	=	244 plf	444 plf
		31 psf	56 psf
V/S	=	1.25 in.	

8'-0"
5 3/4"
2"   2"
14"
4'-0"
3 3/4"

f'_c = 5000 psi
f_{pu} = 270,000 psi
Low-relaxation strand

Key

176 — Safe superimposed service load, psf
0.3 — Estimated camber at erection, in.
0.3 — Estimated long-time camber, in.

8LDT14
No Topping

Table of safe superimposed service load (psf) and cambers

Strand Pattern	e_e / e_c	14	16	18	20	22	24	26	28	30	32	34	36	38	40	42	44	46
28-S	8.51	176	142	109	84	66	51	40	32									
	8.51	0.3	0.3	0.4	0.4	0.5	0.5	0.6	0.6									
		0.3	0.4	0.5	0.5	0.6	0.6	0.6	0.5									
48-S	7.51				170	137	111	91	75	63	52	44	36	30				
	7.51				0.8	1.0	1.1	1.2	1.4	1.5	1.6	1.7	1.7	1.7				
					1.1	1.2	1.4	1.5	1.6	1.7	1.7	1.7	1.6	1.5				
68-S	4.51				183	147	120	99	82	68	57	48	40	34				
	4.51				0.7	0.9	1.0	1.1	1.2	1.3	1.3	1.4	1.4	1.3				
					1.0	1.1	1.2	1.3	1.4	1.4	1.4	1.3	1.2	1.0				
68-D1	4.51									107	91	78	67	58	50	43	37	32
	8.01									2.1	2.3	2.5	2.7	2.8	2.9	3.0	3.1	3.0
										2.6	2.7	2.8	2.9	2.9	2.8	2.7	2.4	2.1

8LDT14+2
2" Normal Weight Topping

Table of safe superimposed service load (psf) and cambers

Strand Pattern	e_e / e_c	14	16	18	20	22	24	26	28	30	32	34	36	38	40	42
28-S	8.51	189	149	111	82	60	44	31								
	8.51	0.2	0.3	0.4	0.4	0.5	0.5	0.5								
		0.2	0.3	0.3	0.3	0.2	0.2	0.0								
48-S	7.51				183	144	114	91	72	57	45	35				
	7.51				0.8	1.0	1.1	1.2	1.4	1.5	1.6	1.7				
					0.8	0.9	1.0	1.0	1.0	0.9	0.7	0.5				
68-S	4.51				167	134	107	87	70	56	45					
	4.51				0.9	1.0	1.1	1.2	1.3	1.3	1.4					
					0.8	0.8	0.8	0.8	0.6	0.4	0.1					
68-D1	4.51									110	91	76	63	52	43	33
	8.01									2.1	2.3	2.5	2.7	2.8	2.9	3.0
										1.6	1.6	1.5	1.3	1.0	0.5	0.0

Strength based on strain compatibility; bottom tension limited to $12\sqrt{f'_c}$; see pages 2-3–2-5 for explanation

Values below heavy line require release strengths higher than 3500 psi.

DOUBLE TEE

8'-0" x 16"
Normal Weight Concrete

Strand Pattern Designation

No of strand (6)
S = straight   D = depressed

68-D1

No. of depression points
Diameter of strand in 16ths

Safe loads shown include dead load of 10 psf for untopped members and 15 psf for topped members. Remainder is live load. Long-time cambers include superimposed dead load but do not include live load.

Key

200 — Safe superimposed service load, psf
0.1 — Estimated camber at erection, in.
0.2 — Estimated long-time camber, in.

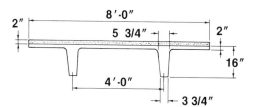

f'_c = 5000 psi
f_{pu} = 270,000 psi
Low-relaxation strand

Section Properties

		Untopped	Topped
A	=	325 in.²	—
I	=	6634 in.⁴	9306 in.⁴
Y_b	=	11.93 in.	13.52 in.
Y_t	=	4.07 in.	4.48 in.
Z_b	=	556 in.³	688 in.³
Z_t	=	1630 in.³	2077 in.³
wt	=	339 plf	539 plf
		42 psf	67 psf
V/S	=	1.29 in.	

8DT16
No Topping

Table of safe superimposed service load (psf) and cambers

Strand Pattern	e_e / e_c	14	16	18	20	22	24	26	28	30	32	34	36	38	40	42	44	46	48	50	52	54	56
28-S	9.93 / 9.93	200 0.1 0.2	160 0.2 0.2	122 0.2 0.2	93 0.2 0.2	71 0.2 0.2	54 0.2 0.2	41 0.2 0.2	31 0.2 0.1														
48-S	8.93 / 8.93				197 0.5 0.6	157 0.5 0.7	127 0.6 0.8	103 0.7 0.8	84 0.7 0.9	69 0.8 0.9	57 0.8 0.9	47 0.8 0.9	38 0.8 0.8	31 0.8 0.7									
68-S	5.93 / 5.93					182 0.5 0.7	147 0.6 0.8	121 0.6 0.8	100 0.7 0.9	82 0.8 0.9	68 0.8 0.9	57 0.8 0.8	47 0.8 0.8	39 0.7 0.6	32 0.7 0.4								
68-D1	5.93 / 9.43							173 0.9 1.2	145 1.0 1.3	122 1.1 1.4	103 1.3 1.5	88 1.3 1.6	75 1.4 1.6	64 1.5 1.6	54 1.5 1.5	46 1.5 1.4	39 1.5 1.3	33 1.4 1.0					
88-D1	3.93 / 9.18													77 1.9 2.1	67 2.0 2.1	58 2.0 2.0	50 2.0 1.9	43 2.0 1.7	37 2.0 1.4	32 1.8 1.1			
108-D1	1.93 / 8.93																			43 2.2 1.5	38 2.1 1.2	33 2.0 0.7	

8DT16+2
2" Normal Weight Topping

Table of safe superimposed service load (psf) and cambers

Strand Pattern	e_e / e_c	16	18	20	22	24	26	28	30	32	34	36	38	40	42	44
28-S	9.93 / 9.93	167 0.2 0.1	124 0.2 0.1	90 0.2 0.1	65 0.2 0.1	47 0.2 0.0	32 0.2 0.0									
48-S	8.93 / 8.93				164 0.5 0.5	129 0.6 0.6	102 0.7 0.6	81 0.7 0.6	64 0.8 0.5	50 0.8 0.4	38 0.8 0.3					
68-S	5.93 / 5.93				161 0.6 0.6	130 0.6 0.6	104 0.7 0.5	84 0.8 0.5	68 0.8 0.4	54 0.8 0.2	42 0.8 0.0					
68-D1	5.93 / 9.43						183 0.9 0.9	151 1.0 0.9	125 1.1 1.0	103 1.3 1.0	85 1.3 0.9	70 1.4 0.8	58 1.5 0.6	47 1.5 0.4	38 1.5 0.1	
88-D1	3.93 / 9.18												73 1.9 0.9	61 2.0 0.7	51 2.0 0.3	

Strength based on strain compatibility; bottom tension limited to $12\sqrt{f'_c}$; see pages 2-3—2-5 for explanation

Values below heavy line require release strengths higher than 3500 psi.

PCI Design Handbook

DOUBLE TEE

8'-0" x 16"
Lightweight Concrete

Strand Pattern Designation

No of strand (6)
S = straight D = depressed

68-D1

No. of depression points
Diameter of strand in 16ths

Safe loads shown include dead load of 10 psf for untopped members and 15 psf for topped members. Remainder is live load. Long-time cambers include superimposed dead load but do not include live load.

Key
168 — Safe superimposed service load, psf
0.3 — Estimated camber at erection, in.
0.3 — Estimated long-time camber, in.

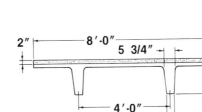

f'_c = 5000 psi
f_{pu} = 270,000 psi
Low-relaxation strand

Section Properties

	Untopped	Topped
A =	325 in.²	—
I =	6634 in.⁴	10,094 in.⁴
Y_b =	11.93 in.	13.99 in.
Y_t =	4.07 in.	4.01 in.
Z_b =	556 in.³	721 in.³
Z_t =	1630 in.³	2517 in.³
wt =	260 plf	460 plf
	33 psf	58 psf
V/S =	1.29 in.	

8LDT16
No Topping

Table of safe superimposed service load (psf) and cambers

Strand Pattern	e_e / e_c	16	18	20	22	24	26	28	30	32	34	36	38	40	42	44	46	48	50	52	54	56
28-S	9.93 / 9.93	168	130	101	79	62	50	39	31													
		0.3	0.3	0.4	0.4	0.4	0.5	0.5	0.5													
		0.3	0.4	0.5	0.5	0.5	0.5	0.4	0.4													
48-S	8.93 / 8.93				165	135	111	92	77	65	55	46	39	33								
					0.8	0.9	1.1	1.2	1.3	1.4	1.5	1.5	1.6	1.6								
					1.0	1.2	1.3	1.4	1.5	1.6	1.6	1.6	1.6	1.5								
68-S	5.93 / 5.93				190	156	129	108	91	77	65	55	47	40	34							
					0.8	0.9	1.0	1.2	1.3	1.4	1.5	1.5	1.5	1.5	1.5							
					1.0	1.2	1.3	1.4	1.5	1.6	1.6	1.6	1.5	1.4	1.2							
68-D1	5.93 / 9.43						181	153	130	111	96	83	72	62	54	47	41	36	31			
							1.4	1.6	1.8	2.0	2.2	2.4	2.5	2.7	2.8	2.8	2.9	2.9	2.8			
							1.9	2.1	2.3	2.4	2.6	2.7	2.8	2.8	2.8	2.6	2.5	2.3	1.9			
88-D1	3.93 / 9.18																58	51	45	40	35	31
																	3.7	3.8	3.9	3.9	3.9	3.8
																	3.7	3.5	3.3	3.0	2.5	2.0

8LDT16+2
2" Normal Weight Topping

Table of safe superimposed service load (psf) and cambers

Strand Pattern	e_e / e_c	16	18	20	22	24	26	28	30	32	34	36	38	40	42	44	46	48
28-S	9.93 / 9.93	175	132	98	74	55	40											
		0.2	0.3	0.3	0.4	0.4	0.4											
		0.2	0.2	0.2	0.2	0.2	0.1											
48-S	8.93 / 8.93				172	137	111	89	72	58	46	36						
					0.8	0.9	1.1	1.2	1.3	1.4	1.5	1.5						
					0.8	0.9	0.9	0.9	0.9	0.9	0.8	0.7						
68-S	5.93 / 5.93					169	138	113	92	76	62	50	41					
						0.9	1.0	1.2	1.3	1.4	1.4	1.5	1.5					
						0.9	0.9	0.9	0.9	0.8	0.7	0.5	0.2					
68-D1	5.93 / 9.43						192	159	133	111	93	79	66	55	46	38		
							1.4	1.6	1.8	2.0	2.2	2.4	2.5	2.7	2.8	2.8		
							1.4	1.5	1.6	1.6	1.6	1.5	1.4	·1.1	0.8	0.3		
88-D1	3.93 / 9.18															51	41	
															3.7	3.8	
															0.8	0.2	

Strength based on strain compatibility; bottom tension limited to $12\sqrt{f'_c}$; see pages 2-3–2-5 for explanation

Values below heavy line require release strengths higher than 3500 psi.

DOUBLE TEE

8'-0" x 18"
Normal Weight Concrete

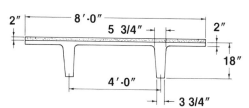

f'_c = 5000 psi
f_{pu} = 270,000 psi
Low-relaxation strand

Strand Pattern Designation

No of strand (10)
S = straight   D = depressed
108-D1
No. of depression points
Diameter of strand in 16ths

Safe loads shown include dead load of 10 psf for untopped members and 15 psf for topped members. Remainder is live load. Long-time cambers include superimposed dead load but do not include live load.

Key

186 — Safe superimposed service load, psf
0.1 — Estimated camber at erection, in.
0.1 — Estimated long-time camber, in.

Section Properties

		Untopped	Topped
A	=	344 in.²	—
I	=	9300 in.⁴	12,749 in.⁴
Y_b	=	13.27 in.	15.00 in.
Y_t	=	4.73 in.	5.00 in.
Z_b	=	701 in.³	850 in.³
Z_t	=	1966 in.³	2550 in.³
wt	=	358 plf	558 plf
		45 psf	70 psf
V/S	=	1.32 in.	

8DT18
No Topping

Table of safe superimposed service load (psf) and cambers

Strand Pattern	e_e / e_c	16	18	20	22	24	26	28	30	32	34	36	38	40	42	44	46	48	50	52	54	56	58	60
28-S	11.27 / 11.27	186 0.1 0.1	143 0.1 0.2	109 0.2 0.2	84 0.2 0.2	65 0.2 0.2	50 0.2 0.1	38 0.2 0.1																
48-S	10.27 / 10.27				185 0.5 0.6	150 0.5 0.7	123 0.6 0.7	101 0.6 0.8	83 0.7 0.8	69 0.7 0.8	57 0.7 0.8	47 0.7 0.7	39 0.7 0.7	31 0.7 0.6										
68-S	7.27 / 7.27				183 0.5 0.7	150 0.6 0.8	125 0.7 0.8	104 0.7 0.9	87 0.8 0.9	73 0.8 0.9	62 0.8 0.9	52 0.8 0.8	43 0.8 0.7	36 0.7 0.6	30 0.7 0.4									
68-D1	7.27 / 10.77							172 0.9 1.2	145 1.0 1.3	123 1.1 1.4	105 1.2 1.4	90 1.3 1.5	77 1.3 1.5	66 1.4 1.5	57 1.4 1.5	49 1.4 1.4	41 1.4 1.2	35 1.3 1.0	30 1.2 0.7					
88-D1	5.27 / 10.52												107 1.7 2.0	93 1.8 2.1	81 1.8 2.1	71 1.9 2.1	62 1.9 2.0	54 2.0 1.9	47 2.0 1.8	41 1.9 1.5	35 1.8 1.2	30 1.6 0.9		
108-D1	3.27 / 10.27																		62 2.3 2.3	55 2.3 2.1	48 2.3 1.9	43 2.2 1.6	37 2.1 1.2	33 2.0 0.8

8DT18 + 2
2" Normal Weight Topping

Table of safe superimposed service load (psf) and cambers

Strand Pattern	e_e / e_c	16	18	20	22	24	26	28	30	32	34	36	38	40	42	44	46	48
28-S	11.27 / 11.27	193 0.1 0.1	144 0.1 0.1	106 0.2 0.1	78 0.2 0.1	57 0.2 0.1	41 0.2 0.0											
48-S	10.27 / 10.27				192 0.5 0.5	153 0.5 0.5	122 0.6 0.5	98 0.6 0.5	78 0.7 0.5	62 0.7 0.4	49 0.7 0.2	37 0.7						
68-S	7.27 / 7.27				196 0.5 0.6	159 0.6 0.6	130 0.7 0.6	106 0.7 0.6	86 0.8 0.5	70 0.8 0.5	57 0.8 0.3	45 0.8 0.1						
68-D1	7.27 / 10.77							177 0.9 0.9	147 1.0 0.9	123 1.1 1.0	103 1.2 0.9	86 1.3 0.9	71 1.3 0.8	59 1.4 0.6	48 1.4 0.4	39 1.4 0.2		
88-D1	5.27 / 10.52												105 1.7 1.2	90 1.8 1.1	76 1.8 1.0	65 1.9 0.8	54 1.9 0.5	46 2.0 0.2

Strength based on strain compatibility; bottom tension limited to $12\sqrt{f'_c}$; see pages 2-3—2-5 for explanation

Values below heavy line require release strengths higher than 3500 psi.

DOUBLE TEE

8'-0" x 18"
Lightweight Concrete

No of strand (10)

S = straight   D = depressed

108-D1

No. of depression points

Diameter of strand in 16ths

Safe loads shown include dead load of 10 psf for untopped members and 15 psf for topped members. Remainder is live load. Long-time cambers include superimposed dead load but do not include live load.

Key

194 — Safe superimposed service load, psf
0.2 — Estimated camber at erection, in.
0.3 — Estimated long-time camber, in.

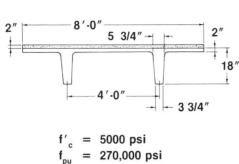

f'_c = 5000 psi
f_{pu} = 270,000 psi
Low-relaxation strand

Section Properties

	Untopped	Topped
A =	344 in.²	—
I =	9300 in.⁴	13,799 in.⁴
Y_b =	13.27 in.	15.51 in.
Y_t =	4.73 in.	4.49 in.
Z_b =	701 in.³	890 in.³
Z_t =	1966 in.³	3073 in.³
wt =	275 plf	475 plf
	34 psf	59 psf
V/S =	1.32 in.	

8LDT18

Table of safe superimposed service load (psf) and cambers
No Topping

Strand Pattern	e_e / e_c	Span, ft.																								
		16	18	20	22	24	26	28	30	32	34	36	38	40	42	44	46	48	50	52	54	56	58	60	62	64
28-S	11.27 / 11.27	194 / 0.2 / 0.3	151 / 0.2 / 0.3	118 / 0.3 / 0.3	93 / 0.3 / 0.4	74 / 0.4 / 0.4	59 / 0.4 / 0.4	47 / 0.4 / 0.4	38 / 0.4 / 0.4	30 / 0.4 / 0.3																
48-S	10.27 / 10.27				194 / 0.7 / 0.9	159 / 0.8 / 1.0	131 / 0.9 / 1.1	109 / 1.0 / 1.2	92 / 1.1 / 1.3	78 / 1.2 / 1.4	66 / 1.3 / 1.5	56 / 1.4 / 1.5	47 / 1.4 / 1.5	40 / 1.4 / 1.4	34 / 1.4 / 1.3											
68-S	7.27 / 7.27				191 / 0.8 / 1.1	159 / 0.9 / 1.2	133 / 1.1 / 1.3	113 / 1.2 / 1.4	96 / 1.3 / 1.5	82 / 1.4 / 1.6	70 / 1.4 / 1.6	60 / 1.5 / 1.6	52 / 1.5 / 1.6	45 / 1.6 / 1.5	38 / 1.6 / 1.4	33 / 1.5 / 1.2										
68-D1	7.27 / 10.77							180 / 1.4 / 1.8	154 / 1.6 / 2.0	132 / 1.7 / 2.2	114 / 1.9 / 2.3	99 / 2.1 / 2.5	86 / 2.2 / 2.6	75 / 2.4 / 2.6	65 / 2.5 / 2.6	57 / 2.6 / 2.6	50 / 2.6 / 2.5	44 / 2.6 / 2.4	38 / 2.6 / 2.2	33 / 2.6 / 2.0						
88-D1	5.27 / 10.52													90 / 3.1 / 3.6	80 / 3.2 / 3.6	71 / 3.4 / 3.7	63 / 3.5 / 3.6	56 / 3.6 / 3.5	49 / 3.7 / 3.4	44 / 3.8 / 3.1	39 / 3.7 / 2.8	35 / 3.7 / 2.4	31 / 3.5 / 1.9			
108-D1	3.27 / 10.27																							46 / 4.5 / 3.6	41 / 4.5 / 3.1	37 / 4.4 / 2.5	33 / 4.2 / 1.8

8LDT18+2

Table of safe superimposed service load (psf) and cambers
2" Normal Weight Topping

Strand Pattern	e_e / e_c	Span, ft.																	
		16	18	20	22	24	26	28	30	32	34	36	38	40	42	44	46	48	50	52
28-S	11.27 / 11.27	201 / 0.2 / 0.2	153 / 0.2 / 0.2	115 / 0.3 / 0.2	87 / 0.3 / 0.2	66 / 0.3 / 0.2	49 / 0.4 / 0.1	36 / 0.4 / 0.0												
48-S	10.27 / 10.27					161 / 0.8 / 0.8	130 / 0.9 / 0.8	106 / 1.0 / 0.9	87 / 1.1 / 0.9	70 / 1.2 / 0.9	57 / 1.3 / 0.8	46 / 1.4 / 0.7	37 / 1.4 / 0.5							
68-S	7.27 / 7.27					168 / 0.9 / 0.9	138 / 1.1 / 1.0	115 / 1.2 / 1.0	95 / 1.3 / 1.0	79 / 1.4 / 0.9	65 / 1.4 / 0.8	54 / 1.5 / 0.6	44 / 1.5 / 0.4	36 / 1.6 / 0.1						
68-D1	7.27 / 10.77							186 / 1.4 / 1.4	156 / 1.6 / 1.5	132 / 1.7 / 1.5	111 / 1.9 / 1.6	94 / 2.1 / 1.6	80 / 2.2 / 1.5	68 / 2.4 / 1.4	57 / 2.5 / 1.1	48 / 2.6 / 0.8	40 / 2.6 / 0.5	33 / 2.6 / 0.0		
88-D1	5.27 / 10.52													85 / 3.1 / 2.0	73 / 3.2 / 1.8	63 / 3.4 / 1.5	54 / 3.5 / 1.1	46 / 3.6 / 0.7	39 / 3.7 / 0.1	

Strength based on strain compatibility; bottom tension limited to $12\sqrt{f'_c}$; see pages 2-3–2-5 for explanation

Values below heavy line require release strengths higher than 3500 psi.

DOUBLE TEE

8'-0" x 20"
Normal Weight Concrete

No of strand (10)
S = straight   D = depressed
108-D1
No. of depression points
Diameter of strand in 16ths

Safe loads shown include dead load of 10 psf for untopped members and 15 psf for topped members. Remainder is live load. Long-time cambers include superimposed dead load but do not include live load.

Key

173 — Safe superimposed service load, psf
0.4 — Estimated camber at erection, in.
0.6 — Estimated long-time camber, in.

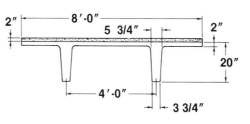

f'_c = 5000 psi
f_{pu} = 270,000 psi
Low-relaxation strand

Section Properties

	Untopped	Topped
A =	363 in.²	—
I =	12,551 in.⁴	16,935 in.⁴
Y_b =	14.59 in.	16.45 in.
Y_t =	5.41 in.	5.55 in.
Z_b =	860 in.³	1029 in.³
Z_t =	2320 in.³	3051 in.³
wt =	378 plf	578 plf
	47 psf	72 psf
V/S =	1.35 in.	

8DT20

Table of safe superimposed service load (psf) and cambers

No Topping

Strand Pattern	e_e / e_c	24	26	28	30	32	34	36	38	40	42	44	46	48	50	52	54	56	58	60	62	64	66	68
48-S	11.59 / 11.59	173 0.4 0.6	142 0.5 0.6	117 0.5 0.7	97 0.6 0.7	81 0.6 0.8	68 0.6 0.8	56 0.7 0.7	47 0.7 0.7	39 0.6 0.6	32 0.6 0.5													
68-S	8.59 / 8.59		180 0.5 0.7	150 0.6 0.8	126 0.7 0.8	106 0.7 0.9	90 0.8 0.9	76 0.8 0.9	65 0.8 0.9	55 0.8 0.8	46 0.8 0.8	39 0.7 0.6	32 0.7 0.5											
68-D1	8.59 / 12.09			199 0.8 1.0	168 0.9 1.1	143 1.0 1.2	123 1.1 1.3	106 1.1 1.4	91 1.2 1.4	78 1.2 1.4	68 1.3 1.4	58 1.3 1.4	50 1.3 1.3	43 1.2 1.2	37 1.2 1.0	31 1.1 0.7								
88-D1	6.59 / 11.84						166 1.3 1.7	144 1.4 1.8	126 1.5 1.9	110 1.6 1.9	96 1.7 2.0	84 1.8 2.0	74 1.8 2.0	65 1.8 1.9	57 1.9 1.8	50 1.8 1.7	44 1.8 1.5	38 1.7 1.3	33 1.6 1.0					
108-D1	4.59 / 11.59														75 2.2 2.4	67 2.3 2.4	59 2.3 2.2	52 2.3 2.0	46 2.2 1.8	41 2.2 1.5	36 2.0 1.2	31 1.8 0.7		
128-D1	2.92 / 11.34																			53 2.6 2.1	47 2.5 1.8	42 2.4 1.4	37 2.3 0.9	33 2.0 0.4

8DT20 + 2

Table of safe superimposed service load (psf) and cambers

2" Normal Weight Topping

Strand Pattern	e_e / e_c	24	26	28	30	32	34	36	38	40	42	44	46	48	50	52	54
48-S	11.59 / 11.59	176 0.4 0.5	141 0.5 0.5	114 0.5 0.5	92 0.6 0.5	74 0.6 0.4	59 0.6 0.3	47 0.6 0.2	36 0.6 0.1								
68-S	8.59 / 8.59		189 0.6 0.6	155 0.6 0.6	128 0.7 0.6	105 0.7 0.6	87 0.8 0.6	71 0.8 0.5	58 0.8 0.4	47 0.8 0.2	37 0.8 0.0						
68-D1	8.59 / 12.09				169 0.9 0.8	142 1.0 0.9	119 1.1 0.9	100 1.1 0.9	84 1.2 0.8	70 1.2 0.7	59 1.3 0.6	48 1.3 0.4	39 1.3 0.2				
88-D1	6.59 / 11.84						169 1.3 1.2	145 1.4 1.3	124 1.5 1.3	106 1.6 1.2	91 1.7 1.1	78 1.8 1.0	66 1.8 0.8	56 1.9 0.6	48 1.9 0.3	40 1.8 0.0	
108-D1	4.59 / 11.59														68 2.2 0.9	59 2.2 0.5	50 2.3 0.1

Strength based on strain compatibility; bottom tension limited to $12\sqrt{f'_c}$; see pages 2-3–2-5 for explanation

Values below heavy line require release strengths higher than 3500 psi.

DOUBLE TEE

8'-0" x 20"
Lightweight Concrete

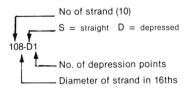

Strand Pattern Designation

108-D1

- No of strand (10)
- S = straight D = depressed
- No. of depression points
- Diameter of strand in 16ths

Safe loads shown include dead load of 10 psf for untopped members and 15 psf for topped members. Remainder is live load. Long-time cambers include superimposed dead load but do not include live load.

Key

- 182 — Safe superimposed service load, psf
- 0.7 — Estimated camber at erection, in.
- 0.9 — Estimated long-time camber, in.

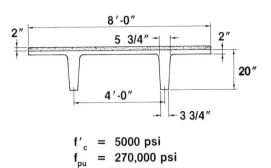

f'_c = 5000 psi
f_{pu} = 270,000 psi
Low-relaxation strand

Section Properties

	Untopped	Topped
A =	363 in.²	—
I =	12,551 in.⁴	18,278 in.⁴
Y_b =	14.59 in.	17.02 in.
Y_t =	5.41 in.	4.98 in.
Z_b =	860 in.³	1074 in.³
Z_t =	2320 in.³	3670 in.³
wt =	290 plf	490 plf
	36 psf	61 psf
V/S =	1.35 in.	

8LDT20

Table of safe superimposed service load (psf) and cambers — No Topping

Strand Pattern	e_e	e_c	24	26	28	30	32	34	36	38	40	42	44	46	48	50	52	54	56	58	60	62	64	66	68
48-S	11.59	11.59	182	151	126	107	90	77	65	56	48	41	35												
			0.7	0.8	0.9	1.0	1.0	1.1	1.2	1.2	1.3	1.3	1.3												
			0.9	1.0	1.1	1.2	1.3	1.3	1.4	1.4	1.4	1.3	1.2												
68-S	8.59	8.59		189	159	135	115	99	85	74	64	55	48	41	36	31									
				0.9	1.0	1.1	1.2	1.3	1.3	1.5	1.5	1.5	1.5	1.5	1.5	1.4									
				1.1	1.2	1.3	1.4	1.5	1.6	1.6	1.6	1.6	1.6	1.5	1.3	1.0									
68-D1	8.59	12.09			177	152	132	115	100	87	77	67	59	52	46	40	35	31							
					1.4	1.5	1.6	1.8	2.0	2.1	2.2	2.3	2.4	2.4	2.4	2.4	2.4	2.3							
					1.8	1.9	2.0	2.2	2.3	2.4	2.5	2.5	2.4	2.4	2.3	2.1	1.9	1.6							
88-D1	6.59	11.84							135	119	105	93	83	74	66	59	53	47	42	37	33				
									2.4	2.6	2.8	2.9	3.1	3.2	3.4	3.5	3.5	3.5	3.5	3.5	3.4				
									3.0	3.2	3.3	3.4	3.5	3.5	3.5	3.5	3.3	3.1	2.8	2.5	2.1				
108-D1	4.59	11.59																	62	55	50	45	41	36	33
																		4.3	4.3	4.4	4.4	4.4	4.3	4.2
																		4.2	4.1	3.8	3.5	3.0	2.5	1.9

8LDT20+2

Table of safe superimposed service load (psf) and cambers — 2" Normal Weight Topping

Strand Pattern	e_e	e_c	24	26	28	30	32	34	36	38	40	42	44	46	48	50	52	54	56	58
48-S	11.59	11.59	185	150	123	101	83	68	56	45	36									
			0.7	0.8	0.9	1.0	1.0	1.1	1.2	1.2	1.2									
			0.7	0.8	0.8	0.8	0.8	0.8	0.7	0.6	0.3									
68-S	8.59	8.59		198	164	137	114	96	80	67	56	46	38							
				0.9	1.0	1.1	1.2	1.3	1.4	1.4	1.5	1.5	1.5							
				0.9	0.9	1.0	1.0	1.0	0.9	0.8	0.7	0.5	0.3							
68-D1	8.59	12.09			178	151	128	109	93	79	68	57	48	41	34					
					1.4	1.5	1.6	1.8	2.0	2.1	2.2	2.3	2.4	2.4	2.4					
					1.4	1.4	1.4	1.5	1.5	1.4	1.3	1.1	0.9	0.6	0.2					
88-D1	6.59	11.84							133	115	100	87	76	66	57	49	42			
									2.4	2.6	2.8	2.9	3.1	3.2	3.4	3.5	3.5			
									2.1	2.1	2.0	2.0	1.8	1.6	1.3	0.9	0.4			
108-D1	4.59	11.59															52	45		
																4.3	4.3		
																0.8	-0.2		

Strength based on strain compatibility; bottom tension limited to $12\sqrt{f'_c}$; see pages 2-3—2-5 for explanation

Values below heavy line require release strengths higher than 3500 psi.

DOUBLE TEE

8'-0" x 24"
Normal Weight Concrete

Strand Pattern Designation

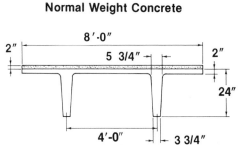

- No of strand (10)
- S = straight   D = depressed
- 108-D1
- No. of depression points
- Diameter of strand in 16ths

Safe loads shown include dead load of 10 psf for untopped members and 15 psf for topped members. Remainder is live load. Long-time cambers include superimposed dead load but do not include live load.

Key
171 — Safe superimposed service load, psf
0.6 — Estimated camber at erection, in.
0.7 — Estimated long-time camber, in.

f'_c = 5000 psi
f_{pu} = 270,000 psi
Low-relaxation strand

Section Properties

	Untopped	Topped
A =	401 in.²	—
I =	20,985 in.⁴	27,720 in.⁴
Y_b =	17.15 in.	19.27 in.
Y_t =	6.85 in.	6.73 in.
Z_b =	1224 in.³	1438 in.³
Z_t =	3063 in.³	4119 in.³
wt =	418 plf	618 plf
	52 psf	77 psf
V/S =	1.41 in.	

8DT24

Table of safe superimposed service load (psf) and cambers — No Topping

Strand Pattern	e_e / e_c	30	32	34	36	38	40	42	44	46	48	50	52	54	56	58	60	62	64	66	68	70	72	74
68-S	11.15 / 11.15	171	145	124	106	91	78	67	57	49	41	35												
		0.6	0.6	0.7	0.7	0.7	0.7	0.7	0.7	0.7	0.7	0.6												
		0.7	0.8	0.8	0.9	0.9	0.9	0.8	0.7	0.7	0.6	0.4												
88-S	9.15 / 9.15		176	152	131	113	98	85	74	64	56	48	41	35										
			0.7	0.7	0.8	0.8	0.8	0.9	0.9	0.9	0.8	0.8	0.7	0.6										
			0.9	0.9	1.0	1.0	1.0	1.0	1.0	0.9	0.8	0.7	0.5	0.3										
88-D1	9.15 / 14.40				187	163	143	126	111	98	87	77	68	60	53	47	41	36	31					
					1.1	1.2	1.3	1.4	1.5	1.5	1.6	1.6	1.6	1.6	1.6	1.5	1.4	1.3	1.2					
					1.5	1.5	1.6	1.7	1.8	1.8	1.8	1.8	1.7	1.6	1.5	1.4	1.2	0.9	0.5					
108-D1	7.15 / 14.15								142	127	113	101	90	81	72	64	57	51	46	40	36			
									1.7	1.8	1.9	2.0	2.0	2.1	2.1	2.1	2.1	2.0	2.0	1.8	1.7			
									2.2	2.2	2.3	2.3	2.3	2.3	2.2	2.1	2.0	1.8	1.6	1.3	1.0			
128-D1	5.48 / 13.90															81	73	66	59	53	48	43	38	34
																2.5	2.5	2.5	2.5	2.5	2.4	2.3	2.1	1.9
																2.7	2.7	2.5	2.3	2.1	1.8	1.5	1.1	0.6
148-D1	4.29 / 13.65																				60	54	49	44
																					2.9	2.9	2.8	2.6
																					2.6	2.3	1.9	1.5

8DT24+2

Table of safe superimposed service load (psf) and cambers — 2" Normal Weight Topping

Strand Pattern	e_e / e_c	26	28	30	32	34	36	38	40	42	44	46	48	50	52	54	56	58	60	62	64	66
48-S	14.15 / 14.15	180	147	120	98	80	65	52	41	32												
		6.4	0.4	0.4	0.5	0.5	0.5	0.5	0.5	0.5												
		0.4	0.4	0.4	0.4	0.3	0.3	0.2	0.1	0.0												
68-S	11.15 / 11.15			171	143	120	100	84	70	58	47	38										
				0.6	0.6	0.7	0.7	0.7	0.7	0.7	0.7	0.7										
				0.6	0.6	0.6	0.6	0.5	0.5	0.4	0.2	0.0										
68-D1	11.15 / 14.65					180	153	130	110	93	79	67	56	46	38							
						0.8	0.8	0.9	1.0	1.0	1.0	1.1	1.1	1.1	1.1							
						0.8	0.8	0.8	0.8	0.8	0.7	0.6	0.5	0.4	0.1							
88-D1	9.15 / 14.40							186	161	139	121	105	91	78	67	58	49	41	34			
								1.1	1.2	1.3	1.4	1.5	1.5	1.6	1.6	1.6	1.6	1.6	1.5			
								1.1	1.1	1.2	1.2	1.1	1.1	0.9	0.9	0.6	0.4	0.1	-0.2			
108-D1	7.15 / 14.15										139	122	107	94	82	72	62	54	46	40		
											1.7	1.8	1.9	2.0	2.0	2.1	2.1	2.1	2.1	2.0		
											1.5	1.5	1.4	1.3	1.2	1.0	1.0	0.5	0.1	-0.3		
128-D1	5.48 / 13.90																	72	64	56	48	41
																		2.5	2.5	2.5	2.5	2.5
																		1.0	0.7	0.3	-0.1	-0.6

Strength based on strain compatibility; bottom tension limited to $12\sqrt{f'_c}$; see pages 2-3—2-5 for explanation
Values below heavy line require release strengths higher than 3500 psi.

PCI Design Handbook

DOUBLE TEE

8'-0" x 24"
Normal Weight Concrete

Strand Pattern Designation

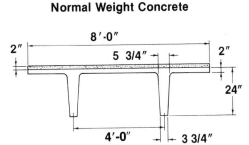

No of strand (10)

S = straight D = depressed

108-D1

No. of depression points

Diameter of strand in 16ths

Safe loads shown include dead load of 10 psf for untopped members and 15 psf for topped members. Remainder is live load. Long-time cambers include superimposed dead load but do not include live load.

Key

171 — Safe superimposed service load, psf
0.5 — Estimated camber at erection, in.
0.6 — Estimated long-time camber, in.

f'_c = 5000 psi
f_{pu} = 270,000 psi
Stress-relieved strand

Section Properties

	Untopped	Topped
A =	401 in.²	—
I =	20,985 in.⁴	27,720 in.⁴
Y_b =	17.15 in.	19.27 in.
Y_t =	6.85 in.	6.73 in.
Z_b =	1224 in.³	1438 in.³
Z_t =	3063 in.³	4119 in.³
wt =	418 plf	618 plf
	52 psf	77 psf
V/S =	1.41 in.	

8DT24

Table of safe superimposed service load (psf) and cambers — No Topping

Strand Pattern	e_e / e_c	30	32	34	36	38	40	42	44	46	48	50	52	54	56	58	60	62	64	66	68	70	72	74
68-S	11.15 / 11.15	171 0.5 0.6	145 0.5 0.7	124 0.6 0.7	106 0.6 0.7	91 0.6 0.7	78 0.6 0.7	67 0.6 0.6	57 0.6 0.6	49 0.6 0.5	41 0.5 0.3	35 0.5 0.2												
88-S	9.15 / 9.15		176 0.6 0.7	152 0.6 0.8	131 0.7 0.8	113 0.7 0.8	98 0.7 0.8	85 0.7 0.8	74 0.7 0.7	64 0.7 0.7	56 0.7 0.6	48 0.7 0.4	41 0.6 0.2	35 0.5 0.0										
88-D1	9.15 / 14.40			187 1.0 1.2	163 1.1 1.3	143 1.2 1.4	126 1.2 1.4	111 1.3 1.4	98 1.3 1.4	87 1.4 1.4	77 1.4 1.4	68 1.4 1.3	60 1.3 1.3	53 1.3 1.1	47 1.2 1.0	41 1.1 0.8	36 1.0 0.5	0.3						
108-D1	7.15 / 14.15					181 1.4 1.7	160 1.5 1.7	142 1.5 1.8	127 1.6 1.8	113 1.7 1.8	101 1.7 1.8	89 1.8 1.8	79 1.8 1.7	70 1.8 1.6	62 1.8 1.5	56 1.7 1.3	50 1.6 1.0	45 1.5 0.8	40 1.4 0.4	35 1.2 0.0				
128-D1	5.48 / 13.90												106 2.1 2.2	95 2.1 2.2	84 2.2 2.1	75 2.2 2.0	67 2.2 1.8	59 2.1 1.6	53 2.1 1.4	47 2.0 1.1	42 1.9 0.8	38 1.7 0.4		
148-D1	4.29 / 13.65																		63 2.5 2.0	56 2.5 1.8	50 2.4 1.5	44 2.3 1.1	40 2.2 0.7	36 2.0 0.2

8DT24+2

Table of safe superimposed service load (psf) and cambers — 2" Normal Weight Topping

Strand Pattern	e_e / e_c	26	28	30	32	34	36	38	40	42	44	46	48	50	52	54	56	58	60	62	64	66
48-S	14.15 / 14.15	180 0.3 0.3	147 0.4 0.3	120 0.4 0.3	98 0.4 0.3	80 0.4 0.3	65 0.5 0.2	52 0.5 0.2	41 0.4 0.1													
68-S	11.15 / 11.15			171 0.5 0.5	143 0.5 0.5	120 0.6 0.5	100 0.6 0.4	84 0.6 0.4	70 0.6 0.3	58 0.6 0.2	47 0.6 0.0	38 0.6 -0.2										
68-D1	11.15 / 14.65			181 0.6 0.5	153 0.6 0.5	130 0.7 0.5	111 0.7 0.5	94 0.7 0.4	80 0.7 0.3	67 0.7 0.2	56 0.7 0.0	47 0.7 -0.2	38 0.6 -0.5									
88-D1	9.15 / 14.40				186 1.0 0.9	161 1.1 1.0	139 1.2 0.9	121 1.2 0.9	105 1.3 0.8	91 1.3 0.7	78 1.4 0.6	67 1.4 0.4	58 1.4 0.2	48 1.3 -0.1	39 1.3 -0.4							
108-D1	7.15 / 14.15						181 1.4 1.2	158 1.5 1.2	139 1.5 1.2	121 1.6 1.1	104 1.7 1.0	89 1.7 0.9	75 1.8 0.7	63 1.8 0.5	53 1.8 0.2	45 1.8 -0.1	37 1.7 -0.5					
128-D1	5.48 / 13.90												95 2.1 1.1	82 2.1 0.9	70 2.2 0.7	59 2.2 0.4	49 2.2 0.0	40 2.1 -0.4				

Strength based on strain compatibility; bottom tension limited to $12\sqrt{f'_c}$; see pages 2-3–2-5 for explanation
Values below heavy line require release strengths higher than 3500 psi.

DOUBLE TEE

8'-0" x 24"
Lightweight Concrete

Strand Pattern Designation

```
        ┌── No of strand (10)
        │  ┌── S = straight  D = depressed
        ▼  ▼
      108-D1
        ▲  ▲
        │  └── No. of depression points
        └───── Diameter of strand in 16ths
```

Safe loads shown include dead load of 10 psf for untopped members and 15 psf for topped members. Remainder is live load. Long-time cambers include superimposed dead load but do not include live load.

Key

116 — Safe superimposed service load, psf
1.1 — Estimated camber at erection, in.
1.4 — Estimated long-time camber, in.

f'_c = 5000 psi
f_{pu} = 270,000 psi
Low-relaxation strand

Section Properties

		Untopped	Topped
A	=	401 in.²	—
I	=	20,985 in.⁴	29,853 in.⁴
Y_b	=	17.15 in.	19.94 in.
Y_t	=	6.85 in.	6.06 in.
Z_b	=	1224 in.³	1497 in.³
Z_t	=	3063 in.³	4926 in.³
wt	=	320 plf	520 plf
		40 psf	65 psf
V/S	=	1.41 in.	

8LDT24

Table of safe superimposed service load (psf) and cambers

No Topping

Strand Pattern	e_e / e_c	36	38	40	42	44	46	48	50	52	54	56	58	60	62	64	66	68	70	72	74	76	78	80
														Span, ft.										
68-S	11.15	116	101	88	77	67	59	52	45	39	34	30												
	11.15	1.1	1.2	1.3	1.4	1.4	1.4	1.4	1.4	1.4	1.3	1.3												
		1.4	1.5	1.5	1.6	1.6	1.6	1.5	1.4	1.3	1.1	0.8												
88-S	9.15	141	123	108	95	84	74	66	58	51	45	40	35	31										
	9.15	1.2	1.3	1.4	1.5	1.6	1.6	1.6	1.7	1.7	1.6	1.6	1.5	1.4										
		1.6	1.6	1.7	1.8	1.8	1.8	1.8	1.7	1.6	1.5	1.3	1.0	0.7										
88-D1	9.15	197	173	153	136	121	101	97	87	78	70	63	57	51	46	41	37	33	29					
	14.40	1.8	1.9	2.1	2.2	2.4	2.5	2.7	2.8	2.9	3.0	3.0	3.1	3.1	3.0	3.0	2.9	2.8	2.6					
		2.3	2.4	2.6	2.8	2.9	3.0	3.1	3.2	3.2	3.2	3.1	2.9	2.8	2.6	2.4	2.1	1.8	1.3					
108-D1	7.15								111	100	91	82	74	68	61	56	50	46	41	37	34	30		
	14.15								3.3	3.4	3.6	3.7	3.8	3.9	4.0	4.1	4.1	4.0	3.9	3.8	3.6	3.4		
									3.9	4.0	4.1	4.1	4.1	4.0	3.9	3.8	3.5	3.2	2.8	2.4	1.9	1.3		
128-D1	5.48													63	58	53	48	44	40	37	33			
	13.90													4.8	4.9	4.9	4.9	4.8	4.7	4.5	4.3			
														4.7	4.5	4.2	3.8	3.4	2.8	2.2	1.5			
148-D1	4.29																						46	42
	13.65																						5.6	5.5
																							3.8	3.2

8LDT24+2

Table of safe superimposed service load (psf) and cambers

2" Normal Weight Topping

Strand Pattern	e_e / e_c	26	28	30	32	34	36	38	40	42	44	46	48	50	52	54	56	58	60	62
											Span, ft.									
48-S	14.15	190	157	130	108	90	75	62	51	42	33									
	14.15	0.6	0.7	0.7	0.8	0.8	0.9	0.9	1.0	1.0	1.0									
		0.6	0.7	0.7	0.7	0.7	0.6	0.6	0.5	0.4	0.2									
68-S	11.15			181	153	130	110	94	80	68	57	48	40	33						
	11.15			0.9	1.0	1.1	1.2	1.2	1.3	1.4	1.4	1.4	1.4	1.4						
				0.9	0.9	1.0	1.0	1.0	0.9	0.9	0.7	0.6	0.4	0.1						
88-S	9.15			191	163	140	121	104	90	77	66	57	48	41						
	9.15			1.1	1.2	1.2	1.3	1.4	1.5	1.6	1.6	1.6	1.7	1.7						
				1.0	1.1	1.1	1.1	1.1	1.0	0.9	0.8	0.6	0.4	0.1						
68-D1	11.15			190	163	140	120	103	89	77	66	56	48	40	34					
	14.65			1.2	1.2	1.4	1.6	1.7	1.8	1.9	1.9	2.0	2.0	2.1	2.1					
				1.2	1.2	1.3	1.4	1.4	1.3	1.2	1.1	1.0	0.8	0.6	0.2					
88-D1	9.15						196	171	149	131	115	101	88	77	68	59	51	44	38	
	14.40						1.8	1.9	2.1	2.2	2.4	2.5	2.7	2.8	2.9	3.0	3.1	3.1		
							1.8	1.9	1.9	2.0	2.0	1.9	1.9	1.7	1.6	1.3	1.0	0.6	0.2	
108-D1	7.15												104	92	82	72	64	56	50	43
	14.15												3.3	3.4	3.6	3.7	3.8	3.9	4.0	4.1
													2.4	2.3	2.1	1.9	1.6	1.3	0.8	0.3

Strength based on strain compatibility; bottom tension limited to $12\sqrt{f'_c}$; see pages 2-3—2-5 for explanation

Values below heavy line require release strengths higher than 3500 psi.

DOUBLE TEE

8'-0" x 32"
Normal Weight Concrete

Strand Pattern Designation

No of strand (10)

S = straight   D = depressed

108-D1

No. of depression points

Diameter of strand in 16ths

Safe loads shown include dead load of 10 psf for untopped members and 15 psf for topped members. Remainder is live load. Long-time cambers include superimposed dead load but do not include live load.

Key

182 — Safe superimposed service load, psf
1.4 — Estimated camber at erection, in.
1.8 — Estimated long-time camber, in.

f'_c = 5000 psi
f_{pu} = 270,000 psi
Low-relaxation strand

Section Properties

	Untopped	Topped
A =	567 in.²	—
I =	55,464 in.⁴	71,886 in.⁴
Y_b =	21.21 in.	23.66 in.
Y_t =	10.79 in.	10.34 in.
Z_b =	2615 in.³	3038 in.³
Z_t =	5140 in.³	6952 in.³
wt =	591 plf	791 plf
	74 psf	99 psf
V/S =	1.79 in.	

8DT32
No Topping

Table of safe superimposed service load (psf) and cambers

Strand Pattern	e_e / e_c	50	52	54	56	58	60	62	64	66	68	70	72	74	76	78	80	82	84	86	88	90	92	94
128-D1	12.04 / 17.96	182	164	147	133	120	108	98	88	79	71	64	57	51	45	40								
		1.4	1.5	1.5	1.5	1.6	1.6	1.6	1.5	1.5	1.5	1.4	1.3	1.2	1.0	0.8								
		1.8	1.8	1.8	1.8	1.8	1.8	1.8	1.7	1.6	1.5	1.3	1.1	0.8	0.5	0.2								
148-D1	9.71 / 17.71		193	175	159	144	131	119	108	98	89	80	73	66	59	53	48	42						
			1.6	1.7	1.7	1.8	1.8	1.9	1.9	1.8	1.8	1.8	1.7	1.6	1.5	1.3	1.2	1.0						
			2.0	2.1	2.1	2.2	2.2	2.1	2.1	2.0	1.9	1.8	1.6	1.5	1.2	0.9	0.6	0.1						
168-D1	8.21 / 17.46				184	168	153	139	127	116	106	97	88	80	73	66	60	54	49	44				
					1.9	2.0	2.0	2.1	2.1	2.1	2.2	2.1	2.1	2.0	2.0	1.8	1.7	1.5	1.3	1.1				
					2.4	2.5	2.5	2.5	2.5	2.4	2.4	2.3	2.1	2.0	1.8	1.5	1.3	1.0	0.6	0.1				
188-D1	6.82 / 17.21						174	159	146	133	122	112	103	94	86	79	72	66	60	54	49			
							2.2	2.3	2.3	2.4	2.4	2.4	2.4	2.4	2.3	2.3	2.2	2.0	1.8	1.6	1.4			
							2.8	2.8	2.8	2.8	2.7	2.7	2.6	2.4	2.2	2.0	1.8	1.5	1.2	0.8	0.4			
208-D1	5.71 / 16.96											127	117	107	99	91	83	77	70	64	59	54	49	
												2.6	2.7	2.7	2.6	2.6	2.5	2.4	2.3	2.1	1.9	1.7	1.4	
												3.0	3.0	2.9	2.7	2.5	2.3	2.0	1.7	1.4	1.0	0.5	1.0	
228-D1	4.80 / 16.71															103	95	87	80	74	67	62	57	52
																2.9	2.8	2.8	2.7	2.5	2.4	2.2	2.0	1.7
																3.0	2.8	2.5	2.2	1.9	1.5	1.1	0.6	0.1

8DT32 + 2
2" Normal Weight Topping

Table of safe superimposed service load (psf) and cambers

Strand Pattern	e_e / e_c	44	46	48	50	52	54	56	58	60	62	64	66	68	70	72	74	76	78	80
108-D1	15.31 / 18.21	200	176	155	137	121	106	93	81	71	61	53	45							
		1.1	1.1	1.2	1.2	1.2	1.3	1.3	1.3	1.3	1.2	1.2	1.1							
		1.1	1.1	1.1	1.1	1.0	1.0	0.9	0.8	0.7	0.5	0.3	0.0							
128-D1	12.04 / 17.96			198	176	157	140	125	111	98	87	77	68	59						
				1.3	1.4	1.5	1.5	1.5	1.6	1.6	1.6	1.5	1.5	1.5						
				1.3	1.3	1.3	1.3	1.2	1.1	1.0	0.9	0.7	0.5	0.3						
148-D1	9.71 / 17.71					188	169	152	136	122	109	98	87	78	69	61				
						1.6	1.7	1.7	1.8	1.8	1.9	1.9	1.8	1.8	1.8	1.7				
						1.5	1.5	1.5	1.4	1.3	1.2	1.1	0.9	0.7	0.4	0.1				
168-D1	8.21 / 17.46					198	178	161	145	131	118	106	96	86	77	68				
						1.8	1.9	2.0	2.0	2.1	2.1	2.1	2.2	2.1	2.1	2.1				
						1.8	1.7	1.7	1.6	1.6	1.4	1.3	1.1	0.9	0.6	0.3				
188-D1	6.82 / 17.21								167	151	137	124	112	101	92	83	74	66		
									2.2	2.3	2.3	2.4	2.4	2.4	2.4	2.4	2.3	2.3		
									1.9	1.8	1.7	1.6	1.4	1.2	1.0	0.7	0.4	0.0		
208-D1	5.71 / 16.96														117	106	96	87	79	71
															2.6	2.7	2.7	2.6	2.6	2.5
															1.7	1.4	1.1	0.8	0.4	0.0

Strength based on strain compatibility; bottom tension limited to $12\sqrt{f'_c}$; see pages 2-3–2-5 for explanation

Values below heavy line require release strengths higher than 3500 psi.

DOUBLE TEE

8'-0" x 32"
Lightweight Concrete

Strand Pattern Designation

No of strand (10)
S = straight   D = depressed

108-D1

No. of depression points
Diameter of strand in 16ths

Safe loads shown include dead load of 10 psf for untopped members and 15 psf for topped members. Remainder is live load. Long-time cambers include superimposed dead load but do not include live load.

Key
147 — Safe superimposed service load, psf
2.7 — Estimated camber at erection, in.
3.2 — Estimated long-time camber, in.

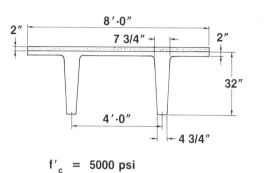

f'_c = 5000 psi
f_{pu} = 270,000 psi
Low-relaxation strand

Section Properties

	Untopped	Topped
A =	567 in.²	—
I =	55,464 in.⁴	77,617 in.⁴
Y_b =	21.21 in.	24.55 in.
Y_t =	10.79 in.	9.45 in.
Z_b =	2615 in.³	3167 in.³
Z_t =	5140 in.³	8213 in.³
wt =	453 plf	653 plf
	57 psf	82 psf
V/S =	1.79 in.	

8LDT32

Table of safe superimposed service load (psf) and cambers

No Topping

Strand Pattern	e_e / e_c	56	58	60	62	64	66	68	70	72	74	76	78	80	82	84	86	88	90	92	94	96	98	100
128-D1	12.04	147	134	122	112	102	93	85	78	71	65	59	54	49	45	40	37	33						
	17.96	2.7	2.7	2.8	2.9	2.9	3.0	3.0	3.0	3.0	2.9	2.8	2.7	2.6	2.5	2.3	2.0	1.8						
		3.2	3.3	3.3	3.3	3.3	3.2	3.1	3.0	2.9	2.8	2.6	2.3	2.0	1.7	1.3	0.8	0.2						
148-D1	9.71	173	158	145	133	122	112	103	94	87	80	73	67	62	57	52	47	43	39	36				
	17.71	2.9	2.9	3.1	3.2	3.3	3.4	3.5	3.6	3.6	3.5	3.5	3.4	3.3	3.2	3.1	2.9	2.7	2.4	2.1				
		3.6	3.5	3.8	3.8	3.9	3.9	3.9	3.8	3.7	3.5	3.3	3.1	2.8	2.5	2.2	1.8	1.4	0.8	0.3				
168-D1	8.21			182	167	153	141	130	120	111	102	94	87	80	74	69	63	58	54	49	45	41	38	
	17.46			3.3	3.4	3.6	3.7	3.8	3.9	4.0	4.0	4.1	4.1	4.0	4.1	4.0	3.9	3.7	3.6	3.4	3.1	2.8	2.5	
				4.1	4.2	4.3	4.4	4.5	4.5	4.5	4.4	4.3	4.2	4.0	3.8	3.5	3.1	2.7	2.3	1.9	1.4	0.8	0.1	
188-D1	6.82								126	117	108	100	93	86	80	74	68	63	59	54	50	46	42	
	17.21								4.3	4.4	4.5	4.6	4.6	4.6	4.6	4.5	4.5	4.4	4.2	4.0	3.7	3.4	3.1	
									5.0	5.0	5.0	4.9	4.7	4.6	4.3	4.1	3.7	3.3	2.8	2.2	1.7	1.1	0.4	
208-D1	5.71													98	91	84	78	73	67	62	57	52	48	
	16.96													5.1	5.1	5.1	5.0	5.0	4.9	4.7	4.5	4.3	4.0	
														5.3	5.1	4.9	4.6	4.2	3.8	3.3	2.8	2.1	1.4	
228-D1	4.80																	81	75	69	64	59	54	50
	16.71																	5.5	5.5	5.4	5.2	5.0	4.8	4.5
																		5.1	4.7	4.3	3.8	3.2	2.6	1.8

8LDT32 + 2

Table of safe superimposed service load (psf) and cambers

2" Normal Weight Topping

Strand Pattern	e_e / e_c	46	48	50	52	54	56	58	60	62	64	66	68	70	72	74	76	78	80	82
108-D1	15.31	190	169	151	135	135	120	107	96	85	76	67	59	52						
	18.21	1.8	1.8	2.0	2.1	2.1	2.2	2.3	2.3	2.4	2.4	2.4	2.4	2.4						
		1.8	1.7	1.9	1.9	1.8	1.8	1.7	1.7	1.6	1.4	1.2	1.0	0.8						
128-D1	12.04			191	171	154	139	125	113	101	91	82	73	66	58					
	17.96			2.3	2.4	2.4	2.7	2.7	2.8	2.9	2.9	3.0	3.0	3.0	3.0					
				2.2	2.2	2.2	2.3	2.2	2.1	2.0	1.8	1.6	1.4	1.1	0.8					
148-D1	9.71					183	166	150	136	123	112	101	92	83	75	67				
	17.71					2.8	2.9	3.1	3.2	3.3	3.4	3.4	3.5	3.6	3.6	3.5				
						2.6	2.6	2.4	2.5	2.5	2.4	2.2	2.0	1.8	1.5	1.1				
168-D1	8.21						175	159	145	132	120	110	100	91	83	75	68			
	17.46						3.3	3.4	3.6	3.7	3.8	3.9	4.0	4.0	4.1	4.1	4.1			
							3.0	3.0	2.9	2.8	2.7	2.6	2.4	2.1	1.8	1.5	1.1			
188-D1	6.82											116	106	97	88	81	73			
	17.21											4.3	4.4	4.5	4.6	4.6	4.6			
												2.9	2.7	2.4	2.1	1.7	1.3			
208-D1	5.71																		85	78
	16.96																		5.1	5.1
																			1.9	1.5

Strength based on strain compatibility; bottom tension limited to $12\sqrt{f'_c}$; see pages 2-3-2-5 for explanation

Values below heavy line require release strengths higher than 3500 psi.

DOUBLE TEE

10'-0'' x 24''
Normal Weight Concrete

Strand Pattern Designation

- No of strand (10)
- S = straight   D = depressed
- 108-D1
- No. of depression points
- Diameter of strand in 16ths

Safe loads shown include dead load of 10 psf for untopped members and 15 psf for topped members. Remainder is live load. Long-time cambers include superimposed dead load but do not include live load.

Key

- 134 — Safe superimposed service load, psf
- 0.5 — Estimated camber at erection, in.
- 0.7 — Estimated long-time camber, in.

f'_c = 5000 psi
f_{pu} = 270,000 psi
Low-relaxation strand

Section Properties

	Untopped	Topped
A =	449 in.²	—
I =	22,469 in.⁴	29,404 in.⁴
Y_b =	17.77 in.	19.89 in.
Y_t =	6.23 in.	6.11 in.
Z_b =	1264 in.³	1478 in.³
Z_t =	3607 in.³	4812 in.³
wt =	468 plf	718 plf
	47 psf	72 psf
V/S =	1.35 in.	

10DT24

Table of safe superimposed service load (psf) and cambers — No Topping

Strand Pattern	e_e / e_c	30	32	34	36	38	40	42	44	46	48	50	52	54	56	58	60	62	64	66	68	70	72	74
68-S	11.77	134	113	96	82	70	59	50	43	36	30													
	11.77	0.5	0.6	0.6	0.7	0.7	0.7	0.7	0.7	0.6	0.6													
		0.7	0.7	0.8	0.8	0.8	0.8	0.7	0.7	0.6	0.4													
88-S	9.77	163	139	119	102	88	75	65	56	48	41	35	30											
	9.77	0.6	0.7	0.7	0.7	0.8	0.8	0.8	0.8	0.8	0.8	0.7	0.6											
		0.8	0.8	0.9	0.9	0.9	0.9	0.9	0.9	0.8	0.7	0.5	0.3											
88-D1	9.77		196	169	147	128	112	98	86	76	67	58	51	45	39	34								
	15.02		0.9	1.0	1.1	1.2	1.3	1.3	1.4	1.4	1.5	1.5	1.5	1.5	1.4	1.3								
			1.2	1.3	1.4	1.5	1.5	1.6	1.6	1.6	1.6	1.6	1.5	1.4	1.2	1.0								
108-D1	7.77						142	125	111	99	87	78	69	61	55	50	43	38	33					
	14.77						1.5	1.6	1.7	1.7	1.8	1.9	1.9	2.0	2.0	1.9	1.9	1.8	1.7					
							1.9	2.0	2.0	2.1	2.1	2.1	2.1	2.0	1.9	1.8	1.6	1.4	1.1					
128-D1	6.10													77	69	62	55	50	44	39	35	31		
	14.52													2.3	2.4	2.4	2.4	2.4	2.3	2.3	2.1	2.0		
														2.6	2.5	2.4	2.3	2.1	1.9	1.6	1.3	0.9		
148-D1	4.91																			50	45	40	36	32
	14.27																			2.8	2.7	2.6	2.5	2.4
																				2.3	2.0	1.7	1.3	0.9

10DT24 + 2

Table of safe superimposed service load (psf) and cambers — 2" Normal Weight Topping

Strand Pattern	e_e / e_c	24	26	28	30	32	34	36	38	40	42	44	46	48	50	52	54	56	58	60	62
48-S	14.77	171	138	111	89	72	57	45	34												
	14.77	0.3	0.3	0.4	0.4	0.4	0.4	0.5	0.5												
		0.3	0.3	0.3	0.3	0.2	0.2	0.1													
68-S	11.77		193	158	131	108	89	74	60	49	39	31									
	11.77		0.4	0.5	0.5	0.6	0.6	0.7	0.7	0.7	0.7	0.7									
			0.5	0.5	0.5	0.5	0.5	0.5	0.4	0.3	0.2	0.0									
68-D1	11.77			198	165	138	116	97	81	68	56	46	38	30							
	15.27			0.6	0.7	0.7	0.8	0.9	0.9	0.9	1.0	1.0	1.0	1.0							
				0.6	0.7	0.7	0.7	0.7	0.7	0.6	0.5	0.4	0.3	0.0							
88-D1	9.77					196	167	143	122	105	90	77	65	56	47	39					
	15.02					0.9	1.0	1.1	1.2	1.3	1.3	1.4	1.4	1.5	1.5	1.5					
						0.9	1.0	1.0	1.0	1.0	1.0	0.9	0.8	0.7	0.5	0.2					
108-D1	7.77									138	120	105	91	79	68	59	50	43			
	14.77									1.5	1.6	1.7	1.7	1.8	1.9	1.9	2.0	2.0			
										1.3	1.3	1.3	1.2	1.1	1.0	0.8	0.5	0.2			
128-D1	6.10																68	59	51	44	38
	14.52																2.3	2.4	2.4	2.4	2.4
																	1.0	0.8	0.4	0.0	-0.4

Strength based on strain compatibility; bottom tension limited to $12\sqrt{f'_c}$; see pages 2-3–2-5 for explanation

Values below heavy line require release strengths higher than 3500 psi.

DOUBLE TEE

10'-0'' x 24''
Lightweight Concrete

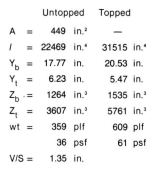

No of strand (10)
S = straight   D = depressed
108-D1
No. of depression points
Diameter of strand in 16ths

Safe loads shown include dead load of 10 psf for untopped members and 15 psf for topped members. Remainder is live load. Long-time cambers include superimposed dead load but do not include live load.

Key
122 — Safe superimposed service load, psf
1.0 — Estimated camber at erection, in.
1.2 — Estimated long-time camber, in.

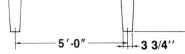
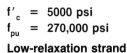

2'' |—— 10'-0'' ——| 2''
5 3/34''
24''
5'-0''   3 3/4''

f'_c = 5000 psi
f_{pu} = 270,000 psi
Low-relaxation strand

Section Properties

	Untopped	Topped
A =	449 in.²	—
I =	22469 in.⁴	31515 in.⁴
Y_b =	17.77 in.	20.53 in.
Y_t =	6.23 in.	5.47 in.
Z_b =	1264 in.³	1535 in.³
Z_t =	3607 in.³	5761 in.³
wt =	359 plf	609 plf
	36 psf	61 psf
V/S =	1.35 in.	

10LDT24

Table of safe superimposed service load (psf) and cambers — No Topping

Strand Pattern	e_e / e_c	32	34	36	38	40	42	44	46	48	50	52	54	56	58	60	62	64	66	68	70	72	74	76
68-S	11.77	122	105	91	79	68	59	52	45	39	34													
	11.77	1.0	1.0	1.1	1.2	1.2	1.3	1.3	1.3	1.3	1.3													
		1.2	1.3	1.3	1.4	1.4	1.4	1.4	1.4	1.3	1.1													
88-S	9.77	148	128	111	97	84	74	65	57	50	44	39	34											
	9.77	1.0	1.1	1.2	1.3	1.4	1.5	1.5	1.6	1.6	1.6	1.6	1.5											
		1.3	1.4	1.5	1.6	1.6	1.7	1.7	1.7	1.6	1.5	1.4	1.2											
88-D1	9.77			178	156	137	121	107	95	85	76	67	60	54	48	43	38	34	30					
	15.02			1.6	1.7	1.9	2.0	2.2	2.3	2.4	2.6	2.7	2.7	2.8	2.8	2.8	2.8	2.8	2.7					
				2.0	2.2	2.3	2.5	2.6	2.7	2.8	2.9	2.9	2.8	2.7	2.6	2.5	2.3	2.1	1.8					
108-D1	7.77								107	96	87	78	70	63	57	52	47	42	34	31				
	14.77								2.8	3.0	3.2	3.3	3.4	3.5	3.6	3.7	3.8	3.8	2.4	2.0				
									3.4	3.6	3.6	3.7	3.7	3.7	3.6	3.5	3.3	3.1	2.4	2.0				
128-D1	6.10																59	53	48	44	40	36	33	30
	14.52																4.4	4.5	4.6	4.6	4.6	4.5	4.4	4.2
																	4.4	4.2	4.0	3.7	3.3	2.8	2.3	1.7
148-D1	4.91																						41	37
	14.27																						5.4	5.3
																							3.8	3.2

10LDT24+2

Table of safe superimposed service load (psf) and cambers — 2'' Normal Weight Topping

Strand Pattern	e_e / e_c	24	26	28	30	32	34	36	38	40	42	44	46	48	50	52	54	56	58	60	62	64
48-S	14.77	180	147	120	98	81	66	54	43	35												
	14.77	0.5	0.6	0.6	0.7	0.7	0.8	0.8	0.9	0.9												
		0.5	0.5	0.6	0.6	0.6	0.5	0.5	0.4	0.3												
68-S	11.77			167	140	117	98	83	69	58	48	40	32									
	11.77			0.8	0.9	1.0	1.0	1.1	1.2	1.3	1.3	1.3	1.3									
				0.8	0.8	0.9	0.9	0.9	0.8	0.7	0.6	0.5	0.3									
68-D1	11.77				174	147	125	106	90	77	65	55	47	39	32							
	15.27				1.0	1.1	1.3	1.4	1.5	1.6	1.7	1.8	1.8	1.9	1.9							
					1.0	1.1	1.2	1.2	1.2	1.1	1.1	1.0	0.8	0.6	0.3							
88-D1	9.77						176	152	131	114	99	86	74	65	56	48	41	35				
	15.02						1.6	1.7	1.9	2.0	2.2	2.3	2.4	2.6	2.7	2.7	2.8	2.8				
							1.5	1.6	1.7	1.7	1.7	1.7	1.6	1.5	1.3	1.0	0.6	0.3				
108-D1	7.77												100	88	77	68	59	52	45	39		
	14.77												2.8	3.0	3.2	3.3	3.4	3.5	3.6	3.7		
													2.2	2.1	2.0	1.8	1.5	1.2	0.8	0.4		
128-D1	6.10																				47	41
	14.52																				4.4	4.5
																					0.7	0.1

Strength based on strain compatibility; bottom tension limited to $12\sqrt{f'_c}$; see pages 2-3-2-5 for explanation

Values below heavy line require release strengths higher than 3500 psi.

DOUBLE TEE

10'-0'' x 32''
Normal Weight Concrete

Strand Pattern Designation

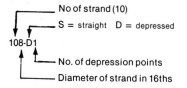

No of strand (10)

S = straight D = depressed

108-D1

No. of depression points

Diameter of strand in 16ths

Safe loads shown include dead load of 10 psf for untopped members and 15 psf for topped members. Remainder is live load. Long-time cambers include superimposed dead load but do not include live load.

Key

200 — Safe superimposed service load, psf
1.1 — Estimated camber at erection, in.
1.4 — Estimated long-time camber, in.

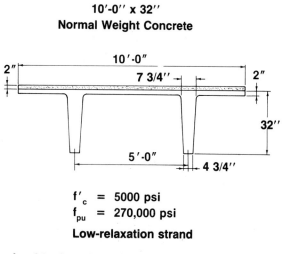

f'_c = 5000 psi
f_{pu} = 270,000 psi
Low-relaxation strand

Section Properties

		Untopped	Topped
A	=	615 in.^{2}	—
I	=	59,720 in.^{4}	77,118 in.^{4}
Y_b	=	21.98 in.	24.54 in.
Y_t	=	10.02 in.	9.46 in.
Z_b	=	2717 in.^{3}	3142 in.^{3}
Z_t	=	5960 in.^{3}	8152 in.^{3}
wt	=	641 plf	891 plf
		64 psf	89 psf
V/S	=	1.69 in.	

10DT32

Table of safe superimposed service load (psf) and cambers

No Topping

Strand Pattern	e_e / e_c	44	46	48	50	52	54	56	58	60	62	64	66	68	70	72	74	76	78	80	82	84	86
128-D1	12.81 18.73	200 1.1 1.4	179 1.2 1.6	160 1.3 1.6	143 1.3 1.7	129 1.4 1.7	116 1.4 1.7	104 1.5 1.7	93 1.5 1.7	84 1.5 1.7	75 1.5 1.6	68 1.4 1.5	61 1.4 1.4	54 1.3 1.2	48 1.2 1.0	43 1.1 0.8							
148-D1	10.48 18.48			187 1.4 1.8	169 1.5 1.9	152 1.6 1.9	137 1.6 2.0	124 1.7 2.0	112 1.7 2.0	102 1.8 2.0	92 1.8 2.0	83 1.8 1.9	75 1.7 1.8	68 1.7 1.7	61 1.6 1.6	55 1.6 1.4	49 1.5 1.2						
168-D1	8.98 18.23			194 1.6 2.1	176 1.7 2.2	159 1.8 2.2	145 1.9 2.3	131 1.9 2.3	119 2.0 2.3	109 2.0 2.3	99 2.0 2.3	90 2.0 2.2	82 2.1 2.1	74 2.1 2.0	67 2.0 1.9	61 2.0 1.7	55 1.8 1.5	50 1.7 1.2					
188-D1	7.59 17.98								150 2.1 2.6	137 2.2 2.6	125 2.2 2.6	114 2.3 2.6	104 2.3 2.6	95 2.3 2.5	87 2.3 2.4	79 2.3 2.3	72 2.1 2.1	66 2.0 2.0	60 2.1 1.8	55 2.0 1.5	49 1.9 1.2		
208-D1	6.48 17.73													108 2.5 2.9	99 2.6 2.8	91 2.6 2.7	83 2.6 2.6	76 2.5 2.4	70 2.5 2.2	64 2.4 2.0	58 2.3 1.7	53 2.2 1.4	
228-D1	5.57 17.48																	86 2.8 2.9	79 2.8 2.7	73 2.7 2.5	67 2.6 2.2	62 2.5 1.9	56 2.4 1.5

10DT32+2

Table of safe superimposed service load (psf) and cambers

2'' Normal Weight Topping

Strand pattern	e_e / e_c	42	44	46	48	50	52	54	56	58	60	62	64	66	68	70	72	74
108-D1	16.08 18.98	175 1.0 1.0	154 1.0 1.0	134 1.1 1.0	118 1.1 1.0	103 1.1 0.9	90 1.2 0.9	78 1.2 0.8	68 1.2 0.7	58 1.2 0.6	50 1.2 0.4							
128-D1	12.81 18.73		195 1.1 1.1	172 1.2 1.2	152 1.3 1.2	135 1.3 1.2	119 1.4 1.1	105 1.4 1.1	93 1.5 1.0	82 1.5 0.9	72 1.5 0.8	63 1.5 0.6						
148-D1	10.48 18.48			182 1.4 1.3	162 1.5 1.4	145 1.6 1.4	129 1.6 1.3	115 1.7 1.3	103 1.7 1.2	91 1.8 1.1	81 1.8 0.9	72 1.8 0.8	63 1.7 0.6					
168-D1	8.98 18.23				190 1.6 1.6	170 1.7 1.6	153 1.8 1.6	137 1.9 1.5	123 1.9 1.5	110 2.0 1.4	99 2.0 1.3	88 2.0 1.1	79 2.1 0.9	70 2.1 0.7				
188-D1	7.59 17.98								142 2.1 1.7	128 2.2 1.7	116 2.2 1.6	104 2.3 1.4	94 2.3 1.3	84 2.3 1.1	75 2.3 0.8	67 2.3 0.5		
208-D1	6.48 17.73											98 2.5 1.4	88 2.6 1.2	80 2.6 0.9	72 2.6 0.6			

Strength based on strain compatibility; bottom tension limited to $12\sqrt{f'_c}$; see pages 2-3–2-5 for explanation

Values below heavy line require release strengths higher than 3500 psi.

Strand Pattern Designation

108-D1
- 10 — No of strand (10)
- D — S = straight D = depressed
- 8 — Diameter of strand in 16ths
- 1 — No. of depression points

Safe loads shown include dead load of 10 psf for untopped members and 15 psf for topped members. Remainder is live load. Long-time cambers include superimposed dead load but do not include live load.

Key

128 — Safe superimposed service load, psf
2.5 — Estimated camber at erection, in.
2.9 — Estimated long-time camber, in.

DOUBLE TEE

10′-0″ x 32″
Lightweight Concrete

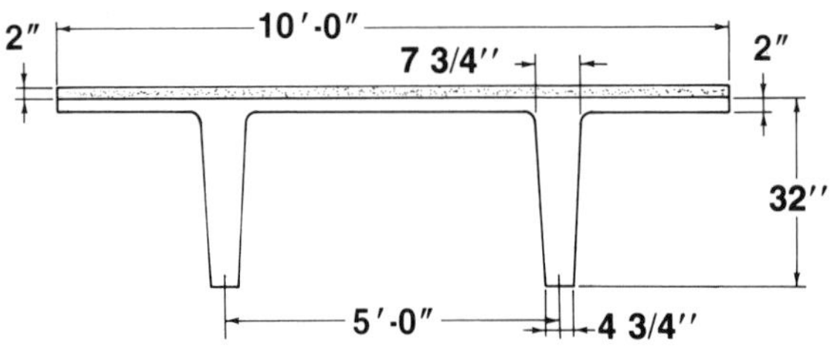

f'_c = 5000 psi
f_{pu} = 270,000 psi
Low-relaxation strand

Section Properties

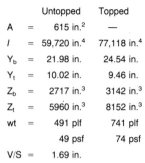

		Untopped	Topped
A	=	615 in.²	—
I	=	59,720 in.⁴	77,118 in.⁴
Y_b	=	21.98 in.	24.54 in.
Y_t	=	10.02 in.	9.46 in.
Z_b	=	2717 in.³	3142 in.³
Z_t	=	5960 in.³	8152 in.³
wt	=	491 plf	741 plf
		49 psf	74 psf
V/S	=	1.69 in.	

10LDT32

Table of safe superimposed service load (psf) and cambers — No Topping

Strand Pattern	e_e / e_c	54	56	58	60	62	64	66	68	70	72	74	76	78	80	82	84	86	88	90	92	94	96	98
128-D1	12.81	128	116	106	96	88	80	73	66	60	55	50	45	41	37	34								
	18.73	2.5	2.5	2.6	2.7	2.7	2.8	2.8	2.8	2.8	2.8	2.7	2.6	2.5	2.4	2.2								
		2.9	3.0	3.0	3.0	3.0	3.0	2.9	2.8	2.7	2.6	2.4	2.1	1.9	1.6	1.2								
148-D1	10.48	150	137	125	114	104	95	87	80	73	67	62	56	52	47	43	39	36						
	18.48	2.7	2.8	2.9	3.0	3.1	3.2	3.3	3.4	3.4	3.4	3.3	3.3	3.2	3.1	3.0	2.8	2.6						
		3.3	3.4	3.5	3.5	3.6	3.6	3.6	3.5	3.2	3.2	3.1	2.9	2.7	2.4	2.1	1.7	1.2						
168-D1	8.98	172	157	144	132	121	111	102	94	87	80	73	68	62	57	53	48	44	41	37				
	18.23	2.9	3.1	3.2	3.3	3.5	3.6	3.7	3.8	3.8	3.9	3.9	4.0	3.9	3.9	3.8	3.6	3.5	3.3	3.1				
		3.7	3.8	3.9	4.0	4.1	4.1	4.2	4.2	4.1	4.0	3.9	3.8	3.5	3.2	2.9	2.6	2.2	1.8	1.3				
188-D1	7.59							116	107	99	92	85	78	72	67	62	57	53	49	45	41	38		
	17.98							4.0	4.1	4.2	4.3	4.4	4.4	4.4	4.4	4.4	4.4	4.3	4.1	3.9	3.7	3.4		
								4.7	4.7	4.7	4.7	4.6	4.5	4.3	4.1	3.8	3.5	3.1	2.6	2.1	1.6	1.0		
208-D1	6.48												89	82	76	71	66	61	56	52	48	44	41	
	17.73												4.8	4.9	4.9	4.9	4.9	4.9	4.8	4.7	4.5	4.3	4.0	
													5.1	5.0	4.8	4.6	4.3	4.0	3.6	3.1	2.6	2.0	1.3	
228-D1	5.57																	69	64	59	55	50	46	42
	17.48																	5.4	5.4	5.3	5.2	5.0	4.8	4.6
																		4.8	4.5	4.1	3.6	3.0	2.4	1.6

(Span, ft.)

10LDT32+2

Table of safe superimposed service load (psf) and cambers — 2″ Normal Weight Topping

Strand Pattern	e_e / e_c	42	44	46	48	50	52	54	56	58	60	62	64	66	68	70	72	74	76	78
108-D1	16.08	188	166	147	130	115	102	90	80	71	62	55	48							
	18.98	1.5	1.6	1.8	1.8	1.9	2.0	2.1	2.2	2.2	2.3	2.3	2.3							
		1.6	1.5	1.6	1.6	1.6	1.6	1.5	1.5	1.3	1.2	1.0	0.8							
128-D1	12.81			184	165	147	132	118	105	94	84	75	67	60	53					
	18.73			2.0	2.1	2.1	2.3	2.5	2.5	2.6	2.7	2.7	2.8	2.8	2.8					
				1.9	2.0	1.9	2.0	2.0	1.9	1.8	1.7	1.5	1.3	1.1	0.8					
148-D1	10.48				195	175	157	142	127	115	104	93	84	75	68	61	54			
	18.48				2.3	2.4	2.5	2.7	2.8	2.9	3.0	3.1	3.2	3.3	3.4	3.4	3.4			
					2.2	2.3	2.3	2.3	2.3	2.3	2.2	2.1	1.9	1.7	1.5	1.1	0.8			
168-D1	8.98							165	149	135	122	111	101	91	82	74	67	61		
	18.23							2.9	3.1	3.2	3.3	3.5	3.6	3.7	3.8	3.8	3.9	3.9		
								2.7	2.7	2.7	2.6	2.5	2.4	2.2	2.0	1.8	1.5	1.1		
188-D1	7.59													106	96	88	80	72	60	59
	17.98													4.0	4.1	4.2	4.3	4.4	4.4	4.4
														2.7	2.5	2.3	2.0	1.7	1.3	0.9
208-D1	6.48																		76	70
	17.73																		4.8	4.9
																			1.9	1.5

(Span, ft.)

Strength based on strain compatibility; bottom tension limited to $12\sqrt{f'_{ci}}$; see pages 2-3–2-5 for explanation
Values below heavy line require release strengths higher than 3500 psi.

SINGLE TEE

8'-0" x 36"
Normal Weight Concrete

Strand Pattern Designation

- No of strand (10)
- S = straight   D = depressed
- 108-D1
- No. of depression points
- Diameter of strand in 16ths

Safe loads shown include dead load of 10 psf for untopped members and 15 psf for topped members. Remainder is live load. Long-time cambers include superimposed dead load but do not include live load.

Key
- 156 — Safe superimposed service load, psf
- 1.6 — Estimated camber at erection, in.
- 2.0 — Estimated long-time camber, in.

f'_c = 5000 psi
f_{pu} = 270,000 psi
Low-relaxation strand

Section Properties

		Untopped	Topped
A	=	570 in.²	—
I	=	68,917 in.⁴	83,212 in.⁴
Y_b	=	26.01 in.	28.28 in.
Y_t	=	9.99 in.	9.72 in.
Z_b	=	2650 in.³	2942 in.³
Z_t	=	6899 in.³	8561 in.³
wt	=	594 plf	794 plf
		74 psf	99 psf
V/S	=	2.16 in.	

8ST36

Table of safe superimposed service load (psf) and cambers   No Topping

Strand Pattern	e_e / e_c	56	58	60	62	64	66	68	70	72	74	76	78	80	82	84	86	88	90	92	94	96	98	100
128-D1	12.68 / 22.76	156	141	128	116	105	95	86	78	71	64	57	51	46	41									
		1.6	1.7	1.7	1.8	1.8	1.8	1.8	1.8	1.8	1.7	1.6	1.6	1.4	1.3									
		2.0	2.1	2.1	2.1	2.1	2.0	2.0	1.9	1.8	1.7	1.6	1.4	1.3	0.9									
148-D1	10.29 / 22.51	185	168	153	140	127	116	106	97	88	80	73	66	60	54	49	44							
		1.8	1.9	1.9	2.0	2.0	2.1	2.1	2.1	2.2	2.1	2.1	2.0	1.8	1.6	1.4	1.1							
		2.3	2.4	2.4	2.5	2.5	2.5	2.5	2.4	2.4	2.3													
168-D1	8.51 / 22.26							125	115	105	97	89	81	74	68	62	56	51	46					
								2.4	2.4	2.5	2.5	2.5	2.5	2.4	2.4	2.3	2.2	2.1	1.9					
								2.9	2.9	2.8	2.8	2.7	2.6	2.4	2.2	2.0	1.8	1.5	1.2					
188-D1	7.12 / 22.01													88	81	74	68	62	57	52				
														2.8	2.8	2.7	2.7	2.6	2.5	2.3				
														3.0	2.9	2.7	2.4	2.1	1.9	1.6				
208-D1	6.01 / 21.76																	73	67	62	57	52	48	
																		3.0	3.0	2.8	2.7	2.5	2.3	
																		2.9	2.6	2.3	1.9	1.5	1.1	
228-D1	5.11 / 21.51																					61	56	52
																						3.1	2.9	2.7
																						2.3	1.9	1.4

8ST36 + 2

Table of safe superimposed service load (psf) and cambers   2" Normal Weight Topping

Strand Pattern	e_e / e_c	50	52	54	56	58	60	62	64	66	68	70	72	74	76	78	80	82
108-D1	16.41 / 23.01	165	147	130	116	102	90	80	70	61	53							
		1.3	1.3	1.4	1.4	1.4	1.4	1.5	1.5	1.4	1.4							
		1.2	1.2	1.2	1.2	1.1	1.1	1.0	0.8	0.7	0.5							
128-D1	12.68 / 27.76		183	164	147	132	118	105	94	84	74	66	58	50				
			1.5	1.6	1.6	1.7	1.7	1.8	1.8	1.8	1.8	1.8	1.8	1.7				
			1.5	1.5	1.5	1.5	1.4	1.3	1.2	1.1	1.0	0.8	0.6	1.3				
148-D1	10.29 / 22.51			178	160	144	130	117	106	95	85	76	68	60				
				1.8	1.9	1.9	2.0	2.0	2.1	2.1	2.1	2.2	2.2	2.1				
				1.8	1.8	1.7	1.7	1.6	1.5	1.4	1.2	1.0	0.8	0.6				
168-D1	8.51 / 22.26								115	104	94	85	76	69	61			
									2.4	2.4	2.5	2.5	2.5	2.5	2.4			
									1.8	1.7	1.5	1.3	1.0	0.8	0.4			
188-D1	7.13 / 22.01															75	68	
																2.8	2.8	
																1.0	0.7	

Strength based on strain compatibility; bottom tension limited to $12\sqrt{f'_c}$; see pages 2-3–2-5 for explanation

Values below heavy line require release strengths higher than 3500 psi.

SINGLE TEE

8'-0" x 36"
Lightweight Concrete

Strand Pattern Designation

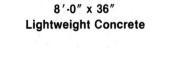

- No of strand (10)
- S = straight   D = depressed
- 108-D1
- No. of depression points
- Diameter of strand in 16ths

Safe loads shown include dead load of 10 psf for untopped members and 15 psf for topped members. Remainder is live load. Long-time cambers include superimposed dead load but do not include live load.

Key

130 — Safe superimposed service load, psf
3.0 — Estimated camber at erection, in.
3.6 — Estimated long-time camber, in.

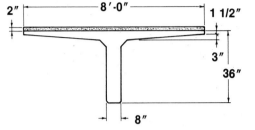

f'_c = 5000 psi
f_{pu} = 270,000 psi
Low-relaxation strand

Section Properties

	Untopped	Topped
A =	570 in.²	—
I =	68,917 in.⁴	88,260 in.⁴
Y_b =	26.01 in.	29.09 in.
Y_t =	9.99 in.	8.91 in.
Z_b =	2650 in.³	3034 in.³
Z_t =	6899 in.³	9906 in.³
wt =	455 plf	655 plf
	57 psf	82 psf
V/S =	2.16 in.	

8LST36

Table of safe superimposed service load (psf) and cambers
No Topping

Strand Pattern	e_e / e_c	62	64	66	68	70	72	74	76	78	80	82	84	86	88	90	92	94	96	98	100	102	104	106	108	110
128-D1	12.68 / 22.76	130 / 3.0 / 3.6	120 / 3.1 / 3.7	110 / 3.2 / 3.8	101 / 3.3 / 3.8	93 / 3.4 / 3.8	85 / 3.4 / 3.8	78 / 3.5 / 3.7	72 / 3.5 / 3.5	66 / 3.4 / 3.4	60 / 3.4 / 3.2	55 / 3.3 / 3.0	50 / 3.3 / 2.8	46 / 3.2 / 2.5	42 / 3.0 / 2.2	38 / 2.9 / 1.9	35 / 2.7 / 1.5									
148-D1	10.29 / 22.51		142 / 3.5 / 4.3	130 / 3.6 / 4.3	120 / 3.7 / 4.4	111 / 3.8 / 4.5	102 / 3.9 / 4.5	95 / 4.0 / 4.5	87 / 4.1 / 4.4	81 / 4.1 / 4.3	74 / 4.1 / 4.2	69 / 4.1 / 4.1	63 / 4.1 / 3.9	58 / 4.1 / 3.6	54 / 4.0 / 3.3	49 / 3.8 / 2.9	45 / 3.7 / 2.6	41 / 3.5 / 2.2	38 / 3.3 / 1.8							
168-D1	8.51 / 22.26									95 / 4.6 / 5.2	88 / 4.7 / 5.1	82 / 4.7 / 5.0	76 / 4.7 / 4.8	70 / 4.7 / 4.7	65 / 4.7 / 4.4	60 / 4.7 / 4.1	56 / 4.6 / 3.8	52 / 4.5 / 3.4	48 / 4.4 / 2.9	44 / 4.2 / 2.4	40 / 3.9 / 1.9					
188-D1	7.12 / 22.01															71 / 5.4 / 5.2	66 / 5.4 / 4.9	62 / 5.3 / 4.6	57 / 5.2 / 4.2	53 / 5.1 / 3.8	49 / 4.9 / 3.2	45 / 4.7 / 2.6	42 / 4.5 / 2.0			
208-D1	6.01 / 21.76																			58 / 5.8 / 4.6	54 / 5.7 / 4.1	50 / 5.5 / 3.5	46 / 5.3 / 2.8	43 / 5.0 / 2.0		
228-D1	5.11 / 21.51																									46 / 4.8 / 2.9

8LST36 + 2

Table of safe superimposed service load (psf) and cambers
2" Normal Weight Topping

Strand Pattern	e_e / e_c	50	52	54	56	58	60	62	64	66	68	70	72	74	76	78	80	82	84	86
108-D1	17.61 / 23.01	179 / 2.0 / 2.0	161 / 2.2 / 2.1	145 / 2.3 / 2.1	130 / 2.4 / 2.1	117 / 2.5 / 2.1	105 / 2.6 / 2.0	94 / 2.6 / 2.0	84 / 2.7 / 1.8	75 / 2.7 / 1.7	67 / 2.8 / 1.5	60 / 2.8 / 1.3	53 / 2.8 / 1.0	47 / 2.8 / 0.8						
128-D1	13.68 / 22.76		197 / 2.4 / 2.4	178 / 2.5 / 2.5	161 / 2.6 / 2.5	146 / 2.8 / 2.5	132 / 2.9 / 2.5	120 / 3.0 / 2.5	108 / 3.1 / 2.4	98 / 3.2 / 2.3	88 / 3.3 / 2.2	80 / 3.4 / 2.0	72 / 3.4 / 1.8	65 / 3.5 / 1.6	58 / 3.5 / 1.2	52 / 3.4 / 0.8				
148-D1	11.15 / 22.51								132 / 3.5 / 2.9	120 / 3.6 / 2.9	109 / 3.7 / 2.8	99 / 3.8 / 2.7	90 / 3.9 / 2.5	82 / 4.0 / 2.3	75 / 4.0 / 2.0	67 / 4.1 / 1.7	61 / 4.1 / 1.4	55 / 4.1 / 1.0		
168-D1	9.26 / 22.26															83 / 4.6 / 2.5	76 / 4.7 / 2.2	69 / 4.7 / 1.8	63 / 4.7 / 1.4	57 / 4.7 / 0.9

Strength based on strain compatibility; bottom tension limited to $12\sqrt{f'_c}$; see pages 2-3–2-5 for explanation

Values below heavy line require release strengths higher than 3500 psi.

PCI Design Handbook

SINGLE TEE

10'-0'' x 36''

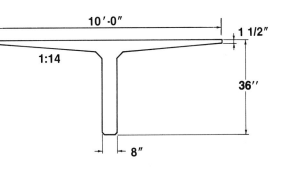

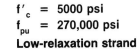

Strand Pattern Designation

208-D1
- No of strand (20)
- S = straight   D = depressed
- No. of depression points
- Diameter of strand in 16ths

Safe loads shown include dead load of 10 psf for untopped members and 15 psf for topped members. Remainder is live load. Long-time cambers include superimposed dead load but do not include live load.

Key
131 — Safe superimposed service load, psf
1.5 — Estimated camber at erection, in.
1.8 — Estimated long-time camber, in.

f'_c = 5000 psi
f_{pu} = 270,000 psi
Low-relaxation strand

Section Properties

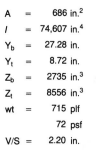

A =	686 in.^{2}
I =	74,607 in.^{4}
Y_b =	27.28 in.
Y_t =	8.72 in.
Z_b =	2735 in.^{3}
Z_t =	8556 in.^{3}
wt =	715 plf
	72 psf
V/S =	2.20 in.

10ST36

Table of safe superimposed service load (psf) and cambers — Normal Weight Concrete — No Topping

Strand Pattern	e_e / e_c	54	56	58	60	62	64	66	68	70	72	74	76	78	80	82	84	86	88	90	92	94	96	98
128-D1	14.95	131	118	106	95	86	77	69	62	55	49	43	38	33										
		1.5	1.5	1.6	1.6	1.6	1.6	1.6	1.6	1.5	1.5	1.4	1.3	1.1										
	24.03	1.8	1.8	1.8	1.8	1.8	1.7	1.7	1.6	1.5	1.3	1.1	0.8	0.5										
148-D1	12.42	156	141	128	116	105	95	86	77	70	63	57	51	45	40	36								
		1.7	1.7	1.8	1.9	1.9	1.9	2.0	2.0	1.9	1.9	1.9	1.8	1.7	1.6	1.4								
	23.78	2.1	2.1	2.2	2.2	2.2	2.2	2.1	2.0	2.0	1.8	1.7	1.5	1.3	1.1	0.7								
168-D1	10.53					123	112	102	93	84	77	69	63	57	51	46	41	37	33					
						2.1	2.2	2.3	2.3	2.4	2.3	2.3	2.2	2.1	1.9	1.7	1.5	1.2	0.8	0.4				
	22.33					2.6	2.6	2.5	2.5	2.4	2.3	2.2	2.1	1.9	1.7	1.5	1.2	0.8	0.4					
188-D1	9.06									90	82	75	69	62	57	51	46	43	37	33				
										2.6	2.6	2.6	2.6	2.5	2.5	2.4	2.3	2.1	1.9	1.7				
	23.28									2.8	2.7	2.6	2.4	2.2	2.0	1.8	1.5	1.3	0.8	0.3				
208-D1	7.88														67	61	56	51	46	41	37	33		
															2.9	2.8	2.7	2.6	2.5	2.3	2.1	1.9		
	23.03														2.6	2.4	2.1	1.8	1.5	1.1	0.7	0.2		
228-D1	6.92																		59	54	49	45	41	37
																			3.1	3.0	2.8	2.7	2.5	2.3
	22.78																		2.5	2.1	1.7	1.3	0.9	0.4

10LST36

Table of safe superimposed service load (psf) and cambers — Lightweight Concrete — No Topping

Strand Pattern	e_e / e_c	62	64	66	68	70	72	74	76	78	80	82	84	86	88	90	92	94	96	98	100	102	104	106
128-D1	14.95	99	91	83	75	69	63	57	52	47	42	38	35	31										
		2.9	3.0	3.0	3.1	3.1	3.1	3.1	3.0	3.0	2.9	2.8	2.7	2.5										
	24.03	3.3	3.3	3.3	3.2	3.1	2.9	2.8	2.7	2.5	2.3	2.0	1.7	1.3										
148-D1	12.42	108	99	91	84	77	70	64	59	54	49	45	41	37	34	30								
		3.3	3.4	3.5	3.6	3.7	3.7	3.8	3.7	3.7	3.6	3.5	3.4	3.3	3.1	2.9								
	23.78	3.9	3.9	3.9	3.9	3.8	3.7	3.6	3.4	3.1	2.9	2.6	2.3	2.0	1.6	1.1								
168-D1	10.53						90	83	77	71	65	60	55	50	46	42	39	35	32	29				
							4.1	4.2	4.3	4.3	4.4	4.4	4.3	4.3	4.2	4.1	3.9	3.7	3.5	3.2				
	23.53						4.5	4.5	4.4	4.3	4.1	3.9	3.6	3.3	2.9	2.5	2.1	1.7	1.2	0.6				
188-D1	9.06											70	65	60	55	51	47	43	40	36	33	30		
												5.0	5.0	5.0	4.9	4.9	4.8	4.7	4.5	4.3	4.0	3.7		
	23.28											4.8	4.6	4.4	4.0	3.7	3.2	2.7	2.2	1.7	1.1	0.5		
208-D1	7.88																55	51	47	44	40	37	34	
																	5.5	5.5	5.3	5.2	5.0	4.8	4.6	
	23.03																4.4	3.9	3.4	2.9	2.2	1.5	0.9	
228-D1	6.92																				47	43	40	37
																					5.9	5.7	5.5	5.2
	22.78																				3.6	2.9	2.1	1.3

Strength based on strain compatibility; bottom tension limited to $12\sqrt{f'_c}$; see pages 2-3–2-5 for explanation

Values below heavy line require release strengths higher than 3500 psi.

SINGLE TEE

10'-0" x 48"

Strand Pattern Designation

No of strand (20)
S = straight   D = depressed

208-D1

No. of depression points
Diameter of strand in 16ths

Safe loads shown include dead load of 10 psf for untopped members and 15 psf for topped members. Remainder is live load. Long-time cambers include superimposed dead load but do not include live load.

Key
126 — Safe superimposed service load, psf
1.5 — Estimated camber at erection, in.
1.8 — Estimated long-time camber, in.

f'_c = 5000 psi
f_{pu} = 270,000 psi
Low-relaxation strand

Section Properties

A = 782 in.^{2}
I = 169,020 in.^{4}
Y_b = 35.19 in.
Y_t = 12.81 in.
Z_b = 4803 in.^{3}
Z_t = 13,194 in.^{3}
wt = 815 plf
     81 psf
V/S = 2.33 in.

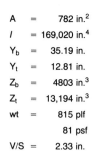

10ST48

Table of safe superimposed service load (psf) and cambers — Normal Weight Concrete — No Topping

Strand Pattern	e_e / e_c	68	70	72	74	76	78	80	82	84	86	88	90	92	94	96	98	100	102	104	106	108	110	112
148-D1	23.78	126	115	106	96	88	80	73	66	60	55	49	44	39										
	33.14	1.5	1.6	1.6	1.6	1.6	1.6	1.5	1.5	1.5	1.4	1.3	1.2	1.1										
		1.8	1.8	1.8	1.8	1.8	1.7	1.6	1.6	1.4	1.3	1.1	0.9	0.6										
168-D1	20.39	148	136	125	115	105	97	89	82	75	68	62	57	51	47	42								
	32.89	1.8	1.8	1.8	1.8	1.9	1.9	1.8	1.8	1.8	1.8	1.7	1.6	1.6	1.4	1.3								
		2.1	2.2	2.2	2.1	2.1	2.1	2.0	1.9	1.9	1.8	1.6	1.5	1.3	1.0	0.8								
188-D1	17.75	169	156	144	133	123	113	104	96	89	82	75	69	63	58	53	48	43						
	32.64	1.9	2.0	2.0	2.1	2.1	2.1	2.2	2.2	2.2	2.1	2.1	2.0	2.0	1.9	1.8	1.7	1.5						
		2.4	2.4	2.5	2.5	2.5	2.5	2.4	2.3	2.2	2.2	2.1	1.9	1.8	1.6	1.4	1.2	0.9						
208-D1	15.44							119	111	102	95	87	81	74	69	63	58	53	48	44				
	32.39							2.4	2.4	2.4	2.4	2.4	2.4	2.3	2.3	2.2	2.1	2.0	1.9	1.7				
								2.7	2.7	2.7	2.6	2.5	2.4	2.2	2.1	1.9	1.7	1.5	1.2	0.9				
228-D1	13.73											100	92	86	79	73	68	63	58	53	49	44		
	32.14											2.7	2.7	2.7	2.6	2.6	2.5	2.4	2.3	2.2	2.0	1.9		
												2.9	2.8	2.7	2.5	2.4	2.2	2.0	1.7	1.5	1.2	0.9		
248-D1	12.14															84	78	72	67	62	57	52	48	44
	31.89															2.9	2.8	2.8	2.7	2.6	2.5	2.4	2.2	2.0
																2.8	2.7	2.5	2.2	2.0	1.7	1.4	1.1	0.7

10LST48

Table of safe superimposed service load (psf) and cambers — Lightweight Concrete — No Topping

Strand Pattern	e_e / e_c	76	78	80	82	84	86	88	90	92	94	96	98	100	102	104	106	108	110	112	114	116	118	120
148-D1	23.78	104	96	89	82	76	70	65	60	55	51	47	43	39	36	32								
	33.14	2.9	2.9	3.0	3.0	3.0	3.0	3.0	3.0	2.9	2.9	2.8	2.7	2.6	2.4	2.2								
		3.3	3.3	3.2	3.1	3.1	3.0	2.9	2.8	2.6	2.5	2.3	2.0	1.8	1.4	1.0								
168-D1	20.39	121	113	105	97	90	84	78	72	67	62	58	53	49	45	42	38	35						
	32.89	3.3	3.4	3.5	3.5	3.5	3.5	3.5	3.5	3.5	3.5	3.4	3.4	3.3	3.2	3.0	2.9	2.7						
		3.9	3.9	3.9	3.9	3.8	3.7	3.6	3.4	3.3	3.1	3.0	2.8	2.5	2.3	2.0	1.7	1.3						
188-D1	17.75		129	120	112	104	97	91	85	79	73	68	64	59	55	51	47	44	40	37				
	32.64		3.7	3.8	3.9	3.9	4.0	4.1	4.1	4.1	4.1	4.1	4.0	3.9	3.9	3.8	3.6	3.5	3.3	3.1				
			4.4	4.4	4.4	4.4	4.4	4.3	4.3	4.2	4.0	3.8	3.5	3.3	3.0	2.8	2.5	2.2	1.8	1.4				
208-D1	15.44								96	90	84	79	74	69	64	60	56	52	48	45	41	38		
	32.39								4.5	4.5	4.6	4.6	4.6	4.6	4.6	4.5	4.4	4.2	4.1	3.9	3.8	3.5		
									4.9	4.8	4.7	4.6	4.4	4.2	4.0	3.7	3.3	2.9	2.6	2.2	1.8	1.3		
228-D1	13.73													78	73	69	64	60	56	52	49	45	42	
	32.14													5.1	5.1	5.1	5.0	5.0	4.9	4.8	4.6	4.4	4.2	
														5.0	4.8	4.6	4.3	4.0	3.6	3.2	2.7	2.2	1.8	
248-D1	12.14																		64	60	56	52	49	46
	31.89																		5.5	5.4	5.3	5.2	5.0	4.8
																			4.5	4.1	3.7	3.3	2.7	2.2

Strength based on strain compatibility; bottom tension limited to $12\sqrt{f'_c}$; see pages 2-3–2-5 for explanation

Values below heavy line require release strengths higher than 3500 psi.

HOLLOW-CORE

4'-0" x 6"
Normal Weight Concrete

Strand Pattern Designation

76-S

└─ S = straight
└─ Diameter of strand in 16ths
└─ Number of strand (7)

Safe loads shown include dead load of 10 psf for untopped members and 15 psf for topped members. Remainder is live load. Long-time cambers include superimposed dead load but do not include live load.

Capacity of sections of other configurations are similar. For precise values, see local hollow-core manufacturer.

Key

257 — Safe superimposed service load, psf
0.2 — Estimated camber at erection, in.
0.2 — Estimated long-time camber, in.

f'_c = 5000 psi
f'_{ci} = 3500 psi

Low-relaxation strand

Section Properties

	Untopped	Topped
A =	187 in.²	—
I =	763 in.⁴	1640 in.⁴
Y_b =	3.00 in.	4.14 in.
Y_t =	3.00 in.	3.86 in.
Z_b =	254 in.³	396 in.³
Z_t =	254 in.³	425 in.³
b_w =	16.00 in.	16.00 in.
wt =	195 plf	295 plf
	49 psf	74 psf
V/S =	1.73 in.	

4HC6
No Topping

Table of safe superimposed service load (psf) and cambers

Strand Designation Code	Span, ft.													
	12	13	14	15	16	17	18	19	20	21	22	23	24	25
66-S	257	224	197	174	154	132	113	98	84	73	63	55	47	40
	0.2	0.2	0.2	0.2	0.2	0.2	0.2	0.2	0.2	0.2	0.1	0.1	0.0	-0.1
	0.2	0.3	0.3	0.3	0.3	0.3	0.2	0.2	0.1	0.1	0.0	-0.2	-0.3	-0.5
76-S		264	233	207	182	157	136	118	102	89	78	68	59	52
		0.2	0.3	0.3	0.3	0.3	0.3	0.3	0.3	0.3	0.3	0.2	0.1	0.1
		0.3	0.3	0.4	0.4	0.4	0.4	0.3	0.3	0.2	0.1	0.0	-0.1	-0.3
96-S			269	235	204	177	155	136	120	106	94	83	73	
			0.4	0.4	0.5	0.5	0.5	0.5	0.5	0.5	0.5	0.5	0.4	
			0.5	0.6	0.6	0.6	0.6	0.6	0.6	0.5	0.5	0.4	0.2	
87-S				278	242	212	186	164	145	129	115	102	91	
				0.6	0.6	0.7	0.7	0.7	0.7	0.8	0.8	0.7	0.7	
				0.7	0.8	0.8	0.8	0.9	0.9	0.8	0.8	0.7	0.7	
97-S					267	234	206	183	162	144	129	115	103	
					0.7	0.8	0.8	0.9	0.9	0.9	0.9	1.0	0.9	
					0.9	1.0	1.0	1.1	1.1	1.1	1.1	1.0	1.0	

4HC6 + 2
2" Normal Weight Topping

Table of safe superimposed service load (psf) and cambers

Strand Designation Code	Span, ft.													
	14	15	16	17	18	19	20	21	22	23	24	25	26	27
66-S	278	245	215	184	153	126	102	82	65	49				
	0.2	0.2	0.2	0.2	0.2	0.2	0.2	0.1	0.1	0.0				
	0.2	0.2	0.1	0.1	0.0	0.0	-0.1	-0.2	-0.4	-0.5				
76-S	291	254	219	189	164	142	122	101	82	66				
	0.3	0.3	0.3	0.3	0.3	0.3	0.3	0.3	0.3	0.1				
	0.2	0.2	0.2	0.2	0.1	0.1	0.0	-0.2	-0.3	-0.5				
96-S		284	247	215	189	166	146	123	104	87	72			
		0.5	0.5	0.5	0.5	0.5	0.5	0.5	0.5	0.4	0.3			
		0.4	0.4	0.4	0.3	0.3	0.2	0.1	-0.1	-0.3	-0.5			
87-S			292	256	226	199	176	156	136	117	100	85		
			0.7	0.7	0.7	0.7	0.8	0.8	0.7	0.7	0.7	0.6		
			0.6	0.6	0.5	0.5	0.4	0.4	0.2	0.1	-0.1	-0.3		
97-S				283	250	221	196	175	156	137	118	102		
				0.8	0.9	0.9	0.9	1.0	1.0	0.9	0.9	0.9		
				0.7	0.7	0.7	0.6	0.6	0.5	0.4	0.2	0.0		

Strength based on strain compatibility; bottom tension limited to $6\sqrt{f'_c}$; see pages 2-3–2-5 for explanation

HOLLOW-CORE

4'-0" x 8"
Normal Weight Concrete

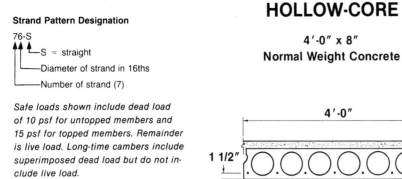

f'_c = 5000 psi
f'_{ci} = 3500 psi

Low-relaxation strand

Strand Pattern Designation

76-S
- S = straight
- Diameter of strand in 16ths
- Number of strand (7)

Safe loads shown include dead load of 10 psf for untopped members and 15 psf for topped members. Remainder is live load. Long-time cambers include superimposed dead load but do not include live load.

Capacity of sections of other configurations are similar. For precise values, see local hollow-core manufacturer.

Section Properties

		Untopped	Topped
A	=	215 in.²	—
I	=	1666 in.⁴	3071 in.⁴
Y_b	=	4.00 in.	5.29 in.
Y_t	=	4.00 in.	4.71 in.
Z_b	=	416 in.³	580 in.³
Z_t	=	416 in.³	652 in.³
b_w	=	12.00 in.	12.00 in.
wt	=	224 plf	323 plf
		56 psf	81 psf
V/S	=	1.92 in.	

Key

- 269 — Safe superimposed service load, psf
- 0.2 — Estimated camber at erection, in.
- 0.3 — Estimated long-time camber, in.

4HC8

Table of safe superimposed service load (psf) and cambers — No Topping

Strand Designation Code	14	15	16	17	18	19	20	21	22	23	24	25	26	27	28	29	30	31	32	33	34	35
66-S	269	242	209	182	159	139	122	107	94	83	73	64	56	49								
	0.2	0.2	0.2	0.2	0.3	0.3	0.3	0.3	0.2	0.2	0.2	0.2	0.1	0.0								
	0.3	0.3	0.3	0.3	0.3	0.3	0.3	0.3	0.3	0.2	0.1	0.0	-0.1	-0.2								
76-S		286	248	216	190	167	147	130	115	102	91	81	72	63	56	49						
		0.3	0.3	0.3	0.3	0.3	0.4	0.4	0.4	0.4	0.3	0.3	0.3	0.2	0.2	0.1						
		0.4	0.4	0.4	0.4	0.4	0.4	0.4	0.4	0.3	0.3	0.2	0.1	0.0	-0.1	-0.2						
58-S			296	275	249	221	196	175	156	140	125	112	101	91	82	73	66	59	53			
			0.4	0.5	0.5	0.5	0.5	0.6	0.6	0.6	0.6	0.6	0.6	0.5	0.5	0.5	0.4	0.3	0.1			
			0.6	0.6	0.6	0.7	0.7	0.7	0.7	0.7	0.7	0.6	0.5	0.4	0.3	0.2	0.0	-0.2	-0.4			
68-S				284	266	250	236	213	191	172	155	140	126	114	104	94	85	77	70	63	57	
				0.6	0.6	0.7	0.7	0.7	0.8	0.8	0.8	0.8	0.8	0.8	0.8	0.8	0.8	0.8	0.7	0.7	0.5	
				0.8	0.8	0.9	0.9	0.9	1.0	1.0	1.0	1.0	1.0	0.9	0.9	0.8	0.7	0.6	0.4	0.2	0.0	
78-S					290	272	256	242	229	215	202	183	166	151	139	125	113	103	94	86	79	72
					0.7	0.7	0.8	0.9	0.9	1.0	1.0	1.0	1.1	1.1	1.1	1.1	1.1	1.1	1.1	1.1	1.0	0.9
					0.9	1.0	1.0	1.1	1.2	1.2	1.3	1.3	1.3	1.3	1.3	1.3	1.2	1.2	1.1	1.0	0.9	0.7

4HC8 + 2

Table of safe superimposed service load (psf) and cambers — 2" Normal Weight Topping

Strand Designation Code	16	17	18	19	20	21	22	23	24	25	26	27	28	29	30	31	32	33	34	35
66-S	262	226	197	171	148	125	105	87	72	58	45									
	0.2	0.2	0.2	0.2	0.2	0.2	0.2	0.2	0.2	0.1	0.1									
	0.2	0.2	0.1	0.1	0.1	0.0	-0.1	-0.1	-0.2	-0.4	-0.5									
76-S		270	236	206	181	160	141	120	102	86	72	59	47							
		0.3	0.3	0.3	0.3	0.3	0.3	0.3	0.3	0.2	0.2	0.2	0.1							
		0.3	0.2	0.2	0.2	0.2	0.1	0.0	-0.1	-0.2	-0.3	-0.4	-0.6							
58-S			273	242	215	191	170	152	136	121	108	96	84	72	60					
			0.5	0.5	0.6	0.6	0.6	0.6	0.6	0.6	0.5	0.5	0.5	0.4	0.3					
			0.5	0.5	0.5	0.4	0.4	0.3	0.3	0.2	0.1	-0.1	-0.2	-0.4	-0.6					
68-S				293	261	234	209	188	169	152	137	123	111	100	87	75	64			
				0.7	0.7	0.8	0.8	0.8	0.8	0.8	0.8	0.8	0.8	0.8	0.7	0.7	0.6			
				0.7	0.7	0.7	0.6	0.6	0.5	0.4	0.3	0.2	0.0	-0.1	-0.2	-0.4	-0.6			
78-S					286	271	245	221	199	180	163	148	134	121	110	98	87	76		
					0.9	1.0	1.0	1.0	1.1	1.1	1.1	1.1	1.1	1.1	1.1	1.1	1.0	0.9		
					0.9	0.9	0.9	0.9	0.9	0.8	0.8	0.7	0.6	0.5	0.3	0.1	-0.1	-0.3		

Strength based on strain compatibility; bottom tension limited to $6\sqrt{f'_c}$; see pages 2-3–2-5 for explanation

HOLLOW-CORE

4'-0" x 8"
Lightweight Concrete

Strand Pattern Designation

76-S
- S = straight
- Diameter of strand in 16ths
- Number of strand (7)

Safe loads shown include dead load of 10 psf for untopped members and 15 psf for topped members. Remainder is live load. Long-time cambers include superimposed dead load but do not include live load.

Capacity of sections of other configurations are similar. For precise values, see local hollow-core manufacturer.

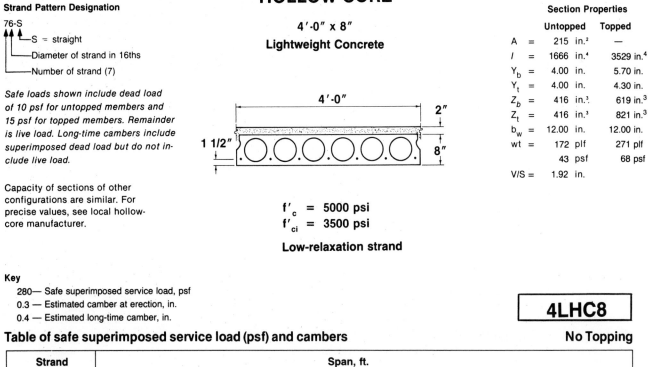

f'_c = 5000 psi
f'_{ci} = 3500 psi

Low-relaxation strand

Section Properties		
	Untopped	**Topped**
A =	215 in.²	—
I =	1666 in.⁴	3529 in.⁴
Y_b =	4.00 in.	5.70 in.
Y_t =	4.00 in.	4.30 in.
Z_b =	416 in.³	619 in.³
Z_t =	416 in.³	821 in.³
b_w =	12.00 in.	12.00 in.
wt =	172 plf	271 plf
	43 psf	68 psf
V/S =	1.92 in.	

Key
280— Safe superimposed service load, psf
0.3 — Estimated camber at erection, in.
0.4 — Estimated long-time camber, in.

4LHC8

Table of safe superimposed service load (psf) and cambers
No Topping

Strand Designation Code	Span, ft.																						
	14	15	16	17	18	19	20	21	22	23	24	25	26	27	28	29	30	31	32	33	35	36
66-S	280	252	220	193	169	150	133	118	105	94	84	75	67	60	54	48						
	0.3	0.4	0.4	0.4	0.5	0.5	0.5	0.5	0.5	0.5	0.5	0.5	0.5	0.4	0.4	0.3	0.3					
	0.4	0.5	0.5	0.5	0.6	0.6	0.6	0.6	0.6	0.5	0.5	0.4	0.3	0.2	0.0	0.1						
76-S		296	259	227	200	178	158	141	126	113	102	91	82	74	67	60	54	49				
		0.4	0.5	0.5	0.6	0.6	0.6	0.7	0.7	0.7	0.7	0.7	0.7	0.7	0.6	0.6	0.6	0.5	0.4			
		0.6	0.6	0.7	0.7	0.7	0.8	0.8	0.8	0.8	0.8	0.7	0.7	0.6	0.5	0.4	0.3	0.1	-0.1			
58-S			286	260	231	207	185	167	150	136	123	112	102	92	84	77	70	64	58	53		
			0.7	0.8	0.8	0.9	0.9	1.0	1.0	1.1	1.1	1.1	1.1	1.1	1.1	1.1	1.0	1.0	0.9	0.8		
			0.9	1.0	1.1	1.1	1.2	1.2	1.2	1.2	1.2	1.2	1.2	1.1	1.1	0.9	0.8	0.7	0.5	0.2		
68-S			292	274	258	243	224	202	183	166	151	137	125	114	105	96	88	81	74	68		
			0.9	1.0	1.1	1.1	1.2	1.3	1.3	1.4	1.5	1.5	1.5	1.6	1.6	1.6	1.6	1.6	1.6	1.5	1.5	
			1.2	1.2	1.3	1.4	1.5	1.6	1.6	1.7	1.7	1.7	1.7	1.7	1.7	1.6	1.5	1.4	1.3	1.1		
78-S				283	267	249	237	225	212	194	177	161	148	135	124	114	105	97	89	82		
				1.2	1.3	1.4	1.5	1.6	1.7	1.7	1.8	1.9	2.0	2.0	2.1	2.1	2.1	2.2	2.2	2.1		
				1.5	1.6	1.8	1.9	2.0	2.0	2.1	2.2	2.2	2.3	2.3	2.3	2.3	2.2	2.2	2.1	2.0		

4LHC8+2

Table of safe superimposed service load (psf) and cambers
2" Normal Weight Topping

Strand Designation Code	Span, ft.																			
	16	17	18	19	20	21	22	23	24	25	26	27	28	29	30	31	32	33	34	35
66-S	272	237	207	182	160	141	125	110	95	80	67	55								
	0.4	0.4	0.4	0.4	0.5	0.5	0.5	0.5	0.5	0.4	0.4	0.4	0.3							
	0.3	0.3	0.3	0.3	0.2	0.2	0.1	0.1	0.0	-0.2	-0.3	-0.5								
76-S		281	246	217	192	170	151	135	120	107	95	81	69	58						
		0.5	0.5	0.6	0.6	0.6	0.6	0.6	0.7	0.6	0.6	0.6	0.5							
		0.4	0.5	0.4	0.4	0.4	0.4	0.3	0.2	0.1	0.0	-0.2	-0.3	-0.5						
58-S			284	253	225	202	181	163	146	132	119	107	97	87	79	71				
			0.8	0.9	0.9	1.0	1.0	1.1	1.1	1.1	1.1	1.1	1.1	1.1	1.0	1.0				
			0.8	0.8	0.8	0.8	0.8	0.7	0.7	0.6	0.5	0.4	0.2	0.1	-0.1	-0.4				
68-S					272	244	220	199	180	163	148	134	122	111	101	91	83	73		
					1.2	1.3	1.3	1.4	1.5	1.5	1.5	1.6	1.6	1.6	1.6	1.6	1.5	1.5		
					1.1	1.1	1.1	1.1	1.1	1.1	1.0	0.9	0.8	0.7	0.5	0.3	0.0	-0.2		
78-S						297	279	256	232	210	191	174	159	145	132	121	110	101	91	
						1.5	1.6	1.7	1.7	1.8	1.9	2.0	2.0	2.1	2.1	2.1	2.2	2.2	2.2	
						1.4	1.5	1.5	1.5	1.5	1.5	1.5	1.4	1.3	1.2	1.1	0.9	0.7	0.5	

Strength based on strain compatibility; bottom tension limited to $6\sqrt{f'_c}$; see pages 2-3–2-5 for explanation

HOLLOW-CORE

4'-0" x 10"
Normal Weight Concrete

Safe loads shown include dead load of 10 psf for untopped members and 15 psf for topped members. Remainder is live load. Long-time cambers include superimposed dead load but do not include live load.

Capacity of sections of other configurations are similar. For precise values, see local hollow-core manufacturer.

Key
- 235 — Safe superimposed service load, psf
- 0.3 — Estimated camber at erection, in.
- 0.4 — Estimated long-time camber, in.

f'_c = 5000 psi
f'_{ci} = 3500 psi
Low-relaxation strand

Section Properties

	Untopped	Topped
A =	259 in.²	—
I =	3223 in.⁴	5328 in.⁴
Y_b =	5.00 in.	6.34 in.
Y_t =	5.00 in.	5.66 in.
Z_b =	645 in.³	840 in.³
Z_t =	645 in.³	941 in.³
b_w =	10.50 in.	10.50 in.
wt =	270 plf	370 plf
	68 psf	93 psf
V/S =	2.23 in.	

4HC10

Table of safe superimposed service load (psf) and cambers — No Topping

Strand Designation Code	20	21	22	23	24	25	26	27	28	29	30	31	32	33	34	35	36	37	38	39	40	41	42
48-S	235	208	185	165	147	131	117	105	94	84	75	67	59										
	0.3	0.3	0.3	0.3	0.3	0.3	0.3	0.3	0.3	0.2	0.2	0.2	0.1										
	0.4	0.4	0.4	0.4	0.4	0.4	0.3	0.3	0.2	0.2	0.1	0.0	-0.1										
58-S	280	263	240	215	193	174	157	141	128	115	104	94	85	77	69	62	56						
	0.4	0.4	0.4	0.5	0.5	0.5	0.5	0.5	0.5	0.5	0.5	0.4	0.4	0.4	0.3	0.2	0.2						
	0.5	0.5	0.6	0.6	0.6	0.6	0.6	0.6	0.6	0.5	0.5	0.4	0.3	0.2	0.1	0.1	-0.1						
68-S	289	272	255	242	231	215	195	176	160	146	133	121	110	100	91	83	76	69	62	51			
	0.5	0.5	0.6	0.6	0.6	0.7	0.7	0.7	0.7	0.7	0.7	0.7	0.7	0.7	0.6	0.6	0.5	0.5	0.4	0.3			
	0.7	0.7	0.8	0.8	0.8	0.8	0.9	0.9	0.9	0.9	0.9	0.8	0.8	0.7	0.7	0.6	0.5	0.4	0.2	0.1	-0.1		
78-S	298	278	264	248	237	223	214	203	192	175	160	147	134	123	113	103	95	87	80	73	67	61	55
	0.6	0.7	0.7	0.7	0.8	0.8	0.9	0.9	0.9	0.9	0.9	1.0	1.0	1.0	0.9	0.9	0.9	0.8	0.8	0.7	0.6	0.5	0.4
	0.8	0.9	0.9	1.0	1.0	1.1	1.1	1.1	1.1	1.2	1.2	1.1	1.1	1.0	1.0	0.9	0.8	0.6	0.5	0.3	0.1	-0.1	
88-S		287	270	257	243	229	220	209	199	189	183	172	158	145	134	123	113	104	96	89	82	75	69
		0.8	0.8	0.9	0.9	1.0	1.0	1.1	1.1	1.2	1.2	1.2	1.2	1.2	1.2	1.2	1.2	1.2	1.2	1.1	1.1	1.0	0.9
		1.0	1.1	1.2	1.2	1.3	1.3	1.4	1.4	1.4	1.5	1.5	1.5	1.5	1.4	1.4	1.3	1.3	1.2	1.1	0.9	0.8	0.6

4HC10 + 2

Table of safe superimposed service load (psf) and cambers — 2" Normal Weight Topping

Strand Designation Code	20	21	22	23	24	25	26	27	28	29	30	31	32	33	34	35	36	37	38	39	40
48-S	276	244	215	191	169	150	133	114	98	83	70	58	47								
	0.3	0.3	0.3	0.3	0.3	0.3	0.3	0.2	0.2	0.2	0.2	0.1	0.1								
	0.2	0.2	0.2	0.2	0.1	0.1	0.0	0.0	-0.1	-0.2	-0.3	-0.4	-0.6								
58-S			280	250	223	200	179	161	145	130	116	102	88	76	64	54					
			0.4	0.4	0.5	0.5	0.5	0.5	0.5	0.5	0.4	0.4	0.4	0.3	0.3	0.2					
			0.4	0.4	0.4	0.3	0.3	0.3	0.2	0.1	0.0	-0.1	-0.2	-0.3	-0.5	-0.6					
68-S				286	272	248	223	202	182	165	149	135	122	111	100	90	81	73			
				0.6	0.6	0.7	0.7	0.7	0.7	0.7	0.7	0.7	0.7	0.7	0.6	0.6	0.5	0.5			
				0.6	0.6	0.6	0.6	0.5	0.5	0.4	0.4	0.3	0.2	0.0	-0.1	-0.3	-0.5				
78-S					295	278	265	250	239	218	199	181	165	150	137	125	113	103	94	85	77	67
					0.7	0.8	0.8	0.9	0.9	0.9	0.9	0.9	1.0	1.0	1.0	0.9	0.9	0.9	0.8	0.8	0.7	0.6
					0.8	0.8	0.8	0.8	0.8	0.8	0.8	0.7	0.7	0.6	0.5	0.4	0.3	0.2	0.0	-0.2	-0.4	-0.6
88-S						287	271	259	245	232	224	211	193	176	161	148	135	124	113	104	95	86
						0.9	1.0	1.0	1.1	1.1	1.2	1.2	1.2	1.2	1.2	1.2	1.2	1.2	1.2	1.2	1.1	1.1
						1.0	1.0	1.0	1.0	1.0	1.0	1.0	1.0	0.9	0.9	0.8	0.7	0.6	0.5	0.3	+0.1	-0.1

Strength based on strain compatibility; bottom tension limited to $6\sqrt{f'_c}$; see pages 2-3–2-5 for explanation

HOLLOW-CORE

4'-0" x 12"
Normal Weight Concrete

Strand Pattern Designation

76-S
- S = straight
- Diameter of strand in 16ths
- Number of strand (7)

Safe loads shown include dead load of 10 psf for untopped members and 15 psf for topped members. Remainder is live load. Long-time cambers include superimposed dead load but do not include live load.

Capacity of sections of other configurations are similar. For precise values, see local hollow-core manufacturer.

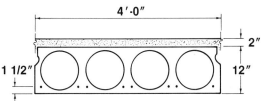

f'_c = 5000 psi
f'_{ci} = 3500 psi

Low-relaxation strand

Section Properties

		Untopped	Topped
A	=	262 in.²	—
I	=	4949 in.⁴	7811 in.⁴
Y_b	=	6.00 in.	7.55 in.
Y_t	=	6.00 in.	6.45 in.
Z_b	=	825 in.³	1035 in.³
Z_t	=	825 in.³	1211 in.³
b_w	=	8.00 in.	8.00 in.
wt	=	273 plf	373 plf
		68 psf	93 psf
V/S	=	2.18 in.	

Key
125 — Safe superimposed service load, psf
0.3 — Estimated camber at erection, in.
0.3 — Estimated long-time camber, in.

4HC12

No Topping

Table of safe superimposed service load (psf) and cambers

Strand Designation Code	Span, ft.																						
	28	29	30	31	32	33	34	35	36	37	38	39	40	41	42	43	44	45	46	47	48	49	50
76-S	125	113	102	92	83	75	67	60	54	48	43												
	0.3	0.3	0.3	0.3	0.2	0.2	0.2	0.1	0.1	0.0	-0.1												
	0.3	0.3	0.2	0.2	0.1	0.0	-0.0	-0.1	-0.2	-0.3	-0.5												
58-S	173	158	144	131	120	109	100	91	83	76	69	63	57	52	47	42							
	0.5	0.5	0.5	0.5	0.5	0.5	0.5	0.5	0.4	0.4	0.3	0.3	0.2	0.1	0.1	0.0							
	0.6	0.6	0.6	0.6	0.6	0.5	0.5	0.4	0.4	0.3	0.2	0.1	-0.1	0.2	-0.4	-0.6							
68-S	182	173	165	157	150	140	128	118	109	100	92	84	78	71	65	60	55	50	45				
	0.7	0.7	0.7	0.7	0.8	0.8	0.8	0.7	0.7	0.7	0.7	0.6	0.6	0.5	0.5	0.4	0.3	0.2	0.1				
	0.9	0.9	0.9	0.9	0.9	0.9	0.9	0.8	0.8	0.7	0.7	0.6	0.5	0.3	0.2	0.1	-0.1	-0.3	-0.5				
78-S	188	179	171	163	156	149	145	139	133	123	114	105	98	90	83	77	71	66	61	56	51	47	
	0.9	0.9	0.9	1.0	1.0	1.0	1.0	1.0	1.0	1.0	1.0	1.0	1.0	0.9	0.9	0.8	0.7	0.7	0.6	0.5	0.4	0.2	
	1.1	1.2	1.2	1.2	1.2	1.2	1.2	1.2	1.2	1.2	1.1	1.1	1.0	0.9	0.8	0.7	0.5	0.3	0.2	-0.1	-0.3	-0.5	
88-S	194	185	177	169	162	155	148	142	137	131	126	121	117	109	101	94	87	81	75	70	65	60	55
	1.0	1.1	1.1	1.2	1.2	1.2	1.3	1.3	1.3	1.3	1.3	1.3	1.3	1.3	1.3	1.2	1.2	1.1	1.1	1.0	0.9	0.8	0.6
	1.3	1.4	1.4	1.5	1.5	1.5	1.6	1.6	1.6	1.6	1.5	1.5	1.5	1.4	1.3	1.2	1.1	1.0	0.9	0.7	0.5	0.3	0.0

4HC12 + 2

2" Normal Weight Topping

Table of safe superimposed service load (psf) and cambers

Strand Designation Code	Span, ft.																						
	23	24	25	26	27	28	29	30	31	32	33	34	35	36	37	38	39	40	41	42	43	44	45
76-S	235	211	189	169	151	135	120	105	91	78	66	56	46										
	0.3	0.3	0.3	0.3	0.3	0.3	0.3	0.3	0.2	0.2	0.2	0.1	0.1										
	0.2	0.2	0.2	0.1	0.1	0.1	0.0	0.0	-0.1	-0.2	-0.3	-0.4	-0.5										
58-S	250	234	222	209	199	188	172	156	141	128	116	105	93	82	71	62	53						
	0.4	0.4	0.5	0.5	0.5	0.5	0.5	0.5	0.5	0.5	0.5	0.4	0.4	0.4	0.3	0.3	0.2						
	0.4	0.4	0.4	0.4	0.4	0.4	0.3	0.3	0.2	0.2	0.1	0.0	-0.1	-0.2	-0.3	-0.5	-0.6						
68-S	262	249	234	224	211	200	189	183	173	164	150	137	125	114	104	95	86	78	71	64			
	0.6	0.6	0.6	0.6	0.7	0.7	0.7	0.7	0.7	0.8	0.8	0.8	0.7	0.7	0.7	0.7	0.6	0.6	0.5	0.5			
	0.6	0.6	0.6	0.6	0.6	0.6	0.6	0.6	0.6	0.5	0.5	0.4	0.3	0.2	0.1	-0.1	-0.2	-0.4	-0.6				
78-S	271	255	243	230	217	206	195	189	179	171	163	155	148	141	130	119	110	101	92	84	77	70	
	0.7	0.7	0.8	0.8	0.8	0.9	0.9	0.9	1.0	1.0	1.0	1.0	1.0	1.0	1.0	1.0	1.0	1.0	0.9	0.9	0.8	0.7	
	0.7	0.8	0.8	0.8	0.8	0.9	0.9	0.9	0.9	0.8	0.8	0.8	0.7	0.7	0.6	0.5	0.4	0.3	0.1	-0.1	-0.2	-0.5	
88-S	280	264	249	236	223	212	201	195	185	177	169	161	154	147	141	135	129	122	112	104	95	88	81
	0.8	0.8	0.9	0.9	1.0	1.1	1.2	1.2	1.2	1.3	1.3	1.3	1.3	1.3	1.3	1.3	1.3	1.3	1.2	1.2	1.1	1.1	1.0
	0.9	0.9	1.0	1.0	1.1	1.1	1.1	1.1	1.1	1.1	1.0	1.0	0.9	0.9	0.8	0.7	0.6	0.4	0.3	0.1	-0.1	-0.3	-0.6

Strength based on strain compatibility; bottom tension limited to $6\sqrt{f'_c}$; see pages 2-3–2-5 for explanation

HOLLOW-CORE

4'-0" x 12"
Lightweight Concrete

Strand Pattern Designation

76-S
- S = straight
- Diameter of strand in 16ths
- Number of strand (7)

Safe loads shown include dead load of 10 psf for untopped members and 15 psf for topped members. Remainder is live load. Long-time cambers include superimposed dead load but do not include live load.

Capacity of sections of other configurations are similar. For precise values, see local hollow-core manufacturer.

Key

138 — Safe superimposed service load, psf
0.6 — Estimated camber at erection, in.
0.7 — Estimated long-time camber, in.

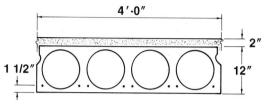

f'_c = 5000 psi
f'_{ci} = 3500 psi

Low-relaxation strand

Section Properties

	Untopped	Topped
A =	262 in.²	—
I =	4949 in.⁴	8800 in.⁴
Y_b =	6.00 in.	8.08 in.
Y_t =	6.00 in.	5.92 in.
Z_b =	825 in.³	1089 in.³
Z_t =	825 in.³	1486 in.³
b_w =	8.00 in.	8.00 in.
wt =	209 plf	309 plf
	52 psf	77 psf
V/S =	2.18 in.	

4LHC12 — No Topping

Table of safe superimposed service load (psf) and cambers

Strand Designation Code	28	29	30	31	32	33	34	35	36	37	38	39	40	41	42	43	44	45	46	47	48	49	50
76-S	138	126	115	105	96	88	80	73	67	61	56	51	46	42									
	0.6	0.6	0.6	0.6	0.6	0.6	0.6	0.5	0.5	0.5	0.4	0.4	0.3	0.2									
	0.7	0.7	0.7	0.7	0.6	0.6	0.5	0.5	0.4	0.3	0.2	0.0	−0.1	−0.3									
58-S	186	171	157	144	133	123	113	104	96	89	82	76	70	65	60	55	51	47	43				
	0.9	0.9	0.9	1.0	1.0	1.0	1.0	1.0	1.0	1.0	1.0	0.9	0.9	0.8	0.8	0.7	0.6	0.5	0.4				
	1.1	1.1	1.2	1.2	1.2	1.2	1.1	1.1	1.1	1.0	0.9	0.8	0.7	0.6	0.4	0.3	0.1	0.1	−0.4				
68-S	192	183	175	170	163	153	141	131	122	113	105	98	91	84	78	73	68	63	59	54	50	47	43
	1.1	1.2	1.2	1.3	1.3	1.4	1.4	1.4	1.4	1.4	1.4	1.4	1.4	1.4	1.9	1.3	1.3	1.2	1.1	1.0	0.9	0.8	0.6
	1.4	1.5	1.5	1.6	1.6	1.6	1.6	1.6	1.6	1.6	1.6	1.5	1.4	1.4	1.3	1.1	1.0	0.8	0.6	0.4	0.2	−0.1	−0.4
78-S	198	189	181	173	166	162	155	149	144	136	127	119	111	103	97	90	84	79	74	69	64	60	56
	1.4	1.5	1.6	1.7	1.7	1.8	1.8	1.8	1.9	1.9	1.9	1.9	1.9	2.0	1.9	1.9	·1.9	1.9	1.8	1.7	1.7	1.6	1.5
	1.8	1.8	1.9	2.0	2.0	2.1	2.1	2.1	2.2	2.2	2.2	2.1	2.1	2.0	1.9	1.8	1.7	1.6	1.4	1.2	1.0	0.7	
88-S	204	195	187	179	172	165	158	152	147	144	139	134	127	119	114	107	100	94	88	83	78	73	68
	1.7	1.8	1.8	1.9	2.0	2.1	2.2	2.2	2.3	2.3	2.4	2.4	2.5	2.5	2.5	2.5	2.5	2.5	2.5	2.5	2.4	2.4	2.3
	2.1	2.2	2.3	2.4	2.5	2.6	2.6	2.7	2.7	2.7	2.8	2.8	2.8	2.8	2.7	2.7	2.6	2.5	2.4	2.3	2.1	2.0	1.8

4LHC12+2 — 2" Normal Weight Topping

Table of safe superimposed service load (psf) and cambers

Strand Designation Code	23	24	25	26	27	28	29	30	31	32	33	34	35	36	37	38	39	40	41	42	43	44	45
76-S	250	224	202	182	165	149	134	122	110	100	90	81	71	62	53								
	0.5	0.5	0.5	0.5	0.5	0.6	0.6	0.6	0.6	0.6	0.5	0.5	0.5	0.4	0.4								
	0.4	0.4	0.4	0.4	0.3	0.3	0.3	0.2	0.1	0.0	−0.0	−0.2	−0.3	−0.4	−0.6								
58-S	263	247	235	222	213	201	185	169	154	141	129	118	108	99	90	82	75	68	62				
	0.7	0.7	0.8	0.8	0.8	0.9	0.9	0.9	0.9	1.0	1.0	1.0	1.0	1.0	0.9	0.9	0.9	0.8	0.8				
	0.7	0.7	0.7	0.7	0.7	0.7	0.7	0.7	0.6	0.6	0.5	0.4	0.3	0.1	0.0	−0.2	−0.3	−0.6					
68-S	275	259	244	234	222	210	203	193	183	175	163	150	138	127	117	108	99	91	84	77	71	64	
	0.9	0.9	1.0	1.0	1.1	1.1	1.2	1.2	1.3	1.3	1.4	1.4	1.4	1.4	1.4	1.4	1.4	1.4	1.4	1.4	1.3	1.2	
	0.9	1.0	1.0	1.0	1.1	1.1	1.1	1.1	1.1	1.1	1.1	1.1	1.0	0.9	0.8	0.7	0.6	0.4	0.3	0.1	−0.1	−0.5	
78-S	281	265	253	240	228	216	209	199	189	181	173	165	161	154	143	132	123	114	105	97	90	83	77
	1.0	1.1	1.2	1.3	1.3	1.4	1.5	1.5	1.6	1.7	1.7	1.8	1.8	1.9	1.9	1.9	1.9	2.0	2.0	1.9	1.9	1.9	1.9
	1.1	1.2	1.2	1.3	1.3	1.4	1.4	1.4	1.5	1.5	1.5	1.4	1.4	1.4	1.3	1.2	1.2	1.1	0.9	0.8	0.6	0.4	0.2
88-S		274	259	246	237	225	215	205	195	187	179	171	164	157	151	145	142	135	125	117	108	101	94
		1.3	1.4	1.5	1.6	1.7	1.8	1.8	1.9	2.0	2.1	2.2	2.2	2.3	2.3	2.4	2.4	2.5	2.5	2.5	2.5	2.5	2.5
		1.4	1.5	1.6	1.6	1.7	1.8	1.8	1.8	1.9	1.9	1.9	1.9	1.9	1.8	1.8	1.7	1.6	1.5	1.4	1.3	1.1	0.9

Strength based on strain compatibility; bottom tension limited to $6\sqrt{f'_c}$; see pages 2-3—2-5 for explanation

HOLLOW-CORE SLABS

Fig. 2.4.1   Section properties — normal weight concrete

Trade name: Dy-Core®
Licensing Organization: Dy-Core Systems, Inc., Vancouver, British Columbia

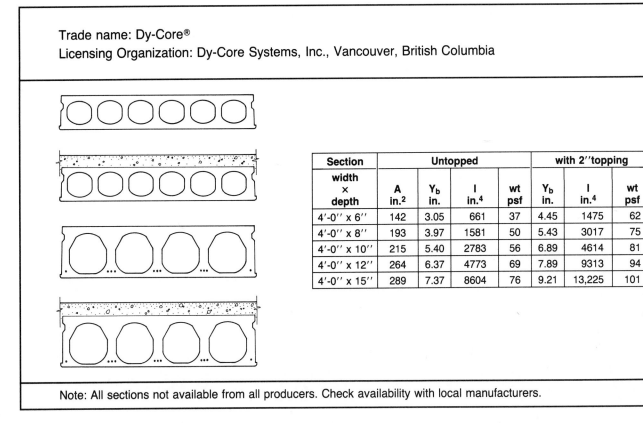

Section	Untopped				with 2"topping		
width × depth	A in.²	Y_b in.	I in.⁴	wt psf	Y_b in.	I in.⁴	wt psf
4'-0'' x 6''	142	3.05	661	37	4.45	1475	62
4'-0'' x 8''	193	3.97	1581	50	5.43	3017	75
4'-0'' x 10''	215	5.40	2783	56	6.89	4614	81
4'-0'' x 12''	264	6.37	4773	69	7.89	9313	94
4'-0'' x 15''	289	7.37	8604	76	9.21	13,225	101

Note: All sections not available from all producers. Check availability with local manufacturers.

Fig. 2.4.2   Section properties — normal weight concrete

Trade name: Dynaspan®
Equipment Manufacturers: Dynamold Corporation, Salina, Kansas

Section	Untopped				with 2"topping		
width × depth	A in.²	Y_b in.	I in.⁴	wt psf	Y_b in.	I in.⁴	wt psf
4'-0'' x 4''	133	2.00	235	35	3.08	689	60
4'-0'' x 6''	165	3.02	706	43	4.25	1543	68
4'-0'' x 8''	233	3.93	1731	61	5.16	3205	86
4'-0'' x 10''	260	4.91	3145	68	6.26	5314	93
8'-0'' x 6''	338	3.05	1445	44	4.26	3106	69
8'-0'' x 8''	470	3.96	3525	61	5.17	6444	86
8'-0'' x 10''	532	4.96	6422	69	6.28	10712	94
8'-0'' x 12''	615	5.95	10505	80	7.32	16507	105

Note: All sections not available from all producers. Check availability with local manufacturers.

HOLLOW-CORE SLABS

Fig. 2.4.3 Section properties — normal weight concrete   **Flexicore**

Trade name: Flexicore®

Licensing Organization: The Flexicore Co. Inc., Dayton, Ohio

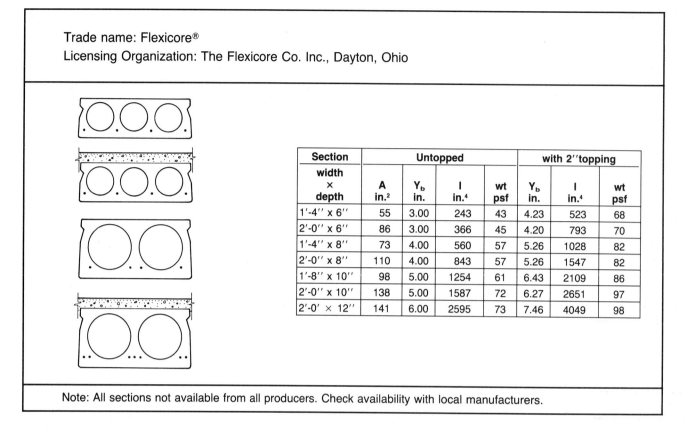

Section	Untopped				with 2″topping		
width × depth	A in.²	Y_b in.	I in.⁴	wt psf	Y_b in.	I in.⁴	wt psf
1′-4″ x 6″	55	3.00	243	43	4.23	523	68
2′-0″ x 6″	86	3.00	366	45	4.20	793	70
1′-4″ x 8″	73	4.00	560	57	5.26	1028	82
2′-0″ x 8″	110	4.00	843	57	5.26	1547	82
1′-8″ x 10″	98	5.00	1254	61	6.43	2109	86
2′-0″ x 10″	138	5.00	1587	72	6.27	2651	97
2′-0′ × 12″	141	6.00	2595	73	7.46	4049	98

Note: All sections not available from all producers. Check availability with local manufacturers.

Fig. 2.4.4 Section properties — normal weight concrete   **Spancrete**

Trade name: Spancrete®

Licensing Organization: Spancrete Machinery Corp., Milwaukee, Wisconsin

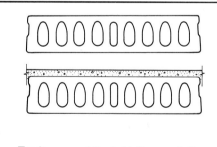

Section	Untopped				with 2″topping		
width × depth	A in.²	Y_b in.	I in.⁴	wt psf	Y_b in.	I in.⁴	wt psf
3′-4″ x 4″	119	1.99	199	37	3.20	659	62
3′-4″ x 6″	160	2.92	635	50	4.28	1585	75
3′-4″ x 8″	218	3.98	1515	68	5.33	3000	93
3′-4″ x 10″	257	5.19	2933	80	6.56	5015	105
3′-4″ x 12″	325	6.20	4981	102	7.54	7976	127

Trade name: Ultralight Spancrete®

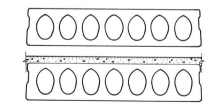

Section	Untopped				with 2″topping		
width × depth	A in.²	Y_b in.	I in.⁴	wt psf	Y_b in.	I in.⁴	wt psf
3′-4″ x 8″	204	4.18	1438	64	5.42	2684	89
3′-4″ x 10″	228	5.22	2641	71	6.59	4466	96
3′-4″ x 12″	254	6.26	4307	79	7.73	6839	104

Note: Spancrete is also available in 48″ and 60″ widths. All sections are not available from all producers. Check availability with local manufacturer.

HOLLOW-CORE SLABS

Fig. 2.4.5 Section properties   **Span-Deck**

Trade name: Span-Deck®
Licensing Organization: Span Deck, Incorporated, Franklin, Tennessee

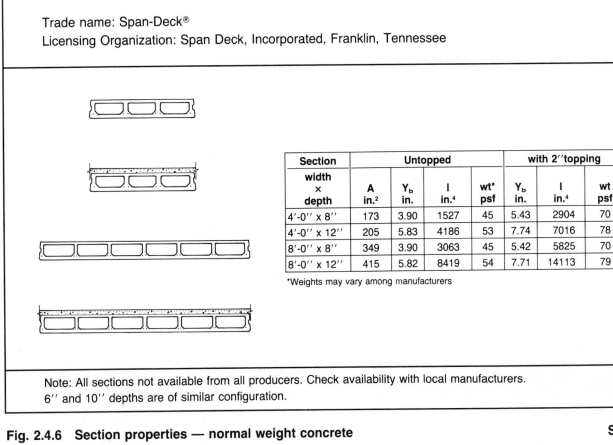

Section	Untopped				with 2''topping		
width × depth	A in.²	Y_b in.	I in.⁴	wt* psf	Y_b in.	I in.⁴	wt psf
4'-0'' x 8''	173	3.90	1527	45	5.43	2904	70
4'-0'' x 12''	205	5.83	4186	53	7.74	7016	78
8'-0'' x 8''	349	3.90	3063	45	5.42	5825	70
8'-0'' x 12''	415	5.82	8419	54	7.71	14113	79

*Weights may vary among manufacturers

Note: All sections not available from all producers. Check availability with local manufacturers.
6'' and 10'' depths are of similar configuration.

Fig. 2.4.6 Section properties — normal weight concrete   **Spiroll**

Trade name: Spiroll,® Corefloor
Licensing Organization: Spiroll Corporation Limited, Winnipeg, Manitoba

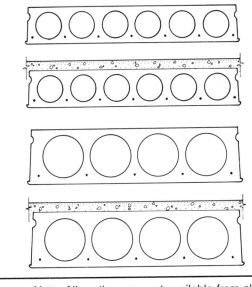

Section	Untopped				with 2''topping		
width × depth	A in.²	Y_b in.	I in.⁴	wt psf	Y_b in.	I in.⁴	wt psf
4'-0'' x 4''	154	2.00	247	40	2.98	723	65
4'-0'' x 6''	188	3.00	764	49	4.13	1641	74
4'-0'' x 8''	214	4.00	1666	56	5.29	3070	81
4'-0'' x 10''	259	5.00	3223	67	6.34	5328	92
4'-0'' x 12''	289	6.00	5272	75	7.43	8195	100

Note: All sections are not available from all producers. Check availability with local manufacturers.

SOLID FLAT SLAB

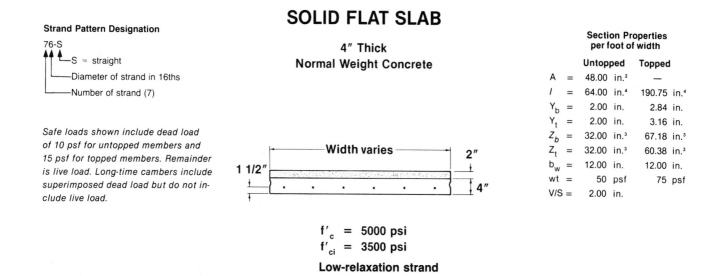

Strand Pattern Designation

76-S

— S = straight
— Diameter of strand in 16ths
— Number of strand (7)

4″ Thick
Normal Weight Concrete

Safe loads shown include dead load of 10 psf for untopped members and 15 psf for topped members. Remainder is live load. Long-time cambers include superimposed dead load but do not include live load.

← Width varies →   2″

1 1/2″   4″

f'_c = 5000 psi
f'_{ci} = 3500 psi

Low-relaxation strand

Section Properties per foot of width

		Untopped	Topped
A	=	48.00 in.²	—
I	=	64.00 in.⁴	190.75 in.⁴
Y_b	=	2.00 in.	2.84 in.
Y_t	=	2.00 in.	3.16 in.
Z_b	=	32.00 in.³	67.18 in.³
Z_t	=	32.00 in.³	60.38 in.³
b_w	=	12.00 in.	12.00 in.
wt	=	50 psf	75 psf
V/S	=	2.00 in.	

Key

167 — Safe superimposed service load, psf
0.1 — Estimated camber at erection, in.
0.1 — Estimated long-time camber, in.

FS4

No Topping

Table of safe superimposed service load (psf) and cambers

Strand Designation Code	Span, ft.							
	10	11	12	13	14	15	16	17
66-S	167	138	116	98	83	66		
	0.1	0.1	0.1	0.1	0.0	0.0		
	0.1	0.1	0.1	0.0	0.0	− 0.1		
76-S	197	164	139	118	98	79		
	0.1	0.1	0.1	0.1	0.1	0.0		
	0.1	0.1	0.1	0.1	0.0	− 0.1		
58-S	253	212	180	154	127	104	86	70
	0.1	0.2	0.2	0.2	0.2	0.1	0.1	0.0
	0.2	0.2	0.2	0.2	0.2	0.1	0.0	− 0.1
68-S	291	234	190	156	129	108	90	75
	0.2	0.2	0.2	0.2	0.2	0.2	0.2	0.1
	0.3	0.3	0.3	0.3	0.2	0.2	0.1	− 0.1

FS4+2

2″ Normal Weight Topping

Table of safe superimposed service load (psf) and cambers

Strand Designation Code	Span, ft.							
	10	11	12	13	14	15	16	17
66-S	315	264	224	167	123	87		
	0.1	0.1	0.1	0.0	0.0	− 0.1		
	0.0	0.0	0.0	− 0.1	− 0.2	− 0.3		
76-S		311	265	203	153	113	80	
		0.1	0.1	0.1	0.0	0.0	− 0.1	
		0.0	0.0	− 0.1	− 0.1	− 0.3	− 0.4	
58-S			274	214	166	127	95	
			0.2	0.1	0.1	0.0	0.0	
			0.0	0.0	− 0.1	− 0.3	− 0.4	
68-S			268	213	169	132		
			0.2	0.2	0.2	0.1		
			0.1	0.0	− 0.1	− 0.3		

Strength based on strain compatibility; bottom tension limited to $6\sqrt{f'_c}$; see pages 2-3–2-5 for explanation

SOLID FLAT SLAB

4″ Thick
Lightweight Concrete

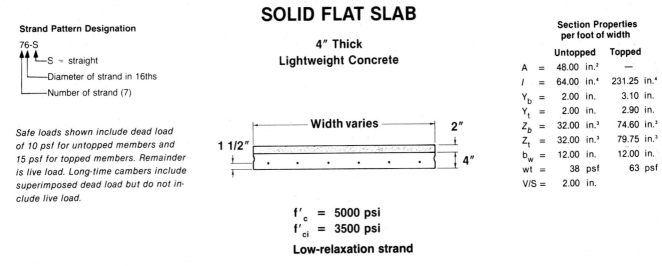

Strand Pattern Designation

76-S
└─ S = straight
└─── Diameter of strand in 16ths
└───── Number of strand (7)

Safe loads shown include dead load of 10 psf for untopped members and 15 psf for topped members. Remainder is live load. Long-time cambers include superimposed dead load but do not include live load.

f'_c = 5000 psi
f'_{ci} = 3500 psi

Low-relaxation strand

Section Properties per foot of width

		Untopped	Topped
A	=	48.00 in.²	—
I	=	64.00 in.⁴	231.25 in.⁴
Y_b	=	2.00 in.	3.10 in.
Y_t	=	2.00 in.	2.90 in.
Z_b	=	32.00 in.³	74.60 in.³
Z_t	=	32.00 in.³	79.75 in.³
b_w	=	12.00 in.	12.00 in.
wt	=	38 psf	63 psf
V/S	=	2.00 in.	

Key

176 — Safe superimposed service load, psf
0.1 — Estimated camber at erection, in.
0.2 — Estimated long-time camber, in.

LFS4

No Topping

Table of safe superimposed service load (psf) and cambers

Strand Designation Code	Span, ft.							
	10	11	12	13	14	15	16	17
66-S	176	148	126	108	93	78		
	0.1	0.1	0.1	0.1	0.1	0.1		
	0.2	0.2	0.2	0.1	0.1	0.0		
76-S	206	174	148	128	110	91	75	
	0.2	0.2	0.2	0.2	0.2	0.2	0.1	
	0.2	0.2	0.2	0.2	0.2	0.1	0.0	
58-S	263	222	190	163	136	113	95	80
	0.3	0.3	0.3	0.3	0.3	0.3	0.3	0.2
	0.3	0.4	0.4	0.3	0.3	0.2	0.1	– 0.1
68-S	301	244	200	166	139	117	99	85
	0.3	0.4	0.4	0.4	0.5	0.5	0.5	0.4
	0.4	0.5	0.5	0.5	0.5	0.4	0.4	0.2

LFS4 + 2

2″ Normal Weight Topping

Table of safe superimposed service load (psf) and cambers

Strand Designation Code	Span, ft.							
	11	12	13	14	15	16	17	18
66-S	273	233	201	163	123	91		
	0.1	0.1	0.1	0.1	0.0	0.0		
	0.1	0.0	– 0.1	– 0.1	– 0.3	– 0.5		
76-S		274	237	197	153	117	86	
		0.2	0.2	0.2	0.1	0.1	0.0	
		0.1	0.0	– 0.1	– 0.2	– 0.3	– 0.6	
58-S		304	258	206	164	129	100	
		0.3	0.3	0.3	0.3	0.2	0.1	
		0.2	0.1	0.0	– 0.1	– 0.3	– 0.5	
68-S			279	229	184	147	116	
			0.5	0.5	0.4	0.4	0.3	
			0.2	0.2	0.0	– 0.1	– 0.3	

Strength based on strain compatibility; bottom tension limited to $6\sqrt{f'_c}$; see pages 2-3–2-5 for explanation

SOLID FLAT SLAB

6″ Thick
Normal Weight Concrete

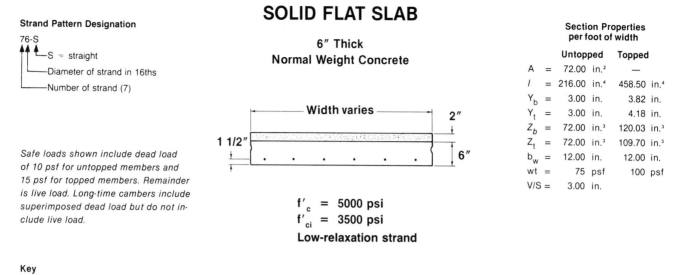

Strand Pattern Designation

76-S

- S = straight
- Diameter of strand in 16ths
- Number of strand (7)

Safe loads shown include dead load of 10 psf for untopped members and 15 psf for topped members. Remainder is live load. Long-time cambers include superimposed dead load but do not include live load.

$f'_c = 5000$ psi
$f'_{ci} = 3500$ psi
Low-relaxation strand

Section Properties per foot of width

		Untopped	Topped
A	=	72.00 in.²	—
I	=	216.00 in.⁴	458.50 in.⁴
Y_b	=	3.00 in.	3.82 in.
Y_t	=	3.00 in.	4.18 in.
Z_b	=	72.00 in.³	120.03 in.³
Z_t	=	72.00 in.³	109.70 in.³
b_w	=	12.00 in.	12.00 in.
wt	=	75 psf	100 psf
V/S	=	3.00 in.	

Key

- 276 — Safe superimposed service load, psf
- 0.1 — Estimated camber at erection, in.
- 0.2 — Estimated long-time camber, in.

FS6

No Topping

Table of safe superimposed service load (psf) and cambers

Strand Designation Code	Span, ft.														
	11	12	13	14	15	16	17	18	19	20	21	22	23	24	25
66-S	276	235	202	175	152	132	110	92	76						
	0.1	0.1	0.1	0.1	0.1	0.1	0.1	0.1	0.0						
	0.2	0.2	0.2	0.2	0.1	0.1	0.1	0.0	– 0.1						
76-S	302	269	242	211	185	160	135	114	96	81					
	0.1	0.2	0.2	0.2	0.2	0.2	0.2	0.1	0.1	0.0					
	0.2	0.2	0.2	0.2	0.2	0.2	0.2	0.1	0.0	– 0.1					
58-S	308	275	248	227	207	187	174	156	134	113	96	81			
	0.2	0.2	0.3	0.3	0.3	0.3	0.3	0.3	0.3	0.2	0.2	0.1			
	0.3	0.3	0.3	0.4	0.4	0.4	0.4	0.3	0.3	0.2	0.1	0.0			
68-S	314	281	254	230	210	193	177	165	152	141	121	103	88	75	
	0.3	0.3	0.3	0.4	0.4	0.4	0.4	0.4	0.4	0.4	0.4	0.3	0.3	0.2	
	0.4	0.4	0.4	0.5	0.5	0.5	0.5	0.5	0.5	0.5	0.4	0.3	0.2	0.0	
78-S	320	287	257	236	213	196	183	168	155	144	134	126	109	94	81
	0.3	0.4	0.4	0.4	0.5	0.5	0.6	0.6	0.6	0.6	0.6	0.6	0.5	0.4	0.4
	0.4	0.5	0.5	0.6	0.7	0.7	0.7	0.7	0.7	0.7	0.7	0.6	0.5	0.4	0.2

FS6 + 2

2″ Normal Weight Topping

Table of safe superimposed service load (psf) and cambers

Strand Designation Code	Span, ft.												
	13	14	15	16	17	18	19	20	21	22	23	24	25
66-S	295	256	213	167	129	99	70						
	0.1	0.1	0.1	0.1	0.1	0.0	0.0						
	0.1	0.1	0.0	0.0	– 0.1	– 0.2	– 0.3						
76-S		303	269	221	177	140	108						
		0.2	0.2	0.2	0.2	0.1	0.1						
		0.1	0.1	0.1	0.0	0.0	– 0.1						
58-S		312	286	261	240	221	181	147	118	93	71		
		0.3	0.3	0.3	0.3	0.3	0.3	0.2	0.2	0.1	0.0		
		0.3	0.3	0.2	0.2	0.2	0.1	0.0	– 0.1	– 0.3	– 0.4		
68-S		318	292	267	246	228	210	193	160	131	105		
		0.4	0.4	0.4	0.4	0.4	0.4	0.4	0.4	0.3	0.3		
		0.4	0.4	0.4	0.4	0.3	0.3	0.2	0.1	0.0	– 0.2		
78-S			298	273	252	231	216	199	185	169	140	115	93
			0.5	0.5	0.6	0.6	0.6	0.6	0.6	0.6	0.5	0.4	0.4
			0.5	0.5	0.5	0.5	0.5	0.4	0.4	0.3	0.1	0.0	– 0.2

Strength based on strain compatibility; bottom tension limited to $6\sqrt{f'_c}$; see pages 2-3–2-5 for explanation

SOLID FLAT SLAB

6" Thick
Lightweight Concrete

Strand Pattern Designation

76-S
- S = straight
- Diameter of strand in 16ths
- Number of strand (7)

Safe loads shown include dead load of 10 psf for untopped members and 15 psf for topped members. Remainder is live load. Long-time cambers include superimposed dead load but do not include live load.

Width varies — 2"

1 1/2" 6"

f'_c = 5000 psi
f'_{ci} = 3500 psi
Low-relaxation strand

Section Properties per foot of width

	Untopped	Topped
A =	72.00 in.²	—
I =	216.00 in.⁴	545.25 in.⁴
Y_b =	3.00 in.	4.11 in.
Y_t =	3.00 in.	3.89 in.
Z_b =	72.00 in.³	132.68 in.³
Z_t =	72.00 in.³	140.18 in.³
b_w =	12.00 in.	12.00 in.
wt =	58 psf	83 psf
V/S =	3.00 in.	

Key
- 291 — Safe superimposed service load, psf
- 0.2 — Estimated camber at erection, in.
- 0.3 — Estimated long-time camber, in.

LFS6

No Topping

Table of safe superimposed service load (psf) and cambers

Strand Designation Code	11	12	13	14	15	16	17	18	19	20	21	22	23	24	25
66-S	291	250	216	189	167	146	124	106	90	77	66				
	0.2	0.2	0.2	0.3	0.3	0.3	0.3	0.2	0.2	0.1	0.1				
	0.3	0.3	0.3	0.3	0.3	0.3	0.2	0.1	0.0	-0.1					
76-S	316	283	257	225	199	174	149	128	110	95	82	71			
	0.2	0.3	0.3	0.3	0.3	0.4	0.4	0.3	0.3	0.3	0.2	0.2			
	0.3	0.4	0.4	0.4	0.4	0.4	0.4	0.4	0.3	0.2	0.1	-0.1			
58-S		289	263	239	222	202	188	170	148	129	113	98	85	73	
		0.4	0.4	0.5	0.5	0.5	0.6	0.6	0.6	0.6	0.6	0.5	0.5	0.4	
		0.5	0.6	0.6	0.7	0.7	0.7	0.7	0.7	0.6	0.6	0.5	0.3	0.1	
68-S		295	269	245	225	208	191	176	167	155	137	120	105	92	81
		0.5	0.5	0.6	0.6	0.7	0.7	0.8	0.8	0.8	0.8	0.8	0.8	0.8	0.7
		0.6	0.7	0.8	0.8	0.9	0.9	1.0	1.0	1.0	1.0	0.9	0.8	0.7	0.5
78-S		298	272	248	228	211	194	179	170	158	149	137	123	109	96
		0.6	0.6	0.7	0.8	0.9	0.9	1.0	1.0	1.1	1.1	1.1	1.1	1.1	1.1
		0.7	0.8	0.9	1.0	1.1	1.2	1.2	1.3	1.3	1.3	1.3	1.2	1.2	1.0

LFS6 + 2

2" Normal Weight Topping

Table of safe superimposed service load (psf) and cambers

Strand Designation Code	14	15	16	17	18	19	20	21	22	23	24	25	26	27
66-S	270	238	202	162	128	99	75							
	0.2	0.2	0.2	0.2	0.2	0.2	0.1							
	0.2	0.1	0.1	0.0	0.0	-0.2	-0.3							
76-S		284	247	209	170	137	109	85						
		0.3	0.3	0.3	0.3	0.3	0.3	0.3						
		0.3	0.2	0.2	0.2	0.1	0.0	-0.1						
58-S	297	272	251	233	208	173	142	116	93	73				
	0.5	0.5	0.5	0.6	0.6	0.6	0.5	0.5	0.5	0.3				
	0.4	0.4	0.4	0.4	0.3	0.3	0.1	0.0	-0.2	-0.4				
68-S	303	281	260	239	224	208	192	169	130	107	87	69		
	0.6	0.7	0.8	0.8	0.8	0.8	0.8	0.8	0.8	0.7	0.6	0.5		
	0.6	0.7	0.7	0.7	0.6	0.6	0.5	0.4	0.2	0.0	-0.2	-0.5		
78-S	309	284	263	245	227	214	200	187	172	138	117	97	79	
	0.8	0.9	0.9	1.0	1.0	1.1	1.1	1.1	1.1	1.1	1.0	1.0	0.9	
	0.8	0.8	0.9	0.9	0.9	0.9	0.8	0.8	0.7	0.4	-0.3	0.0	-0.3	

Strength based on strain compatibility; bottom tension limited to $6\sqrt{f'_c}$; see pages 2-3–2-5 for explanation

SOLID FLAT SLAB

8″ Thick
Normal Weight Concrete

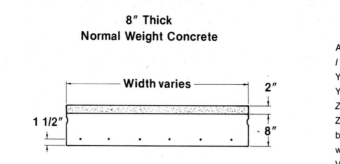

f'_c = 5000 psi
f'_{ci} = 3500 psi

Low-relaxation strand

Strand Pattern Designation

76-S

└─ S = straight
└─── Diameter of strand in 16ths
└───── Number of strand (7)

Safe loads shown include dead load of 10 psf for untopped members and 15 psf for topped members. Remainder is live load. Long-time cambers include superimposed dead load but do not include live load.

Key

266 — Safe superimposed service load, psf
0.1 — Estimated camber at erection, in.
0.2 — Estimated long-time camber, in.

Section Properties per foot of width

	Untopped	Topped
A =	96.00 in.²	—
I =	512.00 in.⁴	907.50 in.⁴
Y_b =	4.00 in.	4.81 in.
Y_t =	4.00 in.	5.19 in.
Z_b =	128.00 in.³	188.68 in.³
Z_t =	128.00 in.³	174.85 in.³
b_w =	12.00 in.	12.00 in.
wt =	100 psf	125 psf
V/S =	4.00 in.	

FS8

No Topping

Table of safe superimposed service load (psf) and cambers

Strand Designation Code	Span, ft.															
	14	15	16	17	18	19	20	21	22	23	24	25	26	27	28	29
66-S	266	233	205	173	146	122	103	86	71							
	0.1	0.1	0.1	0.1	0.1	0.1	0.1	0.0	0.0							
	0.2	0.1	0.1	0.1	0.1	0.1	0.0	0.0	-0.1							
76-S	306	279	249	212	180	153	131	111	94	79	66					
	0.1	0.2	0.2	0.2	0.2	0.1	0.1	0.1	0.1	0.0	-0.1					
	0.2	0.2	0.2	0.2	0.2	0.2	0.1	0.1	0.0	-0.1	-0.2					
58-S	312	285	260	239	221	206	184	160	138	120	103	89	76	65		
	0.2	0.2	0.2	0.3	0.3	0.3	0.3	0.2	0.2	0.2	0.1	0.1	0.0	-0.1		
	0.3	0.3	0.3	0.3	0.3	0.3	0.3	0.3	0.2	0.2	0.1	-0.0	-0.1	-0.3		
68-S	318	291	266	245	227	209	196	181	169	155	136	119	103	88	75	63
	0.3	0.3	0.3	0.3	0.4	0.4	0.4	0.4	0.4	0.4	0.4	0.3	0.3	0.2	0.2	0.1	0.0
	0.4	0.4	0.4	0.5	0.5	0.5	0.5	0.5	0.5	0.4	0.4	0.4	0.3	0.2	-0.1	-0.3
78-S		297	272	251	230	215	199	184	172	160	150	141	128	112	97	83
		0.4	0.4	0.4	0.5	0.5	0.5	0.5	0.5	0.5	0.5	0.4	0.4	0.3	0.2	0.1
		0.5	0.6	0.6	0.6	0.6	0.7	0.7	0.7	0.6	0.6	0.5	0.3	0.2	0.1	-0.1

FS8 + 2

2″ Normal Weight Topping

Table of safe superimposed service load (psf) and cambers

Strand Designation Code	Span, ft.												
	17	18	19	20	21	22	23	24	25	26	27	28	29
66-S	203	161	126	95	69								
	0.1	0.1	0.1	0.0	0.0								
	0.0	0.0	-0.1	-0.1	-0.2								
76-S	270	221	179	144	113	87	64						
	0.1	0.1	0.1	0.1	0.1	0.0	0.0						
	0.1	0.1	0.0	0.0	-0.1	-0.2	-0.3						
58-S		298	273	237	205	178	152	124	100	79			
		0.3	0.3	0.3	0.2	0.2	0.2	0.1	0.1	0.0			
		0.2	0.2	0.2	0.1	0.1	0.0	-0.1	-0.2	-0.4			
68-S		304	283	261	245	225	197	167	140	115	94	74	
		0.4	0.4	0.4	0.4	0.4	0.4	0.3	0.3	0.2	0.2	0.1	
		0.4	0.4	0.3	0.3	0.3	0.2	0.1	0.0	-0.1	-0.3	-0.4	

Strength based on strain compatibility; bottom tension limited to $6\sqrt{f'_c}$; see pages 2-3—2-5 for explanation

SOLID FLAT SLAB

8″ Thick
Lightweight Concrete

Strand Pattern Designation

76-S
- S = straight
- Diameter of strand in 16ths
- Number of strand (7)

Safe loads shown include dead load of 10 psf for untopped members and 15 psf for topped members. Remainder is live load. Long-time cambers include superimposed dead load but do not include live load.

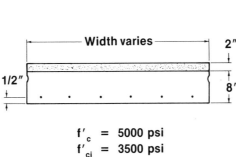

$f'_c = 5000$ psi
$f'_{ci} = 3500$ psi

Low-relaxation strand

Section Properties per foot of width

	Untopped	Topped
A =	96.00 in.²	—
I =	512.00 in.⁴	1058.50 in.⁴
Y_b =	4.00 in.	5.12 in.
Y_t =	4.00 in.	4.88 in.
Z_b =	128.00 in.³	206.75 in.³
Z_t =	128.00 in.³	216.90 in.³
b_w =	12.00 in.	12.00 in.
wt =	77 psf	102 psf
V/S =	4.00 in.	

Key
- 285 — Safe superimposed service load, psf
- 0.2 — Estimated camber at erection, in.
- 0.3 — Estimated long-time camber, in.

LFS8

No Topping

Table of safe superimposed service load (psf) and cambers

Strand Designation Code	14	15	16	17	18	19	20	21	22	23	24	25	26	27	28	29	30	31	32	33
66-S	285	252	225	192	165	142	122	105	90	77	66									
	0.2	0.2	0.2	0.2	0.2	0.2	0.2	0.2	0.1	0.1	0.1									
	0.3	0.3	0.3	0.3	0.2	0.2	0.2	0.1	0.1	0.0	-0.1									
76-S		298	269	231	199	173	150	130	113	98	85	74	64							
		0.3	0.3	0.3	0.3	0.3	0.3	0.3	0.3	0.2	0.2	0.1	0.0							
		0.4	0.4	0.4	0.4	0.4	0.3	0.3	0.3	0.3	0.3	0.0	-0.1							
58-S			279	258	240	225	204	179	158	139	123	108	95	84	74	65				
			0.4	0.5	0.5	0.5	0.5	0.5	0.5	0.5	0.5	0.5	0.4	0.3	0.2	0.1				
			0.6	0.6	0.6	0.6	0.6	0.6	0.6	0.6	0.5	0.4	0.3	0.2	0.0	-0.2				
68-S				243	228	212	200	188	174	155	138	123	109	97	87	76	65			
				0.6	0.7	0.7	0.7	0.7	0.7	0.7	0.7	0.7	0.7	0.7	0.6	0.5	0.4	0.3		
				0.8	0.9	0.9	0.9	0.9	0.9	0.9	0.8	0.7	0.6	0.5	0.3	0.1	-0.1			
78-S						231	218	203	191	180	170	161	149	133	120	107	95	84	74	65
						0.8	0.9	0.9	1.0	1.0	1.0	1.0	1.0	1.0	1.0	0.9	0.8	0.7	0.6	0.5
						1.1	1.1	1.2	1.2	1.2	1.2	1.2	1.2	1.1	1.0	0.9	0.7	0.5	0.3	0.0

LFS8 + 2

2″ Normal Weight Topping

Table of safe superimposed service load (psf) and cambers

Strand Designation Code	17	18	19	20	21	22	23	24	25	26	27	28	29	30	31
66-S	244	208	170	137	109	85	64								
	0.2	0.2	0.2	0.2	0.1	0.1	0.0								
	0.1	0.1	0.0	0.0	-0.1	-0.2	-0.3								
76-S		253	218	185	153	125	100	78							
		0.3	0.3	0.3	0.3	0.2	0.2	0.1							
		0.2	0.2	0.1	0.0	0.1	-0.2	-0.3							
58-S			290	256	225	197	168	141	117	96	77	60			
			0.5	0.5	0.5	0.5	0.5	0.4	0.4	0.3	0.3	0.2			
			0.4	0.4	0.3	0.3	0.2	0.1	0.0	-0.2	-0.3	-0.5			
68-S				278	262	244	216	192	167	142	119	100	82	66	
				0.7	0.7	0.7	0.7	0.7	0.7	0.7	0.6	0.5	0.5	0.4	
				0.6	0.6	0.6	0.5	0.4	0.4	0.2	0.1	-0.1	-0.3	-0.5	
78-S					268	251	236	222	204	182	163	134	113	95	79
					0.9	1.0	1.0	1.0	1.0	1.0	1.0	0.9	0.9	0.8	0.7
					0.9	0.9	0.9	0.8	0.8	0.7	0.6	0.3	0.2	0.0	-0.3

Strength based on strain compatibility; bottom tension limited to $6\sqrt{f'_c}$; see pages 2-3–2-5 for explanation

RECTANGULAR BEAMS

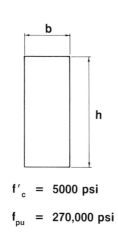

f'_c = 5000 psi

f_{pu} = 270,000 psi

Key

3246 — Safe superimposed service load, plf
0.4 — Estimated camber at erection, in.
0.1 — Estimated long-time camber, in.

Section Properties

Designation	b (in.)	h (in.)	A (in.²)	I (in.⁴)	y_b (in.)	Z (in.³)	wt (plf)
12RB16	12	16	192	4096	8.00	512	200
12RB20	12	20	240	8000	10.00	800	250
12RB24	12	24	288	13,824	12.00	1152	300
12RB28	12	28	336	21,952	14.00	1568	350
12RB32	12	32	384	32,768	16.00	2048	400
12RB36	12	36	432	46,656	18.00	2592	450
16RB24	16	24	384	18,432	12.00	1536	400
16RB28	16	28	448	29,269	14.00	2091	467
16RB32	16	32	512	43,691	16.00	2731	533
16RB36	16	36	576	62,208	18.00	3456	600
16RB40	16	40	640	85,333	20.00	4267	667

Safe loads shown include 50% dead load and 50% live load. 800 psi top tension has been allowed, therefore additional top reinforcement is required.

Table of safe superimposed service load (plf) and cambers

Designation	No. Strand	e	16	18	20	22	24	26	28	30	32	34	36	38	40	42	44	46	48	50
12RB16	5	5.67	3246 0.4 0.1	2527 0.5 0.1	2012 0.6 0.2	1632 0.7 0.2	1342 0.7 0.2	1117 0.8 0.2												
12RB20	8	6.60	5816 0.3 0.1	4548 0.4 0.1	3641 0.5 0.1	2970 0.6 0.2	2459 0.7 0.2	2062 0.8 0.2	1747 0.9 0.2	1493 1.0 0.2	1285 1.1 0.2	1112 1.2 0.2								
12RB24	10	7.76	8585 0.3 0.1	6726 0.4 0.1	5397 0.4 0.1	4413 0.5 0.1	3665 0.6 0.1	3083 0.7 0.2	2621 0.8 0.2	2248 0.9 0.2	1940 1.0 0.2	1684 1.1 0.2	1470 1.1 0.2	1288 1.2 0.2	1133 1.3 0.2	1000 1.4 0.1				
12RB28	12	8.89		9074 0.3 0.1	7290 0.4 0.1	5970 0.5 0.1	4966 0.5 0.2	4184 0.6 0.2	3564 0.7 0.2	3064 0.8 0.2	2655 0.9 0.2	2316 1.0 0.2	2031 1.1 0.2	1791 1.1 0.2	1585 1.2 0.2	1409 1.3 0.2	1255 1.4 0.2	1122 1.4 0.1	1002 1.5 0.1	
12RB32	13	10.48			9584 0.3 0.1	7858 0.4 0.1	6545 0.5 0.1	5524 0.5 0.1	4713 0.6 0.2	4059 0.7 0.2	3524 0.8 0.2	3080 0.9 0.2	2708 0.9 0.2	2394 1.0 0.2	2125 1.1 0.2	1894 1.2 0.2	1694 1.2 0.2	1519 1.3 0.1	1365 1.4 0.1	1230 1.4 0.1
12RB36	15	11.64					8450 0.4 0.1	7140 0.5 0.1	6100 0.6 0.1	5261 0.6 0.2	4575 0.7 0.2	4006 0.8 0.2	3530 0.9 0.2	3123 0.9 0.2	2775 1.0 0.2	2475 1.1 0.2	2215 1.2 0.2	1989 1.2 0.2	1790 1.3 0.2	1614 1.4 0.1
16RB24	13	7.86		8847 0.4 0.1	7098 0.4 0.1	5803 0.5 0.1	4819 0.6 0.1	4052 0.7 0.1	3444 0.8 0.2	2954 0.9 0.2	2552 1.0 0.2	2220 1.1 0.2	1941 1.2 0.1	1705 1.2 0.1	1503 1.3 0.1	1330 1.4 0.1	1180 1.5 0.0			
16RB28	13	8.89			9720 0.4 0.1	7959 0.5 0.1	6621 0.5 0.1	5579 0.6 0.2	4752 0.7 0.2	4086 0.8 0.2	3540 0.9 0.2	3087 1.0 0.2	2708 1.1 0.2	2388 1.1 0.2	2114 1.2 0.2	1878 1.3 0.2	1674 1.4 0.2	1496 1.4 0.1	1335 1.5 0.1	1194 1.5 0.1
16RB32	18	10.29					8808 0.5 0.1	7434 0.5 0.1	6343 0.6 0.2	5464 0.7 0.2	4744 0.8 0.2	4147 0.9 0.2	3647 0.9 0.2	3224 1.0 0.2	2863 1.1 0.2	2549 1.2 0.2	2275 1.3 0.2	2036 1.3 0.2	1827 1.4 0.1	1642 1.5 0.1
16RB36	20	11.64						9519 0.5 0.1	8133 0.6 0.1	7015 0.6 0.2	6100 0.7 0.2	5342 0.8 0.2	4706 0.9 0.2	4165 0.9 0.2	3700 1.0 0.2	3300 1.1 0.2	2954 1.2 0.2	2651 1.2 0.2	2386 1.3 0.2	2152 1.4 0.1
16RB40	22	13.00							8647 0.6 0.1	7527 0.6 0.1	6599 0.7 0.1	5821 0.8 0.2	5163 0.9 0.2	4601 0.9 0.2	4117 1.0 0.2	3698 1.1 0.2	3332 1.1 0.1	3011 1.2 0.1	2728 1.3 0.1	

L-SHAPED BEAMS

Normal Weight Concrete

1'-0" 6"

f'_c = 5000 psi

f_{pu} = 270,000 psi

1'-6"

Section Properties

Designation	h (in.)	h_1/h_2 (in.)	A (in.²)	I (in.⁴)	y_b (in.)	Z_b (in.³)	Z_t (in.³)	wt (plf)
18LB20	20	12/8	288	9696	9.00	1077	882	300
18LB24	24	12/12	360	16,762	10.80	1552	1270	375
18LB28	28	16/12	408	26,611	12.59	2114	1727	425
18LB32	32	20/12	456	39,695	14.42	2753	2258	475
18LB36	36	24/12	504	56,407	16.29	3463	2862	525
18LB40	40	24/16	576	77,568	18.00	4309	3526	600
18LB44	44	28/16	624	103,153	19.85	5197	4271	650
18LB48	48	32/16	672	133,705	21.71	6159	5086	700
18LB52	52	36/16	720	169,613	23.60	7187	5972	750
18LB56	56	40/16	768	211,264	25.50	8285	6927	800
18LB60	60	44/16	816	259,046	27.41	9451	7949	850

Safe loads shown include 50% dead load and 50% live load. 800 psi top tension has been allowed, therefore additional top reinforcement is required.

Key

6486 — Safe superimposed service load, plf
0.3 — Estimated camber at erection, in.
0.1 — Estimated long-time camber, in.

Table of safe superimposed service load (plf) and cambers

Desig- nation	No. Strand	e	16	18	20	22	24	26	28	30	32	34	36	38	40	42	44	46	48	50
18LB20	9	6.26	6486 0.3 0.1	5068 0.4 0.1	4053 0.5 0.1	3303 0.6 0.2	2732 0.7 0.2	2288 0.8 0.2	1935 0.9 0.2	1650 0.9 0.2	1414 1.0 0.2	1218 1.1 0.2								
18LB24	10	7.67	9179 0.3 0.1	7182 0.3 0.1	5753 0.4 0.1	4696 0.5 0.1	3891 0.5 0.1	3266 0.6 0.1	2769 0.7 0.1	2369 0.7 0.1	2041 0.8 0.1	1769 0.9 0.1	1541 1.0 0.1	1349 1.0 0.0	1184 1.1 0.0					
18LB28	12	8.93		8039 0.3 0.1	6578 0.4 0.1	5466 0.5 0.1	4600 0.6 0.1	3914 0.6 0.2	3360 0.7 0.2	2906 0.8 0.2	2531 0.8 0.2	2216 0.9 0.2	1949 1.0 0.1	1722 1.0 0.1	1524 1.1 0.1	1351 1.1 0.1	1200 1.2 0.2			
18LB32	14	10.22			8814 0.4 0.1	7331 0.4 0.1	6176 0.5 0.1	5260 0.6 0.2	4521 0.7 0.2	3916 0.7 0.2	3414 0.8 0.2	2994 0.9 0.2	2639 0.9 0.2	2335 1.0 0.2	2074 1.1 0.2	1847 1.1 0.1	1650 1.2 0.1	1476 1.2 0.1	1323 1.3 0.0	
18LB36	16	11.52				9358 0.4 0.1	7903 0.5 0.1	6744 0.5 0.1	5807 0.6 0.1	5040 0.7 0.2	4405 0.7 0.2	3872 0.8 0.2	3422 0.9 0.2	3037 0.9 0.2	2706 1.0 0.2	2419 1.1 0.2	2168 1.1 0.1	1948 1.2 0.1	1755 1.2 0.1	
18LB40	18	12.52					9693 0.4 0.1	8284 0.5 0.1	7146 0.5 0.2	6215 0.6 0.2	5443 0.7 0.2	4797 0.7 0.2	4250 0.8 0.2	3783 0.9 0.2	3380 0.9 0.2	3026 1.0 0.2	2718 1.0 0.2	2447 1.1 0.2	2208 1.1 0.2	
18LB44	19	14.19							8729 0.5 0.1	7601 0.6 0.2	6666 0.6 0.2	5883 0.7 0.2	5219 0.7 0.2	4653 0.8 0.2	4166 0.9 0.2	3743 0.9 0.2	3370 1.0 0.2	3042 1.0 0.2	2752 1.1 0.2	
18LB48	21	15.48								9166 0.5 0.1	8048 0.6 0.1	7110 0.6 0.1	6313 0.7 0.2	5629 0.8 0.2	5041 0.8 0.2	4531 0.9 0.2	4086 0.9 0.2	3695 1.0 0.2	3351 1.0 0.2	
18LB52	23	16.78									9538 0.5 0.2	8427 0.6 0.2	7486 0.7 0.2	6683 0.7 0.2	5992 0.8 0.2	5393 0.8 0.2	4871 0.9 0.2	4412 0.9 0.2	4007 1.0 0.2	
18LB56	25	18.07									9842 0.6 0.2	8752 0.6 0.2	7820 0.7 0.2	7019 0.7 0.2	6324 0.8 0.2	5718 0.8 0.2	5186 0.9 0.2	4717 1.0 0.2		
18LB60	27	19.36										9026 0.6 0.2	8116 0.7 0.2	7326 0.7 0.2	6630 0.8 0.2	6020 0.9 0.2	5481 0.9 0.2			

INVERTED TEE BEAMS

$f'_c = 5000$ psi

$f_{pu} = 270,000$ psi

Normal Weight Concrete

Section Properties

Designation	h (in.)	h_1/h_2 (in.)	A (in.²)	I (in.⁴)	y_b (in.)	Z_b (in.³)	Z_t (in.³)	wt (plf)
24IT20	20	12/8	336	10,981	8.29	1325	938	350
24IT24	24	12/12	432	19,008	10.00	1901	1358	450
24IT28	28	16/12	480	30,131	11.60	2598	1837	500
24IT32	32	20/12	528	44,969	13.27	3388	2401	550
24IT36	36	24/12	576	63,936	15.00	4262	3045	600
24IT40	40	24/16	672	87,845	16.57	5301	3749	700
24IT44	44	28/16	720	116,877	18.27	6397	4542	750
24IT48	48	32/16	768	151,552	20.00	7578	5413	800
24IT52	52	36/16	816	192,275	21.76	8836	6358	850
24IT56	56	40/16	864	239,445	23.56	10,163	7381	900
24IT60	60	44/16	912	293,460	25.37	11,567	8474	950

Key
6888 — Safe superimposed service load, plf
0.3 — Estimated camber at erection, in.
0.1 — Estimated long-time camber, in.

Safe loads shown include 50% dead load and 50% live load. 800 psi top tension has been allowed, therefore additional top reinforcement is required.

Table of safe superimposed service load (plf) and cambers

Desig-nation	No. Strand	e	16	18	20	22	24	26	28	30	32	34	36	38	40	42	44	46	48	50
24IT20	9	6.20	6888	5376	4294	3494	2886	2412	2033	1726	1474	1266								
			0.3	0.3	0.4	0.5	0.6	0.6	0.7	0.8	0.8	0.9								
			0.1	0.1	0.1	0.1	0.1	0.1	0.1	0.1	0.1	0.0								
24IT24	11	7.17	9759	7625	6099	4970	4111	3443	2913	2485	2135	1845	1601							
			0.2	0.3	0.3	0.4	0.5	0.5	0.6	0.7	0.7	0.8	0.8							
			0.1	0.1	0.1	0.1	0.1	0.1	0.1	0.1	0.0	0.0	0.0							
24IT28	13	8.44			8505	6951	5768	4848	4118	3529	3047	2648	2313	2030	1786					
					0.3	0.4	0.4	0.5	0.6	0.6	0.7	0.7	0.8	0.8	0.9					
					0.1	0.1	0.1	0.1	0.1	0.1	0.1	0.1	0.1	0.0	0.0					
24IT32	15	9.77				9248	7691	6480	5519	4744	4109	3583	3138	2760	2437	2159	1919	1709		
						0.3	0.4	0.5	0.5	0.6	0.6	0.7	0.8	0.8	0.9	0.9	1.0	1.0		
						0.1	0.1	0.1	0.1	0.1	0.1	0.1	0.1	0.1	0.1	0.1	0.0	0.0		
24IT36	16	11.50				9879	8337	7114	6127	5320	4644	4077	3598	3189	2836	2531	2265	2031	1825	
						0.4	0.4	0.5	0.5	0.6	0.6	0.6	0.7	0.7	0.8	0.9	0.9	0.9	1.0	1.0
						0.1	0.1	0.1	0.1	0.1	0.1	0.1	0.1	0.1	0.1	0.1	0.0	0.0	0.0
24IT40	19	12.02							8675	7475	6494	5680	4998	4421	3928	3504	3137	2816	2535	2286
									0.4	0.5	0.5	0.6	0.6	0.7	0.7	0.8	0.8	0.9	0.9	1.0
									0.1	0.1	0.1	0.1	0.1	0.1	0.1	0.1	0.1	0.0	0.0
24IT44	20	13.73							9300	8083	7075	6230	5514	4903	4378	3922	3525	3176	2868
									0.4	0.5	0.5	0.6	0.6	0.7	0.7	0.8	0.8	0.9	0.9
									0.1	0.1	0.1	0.1	0.1	0.1	0.1	0.1	0.1	0.1
24IT48	22	15.08								9723	8522	7515	6663	5935	5309	4766	4293	3877	3510
										0.5	0.5	0.6	0.6	0.7	0.7	0.8	0.8	0.9	0.9
										0.1	0.1	0.1	0.1	0.1	0.1	0.1	0.1	0.1	0.1
24IT52	24	16.44									8917	7916	7061	6326	5688	5132	4644	4213
											0.5	0.6	0.6	0.7	0.7	0.8	0.8	0.9
											0.1	0.1	0.1	0.1	0.1	0.1	0.1	0.1
24IT56	26	17.82										9279	8287	7433	6692	6046	5480	4979
											0.6	0.6	0.7	0.7	0.8	0.8	0.9
											0.1	0.1	0.1	0.1	0.1	0.1	0.1
24IT60	28	19.18											9597	8616	7766	7025	6374	5800
												0.6	0.6	0.7	0.7	0.8	0.8
												0.1	0.1	0.1	0.1	0.2	0.1

PCI Design Handbook

PRECAST, PRESTRESSED COLUMNS

Fig. 2.6.1 Design strength interaction curves for precast, prestressed concrete columns

Criteria

1. Minimum prestress = 225 psi
2. All strand assumed 1/2 in. diameter, f_{pu} = 270 ksi
3. Curves shown for partial development of strand near member end, where $f_{ps} \approx f_{se}$
4. Horizontal portion of curve is the maximum for tied columns = $0.80\phi P_o$
5. ϕ = 0.9 for ϕP_n = 0
   = 0.7 for $\phi P_n \geqslant 0.10f'_c A_g$

   Varies from 0.9 to 0.7 for points between

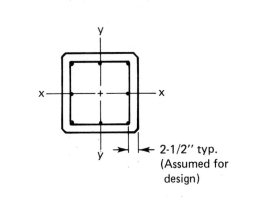

2-1/2'' typ.
(Assumed for design)

Use of curves

1. Enter at left with applied factored axial load, P_u
2. Enter at bottom with applied magnified factored moment, δM_u
3. Intersection point must be to the left of curve indicating required concrete strength.

Notation

ϕP_n = Design axial strength
ϕM_n = Design flexural strength
ϕP_o = Design axial strength at zero ecentricity
A_g = Gross area of the column
δ = Moment magnifier (Sect. 10.11, ACI 318-83)

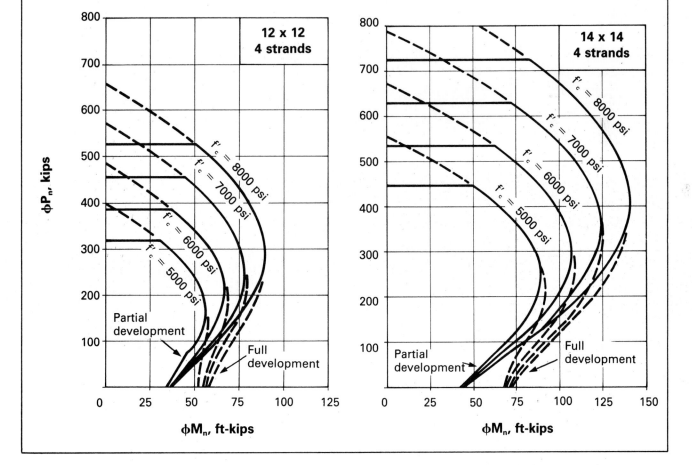

PRECAST, PRESTRESSED COLUMNS

Fig. 2.6.1 (cont.) Design strength interaction curves for precast, prestressed concrete columns

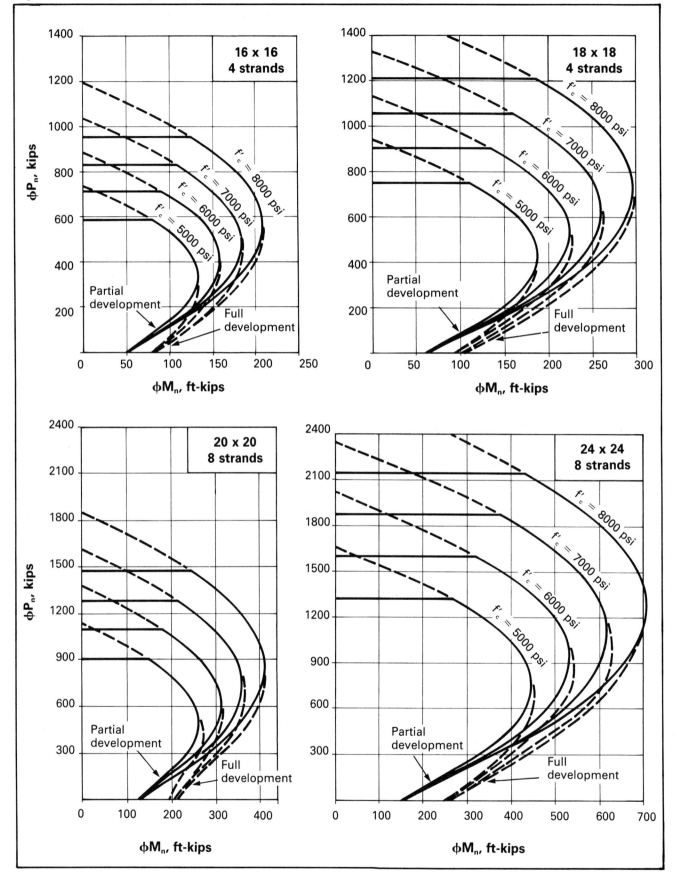

PRECAST, REINFORCED COLUMNS

Fig. 2.6.2 Design strength interaction curves for precast, reinforced concrete columns

Criteria

1. Concrete f'_c = 5000 psi
2. Reinforcement f_y = 60,000 psi
3. Curves shown for full development of reinforcement
4. Horizontal portion of curve is the maximum for tied columns = $0.80\phi P_o$.
5. ϕ = 0.9 for $\phi P_n = 0$
   = 0.7 for $\phi P_n \geqslant 0.10 f'_c A_g$

   Varies from 0.9 to 0.7 for points between

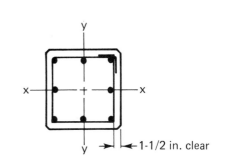

1-1/2 in. clear

Use of curves

1. Enter at left with applied factored axial load, P_u
2. Enter at bottom with applied magnified factored moment, δM_u
3. Intersection point must be to the left of curve indicating required reinforcement.

Notation

ϕP_n = Design axial strength
ϕM_n = Design flexural strength
ϕP_o = Design axial strength at zero eccentricity
A_g = Gross area of the column
δ = Moment magnifier (Sect. 10.11, ACI 318-83)

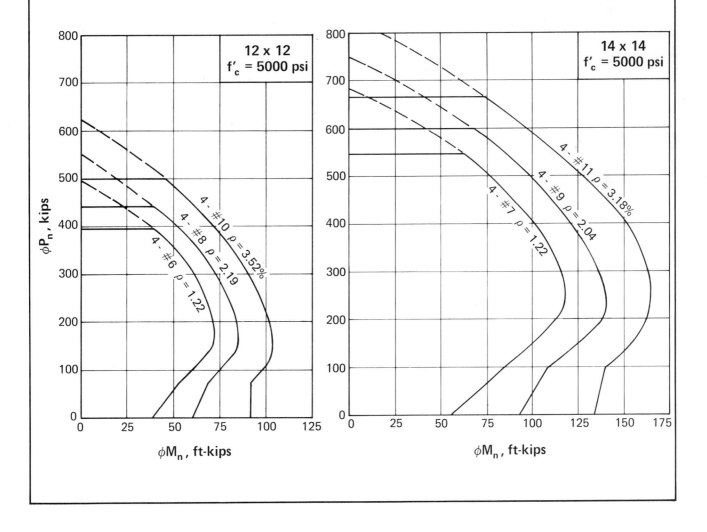

PRECAST, REINFORCED COLUMNS

Fig. 2.6.2 (Cont.) Design strength interaction curves for precast, reinforced concrete columns

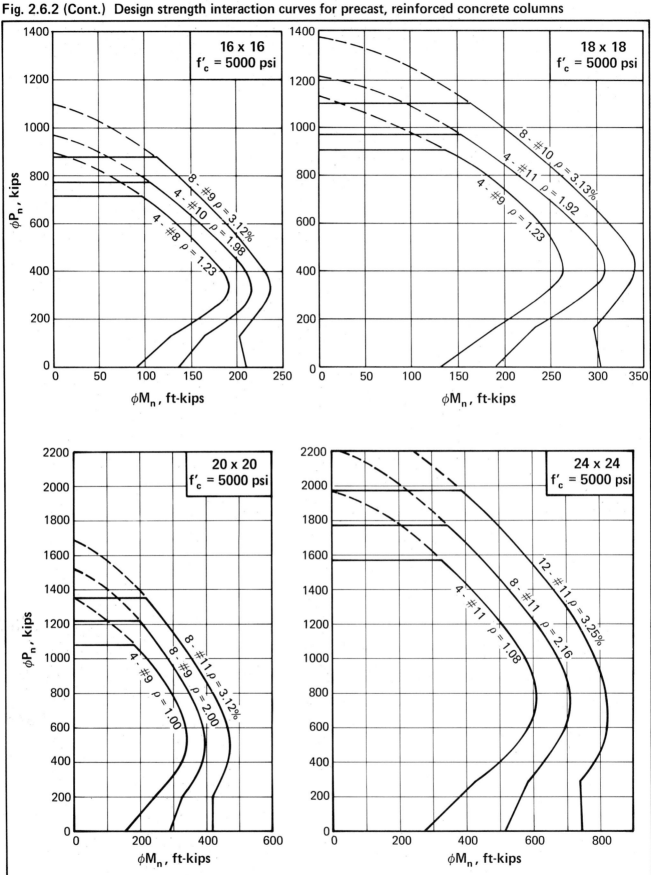

DOUBLE TEE WALL PANELS

Fig. 2.6.3 Partial interaction curve for prestressed double tee wall panels

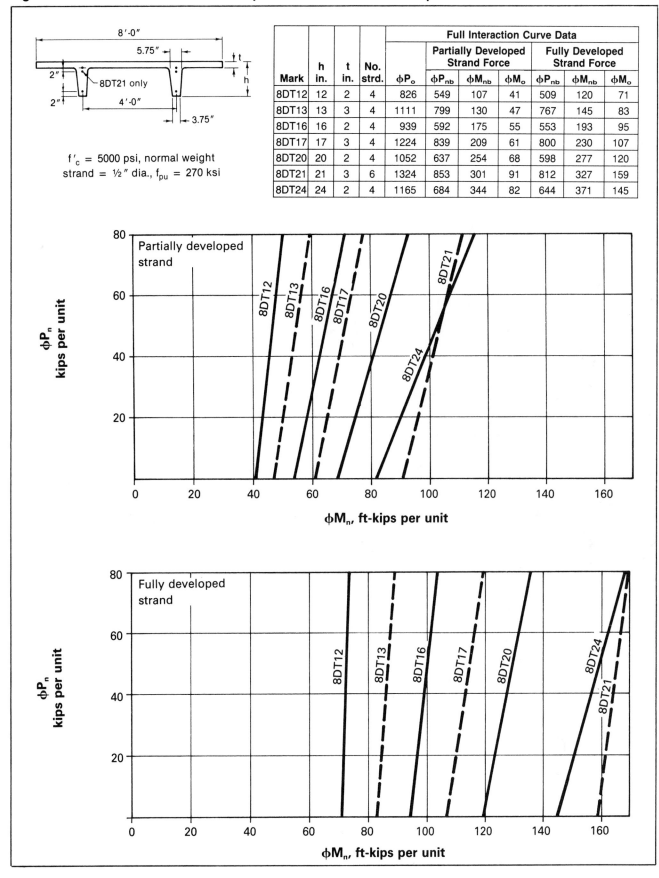

$f'_c = 5000$ psi, normal weight
strand = ½" dia., $f_{pu} = 270$ ksi

| | | | | | Full Interaction Curve Data | | | | | |
| Mark | h in. | t in. | No. strd. | ϕP_o | Partially Developed Strand Force | | | Fully Developed Strand Force | | |
					ϕP_{nb}	ϕM_{nb}	ϕM_o	ϕP_{nb}	ϕM_{nb}	ϕM_o
8DT12	12	2	4	826	549	107	41	509	120	71
8DT13	13	3	4	1111	799	130	47	767	145	83
8DT16	16	2	4	939	592	175	55	553	193	95
8DT17	17	3	4	1224	839	209	61	800	230	107
8DT20	20	2	4	1052	637	254	68	598	277	120
8DT21	21	3	6	1324	853	301	91	812	327	159
8DT24	24	2	4	1165	684	344	82	644	371	145

HOLLOW-CORE WALL PANELS

Fig. 2.6.4 Partial interaction curve for prestressed hollow-core wall panels

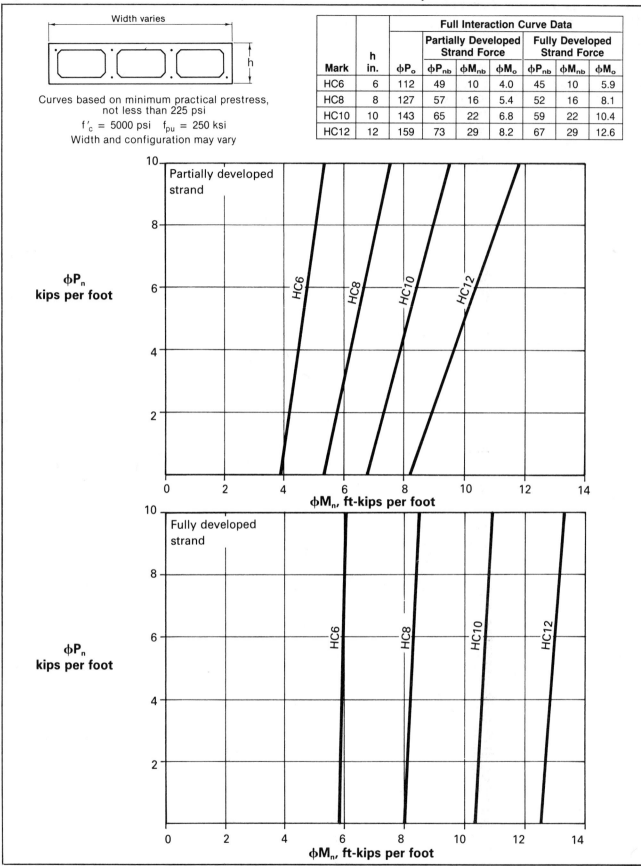

Width varies

Curves based on minimum practical prestress,
not less than 225 psi

f'_c = 5000 psi f_{pu} = 250 ksi

Width and configuration may vary

Mark	h in.	ϕP_o	Full Interaction Curve Data					
			Partially Developed Strand Force			Fully Developed Strand Force		
			ϕP_{nb}	ϕM_{nb}	ϕM_o	ϕP_{nb}	ϕM_{nb}	ϕM_o
HC6	6	112	49	10	4.0	45	10	5.9
HC8	8	127	57	16	5.4	52	16	8.1
HC10	10	143	65	22	6.8	59	22	10.4
HC12	12	159	73	29	8.2	67	29	12.6

PCI Design Handbook

PRECAST, PRESTRESSED SOLID WALL PANELS

Fig. 2.6.5 Partial interaction curve for prestressed solid wall panels

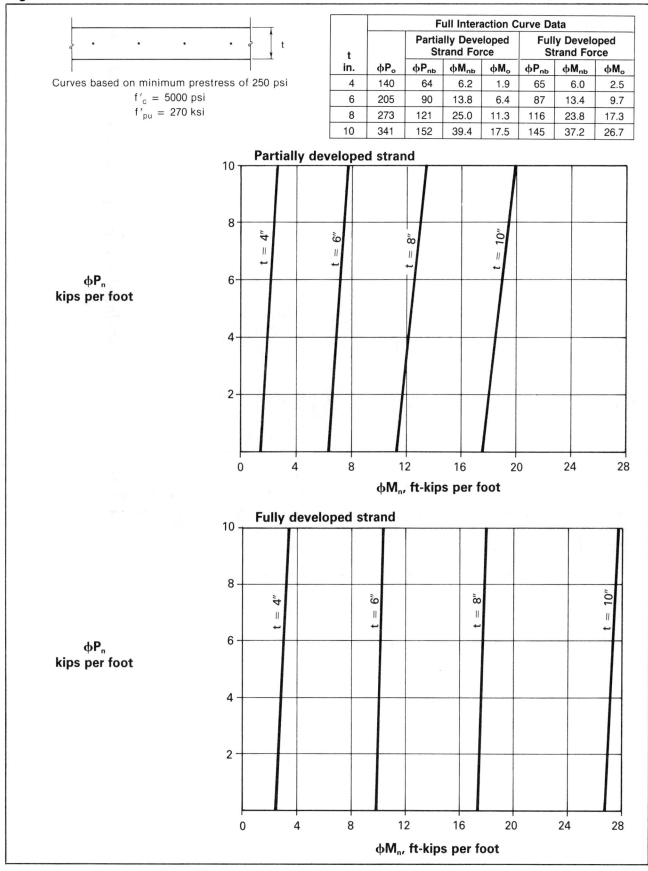

Curves based on minimum prestress of 250 psi
$f'_c = 5000$ psi
$f'_{pu} = 270$ ksi

t in.	ϕP_o	Full Interaction Curve Data					
		Partially Developed Strand Force			Fully Developed Strand Force		
		ϕP_{nb}	ϕM_{nb}	ϕM_o	ϕP_{nb}	ϕM_{nb}	ϕM_o
4	140	64	6.2	1.9	65	6.0	2.5
6	205	90	13.8	6.4	87	13.4	9.7
8	273	121	25.0	11.3	116	23.8	17.3
10	341	152	39.4	17.5	145	37.2	26.7

PRECAST, REINFORCED SOLID WALL PANELS

Fig. 2.6.6 Partial interaction curve for precast, reinforced concrete wall panels

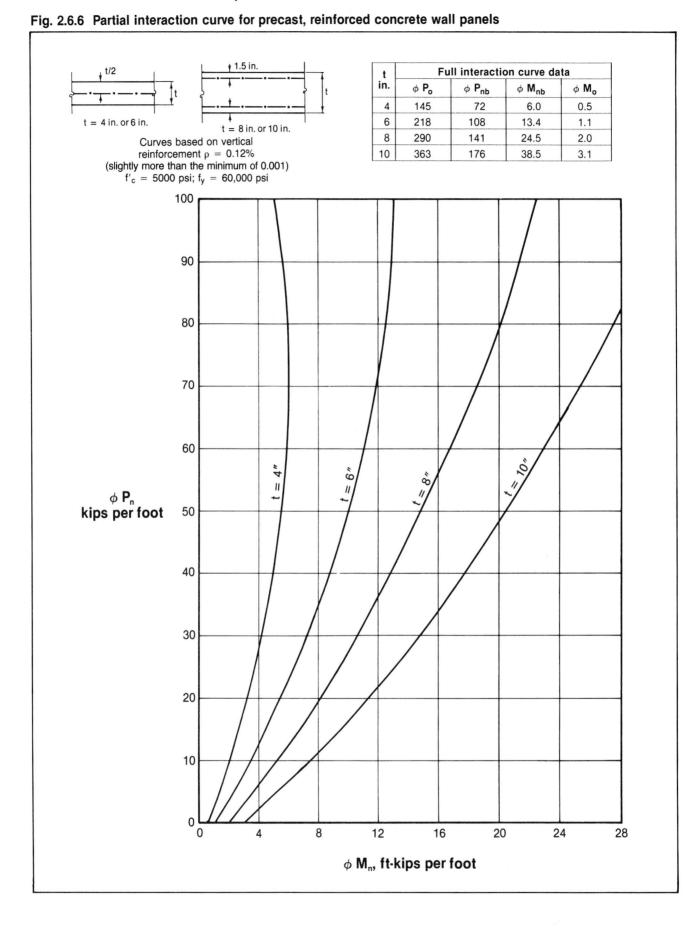

Curves based on vertical
reinforcement ρ = 0.12%
(slightly more than the minimum of 0.001)
f'_c = 5000 psi; f_y = 60,000 psi

t in.	Full interaction curve data			
	ϕP_o	ϕP_{nb}	ϕM_{nb}	ϕM_o
4	145	72	6.0	0.5
6	218	108	13.4	1.1
8	290	141	24.5	2.0
10	363	176	38.5	3.1

PILES

Table 2.7.1 Section properties and allowable loads of prestressed concrete piles

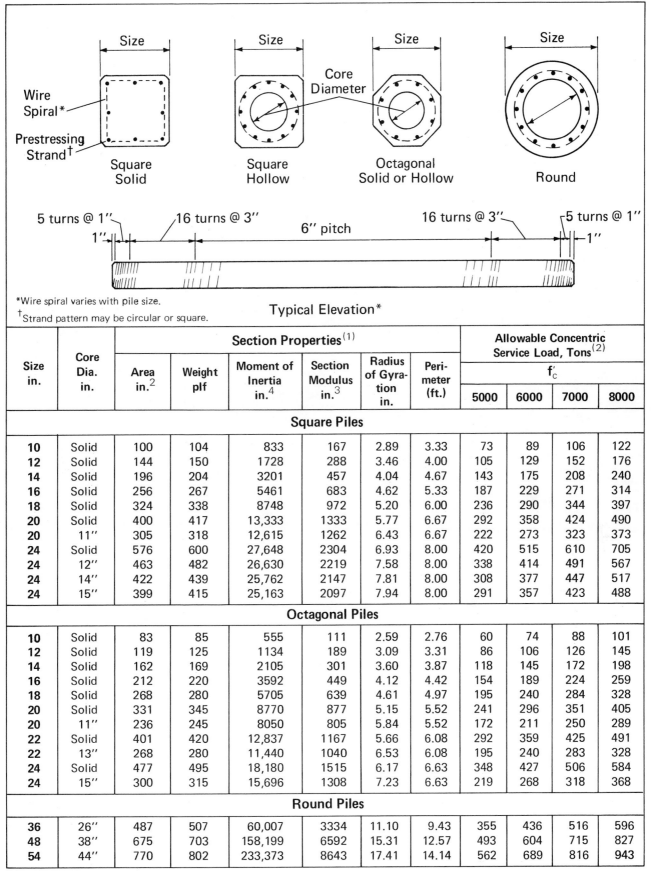

*Wire spiral varies with pile size.

†Strand pattern may be circular or square.

Typical Elevation*

Size in.	Core Dia. in.	Section Properties[1]						Allowable Concentric Service Load, Tons[2]			
		Area in.^{2}	Weight plf	Moment of Inertia in.^{4}	Section Modulus in.^{3}	Radius of Gyration in.	Perimeter (ft.)	f'_c			
								5000	6000	7000	8000
Square Piles											
10	Solid	100	104	833	167	2.89	3.33	73	89	106	122
12	Solid	144	150	1728	288	3.46	4.00	105	129	152	176
14	Solid	196	204	3201	457	4.04	4.67	143	175	208	240
16	Solid	256	267	5461	683	4.62	5.33	187	229	271	314
18	Solid	324	338	8748	972	5.20	6.00	236	290	344	397
20	Solid	400	417	13,333	1333	5.77	6.67	292	358	424	490
20	11″	305	318	12,615	1262	6.43	6.67	222	273	323	373
24	Solid	576	600	27,648	2304	6.93	8.00	420	515	610	705
24	12″	463	482	26,630	2219	7.58	8.00	338	414	491	567
24	14″	422	439	25,762	2147	7.81	8.00	308	377	447	517
24	15″	399	415	25,163	2097	7.94	8.00	291	357	423	488
Octagonal Piles											
10	Solid	83	85	555	111	2.59	2.76	60	74	88	101
12	Solid	119	125	1134	189	3.09	3.31	86	106	126	145
14	Solid	162	169	2105	301	3.60	3.87	118	145	172	198
16	Solid	212	220	3592	449	4.12	4.42	154	189	224	259
18	Solid	268	280	5705	639	4.61	4.97	195	240	284	328
20	Solid	331	345	8770	877	5.15	5.52	241	296	351	405
20	11″	236	245	8050	805	5.84	5.52	172	211	250	289
22	Solid	401	420	12,837	1167	5.66	6.08	292	359	425	491
22	13″	268	280	11,440	1040	6.53	6.08	195	240	283	328
24	Solid	477	495	18,180	1515	6.17	6.63	348	427	506	584
24	15″	300	315	15,696	1308	7.23	6.63	219	268	318	368
Round Piles											
36	26″	487	507	60,007	3334	11.10	9.43	355	436	516	596
48	38″	675	703	158,199	6592	15.31	12.57	493	604	715	827
54	44″	770	802	233,373	8643	17.41	14.14	562	689	816	943

(1) Form dimensions may vary with producers, with corresponding variations in section properties.
(2) Allowable loads based on $N = A_c (0.33 f'_c - 0.27 f_{pc})$; f_{pc} = 700 psi. Check local producer for available concrete strengths.

SHEET PILES

Table 2.7.2 Section properties and allowable moments of prestressed sheet piles

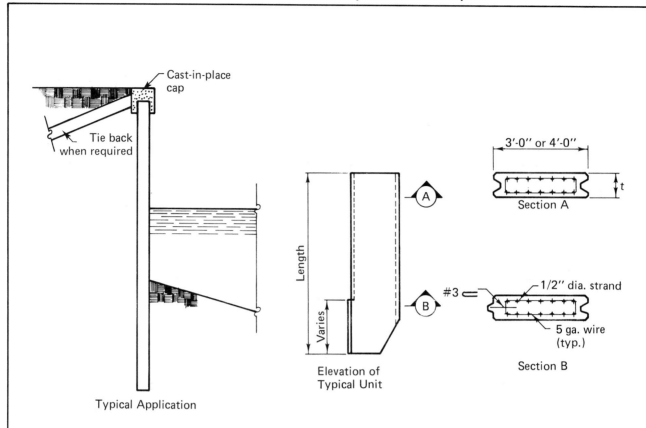

Cast-in-place cap

Tie back when required

Typical Application

Length

Varies

Elevation of Typical Unit

Section A

3'-0" or 4'-0"

t

#3

1/2" dia. strand

5 ga. wire (typ.)

Section B

Thickness t in.	Section Properties per Foot of Width				Maximum Allowable Service Load Moment(2) ft-kips per foot	
	Area in.²	Weight(1) psf	Moment of Inertia in.⁴	Section Modulus in.³	f'_c = 5000 psi	f'_c = 6000 psi
6(3)	72	75	216	72	6.0	7.2
8(3)	96	100	512	128	10.6	12.8
10	120	125	1000	200	16.6	20.0
12	144	150	1728	288	24.0	28.8
16	192	200	4096	512	42.7	51.2
18	216	225	5832	648	54.0	64.8
20	240	250	8000	800	66.7	80.0
24	288	300	13,824	1152	96.0	115.2

(1) Normal weight concrete
(2) Based on zero tension and maximum $0.4f'_c$ compression
(3) Strand can be placed in a single layer in thin sections. Where site conditions require it, strand may be placed eccentrically.

PART 3
ANALYSIS AND DESIGN OF PRECAST, PRESTRESSED CONCRETE STRUCTURES

		Page No.
3.1	General	3–3
	3.1.1 Notation	3–3
	3.1.2 Introduction	3–4
3.2	Preliminary Analysis	3–4
	3.2.1 Framing Dimensions	3–5
	3.2.2 Span to Depth Ratios	3–5
	3.2.3 Lateral Load Resisting Systems	3–5
	3.2.4 Control of Volume Change Deformations and Restraint Forces	3–5
	3.2.5 Connection Concepts	3–6
3.3	Volume Changes	3–6
	3.3.1 Axial Volume Change Strains	3–6
	3.3.2 Thermal Bowing	3–11
	3.3.3 Expansion Joints	3–14
3.4	Component Analysis	3–16
	3.4.1 Non-Bearing Wall Panels	3–16
	3.4.2 Load Bearing Wall Panels	3–17
	3.4.3 Non-Bearing Spandrels	3–18
	3.4.4 Load Bearing Spandrels	3–19
	3.4.5 Eccentrically Loaded Columns	3–20
3.5	Slenderness Effects in Columns and Wall Panels	3–22
	3.5.1 Second-Order (P-Δ) Analysis	3–22
	3.5.2 Moment Magnification Method	3–23
3.6	Diaphragm Design	3–27
	3.6.1 Method of Analysis	3–27
	3.6.2 Shear Transfer between Members	3–27
	3.6.3 Chord Forces	3–29
3.7	Shear Wall Buildings	3–29
	3.7.1 General	3–29
	3.7.2 Rigidity of Solid Shear Walls	3–30
	3.7.3 Distribution of Lateral Loads	3–31
	3.7.4 Unsymmetrical Shear Walls	3–31
	3.7.5 Coupled Shear Walls	3–32
	3.7.6 Shear Walls with Large Openings	3–33
	3.7.7 Architectural Panels as Shear Walls	3–42

3.8   Buildings with Moment-Resisting Frames 3–43
    3.8.1   General... 3–43
    3.8.2   Moment Resistance of Column Bases....................... 3–43
    3.8.3   Fixity of Column Bases................................. 3–47
    3.8.4   Modeling Partially Fixed Bases......................... 3–48
    3.8.5   Volume Change Effects in Moment-Resisting Frames 3–48
    3.8.6   Computer Models for Frame Analyses..................... 3–55

3.9   Shear Wall-Frame Interaction 3–55

3.10 Earthquake Analysis 3–56
    3.10.1   Notation.. 3–56
    3.10.2   General... 3–57
    3.10.3   Building Code Requirements.......................... 3–57
    3.10.4   Design Guidelines for Wall Panels 3–58
    3.10.5   Concept of Box-Type Buildings....................... 3–59
    3.10.6   Structural Layout and Connections 3–59
    3.10.7   Example – One-Story Building........................ 3–60
    3.10.8   Example – 4-Story Building 3–65
    3.10.9   Example – 12-Story Building......................... 3–72
    3.10.10   Example – 23-Story Building 3–72

3.11 References 3–74

ANALYSIS AND DESIGN OF
PRECAST, PRESTRESSED CONCRETE STRUCTURES

3.1 General

3.1.1 Notation

Note: Notation for earthquake analysis is in Sect. 3.10.1

A	=	area (with subscripts)
A_{vf}	=	area of shear-friction reinforcement
b	=	width of a section or structure
C	=	coefficient of thermal expansion
C	=	compressive force
C_m	=	a factor relating actual moment to equivalent uniform moment
C_u	=	factored compressive force
D	=	dead load
e	=	eccentricity of axial load
E	=	modulus of elasticity (with subscripts)
E_t	=	modulus of elasticity modified for time-dependent effects
f'_c	=	concrete compressive stress
f'_m	=	compressive strength of masonry
f_r	=	modulus of rupture of concrete
f_t	=	unfactored tensile stress
f_{ut}	=	factored tensile stress
f_y	=	yield strength of non-prestressed reinforcement
F_b	=	degree of base fixity (decimal)
F_i	=	lateral force at bay i or in shear wall i
F_u	=	factored force
g	=	assumed length over which elongation of the anchor bolt takes place
G	=	shear modulus of elasticity (modulus of rigidity)
h	=	column width in direction of bending
h_s	=	story height
H_u	=	total factored lateral force within a story
I	=	moment of inertia
I_b	=	moment of inertia of a beam
I_{bp}	=	moment of inertia of base plate (vertical cross-section dimensions)
I_c	=	moment of inertia of a column
I_{eq}	=	approximate moment of inertia that results in a flexural deflection equal to the combined shear and flexural deflections of a wall
I_f	=	moment of inertia of the footing (plan dimensions)
I_g	=	uncracked moment of inertia
I_p	=	polar moment of inertia
k	=	effective length factor
k_b, k_f, k_m	=	coefficients used to determine forces and moments in beams and columns
k_s	=	coefficient of subgrade reaction
K	=	stiffnesses (with subscripts)
K_ℓ	=	constant used for the calculation of equivalent creep and shrinkage shortening
K_r	=	relative stiffness
K_t	=	constant used for the calculation of equivalent temperature shortening
ℓ	=	length of span or structure
ℓ_n	=	clear span
ℓ_s	=	distance from column to center of stiffness
ℓ_u	=	unbraced length
ℓ_w	=	length of weld
M	=	unfactored moment
M_c	=	factored moment to be used for design of compression member
M_R	=	resisting moment
M_T	=	torsional moment
M_u	=	factored moment
M_{2b}	=	value of larger factored end moment on compression member due to loads that result in no appreciable sidesway
M_{2s}	=	value of larger factored end moment on compression member due to loads that result in appreciable sidesway
N	=	unfactored horizontal force
N_u	=	factored horizontal force
P	=	applied axial load
P_c	=	critical load
P_o	=	axial load nominal strength of a compression member with zero eccentricity
P_o	=	prestressing force after assumed initial loss
P_u	=	factored axial load
Q	=	stability index
Q	=	statical moment

r	=	radius of gyration
r	=	rigidity
R_{du}	=	factored dead load reaction
t	=	thickness
T	=	tensile force
T_u	=	factored tensile force
T_1, T_2	=	outside, inside temperature
v_r	=	unit shear on panel edge
v_u	=	factored unit shear
V_n	=	nominal shear strength
$V_{R,L}$	=	shear at right, left support
V_u	=	factored shear force
V_w	=	total wind shear
w	=	uniform load
W	=	total load
x_1	=	distance from face of column to center of anchor bolts
x_2	=	distance from face of column to base plate anchorage
Z	=	section modulus
α	=	rotation
β_d	=	dead load/total load ratio
γ	=	flexibility coefficient (with subscripts)
δ	=	moment magnifier
δ	=	volume change shortening (with subscripts)
Δ	=	total equivalent shortening or column deflection
Δ_u	=	deflection due to factored loads
η	=	see Fig. 3.5.1
θ	=	see Fig. 3.5.1
λ	=	see Fig. 3.5.1
μ_e	=	effective shear-friction coefficient
ϕ	=	strength reduction factor
ϕ	=	rotation (with subscripts)
ψ	=	ratio of column to beam stiffnesses

3.1.2 Introduction

This chapter provides guidelines for the analysis and design of structures that are comprised wholly or partially of precast and/or precast, prestressed components. The primary advantages of precast concrete include:

1. Construction speed
2. Plant-controlled quality control
3. Fire resistance and durability
4. With prestressing: greater span-depth ratios,

more controllable performance, less material usage

5. Architectural precast concrete provides a wide variety of highly attractive surfaces and shapes

6. Thermal and acoustical control.

To fully realize these benefits and thereby gain the most economical and effective use of the material, the following general principles are offered:

1. Precast concrete is basically a "simple-span" material. While continuity can be, and often is, effectively achieved with properly concieved connection details, this is not an inherent quality of the method.

2. Sizes and shapes of members are often limited by production, hauling and erection considerations.

3. Concrete is a massive material. This is an advantage for such things as stability under wind loads, acoustical and vibration control and fire resistance. Also, the high dead to live load ratio will provide a greater safety factor against gravity overloads. However, framing details, or loading conditions such as earthquake, which result in eccentrically loaded supports need careful attention.

4. Maximum economy is achieved with maximum repetition, and standard sections should be used whenever possible.

5. Successful use is largely dependent on carefully conceived connection design and details.

6. The effects of restraint of volume changes caused by creep, shrinkage and temperature change must be considered in every structure.

7. While architectural panels are often used only as cladding, the inherent load-carrying capacity of these products should not be overlooked.

8. Prestressing improves the economy and performance of precast members, but is usually only feasible when they are standard shapes and capable of being cast in "long-line" beds.

3.2 Preliminary Analysis

Maximum economy occurs when the building is laid out to take advantage of the principles discussed above. The primary considerations in preliminary analysis of the total structure are:

1. Framing dimensions
2. Span to depth ratios

3. Lateral load resisting systems
4. Control of volume change deformations and restraint forces
5. Connection concepts.

3.2.1 Framing Dimensions

When possible, bays should be laid out to fit the module of the components selected. Standard section dimensions are shown in Part 2, but others may be available in a given area. Width of wall and deck units may be limited by over the road hauling regulations.

It is often feasible to cast wall panels and columns in multi-story units, and economy is achieved with fewer pieces to handle. Length and weight limitations for hauling and erection, and stability during erection, are items to be considered in this decision.

3.2.2 Span to Depth Ratios

Selection of floor to floor dimensions should consider the practical span-depth ratio of the horizontal framing members, allowing adequate space for mechanical ductwork.

Typical span to depth ratios of flexural precast, prestressed concrete members are:

Hollow-core floor slabs	30 to 40
Hollow-core roof slabs	40 to 50
Stemmed floor slabs	25 to 35
Stemmed roof slabs	35 to 40
Beams	10 to 20

These values are intended as guidelines, not limits. The required depth of a beam or slab is influenced by the ratio of live load to total load. Where this ratio is high, deeper sections may be needed.

For non-prestressed flexural members, span-depth ratios are given in ACI 318-83, Sect. 9.5.2.1.

3.2.3 Lateral Load Resisting Systems

Often the most time consuming task in the preliminary analysis is the selection of the lateral load resisting system. Methods used to resist lateral loads, in the approximate order of economy, include:

1. Shear walls. These can be of precast concrete, cast-in-place concrete, or masonry. These are discussed in more detail in Sect. 3.7. When architectural or structural precast members are used for the exterior cladding, they can often be used as shear walls.
2. Cantilevered columns or wall panels. This is usually only feasible in low-rise buildings.

Base fixity can be attained through a moment couple between the footing and ground-floor slab, or by fixing the column to the footing. In the latter case, a detailed analysis of the footing rotation can be made as described in Sect. 3.8.2.

3. Steel or concrete X-bracing. This has been used effectively in some mid-rise buildings, and is largely dependent on connection details. A related resistance system, in terms of analysis procedures, occurs naturally in parking structures with sloped decks in the direction of traffic flow.
4. Moment resisting frames. This is the least desirable system, because of the volume change restraint forces that can build up. However, building function sometimes makes it necessary. It is sometimes feasible to provide a moment connection at only one end of a member, or a connection that will resist moments with lateral forces in one direction but not in the other, in order to reduce the buildup of restraint forces. To reduce the number of moment frames required, a combined shear wall-frame system may be used. Moment resisting frames are discussed in more detail in Sect. 3.8.

All of the above systems depend on distribution of lateral loads through diaphragm action of the roof and floor systems (see Sect. 3.6).

3.2.4 Control of Volume Change Deformations and Restraint Forces

Volume changes of concrete are those resulting from creep, shrinkage and temperature change. Creep and shrinkage cause a shortening of the member, so the critical temperature change condition is nearly always the result of temperature drop.

It is usually desirable to design connections so that volume change shortening is accommodated. Sect. 3.3 provides data and guidelines for estimating the amount of shortening that may take place.

Most of the severe problems that have been caused by restraint of volume change movements have appeared when relatively long members, usually stemmed deck units, were welded to their supports at the bottom on both ends. When such members are connected only at the top, experience has shown that volume changes are adequately accommodated. An unyielding top connection may attract negative moments if compression resistance is encountered at the bottom. This may be difficult to accommodate.

Connections using cast-in-place concrete have

exhibited few volume change problems. This is probably because micro-cracking and creep in the cast-in-place portion effectively relieve the restraint.

Long buildings may require full height expansion joints. This is discussed in Sect. 3.3.3.

3.2.5 Connection Concepts

The types of connections to be used should be determined during the preliminary analysis, as this may have an effect on the component dimensions, the overall structural behavior, as discussed above, and on the erection procedure. Part 6 and some PCI publications are devoted entirely to connections.

3.3 Volume Changes

The strains resulting from creep, shrinkage and temperature change, and the forces caused by restraining these strains have important effects on connections, service load behavior and ultimate capacity of precast, prestressed structures. Consequently, these strains and forces must be considered in the design.

Vertical members, such as load bearing wall panels, are also subject to volume change strains. The approximate magnitude can be calculated using Tables 3.3.1 through 3.3.5, adding the dead load stress to the prestress. The effects will only be significant in high rise buildings, and then only differential movements between elements will significantly affect performance of the structure. This can occur, for example, at the corner of a building where load bearing and non-load bearing panels meet.

3.3.1 Axial Volume Change Strains

Tables 3.3.1 through 3.3.5 and Figs. 3.3.1 and 3.3.2 provide the data needed to determine volume change strains.[1,2] These values can be reduced in rigid frames (see Sect. 3.8.5).

Example 3.3.1 Calculation of volume change shortening

Given:

Heated structure in Denver, Colorado
Normal weight concrete beam – 12RB28
8-1/2-in. diameter, 270K, stress-relieved strands
Initial tension = $0.70f_{pu}$
Assume initial prestress loss = 10%
Release strength = 4500 psi (accelerated cure)
Length = 24 ft

Problem:

Determine the actual shortening that can be anticipated from:
a. Casting to erection at 60 days
b. Erection to the end of service life.

Solution:

From Figs. 3.3.1 and 3.3.2:

Design temperature = 70° F
Average ambient relative humidity = 55%

Prestress force:

A_{ps} = 8 (0.153) = 1.224 in.²
P_o = 1.224 (270) (0.70) (0.90)
    = 208.2 kips
P_o/A = 208.2 (1000) / (12 × 28)
      = 620 psi

Volume/surface ratio
    = (12 × 28) / [(2 × 12) + (2 × 28)]
    = 4.2 in.

a. At 60 days:

From Table 3.3.1:

Creep strain = 169 × 10⁻⁶ in./in.
Shrinkage strain = 292 × 10⁻⁶ in./in.

From Table 3.3.2:

Creep correction factor
    = 0.88 + (20/200) (1.18 − 0.88) = 0.91

From Table 3.3.3:

Creep correction = 1.17 − 0.5 (1.17 − 1.08)
               = 1.13
Shrinkage correction = 1.29 − 0.5 (1.29 − 1.14)
                  = 1.22

From Table 3.3.4:

Creep correction = 0.48 − 0.2 (0.48 − 0.36)
               = 0.46
Shrinkage correction
    = 0.46 − 0.2 (0.46 − 0.31) = 0.43

(Note: Temperature shortening is not significant for this calculation.)

Total strain:

Creep = 169 × 10⁻⁶ (0.91) (1.13) (0.46)
       = 80 × 10⁻⁶ in./in.
Shrinkage = 292 × 10⁻⁶ (1.22) (0.43)
            = 153 × 10⁻⁶ in./in.

Fig. 3.3.1 Maximum seasonal climatic temperature change, deg F

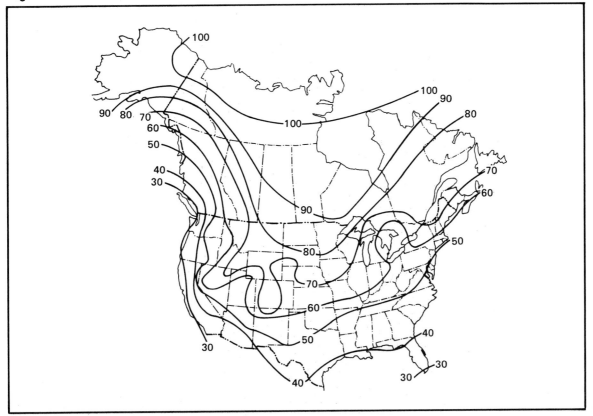

Fig. 3.3.2 Annual average ambient relative humidity, percent

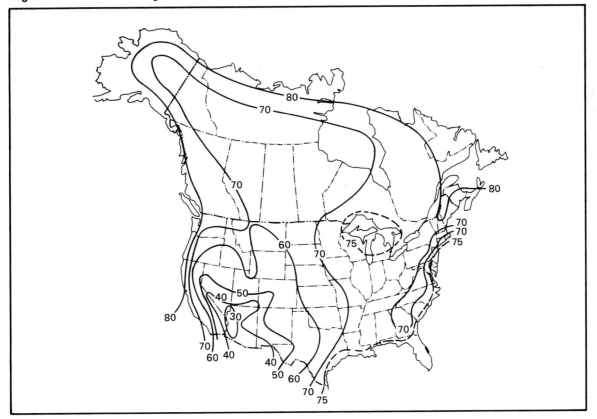

Table 3.3.1 Creep and shrinkage strains (millionths)

Concrete Release Strength = 3500 psi
Average Prestress = 600 psi
Relative Humidity = 70%
Volume / Surface Ratio = 1.5 in.

Time, days	Creep		Shrinkage	
	Normal weight	Lightweight	Accelerated cure	Moist cure
1	29	43	10	14
3	51	76	29	40
5	65	97	47	64
7	76	114	63	85
9	86	127	79	104
10	90	133	86	113
20	118	176	149	185
30	137	204	198	235
40	150	224	236	272
50	161	239	267	300
60	169	252	292	322
70	177	263	314	340
80	183	272	332	355
90	188	280	348	367
100	193	287	361	378
200	222	331	439	434
1 Yr	244	363	487	465
3 Yr	273	407	533	494
5 Yr	283	422	544	500
Final	315	468	560	510

Table 3.3.2 Correction factors for prestress and concrete strength (creep only)

Ave. P/A (psi)	Release Strength, f_{ci} (psi)						
	2500	3000	3500	4000	4500	5000	6000
0	0.00	0.00	0.00	0.00	0.00	0.00	0.00
200	0.39	0.36	0.33	0.31	0.29	0.28	0.25
400	0.79	0.72	0.67	0.62	0.59	0.56	0.51
600	1.18	1.08	1.00	0.94	0.88	0.84	0.76
800	1.58	1.44	1.33	1.25	1.18	1.12	1.02
1000	1.97	1.80	1.67	1.56	1.47	1.39	1.27
1200	2.37	2.16	2.00	1.87	1.76	1.67	1.53
1400	2.76	2.52	2.33	2.18	2.06	1.95	1.78
1600		2.88	2.67	2.49	2.35	2.23	2.04
1800		3.24	3.00	2.81	2.65	2.51	2.29
2000			3.33	3.12	2.94	2.79	2.55
2200				3.43	3.23	3.07	2.80
2400				3.74	3.53	3.35	3.06
2600					3.82	3.63	3.31
2800						3.90	3.56
3000						4.18	3.82

Table 3.3.3 Correction factors for relative humidity

Ave. ambient R.H. (from Fig. 3.3.2)	Creep	Shrinkage
40	1.25	1.43
50	1.17	1.29
60	1.08	1.14
70	1.00	1.00
80	0.92	0.86
90	0.83	0.43
100	0.75	0.00

Table 3.3.4 Correction factors for volume/surface ratio

Time, days	Creep V/S 1	2	3	4	5	6	Shrinkage V/S 1	2	3	4	5	6
1	1.30	0.78	0.49	0.32	0.21	0.15	1.25	0.80	0.50	0.31	0.19	0.11
3	1.29	0.78	0.50	0.33	0.22	0.15	1.24	0.80	0.51	0.31	0.19	0.11
5	1.28	0.79	0.51	0.33	0.23	0.16	1.23	0.81	0.52	0.32	0.20	0.12
7	1.28	0.79	0.51	0.34	0.23	0.16	1.23	0.81	0.52	0.33	0.20	0.12
9	1.27	0.80	0.52	0.35	0.24	0.17	1.22	0.82	0.53	0.34	0.21	0.12
10	1.26	0.80	0.52	0.35	0.24	0.17	1.21	0.82	0.53	0.34	0.21	0.13
20	1.23	0.82	0.56	0.39	0.27	0.19	1.19	0.84	0.57	0.37	0.23	0.14
30	1.21	0.83	0.58	0.41	0.30	0.21	1.17	0.85	0.59	0.40	0.26	0.16
40	1.20	0.84	0.60	0.44	0.32	0.23	1.15	0.86	0.62	0.42	0.28	0.17
50	1.19	0.85	0.62	0.46	0.34	0.25	1.14	0.87	0.63	0.44	0.29	0.19
60	1.18	0.86	0.64	0.48	0.36	0.26	1.13	0.88	0.65	0.46	0.31	0.20
70	1.17	0.86	0.65	0.49	0.37	0.28	1.12	0.88	0.66	0.48	0.32	0.21
80	1.16	0.87	0.66	0.51	0.39	0.29	1.12	0.89	0.67	0.49	0.34	0.22
90	1.16	0.87	0.67	0.52	0.40	0.31	1.11	0.89	0.68	0.50	0.35	0.23
100	1.15	0.87	0.68	0.53	0.42	0.32	1.11	0.89	0.69	0.51	0.36	0.24
200	1.13	0.90	0.74	0.61	0.51	0.42	1.08	0.92	0.75	0.59	0.44	0.31
1 Yr	1.11	0.91	0.77	0.67	0.58	0.50	1.07	0.93	0.79	0.64	0.50	0.38
3 Yr	1.10	0.92	0.81	0.73	0.67	0.62	1.06	0.94	0.82	0.71	0.59	0.47
5 Yr	1.10	0.92	0.82	0.75	0.70	0.66	1.06	0.94	0.83	0.72	0.61	0.49
Final	1.09	0.93	0.83	0.77	0.74	0.72	1.05	0.95	0.85	0.75	0.64	0.54

Table 3.3.5 Design temperature strains* (millionths)

Temperature zone (from Fig. 3.3.1)	Normal weight Heated	Unheated	Lightweight Heated	Unheated
10	30	45	25	38
20	60	90	50	75
30	90	135	75	113
40	120	180	100	150
50	150	225	125	188
60	180	270	150	225
70	210	315	175	263
80	240	360	200	300
90	270	405	225	338
100	300	450	250	375

* Based on accepted coefficients of thermal expansion, reduced to account for thermal lag.
(See referenced committee report,[2] *PCI Journal,* September-October 1977)

Table 3.3.6 Volume change strains for typical building elements (millionths)

Temp. zone (from map)	Normal weight concrete Ave. R.H. (from map)					Lightweight concrete Ave. R.H. (from map)				
Prestressed members (P/A = 600 psi)										
	40	50	60	70	80	40	50	60	70	80
Heated Buildings										
0	584	533	483	432	382	617	564	512	459	407
10	614	563	513	462	412	642	589	537	484	432
20	644	593	543	492	442	667	614	562	509	457
30	674	623	573	522	472	692	639	587	534	482
40	704	653	603	552	502	717	664	612	559	507
50	734	683	633	582	532	742	689	637	584	532
60	764	713	663	612	562	767	714	662	609	557
70	794	743	693	642	592	792	739	687	634	582
80	824	773	723	672	622	817	764	712	659	607
90	854	803	753	702	652	842	789	737	684	632
100	884	833	783	732	682	867	814	762	709	657
Unheated Structures										
0	584	533	483	432	382	617	564	512	459	407
10	629	578	528	477	427	654	602	549	497	444
20	674	623	573	522	472	692	639	587	534	482
30	719	668	618	567	517	729	677	624	572	519
40	764	713	663	612	562	767	714	662	609	557
50	809	758	708	657	607	804	752	699	647	594
60	854	803	753	702	652	842	789	737	684	632
70	899	848	798	747	697	879	827	774	722	669
80	944	893	843	792	742	917	864	812	759	707
90	989	938	888	837	787	954	902	849	797	744
100	1034	983	933	882	832	992	939	887	834	782

Table 3.3.7 Volume change strains for typical building elements (millionths)

Temp. zone (from map)	Normal weight concrete Ave. R.H. (from map)					Lightweight concrete Ave. R.H. (from map)				
Non-prestressed members										
	40	50	60	70	80	40	50	60	70	80
Heated Buildings										
0	269	242	215	188	161	269	242	215	188	161
10	299	272	245	218	191	294	267	240	213	186
20	329	302	275	248	221	319	292	265	238	211
30	359	332	305	278	251	344	317	290	263	236
40	389	362	335	308	281	369	342	315	288	261
50	419	392	365	338	311	394	367	340	313	286
60	449	422	395	368	341	419	392	365	338	311
70	479	452	425	398	371	444	417	390	363	336
80	509	482	455	428	401	469	442	415	388	361
90	539	512	485	458	431	494	467	440	413	386
100	569	542	515	488	461	519	492	465	438	411
Unheated Structures										
0	269	242	215	188	161	269	242	215	188	161
10	314	287	260	233	206	306	279	252	226	199
20	359	332	305	278	251	344	317	290	263	236
30	404	377	350	323	296	381	354	327	301	274
40	449	422	395	368	341	419	392	365	338	311
50	494	467	440	413	386	456	429	402	376	349
60	539	512	485	458	431	494	467	440	413	386
70	584	557	530	503	476	531	504	477	451	424
80	629	602	575	548	521	569	542	515	488	461
90	674	647	620	593	566	606	579	552	526	499
100	719	692	665	638	611	644	617	590	563	536

Total strain = 233×10^{-6} in./in.

Total shortening = 233×10^{-6} (24) (12)
= 0.07 in.

b. At final:

From Table 3.3.1:

Creep strain = 315×10^{-6} in./in.
Shrinkage strain = 560×10^{-6} in./in.

Factors from Tables 3.3.2 and 3.3.3 same as for 60 days.

From Table 3.3.4:

Creep correction = 0.77 − 0.2 (0.77 − 0.74)
= 0.76

Shrinkage correction = 0.75 − 0.2 (0.75 − 0.64)
= 0.73

From Table 3.3.5:

Temperature strain = 210×10^{-6} in./in.

Total creep and shrinkage strain:

Creep = 315×10^{-6} (0.91) (1.13) (0.76)
= 246×10^{-6} in./in.

Shrinkage = 560×10^{-6} (1.22) (0.73)
= 499×10^{-6} in./in.

Total = 745×10^{-6} in./in.

Difference from 60 days to final
= 745 − 233
= 512×10^{-6} in./in.

Total strain
= 512 + 210
= 722×10^{-6} in./in.

Total shortening
= 722×10^{-6} (24) (12)
= 0.21 in.

For the effects of shortening in frame structures, see Sect. 3.8.5.

The behavior of actual structures indicates that reasonable estimates of volume change characteristics are satisfactory for the design of most structures even though test data relating volume changes to the variables shown in Tables 3.3.1 through 3.3.5 exhibit a considerable scatter. Therefore, it is possible to reduce the variables and use approximate values as shown in Tables 3.3.6 and 3.3.7.

Example 3.3.2 Determine volume change shortening by Tables 3.3.6 and 3.3.7

Given:
Same as Example 3.3.1

Solution:
For prestressed, normal weight concrete, in a heated building, use Table 3.3.6.
For 55% relative humidity and 70°F temperature, interpolating from Table 3.3.6:

Actual strain = 718×10^{-6} in./in.

This result compares with the value of 722×10^{-6} calculated from Tables 3.3.1 through 3.3.5.

3.3.2 Thermal Bowing

Temperature difference between the inside and outside of a wall panel, especially composite insulated sandwich panels, or between the top and underside of an uninsulated roof deck, can cause the members to bow. The theoretical magnitude of bowing (see Fig. 3.3.3) can be determined by:

$$\Delta = \frac{C(T_1 - T_2)\,\ell^2}{8h} \qquad \text{(Eq. 3.3.1)}$$

where:

C = coef. of thermal expansion
T_2, T_1 = inside and outside temperature
ℓ = distance between supports
h = member thickness

Limited records of temperature measurements indicate that in open structures, such as the roofs

Fig. 3.3.3 Thermal bow of wall panel

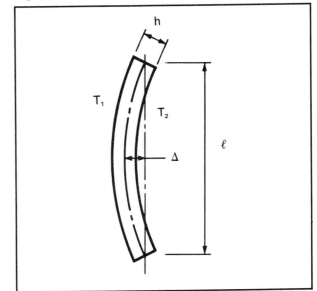

of parking decks, the temperature differential $(T_1 - T_2)$ seldom exceeds 30 to 40°F. In an insulated sandwich wall panel, the theoretical difference can be higher, but this is tempered by "thermal lag" due to the mass of the concrete (see Sect. 9.1).

Moisture differences between the inside and outside of an enclosed building can also cause bowing, however, calculation is much less precise and involves more variables. The exterior layer of the concrete panel absorbs moisture from the atmosphere and periodic precipitation, while the interior layer is relatively dry, especially when the building is heated. This causes the inside layer to shrink more than the outside, causing an outward bow. This would tend to balance the theoretical inward thermal bowing in cold weather, which is believed to explain the observation that "wall panels always bow out".

Example 3.3.3 Thermal bow in a wall panel

Given:

A 20 ft high, 6 in. thick wall panel as shown below. Assume a coefficient of thermal expansion $C = 6 \times 10^{-6}$ in./in./°F; a temperature differential $T_1 - T_2 = 35°F$; $E_c = 4300$ ksi.

Problem:

Determine the potential thermal bow Δ_1, the force, P, required at midheight to restrain the bowing, the stress in the panel caused by the restraint, and the residual bow, Δ_2.

Solution:

From Eq. 3.3.1:

$$\Delta_1 = \frac{(6 \times 10^{-6})(35)(20 \times 12)^2}{8(6)} = 0.25 \text{ in.}$$

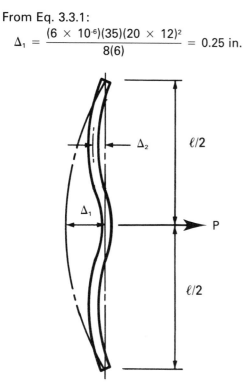

From Table 3.3.8:

$$E_t = 0.75(4300) = 3225 \text{ ksi}$$

$$I = \frac{bh^3}{12} = \frac{12(6)^3}{12} = 216 \text{ in.}^4/\text{ft}$$

From Table 3.3.8, Case 1:

$$P = \frac{48E_t I \Delta}{\ell^3} = \frac{48(3225)(216)(0.25)}{(20 \times 12)^3}$$
$$= 0.605 \text{ k/ft width}$$

$$M = \frac{P\ell}{4} = \frac{0.605(20)}{4} = 3.02 \text{ ft-kips/ft width}$$

$$\text{Panel stress} = \frac{My}{I} = \frac{3.02(12,000)(3)}{216} = 503 \text{ psi}$$

The residual bow can be calculated by adjusting the equation in Table 3.3.8, Case 5, to read:

$$\Delta = \frac{M\ell^2}{16E_t I} \text{ ; and substituting } \ell/2 \text{ for } \ell$$

$$\Delta_2 = \frac{(3.02 \times 12)(10 \times 12)^2}{16(3225)(216)} = 0.05 \text{ in.}$$

While the magnitude of bowing is usually not very significant, in the case of wall panels it may cause unacceptable separation at the corners (see Fig. 3.3.4), and possible damage to joint sealants. It may therefore be desirable to restrain bowing with one or more connectors between panels. Table 3.3.8 gives equations for calculating the required restraint and the moments this would cause in the panel.

Similarly, differential temperature can cause upward bowing in roof members, especially in open structures such as parking decks. If these members are restrained from rotations at the ends, positive moments (bottom tension) can develop at the support, as shown in Cases 4 and 5, Table 3.3.8. The bottom tension can cause severe crack-

Fig. 3.3.4 Corner separation due to thermal bow

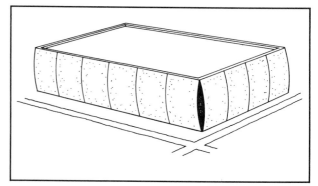

Table 3.3.8 Forces required to restrain bowing

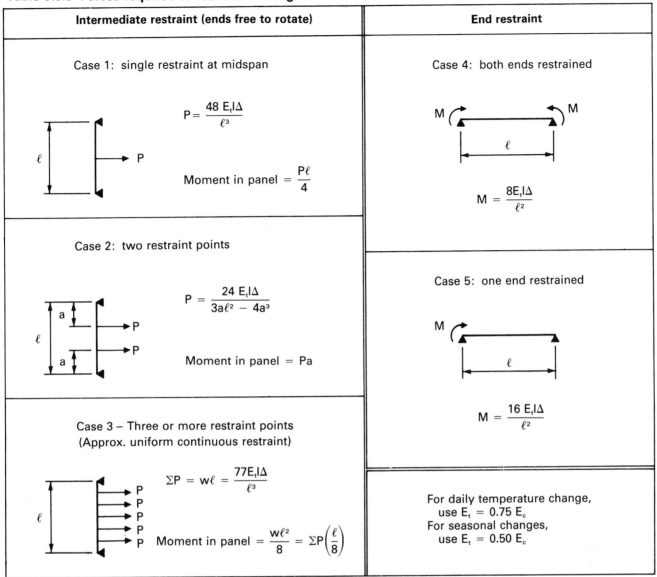

Intermediate restraint (ends free to rotate)	End restraint
Case 1: single restraint at midspan $P = \dfrac{48\,E_t I\Delta}{\ell^3}$ Moment in panel $= \dfrac{P\ell}{4}$	Case 4: both ends restrained $M = \dfrac{8E_t I\Delta}{\ell^2}$
Case 2: two restraint points $P = \dfrac{24\,E_t I\Delta}{3a\ell^2 - 4a^3}$ Moment in panel $= Pa$	Case 5: one end restrained $M = \dfrac{16\,E_t I\Delta}{\ell^2}$
Case 3 – Three or more restraint points (Approx. uniform continuous restraint) $\Sigma P = w\ell = \dfrac{77E_t I\Delta}{\ell^3}$ Moment in panel $= \dfrac{w\ell^2}{8} = \Sigma P\left(\dfrac{\ell}{8}\right)$	For daily temperature change, use $E_t = 0.75\,E_c$ For seasonal changes, use $E_t = 0.50\,E_c$

ing, but once the cracks occur, the tension is relieved. Examination of the equations shows that the thermal induced positive moments are independent of the span length. (For example, substitute Eq. 3.3.1 for Δ in Table 3.3.8, Case 4). Note from Table 3.3.8 that if only one end is restrained, as is sometimes done to relieve axial volume change force, the restraint moment is doubled. Also note that, since thermal bow occurs with daily temperature changes, the cyclical effects could magnify the potential damage.

Example 3.3.4 Thermal bow in a roof member

Given:
An inverted tee beam supporting the double tees of the upper level of a parking deck, as shown in Fig. 3.3.5, is welded at both ends at the bottom as part of a moment-resisting frame. Coefficient of thermal expansion $C = 6 \times 10^{-6}$ in./in./°F; $T_1 - T_2 = 35$ °F; $E_c = 4300$ ksi; composite $I = 49,700$ in.^{4}

Problem:
Find the tensile force developed at the support.

Solution:
From Eq. 3.3.1:

$$\Delta = \frac{6 \times 10^{-6}(35)(24 \times 12)^2}{8(27)} = 0.081 \text{ in.}$$

From Table 3.3.8, Case 4:

$$E_t = 0.75\,(4300) = 3225 \text{ ksi}$$

$$M = \frac{8(3225)(49,700)(0.081)}{(24 \times 12)^2} = 1252 \text{ in.-kips}$$

$$T = \frac{1252}{27} = 46 \text{ kips}$$

Fig. 3.3.5 Inverted tee beam for Example 3.3.4

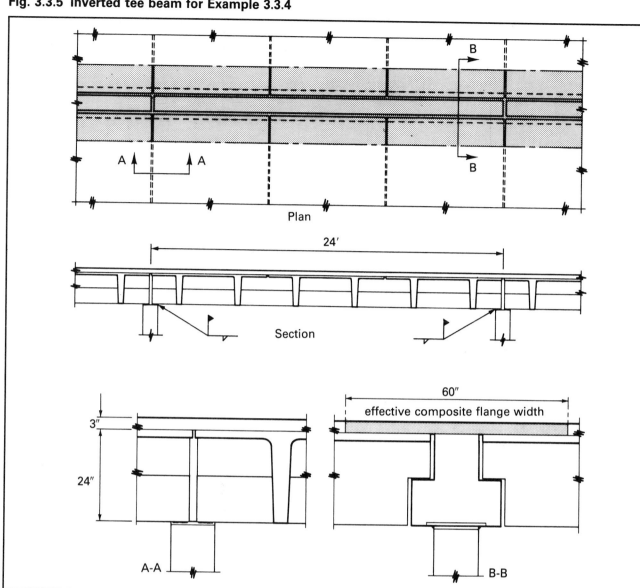

3.3.3 Expansion Joints

Joints are placed in structures to limit the magnitude of forces which result from volume change deformations (temperature changes, shrinkage and creep), and to permit movements (volume change deformations) of structural elements. If the forces generated by temperature rise are significantly greater than shrinkage and creep forces, a true "expansion joint" is needed. However, in concrete structures, true expansion joints are seldom required. Instead, joints that permit contraction of the structure are needed to relieve the strains caused by temperature drop and restrained creep and shrinkage, which are additive. Such joints are properly called contraction or control joints but are commonly referred to as expansion joints.

It is desirable to have as few expansion joints as possible. Expansion joints are often located by "rule of thumb" without considering the structural framing method. The purpose of this section is to present guidelines for determining the spacing and width of expansion joints.

Spacing of expansion joints

There is a wide divergence of opinion concerning the spacing of expansion joints. Typical practice in concrete structures, prestressed or non-prestressed, is to locate expansion joints at distances between 150 and 200 ft. However, reinforced concrete buildings exceeding these limits have performed well without expansion joints. Recommended joint spacings for precast concrete buildings are generally based on experience and may not consider several important items. Among these items are the types of connections used, the column stiffnesses in simple span structures, the relative stiffness between beams and columns in framed structures, and the weather exposure conditions. Non-heated structures, such as parking garages, are subjected to greater temperature changes than occupied structures, so shorter distances between expansion joints are warranted.

Sects. 3.3 and 3.8 present methods for analyzing the potential movement of framed structures, and the effect of restraint of movement on the connections and structural frame. This information along with the connection design methods in Part 6 can aid in determining spacing of expansion joints.

Fig. 3.3.6 shows joint spacing as recommended by the Federal Construction Council, and is adapted from *Expansion Joints in Buildings*, Technical Report No. 65, prepared by the Standing Committee on Structural Engineering of the Federal Construction Council, Building Research Advisory Board, Division of Engineering, National Research Council, National Academy of Sciences, 1974. Note that the spacings obtained from the graph in Fig. 3.3.6 should be modified for various conditions as shown in the notes below the graph. Values for the design temperature change can be obtained from Fig. 3.3.1.

When expansion joints are required in non-rectangular structures, they should be located at places where the plan or elevation dimensions change radically.

Width of expansion joints

The width of the joint can be calculated theoretically using a coefficient of expansion of 6×10^{-6} in./in./°F for normal weight and 5×10^{-6} in./in./°F for sand-lightweight concrete. The Federal Construction Council report, referenced above, recommends a minimum width of 1 in. However, since the primary problem in concrete buildings is contraction rather than expansion, joints that are too wide may result in problems with reduced bearing or loss of filler material.

Fig. 3.3.6 Maximum building length without use of expansion joints

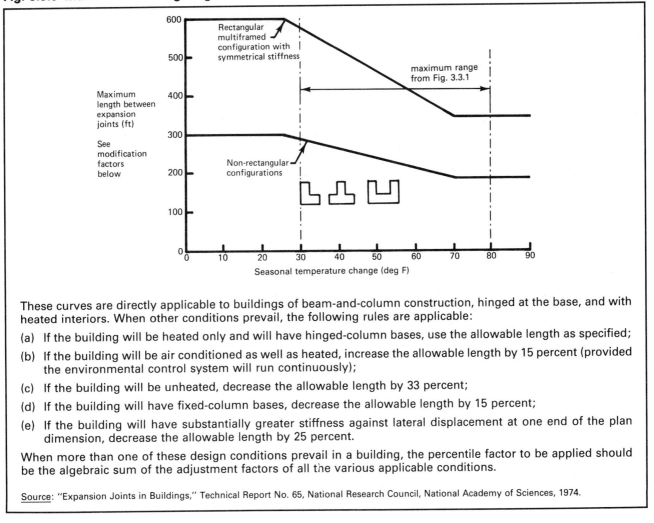

These curves are directly applicable to buildings of beam-and-column construction, hinged at the base, and with heated interiors. When other conditions prevail, the following rules are applicable:

(a) If the building will be heated only and will have hinged-column bases, use the allowable length as specified;

(b) If the building will be air conditioned as well as heated, increase the allowable length by 15 percent (provided the environmental control system will run continuously);

(c) If the building will be unheated, decrease the allowable length by 33 percent;

(d) If the building will have fixed-column bases, decrease the allowable length by 15 percent;

(e) If the building will have substantially greater stiffness against lateral displacement at one end of the plan dimension, decrease the allowable length by 25 percent.

When more than one of these design conditions prevail in a building, the percentile factor to be applied should be the algebraic sum of the adjustment factors of all the various applicable conditions.

Source: "Expansion Joints in Buildings," Technical Report No. 65, National Research Council, National Academy of Sciences, 1974.

3.4 Component Analysis

The design of components is covered in Part 4. This section is intended to assist in establishing design parameters for some components, and to discuss the effects of non-frame components on the frame.

3.4.1 Non-Bearing Wall Panels

Non-bearing panels are designed to resist wind, seismic forces generated from the self weight, and forces required to transfer the weight of the panel to the support. It is rare that these externally applied loads will produce the maximum stresses; the forces imposed during manufacturing and erection will usually govern the design, except for the connections.

Deformations

The relationship of the deformations of the panel and the supporting structure must be evaluated, and care taken to prevent unintended restraints from imposing additional loads. Such deformation may be caused by the weight of the panel, volume changes of concrete frames, and rotation of spandrel beams. To prevent imposing loads on the panel, the connections must be designed and installed to permit these deformations to freely occur.

For example, the tendency for the panels to follow the beam shown in Fig. 3.4.1 may cause restraining forces to develop at panel joints. The connections should be designed to allow the supporting beam to deflect, but the beam should be stiff enough that panel joint widths are within specified tolerances. These effects can sometimes

Fig. 3.4.1 Deformation of panels on flexible beam

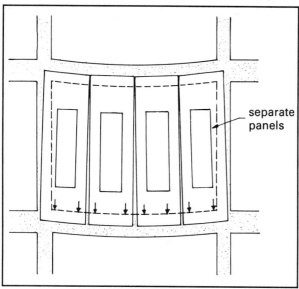

separate panels

be controlled by adjusting the erection sequence.

The most prevalent cause of panel deformation after placement in the structure is bowing due to thermal gradients. If supported in a manner that will permit bowing, the panel will not be subjected to stress. However, if the panel is restrained laterally between supports, stresses in the panel and forces in the structure will occur. This is discussed in Sect. 3.3.2.

Non-bearing panels should be designed and installed so that they do not restrain frames from lateral translation. If such restraint occurs, the panels may tend to act as shear walls and become overstressed (Fig. 3.4.2). Panels which are installed on a frame should be connected in a manner to allow frame distortion. In some cases, especially in high seismic regions, special connections which allow movement may be required.

Fig. 3.4.2 Panel forces induced by frame distortion

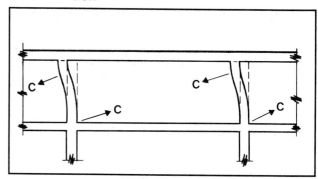

The vertical shortening of concrete columns from elastic and plastic deformation should be considered in very tall structures. At intermediate levels, the differential shortening between two adjacent floors will be negligible, and the panel will follow the frame movement. At the lowest level, if the panel is rigidly supported at the base (such as foundation or transfer girder), the accumulated shortening of the structure above may induce unintended loading on the panel. In such cases, the panel connection should be designed to permit the calculated deformation.

A similar situation will result when two adjacent columns have significantly different loads. For example, the corner column of a structure will usually be subjected to a smaller load than the adjacent columns. If both columns are the same size (as is often the situation for architectural reasons) and reinforced approximately the same, they will undergo different shortening.

Wind Load

Building codes specify the wind pressure for

which a building is to be designed. These loads may not be adequate for localized portions of a tall structure, due to gusting or funnel effects produced by adjacent structures.

The lateral deflection of thin panels when subjected to wind should be determined, particularly if they are attached to, or include, windows. Panels with deep protruding ribs may require analysis for shearing winds, as indicated in Fig. 3.4.3, and the connections designed for the twist produced. Panels and their connections should also be examined for suction. Although the design of the panel itself will generally not be critical for wind, the connections may be. This is particularly true for the tension connection which resists eccentric gravity loads, as indicated in Fig. 3.4.4.

Fig. 3.4.3 Shearing wind on ribbed panels

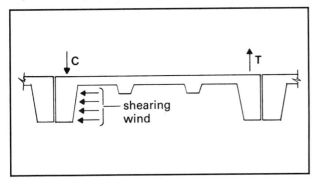

Fig. 3.4.4 Forces on a panel subjected to wind suction

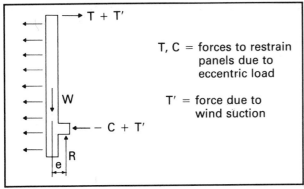

Panels with Openings

Non-bearing panels which contain openings may develop stress concentrations at these openings, resulting from unintended loading or restrained bowing. Hairline cracks radiating from the corners may result (Fig. 3.4.5). While these stress concentrations may be partially resisted by reinforcement, the designer should consider methods of eliminating imposed restraints. Areas of abrupt change in cross-section should be reinforced.

Loads from adjacent floors can be imposed on non-load bearing panels by methods of joinery,

and these loads can cause excessive stresses at the "beam" portion of an opening (Fig. 3.4.6). This can be prevented by locating connections away from critical sections.

Unless a method of preventing load transfer can be developed and permanently maintained, the "beam" should be designed for some loads from the floor. The magnitude of such loads requires engineering judgment.

Fig. 3.4.5 Corner cracking due to restrained bowing

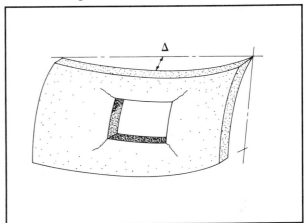

Fig. 3.4.6 Unanticipated loading on a non-load bearing panel

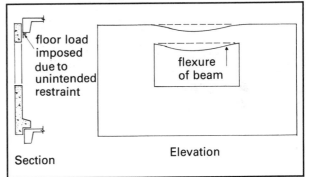

3.4.2 Load Bearing Wall Panels

Most of the items in the previous section must also be considered in the analysis of load bearing wall panels. Panels may be designed to span horizontally between columns, or vertically. When spanning horizontally, they are designed as beams, or, if they have frequent, regularly spaced window openings as shown in Fig. 3.4.7(a), as Vierendeel trusses. When so designed there must be a space or joint horizontally between panels, to insure that they will not transfer loads to panels below.

When the panels are placed vertically, they are usually designed as columns, and slenderness, as described in Sect. 3.5 should be considered. If a large portion of the panel is window opening, as in Fig. 3.4.7(c), it may be necessary to analyze it as a rigid frame.

Fig. 3.4.7 shows architectural wall panels, generally used with relatively short vertical spans, although they will sometimes span continuously over two or more floors. They are usually custom-made for each project, and reinforced with mild steel. Standard flat, hollow-core and stemmed members are also used as wall panels, and are frequently prestressed.

Dimensions of architectural panels are usually selected based on a desired appearance. When these panels are also used to carry loads, or act as shear walls, it is obviously important to have some engineering input early in the preliminary stages of the project.

Fig. 3.4.7 Horizontal and vertical rib panels

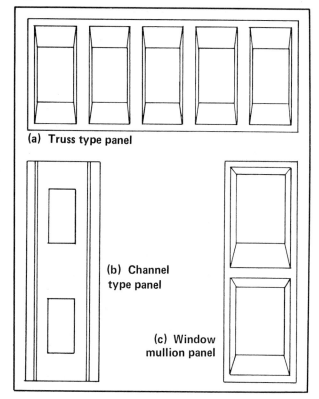

(a) Truss type panel

(b) Channel type panel

(c) Window mullion panel

3.4.3 Non-Bearing Spandrels

These are precast elements which are less than story height, made up either as a series of individual units or as one unit extending between columns. Support for spandrel weight may be the floor or the column, and stability against eccentric loading is achieved by connections to the underside of the floor or to the column (see Fig. 3.4.8). Spandrels are usually part of a window wall, so consideration should be given to the effect of deflections and rotations of the spandrel on the win-

dow. Deformation calculations should be based on gross concrete section since the stresses will generally be less than those which cause cracking. For elements which extend in one piece between columns, it is preferable that the connections which provide vertical support be located close to the ends. This arrangement will minimize interaction and load transfer between floor and spandrel.

Consideration should also be given to spandrels which are supported at the ends of long cantilevers. The designer must determine the effect of deflection and rotation of the support, including the effects of creep, and arrange the details of all attachments to accommodate this condition (Fig. 3.4.9). A particularly critical condition can occur at corners of buildings, especially when there is a cantilever on both faces.

Fig. 3.4.8 Typical spandrel connections

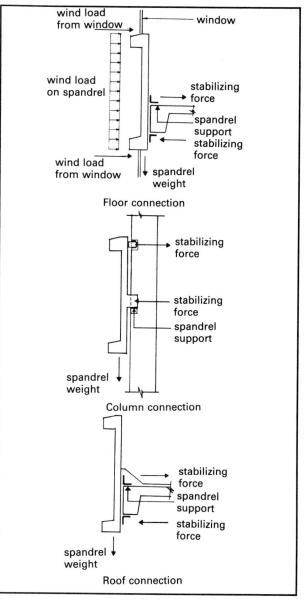

Floor connection

Column connection

Roof connection

Fig. 3.4.9 Effect of cantilever supports

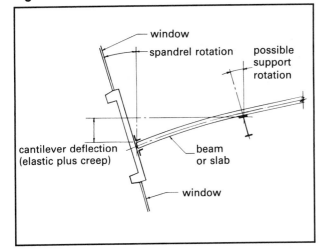

Example 3.4.1 Spandrel panel rotation

Given:
The spandrel panel shown.

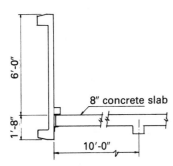

Panel weight = 417 plf
Normal weight concrete slab
E = 4000 ksi
Assume creep reduces effective E to 2000 ksi

Superimposed dead load = 10 psf

Problem:
Determine rotation and displacement at top of spandrel.

Solution:
Rotation of cantilever due to slab weight plus superimposed loads (neglecting support rotation):

w = 8(150)/12 + 10 = 110 psf

I = bh³/12 = 12(8)³/12 = 512 in.

$$\theta = \frac{w\ell^3}{6EI} = \frac{(0.110/12)(10 \times 12)^3}{6(2000)(512)}$$

$$= 0.00258 \text{ radians}$$

Rotation of cantilever due to weight of spandrel:

$$\theta = \frac{w\ell^2}{2EI} = \frac{(0.417)(10 \times 12)^2}{2(2000)(512)}$$

$$= 0.00293 \text{ radians}$$

Total rotation of cantilever

$$= 0.00258 + 0.00293 = 0.00551 \text{ radians}$$

Displacement of top of spandrel

$$= (0.00551)(72 + 8/2) = 0.42 \text{ in.}$$

3.4.4 Load Bearing Spandrels

Load bearing spandrels are panels which support floor or roof loads. Except for the magnitude and location of these additional loads, the design is the same as for non-bearing spandrels.

Load bearing spandrels support structural loads which are usually applied eccentrically with respect to the support. A typical arrangement of spandrel and supported floor is shown in Fig. 3.4.10.

Torsion due to eccentricity must be resisted by the spandrel, or resisted by a horizontal couple developed in the floor construction. In order to take torsion in the floor construction, the details must provide for a compressive force transfer at the top of the floor, and a tensile force transfer at the bearing of the precast floor element. The load path of these floor forces must be followed through the structure, and considered in the design of other members in the building. Even when torsion is resisted in this manner in the completed structure, twisting on the spandrel prior to completion must be considered.

If torsion cannot be removed by floor connections, the spandrel panel should be designed for induced stresses. Torsion design is illustrated in Part 4.

Fig. 3.4.10 Load bearing spandrel

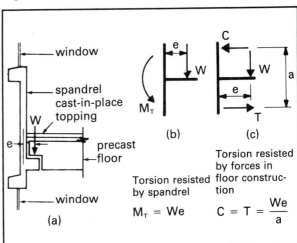

3.4.5 Eccentrically Loaded Columns

Many precast concrete structures utilize multi-story columns with simple-span beams resting on haunches. Fig. 3.4.11 and Table 3.4.1 are provided as aids for determining the various combinations of load and moment that can occur with such columns.

The following conditions and limitations apply to Fig. 3.4.11 and Table 3.4.1.

1. The coefficients are only valid for braced columns. Lateral stability must be achieved by other shear walls or moment-resisting frames.

2. For partially fixed column bases (see Sect. 3.8.3), a straight line interpolation between the coefficients for pinned and fixed bases can be used with small error.

3. For higher columns, the coefficients for the 4-story columns can be used with small error.

4. The coefficients in the "Σ Max" line will yield the maximum required restraining force, F_i, and column moments caused by loads (equal at each level) which can occur on either side of the column, for example, live loads on interior columns. The maximum force will not necessarily occur with the same loading pattern that causes the maximum moment.

5. The coefficients in "Σ One Side" line will yield the maximum moments which can occur if the column is loaded on only one side, such as the end column in a bay.

Example 3.4.2   Use of Fig. 3.4.11 and Table 3.4.1

Using Table 3.4.1, determine the maximum restraining force and moment in the lowest story of a 3-story frame for:

a. An interior column in a multi-bay frame

b. An exterior column

Beam reactions to column haunch at each level:

D.L. = 50 kips
L.L. = 20 kips

Eccentricity e = 14 in.
Story height h_s = 16 ft
Column base is determined to be 65% fixed

Solution:

Factored loads: D.L. = 1.4(50) =   70.0 kips
               L.L. = 1.7(20) =   34.0 kips
                                   _____
                                    104.0 kips

Fig. 3.4.11 Use of Table 3.4.1

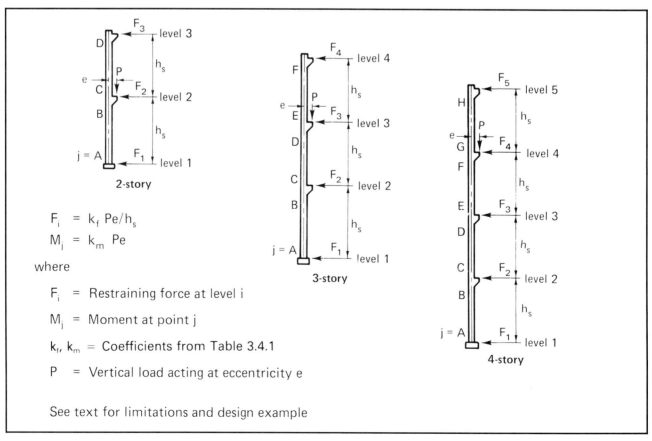

$$F_i = k_f Pe/h_s$$
$$M_j = k_m Pe$$

where

F_i = Restraining force at level i

M_j = Moment at point j

k_f, k_m = Coefficients from Table 3.4.1

P = Vertical load acting at eccentricity e

See text for limitations and design example

Table 3.4.1 Coefficients k_f and k_m for determining moments and restraining forces on eccentrically loaded columns braced against sidesway

See Fig. 3.4.11 for explanation of terms

+ Indicates clockwise moments on the columns and compression in the restraining beam															
No. of stories	Base Fixity	P acting at level	k_f at level					k_m at point							
			1	2	3	4	5	A	B	C	D	E	F	G	H
2	PINNED	3	+0.25	−1.50	+1.25			0	−0.25	+0.25	+1.0				
		2	−0.50	0	+0.50			0	+0.50	+0.50	0				
		Σ Max	±0.75	±1.50	±1.75			0	±0.75	±0.75	±1.0				
		Σ One Side	−0.25	−1.50	+1.75			0	+0.25	+0.75	+1.0				
	FIXED	3	+0.43	−1.72	+1.29			−0.14	−0.29	+0.29	+1.0				
		2	−0.86	+0.43	+0.43			+0.29	+0.57	+0.43	0				
		Σ Max	±1.29	±2.15	±1.72			±0.43	±0.86	±0.72	±1.0				
		Σ One Side	−0.43	−1.29	+1.72			+0.15	+0.28	+0.72	+1.0				
3	PINNED	4	−0.07	+0.40	−1.60	+1.27		0	+0.07	−0.07	−0.27	+0.27	+1.0		
		3	+0.13	−0.80	+0.20	+0.47		0	−0.13	+0.13	+0.53	+0.47	0		
		2	−0.47	−0.20	+0.80	−0.13		0	+0.47	+0.53	+0.13	−0.13	0		
		Σ Max	±0.67	±1.40	±2.60	±1.87		0	±0.67	±0.73	±0.93	±0.87	±1.0		
		Σ One Side	−0.41	−0.60	−0.60	+1.61		0	+0.40	+0.60	+0.40	+0.60	+1.0		
	FIXED	4	−0.12	+0.47	−1.62	+1.27		+0.04	+0.08	−0.08	−0.27	+0.27	+1.0		
		3	+0.23	−0.92	+0.23	+0.46		−0.08	−0.15	+0.15	+0.54	+0.46	0		
		2	−0.81	+0.23	+0.70	−0.12		+0.27	+0.54	+0.46	+0.12	−0.12	0		
		Σ Max	±1.16	±1.62	±2.55	±1.85		±0.38	±0.77	±0.69	±0.92	±0.85	±1.0		
		Σ One Side	−0.70	−0.22	−0.69	+1.61		+0.23	+0.46	+0.54	+0.38	+0.62	+1.0		
4	PINNED	5	+0.02	−0.11	+0.43	−1.61	+1.27	0	−0.02	+0.02	+0.07	−0.07	−0.27	+0.27	+1.0
		4	−0.04	+0.22	−0.86	+0.22	+0.46	0	+0.04	−0.04	−0.14	+0.14	+0.54	+0.46	0
		3	+0.13	−0.75	0	+0.75	−0.12	0	−0.13	+0.13	+0.50	+0.50	+0.12	−0.12	0
		2	−0.46	−0.22	+0.86	−0.22	+0.04	0	+0.46	+0.54	+0.14	−0.14	−0.04	+0.04	0
		Σ Max	±0.65	±1.30	±2.15	±2.80	±1.89	0	±0.64	±0.72	±0.86	±0.86	±0.97	±0.89	±1.0
		Σ One Side	−0.35	−0.86	+0.43	−0.86	+1.65	0	+0.35	+0.65	+0.57	+0.43	+0.35	+0.65	+1.0
	FIXED	5	+0.03	−0.12	+0.43	−1.61	+1.27	−0.01	−0.02	+0.02	+0.07	−0.07	−0.27	+0.27	+1.0
		4	−0.06	+0.25	−0.87	+0.22	+0.46	+0.02	+0.04	−0.04	−0.14	+0.14	+0.54	+0.46	0
		3	+0.22	−0.87	+0.03	+0.74	−0.12	−0.07	−0.14	+0.14	+0.51	+0.50	+0.12	−0.12	0
		2	−0.80	+0.21	+0.74	−0.18	+0.03	+0.27	+0.54	+0.46	+0.12	−0.12	−0.03	+0.03	0
		Σ Max	±1.11	±1.45	±2.07	±2.75	±1.88	±0.37	±0.74	±0.67	±0.84	±0.83	±0.96	±0.88	±1.0
		Σ One Side	−0.61	−0.53	+0.33	−0.83	+1.64	+0.21	+0.41	+0.59	+0.56	+0.44	+0.36	+0.64	+1.0

a. For the interior column, the dead load reaction would be the same on either side, thus no moment results. The live load could occur on any one side at any floor, hence use of the coefficients in the "Σ Max" line:

$$P_u e = 34.0(14) = 476 \text{ in.-kips} = 39.7 \text{ ft-kips}$$

To determine the maximum moment at point B:

For a pinned base: $k_m = 0.67$

For a fixed base: $k_m = 0.77$

For 65% fixed: $k_m = 0.67 + 0.65 (0.77 - 0.67)$
$= 0.74$

$$M_u = k_m P_u e = 0.74 (39.7) = 29.4 \text{ ft-kips}$$

Maximum restraining force at level 2:

$$F_u = k_f P_u e / h_s$$

$$k_f = 1.40 + 0.65 (1.62 - 1.40) = 1.54$$

$$F_u = 1.54 (39.7)/16 = 3.82 \text{ kips (tension or compression)}$$

b. For the exterior column, the total load is eccentric on the same side of the column, hence use the coefficients in the "Σ One Side" line:

$$P_u e = 104.0(14) = 1456 \text{ in.-kips} = 121.3 \text{ ft-kips}$$

To determine the maximum moment at point B:

For a pinned base, $k_m = 0.40$

For a fixed base, $k_m = 0.46$

For 65% fixed, $k_m = 0.40 + 0.65 (0.46 - 0.40)$
$= 0.44$

$$M_u = k_m P_u e = 0.44 (121.3) = 53.4 \text{ ft-kips}$$

Maximum restraining force at level 2:

$$F_u = k_f P_u e / h_s$$

$$k_f = -0.60 - 0.65 (-0.60 + 0.22) = -0.35$$

$$F_u = -0.35 (121.3)/16$$
$$= -2.65 \text{ kips (tension)}$$

3.5 Slenderness Effects in Columns and Wall Panels

The term "slenderness effects," used by ACI 318, can be described as the moments in a member, not accounted for in the primary analysis, which are generated when the line of action of the axial force is not coincident with the centroid of the member. These secondary moments arise from changes in the geometry of the structure, and may be caused by one or more of the following:

1. Relative displacement of the ends of the member due to:
   a. Lateral or unbalanced vertical loads in an unbraced frame, usually labeled "translation" or "sidesway", or
   b. Manufacturing and erection tolerances.

2. Deflections away from the end of the member due to:
   a. End moment due to eccentricity of the axial load.
   b. End moments due to frame action — continuity, fixity or partial fixity of the ends.
   c. Applied lateral loads, such as wind.
   d. Thermal bowing from differential temperature (see Sect. 3.3.2).
   e. Manufacturing tolerances.
   f. Camber due to prestressing.

These secondary effects can be considered in the design of the member by either using the Moment Magnification Method, described in Sect. 3.5.2, or by a direct iterative analysis usually termed "second-order" or "P-Δ" analysis.

3.5.1 Second-Order (P-Δ) Analysis

The usual procedure for this analysis is to perform an elastic type analysis using factored loads. Out-of-plumbness (items 1b and 2e above) are initially assumed based on experience and/or specified tolerances. The thermal bowing effect is usually neglected in columns, but may be significant in exterior wall panels (see Sect. 3.3.2).

At each iteration, the lateral deflection is calculated, and the moments caused by the axial load acting at that deflection are accumulated. After three or four iterations, the increase in deflection should be negligible (convergence). If it is not, the member may be approaching stability failure, and the section dimensions should be re-evaluated.

Note that, if the calculated moments indicate that cracking will occur, this needs to be taken into account in the deflection calculations, which may involve iterations within iterations, greatly complicating the procedure, although approximations of cracked section properties are usually satisfactory.

Effects of creep should also be included. The most common method is to divide the stiffness (EI) by the factor $1 + \beta_d$ as specified in the ACI moment magnification method.

In unbraced continuous frames, the joint translations affect and are affected by the total frame response. Thus, a practical analysis, especially when potential cracking is a parameter, usually will require the use of computers.

A good review of second-order analysis, along with an extensive bibliography and an outline of a complete program, is contained in Ref. 3.

An example of P-Δ calculations for a simple yet frequently encountered problem is shown in Example 3.5.1.

Example 3.5.1 Second-order analysis of an uncracked member

Given:

A 6 in. thick, 8 ft wide prestressed wall panel as shown.

Loading assumptions are as follows:

1. Axial load eccentricity = 1 in.

2. Midspan bowing due to temperature = 0.6 in.

3. No lateral loads. (For this example — lateral loads would be considered with different load factors.)

4. Braced frame—no joint translation.

5. Joints assumed pinned top and bottom.

Concrete: $f_c' = 5000$ psi
$E_c = 4300$ ksi

Problem:

Determine if standard panel (see p. 2-53) is adequate.

Solution:

$P_u = 1.4D + 1.7L = 1.4(20) + 1.7(10)$
$\quad = 45$ kips

$I_g = bh^3/12 = 96(6)^3/12 = 1728$ in.^{4}

$\phi = 0.7$ at $0.1f_c'A_g = 288$ kips; 0.9 at $P_u = 0$
$\quad = 0.9 - 0.2 (45/288) = 0.87$

$\beta_d = 28/45 = 0.62$

Note: the ϕ-factor is used to allow for unplanned deviations in the material dimensions and properties. The β_d factor is used to account for sustained load effects. Thus both factors affect the stiffness, EI. For convenience use:

$$EI = \frac{\phi E_c I_g}{1 + \beta_d} = \frac{0.87(4300)(1728)}{1.62}$$

$$= 3.99 \times 10^6 \text{ k-in}^2$$

Moments on panel

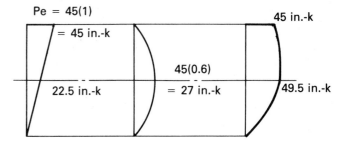

Deflection at midspan due to Pe:

$$\Delta = \frac{Pe\ell^2}{16EI} = \frac{45(1)(360)^2}{16(3.99 \times 10^6)} = 0.09 \text{ in.}$$

Total midspan deflection including thermal:

$$= 0.6 + 0.09 = 0.69 \text{ in.}$$

Deflection due to P-Δ moment at midspan:

$$\Delta = \frac{Pe\ell^2}{8EI} = \frac{45e(360)^2}{8(3.99 \times 10^6)}$$
$$= 0.183e = 0.183(0.69) = 0.13 \text{ in.}$$

Second iteration:

$e = 0.69 + 0.13 = 0.82$ in.
$\Delta = 0.183(0.82) = 0.15$ in.

Third iteration:

$e = 0.69 + 0.15 = 0.84$ in.
$\Delta = 0.183(0.84) = 0.15$ in. (convergence)

M_u at midspan $= 22.5 + 45(0.84)$
$\quad\quad\quad\quad = 60.3$ in.-kips

Check for cracking:

$My/I = 60.3(3)(1000)/1728 = 105$ psi

$f_r = 7.5 \sqrt{f_c'} = 530$ psi > 105 psi
therefore, analysis is valid.

Using interaction curve p. 2−53:

$P_u = 45/8 = 5.63$ kips/ft
$M_u = 60.3/(12 \times 8) = 0.63$ ft-kips/ft

Point is below curve, OK.

3.5.2 Moment Magnification Method

This is an approximate method described in Sect. 10.11 of ACI 318-83, and is applicable to precast and prestressed members. The ACI equations are repeated here for convenience:

$$M_c = \delta_b M_{2b} + \delta_s M_{2s} \quad\quad \text{(Eq. 3.5.1)}$$

Fig. 3.5.1 Coefficients, λ, for modified EI

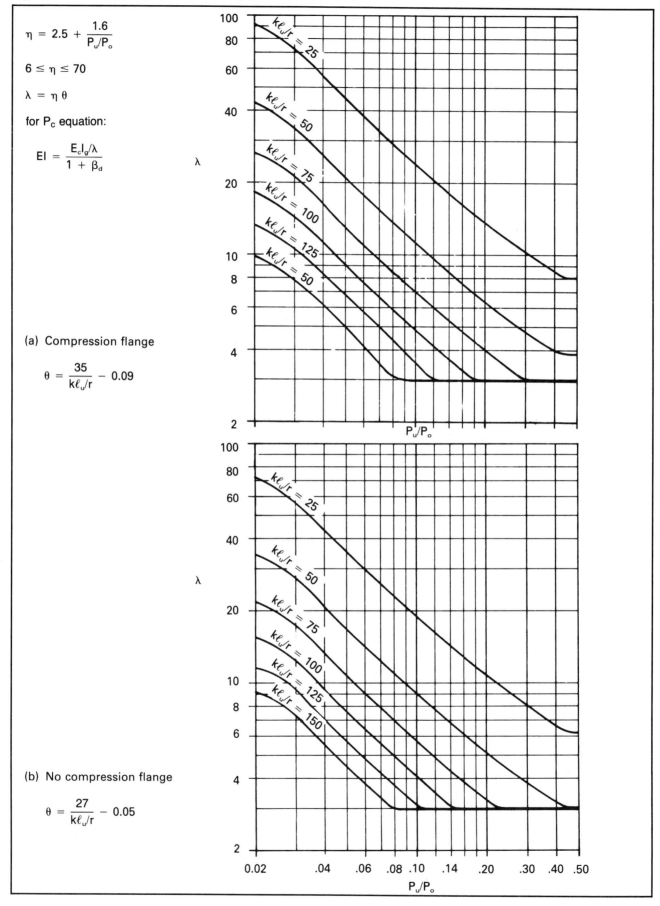

$$\eta = 2.5 + \frac{1.6}{P_u/P_o}$$

$6 \leq \eta \leq 70$

$\lambda = \eta \, \theta$

for P_c equation:

$$EI = \frac{E_c I_g / \lambda}{1 + \beta_d}$$

λ

(a) Compression flange

$$\theta = \frac{35}{k\ell_u/r} - 0.09$$

(b) No compression flange

$$\theta = \frac{27}{k\ell_u/r} - 0.05$$

Fig. 3.5.2 Alignment charts for determining effective length factors

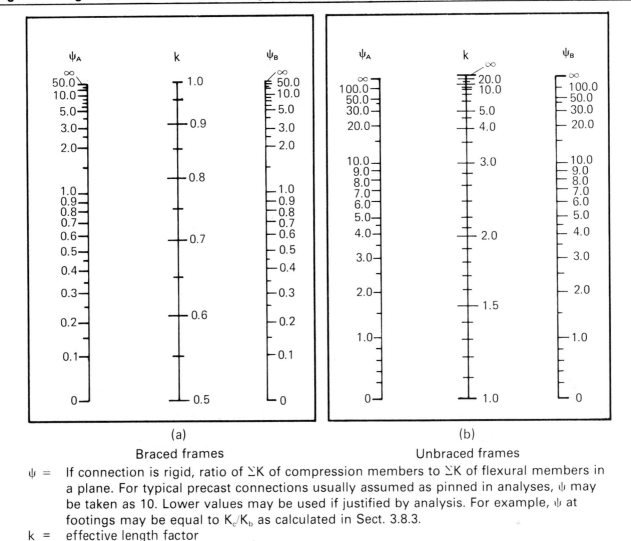

(a)
Braced frames

(b)
Unbraced frames

ψ = If connection is rigid, ratio of ΣK of compression members to ΣK of flexural members in a plane. For typical precast connections usually assumed as pinned in analyses, ψ may be taken as 10. Lower values may be used if justified by analysis. For example, ψ at footings may be equal to K_c/K_b as calculated in Sect. 3.8.3.

k = effective length factor

where

$$\delta_b = \frac{C_m}{1 - \dfrac{P_u}{\phi P_c}} \geq 1.0 \qquad \text{(Eq. 3.5.2)}$$

$$\delta_s = \frac{1}{1 - \dfrac{\Sigma P_u}{\phi \Sigma P_c}} \geq 1.0 \qquad \text{(Eq. 3.5.3)}$$

$$P_c = \frac{\pi^2 EI}{(k\ell_u)^2} \qquad \text{(Eq. 3.5.4)}$$

$$C_m = 0.6 + 0.4\frac{M_{1b}}{M_{2b}} \geq 0.4 \qquad \text{(Eq. 3.5.5)}$$

The Code suggests Eqs. 3.5.6 and 3.5.7 for the value of EI in Eq. 3.5.4. They were developed for reinforced concrete columns with at least 1% reinforcement, and may be used for precast col-

umns which meet that criterion.

$$EI = \frac{(E_c I_g/5) + E_s I_{se}}{1 + \beta_d} \qquad \text{(Eq. 3.5.6)}$$

or

$$EI = \frac{E_c I_g/2.5}{1 + \beta_d} \qquad \text{(Eq. 3.5.7)}$$

Since most prestressed compression members and precast load bearing wall panels have much less than 1% steel, modified equations for EI are recommended. Fig. 3.5.1 presents such equations based on "curve fitting" with theoretically exact procedures, and agrees reasonably well for $k\ell_u/r$ values up to 150. Other modifications have also been used successfully. These approximations are necessarily conservative, so the second-order analysis described in Sect. 3.5.1 is preferred for major structures. In Eq. 3.5.1, the subscript "s" is used to denote loads and moments which contribute to sidesway in a frame, and the subscript "b" those which do not. In a braced frame, the second term becomes zero. Example 3.5.2 illustrates the use of this equation.

In. Eq. 3.5.4, the value of k can be determined from the Jackson-Moreland alignment charts, Fig. 3.5.2. Since most precast members are used in braced frames, and the connections not designed to transfer moment to horizontal members, a k of 1 is usually used.

Example 3.5.2 Slenderness effects using moment magnification

Given:

The structure below is the interior portion of a long building that is isolated from the remaining structure by expansion joints, creating an unbraced frame. A frame analysis of the structure yields the following data:

Each wall panel:

    Axial load: D = 14.4 kips, L = 7.2 kips, W = 0

    Top moment: D = 9.6 ft-kips, L = 4.8 ft-kips, W = 0

    Bott. moment: D = 4.2 ft-kips, L = 2.1 ft-kips, W = 17.0 ft-kips (flange compression)

Each column:

    Axial load: D = 115.2 kips, L = 57.6 kips, W = 0

    Top moment: D = L = W = 0 (pinned)

    Bott. moment: D = L = 0, W = 3.0 ft-kips

Wall panel properties:

    A = 401 in.²; I = 20,985 in.⁴; r = 7.23 in.

    P_o (see p. 2-51) = 1165/ϕ = 1664 kips

    f_c' = 5000 psi, E_c = 4300 ksi

Column properties:

    A = 576 in.²; I = 27,648 in.⁴; r = 7.2 in.

    P_o (see p. 2-48) = 1650/0.7 = 2357 kips

    f_c' = 5000 psi, E_c = 4300 ksi

Problem:

Find magnified moments for wall panels and columns.

Solution:

Wall panel:

Case 1 (Dead + Live)

    P_u = 1.4(14.4) + 1.7(7.2) = 32.4 kips

    M_2 = 1.4(9.6) + 1.7(4.8) = 21.6 ft-kips

    M_1 = 1.4(4.2) + 1.7(2.1) = 9.5 ft-kips

In this case, the larger moment M_2 occurs at the top.

    β_d = 1.4(9.6)/21.6 = 0.62

Since in this structure the axial loads are symmetrical and do not contribute to sidesway, the value of k can be taken as 1 and $\delta_s M_{2s}$ = 0.

    $k\ell_u/r$ = 16(12)/7.23 = 26.6

    P_u/P_o = 32.4/1664 = 0.02

From Fig. 3.5.1(a): λ = 86

$$EI = \frac{E_c I_g/\lambda}{1 + \beta_d} = \frac{4300(20,985)/86}{1.62} = 647,685$$

$$C_m = 0.6 + 0.4(9.5/21.6) = 0.78$$

$$P_c = \pi^2 EI/(k\ell_u)^2 = \pi^2(647,685)/(16 \times 12)^2 = 173 \text{ kips}$$

ϕ = 0.7 at P_u = 0.1$f_c' A_g$ = 200 kips

    = 0.9 at P_u = 0

ϕ = 0.9 − 0.2(32.4/200) = 0.87

$$\delta_b = \frac{C_m}{1 - \dfrac{P_u}{\phi P_c}} = \frac{0.78}{1 - \dfrac{32.4}{0.87(173)}} = 0.99 \text{ Use } 1.0$$

60'   60'

8DT24 P/S wall panels (4 per bay)

1-24 x 24 P/S column each bay

16'

8DT24 wall panels

tension tie

slab isolation joint around column

3'

shear resistance

$M_c = \delta_b M_2 = 1.0(21.6) = 21.6$ ft-kips

Case 2 (Dead + Live + Wind)

Since wind loads contribute to sidesway:

$P_u = 0.75(32.4) = 24.3$ kips

$M_{2b} = 0.75(9.5) = 7.1$ ft-kips

$M_{2s} = 0.75(1.7)(17.0) = 21.7$ ft-kips

In this case, the larger moment M_2 occurs at the bottom.

$\beta_{db} = 0.62$ as before; $\beta_{ds} = 0$

Assume ψ_A for use in Jackson-Moreland alignment charts is approximately the ratio of ℓ_u to length below floor = $16/3 = 5.3$; $\psi_B = 10$ (max.)

$k = 2.6$; $k\ell_u = 2.6(16)(12) = 499.2$ in.

Using Fig. 3.5.1(a):

$P_u/P_o = 24.3/1664 = 0.015$, $k\ell_u/r = 69.0$

$\lambda = 29$

$(EI)_b = \dfrac{4300(20,985)/29}{1.62} = 1.92 \times 10^6$ kips-in.^{2}

$(C_m)_b = 0.78$ as before

$(P_c)_b = \pi^2 \, 1.92 \times 10^6/(16 \times 12)^2 = 514$ kips

$\phi = 0.9 - 0.2(24.3/200) = 0.88$

$\delta_b = \dfrac{0.78}{1 - \dfrac{24.3}{0.88(514)}} = 0.82$ Use 1.0

$(EI)_s = \dfrac{4300(20,985)/29}{1.0} = 3.11 \times 10^6$ kips-in.^{2}

$C_m = 1.0$

$(P_c)_s = \pi^2 \, (3.11 \times 10^6)/(499.2)^2 = 123$ kips

Column (case 2 continued):

Column $P_u = 0.75[1.4(115.2) + 1.7(57.6)]$
$\qquad\qquad = 194.4$ kips

From analysis of column/base relationship (see Sect. 3.8.3), $K_c/K_b = 1.0$

Using alignment charts: $\psi_A = 1$; $\psi_B = 10$
$\quad k = 1.9$

$k\ell_u = 1.9(16)(12) = 364.8$ in.

$k\ell_u/r = 364.8/7.2 = 51$

$P_u/P_o = 194.4/2357 = 0.08$

From Fig. 3.5.1(b): $\lambda = 13$

$(EI)_s = \dfrac{4300(27,648)/13}{1.0} = 9.15 \times 10^6$ kips-in.^{2}

$(P_c)_s = \pi^2 \, (9.15 \times 10^6)/(364.8)^2 = 678$ kips

$\Sigma P_u = 8(24.3) + 194.4 = 389$ kips

$\Sigma(P_c)_s = 8(123) + 678 = 1662$ kips

$0.1 f'_c A_g = 0.1(5)[8(401) + 576] = 1892$ kips

$\phi = 0.9 - 0.2(389/1892) = 0.86$

$\delta_s = \dfrac{1.0}{1 - \dfrac{389}{0.86(1662)}} = 1.37$

Wall panel $M_c = \delta_b M_{2b} + \delta_s M_{2s}$
$\qquad\qquad\quad = 1.0(7.1) + 1.37(21.7)$
$\qquad\qquad\quad = 36.8$ ft-kips

Column $M_{2b} = 0$
$\qquad\quad M_{2s} = 0.75(1.7)(3.0) = 3.82$ ft-kips
$\qquad\quad M_c = 1.37(3.82) = 5.23$ ft-kips

Use interaction diagrams in Part 2 to complete solution.

3.6 Diaphragm Design

Horizontal loads from wind or earthquake are usually transmitted to shear walls or moment-resisting frames through the roof and floors acting as horizontal diaphragms.

3.6.1 Method of Analysis

The diaphragm is analyzed by considering the roof or floor as a deep horizontal beam, analogous to a plate girder or I-beam. The shear walls or structural frames are the supports for this analogous beam. Thus the lateral loads are transmitted to these supports as reactions. As in a beam, tension and compression are induced in the chords or "flanges" of the analogous I-beam as shown in Fig. 3.6.1

When precast concrete members which span parallel to the supporting shear walls or frames are used for the diaphragm, it is apparent that the shear in the analogous beam must be transferred between adjacent members and also to the supporting elements. The "web" shear must also be transferred to the chord elements. Thus the design of a diaphragm is essentially a connection design problem.

3.6.2 Shear Transfer between Members

In floors or roofs without composite topping, the shear transfer between members is usually accomplished by weld plates or grout keys, depending on the member.

Weld plates may be analyzed as illustrated in Fig. 3.6.2. In addition to the hardware details shown, many others are used by precast concrete manufacturers.

For members connected by grout keys, a con-

Fig. 3.6.1 Analogous beam design of a diaphragm

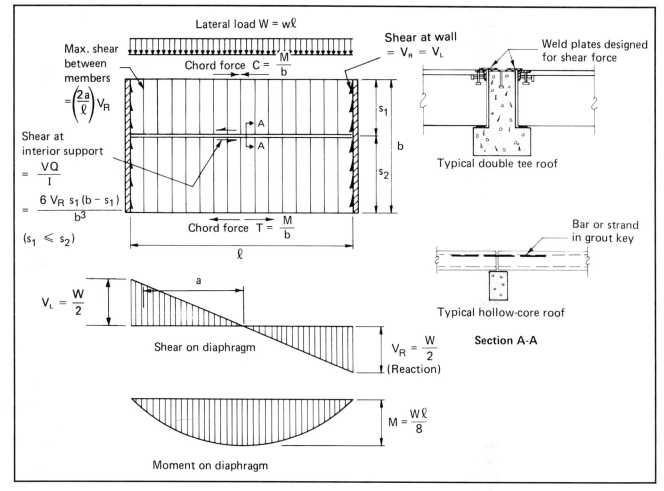

Fig. 3.6.2 Typical flange weld plate details

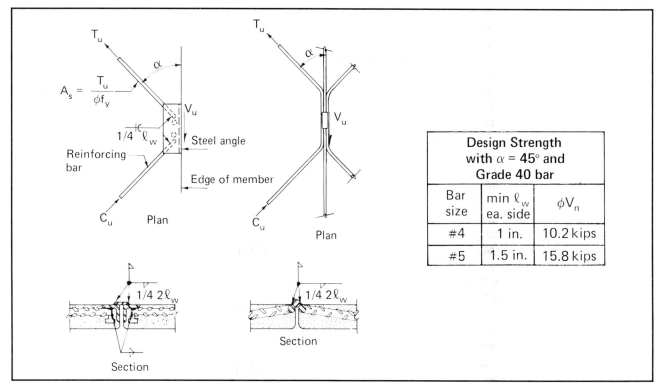

Design Strength with $\alpha = 45°$ and Grade 40 bar		
Bar size	min ℓ_w ea. side	ϕV_n
#4	1 in.	10.2 kips
#5	1.5 in.	15.8 kips

servative value of 80 psi can be used for the design strength of the grouted key. If necessary, reinforcement placed as shown in Fig. 3.6.3 can be used to transfer the shear. This steel is designed by the shear-friction principles discussed in Part 6.

Fig. 3.6.3 Use of perimeter reinforcement as shear-friction steel

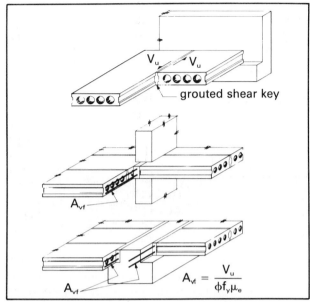

$$A_{vf} = \frac{V_u}{\phi f_y \mu_e}$$

In floors or roofs with composite topping, the topping itself can act as the diaphragm, if it is adequately reinforced. Reinforcement requirements can be determined by shear-friction analysis.

It should be noted that the connections between members often serve functions in addition to the transfer of shear for lateral loads. For example, weld plates in flanged members are often used to adjust differential camber. Grout keys may be called upon to distribute concentrated loads.

Connections which transfer shear from the diaphragm to the shear walls or moment-resisting frame are analyzed in the same manner as the connection between members.

3.6.3 Chord Forces

Chord forces are calculated as shown in Fig. 3.6.1. For roofs with intermediate supports as shown, the shear stress is carried across the beam with weld plates or bars in grout keys as shown in Section A-A. Bars are designed by shear-friction. Stresses are usually quite low, and only as many bars or weld plates as required should be used.

In flanged deck members the chord tension at the perimeter of the building is usually transferred between members by the same type of connection used for shear transfer (Fig. 3.6.2). Between connections, within the member flange, the ten-

sion is usually assumed to be taken by the tensile strength of the concrete.

Static friction as discussed in Sect. 6.6 can be used to transfer wind loads to walls. Note that the static coefficients of friction (Table 6.6.1) should be divided by 5 when used for this purpose. Static friction is not reliable for transferring earthquake forces.

In some bearing wall buildings, a minimum amount of perimeter reinforcement is recommended for resistance to "abnormal loads."[7] When abnormal load design is required by the building code or owner, these minimum requirements may be more than enough to resist the chord tension.

3.7 Shear Wall Buildings

3.7.1 General

In most precast, prestressed concrete buildings, it is desirable to resist lateral loads with shear walls of precast or cast-in-place concrete, or unit masonry. Shear walls are usually the exterior wall system, interior walls, or walls of elevator, stairway, mechanical shafts or cores. The transfer of load from horizontal diaphragm to shear walls, or walls of elevator and stairway cores or mechanical shafts, can be achieved either through connections or by direct bearing.

Shear walls act as vertical cantilever beams which transfer lateral forces from the superstructure to the foundation. Most structures contain a number of walls which resist lateral load in two orthogonal directions. The portion of the total lateral force which each wall resists depends on the bending and shear resistance of the wall, the participation of the floor, and the characteristics of the foundation. It is common practice to assume that floors act as rigid elements for loads in the plane of the floor, and that the deformations of the footings and soil can be neglected. Thus, for most structures, lateral load distribution is based only on the properties of the walls.

If the floor is considered to be a rigid body, it will translate in a direction parallel to the applied load an amount related to the flexural and shear rigidity of the participating shear walls (Fig. 3.7.1). If the center of rigidity is not coincident with the line of action of the applied loads, the floor will tend to rotate about the center of ridigity, introducing additional forces (Fig. 3.7.1b). Therefore, the load on each shear wall will be determined by combining the effects produced by rigid body translation and rotation.

A shear wall need not consist of a single element. It can be composed of independent units such as double tee, hollow-core, or architectural precast wall panels. If such units have adequate

Fig. 3.7.1 Translation and rotation of rigid floors

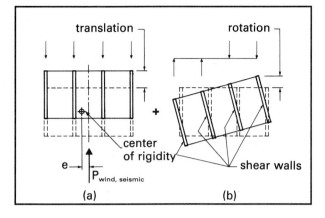

(a)   (b)

Fig. 3.7.2 Effective width of walls perpendicular to shear walls[7]

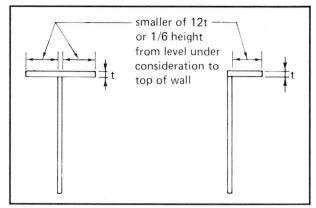

shear ties between them, they can be designed to act as a single unit, greatly increasing their shear resistance. Connecting such units can, however, result in a buildup of volume change forces, so it is usually desirable to connect only as many units as necessary to resist the overturning moment. Connecting as few units as necessary near midlength of the wall will minimize the volume change restraint forces.

Connection of rectangular wall units to form "T" or "L" shaped walls increases their flexural rigidity, but has little effect on shear rigidity. The effective flange width that can be assumed for such walls is illustrated in Fig. 3.7.2 The designer should evaluate whether such connections are worth the extra cost. In some structures, such as Example 3.7.3, it may be desirable to provide shear connections between non-load bearing and load bearing shear walls in order to increase the dead load resistance to moments caused by lateral loads.

3.7.2 Rigidity of Solid Shear Walls

In order to determine the distribution of lateral loads, the relative rigidity of all shear walls must

be established. Rigidity is defined as:

$$r = 1/\Delta \qquad \text{(Eq. 3.7.1)}$$

where

Δ = sum of flexure and shear deflections

For a structure with rectangular shear walls of the same material, flexural deflections can be neglected when the wall height to length ratio is less than about 0.3. The rigidity of the element is then directly proportional to its cross-sectional area. When the wall height to length ratio is greater than about 3.0, shear deflections can be neglected, and the rigidity is proportional to the moment of inertia (plan dimensions). When the height to length ratio is between 0.3 and 3.0, an equivalent moment of inertia, I_{eq}, can be derived for simplifying the calculation of wall rigidity. I_{eq} is an approximation of the moment of inertia that would result in a flexural deflection equal to the combined flexural and shear deflections of the wall. Table 3.7.1 compares the deflections and I_{eq} for several load and restraint conditions.

Table 3.7.1 Shear wall deflections

| Case | Deflection due to | | Equivalent Moment of Inertia, I_{eq} | |
	Flexure	Shear	Single story	Multistory
	$\dfrac{Ph^3}{3EI}$	$\dfrac{2.78\,Ph}{A_w E}$	$\dfrac{I}{1 + \dfrac{8.34\,I}{A_w h^2}}$	$\dfrac{I}{1 + \dfrac{13.4\,I}{A_w h^2}}$
	$\dfrac{Wh^3}{8EI}$	$\dfrac{1.39\,Wh}{A_w E}$	—	$\dfrac{I}{1 + \dfrac{23.6\,I}{A_w h^2}}$
	$\dfrac{Ph^3}{12EI}$	$\dfrac{2.78\,Ph}{A_w E}$	$\dfrac{I}{1 + \dfrac{33.4\,I}{A_w h^2}}$	—

PCI Design Handbook

Connecting or coupling of shear walls and large openings in walls also affect stiffness, as discussed in Sects. 3.7.5 and 3.7.6.

3.7.3 Distribution of Lateral Loads

Lateral loads are distributed to each shear wall in proportion to its rigidity. It is usually considered sufficient to design for lateral loads in only two orthogonal directions.

When the shear walls are symmetrical with respect to the center of load application, the force resisted by any shear wall is:

$$F_i = r_i / \Sigma r \qquad \text{(Eq. 3.7.2)}$$

where: F_i = the force resisted by an individual shear wall, i

r_i = the rigidity of wall i

Σr = sum of rigidities of all shear walls

3.7.4 Unsymmetrical Shear Walls

Structures which have shear walls placed unsymmetrically with respect to the center of the lateral load should take the torsional effect into account. Typical examples are shown in Fig. 3.7.3. For wind loading on most structures, a simplified method of determining the torsional resistance may be used in lieu of more exact design. The method is similar to the design of bolt groups in steel connections, and is illustrated in the follow-

Fig. 3.7.3 Unsymmetrical shear walls

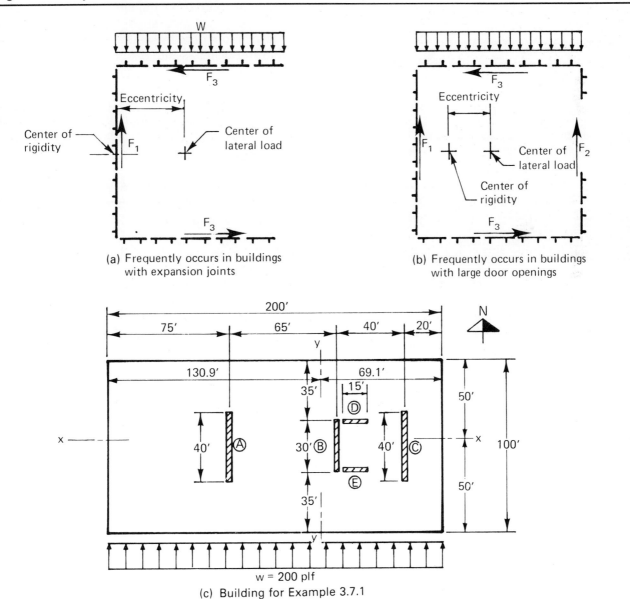

(a) Frequently occurs in buildings with expansion joints

(b) Frequently occurs in buildings with large door openings

(c) Building for Example 3.7.1

ing example.

Example 3.7.1 Design of unsymmetrical shear walls

Given:

The structure of Fig. 3.7.3c. All walls are 8 ft high and 8 in. thick.

Problem:

Determine the shear in each wall, assuming the floors and roof are rigid diaphragms. Walls D and E are not connected to Wall B.

Solution:

Maximum height to length ratio of north-south walls = 8/30 < 0.3. Thus, for distribution of the direct wind shear, neglect flexural stiffness. Since walls are the same thickness and material, distribute in proportion to length.

Total lateral load, $W = 0.20 \times 200 = 40$ kips

Determine center of rigidity:

$$\bar{x} = \frac{40(75) + 30(140) + 40(180)}{40 + 30 + 40}$$

$$= 130.9 \text{ ft from left}$$

$\bar{y}$ = center of building, since walls D and E are placed symmetrically about the center of the building in the north-south direction

Torsional moment, $M_T = 40(130.9 - 200/2)$
$$= 1236 \text{ ft-kips.}$$

Determine the polar moment of inertia of the shear wall group about the center of rigidity:

$$I_p = I_{xx} + I_{yy}$$

$I_{xx} = \Sigma \ell y^2$ of east-west walls
$$= 2(15)(15)^2 = 6750 \text{ ft}^3$$

$I_{yy} = \Sigma \ell x^2$ of north-south walls
$$= 40(130.9 - 75)^2 + 30(140 - 130.9)^2$$
$$+ 40(180 - 130.9)^2$$
$$= 223,909 \text{ ft}^3$$

$I_p = 6750 + 223,909 = 230,659 \text{ ft}^3$

Shear in north-south walls $= \dfrac{W\ell}{\Sigma \ell} + \dfrac{M_T x \ell}{I_p}$

Wall A $= \dfrac{40(40)}{110} + \dfrac{1236(130.9 - 75)(40)}{230,659}$
$$= 14.5 + 12.0 = 26.5 \text{ kips}$$

Wall B $= \dfrac{40(30)}{110} + \dfrac{1236(-9.1)(30)}{230,659}$
$$= 10.9 - 1.5 = 9.4 \text{ kips}$$

Wall C $= \dfrac{40(40)}{110} + \dfrac{1236(-49.1)(40)}{230,659}$
$$= 14.5 - 10.5 = 4.0 \text{ kips}$$

Shear in east-west walls $= \dfrac{M_T y \ell}{I_p}$

$$= \dfrac{1236(15)(15)}{230,659} = 1.2 \text{ kips}$$

3.7.5 Coupled Shear Walls

Fig. 3.7.4 shows two examples of coupled shear walls. The effect of coupling two walls is to increase the stiffness by transfer of shear through the coupling beam. The wall curvatures are altered from that of a cantilever because of the frame action developed. Fig. 3.7.5 shows how the deflected shapes differ in response to lateral loads.

Several approaches may be used to analyze the response of coupled shear walls. A simple approach is to ignore the coupling effect by considering the walls as independent cantilevers. This method results in a conservative wall design. However, if the coupling beam is rigidly connected, significant shears and moments will occur in the beam that may cause unsightly and possibly dangerous cracking. To avoid the problem, the

Fig. 3.7.4 Coupled shear walls

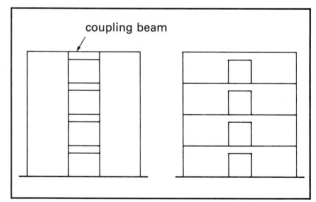

Fig. 3.7.5 Response to lateral loads

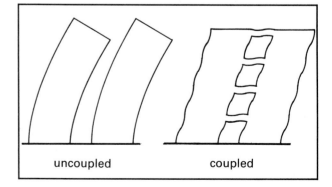

PCI Design Handbook

beam to panel connection can be detailed for little or no rigidity, or the beam can be designed to resist the actual shears and moments.

Finite element analysis may be used to determine the stiffness and the distribution of shears and moments within a coupled shear wall. The accuracy (and cost) of such an analysis is a function of the element size used. This method is usually reserved for very complex structures.

A "plane frame" computer analysis will be sufficiently precise for the great majority of structures. In modelling the coupled shear wall as a frame, the member dimensions must be considered, as a centerline analysis may yield very inaccurate results. A suggested model is shown in Fig. 3.7.6a.

Either a finite element or frame analysis may be used to determine the deflection of a coupled shear wall, and hence its equivalent moment of inertia. This may then be used to determine distribution of shears in a building which contains both solid and coupled shear walls. Most frame analysis programs do not calculate shear deformations so, if significant, shear deformations may have to be manually calculated.

3.7.6 Shear Walls with Large Openings

Window panels and other wall panels with large openings may also be analyzed with a plane frame computer program. Fig. 3.7.6b shows suggested models. Where length to depth ratios for vertical and horizontal segments are similar, a frame model based on segment centerlines will be reasonably accurate. Otherwise, an analysis similar to that described for coupled shear walls may be used.

As with coupled shear walls, the deflections yielded by the computer analysis may be used to determine equivalent stiffness for determining lateral load distribution. Again, shear deflections, if significant, may have to be hand calculated and added to the flexural stiffnesses from the frame analysis.

In very tall structures, vertical shear and axial deformations influence the rigidity of panels with large openings, so a more rigorous analysis may be required. This is beyond the scope of this Handbook.

Example 3.7.2 One-story building

Given:

The wind load analysis and design of a typical one-story industrial building are illustrated by the structure shown in Fig. 3.7.7. 8-ft wide double tees are used for both the roof and walls. The local building code specifies that a wind load of 25 psf be used for buildings of this height.

Fig. 3.7.6 Computer models

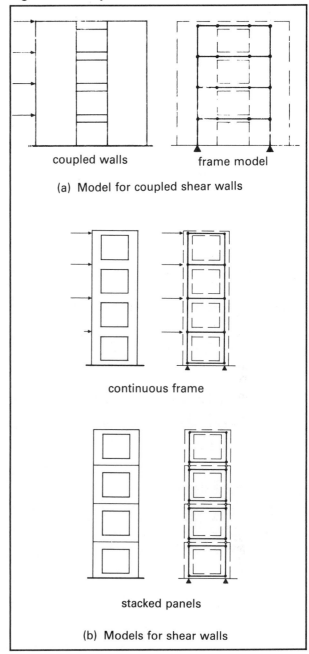

(a) Model for coupled shear walls

continuous frame

stacked panels

(b) Models for shear walls

Solution:

1. Calculate forces, reactions, shears and moments:

   Total wind force to roof:

   $W = 25$ psf $(160$ ft$)(18/2 + 2.5)$

   $\quad = 46,000$ lb $= 46$ kips

   $V_L = V_R = 23$ kips

   Diaphragm moment $= \dfrac{W\ell}{8} = \dfrac{46(160)}{8}$

   $\qquad\qquad\qquad\quad = 920$ ft-kips

Fig. 3.7.7 Example 3.7.2

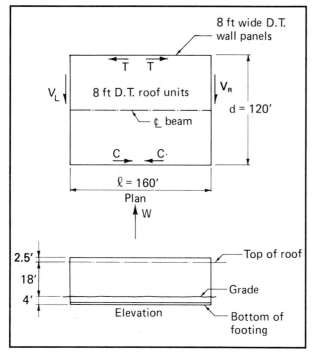

2. Check sliding resistance of the shear wall:

   Determine dead load on the footing:

   8DT12 wall = 37 psf(23.5)(120) = 104,340 lb

   12″ × 18″ footing
   = 1(1.5)(150 pcf)(120)   = 27,000 lb

   Assume 2 ft backfill
   = (100 pcf)(1.5)(120)   = 36,000 lb

   Total = 167,340 lb

   = 167.34 kips

   Assume coefficient of friction against granular soil, μ_s = 0.5.

   Sliding resistance = $\mu_s N$ = 0.5(167.34)

   = 83.67 kips

   Factor of safety = 83.67/23 = 3.64 OK

   (Note: A factor of safety of 1.5 is specified by some building codes.)

3. Check overturning resistance:

   Applied overturning moment = 23 (4 + 18)
   = 506.0 ft-kips

   Resistance to overturning:
   Assume axis of rotation at leeward edge of the building.

   (Note: Some engineers prefer to use the more conservative assumption of an axis at d/5, d/4 or d/3 from the leeward edge, depending on the foundation conditions.)

   Resisting moment = 167.34 (120/2)
   = 10,040 ft-kips

   Factor of safety = 10,040/506
   = 19.8 > 1.5 OK

4. Analyze connections:

   a. Shear ties in double tee roof joint:
   (Maximum load at next to last joint)

   Applied shear = [(80 − 8)/80](23)
   = 20.7 kips

   Load factor by ACI 318-83 = 1.3

   Connection load factor (see Sect. 6.3) = 1.3

   V_u = 20.7(1.3)(1.3) = 35.0 kips

   v_u = 35.0/120 = 0.292 kips/ft

   Use #4 ties as shown in Fig. 3.6.2

   ϕV_n = 10.2 kips

   Required spacing = 10.2/0.292 = 34.9 ft

   (Note: Most engineers and precasters prefer a maximum connection spacing of about 8 to 10 ft.)

   b. Shear ties at the shear walls:

   V_u = 23(1.3)(1.3) = 38.9 kips

   v_u = 38.9/120 = 0.324 kips/ft

   A connection as shown in Fig. 3.6.2 is designed similar to the shear tie between tees. This would require a spacing of 10.2/0.324 = 31.5 ft. In order to distribute the load to the wall panels, at least one connection per panel is required. From Fig. 3.7.8a it is apparent that these connections should occur at the tee stems. Thus a spacing of 4 ft or 8 ft would be used in this case.

   Other types of connections using short welded headed studs are commonly used for this application. Design of studs is shown in Part 6.

   In some cases, the designer may find it necessary to provide a connection that permits vertical movement of the roof member. This is illustrated in Fig. 3.7.8b.

   c. Chord force (see Fig. 3.7.9)

   T = C = M/d = 920/120

   = 7.67 kips

   T_u = 1.3(7.67) = 9.97 kips

   This force can be transmitted between

Fig. 3.7.8 Connection of roof tee to wall

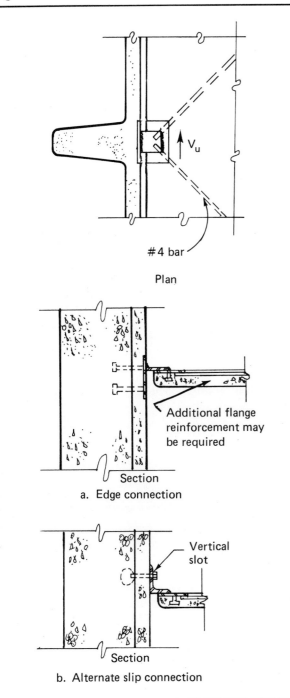

Plan

#4 bar

Section

a. Edge connection

Vertical slot

Section

b. Alternate slip connection

Fig. 3.7.9 Chord forces

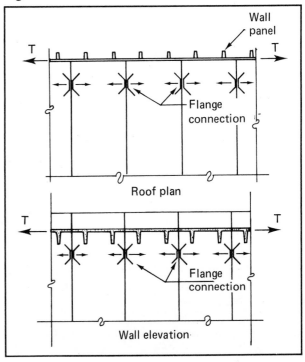

Wall panel

T — T

Flange connection

Roof plan

T — T

Flange connection

Wall elevation

Fig. 3.7.10 Panels acting as individual units in a shear wall

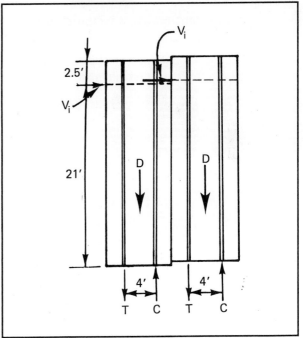

members by ties at the roof tees, wall panels or a combination, as illustrated in Fig. 3.7.9. The force through the member flanges can be transmitted by concrete tension, using reasonable assumptions of effective areas and tensile strength of the concrete. A conservative value of $3 \lambda \sqrt{f_c'}$ is suggested for the tensile strength of the concrete.

For higher forces, or where more ductility is required, reinforcing bars can be placed in the flanges. Design procedures are shown in Sect. 3.10.

d. Wall panel connections:

This shear wall may be designed to act as a series of independent units, without ties between the panels. The shear force is assumed to be distributed equally among the wall panels (see Fig. 3.7.10).

$$n = 120/8 = 15 \text{ panels}$$

$$V = V_R/n = 23/15 = 1.53 \text{ kips}$$

$$D = 37 \text{ psf}(8)(23.5) = 6956 \text{ lb} = 6.96 \text{ kips}$$

Design base connection for $1.3W - 0.9D$

$$T_u = [1.3(1.53)(21) - 0.9(6.96)(2)]/4$$
$$= 7.31 \text{ kips tension}$$

As an alternative, the shear wall may be designed with 2 or more panels connected together, as described in Sect. 3.7.1. The following illustrates the connection design if all panels are connected (see Fig. 3.7.11).

Shear ties between panels:

$$v_r = V_R/d = 23/120 = 0.192 \text{ klf}$$

The unit shear stress, v_r, is equal on all sides of the panel (see Sect. 3.10.7).

$$v_u = 0.192(1.3)(1.3) = 0.324 \text{ klf}$$

Using the same shear ties as on the roof: maximum spacing $= 10.2/0.324 = 31.5$ ft

(Note: Most designers prefer a minimum of 2 or 3 ties per panel, at a maximum spacing of 8 to 10 ft.)

Check for tension using factored loads:

By ACI 318-83, the required load factor equation to use under this condition is $1.3W - 0.9D$. The tensile stress would be:

$$f_t = \frac{M}{Z} - \frac{P}{A}$$

$$Z = \ell^2/6 = 120^2/6 = 2400 \text{ ft}^2$$

$$A = \ell = 120 \text{ ft}$$

$$M = V_R h_s = 23(21) = 483 \text{ ft-kips}$$

$$P = \text{D.L. of wall} = 104.34 \text{ kips}$$

$$f_{ut} = \frac{1.3M}{Z} - \frac{0.9P}{A}$$

Fig. 3.7.11 Panels connected together as a monolithic shear wall

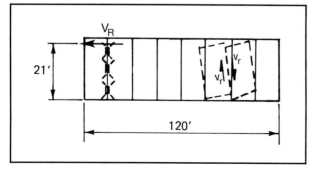

$$= \frac{1.3(483)}{2400} - \frac{0.9(104.34)}{120}$$
$$= -0.521 \text{ klf (compression)}$$

Thus, no tension connections are required.

Example 3.7.3 Four-story building

Given:

The wind load analysis and design of a typical four-story residential building are illustrated by the structure shown in Fig. 3.7.12. 8-in. deep hollow-core units are used for the floors and roof, and 8-in. thick precast concrete walls are used for all walls shown. Unfactored loads are given as follows:

Gravity loads:	L.L.	D.L.
Roof	30	
Roofing, mechanical, etc.		10
Hollow-core slabs		64
	30 psf	74 psf
Typical floor		
Living areas	40	
Corridors & stairs	100	
Partitions		10
Hollow-core slabs		64
		74 psf
Walls		100 psf
Stairs	100	130 psf

Wind loads:

0 to 30 ft above grade = 25 psf
30 to 34 ft 8 in. above grade = 30 psf

Solution:

For wind in the transverse (east-west) direction, normal practice for this structure would be to conservatively neglect the resistance provided by the stair, elevator and longitudinal walls. Thus, two 27 ft long interior bearing walls can be assumed to resist the wind on one 26 ft bay. The wind and gravity loads on the wall are shown in Fig. 3.7.13.

Concentrated loads from the corridor lintels can be assumed to be distributed as shown in Fig. 3.7.13. In this example, these loads are conservatively neglected to simplify the calculations.

Check overturning of shear wall:

D.L. resisting moment about toe of wall

$$= 27 (27/2) [1.92 + 3(2.72) + 0.8]$$

$$= 3966 \text{ ft-kips}$$

Factor of safety $= 3966/205.1$

$$= 19.3 > 1.5 \text{ OK}$$

PCI Design Handbook

Fig. 3.7.12 Example 3.7.3: Four-story building design

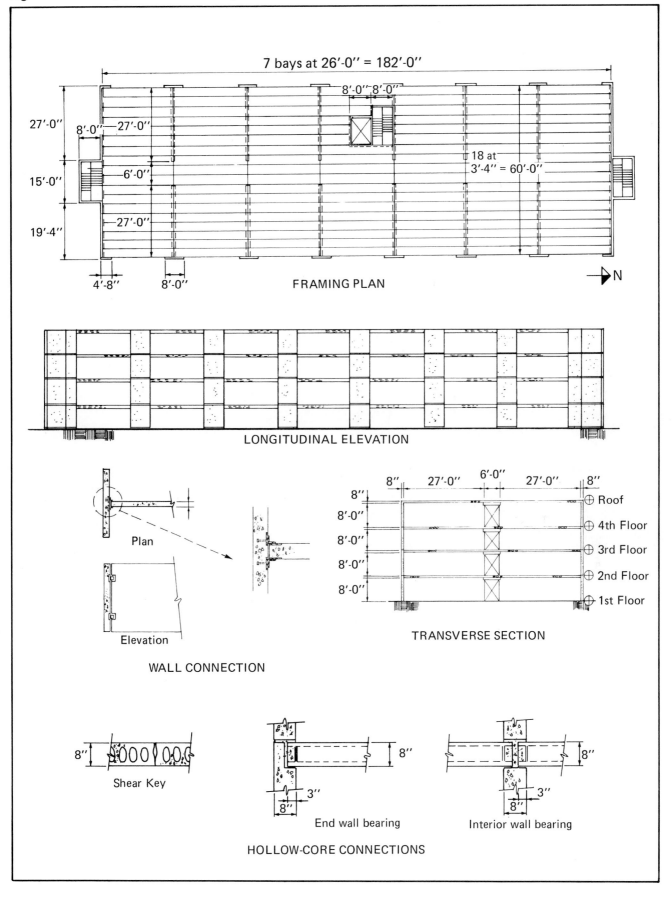

FRAMING PLAN

LONGITUDINAL ELEVATION

WALL CONNECTION

Plan

Elevation

TRANSVERSE SECTION

Shear Key

End wall bearing

Interior wall bearing

HOLLOW-CORE CONNECTIONS

Fig. 3.7.13 Loads to transverse walls — 4-story design example

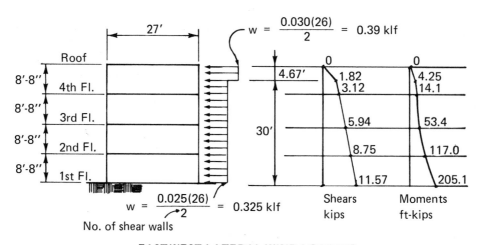

$$w = \frac{0.030(26)}{2} = 0.39 \text{ klf}$$

$$w = \frac{0.025(26)}{2} = 0.325 \text{ klf}$$

No. of shear walls

EAST-WEST LATERAL WIND LOADING

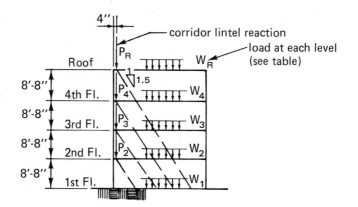

— corridor lintel reaction
— load at each level (see table)

GRAVITY LOADS ON BEARING WALL

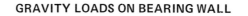

Summary of gravity loads

Load Mark	Tributary Area	Unit loads, psf		Wall weight, klf	Total unfactored loads		
		L.L.	D.L.		L.L.	D.L.	T.L.
P_R	78 sq. ft.	30	74	—	2.3 kips	5.8 kips	8.1 kips
P_4	78 sq. ft.	100	64	—	7.8 kips	5.0 kips	12.8 kips
P_3	78 sq. ft.	100	64	—	7.8 kips	5.0 kips	12.8 kips
P_2	78 sq. ft.	100	64	—	7.8 kips	5.0 kips	12.8 kips
W_R	26 lin. ft.	30	74	—	0.78 klf	1.92 klf	2.70 klf
W_4	26 lin. ft.	16*	74	0.8	0.42 klf	2.72 klf	3.14 klf
W_3	26 lin. ft.	16*	74	0.8	0.42 klf	2.72 klf	3.14 klf
W_2	26 lin. ft.	16*	74	0.8	0.42 klf	2.72 klf	3.14 klf
W_1	N / A	—	—	0.8	0	0.80 klf	0.80 klf

*Includes live load reduction allowed by codes

Check for tension using factored loads:
Dead weight on wall

$$P = [1.92 + 3(2.72) + 0.8](27)$$
$$= 293.8 \text{ kips}$$

Maximum moment at foundation

$$= 205.1 \text{ ft-kips}$$

$$f_{ut} = \frac{1.3M}{(\ell^2/6)} - \frac{0.9(293.8)}{27}$$

$$= \frac{1.3(205.1)}{(27^2/6)} - \frac{0.9(326.2)}{27}$$

$$= -7.60 \text{ klf (compression)}$$

No tension connections required between panels and the foundation. Thus the building is stable under wind loads in east-west direction. (Note: Other design considerations may dictate the use of minimum vertical ties.[7])

For wind in the longitudinal (north-south) direction, the shear walls will be connected to the load bearing walls. The assumed resisting elements are shown in Fig. 3.7.14; a summary of the properties is shown in Table 3.7.2. Sample calculations of these properties are given below for element A.

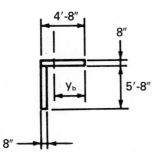

Effective width of perpendicular wall (see Fig. 3.7.2) is the smaller of:

$12t = 12(8) = 96$ in., or
$1/6(34.67 \times 12) = 69.3$ in. Use 5 ft 8 in.

Area of web $= 4.67 \times 0.67 = 3.11 \text{ ft}^2$
Area of flange $= 5.67 \times 0.67 = \underline{3.78}$
$ 6.89 \text{ ft}^2$

$$y_b = \frac{3.11(4.67/2) + 3.78(4.67 - 0.33)}{6.89}$$

$$= 3.43 \text{ ft}$$

$$y_t = 4.67 - 3.43 = 1.24 \text{ ft}$$

$$I = \frac{0.67(4.67)^3}{12} + 3.11(3.43 - 2.33)^2$$
$$+ 3.78(1.24 - 0.33)^2$$
$$= 12.58 \text{ ft}^4$$

Equivalent stiffness is calculated using the Case 1 multistory formula from Table 3.7.1.

$$I_{eq} = \frac{I}{1 + \dfrac{13.4\,I}{A_w h_s^2}}$$

$$= \frac{12.58}{1 + \dfrac{13.4(12.58)}{3.11(8.67)^2}} = 7.31 \text{ ft}^4$$

I_{eq} is essentially a relative stiffness:

$$K_r = 1/\Delta; \quad \Delta = \frac{Ph^3}{3EI_{eq}}$$

$$K_r = \frac{3EI_{eq}}{Ph^3}$$

Since 3, E, P, and h are all constants when comparing stiffnesses, $K_r = I_{eq}$.

Distribution of load to element A based on its relative stiffness is (See Table 3.7.2):

$$\frac{I_{eq}}{\Sigma nI_{eq}} = \frac{7.31(100)}{368.54} = 1.98\%$$

The shears and moments in the north-south direction are shown in Fig. 3.7.15, and the distributions are shown in Table 3.7.3.

To check overturning, consider element B at the first floor. From Fig. 3.7.13 the dead load on the 6'-4" portion of element B:

$$= 1.92 + 3(2.7) + 0.8 = 10.88 \text{ kips/ft}$$

The dead load on the 8'-0" portion of element B is the weight of the wall:

$$= 34.67 \times 0.1 = 3.47 \text{ kips/ft}$$

The resisting moment is then:

$$M_R = 10.88(5.67)(4) + 3.47(8)(4)$$
$$= 358 \text{ ft-kips} \times 11 \text{ elements}$$
$$= 3938 \text{ ft-kips}$$

Factor of safety = 3938/966.9 = 4.1 > 1.5 OK

(Note: This conservatively neglects the contribution of the other elements.)

To check for tension, also consider element B:

Total dead weight on the wall
$$= 10.88(5.67) + 3.47(8) = 89.45 \text{ kips}$$

Total wall area
$$= (8.0 + 5.67)\,0.67 = 9.16 \text{ ft}^2$$

$M = 38.5$ ft-kips (see Table 3.7.3)

$$f_{ut} = \frac{1.3M(d/2)}{I} - \frac{0.9P}{A}$$

$$= \frac{1.3(38.5)(4.0)}{28.7} - \frac{0.9(89.45)}{9.16}$$

$$= -1.81 \text{ ksf (compression)}$$

Fig. 3.7.14 Wind resisting elements for north-south wind

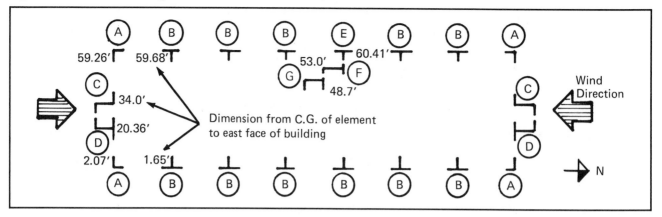

Table 3.7.2 Properties of resisting elements for wind in longitudinal direction

Element	A_w	I	y_b	I_{eq}	No. of elements	nI_{eq}	$\dfrac{I_{eq}}{\Sigma nI_{eq}}(100)$	$\Sigma \bar{y}$	$I_{eq}(\Sigma \bar{y})$
Ⓐ	3.11	12.6	3.43	7.31	4	29.24	1.98	123	899
Ⓑ	5.36	28.7	4.0	14.68	11	161.5	3.98	308	4521
Ⓒ	5.81	158.1	4.34	27.02	2	54.04	7.33	68	1837
Ⓓ	5.81	205.6	3.45	28.13	2	56.26	7.63	41	1153
Ⓔ	5.36	29.0	4.0	14.76	1	14.76	4.01	60	886
Ⓕ	5.81	114.1	2.72	25.35	1	25.35	6.88	53	1344
Ⓖ	5.81	171.6	4.09	27.39	1	27.39	7.43	49	1342

$\Sigma nI_{eq} = 368.54$ $\Sigma = 11,982$

Center of rigidity = 11,982/368.54 = 32.51 ft from east
Note: The north-south wind load is slightly eccentric by 32.51 − 61.33/2 = 1.85 ft.
Torsion due to this eccentricity is neglected in calculating shears and moments in Table 3.7.3.

PCI Design Handbook

Table 3.7.3 Distribution of wind shears and moments (north-south direction)

Element	% Dist.	4th floor Shear 14.71 kips	Moment 66.7 ft-kips	3rd floor Shear 27.98 kips	Moment 251.7 ft-kips	2nd floor Shear 41.24 kips	Moment 551.8 ft-kips	1st floor Shear 54.51 kips	Moment 966.9 ft-kips
A	1.98	0.29	1.32	0.55	4.98	0.82	10.9	1.08	19.1
B	3.98	0.59	2.65	1.11	10.0	1.64	22.0	2.17	38.5
C	7.33	1.08	4.89	2.05	18.4	3.02	40.4	4.00	70.9
D	7.63	1.12	5.09	2.13	19.2	3.15	42.1	4.16	73.8
E	4.01	0.59	2.67	1.12	10.1	1.65	22.1	2.19	38.8
F	6.88	1.01	4.59	1.93	17.3	2.84	38.0	3.75	66.5
G	7.43	1.09	4.96	2.08	18.7	3.06	41.0	4.05	71.8

The relative stiffness and percent distribution for the elements in this table are assumed the same for all stories. The exact values may be slightly different for each story because the values change due to the reduced flange width (see Fig. 3.7.2).

Fig. 3.7.15 Wind load in north-south direction

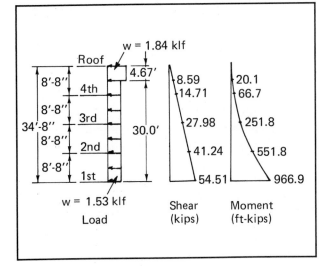

Fig. 3.7.16 Diaphragm analysis

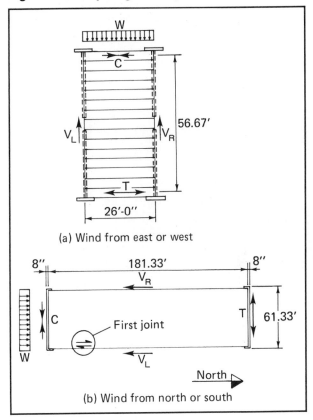

(a) Wind from east or west

(b) Wind from north or south

No tension connections are required between panels and the foundation. Thus the building is stable under wind loads in the north-south direction.

The connections required to assure that the elements will act in a composite manner as assumed can be designed by considering element A. The unit stress at the interface is determined using the classic equation for horizontal shear:

$$v_h = \frac{VQ}{I}$$

$$Q = 5.67(0.67)(1.24 - 0.33) = 3.46 \text{ ft}^3$$

$$v_h = \frac{1.08(3.46)}{12.6} = 0.297 \text{ kips/ft}$$

Total shear = 0.297(8.0) = 2.37 kips

Connections similar to those shown in Fig. 3.7.12 can be designed using the principles outlined in Part 6.

Design of floor diaphragm:

Analysis procedures for the floor diaphragm are described in Sect. 3.6. For this example refer to Fig. 3.7.16.

The factored wind load for a typical floor is:

$$W_u = 1.3(25 \text{ psf})(8.67 \text{ ft}) = 282 \text{ plf}$$

For wind from the east or west:

$$V_{Ru} = \frac{0.282(26)}{2} = 3.67 \text{ kips}$$

$$C_u = T_u = \frac{M_u}{\ell} = \frac{0.282(26)^2}{8(56.67)}$$

$$= 0.42 \text{ kips}$$

The reaction V_{Ru} is transferred to the shear wall by static friction:

Dead load of floor
$\quad = (26/2)(64 + 10)(60) \qquad = 57,720$
Dead load of wall $= 800/2(54) \qquad = \underline{21,600}$
$\qquad\qquad\qquad\qquad\qquad\qquad\qquad 79,320$

Static coefficient of friction from Table 6.6.1 (hardboard to concrete) = 0.5. Reduce by factor of 5 as recommended in Sect. 6.6.

$\quad \mu = 0.5/5 = 0.10$
Resisting force $= 0.10 (79.3)$
$\qquad\qquad\qquad = 7.93 > 3.67$ OK

The chord tension, T_u, is resisted by the tensile strength of the floor slab. The grout key between slabs must also resist approximately the same force.

Assume area of exterior slab = 218 in.²
Grout key = 3 in. deep
Concrete f'_c = 5000 psi
Use a resisting tensile strength of:
$\quad 3\lambda \sqrt{f'_c} = 212$ psi
Grout key resisting strength (see Sect. 3.6.2).
$\quad = 80$ psi
Resisting tensile strength of slab
$\quad = 218(0.212) = 46.2$ kips > 0.42 OK

Resisting strength of grout key
$\quad = 26 (12) (3) (0.080)$
$\quad = 74.9$ kips > 0.42 OK

For wind from the north or south:

$$V_{Ru} = \frac{0.282(61.33)}{2} = 8.65 \text{ kips}$$

Resisting force in the first joint
$\quad = 181.33(12)(3)(0.080) = 522$ kips OK

$$C_u = T_u = \frac{0.282(61.33)^2}{8(181.33)}$$

$$= 0.73 \text{ kips} < 7.93 \text{ OK}$$

In this example, only the resistance to wind loading was analyzed. Any other required loading (including "abnormal" loads) must be reviewed for a complete analysis.

3.7.7 Architectural Panels as Shear Walls

In many structures it is economical to take advantage of the inherent strength and rigidity of exterior panels, and design them to serve as the lateral load resisting system. The effectiveness of such a system is largely dependent on the panel-to-panel connections.

Fig. 3.7.17 illustrates the foundation reaction distributions of exterior architectural precast shear wall systems under the action of lateral load, with and without connections between the shear walls and the windward or leeward walls. The structure with corner connections is structurally more efficient for resisting lateral loads.

The lateral load resisting system shown in Fig. 3.7.17b is frequently labeled a "tube". However, because the components and the connections are not perfectly rigid, full tube behavior does not develop. Fig. 3.7.18 illustrates the difference. The peaking of the foundation reaction at the corner results from shear lag, which limits the effective width of the "flange". Accurate evaluation of shear lag is difficult, but analytical and research studies have indicated the following limitations on the effective flange are sufficiently accurate for most structures:

1. One-half the length of the shear wall
2. One-third the length of the windward or leeward wall
3. One-tenth the height of the building
4. Six times the thickness of the "flange" wall
5. Distance to nearest major opening
6. One-half the distance to the nearest shear wall.

Fig 3.7.17 Foundation reaction distributions resulting from lateral loads

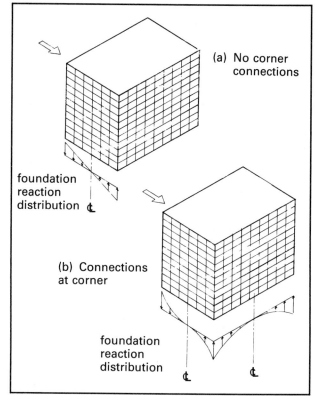

(a) No corner connections

foundation reaction distribution ₵

(b) Connections at corner

foundation reaction distribution ₵ ₵

Fig 3.7.18 Influence of shear lag on tube behavior

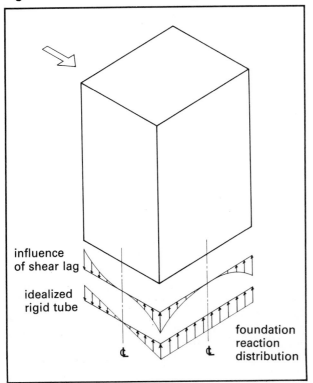

influence of shear lag

idealized rigid tube

foundation reaction distribution

3.8 Buildings With Moment-Resisting Frames

3.8.1 General

Precast, prestressed concrete beams and deck members are usually most economical when they can be designed and connected into a structure as simple-span members. This is because:

1. Positive moment-resisting capacity is much easier and less expensive to attain with pretensioned members than negative moment capacity at supports.

2. Connections which achieve continuity at the supports are usually complex and costly.

3. The restraint to volume changes that occurs in rigid connections may cause serious cracking and unsatisfactory performance or, in extreme cases, even structural failure.

Therefore, it is most desirable when designing precast, prestressed concrete structures to have connections which allow lateral movement and rotation, and achieve lateral stability through use of floor and roof diaphragms and shear walls.

However, in some structures, adequate shear walls interfere with the function of the building, or are more expensive than alternate solutions. In these cases, the lateral stability of the structure depends on the moment-resisting capacity of either the column bases, a beam-column frame, or both.

When moment connections between beams and columns are required to resist lateral loads, it is desirable, when possible, to make the moment connection after most of the dead loads have been applied. This requires careful detailing, specification of the construction process, and inspection. If such details are possible, the moment connections need only resist the negative moments from live load, lateral loads and volume changes, and will then be less costly.

3.8.2 Moment Resistance of Column Bases

Single-story and some low-rise buildings without shear walls may depend on the fixity of the column base to resist lateral loads. The ability of a spread footing to resist moments caused by lateral loads is dependent on the rotational characteristics of the base. The total rotation of the column base is a function of rotation between the footing and soil, bending in the base plate, and elongation of the anchor bolts, as shown in Fig. 3.8.1.

The total rotation of the base is:

$$\phi_b = \phi_f + \phi_{bp} + \phi_{ab}$$

If the axial load is large enough so that there is no tension in the anchor bolts, ϕ_{bp} and ϕ_{ab} are zero, and:

$$\phi_b = \phi_f$$

Fig. 3.8.1 Assumptions used in derivation of rotational coefficients for column bases

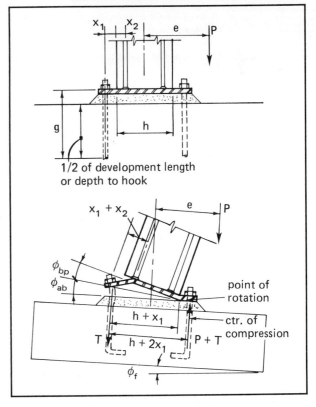

1/2 of development length or depth to hook

point of rotation

ctr. of compression

Rotational characteristics can be expressed in terms of flexibility or stiffness coefficients:

$$\phi = \gamma M = M/K$$

where

M = applied moment = Pe
e = eccentricity of the applied load, P
γ = flexibility coefficient
K = stiffness coefficient = $1/\gamma$

If bending of the base plate and strain in the anchor bolts are assumed as shown in Fig. 3.8.1, the flexibility coefficients for the base can be derived, and the total rotation of the base becomes:

$$\phi_b = M (\gamma_f + \gamma_{bp} + \gamma_{ab}) = Pe (\gamma_f + \gamma_{bp} + \gamma_{ab})$$

$$\gamma_f = 1/k_s I_f \qquad \text{(Eq. 3.8.1)}$$

$$\gamma_{bp} = \frac{(x_1 + x_2)^3 [2e/(h + 2x_1) - 1]}{6e \, E_s \, I_{bp} \, (h + x_1)} \geq 0 \qquad \text{(Eq. 3.8.2)}$$

$$\gamma_{ab} = \frac{g [2e/(h + 2x_1) - 1]}{2e \, A_b \, E_s \, (h + x_1)} \geq 0 \qquad \text{(Eq. 3.8.3)}$$

where:

$\gamma_f, \gamma_{bp}, \gamma_{ab}$ = flexibility coefficients of the footing/soil interaction, the base plate and the anchor bolts, respectively

k_s = coefficient of subgrade reaction from Fig. 3.8.2

I_f = moment of inertia of the footing (plan dimensions)

E_s = modulus of elasticity of steel

I_{bp} = moment of inertia of the base plate (vertical cross-section dimensions)

A_b = total area of anchor bolts which are in tension

h = width of the column in the direction of bending

x_1 = distance from face of column to the center of the anchor bolts, positive when anchor bolts are outside the column, and negative when anchor bolts are inside the column

x_2 = distance from the face of the column to base plate anchorage

g = assumed length over which elongation of the anchor bolt takes place = 1/2 of development length + projection for anchor bolts made from reinforcing bars or the length to the hook + projection for smooth anchor bolts (see Fig. 3.8.1)

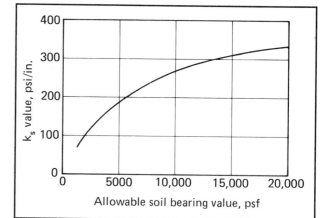

Fig. 3.8.2 Approximate relationship between allowable soil bearing value and coefficient of subgrade reaction, k_s

Rotation of the base may cause an additional eccentricity of the loads on the columns, causing moments which must be added to the moments induced by the lateral loads.

Note that in Eqs. 3.8.2 and 3.8.3, if the eccentricity, e, is less than $h/2 + x_1$ (inside the center of compression), γ_{bp} and γ_{ab} are less than zero, meaning that there is no rotation between the column and the footing, and only the rotation from soil deformation (Eq. 3.8.1) need be considered.

Values of Eqs. 3.8.1 through 3.8.3 are tabulated for typical cases in Tables 3.8.1 and 3.8.2.

Example 3.8.1 Stability analysis of an unbraced frame

Given:

The column shown in Fig. 3.8.3
Soil bearing capacity = 5000 psf
P = 80 kips dead load, 30 kips live load
W = 2 kips wind load

Determine the column design loads and moments for stability as an unbraced frame.

Solution:

ACI 318-83 requires that the column be designed for the following conditions:

1. $1.4 D + 1.7 L$
2. $0.75 (1.4 D + 1.7 L + 1.7 W)$
3. $0.9 D + 1.3 W$

The maximum eccentricity would occur when 3 is applied. Moment at base of column
= 2 (16) = 32 ft-kips = 384 in.-kips

$0.9 D = 0.9 (80) = 72$ kips
$1.3 W = 1.3 (384) = 499.2$ in.-kips

PCI Design Handbook

Table 3.8.1 Flexibility coefficients for footing/soil interaction

Flexibility of base $= \gamma_b = \gamma_f + \gamma_{ab} + \gamma_{bp}$

Rotation of base $= \gamma_b \times Pe$ in inch-lbs

Stiffness of base $= K_b = 1/\gamma_b$

Fixity of base $= K_b/(K_c + K_b)$

$K_c =$ Column stiffness $= 4E_c I_c/h_s$
$E_c =$ Modulus of elasticity of column concrete, psi
$I_c =$ Moment of inertia of column, in.^{4}
$h_s =$ Story height, in.

γ_f, 1/in-lb $\times 10^{-10}$ for square footings

Footing Size (ft)	k_s				
	100	150	200	250	300
2.0 X 2.0	3616.9	2411.3	1808.4	1446.8	1205.6
2.5 X 2.5	1481.5	987.7	740.7	592.6	493.8
3.0 X 3.0	714.4	476.3	357.2	285.8	238.1
3.5 X 3.5	385.6	257.1	192.8	154.3	128.5
4.0 X 4.0	226.1	150.7	113.0	90.4	75.4
4.5 X 4.5	141.1	94.1	70.6	56.5	47.0
5.0 X 5.0	92.6	61.7	46.3	37.0	30.9
5.5 X 5.5	63.2	42.2	31.6	25.3	21.1
6.0 X 6.0	44.7	29.8	22.3	17.9	14.9
6.5 X 6.5	32.4	21.6	16.2	13.0	10.8
7.0 X 7.0	24.1	16.1	12.1	9.6	8.0
7.5 X 7.5	18.3	12.2	9.1	7.3	6.1
8.0 X 8.0	14.1	9.4	7.1	5.7	4.7
9.0 X 9.0	8.8	5.9	4.4	3.5	2.9
10.0 X 10.0	5.8	3.9	2.9	2.3	1.9
11.0 X 11.0	4.0	2.6	2.0	1.6	1.3
12.0 X 12.0	2.8	1.9	1.4	1.1	0.9

Eccentricity due to wind load

$$= \frac{M_u}{P_u} = \frac{499.2}{72} = 6.93 \text{ in.}$$

To determine the moments caused by base rotation, an iterative procedure is required.

Estimate eccentricity due to rotation = 0.25 in.

$e = 6.93 + 0.25 = 7.18$ in.

Check rotation between column and footing:

$h/2 + x_1 = 20/2 + (-2) = 8$ in. > 7.18,
thus there is no tension in the anchor bolts and no rotation between the column and footing.

$I_f = (6 \times 12)^4/12 = 2.24 \times 10^6$ in.^{4}

From Fig. 3.8.2: $k_s \approx 200$ psi/in.

$\gamma_f = 1/k_s I_f = 1/[200 (2.24 \times 10^6)]$
$= 2.23 \times 10^{-9}$

(Note: This could also be read from Table 3.8.1)

$M_u = 72 (7.18) = 517$ in.-kips
$\phi_b = \gamma_f M_u = (2.23 \times 10^{-9})(517 \times 10^3)$
$= 0.00115$ radians

Eccentricity caused by rotation:

$\phi_b h_s = 0.00115 (16 \times 12) = 0.22$ in.
≈ 0.25
no further trial is required

Design requirements for 0.9 D + 1.3 W:

$P_u = 72$ kips
$M_u = 517$ in.-kips $= 43.1$ ft-kips

Table 3.8.2 Flexibility coefficients for anchor bolts and bare plates

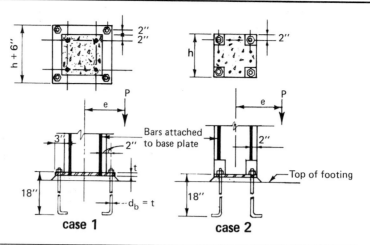

case 1   case 2

Column size, h (in)	e (in)	$\gamma_{ab} + \gamma_{bp}$, 1/in.-lb × 10⁻¹⁰ for typical details (see Eq. 3.8.2 and 3.8.3)							
		Case 1: Exterior anchor bolts				**Case 2: Interior anchor bolts**			
		Base plate thickness & anc. bolt diameter				Base plate thickness & anc. bolt diameter			
		.75	1.00	1.25	1.50	.75	1.00	1.25	1.50
12 × 12	4	.0	.0	.0	.0	.0	.0	.0	.0
	6	.0	.0	.0	.0	29.3	16.5	10.5	7.3
	8	.0	.0	.0	.0	43.9	24.7	15.8	11.0
	10	16.6	7.9	4.5	2.9	52.7	29.6	19.0	13.2
	12	27.7	13.2	7.5	4.8	58.5	32.9	21.1	14.6
	14	35.7	16.9	9.6	6.1	62.7	35.3	22.6	15.7
	16	41.6	19.8	11.2	7.2	65.8	37.0	23.7	16.5
	18	46.2	22.0	12.5	8.0	68.3	38.4	24.6	17.1
16 × 16	6	.0	.0	.0	.0	.0	.0	.0	.0
	8	.0	.0	.0	.0	10.5	5.9	3.8	2.6
	10	.0	.0	.0	.0	16.7	9.4	6.0	4.2
	12	7.7	3.7	2.1	1.4	20.9	11.8	7.5	5.2
	14	13.1	6.3	3.6	2.3	23.9	13.4	8.6	6.0
	16	17.2	8.3	4.8	3.1	26.1	14.7	9.4	6.5
	18	20.4	9.8	5.7	3.6	27.9	15.7	10.0	7.0
	20	23.0	11.1	6.4	4.1	29.3	16.5	10.5	7.3
20 × 20	8	.0	.0	.0	.0	.0	.0	.0	.0
	10	.0	.0	.0	.0	4.9	2.7	1.8	1.2
	12	.0	.0	.0	.0	8.1	4.6	2.9	2.0
	14	4.1	2.0	1.2	.7	10.5	5.9	3.8	2.6
	16	7.1	3.5	2.0	1.3	12.2	6.9	4.4	3.0
	18	9.5	4.6	2.7	1.7	13.5	7.6	4.9	3.4
	20	11.4	5.6	3.2	2.1	14.6	8.2	5.3	3.7
	22	13.0	6.3	3.7	2.4	15.5	8.7	5.6	3.9
24 × 24	10	.0	.0	.0	.0	.0	.0	.0	.0
	12	.0	.0	.0	.0	2.7	1.5	1.0	.7
	14	.0	.0	.0	.0	4.6	2.6	1.6	1.1
	16	2.4	1.2	.7	.5	6.0	3.4	2.2	1.5
	18	4.3	2.1	1.2	.8	7.1	4.0	2.6	1.8
	20	5.8	2.8	1.7	1.1	8.0	4.5	2.9	2.0
	22	7.0	3.4	2.0	1.3	8.7	4.9	3.1	2.2
	24	8.0	3.9	2.3	1.5	9.3	5.2	3.4	2.3

Fig. 3.8.3 Examples 3.8.1 and 3.8.2

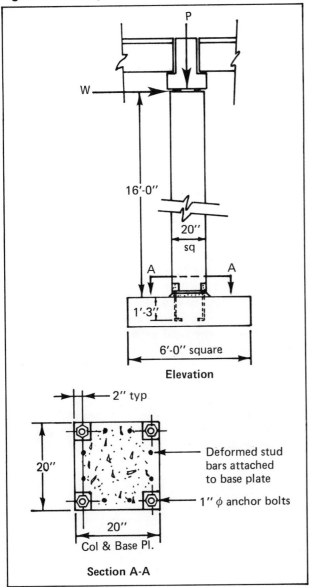

Elevation

Section A-A

2" typ

20"

20"
Col & Base Pl.

Deformed stud
bars attached
to base plate

1" ϕ anchor bolts

Check for 0.75 (1.4 D + 1.7 L + 1.7 W):

P_u = 0.75 (1.4 D + 1.7 L)
= 0.75 [1.4 (80) + 1.7 (30)]
= 122.3 kips

M_u = 0.75 (1.7 W) = 0.75 [1.7 (384)]
= 489.6 in.kips

e = $\frac{489.6}{122.3}$ = 4.0 in.

Estimate eccentricity due to rotation = 0.22 in.

M_u = 122.3 (4.22) = 516.1 in.-kips

ϕ_b = $\gamma_f M_u$ = (2.23 × 10⁻⁹) (516.1 × 10³)
= 0.00115 radians

$\phi_b h_s$ = 0.00115 (16 × 12) = 0.22 in. OK

Design requirements for
0.75 [1.4 D + 1.7 L + 1.7W]:

P_u = 122.3 kips
M_u = 516.1 in.-kips = 43.0 ft-kips

Section 10.11.5 (ACI 318-83) also requires that
the moment caused by a minimum eccentricity of
0.6 + 0.03h be considered when designing for
1.4 D + 1.7 L.

P_u = 1.4 D + 1.7 L = 1.4 (80) + 1.7 (30)
= 163 kips

e = 0.6 + 0.03h = 0.6 + 0.03 (20)
= 1.2 in.

Estimate eccentricity due to rotation = 0.1 in.

M_u = P_ue = 163 (1.2 + 0.1)
= 211.9 in.-kips

ϕ_b = (2.23 × 10⁻⁹)(211.9 × 10³)
= 0.000473 radians

$\phi_b h_s$ = 0.000473 (16 × 12) = 0.09 in.
≈ 0.1 in. OK

Design requirements for 1.4 D + 1.7 L:

P_u = 163 kips
M_u = 212 in.-kips = 17.7 ft-kips

3.8.3 Fixity of Column Bases

The degree of fixity of a column base is the ratio
of the rotational stiffness of the base to the sum
of the rotational stiffnesses of the column plus the
base:

$$F_b = \frac{K_b}{K_b + K_c} \qquad \text{(Eq. 3.8.4)}$$

where

F_b = degree of base fixity, expressed as a decimal

K_b = $1/\gamma_b$

K_c = $\frac{4 E_c I_c}{h_s}$

E_c = modulus of elasticity of the column concrete

I_c = moment of inertia of the column

h_s = column height

Example 3.8.2 Calculation of degree of fixity

Determine the degree of fixity of the column
base in Example 3.8.1; E_c = 4300 ksi:

$$K_b = 1/\gamma_b = 1/(2.23 \times 10^{-9}) = 4.48 \times 10^8$$

$$I_c = 20^4/12 = 13,333$$

$$K_c = \frac{4\,(4.3 \times 10^6)\,(13,333)}{16 \times 12} = 11.94 \times 10^8$$

$$F_b = \frac{4.48}{4.48 + 11.94} = 0.27$$

3.8.4 Modeling Partially Fixed Bases

A simple way to model base fixity for computer analyses is to incorporate an "imaginary column" below the actual column base. If the bottom of the imaginary column is modeled as pinned, then the expression for its rotational stiffness is:

$$K_{ci} = 3E_{ci}I_{ci}/h_{ci} \qquad \text{(Eq. 3.8.5)}$$

where the subscript "ci" denotes the properties of the imaginary column (Fig. 3.8.4), and $K_{ci} = K_b$ as calculated in Sect. 3.8.2. Or, the degree of base fixity, F_b, can be determined or estimated, and K_b calculated from Eq. 3.8.4.

For the computer model, either I_{ci} or h_{ci} may be varied for different values of K_{ci}, with the other terms left constant for a given problem. It is usually preferable to use $E_{ci} = E_c$. For the assumptions of Fig. 3.8.4:

$$h_{ci} = 3E_{ci}\,I_{ci}/K_{ci} \qquad \text{(Eq. 3.8.5a)}$$

or

$$I_{ci} = K_{ci}\,h_{ci}/(3E_{ci}) \qquad \text{(Eq. 3.8.5b)}$$

Example 3.8.3 Imaginary column for computer model

Determine the length of an imaginary column to model the fixity of the column base of Examples 3.8.1 and 3.8.2; $f'_c = 5000$ psi, $E_c = 4300$ ksi.

Fig. 3.8.4 Model for partially fixed column base

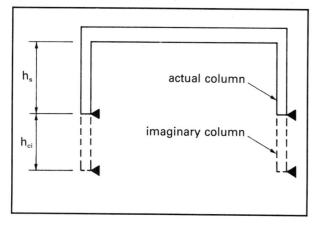

Assume: $E_{ci} = E_c = 4.3 \times 10^6$ psi
$$I_{ci} = I_c = 13,333 \text{ in.}$$
$$K_{ci} = K_b = 4.48 \times 10^8$$

$$h_{ci} = \frac{3\,E_{ci}I_{ci}}{K_{ci}} = \frac{3(4.3 \times 10^6)\,(13,333)}{4.48 \times 10^8}$$

$$= 383 \text{ in.} = 31.9 \text{ ft}$$

3.8.5 Volume Change Effects in Moment-Resisting Frames

The restraint of volume changes in moment-resisting frames causes tension in the girders and deflections and moments in the columns. The magnitude of these tensions, moments and deflections is dependent on the distance from the center of stiffness of the frame.

The center of stiffness is that point of a frame, which is subject to a uniform unit shortening, at which no lateral movement will occur. For frames which are symmetrical with respect to bay sizes, story heights and member stiffnesses, the center of stiffness is located at the midpoint of the frame, as shown in Fig. 3.8.5.

Tensions in girders are maximum in the bay nearest the center of stiffness. Deflections and moments in columns are maximum furthest from the center of stiffness. Thus in Fig. 3.8.5:

$$F_1 < F_2 < F_3$$
$$\Delta_1 > \Delta_2 > \Delta_3$$
$$M_1 > M_2 > M_3$$

The degree of fixity of the column base as described in Sect. 3.8.3 has a great effect on the magnitude of the forces and moments caused by volume change restraint. An assumption of a fully fixed base in the analysis of the structure may result in signficant overestimation of the restraint forces, whereas assuming a pinned base may have the opposite effect. The degree of fixity used in the volume change analysis should be consistent with that used in the analysis of the column for other loadings, and determination of slenderness effects.

3.8.5.1 Equivalent volume change

If a horizontal framing member is connected at the ends, such that the volume change shortening is restrained, a tensile force is built up in the member and transmitted to the supporting elements. However, since the shortening takes place gradually over a period of time, the effect of the shortening on the shears and moments of the support

Fig. 3.8.5 Effect of volume change restraints in building frames

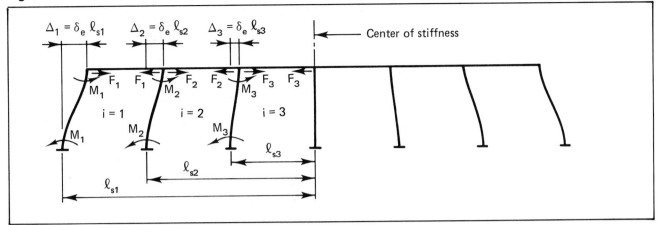

is lessened because of creep and micro-cracking of the member and its support.

For ease of design, the volume change shortenings can be treated in the same manner as short term elastic deformations by using a concept of "equivalent" shortening.

Thus, the following relations can be assumed:

$$\delta_{ec} = \delta_c/K_\ell$$

$$\delta_{es} = \delta_s/K_\ell$$

where

δ_{ec}, δ_{es} = equivalent creep and shrinkage shortenings, respectively

δ_c, δ_s = calculated creep and shrinkage shortenings, respectively

K_ℓ = a constant for design purposes which varies from 4 to 6

The value of K_ℓ will be near the lower end of the range when the members are heavily reinforced, and near the upper end when they are lightly reinforced. For most common structures, a value of $K_\ell = 5$ is sufficiently conservative.

Shortening due to temperature change* will be similarly modified. However, the maximum temperature change will usually occur over a much shorter time, probably within 60 to 90 days. Thus,

$$\delta_{et} = \delta_t/K_t$$

where

δ_{et} and δ_t = the equivalent and calculated temperature shortening, respectively

K_t = a constant
recommended value = 1.5

---

* Temperature change is, of course, a reversible effect; increases cause expansion and are important in design and location of expansion joints (see Sect. 3.3.3). Temperature differentials in roof and wall elements should also be considered.

The total equivalent shortening to be used for design is:

$$\Delta = \delta_{ec} + \delta_{es} + \delta_{et}$$

$$= \frac{\delta_c + \delta_s}{K_\ell} + \frac{\delta_t}{K_t}$$

When the equivalent shortening is used in the frame analysis for determining shears and moments in the supporting elements, the actual modulus of elasticity of the members is used, rather than a reduced modulus as used in other methods.

Tables 3.8.3 and 3.8.4 provide equivalent volume change strains for typical building frames.

Example 3.8.4 Calculation of column moment caused by volume change shortening of a beam

Given:

The beam of Example 3.3.1 is supported and attached to two 16 × 16-in. columns as shown in the sketch.

f'_c (col.) = 5000 psi

$E_c = 4.3 \times 10^6$ psi

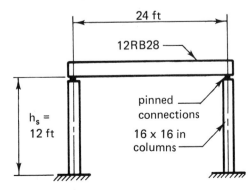

Table 3.8.3 Equivalent volume change strains for typical continuous building frames (millionths)

Prestressed members (P/A = 600 psi)										
	Normal weight concrete					Lightweight concrete				
Temp. zone (from map)	Ave. R.H. (from map)					Ave. R.H. (from map)				
	40	50	60	70	80	40	50	60	70	80
Heated Buildings										
0	117	107	97	86	76	123	113	102	92	81
10	137	127	117	106	96	140	130	119	109	98
20	157	147	137	126	116	157	146	136	125	115
30	177	167	157	146	136	173	163	152	142	131
40	197	187	177	166	156	190	180	169	159	148
50	217	207	197	186	176	207	196	186	175	165
60	237	227	217	206	196	223	213	202	192	181
70	257	247	237	226	216	240	230	219	209	198
80	277	267	257	246	236	257	246	236	225	215
90	297	287	277	266	256	273	263	252	242	231
100	317	307	297	286	276	290	280	269	259	248
Unheated Structures										
0	117	107	97	86	76	123	113	102	92	81
10	147	137	127	116	106	148	138	127	117	106
20	177	167	157	146	136	173	163	152	142	131
30	207	197	187	176	166	198	188	177	167	156
40	237	227	217	206	196	223	213	202	192	181
50	267	257	247	236	226	248	238	227	217	206
60	297	287	277	266	256	273	263	252	242	231
70	327	317	307	296	286	298	288	277	267	256
80	357	347	337	326	316	323	313	302	292	281
90	387	377	367	356	346	348	338	327	317	306
100	417	407	397	386	376	373	363	352	342	331

Table 3.8.4 Equivalent volume change strains for typical continuous building frames (millionths)

Non-prestressed members										
	Normal weight concrete					Lightweight concrete				
Temp. zone (from map)	Ave. R.H. (from map)					Ave. R.H. (from map)				
	40	50	60	70	80	40	50	60	70	80
Heated Buildings										
0	54	48	43	38	32	54	48	43	38	32
10	74	68	63	58	52	70	65	60	54	49
20	94	88	83	78	72	87	82	76	71	66
30	114	108	103	98	92	104	98	93	88	82
40	134	128	123	118	112	120	115	110	104	99
50	154	148	143	138	132	137	132	126	121	116
60	174	168	163	158	152	154	148	143	138	132
70	194	188	183	178	172	170	165	160	154	149
80	214	208	203	198	192	187	182	176	171	166
90	234	228	223	218	212	204	198	193	188	182
100	254	248	243	238	232	220	215	210	204	199
Unheated Structures										
0	54	48	43	38	32	54	48	43	38	32
10	84	78	73	68	62	79	73	68	63	57
20	114	108	103	98	92	104	98	93	88	82
30	144	138	133	128	122	129	123	118	113	107
40	174	168	163	158	152	154	148	143	138	132
50	204	198	193	188	182	179	173	168	163	157
60	234	228	223	218	212	204	198	193	188	182
70	264	258	253	248	242	229	223	218	213	207
80	294	288	283	278	272	254	248	243	238	232
90	324	318	313	308	302	279	273	268	263	257
100	354	348	343	338	332	304	298	293	288	282

Problem:

Determine the horizontal force at the top of the column and the moment at the base of the column caused by volume change shortening of the beam.

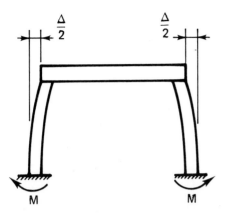

Solution:

$I_c = bh^3/12 = 16^4/12 = 5461 \text{ in.}^4$

From Example 3.3.1:

Total volume change shortening from erection to final is 0.21 in., or 0.11 in. each end.

Calculate the equivalent shortening:

$$\Delta = \frac{\delta_c + \delta_s}{K_\ell} + \frac{\delta_t}{K_t}$$

$$= \left[\frac{(246 - 80 + 499 - 153)}{5} + \frac{210}{1.5}\right] \times (10^{-6})(24)(12)$$

$$= 140 \times 10^{-6}(288) = 0.040 \text{ in.}$$

$\Delta/2 = 0.040/2 = 0.020$ in. each end

$\Delta/2 = Nh_s^3/3E_cI_c$

$N = 3E_cI_c \, (\Delta/2)/h_s^3$
$\quad = 3(4.3 \times 10^6)(5461)(0.020)/(12 \times 12)^3$
$\quad = 472$ lb

$M = Nh_s$
$\quad = 472(144)$
$\quad = 67,968$ in.-lb $= 5.66$ ft-kips

3.8.5.2 Calculating restraint forces

Most "plane frame" computer analysis programs allow the input of shortening strains of members from volume changes. The equivalent strains as described in Sect. 3.8.5.1 can be input directly into such programs.

For frames that are approximately symmetrical, the coefficients from Tables 3.8.5 and 3.8.6 can be used with small error. The use of these tables is described in Fig. 3.8.6.

Table 3.8.5 Build-up of restraint forces in beams (k_b)

Total number of bays (n)	Number of bays from end (i)							
	1	2	3	4	5	6	7	8
2	1.00							
3	1.00	4.00						
4	1.00	3.00						
5	1.00	2.67	9.00					
6	1.00	2.50	6.00					
7	1.00	2.40	5.00	16.00				
8	1.00	2.33	4.50	10.00				
9	1.00	2.29	4.20	8.00	25.00			
10	1.00	2.25	4.00	7.00	15.00			
11	1.00	2.22	3.86	6.40	11.67	36.00		
12	1.00	2.20	3.75	6.00	10.00	21.00		
13	1.00	2.18	3.67	5.71	9.00	16.00	49.00	
14	1.00	2.17	3.60	5.50	8.33	13.50	28.00	
15	1.00	2.15	3.55	5.33	7.86	12.00	21.00	64.00
16	1.00	2.14	3.50	5.20	7.50	11.00	17.50	36.00

Fig. 3.8.6 Use of Table 3.8.6

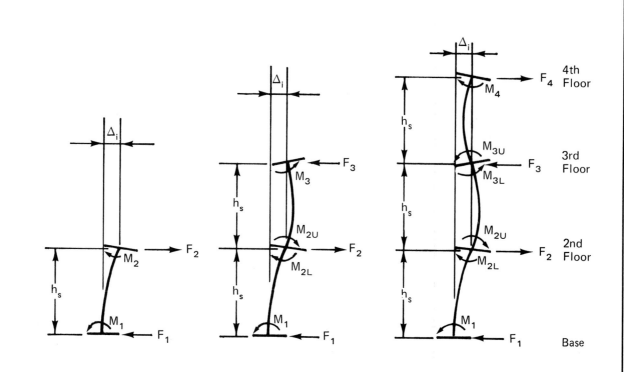

$$\Delta_i = \delta_e \ell_s$$

$$F_i = k_f k_b \Delta_i E_c I_c / h_s^3$$

$$M_i = k_m \Delta_i E_c I_c / h_s^2$$

where: δ_e = equivalent unit strain (see Sect. 3.3)

ℓ_s = distance from column to center of stiffness

F_i = F_1, F_2, etc., as shown above

k_f, k_m = coefficients from Table 3.8.6

$k_b = i\left(\dfrac{n + 1 - i}{n + 2 - 2i}\right)$ (or from Table 3.8.5)

n = no. of bays

i = as shown in Fig. 3.8.5

E_c = modulus of elasticity of the column concrete

I_c = moment of inertia of the column

Table 3.8.6 Coefficients k_f and k_m for forces and moments caused by volume change restraint forces (see Fig. 3.8.6 for notation)

No. of Stories	$K_r = \dfrac{\Sigma E_b I_b / \ell}{\Sigma E_c I_c / h_s}$	Base Fixity	Values of k_f				Values of k_m					
							Base	2nd floor		3rd floor		4th
			F_1	F_2	F_3	F_4	M_1	M_{2L}	M_{2U}	M_{3L}	M_{3U}	M_4
1	0	Fixed	3.0	3.0			3.0	0				
		Pinned	0	0			0	0				
	0.5	Fixed	6.0	6.0			4.0	2.0				
		Pinned	1.2	1.2			0	1.2				
	1.0	Fixed	7.5	7.5			4.5	3.0				
		Pinned	1.7	1.7			0	1.7				
	2.0	Fixed	9.0	9.0			5.0	4.0				
		Pinned	2.2	2.2			0	2.2				
	4.0 or more	Fixed	10.1	10.1			5.4	4.7				
		Pinned	2.5	2.5			0	2.5				
2	0	Fixed	6.8	9.4	2.6		4.3	2.6	2.6	0		
		Pinned	0	3.0	1.5		0	1.5	1.5	0		
	0.5	Fixed	8.1	10.7	2.6		4.7	3.4	2.1	0.4		
		Pinned	1.9	3.4	1.4		0	1.9	1.2	0.2		
	1.0	Fixed	8.9	11.2	2.3		4.9	3.9	1.8	0.5		
		Pinned	2.1	3.4	1.3		0	2.1	1.0	0.3		
	2.0	Fixed	9.7	11.6	1.9		5.2	4.5	1.4	0.5		
		Pinned	2.4	3.4	1.0		0	2.4	0.8	0.3		
	4.0 or more	Fixed	10.4	11.9	1.4		5.5	5.0	1.0	0.4		
		Pinned	2.6	3.4	0.8		0	2.6	0.5	0.2		
3 or more	0	Fixed	7.1	10.6	4.1	0.7	4.4	2.8	2.8	0.7	0.7	0
		Pinned	1.6	3.6	2.4	0.4	0	1.6	1.6	0.4	0.4	0
	0.5	Fixed	8.2	11.1	3.5	0.5	4.7	3.5	2.2	0.7	0.4	0.09
		Pinned	1.9	3.6	1.9	0.3	0	1.9	1.2	0.4	0.2	0.05
	1.0	Fixed	8.9	11.4	2.9	0.4	5.0	3.9	1.9	0.7	0.3	0.09
		Pinned	2.2	3.5	1.6	0.2	0	2.2	1.0	0.4	0.2	0.05
	2.0	Fixed	9.7	11.7	2.2	0.2	5.2	4.7	1.4	0.6	0.2	0.06
		Pinned	2.4	3.5	1.2	0.1	0	2.4	0.8	0.3	0.1	0.03
	4.0 or more	Fixed	10.4	11.9	1.5	0.04	5.5	5.0	1.0	0.5	0.04	0.01
		Pinned	2.6	3.4	0.8	0.02	0	2.6	0.5	0.2	0.02	0.00

Example 3.8.5 Volume change restraint forces

Given:

The 4-bay, 2-story frame shown
Beam modulus of elasticity = E_b = 4300 ksi
Column modulus of elasticity = E_c = 4700 ksi
Column bases 20% fixed (see Sect. 3.8.3)
Design R.H. = 70%
Design temperature change = 70 deg. F

Problem:

Determine the maximum tension in the beams and the maximum moment in the columns caused by volume change restraint.

Solution:

1. Determine relative stiffness between columns and beams:

   $I_b = 12(24)^3/12 = 13,824$ in.[4]
   $E_b I_b/\ell = 4300(13,824) / (24 \times 12)$
   $= 206,400$
   $I_c = 16(16)^3/12 = 5461$ in.[4]
   $E_c I_c/h_s = 4700 (5461) / (16 \times 12)$
   $= 133,681$

   $K_r = \dfrac{E_b I_b/\ell}{E_c I_c/h_s} = \dfrac{206,400}{133,681} = 1.5$

2. Determine deflections:

   From Table 3.8.3: $\delta_e = 226 \times 10^{-6}$ in./in.

   $\Delta_B = \delta_e \ell = 0.000226 (24)(12) = 0.065$ in.

   $\Delta_A = \delta_e(2\ell) = 0.130$ in.

3. Determine maximum beam tension:

   Maximum tension is nearest the center of stiffness, i.e., beams BC and CD, 2nd floor.

From Table 3.8.5:
For n = 4 and i = 2; k_b = 3.00

From Table 3.8.6:
For K_r = 1.0, fixed base; k_f = 11.2
For K_r = 2.0, fixed base; k_f = 11.6
Therefore for K_r = 1.5; k_f = 11.4

For pinned based, k_f = 3.4
(for K_r = 1.0 and 2.0)

For 20% fixed:
$k_f = 3.4 + 0.20 (11.4 - 3.4) = 5.0$
$F_2 = k_f k_b \Delta_i E_c I_c/h_s^3$
$= \dfrac{5.0(3.0)(0.065)(4700)(5461)}{(16 \times 12)^3}$
$= 3.54$ kips

4. Determine maximum column moments:

   For base moment, M_1:
   From Table 3.8.6, by interpolation similar to above:
   k_m (fixed) = $(4.9 + 5.2) / 2 = 5.05$
   k_m (pinned) = 0
   k_m (20% fixed) = $0 + 0.20 (5.05) = 1.0$
   $M_1 = k_m \Delta_i E_c I_c/h_s^2$
   $= 1.0(0.130)(4700)(5461)/(16 \times 12)^2$
   $= 90.5$ in.-kips

   For second floor moment, M_{2L}:

   k_m (fixed) = $(3.9 + 4.5)/2 = 4.2$
   k_m (pinned) = $(2.1 + 2.4)/2 = 2.25$
   k_m (20% fixed) = $2.25 + 0.20 (4.20 - 2.25)$
   $= 2.64$

   $M_{2L} = \dfrac{2.64(0.130)(4700)(5461)}{(16 \times 12)^2}$
   $= 239$ in.-kips

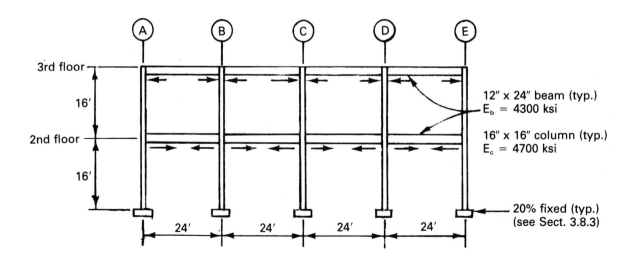

Fig. 3.8.7 Computer models

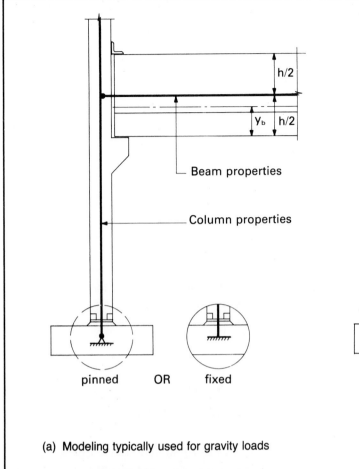

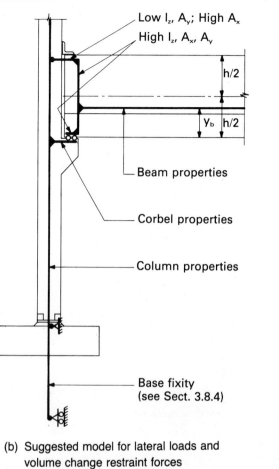

(a) Modeling typically used for gravity loads

(b) Suggested model for lateral loads and volume change restraint forces

3.8.6 Computer Models for Frame Analyses

When precast concrete framing members are modeled for computer analyses as "sticks", as is usually done with steel frames, the results are often very misleading. For example, the structure in Fig. 3.8.7, when assumed connected at the centerline as in (a), will indicate more flexibility than is actually true. Thus, such things as wind drift will be overestimated, and the moments caused by axial shortening may be underestimated. Fig. 3.8.7(b) shows a suggested model that will more nearly estimate the true condition.

3.9 Shear Wall-Frame Interaction

Rigid frames and shear walls exhibit different responses to lateral loads, which may be important, especially in high-rise structures. This difference is illustrated in Fig. 3.9.1.

A frame bends predominantly in a shear mode as shown in Fig. 3.9.1(a), while a shear wall deflects predominantly in a cantilever bending mode, Fig. 3.9.1(b). Elevator shafts, stairwells, and con-

Fig. 3.9.1 Deformation modes

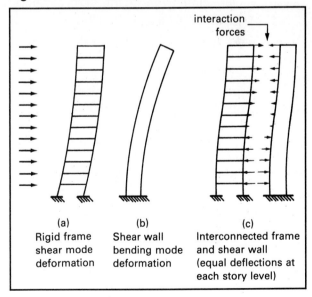

(a) Rigid frame shear mode deformation

(b) Shear wall bending mode deformation

(c) Interconnected frame and shear wall (equal deflections at each story level)

crete walls normally exhibit this behavior.

It is not always easy to differentiate between modes of deformation. For example, a shear wall weakened by a row, or rows of openings may tend to act like a frame, and an infilled frame will tend to deflect in a bending mode. Also, shear deformation of a shear wall can be more important than bending deformation if the height to length ratio is low, as discussed in Sect. 3.7.2.

If all vertical elements of a structure exhibit the same behavior under load, that is, if they are all frames or all shear walls, the load can be distributed to the units in proportion to their stiffnesses (Sect. 3.7.3). However, because of the difference in bending modes, the load distribution in structures with both frames and shear walls is considerably more complex. References 8 through 12 address this problem in more detail.

3.10 Earthquake Analysis

3.10.1 Notation

A_ℓ = cross sectional area in linear measure

A_s = area of non-prestressed reinforcement

A_{vf} = area of shear-friction reinforcement

b = width of panel

C = coefficient for base shear (UBC-82); total compressive force

C_p = coefficient for horizontal force (UBC-82)

C_o = compressive chord force

C_1 = overturning couple force

C_u = factored compressive force

D = dimension of building in direction parallel to applied lateral force (UBC-82)

D_s = plan dimension of the vertical lateral force resisting system (UBC-82)

d = dimension of building; distance from extreme compression fiber to centroid of tension reinforcement

F_p = lateral force on the part of the structure and in the direction under consideration (UBC-82)

F_t = that portion of V considered concentrated at the top of the structure, level n (UBC-82)

F_x = forces in x direction; force at level x

F_y = forces in y direction

f'_c = concrete compressive strength

f_s = stress in steel

f_y = yield strength of non-prestressed reinforcement

H = horizontal force needed to overcome friction

h = height of member

h_i, h_n, h_x = height above base level to level "i", "n", or "x", respectively

I = occupancy importance factor (UBC-82)

K = coefficient relating to type of construction (UBC-82)

ℓ = length of building or member

M = moment

M_1 = overturning moment

M_R = overturning moment resistance

N = force normal to friction plane

n = uppermost level in the structure

$P_{1,2,3,4}$ = forces

R_o, R_1 = reactions

R_s = resistance to sliding

S = coefficient for site-structure resonance (UBC-82)

s = spacing of weld clips

T = fundamental period of vibration of the building in the direction under consideration (UBC-82); total tensile capacity or force

T_o = tensile chord force

T_1 = overturning couple force

T_s = characteristic site period (UBC-82)

T_u = factored tensile force

V = total lateral load or shear at the base (UBC-82); shear force

V_{Ru} = design shear strength

v = unit shear stress

$v_{0,1,2,3}$ = unit shear stresses

v_{ru} = design unit shear strength

v_u = factored shear stress

W = total dead load of building (UBC-82)

W_p = total weight of a part or portion of a structure (UBC-82)

w_i, w_x = that portion of "W" which is located at or is assigned to level "i" or "x" respectively.

Z = coefficient dependent upon the seismic zone (UBC-82)

Z_ℓ = section modulus in linear measure

ϕ = capacity reduction factor (ACI-318-83)

μ = shear-friction coefficient

μ_s = static coefficient of friction

3.10.2 General

Earthquakes generate horizontal and vertical ground movement. When the earthquake passes beneath a structure, the foundation will tend to move with the ground, while the superstructure will tend to remain in its original position. The lag between foundation and superstructure movement will cause distortions and develop forces in the structure. As the ground moves, distortions and forces are produced throughout the height of the structure.

Load tests of prestressed concrete members have consistently shown that large deflections occur as the design strength is approached. Because of prestress, the transition from linear to nonlinear response is gradual and smooth. Cyclic load tests have shown that prestressed concrete beams can undergo several cycles of intense load reversals and still maintain their design strengths.

The current philosophy for the design of earthquake resistant structures permits minor damage for moderate earthquakes, and accepts major damage for severe earthquakes provided that complete collapse is prevented. Seismic performance can be improved by setting limitations on structural deflections. The design details often require large, inelastic deformations to occur in order to absorb the inertia forces. This is achieved by providing member and connection ductility. While this ductility prevents total collapse, the resultant distortions may lead to significant damage to mechanical, electrical, and architectural elements.

To limit damage, three paths are open to the designer. First, the elements may be uncoupled from the structural system, so that these elements are not forced to undergo as much deformation as the supporting structure. Second, the deflection of the support could be reduced in order to minimize deformations of the architectural elements. Third, the connection between individual elements and the supporting frame could be designed to sustain large deformations and rotations without fracture. Generally, the first or third approach is adopted for non-structural architectural wall panels.

Buildings may be designed as either flexible or stiff. Flexible structures will develop large deflections and small inertial forces; conversely, stiff structures will develop large inertial forces but small deflections. Either type may be designed to be safe against total collapse. However, experience demonstrates that a stiff structure, properly designed to account for the large inertia forces, will incur significantly less damage to architectural, mechanical, and electrical elements.

Since ground motion is random in direction, a structure which is shaped so as to be equally resistant in any direction is the optimum solution. Furthermore, closed sections (i.e., boxes or tubes) have demonstrated markedly improved behavior when compared with open sections, because: (1) closed sections provide a high degree of torsional resistance, and (2) the higher axial stresses and resultant deformations in the exterior columns provide significant energy absorption.

3.10.3 Building Code Requirements

In this section, the seismic requirements of the 1982 Uniform Building Code (referred to herein as UBC-82) are used, except as noted.

The response of a structure to the ground motion of an earthquake depends on the structural system with its damping characteristics, and on the distribution of its mass. With mathematical idealization a designer can determine the probable response of the structure to an imposed earthquake. UBC-82 requires a dynamic analysis for structures which have highly irregular shapes or framing systems and allows it for other structures. However, most buildings have structural systems and shapes which are more or less regular, and many designers use the equivalent static load method for these structures. Calculations must be supplemented with engineering judgment.

In its simplest form, UBC-82 requires that a total base shear, V, be applied to the building in any horizontal direction, where

$$V = ZIKCSW$$

in which

Z is based on the expected earthquake intensity,

I is based on the type of occupancy anticipated,

K is based on the type of framing,

C is based on the flexibility of the structure,

S is based on the relative natural vibration periods of the site and the structure, and

W is the dead load of the structure, plus portions of the live load for some occupancies.

This total shear is divided among the story levels, with the upper stories being assigned more of the horizontal load than the lower stories. This is the method of *equivalent static loads.*

Also, UBC-82 requires that parts of the structure such as roofs, floors, walls, and their connections be designed locally for either the distributed base shear or the lateral forces determined by the expression:

$$F_p = ZIC_pSW_p$$

in which the subscript "p" denotes the effects of

the building part. For certain parts of buildings, this latter requirement is sometimes more severe than the distributed base shear.

The occupancy importance factor, I, has the effect of making structures that would be essential during an earthquake disaster (e.g., hospitals, fire stations, etc.), or those that could house large numbers of people, less likely to be severely damaged.

The coefficient for site-structure resonance, S, is a function of the ratio of the fundamental period of vibration of the building, T, and the characteristic site period, T_s. It can be calculated by equations given in UBC-82 or, if T_s is not known, the value of S is taken as 1.5.

3.10.4 Design Guidelines for Wall Panels

The sections that follow deal primarily with totally precast structures. In addition, architectural wall panels, whether connected to precast concrete or other materials, require special considerations:

1. Exterior walls perforated for windows will act somewhere between a solid wall and a flexible frame. For tall buildings, this will result in a non-linear distribution of forces, due to shear lag (see Sect. 3.7.7). When similar to a flexible frame, they must be designed as moment-resisting frames to resist seismic loads.

2. Portions of walls with openings can be subjected to significant axial loads. These portions may require reinforcement with closely spaced ties, in accordance with ACI 318-83, Appendix A.

3. Connected walls may act as coupled walls (see Sect 3.7.5).

4. Walls will be subjected to lateral loads perpendicular to the plane of the wall (wind, seismic) in combination with loads in the plane.

5. Design should consider the eccentricities produced by story drift. These are combined with the eccentricities due to manufacturing and erection tolerances. Drift is defined as the relative movement of one story with respect to the stories immediately above or below the level under consideration. Between points of connection, non-load bearing panels should be separated from the building frame to avoid contact under seismic action. In the immediate area of connections, the panel will tend to distort the same amount as the supporting frame. Internal stresses induced due to a statically indeterminate support system should be

checked. Even in a statically determinate panel there may be some built-in restraint at the connections, so that some allowance for internal stresses should be considered.

6. Under severe earthquake, large deflection may be anticipated. The investigation of individual walls and of the entire structure should include the consideration of deflection (P-Δ effect).

7. Accidental torsion may exist for any member subjected to seismic forces. Arrangement of reinforcement should consider this possibility.

8. Seismic induced forces are reversible. This is particularly important at joints.

9. The best energy absorbing members are those with high moment-rotation capabilities. The energy absorbing capacity of a flexural member is measured by the area under the moment-rotation curve. Correctly reinforced, concrete can exhibit high ductility. Refer to ACI 318-83, Appendix A, for proper methods of reinforcing to achieve ductility.

10. Joints represent discontinuities, and may be the location of stress concentrations. Reinforcement or mechanical anchorage must be provided through the joint to fully transmit the horizontal shear and flexure developed during seismic activity. See Part 6 for a discussion on connections. In zones of high seismicity, cast-in-place reinforced concrete in combination with precast concrete has proved successful in economically transferring seismic forces (see Sect. 7.3).

11. Wherever possible, make panel connections to the supporting structure statically determinate, in order to permit a more accurate determination of force distribution.

12. Choose the number and location of connections to minimize internal stresses and permit movements in the plane of the panel to accommodate story drift and volume changes.

13. Locate connections to minimize torsional moments on supporting spandrel beams, particularly if the beams are structural steel.

14. Provide separation between the panel and the building frame to prevent contact during an earthquake.

A generalized arrangement for connecting a non-load bearing panel to the supporting structure is shown in Fig. 3.10.1.

Fig. 3.10.1 Typical wall panel connection arrangement

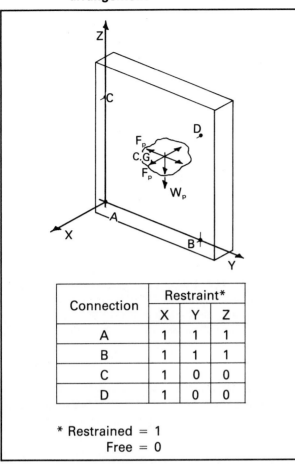

Connection	Restraint*		
	X	Y	Z
A	1	1	1
B	1	1	1
C	1	0	0
D	1	0	0

* Restrained = 1
  Free = 0

3.10.5 Concept of Box-Type Buildings

A box-type building consists of a roof, floor diaphragms, and shear walls. These are connected wherever they meet, and if the arrangement is suitable, the structure is resistant to lateral loads.

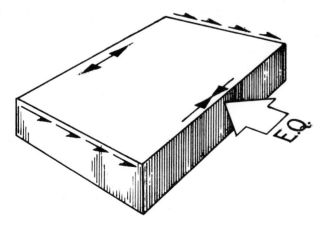

Since an earthquake is a ground motion reacting with the inertia of the building and its parts, the equivalent static loads are applied at the centroids of the parts. The internal forces that link the applied loads and the ground reactions follow the stiffest paths consistent with equilibrium and compatibility of deflection. With box-type buildings of moderate height-to-width ratio, the stiffnesses of diaphragms and walls in their own planes greatly exceed other resistances. The load paths are then along diaphragms and walls rather than through moment-resisting frames.

In multi-story, box-type buildings, the equivalent static loads find resistances in the several diaphragms and walls. If a load originates on a floor, the load path is through that floor to adjacent connected walls, and downward to the footings. After the load path has entered a wall, it does not leave that wall unless the wall is discontinued or its stiffness is reduced substantially. The designer should try to arrange the path of resistance to be direct.

A diaphragm made up of precast elements requires a special effort to be assured that it is strong enough in shear and moment. Walls may be full of windows. The designer must judge if the wall should be considered a shear wall or a moment-resisting frame depending on the number and size of windows and doors. The UBC-82 requirements for a box-type structure are severe, that is, they call for a short period of vibration and high values for the coefficients "C" (≈ 0.1) and "K" ($= 1.33$).

3.10.6 Structural Layout and Connections

A box-type structure may have a large number of precast concrete elements that are assembled into walls, floors, roof, and maybe frames. Proper connections between the many pieces create the diaphragms and shear walls, and the connections between these, in turn, create the box-type structure. In the seismic design of a box-type structure, there are two fundamental and different requirements of the two groups of connections:

1. One group of connections that transmits forces between elements within a horizontal diaphragm or a shear wall, and

2. Another group of connections that transmits forces between a horizontal diaphragm and a shear wall.

In seismic design, forces must be positively transmitted. Load paths must be as direct as possible. Anchors should be attached to or hooked around reinforcing bars or otherwise terminated so as to effectively transfer forces to the bars. Reinforcement in the vicinity of the anchors should be designed to distribute the forces so as to preclude local failure. Concrete dimensions must be ample, so that the hardware of the connection is confined, and the connection thus can transmit accidental forces that are normal to the usual plane of the load path. Finally, a connection should be such that, if it were to yield, it will do so in a ductile manner, i.e., without loss of load-carrying capacity when the concrete cracks.

3.10.7 Example — One-Story Building

General

By taking advantage of walls already present, one-story buildings usually can be designed to resist lateral loads (wind or earthquake) by shear wall and diaphragm action. If a shear wall and diaphragm concept is feasible, it is generally the most economical concept. This section of the Handbook is intended to assist a designer with the shear wall/diaphragm concept for precast prestressed concrete buildings.

To show a fairly complete design, the simple one-story example building in Fig. 3.10.2 will be illustrated. It is 128 ft × 160 ft in plan, and has 16-ft clear height inside. It is entirely precast above the floor, using 16-in. double tees for the walls

and 24-in. double tees for the roof. The double tees are 8 ft wide, with stems 4 ft on centers. The flanges on the roof tees are 2 in. thick, and on the wall tees, 4 in. thick.

Because all loads must funnel through the connections, gravity and lateral loads must be considered together. Thus, this example shows both gravity and lateral load connections. The example emphasizes both free bodies and the concept of load path. Most serious design errors are ones of concept caused by not drawing enough free-body diagrams, and failure to provide a continuous load path.

Load analysis

Earthquakes impose lateral and vertical ground motions upon a structure. The structure responds to these motions with its own deflections. These deflections are accompanied by corresponding strains and the resulting stresses. However, most designers are used to thinking of stress as caused by load rather than as caused by deflection. Con-

Fig. 3.10.2 One-story example building

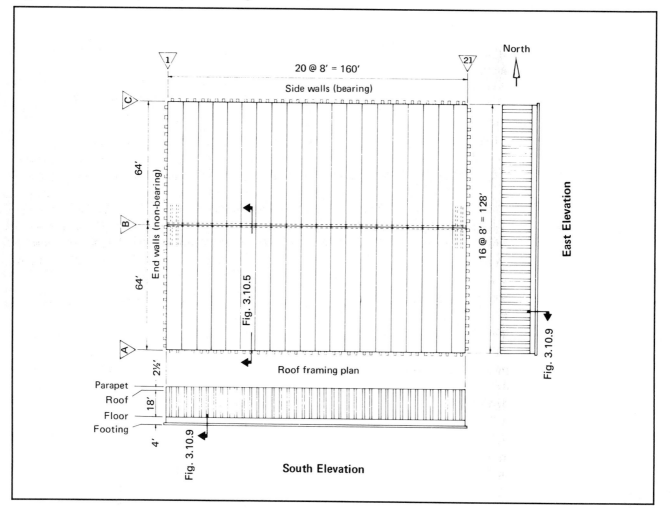

sequently, all the common methods of design use a set of static lateral loads intended to be equivalent to (i.e., produce the same stresses as) the real, dynamic loads caused by deflection. One such method is used in UBC-82. The example building is analyzed here for N-S earthquake only. In a real design situation, it would also have to be analyzed for E-W earthquake.

The example building resists lateral load by diaphragm and shear wall action. Inertia loads (mass × acceleration) are delivered to the roof diaphragm. The diaphragm acts like a plate girder laid flat, spanning between the shear walls. In determining the equivalent static loads on this diaphragm, the mass tributary to the diaphragm must be determined. This is done in terms of dead weight, W, as follows:

N and S walls:
half ht + parapet	= 11.5 ft
total length = (2)(160)	= 320 ft
weight = 11.5(320)(0.080)	= 295 kips

Roof:
weight = 128(160)(0.075)	= <u>1535 kips</u>
Total W	= 1830 kips

The equivalent lateral load, V, is computed by multiplying W by an acceleration. The acceleration is determined by multiplying together the five constants, ZIKCS. The "Z" factor, denoting geographical zones of equal probability of serious earthquake, is taken here as 0.75, representing the second-most earthquake prone areas, Zone 3. The occupancy importance factor, I, is assumed as 1.0 and the "K" factor, used to indicate the performance certain types of framing have shown in actual earthquakes is taken as 1.33. Since the characteristic site period, T_s, is assumed unknown in this example, the value of "S" is taken as 1.5. The "C" factor is used to allow for the fact that flexible long-period (low natural frequency) buildings tend to absorb ground motion with less damage than do stiff buildings. The example building is very stiff and rates the highest "C" value of 0.12, however, the product of CS need not exceed 0.14. The coefficient used in design is thus:

$$ZIKCS = 0.75(1.0)(1.33)(0.14) = 0.14$$

Shear on the diaphragm due to earthquake is:

$$V = ZIKCS(W) = 0.14(1830) = 256 \text{ kips}$$

In comparison, the shear caused by a 20 psf wind load is:

$$V = 11.5(160)(0.020) = 37 \text{ kips}$$

This building is definitely governed by earthquake rather than by wind. A word of caution is in order. The base shear due to earthquake as cal-

culated above is a "service" load. Accelerations in real earthquakes may be many times the "ZIKCS" shown above. This will cause elements of the structure to be stressed to the yield point and beyond in a severe earthquake. The structure must have deformation capability to absorb overloads of short duration without failure.

To guard against accidental horizontal torsion effects, UBC-82 requires that a minimum 5 percent eccentricity be assumed between centroid of resistance and center of mass as shown in Fig. 3.10.3. With this in mind, the forces internal to the roof diaphragm are:

Max. shear reaction, $R_o = 0.55 \text{ V} = 0.55(256)$
$= 141 \text{ kips}$

Max. shear intensity, $v_o = 141/128 = 1.10 \text{ klf}$

Max. bending moment, $M_o = V\ell/8$
$= 256(160/8)$
$= 5120 \text{ ft-kips}$

Max. chord forces, $C_o = T_o = M_o/d = 5120/128$
$= 40.0 \text{ kips}$

The shear forces are analogous to those in the web of a plate girder. The chord forces are analogous to those in the flanges of a plate girder.

Considering the roof as a free body, Fig. 3.10.3 shows that equilibrium is maintained by the reactions R_o from the tops of the shear walls. One must also consider the shear wall as a free body. Fig. 3.10.3 shows that the wall can be in equilibrium only if sufficient sliding resistance and overturning capacity are provided. The forces acting which must be resisted by the shear wall are:

Sliding force, $R_o = 141 \text{ kips} = R_1$

Overturning moment, $M_1 = R_1h = 141 (22)$
$= 3102 \text{ ft-kips}$

Overturning couple, $C_1 = T_1 = M_1/\ell$
$= 3102/128$
$= 24.2 \text{ kips}$

The weight of the end wall, tributary roof, floor, backfill, etc., is about N = 410 kips. The available resistance to overturning, then, is:

$$M_R = N(d/2) = 410(128/2) = 26,240 \text{ ft-kips}$$

The sliding force is resisted by friction in the bottom of the wall footing. Assume a granular soil, the coefficient of sliding friction is about

$$\mu_s N = 0.5(410) = 205 \text{ kips}$$

The factors of safety (load factors) for overturning and sliding are seen to be sufficient.

$$M_R/M_1 = 26,240/3102 = 8.46$$
$$R_s/R_o = 205/141 = 1.45$$

The designer must be careful to follow the loads all the way down into the ground. This is the "load-path" concept. The designer *must* provide for a complete and continuous load path.

Fig. 3.10.3 Forces acting on roof and walls

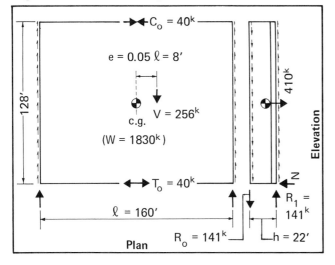

Fig. 3.10.4 Connections between flanges of roof double tees

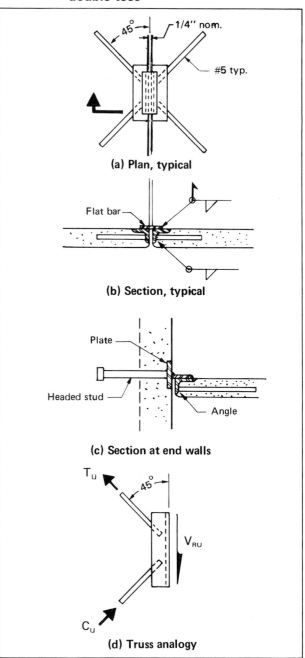

(a) Plan, typical

(b) Section, typical

(c) Section at end walls

(d) Truss analogy

Strength analysis—roof

In this example, strengths of concrete components are analyzed using the load factors of ACI 318-83, and strength of structural steel parts by working-stress methods. Where welded, reinforcing bars are ASTM A615, Grade 40; otherwise Grade 60. (See Part 6 for guides in welding reinforcing bars.)

Following the load path, the diaphragm is first analyzed for shear. The applied design shear in the double tee flanges is:

$$v_u = 1.4v_o = 1.4(1.10) = 1.54 \text{ klf}$$

(The load factor, 1.4, is derived from Sects. 9.2.2 and 9.2.3 of ACI 318-83. From Eq. 9-2, $U = 0.75$ $(1.7 \times 1.1E) = 1.4E$, and from Eq. 9-3, $U = 1.3 \times 1.1E \approx 1.4E$. Also, UBC-82 stipulates a load factor of 1.4.)

The design shear strength of the reinforced concrete in the double tee flanges is:

$$v_{ru} = 2\phi\sqrt{f'_c}\,t = 2(0.85)\sqrt{6000}(2)$$
$$= 263 \text{ lb/in.} = 3.16 \text{ klf}$$

This is greater than required, therefore satisfactory.

The double tee flanges must be connected at their edges to each other and to the end shear walls. This is analogous to providing shear strength along vertical joints in the web of a plate girder, and is done by weld clips as shown in Fig. 3.10.4 (a), (b) and (c). This clip is analyzed by truss analogy, illustrated in Fig. 3.10.4 (d). The design forces in the bars, and their resultant along the double tee edge, are:

$$C_u = T_u = \phi A_s f_y = 0.9(0.31)(40)$$
$$= 11.2 \text{ kips, and}$$
$$V_{Ru} = (C_u + T_u)\cos 45° = (11.2 + 11.2)(0.707)$$
$$= 15.8 \text{ kips}$$

Hence, at the junction with the end walls, the spacing between clips (as limited by horizontal or diaphragm shear) must be no more than:

$$s = V_{Ru}/v_u = 15.8/1.54 = 10.3 \text{ ft, say 10 ft}$$

Ratioing by distance from the center of the roof, the shear between the first and second double tee and the corresponding spacing between clips, is:

$$v_u = (72/80)(1.54) = 1.4 \text{ klf, and}$$
$$s = 15.8/1.4 = 11.3 \text{ ft, say 10 ft on centers}$$

Ten feet is a reasonable maximum spacing. However, in an earthquake, the double-tee next to the end wall is subject to vertical bouncing. This could destroy the essential connection of diaphragm to shear wall, unless the connection is strengthened sufficiently to yield the double tee flange as a cantilever in bending where it joins the first interior stem. The flanges of the double tees used here are reinforced with 12 × 6-W1.4 × W2.5 welded wire fabric (A_s = 0.050 in.²/ft), ½ in. clear top. If standard weld clips are placed at 4 ft on centers (2 per wall panel), they will easily force uniform yield in the flange. Hence, the clips are spaced:

s = 10 ft on centers typically, and

s = 4 ft on centers at end walls

The double tee flanges must also be connected to the north and south walls, in order to transfer the "VQ/Ib" type web shears to the chords. This is analogous to the connection between web and flange of a plate girder. This same connection must also function as a tension tie, by holding the wall panels onto the roof against wind and earthquake, and by holding the building together against shrinkage and related forces.

The connection which holds the wall panel to the roof must be designed for an earthquake force, F_p, which is obtained by multiplying the tributary weight of this part of the building, W_p, by an acceleration, ZIC_p. Z and I are the same zone and occupancy factors as before. C_p is a factor used to correlate for past good and bad experience in earthquakes. The UBC-82 requires a C_p of 0.3 for connections of wall panels, but also requires that the fasteners, such as bolts, inserts, welds, dowels, etc., be designed for four times the load determined from the above formula. Thus, for most parts of the connection, design acceleration is:

ZIC_p = 0.75(1.0)(0.3)(4) = 0.90

Tributary weight of an 8-ft wide wall panel is:

W_p = (half ht + parapet = 11.5)(8)(0.080)
= 7.4 kips

The corresponding lateral earthquake force is:

F_p = ZIC_pW_p = 0.90(7.4) = 6.66 kips,
            or 3.33 kips per double tee stem

As this is equivalent to a wind force of 72 psf, wind is no problem. The connection must transmit diaphragm shear and at the same time allow for the live load end rotation of the simple span roof tees. This is done by welding the roof and wall tees together at the top of the roof tee, but *not* at the bearing (Fig 3.10.5). The bearing will then slide under live load end rotation, and friction at the bearing produces tension in the top connection.

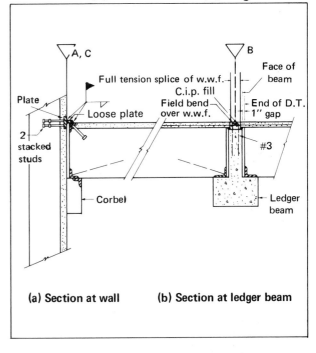

(a) Section at wall   (b) Section at ledger beam

The horizontal friction at each bearing, and the corresponding tension tie strength required, assuming a coefficient of friction of 0.5 is:

H = μ_sN = 0.5 (9.6) = 4.8 kips

This H is larger than the F_p = 3.33 kips required for earthquakes. The wall panel to roof connection is thus designed for a horizontal pullout force of 4.8 kips. The bar anchorage is sized thus:

T_u = 1.4H = 1.4 (4.8) = 6.7 kips

A_s = $T_u/\phi f_y$ = 6.7/(0.9 × 40) = 0.19 in.²

Use 2 - #4 bars (A_s = 0.40 in.²)

The flange, or "chord" reinforcement is sized for tension as follows:

T_u = C_u = 1.4T_o = 1.4 (40.0) = 56 kips

A_s = $T_u/\phi f_y$ = 56/(0.9 × 40) = 1.55 in.²,

Use 2 - #8 bars (A_s = 1.58 in.²)

These bars are located at the top of the wall panels, and must be spliced at the vertical joints as shown in Fig. 3.10.6. The cross-sectional area of the angle should be sufficient to assure prior tensile yield in the bars, and of a size to provide reasonable concentricity between bar and angle. The weld is field-made and is designed as shown in Part 6. The No. 3 hairpins shown provide enough shear-friction capacity to transmit the maximum diaphragm shear into the two No. 8 chord bars. The bars should extend an anchorage length below the double tee-to-wall tension ties. This provides a load path for the diaphragm shears between the chord bars and the roof weld clips.

The diaphragm must be continuous in shear over the centerline ledger beam (see Fig. 3.10.5 (b)). This bearing connection is designed very much like that of Fig. 3.10.5 (a). By shear-friction, the required diaphragm continuity steel area is:

$$A_{vf} = v_u/\phi \, f_y\mu$$
$$= 1.54/[0.85 \, (60) \, (1.4)] = 0.022 \text{ in.}^2/\text{ft}$$

The connection should also have a horizontal tensile capacity, as was required at the exterior wall. Select a 6×6–$W2.9 \times W2.9$, $A_s = 0.058$ in.^{2}/ft. This will provide about the same tensile capacity as provided at the exterior wall.

This completes the strength analysis of the diaphragm for N-S earthquake. For E-W earthquake, chords would also be needed in the east and west walls.

Strength analysis — walls

Following the load path, the shear walls are analyzed next. For simplicity, it is assumed the walls have no openings. Thus, there are interior and exterior (corner) wall panels, as shown in Fig. 3.10.7. The distinction between "interior" and "exterior" is that an interior panel has weld clips on both vertical edges for transfer of shear. An exterior panel lacks the capability for shear transfer on one vertical edge. As will be seen, an interior panel requires no provision for overturning; an exterior panel does.

Fig. 3.10.6 Flange or "chord" reinforcement at top of wall panels

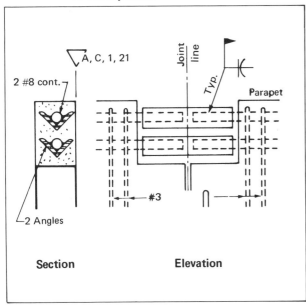

| Section | Elevation |

Note: This example assumes a mild climate. In colder climates, it would be preferable to locate the tensile chord in a location less exposed to the elements — say, under the roof insulation at the roof-to-wall junction.

Fig. 3.10.7 Forces acting on wall panels

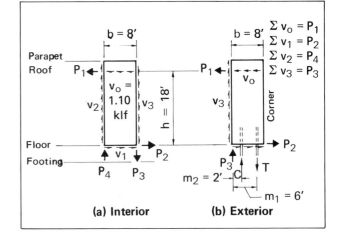

(a) Interior   **(b) Exterior**

An interior panel is analyzed first (see Fig. 3.10.7 (a)). The diaphragm shear $v_o = 1.10$ klf is applied from the roof. The distributed shears v_1, v_2, and v_3 resist v_o. These distributed shears have resultants P_1, P_2, P_3, and P_4. Treating the panel as a free body of zero weight:

$$\Sigma F_x = 0: P_2 = P_1; \, v_1 b = v_o b; \, v_1 = v_o, \text{ and}$$
$$\Sigma F_y = 0: P_3 = P_4; \, v_2 h = v_3 h; \, v_2 = v_3, \text{ and}$$
$$\Sigma M = 0: P_1 h = P_3 b; \, P_3 = P_1 h/b$$
$$v_3 h = v_o bh/b, \text{ so } v_3 = v_o$$
$$v_o = v_1 = v_2 = v_3 = 1.10 \text{ klf}$$

Weld clips similar to Fig. 3.10.4 (b) are used at the vertical joints. The number required at each joint is determined by dividing the factored shear force by the design strength of the clip:

$$P_3 = v_o h = 1.10 \, (18) = 19.8 \text{ kips}$$
$$P_{3u} = 1.4 \, P_3 = 1.4 \, (19.8) = 27.7 \text{ kips}$$

required number $= 27.7/15.8 = 1.8$, use 3 min.

The shear at the panel bottom is taken by projecting dowels. These dowels are anchored into grouted sleeves, sufficient to develop full tensile yield, and act in shear-friction. Per double tee stem, the shear-friction steel required is:

$$v_u = 1.4 v_o = 1.4 \, (1.10) = 1.54 \text{ klf}$$
$$V_u = v_u \ell = 1.54 \, (4.0) = 6.16 \text{ kips}$$
$$A_{vf} = V_u/\phi f_y \mu = 6.16/[0.85 \, (60) \, (0.6)]$$
$$= 0.20 \text{ in.}^2, \text{ or one No. 4 bar}$$

By judgment, this size dowel is too delicate. Use one No. 6 at each double tee stem, typically.

An exterior (corner) panel, is analyzed next. (see Fig. 3.10.7 (b)). To obtain reliable strength, the anchor bars are located at the double tee stems, rather than at the flange edge. Again taking the panel as a free body of zero weight,

$$\Sigma F_x = 0: P_2 = P_1 = v_o b = 1.10 \, (8) = 8.8 \text{ kips}$$
$$\Sigma F_y = 0: C = T - P_3 = T - 19.8$$

PCI Design Handbook

$$\Sigma M = 0: T m_1 = P_1 h + C m_2$$
$$= P_1 h + (T - 19.8) m_2, \text{ so}$$
$$T (m_1 - m_2) = P_1 h - 19.8 m_2$$
$$T = \frac{P_1 h - 19.8 m_2}{m_1 - m_2}$$
$$= \frac{(8.8)(18) - (19.8)(2)}{6 - 2} = 29.7 \text{ kips}$$
$$C = 29.7 - 19.8 = 9.9 \text{ kips}$$

The reinforcement required to take T is determined as follows:

$$T_u = 1.4T = 1.4(29.7) = 41.6 \text{ kips}$$
$$A_s = T_u / \phi f_y = 41.6 / (0.9)(60) = 0.77 \text{ in.}^2, \text{ or}$$

two No. 6 bars ($A_s = 0.88$ in.^{2})

Fig. 3.10.8 Wall double tee connections

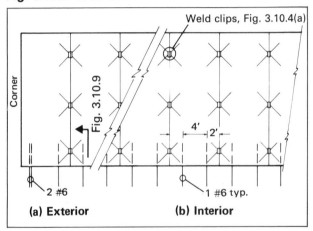

Weld clips, Fig. 3.10.4(a)

Corner

Fig. 3.10.9

4' 2'

2 #6

1 #6 typ.

(a) Exterior   **(b) Interior**

Fig. 3.10.9 Wall-footing connection

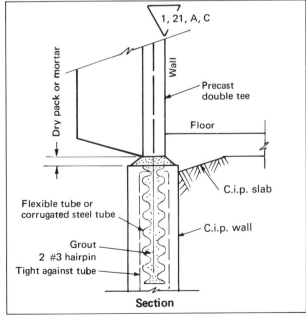

1, 21, A, C

Dry pack or mortar

Wall

Precast double tee

Floor

C.i.p. slab

C.i.p. wall

Flexible tube or corrugated steel tube

Grout
2 #3 hairpin
Tight against tube

Section

Hence, at each double tee stem, use one No. 6 typically and two No. 6's at corners (see Figs. 3.10.8 and 3.10.9).

In this example, the wall panels act together as a unit because of the weld clips. In locations subject to severe volume change deformations, the weld clips might cause local cracking. To avoid this problem, the designer may choose to utilize only those panels located near the middle of the wall to resist shear forces. These panels could be tied together with weld clips, but the other panels would be fastened only at the top and bottom.

3.10.8 Example — 4-Story Building

General

The following is an example analysis of a four-story building in which the structural system resisting lateral forces is of the box type, that is, floors and roof are horizontal diaphragms, and exterior walls are shear walls. The building is rectangular in plan, and dimensions and layout are shown in Figs. 3.10.10 and 3.10.11. Floors and roof are made up of 8 ft wide precast, prestressed concrete double tees, which are supported by exterior bearing walls and a central interior frame of precast, reinforced concrete beams and columns. There are smaller frames at the two elevator-stairway shafts, one at each end of the building. The interior frames are made up of full height columns and short beams. The exterior walls are 8 ft wide precast, prestressed concrete double tees placed on end, and are continuous from top of footing to top of parapet.

Approximations and partial designs

In the example, several approximations are made, but they are all on the conservative side. The designs are not complete in all respects, but in the case of several similar parts the more critical part is analyzed to illustrate procedure. Some items, such as number or spacing of connections, are in some cases determined by judgment.

Weights

The unit dead loads are:

    Roof: 100 psf

    Floors: 110 psf

    Walls: 90 psf max., 67 psf average

The dead load assigned to each level of the building is the weight of the roof or floor plus the walls from midstory to midstory (only walls for first level).

This gives values of w_x as listed in Table 3.10.1 (Column 3).

Fig. 3.10.10 Four-story example building — elevations

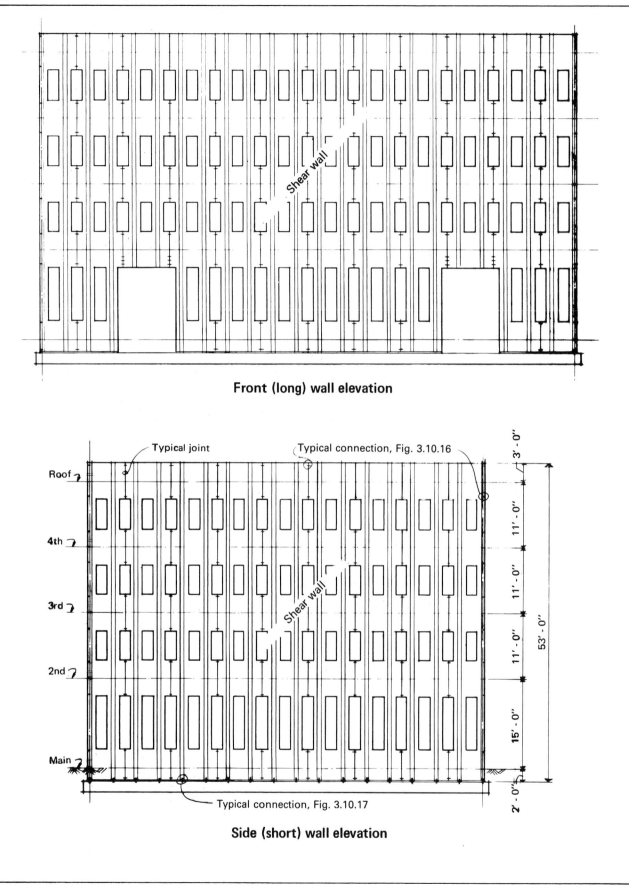

Front (long) wall elevation

Side (short) wall elevation

Fig. 3.10.11 Four-story example building — typical floor plan

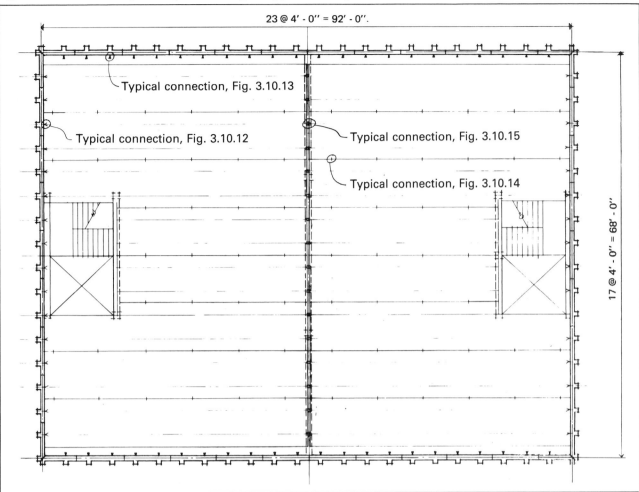

Equivalent static loads

The building is located in a Seismic Zone 2; $Z = 3/8$.

The structural system is of the box type; $K = 1.33$.

The fundamental period of vibration in the longitudinal direction of the building is:

$$T = 0.05\, h_n / \sqrt{D} = 0.05\,(50) / \sqrt{92} = 0.26 \text{ sec}$$

and the corresponding coefficient for base shear is:

$$C = 1/15 \sqrt{T} = 1/15 \sqrt{0.26} = 0.131$$

In the transverse direction, $C = 0.121$. However, C need not be greater than 0.12 nor CS greater than 0.14. $CS = 0.14$ governs. The total dead load of the building is:

$W = 3823$ kips, as shown in Table 3.10.1.

The base shear is then:

$$V = ZIKCSW$$
$$= 0.375\,(1.0)\,(1.33)\,(0.14)\,(3823) = 267 \text{ kips}$$

The base shear is distributed over the height of the building. In this example, the extra lateral force at the top level is $F_t = 0$, since $T < 0.7$ sec. The distribution to the several levels is then

$$F_x = \frac{(V - F_t)\, w_x\, h_x}{\sum\limits_{i=1}^{n} w_i h_i} = \frac{267\, w_x\, h_x}{120,000}$$

which gives the values of F_x listed in Table 3.10.1 (Column 4). These lateral loads are used for design of the building as a whole, like shear in and overturning of walls. For floor and roof diaphragms the following equations are used:

$$F_{px} = \frac{\sum\limits_{\ell=x}^{n} F_\ell}{\sum\limits_{\ell=x}^{n} w_\ell}\, w_{px}$$

Table 3.10.1

(1) Level	(2) h_x, ft	(3) w_x, kips	(4) F_x, kips	(5) F_{px} kips
Roof	50	831	92.4*	59.5
4th	39	926	80.4*	56.1
3rd	28	926	57.7*	48.6
2nd	17	962	36.4	50.5*
1st	2	178	6.8	9.3*
Totals		3823	267.7	

*These are used for design of diaphragms.

$$0.14 ZI \; w_{px} < F_{px} < 0.30 \; ZI \; w_{px}$$

The resulting values of F_{px} are listed in Table 3.10.1 (Column 5). However, at each level the larger of the loads F_x or F_{px} must be used; this is indicated by asterisks in Table 3.10.1. Note that these loads include the effects of the weights of walls that are parallel to the direction of motion, and although this is not really required, the design is somewhat conservative.

Chord forces in diaphragm

The roof is the most heavily loaded diaphragm, with a total lateral load of 92.4 kips. With the diaphragm spanning in the long direction, the moment T (d) = 1/8 Wℓ and the chord force

$$T = \frac{W\ell}{8d} = \frac{92.4 \; (92)}{8 \; (68 - 2)} = 16.1 \text{ kips}$$

Design of the chord reinforcement:
$$A_s = T_u/\phi f_y = 1.4 \; (16.1)/(0.9) \; (60) = 0.42 \text{ in.}^2$$

Use two No. 5 bars placed as shown in Figs. 3.10.12 and 3.10.13.

Requirements at lower levels and for the diaphragm spanning in the short direction are even less, but, for simplicity, the same diaphragm chord reinforcement is used throughout.

Shear stresses in diaphragm

This example building is assumed to be symmetrical, and the centers of mass and rigidity coincide. It is then necessary to include the effect of an arbitrary torsion, that is, the lateral load is offset from the center of rigidity by 5 percent of the maximum dimension of the diaphragm.

The torsional moment is resisted by all four exterior walls, depending on rigidities and distances from the center or rigidity. However, it is conservative to assume that the torsional moment is resisted by the two walls that are parallel to the direction of motion. For earthquake in the transverse direction, the maximum total shear to one wall is:

$$R_o = 0.55 \; (F_x \text{ or } F_{px}) = 0.55 \; (92.4) = 50.8 \text{ kips}$$

Deducting the length of the service shafts, the maximum unit shear is:

$$v_o = R_o/\ell = 50.8/ \; (68 - 20) = 1.06 \text{ klf}$$

For earthquake in the longitudinal direction, the maximum unit shear to the long walls is less. For simplicity in production of elements and erection at the site, the 1.06 klf is used throughout all diaphragms and for their connections to the walls.

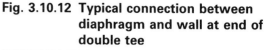
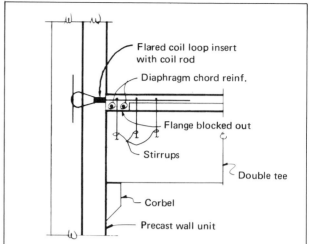

Fig. 3.10.13 Typical connection between diaphragm and wall at side of double tee

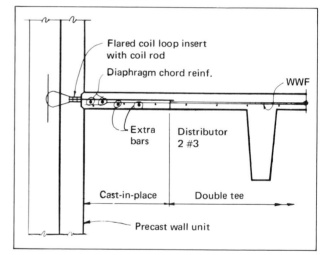

Typical connections for interior of diaphragm

A typical connection transmitting shear between double tees is illustrated in Fig. 3.10.14. In each edge there is a small angle anchored with two bars at 45°. The shear strength is calculated from force components in the anchor bars, as in the previous example:

$$V_{Ru} = 15.8 \text{ kips}$$

To equalize cambers and deflections in neighboring elements, a maximum spacing should not exceed about 8 ft. This provides a resisting unit shear of:

$$v_u = 15.8/8 = 1.98 \text{ klf}$$

which compares to the requirement of:

$$1.4 \; (1.06) = 1.48 \text{ klf}$$

The cast-in-place strip along the longitudinal wall is reinforced with 8 × 4 − W2.9 × W2.9 WWF

Fig. 3.10.14 Typical diaphragm connections between double tees

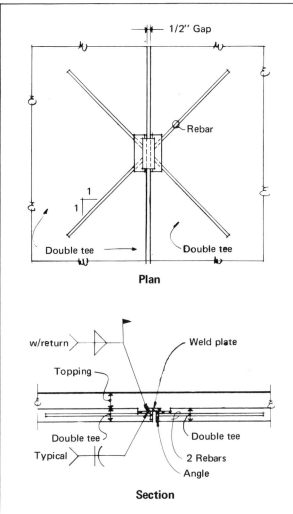

Plan

Section

projecting out of the double tee flange. By shear-friction, this is good for

$$V_{Ru} = \phi A_s f_y \, \mu = 0.85 \, (0.087) \, (60) \, (1.0)$$
$$= 4.43 \text{ klf which is several times that required.}$$

Across the center of the building, the double tee elements are supported by a beam, and they meet end to end. It is desirable to make the shear path stay at the level of the flange, that is, making the path as direct as feasible. The connections to take care of this may be angle splices on bars projecting out of the tops of the ribs. The gap is filled with concrete, and shear-friction is applicable. Such a connection is shown in Fig. 3.10.15.

It should be noted that structural topping containing a rather light welded wire fabric could be used in place of the direct connections between neighboring elements. Even so, it may be desirable to use some of the hardware type connections to make the building resistant to lateral loads during construction.

Connection of diaphragm to wall

First consider the connections between the ends of the double tees and the short wall. The ribs of the double tees and of the wall elements are aligned, and the vertical loads are carried by corbels projecting from the interior wall surface. Two requirements must be satisfied: a) horizontal shear between diaphragm and wall, and b) direct horizontal tension caused by the tendency of the wall to fall away from the building. The two requirements will be superimposed because earthquake motion may occur in any horizontal direction, and is not limited to the principal axes of the building.

It is convenient to space the connections 4 ft apart so that they will occur at wall ribs. At each connection the forces are:

a) Horizontal shear, $V = 1.06 \, (4) = 4.24$ kips

This can be handled by shear-friction. Then the tensile capacity must be:

$$V_u/\mu = 1.4 \times 4.24/1.0 = 5.9 \text{ kips}$$

b) Direct horizontal tension is given by:
$$F_p = Z I C_p W_p$$

UBC-82, Table 23-J, gives $C_p = 0.3$. For the weight of wall contributing to one connection, a unit weight of 90 psf over an area of $4 \times 14 = 56$ ft^2 will be used, so that:

$$W_p = 0.090 \, (56) = 5.04 \text{ kips.}$$
$$F_p = 0.375 \, (1.0) \, (0.3) \, (5.04) = 0.57 \text{ kips.}$$
Design tensile strength $= 1.4 \, (0.57) = 0.79$ kips.

A choice must be made of the type and size of connection. The coil-loop insert with continuous threaded coil rod is a simple and useful type, and there are several proprietary makes. The installation is illustrated in Fig. 3.10.12. A note of caution is in order: the coil-loop insert and the coil rod should each be amply anchored in their respective parts so that failure by yielding in the coil rod is assured.

Fig. 3.10.15 Typical end connection between double tees

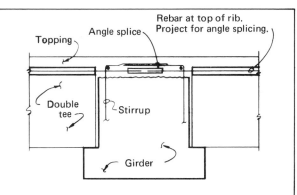

The connections between the edge of the double tee and the long wall are of the same kind, and are calculated in the same way. The primary difference is caused by the cast-in-place strip of diaphragm adjacent to the wall. The threaded coil rods extend into this strip and overlap the welded wire fabric that projects out of the edge of the double tee flange (see Fig. 3.10.13).

Special situation at service shafts

From the typical floor plan, Fig. 3.10.11, it is seen that several wall ribs are not tied to the floor, and the wall, because of vertical joints, is not suited to span horizontally past the shaft openings. This situation is readily alleviated with a strongback type beam placed horizontally at each floor level, and spanning the shaft opening. The adjacent connections at each side of the shaft will have to be stronger than the standard connection designed in the preceding section.

Shear wall

The exterior walls of the building provide the horizontal reactions for the diaphragms. These forces are in the plane of the wall and become horizontal shears, therefore the term shear wall. The end walls in the example building do double duty in that they are also bearing walls.

The design of the individual wall elements can be important, but will not be covered here. For the example, next examine the shear in the joints, and overall stability. The equivalent static loads, which are assumed to act horizontally at each floor level, cause a tendency for the wall to overturn. Gravity loads and footing will stabilize the wall and prevent overturning, but internal shear and direct stresses will develop. In this case, no advantage will be taken of the slightly reduced overturning permitted by UBC-82, but the full value of the cantilever moment will be used. Results of the calculations are shown in Table 3.10.2 for one wall.

Shear in shear wall

The maximum shear in the short end wall due

Table 3.10.2

Level	V_x, kips	M_x, ft-kips
Roof	46.2	0
4th	86.4	508
3rd	115.3	1458
2nd	133.5	2726
1st	133.9	4728
Top of footing		4996

to the equivalent static loads and the arbitrary 5 percent eccentricity is

$$V = 0.55 \ (267) = 147 \text{ kips}$$

$$v = 147/68 = 2.16 \text{ klf}$$

UBC-82 requires a load factor of 2.0 for shear in shear walls, so

$$v_u = 2.0 \ (2.16) = 4.32 \text{ klf}$$

This unit shear occurs near the bottom of the wall, and is both horizontal and vertical.

A typical connection for the vertical joints between wall elements is shown in Fig. 3.10.16. In each edge there is a small angle anchored with two No. 5 bars. With 6-in. thick walls, it is possible to drypack the joints effectively. Apply the shear-friction concept, with the anchor bars placed at 90° to the joint. The shear capacity of one connection is:

$$V_{Ru} = \phi \ A_s \ f_y \ \mu = 0.85 \ (2) \ (0.31) \ (40) \ (1.0)$$
$$= 21.1 \text{ kips}$$

Near the bottom of the wall, the average vertical spacing should not be more than $21.1/4.32 = 4.9$ ft. Use three connections in each story. Since windows cross the joints, be sure to allow for this when locating the connections.

Moment in shear wall

The maximum moment in the short end wall, with allowance for the arbitrary eccentricity, is:

$$M = 4996 \times 1.10 = 5496 \text{ ft-kips}$$

The maximum gravity load is:

W of one end wall, 68 (53) (0.067) = 241

W of ¼ of roof, 68 (92/4) (0.100)   = 156

W of ¼ of floors, 68 (92/4) (0.110) (3) = 516

Total W = 913 kips

With wall ribs neglected, section properties in linear measure are, area $A_\ell = 68$ ft, and section modulus $Z_\ell = (1/6) \ (68)^2 = 771$ ft².
Fiber stresses at bottom of wall are:

$$f = \frac{W}{A} \pm \frac{M}{Z} = \frac{913}{68} \pm \frac{5496}{771} = \begin{matrix} 20.6 \text{ klf} \\ 6.3 \text{ klf} \end{matrix}$$

These values are a good measure of the stresses in the gross wall section, and can also be used to calculate the necessary width of wall footing. Additional refinement of stress calculations are necessary at window openings.

PCI Design Handbook

Fig. 3.10.16 Typical connection, wall to wall

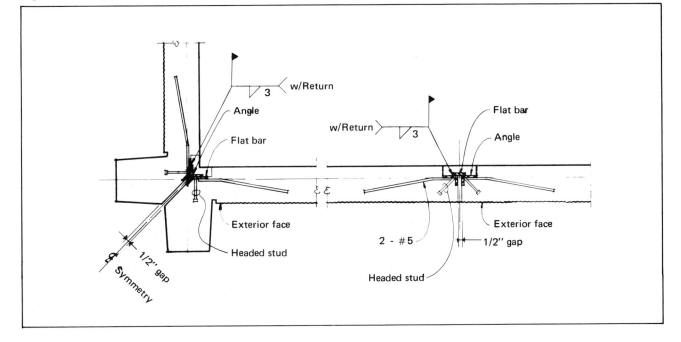

Other shear walls

The long shear walls, which provide the resistance to earthquake in the longitudinal direction of the building, are less critical than the short walls, and they will not be discussed. Details of design will be the same as for the short walls.

There is a special situation at the building's entrance doors. The wall panel above the doorway is supported by connections to neighboring units. The necessary strength can be provided by two additional connections in each joint above the doorway.

Connections of walls to footings

The footings can be proportioned on the basis of reactions from the walls, as indicated above. Each wall element can be connected to the footing by projecting the main reinforcement of the ribs downward into blockouts in the footings, and filling the blockouts with concrete. The space between wall element and footing must be drypacked as shown in Fig. 3.10.17.

Fig. 3.10.17 Typical connection of wall to footing

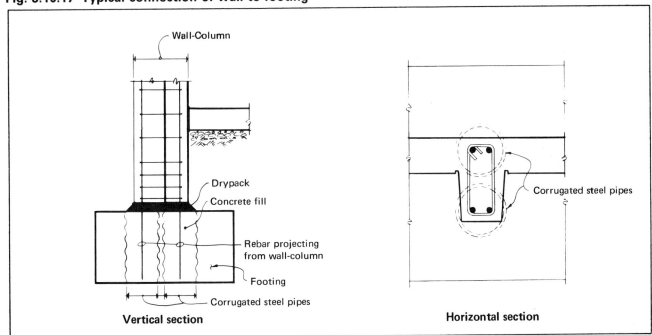

3.10.9 Example — 12-Story Building

Box-type buildings can be built to a height of 160 ft, according to UBC-82. For buildings approaching that height, the structural layout and connection details may differ from those of the preceding example, but the design procedure is the same. The base shear, its distribution as equivalent static loads over the height of the building, and the load paths down to the foundations, are calculated as illustrated in Sect. 3.10.8

Fig. 3.10.18 shows a 12-story building which is square in plan. Wall units are 12 ft high and 30 ft or 15 ft in width. The "running bond" pattern illustrated is not an essential feature of seismic design, but it does help tie the building together, thus reducing the likelihood of progressive collapse in case of accidental overloads.

Floors consist of precast, prestressed hollow-core slabs, 12 in. deep and 4 ft wide. They bear on continuous corbels on the walls and on interior beams.

Buildings of this type are generally erected one story at a time, and locations of joints, temporary bracing, and connection details must be correlated with the erection sequence. Fig. 3.10.19 in-

Fig. 3.10.19 Twelve-story building — isometric view of assembly

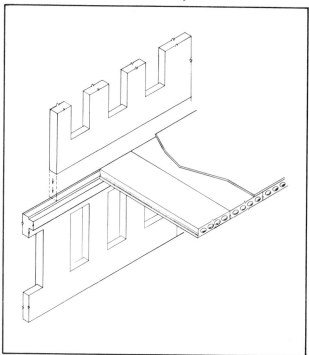

dicates the erection cycle. The walls of the story below have been erected, braced, and fully connected. The hollow-core slabs are then erected, bearing on wall corbels and interior beams. The wall panels of the next story are then erected; these have dowels projecting downward into sleeves filled with grout in the wall panels below (Fig. 3.10.20a). Hardware in the vertical joints for welded connections is shown in Fig. 3.10.20(b). The next step is to place continuous welded wire fabric for the topping, and finally place the topping. The erection cycle can be repeated as soon as the grout and concrete have adequate strength, and the welding of the connections between the wall panels has been completed.

The connections shown in Fig. 3.10.20 are illustrations of methods of providing the diaphragms and continuous load paths discussed in the previous examples. Many other details have been used successfully on similar structures in high-intensity earthquake areas.

3.10.10 Example — 23-Story Building

Figs. 3.10.21 and 3.10.22 show a 23-story building 300 ft high. The building was designed for earthquake resistance with a ductile moment-resisting space frame.

All floors consist of 10-ft wide single-tee units that were precast and prestressed. Adjacent tees were connected by means of hardware in the

Fig. 3.10.18 Twelve-story building — elevation

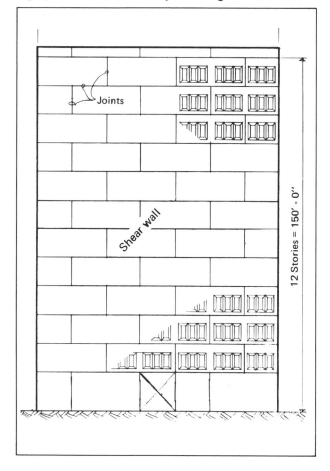

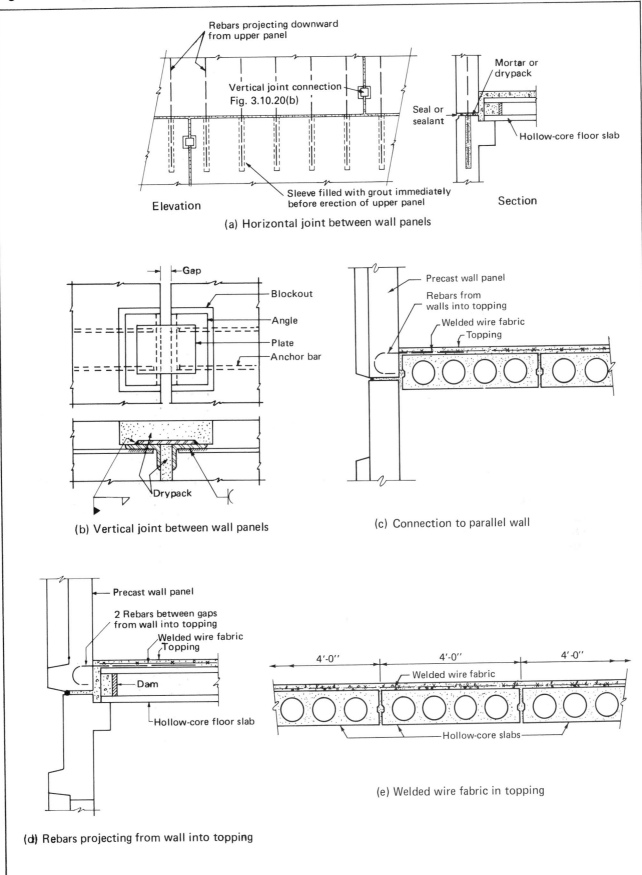

(a) Horizontal joint between wall panels

(b) Vertical joint between wall panels

(c) Connection to parallel wall

(d) Rebars projecting from wall into topping

(e) Welded wire fabric in topping

Fig. 3.10.21 23-story example building — elevation

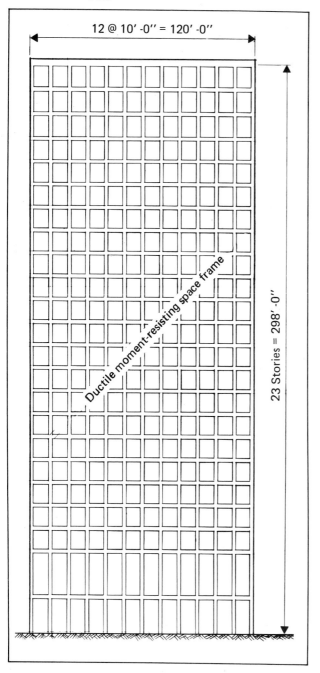

12 @ 10'-0" = 120'-0"

Ductile moment-resisting space frame

23 Stories = 298'-0"

ment must be provided in the diaphragm around the core in the same manner as around any large hole in a slab.

The exterior walls of the building are the moment-resisting frames. They consist of reinforced concrete columns and spandrel beams. For these frames to be ductile, they must have closely spaced spirals in the columns and ties in the spandrels. At the intersections, the congestion of reinforcement is so great that it is not feasible to have the tees align with the columns. Instead they are located midway between columns, but again the connection of the tee stem to the spandrel is immaterial to the earthquake resistance. The vital connection here is between the flange and the spandrel. This was accomplished by reinforcement projecting out of the end of the tee flange, which is later embedded in the cast-in-place concrete of the spandrel.

3.11 References

1. Branson, Dan E., *Deformations of Concrete Structures*, McGraw-Hill International Book Co., New York, 1977.

2. PCI Committee on Design Handbook, "Volume Changes in Precast, Prestressed Concrete Structures", *PCI Journal*, V. 22, No. 5, Sept.-Oct., 1977.

3. Nathan, Noel D., "Rational Analysis of Slender Prestressed Beam-Columns and Walls", *PCI Journal* (scheduled for publication 1985).

4. PCI Committee on Prestressed Concrete Columns, "Recommended Practice for the Design of Prestressed Concrete Columns and Walls", *PCI Journal* (scheduled for publication 1985).

5. Neville, Gerald B. (Editor), "Notes on ACI 318-83", Portland Cement Association, Skokie, Illinois, 1984.

6. "Commentary on Building Code Requirements for Reinforced Concrete (ACI 318-83)", American Concrete Institute, Detroit, Michigan, 1983.

7. Speyer, Irwin J. "Considerations for the Design of Precast Concrete Bearing Wall Buildings to Withstand Abnormal Loads", *PCI Journal*, V.21, No. 2, March-April, 1976.

8. "Response of Multi-Story Concrete Structures to Lateral Force", Special Publication SP-36, American Concrete Institute, Detroit, Michigan, 1973.

flanges, but these connections were not considered as diaphragm connections. The tees have a structural concrete topping reinforced with welded wire fabric. The diaphragm is provided by the topping rather than by the flanges of the tees.

The core of the building has no structural walls or frames that contribute measurably to lateral stability of the building. Each single tee is supported at the core by a cast-in-place column, but the connection is immaterial for earthquake resistance. The core has additional columns and partial slabs, but in effect, the core creates a large square hole in the floor diaphragm. Reinforce-

Fig. 3.10.22 23-story example building — typical floor plan

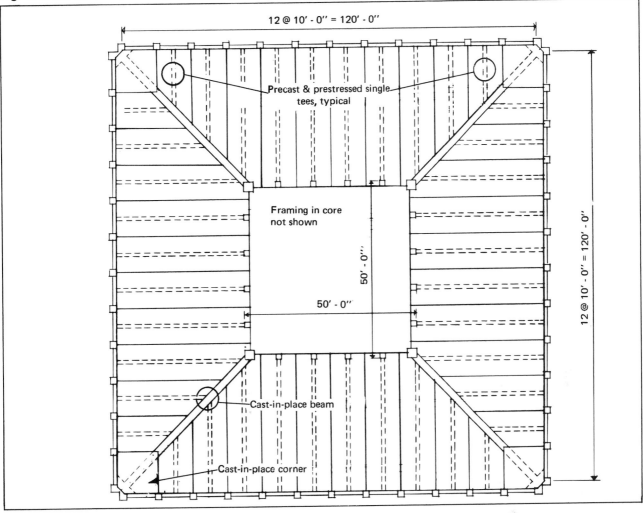

9. ACI Committee 442, "Response of Buildings to Lateral Forces", *Journal of the American Concrete Institute*, V. 68, No. 2, Feb., 1971.

10. "Design of Combined Frames and Shear Walls", Advanced Engineering Bulletin No. 14, Portland Cement Association, Skokie, Illinois, 1965.

11. Fintel, Mark (Editor), "Handbook of Concrete Engineering", 2nd Edition, Van Nostrand Reinhold Company, New York, 1985.

12. MacLeod, I.A., "Shear Wall-Frame Interaction — A Design Aid with Commentary", Portland Cement Association, Skokie, Illinois, April, 1970.

PART 4
DESIGN OF PRECAST AND PRESTRESSED CONCRETE COMPONENTS

Page No.

4.1 General 4-3
    4.1.1 Notation 4–3
    4.1.2 Introduction 4–6

4.2 Flexure 4–6
    4.2.1 Strength Design 4–6
    4.2.2 Service Load 4–14
        4.2.2.1 Non-Prestressed Element Design 4–14
        4.2.2.2 Prestressed Element Design 4–16
    4.2.3 Prestress Transfer and Strand Development 4–20

4.3 Shear 4–20
    4.3.1 Shear Resistance of Non-Prestressed Concrete 4–21
    4.3.2 Shear Resistance of Prestressed Concrete Members 4–21
    4.3.3 Design Using Design Aids 4–23
    4.3.4 Shear Reinforcement 4–26
    4.3.5 Horizontal Shear Transfer in Composite Members 4–27

4.4 Torsion 4–29
    4.4.1 Combined Shear and Torsion Strength — Non-Prestressed
        Members 4–29
    4.4.2 Torsion with Shear and Flexure — Prestressed Members ... 4–31

4.5 Loss of Prestress 4–39
    4.5.1 Sources of Stress Loss 4–39
    4.5.2 Range of Values for Total Losses 4–39
    4.5.3 Estimating Prestress Losses 4–40
    4.5.4 Critical Locations 4–40

4.6 Camber and Deflection 4–42
    4.6.1 Initial Camber 4–43
    4.6.2 Elastic Deflections 4–44
    4.6.3 Bilinear Behavior 4–44
    4.6.4 Effective Moment of Inertia 4–45
    4.6.5 Long-Time Camber/Deflection 4–46

4.7 Compression Members 4–47
    4.7.1 Strength Design of Precast Concrete
        Compression Members 4–47
    4.7.2 Slenderness Effects 4–52
    4.7.3 Service Load Stresses 4–52
    4.7.4 Effective Width of Wall Panels 4–52
    4.7.5 Varying Section Properties of Compression Members 4–52
    4.7.6 Piles 4–54

4.8 Special Considerations 4–57

    4.8.1 Load Distribution 4–57

    4.8.2 Effect of Openings 4–58

    4.8.3 Continuity 4–58

    4.8.4 Cantilevers 4–58

    4.8.5 Composite Topping with Hollow-Core Slabs 4–59

4.9 References 4–60

4.10 Design Aids... 4–61

    Flexure 4–61

    Shear 4–65

    Camber and Deflection... 4–73

DESIGN OF PRECAST AND PRESTRESSED CONCRETE COMPONENTS

4.1 General

4.1.1 Notation

A = cross sectional area

A_c = area enclosed by perimeter of cross section, p_c

A_{cr} = area of crack interface

A_{cs} = area of horizontal shear ties

A_{comp} = cross-sectional area of the equivalent rectangular stress block

A_g = gross area of concrete cross section

A_ℓ = total area of longitudinal reinforcement to resist torsion

A_o = area enclosed by shear flow path

A_{oh} = area enclosed by hoop centerline

A_{ps} = area of prestressed reinforcement

A_s = area of non-prestressed tension reinforcement

A'_s = area of compression reinforcement

A_t = area of one leg of closed stirrup

A_{top} = effective area of cast-in-place composite topping

A_v = area of shear reinforcement

a = depth of equivalent rectangular stress block

a_o = equivalent depth of compression in torsion

b = width of compression face of member

b_v = minimum effective web width (Sect. 4.4.2)

b_v = width of interface surface in a composite member (Sect. 4.3.5)

b_w = web width

C = coefficient as defined in section used (with subscripts)

C = compressive force

C_c = compressive force capacity of composite topping

C_f = coefficient = $bd^2/12,000$

CR = creep of concrete

c = distance from extreme compression fiber to neutral axis

DW = distribution width

d = distance from extreme compression fiber to centroid of non-prestressed tension reinforcement

d' = distance from extreme compression fiber to centroid of compression reinforcement

d_b = nominal diameter of reinforcing bar or prestressing strand

d_p = distance from extreme compression fiber to centroid of prestressed reinforcement

d_v = effective shear depth

d_{ve} = effective shear depth at end of beam

E = modulus of elasticity

E_c = modulus of elasticity of concrete

E_{ci} = modulus of elasticity of concrete at time of initial prestress

ES = elastic shortening

E_s = modulus of elasticity of steel

e = eccentricity of design load or prestress force parallel to axis measured from the centroid of the section

e' = distance between c.g. of strand at end and c.g. of strand at lowest point = $e_c - e_e$

e_c = eccentricity of prestress force from the centroid of the section at the center of the span

e_e = eccentricity of prestress force from the centroid of the section at the end of the span

F = force as defined in section used (with subscripts)

F_h = horizontal shear force

f_b = stress in the bottom fiber of the cross section

f_c = unit stress in concrete

f'_c = specified compressive strength of concrete

f'_{cc} = specified compressive strength of composite topping

f_{cds} = concrete stress at center of gravity of prestressing force due to all permanent (dead) loads not used in computing f_{cir}

f'_{ci} = compressive strength of concrete at time of initial prestress

f_{cir} = concrete stress at center of gravity of prestressing force immediately after transfer

f_{ct} = splitting tensile strength of light-weight concrete

f_d = stress due to sevice dead load

f_e	=	total load stress in excess of f_r
f_ℓ	=	stress due to service live load
f_{pc}	=	compressive stress in concrete at centroid of cross section due to prestress (after allowance for all prestress losses)
f_{pe}	=	compressive stress in concrete due to effective prestress forces only (after allowance for all prestress losses) at extreme fiber of section where tensile stress is caused by externally applied loads
f_{po}	=	stress in prestressing steel when strain in surrounding concrete is zero
f_{ps}	=	stress in prestressed reinforcement at nominal strength
f_{pu}	=	ultimate strength of prestressing steel
f_{py}	=	specified yield strength of prestressing steel
f_r	=	modulus of rupture of concrete
f_s, f'_s	=	stress in non-prestressed reinforcement
f_{se}	=	effective stress in prestressing steel after losses
f_t	=	stress in the top fiber of the cross section
$f_{t\ell}$	=	final calculated stress in the member
f_y	=	specified yield strength of non-prestressed reinforcement
h	=	overall depth of member
I	=	moment of inertia
I_{cr}	=	moment of inertia of cracked section transformed to concrete
I_e	=	effective moment of inertia for computation of deflection
I_g	=	moment of inertia of gross section
$\bar{j}_u$	=	$j_u (f_{ps}/f_{pu})$
j_u	=	for resisting lever arm used in $j_u d$
K_u	=	$[M_u/\phi - A'_s f'_y (d - d'_s)]/f'_c bd^2$
K'_u	=	a coefficient = $\phi M_n (12,000)/bd^2$
k	=	effective length factor for compression members
ℓ	=	span length
ℓ_b	=	effective length of bearing area
ℓ_d	=	development length
ℓ_u	=	unsupported length of a compression member
ℓ_{vh}	=	horizontal shear length as defined in Fig. 4.3.5
M	=	service load moment
M_a	=	total moment at the section

M_{cr}	=	cracking moment
M_d	=	moment due to service dead load (unfactored)
M_ℓ	=	moment due to service live load (unfactored)
M_{max}	=	maximum factored moment at section due to externally applied loads
M_n	=	nominal moment strength of a section
M_{nb}	=	nominal moment strength under balanced condition
M_o	=	nominal moment strength of a compressive member with zero axial load
M_{ocr}	=	pure flexural cracking strength
M_{sd}	=	moment due to superimposed dead load (unfactored)
M_{top}	=	moment due to topping (unfactored)
M_u	=	applied factored moment at a section
M_1	=	smaller factored end moment on a compression member, positive if bent in single curvature, negative if double curvature
M_2	=	larger factored end moment, always positive
N	=	unfactored axial load
N_u	=	factored axial load
ΔN_u	=	equivalent factored axial load caused by shear and torsion
n	=	modular ratio = E_s/E_c
P	=	prestress force after losses
P_i	=	initial prestress force
p_c	=	outside perimeter of concrete cross section
p_h	=	perimeter of hoop centerline
p_o	=	perimeter of shear flow path
P_n	=	axial load nominal strength of a compression member at given eccentricity
P_{nb}	=	axial load nominal strength under balanced conditions
P_o	=	prestress force at transfer
P_o	=	axial load nominal strength of a compression member with zero eccentricity
P_u	=	factored axial load
RE	=	relaxation of tendons
R_t, R_v	=	coefficients used in torsion design
r	=	radius of gyration

SH	=	shrinkage of concrete
s	=	shear or torsion reinforcement spacing in a direction parallel to the longitudinal reinforcement
T	=	tensile force
T_{cr}	=	torsional cracking moment under combined loading
TL	=	total prestress loss
T_n	=	nominal torsional moment strength provided by circulatory shear flow
T_{ocr}	=	pure torsional cracking strength
T_u	=	factored torsional moment on a section
t	=	thickness (used for various parts of members with subscripts)
V_c	=	nominal shear strength provided by the concrete
V_{cr}	=	cracking shear under combined loading
V_d	=	dead load shear (unfactored)
V_i	=	factored shear force at section due to externally applied loads occurring simultaneously with M_{max}
V_ℓ	=	live load shear (unfactored)
V_n	=	nominal shear strength
V_{nh}, V_{nv}	=	nominal shear strength of the connection in the horizontal and vertical directions, respectively
V_{ocr}	=	pure shear cracking strength
V_p	=	vertical component of the effective prestress force at the section considered
V/S	=	volume-surface ratio
V_u	=	factored shear force at section
V_{uh}, V_{uv}	=	applied factored shear loads in horizontal and vertical directions, respectively
V_{se}	=	service load shear
v_c	=	nominal shear stress carried by concrete
v'_c	=	nominal shear stress carried by concrete if no torsion is present
v_{ci}	=	shear stress at diagonal cracking due to all design loads when such cracking is the result of combined shear and moment
v_{cw}	=	shear stress at diagonal cracking due to all design loads when such cracking is the result of excessive principal tensile stresses in the web
v_{max}	=	maximum factored shear stress on a section
v_{tc}	=	nominal torsional stress carried by the concrete

v'_{tc}	=	nominal torsional stress carried by concrete if no shear is present
v_{tu}	=	factored torsional stress
v_u	=	factored shear stress
w	=	maximum crack width at extreme tension fiber
w	=	unfactored load per unit length of beam or per unit area of slab
w_d	=	unfactored dead load per unit length
w_ℓ	=	unfactored live load per unit of length
w_{sd}	=	dead load due to superimposed loading
$w_{t\ell}$	=	unfactored total load per unit of length $= w_d + w_\ell$
x	=	distance from support to point being investigated
x, x_1	=	shorter side of component rectangle and closed stirrup, respectively (torsion design)
y, y_1	=	longer side of component rectangle and closed stirrup, respectively (torsion design)
y'	=	distance from top to c.g. of A_{comp}
y_b	=	distance from bottom fiber to center of gravity of the section
y_t	=	distance from top fiber to center of gravity of the section
Z	=	section modulus
Z_b	=	section modulus with respect to the bottom fiber of a cross section
Z_t	=	section modulus with respect to the top fiber of a cross section
z	=	quantity limiting distribution of flexural reinforcement (Sect. 4.2.2.1)
Δ	=	deflection (with subscripts)
α_t, β	=	coefficients used in torsion design
β_1	=	factor defined in Sect. 4.2.1
γ_t	=	prestress factor used in torsion design
ϵ_{ps}	=	strain in prestressing steel corresponding to f_{ps}
ϵ_{sa}	=	strain in prestressing steel caused by external loads $= \epsilon_{ps} - \epsilon_{se}$
ϵ_{se}	=	strain in prestressing steel after losses
λ	=	a conversion factor for shear in lightweight concrete
μ	=	shear-friction coefficient
μ_e	=	effective shear-friction coefficient
ρ_p	=	A_{ps}/bd = ratio of prestressed reinforcement
ϕ	=	strength reduction factor
ω	=	$\rho f_y / f'_c$
ω'	=	$\rho' f_y / f'_c$
ω_p	=	$\rho_p f_{ps} / f'_c$

$$\omega_{pu} = \rho_p f_{pu}/f'_c$$

$\omega_w, \omega_{pw}, \omega'_w$ = reinforcement indices for flanged sections computed as for ω, ω_p, and ω' except that b shall be the web width, and reinforcement area shall be that required to develop compressive strength of web only

4.1.2 Introduction

This part of the Handbook provides a summary of design theory and procedures used in the design of precast concrete elements reinforced with prestressed and non-prestressed reinforcement. Designs are based on the provisions of the ACI Building Code[1] (referred to as "the Code" or "ACI 318" in the Handbook).

There are two phases which require consideration when designing precast concrete elements: (1) the manufacturing through erection phase and (2) the in-service conditions, those imposed on an element after it is permanently connected to the supporting structure. The designer is referred to Part 5 for the first phase. This part is concerned with the in-service conditions.

Rarely will the load tables in Part 2 provide all the design data necessary. In most cases, the engineer will select a standard section, with the detailed design calculations furnished by the staff or consulting engineer of the local prestressed concrete producer. The engineer of record should verify that the section selected is capable of satisfying both strength and performance criteria for the use intended. Under some circumstances, the engineer of record will furnish complete designs. When complete designs are furnished, some economy may be realized if the producers are permitted to suggest modifications to better fit their own production procedures.

4.2 Flexure

Design for flexure in accordance with the Code requires precast members be checked for both design strength and service load.

4.2.1 Strength Design

Strength design is based on the solution of the equations of equilibrium, normally using the rectangular stress block in accordance with Sect. 10.2.7 of the Code (see Fig. 4.2.1). The steel stress in prestressed design at nominal strength, f_{ps}, can be determined by strain compatibility or by the approximate equation given in the Code (Eq. 18-3). For elements with compression reinforcement, the nominal strength can be calculated by assum-

Fig. 4.2.1 Nominal flexural resistance

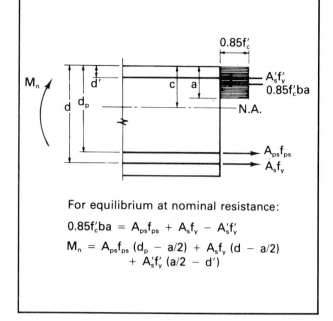

For equilibrium at nominal resistance:

$$0.85f'_c ba = A_{ps}f_{ps} + A_s f_y - A'_s f'_y$$

$$M_n = A_{ps}f_{ps}(d_p - a/2) + A_s f_y (d - a/2) + A'_s f'_y (a/2 - d')$$

ing that the compression reinforcement yields. The designer will normally choose a section and reinforcement and then determine if it has the capacity to meet the design requirements.

$$M_u \leq \phi M_n$$

A flow chart illustrating the application of design calculations for the computation of the nominal strength of flexural elements is given in Fig. 4.2.2.

Depth of stress block

The depth "a" of a rectangular stress block is related to the depth to the neutral axis "c" by the equation:

$$a = \beta_1 c$$

where:

f'_c, psi	β_1
3000	0.85
4000	0.85
5000	0.80
6000	0.75
7000	0.70
8000 and higher	0.65

Flanged elements

The equations for nominal strength given in Fig. 4.2.1 apply only to rectangular cross sections and flanged sections in which the stress block lies entirely within the depth of the flange h_f. To derive the depth of the stress block "a", divide the first equation of equilibrium given in Fig. 4.2.1 by $0.85f'_c b$ giving:

Fig. 4.2.2 Flow chart for nominal strength calculations for flexure

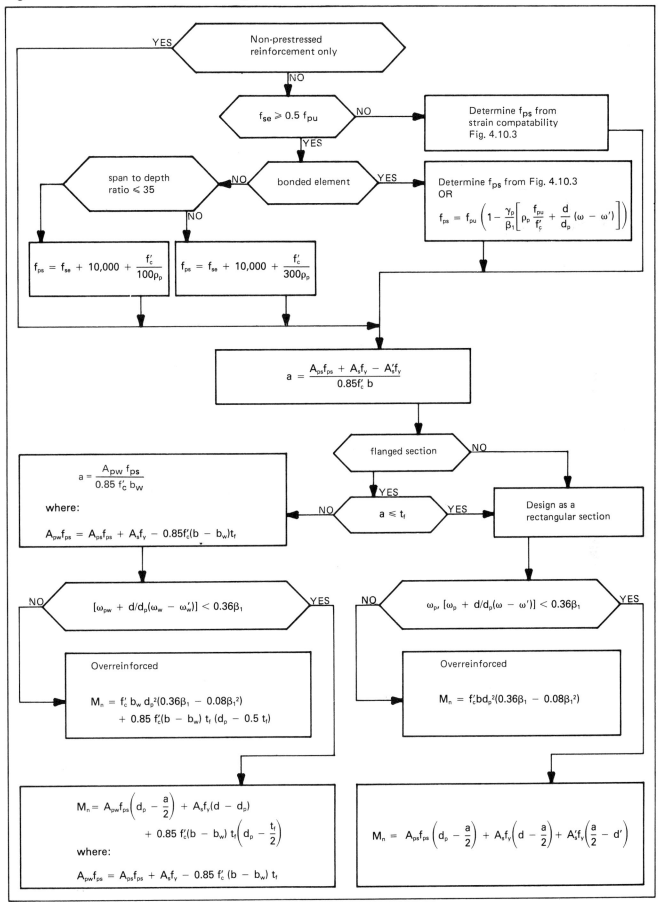

$$a = \frac{A_{ps}f_{ps} + A_s f_y - A'_s f'_y}{0.85 f'_c b} \qquad \text{(Eq. 4.2.1)}$$

If $a > h_f$ and compression reinforcement is present, use fundamental strain compatability or reasonable approximations to find the nominal strength. If compression reinforcement is not present the nominal strength can be found using the code equations shown in Fig. 4.2.2 or by strain compatability.

Limitations on reinforcement

For non-prestressed flexural elements, except slabs of uniform thickness, the minimum reinforcement ratio ρ_{min} is:

$$\rho_{min} = \frac{200}{f_y} \qquad \text{(Eq. 4.2.2)}$$

unless the area of reinforcement provided is 1/3 greater than required by analysis. For flanged sections, ρ is based upon the width of the web. For slabs, the minimum flexural reinforcement is that amount required for shrinkage and temperature reinforcement.

The maximum reinforcement ratio for non-prestressed elements is limited to 0.75 times the balanced reinforcement ratio:

$$\rho_{max} = 0.75 \, \rho_{bal} \qquad \text{(Eq. 4.2.3)}$$

where:

$$\rho_{bal} = \frac{0.85\beta_1 f'_c}{f_y} \left(\frac{87,000}{87,000 + f_y} \right)$$

Substituting the equation for ω yields:

$$\omega_{max} = \frac{\rho_{max} f_y}{f'_c}$$

$$= 0.64 \, \beta_1 \left(\frac{87,000}{87,000 + f_y} \right) \qquad \text{(Eq. 4.2.4)}$$

For prestressed elements, the Code requires that the total prestressed and non-prestressed reinforcement be adequate to develop a factored load at least 1.2 times the cracking load, except for flexural members with shear and flexural strength at least twice that required by analysis. The cracking strength is based on a modulus of rupture of:

$$f_r = 7.5 \sqrt{f'_c} \qquad \text{(Eq. 4.2.5)}$$

No upper limit is placed on the reinforcement for prestressed elements. However when:

$$\omega_p, \, [\omega_p + d/d_p \, (\omega - \omega')]$$
$$\text{or } [\omega_{pw} + d/d_p \, (\omega_w - \omega'_w)] > 0.36\beta_1$$

the nominal strength is calculated as shown in Fig. 4.2.2.

Critical section

For simply supported, uniformly loaded, prismatic non-prestressed elements, the critical section for flexural design will occur at midspan. Provided that reinforcement is properly developed and adequate shear reinforcement is provided, the amount of flexural reinforcement may be reduced in areas of lower moment towards the supports.

For prestressed elements, because of the Code limitation on end stresses at release (see Sect. 4.2.2.2), tendons are often draped or depressed, producing a varying effective depth, d, along the element length. For draped tendons, and also for non-uniform loading, it may be necessary to check the factored moment (M_u) and factored resistance (ϕM_n) at points other than midspan. This is illustrated in Fig. 4.2.3. For uniform loads with straight tendons, it is sufficient to check the midspan condition. For uniform loads with single point depressed tendons, the critical point is usually near

Fig. 4.2.3 Critical section for flexural design

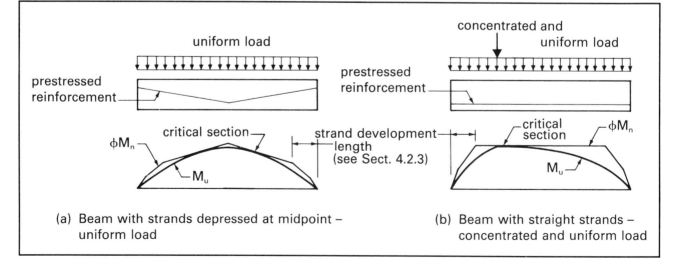

(a) Beam with strands depressed at midpoint – uniform load

(b) Beam with straight strands – concentrated and uniform load

PCI Design Handbook

0.4ℓ, and it is common practice to check only at midspan and 0.4ℓ.

Analysis using Code equations

The equations in ACI 318-83 combine the design of non-prestressed, prestressed, and partially prestressed members. Fig. 4.2.2 essentially outlines the design procedures using the Code equations. However, if compression reinforcement is present, the Code requires certain checks to assure that the stress in the compression reinforcement is at its yield strength. In computing f_{ps}, if any compression reinforcement is taken into account, the term

$$\left[\rho_p \frac{f_{pu}}{f'_c} + \frac{d}{d_p}(\omega - \omega') \right]$$

shall be taken not less than 0.17 and d' shall be no greater than $0.15d_p$. Also in order for the compression reinforcement to be considered effective:

$$\frac{A_{ps}f_{ps} + A_sf_y - A'_sf'_y}{bd}$$

$$\geq 0.85\, \beta_1\, f'_c \frac{d'}{d}\left(\frac{87,000}{87,000 - f_y} \right) \quad \text{(Eq. 4.2.6)}$$

If the above equation is not true the effects of the compression reinforcement are neglected.

Analysis using strain compatibility

The strain compatability approach is recognized by the Code as being an alternate method to the code equations. The procedure consists of assuming the location of the neutral axis, computing the strain in the prestressed and non-prestressed reinforcement, and establishing the depth of the stress block. Knowing the stress-strain relationship for the reinforcement, and assuming that the maximum strain in the concrete is 0.003, the forces in the reinforcement and in the concrete are determined and the sum of compression and tension forces is computed. If necessary, the neutral axis location is moved on a trial and error basis until the sum of forces is zero. The first moment of these forces is then computed to obtain the nominal strength of the section.

Design aids

Figs. 4.10.1 through 4.10.3 are provided to assist in the strength design of flexural members. Fig. 4.10.1 can be used for members with prestressed or non-prestressed reinforcement, or combinations (partial prestressing). Note that to use this aid, it is necessary to determine f_{ps} from some other source, such as Eq. 18-3 of ACI 318-83 or Fig. 4.10.3.

Fig. 4.10.2 is for use only with fully prestressed members. The value of f_{ps} is determined by strain compatibility in a manner similar to that used in Fig. 4.10.3. The reduction factor, ϕ, is included in the values of K'_u. The following examples illustrate the use of these design aids.

Example 4.2.1 Use of Fig. 4.10.1 for determination of non-prestressed reinforcement

Given:

The ledger beam shown.
Applied factored moment, $M_u = 1460$ ft-kips
$f'_c = 5000$ psi, normal weight concrete
$d = 72$ in. $f_y = 60$ ksi

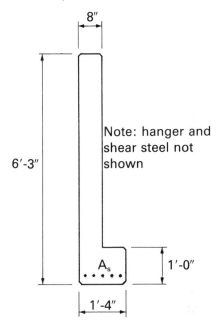

Problem:

Find the amount of mild steel reinforcement, A_s, required.

Solution:

Referring to Fig. 4.10.1, $A_{ps} = 0$; $A'_s = 0$

$$K_u = \frac{M_u/\phi}{f'_c\, bd^2} = \frac{(1460/0.9)(12,000)}{5000(8)(72)^2} = 0.0939$$

Required $\bar{\omega} = 0.100 < \omega_{max} = 0.302$

$$A_s = \frac{\bar{\omega}bdf'_c}{f_y} = \frac{0.100(8)(72)(5)}{60}$$

$$= 4.80 \text{ sq in.}$$

Use 5 - #9 $A_s = 5$ sq in.

$$\rho = A_s/bd = 5/(8)(72) = 0.0087$$
$$\rho_{min} = 200/f_y = 200/60,000$$

$$= 0.0033 < 0.0087$$

Example 4.2.2 Use of Fig.4.10.2 for determination of prestressing steel requirements—bonded strand

Given:

PCI standard rectangular beam 16RB24

Applied factored moment, $M_u = 600$ ft-kips

$f'_c = 6000$ psi normal weight concrete

$f_{pu} = 270$ ksi, low-relaxation strand

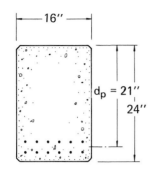

Problem:

Find the required amount of prestressing steel.

Solution:

Referring to Fig. 4.10.2:

$M_u \leq \phi M_n = K'_u \, bd_p^2/12{,}000$

$\text{Req'd } K'_u = \dfrac{M_u (12{,}000)}{bd_p^2} = \dfrac{600(12{,}000)}{16(21)^2}$

$= 1020$

for $\omega_{pu} = 0.20$, $K'_u = 1014$

$= 0.21$, $K'_u = 1052$

therefore

$\omega_{pu} = 0.20 + \dfrac{1020 - 1014}{1052 - 1014}(0.01) = 0.202$

$A_{ps} = \dfrac{\omega_{pu}\, bd_p f'_c}{f_{pu}} = \dfrac{0.202(16)(21)(6)}{270}$

$= 1.51$ sq in.

Use 10–1/2″ diameter strands; $A_{ps} = 1.53$ sq in.

Example 4.2.3 Use of Fig. 4.10.3—values of f_{ps} by stress-strain relationship—bonded strand

Given:

3′-4″ × 8″ hollow-core slab

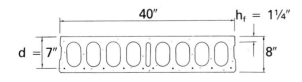

Concrete:

$f'_c = 5000$ psi normal weight concrete

Prestressing steel:

10 - 3/8″ diameter 250K stress-relieved strand

$A_{ps} = 10(0.080) = 0.800$ sq in.

Section properties:

$A \quad = 218$ in²

$Z_b \quad = 381$ in³

$y_b \quad = 3.98$ in.

Problem:

Find design flexural strength, ϕM_n

Determine $C\omega_{pu}$ for the section:

$$C\omega_{pu} = C\,\frac{A_{ps}f_{pu}}{bd_p f'_c} + \frac{d}{d_p}(\omega - \omega')$$

since $\omega = \omega' = 0$

$$C\omega_{pu} = \frac{1.06(0.8)(250)}{(40)(7)(5)} = 0.151$$

Entering Fig.4.10.3 with this parameter and an assumed effective stress, $f_{se} = 150$ ksi gives a value of:

$f_{ps}/f_{pu} = 0.965$ or $f_{ps} = 0.965(250)$

$= 241$ ksi

Determine the flexural strength:

$\phi M_n = \phi[A_{ps}f_{ps}(d_p - a/2) + A_s f_s(d - a/2)]$

$a = (A_{ps}f_{ps} + A_s f_y)/(0.85 f'_c b)$

Since $A_s = 0$:

$a = \dfrac{0.800(241)}{0.85(5)(40)} = 1.134$ in.

$\phi M_n = 0.9[0.8\,(241)\,(7 - 1.134/2) + 0]$

$= 1116$ in-kips $= 93.0$ ft-kips

Check the ductility requirement, $\phi M_n > 1.2 M_{cr}$

$P = f_{se}\,A_{ps} = 150\,(0.80)$

$= 120$ kips

$1.2 M_{cr} = 1.2(P/A + Pe/Z_b + 7.5\sqrt{f'_c}\,)Z_b$

$= 1.2\left(\dfrac{120}{218} + \dfrac{120(2.98)}{381}\right.$

$\left. + \dfrac{7.5\sqrt{5000}}{1000}\right)381$

$= 923$ in.-kips $= 76.9$ ft-kips

< 93.0 ft-kips \qquad OK

PCI Design Handbook

Example 4.2.4 Use of Fig. 4.10.3 and Eq. 18-3 (ACI 318-83) for partial prestressed member

Given:

PCI standard double tee
8DT24 +2

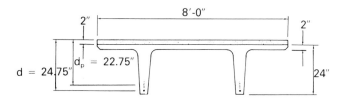

Concrete:
Precast: f'_c = 5000 psi
Topping: f'_c = 3000 psi, normal weight
Reinforcement:
12 - ½ in. diameter 270K low-relaxation strands
(6 each stem)
A_{ps} = 12(0.153) = 1.84 sq in.
A_s = 2 - #9 = 2.0 sq in.

Problem:

Find the design flexural strength of the composite section by the stress-strain relationship, Fig. 4.10.3, and compare using Code Eq. 18-3.

Solution:

Assume f_{se} = 150 ksi,

$$C\omega_{pu} = C\frac{A_{ps} f_{pu}}{bd_p f'_c} + \frac{d}{d_p}(\omega - \omega')$$

$$\omega = A_s f_y / bdf'_c$$

$$= \frac{2(60)}{96(24.75)(3)} = 0.017$$

$$C\omega_{pu} = \frac{1.00(1.84)(270)}{96(22.75)(3)} + \frac{24.75}{22.75}(0.017)$$

$$= 0.094$$

From Fig. 4.10.3, f_{ps}/f_{pu} = 0.97

$$f_{ps} = 0.97(270) = 261.9 \text{ ksi}$$

$$a = \frac{1.84(261.9) + 2(60)}{0.85(3)(96)} = 2.46 \text{ in.}$$

$$M_n = 1.84(261.9)[22.75 - (2.46/2)]$$

$$+ 2(60)[(24.75 - (2.46/2)]$$

$$= 13,193 \text{ in-kips} = 1099 \text{ ft-kips}$$

$$\phi M_n = 0.9(1099) = 989 \text{ ft-kips}$$

Find f_{ps} using Eq. 18-3 (ACI 318-83)

$$f_{ps} = f_{pu}\left(1 - \frac{\gamma_p}{\beta_1}\left[\rho_p\frac{f_{pu}}{f'_c} + \frac{d}{d_p}(\omega - \omega')\right]\right)$$

ω' = 0 in this example

$$f_{ps} = 270\left(1 - \frac{0.28}{0.85}\left[\frac{1.84(270)}{96(22.75)(3)}\right.\right.$$

$$\left.\left. + \frac{24.75}{22.75}(0.017)\right]\right)$$

$$= 261.6 \text{ ksi} \approx 261.9 \text{ ksi}$$

Therefore, as above: ϕM_n = 989 ft-kips

Example 4.2.5 Use of strain compatability analysis of a non-prestressed member

Given:

PCI standard double tee of Example 4.2.4. Topping is reinforced with 6 × 6—W1.4 × W1.4 WWF, and the flange is reinforced with 6 × 6—W4 × W4 WWF.

f'_c topping = 3000 psi; f'_c precast = 5000 psi; f_y = 60,000 psi

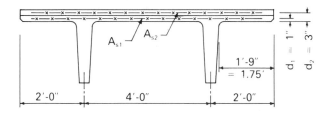

Problem:

Find the uniform live load which the flange can support.

Solution:

A_{s1} (WWF in precast) = 0.080 sq in./ft
A_{s2} (WWF in topping) = 0.029 sq in./ft
The cantilevered flange controls the design since the negative moment over the stem reduces the positive moment between the stems. Construct a strain diagram:

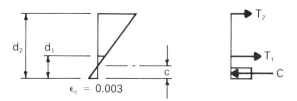

Try c = 1/4 in.

$$\frac{\epsilon_s + 0.003}{0.003} = \frac{d}{c}$$

Therefore:

$$\epsilon_{s1} = \frac{0.003\, d_1}{c} - 0.003$$

$$\epsilon_{s2} = \frac{0.003\, d_2}{c} - 0.003$$

$$\epsilon_{s1} = \frac{0.003 \times 1}{0.25} - 0.003 = 0.009$$

$$\epsilon_{s2} = \frac{0.003 \times 3}{0.25} - 0.003 = 0.033$$

$$\epsilon_y = \frac{f_y}{E_s} = \frac{60}{29,000} = 0.0021 < 0.009$$

Therefore, reinforcement yields, and

$f_s = f_y = 60$ ksi

$a = \beta_1 c = 0.80(0.25) = 0.20$ in.

$C = 0.85\, f'_c\, ba = 0.85(5)(12)(0.20)$
$\quad = 10.2$ kips/ft

$T_1 = A_{s1}f_y = 0.080(60) = 4.80$ kips/ft

$T_2 = A_{s2}f_y = 0.029(60) = 1.74$ kips/ft

$T_1 + T_2 = 6.54 < 10.2$ kips/ft

Therefore, try $a = (T_1 + T_2)/0.85\, f'_c b$

$$= 6.54/0.85(5)(12)$$

$$= 0.13 \text{ in.}$$

$$\text{Check } \epsilon_{s1} = \frac{0.003 \times 1}{0.13} - 0.003$$

$$= 0.020 > 0.0021$$

Therefore, the reinforcement yields, and analysis is valid.

$\phi M_n = \phi\,[T_1\,(d_1 - a/2) + T_2\,(d_2 - a/2)]$

$\quad = 0.9\,[4.80\,(1 - 0.13/2)$
$\quad\quad + 1.74\,(3 - 0.13/2)]$

$\quad = 8.64$ in.-kips/ft $= 720$ ft-lb/ft

Check ρ_{min} and ρ_{max}

$d = (0.029 \times 3 + 0.080 \times 1)/(0.029 + 0.080)$
$\quad = 1.53$ in.

$$\rho = \frac{0.029 + 0.080}{12 \times 1.53} = 0.0059$$

$\rho_{min} = 200/60,000 = 0.0033 < 0.0059$ OK

$\rho_{max} = 0.75\,\rho_{bal}$

$$= 0.75(0.85\beta_1)\left(\frac{f'_c}{f_y}\right)\left(\frac{87,000}{87,000 + f_y}\right)$$

$$= 0.75(0.85)(0.80)\left(\frac{5000}{60,000}\right)\left(\frac{87}{87 + 60}\right)$$

$$= 0.025 > 0.0059 \quad \text{OK}$$

w_d (flange self weight) $= 50$ psf

$$M_d = \frac{\omega_d \ell^2}{2} = \frac{50(1.4)(1.75)^2}{2} = 107.2 \text{ ft-lb/ft}$$

$M_\ell = 720 - 107.2 = 612.8$ ft-lb/ft

$$w_\ell = \frac{612.8(2)}{(1.75)^2(1.7)} = 235 \text{ psf}$$

Example 4.2.6 Design of a partially prestressed flanged section using strain compatability

Given:

Inverted tee beam with 2 in. composite topping as shown

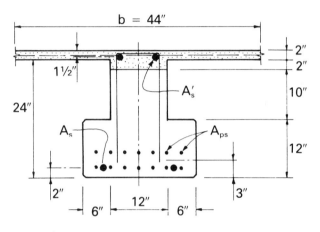

Concrete:

f'_c (precast) $= 5000$ psi

f'_c (topping) $= 3000$ psi

Reinforcement:

14–1/2″ diameter 270K low relaxation strand

$A_{ps} = 14 \times 0.153 = 2.142$ sq in.

$E_{ps} = 28,000$ ksi

$A_s = 2-\#8 = 1.58$ in.^{2} $E_s = 29,000$ ksi

$A'_s = 2-\#9 = 2.0$ in.^{2} $E_s = 29,000$ ksi

Problem:

Find flexural strength, ϕM_n

Solution:

Determine effective flange width, b, from Sect. 8.10.2 of the Code, overhanging width = 8 times thickness

$b = b_w + 2(8t) = 12 + 2(8)(2) = 44$ in.

$d_p = 26 - 3 = 23$ in.

$d = 26 - 2 = 24$ in.

$d' = 1\frac{1}{2}$ in.

Assume 20% loss of prestress

Strand initially tensioned to 75% of f_{pu}

$f_{se} = 0.8(0.75)(270) = 162$ ksi

$$\epsilon_{se} = \frac{f_{se}}{E_{ps}} = \frac{162}{28,000} = 0.0058$$

Construct a strain diagram as above:

$$\frac{\epsilon_{sa} + 0.003}{0.003} = \frac{d_p}{c}$$

$$\epsilon_{sa} = \frac{0.003\, d_p}{c} - 0.003$$

also $\epsilon_s = \dfrac{0.003\, d}{c} - 0.003$

and $\epsilon'_s = 0.003 - \dfrac{0.003\, d'}{c}$

$$\epsilon_{sa} = \frac{0.003(23)}{c} - 0.003 = \frac{0.069}{c} - 0.003$$

$$\epsilon_{ps} = \epsilon_{sa} + \epsilon_{se} = \frac{0.069}{c} - 0.003 + 0.0058$$

$$= \frac{0.069}{c} + 0.0028$$

$$\epsilon_s = \frac{0.003(24)}{c} - 0.003 = \frac{0.072}{c} - 0.003$$

$$\epsilon'_s = 0.003 - \frac{0.003(1.5)}{c} = 0.003 - \frac{0.0045}{c}$$

$$\epsilon_y = \frac{f_y}{E_s} = \frac{60}{29,000} = 0.0021$$

Try c = 11 in.

$$\epsilon_{ps} = \frac{0.069}{11} + 0.0028 = 0.0091$$

From the strand stress-strain curve equations in Fig. 11.2.5

$$f_{ps} = 268 - \left(\frac{0.075}{0.0091 - 0.0065}\right) = 239 \text{ ksi}$$

$$\epsilon_s = \frac{0.072}{11} - 0.003 = 0.0035 > 0.0021$$

$f_s = 60$ ksi

$$\epsilon'_s = 0.003 - \frac{(0.003)(1.5)}{11} = 0.0026 > 0.0021$$

$f'_s = 60$ ksi

$a_3 = \beta_1 c - 4 \text{ in.} = (0.8)(11) - 4 = 4.8 \text{ in.}$

$C_1 = (0.85)(3)(2)(44) = 224.4$ kips

$C_2 = (0.85)(3)(2)(12) = 61.2$ kips

$C_1 + C_2 = 285.6$ kips

$C_3 = (0.85)(5)(4.8)(12) = 244.8$ kips

$C_4 = (2.0)(60) = 120$ kips

$C_1 + C_2 + C_3 + C_4 = 650.4$ kips

$T_1 = A_{ps}f_{ps} = (2.142)(239) = 511.9$ kips

$T_2 = (1.58)(60) = 94.8$ kips

$T_1 + T_2 = 606.7$ kips < 650.4 kips

Try c = 10.25 in.

$$\epsilon_{ps} = \frac{0.069}{10.25} + 0.0028 = 0.0095$$

$$f_{ps} = 268 - \left(\frac{0.075}{0.0095 - 0.0065}\right) = 243 \text{ ksi}$$

$$\epsilon_s = \frac{0.072}{10.25} - 0.003 = 0.004 > 0.0021$$

$f_s = 60$ ksi

$$\epsilon'_s = 0.003 - \frac{(0.003)(1.5)}{10.25} = 0.0026 > 0.0021$$

$f'_s = 60$ ksi

$a_3 = (0.8)(10.25) - 4 = 4.2$ in.

$C_1 + C_2 = 285.6$ kips

$C_3 = (0.85)(5)(4.2)(12) = 214.2$ kips

$C_4 = (2.0)(60) = 120$ kips

$C_1 + C_2 + C_3 + C_4 = 619.8$ kips

$T_1 = (2.142)(243) = 520.5$ kips

$T_2 = (1.58)(60) = 94.8$ kips

$T_1 + T_2 = 615.3$ kips ≈ 619.8 kips OK

Check whether the section is over-reinforced (see ACI 318-83, Sect. 18.8.1):

Determine force to develop overhanging flanges:

$$C_f = 0.85\, f_c'\, (b - b_w) t_f$$

$$= (0.85)(3)(44 - 12)(2) = 163.2 \text{ kips}$$

Using average $f_c' = 4$ ksi find ω_{pw}, ω_w, ω_w'

$$\omega_{pw} = \frac{A_{ps}\, f_{ps}}{b_w\, d_p\, f_c'} = \frac{(2.142)(243)}{(12)(23)(4)} = 0.472$$

$$\omega_w = \frac{A_s\, f_s}{b_w\, d\, f_c'} = \frac{(1.58)(60)}{(12)(24)(4)} = 0.082$$

$$\omega_w' = \frac{A_s'\, f_s'}{b_w\, d\, f_c'} = \frac{(2.0)(60)}{(12)(24)(4)} = 0.104$$

$$\omega_f = \frac{C_f}{b_w\, d\, f_c'} = \frac{163.2}{(12)(24)(4)} = 0.142$$

$$\overline{\omega}_w = \left[\omega_{pw} + \frac{d}{d_p}(\omega_w - \omega_w' - \omega_f) \right] < 0.36\, \beta_1$$

$$\left[0.472 + \frac{24}{23}(0.082 - 0.104 - 0.142) \right]$$
$$< (0.36)(0.85)$$

$$0.301 < 0.306$$

The section is not over-reinforced and tension governs.

$$\begin{aligned}
M_n =\ & 224.4(10.25 - 1) + 61.2(10.25 - 3) \\
& + 214.2\left(10.25 - 4 - \frac{4.2}{2}\right) + 120(10.25 - 1.5) \\
& + 520.5(23 - 10.25) + 94.8(24 - 10.25) \\
=\ & 12{,}398 \text{ in.-kips} \\
=\ & 1033 \text{ ft-kips}
\end{aligned}$$

$$\phi M_n = 0.9(1033) = 930 \text{ ft-kips}$$

Note: this example shows an exact method; approximate methods may be satisfactory.

4.2.2 Service Load Design

Precast members are checked under service load, primarily for meeting performance criteria and to control cracking.

4.2.2.1 Non-Prestressed Element Design

Non-prestressed flexural elements are normally proportioned, and reinforcement selected, on the basis of the procedures described in Sect. 4.2.1. However, depending upon the application and exposure of the member, designers may want to control the degree of cracking. In some applica-

tions, such as architectural concrete panels, they may not want any discernible cracking. In other cases cracking may be permitted, but the crack width must be limited. In addition, the Code requires that the crack width be limited when the yield strength of the reinforcement exceeds 40,000 psi.

If no discernible cracking is the criteria, the flexural tensile stress level should be limited to:

$$f_r' \leqslant 5\, \lambda\, \sqrt{f_c'} \qquad \text{(Eq. 4.2.7)}$$

where:

f_r' = allowable flexural tension, computed using gross concrete section

f_c' = concrete strength at the time considered

λ = 1.0 for normal weight concrete
= 0.85 for sand-lightweight concrete
= 0.75 for all-lightweight concrete

When f_r' exceeds this value, required reinforcement is determined by a Code limitation on the maximum value of the quantity z as shown in Fig. 4.2.5, in which z is calculated from the equation:

$$z = f_s \sqrt[3]{d_c A} \qquad \text{(Eq. 4.2.8)}$$

where:

f_s = reinforcement stress at service load, ksi
= 0.6 f_y unless otherwise determined

d_c = concrete cover to the center of the reinforcement, in.

A = average effective area around one reinforcing bar, in.^{2}
= 2b d_c/n

b = width of tension face, in.

n = number of reinforcing bars

Fig. 4.2.4 Notation for crack-control equations

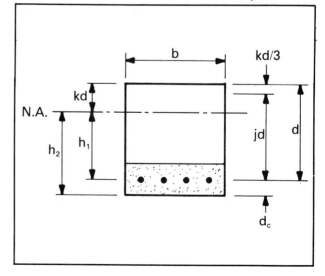

PCI Design Handbook

Fig. 4.2.5 Recommended maximum values of z and crack widths

Type	Not exposed to view		Critical appearance	
Exposure	Not exposed to weather	Exposed to weather	Not exposed to weather	Exposed to weather
Max. value of z (k/in)	175	145	105	53
Corresponding value of w (in)	0.016	0.013	0.010	0.005

Note: Based on $h_2/h_1 = 1.2$

This equation is derived from the Gergely-Lutz[3] expression:

$$w = (7.6 \times 10^{-5}) \frac{h_2}{h_1} f_s \sqrt[3]{d_c A} \quad \text{(Eq. 4.2.9)}$$

with h_2/h_1 taken equal to 1.2

where:

w = maximum crack width at extreme tension fiber, in.

h_1 = distance from centroid of tensile reinforcement to neutral axis, in.

h_2 = distance from extreme tension fiber to neutral axis, in.

If values of f_s under service load conditions are required to be less than $0.6 f_y$ to satisfy crack control requirements, reinforcement should be provided equal to:

$$A_s = \frac{M}{0.9 f_s d} \quad \text{(Eq. 4.2.10)}$$

where M = service load moment.

This equation is based on working stress design principles with the assumption that j = 0.9 and k = 0.3.

Example 4.2.7 Non-prestressed panel design

Given:

A 6 in. thick architectural panel exposed to the weather, with dimensions as shown:

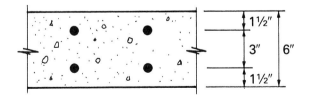

Concrete $f'_c = 5000$ psi

Service load moment, M = 2.4 ft-kips/ft

Problem:

Determine if reinforcement is needed to control cracking, and if so, amount required.

Solution:

For a 12 in. width:

$$f_c = \frac{M}{Z} = \frac{2.4(12)(1000)}{12(6)^2/6} = 400 \text{ psi}$$

From Eq. 4.2.7:

$f'_r = 5(1.0) \sqrt{5000}$

= 354 psi < 400, reinforcement required

For a panel with critical appearance exposed to the weather, the recommended maximum value of w from Fig. 4.2.5 is:

$$w = 0.005 \text{ in.}$$

Assuming j = 0.9 and k = 0.3, calculate:

$$d = 6 - 1.5 = 4.50 \text{ in.}$$
$$kd = (0.3)(4.5) = 1.35 \text{ in.}$$
$$h_1 = 4.5 - 1.35 = 3.15 \text{ in.}$$
$$h_2 = 6 - 1.35 = 4.65 \text{ in.}$$
$$\frac{h_2}{h_1} = \frac{4.65}{3.15} = 1.48$$
$$d_c = 1.5 \text{ in.}$$

Try a bar spacing of 4 in.

$$A = 2(4)(1.5) = 12 \text{ sq in.}$$

From Eq. 4.2.9:

$$f_s = \frac{w}{(7.6 \times 10^{-5}) \frac{h_2}{h_1} \sqrt[3]{d_c A}}$$

$$\frac{0.005}{(7.6 \times 10^{-5})(1.48) \sqrt[3]{1.5(12)}} = 16.96 \text{ ksi}$$

$$A_s = \frac{M}{0.9 f_s d}$$

$$= \frac{2.4(12)}{0.9(16.96)(4.5)} = 0.42 \text{ sq in./ft}$$

Use #4 at 4 in. $A_s = 0.60$ sq in./ft

(Note: This is an unusually high amount of rein-forcement for this type of panel. Ordinarily, span would be reduced to keep the stress below f'_r.)

4.2.2.2 Prestressed Element Design

Unlike concrete reinforced with non-pre-stressed steel, the ACI Code requires that service load stresses be checked at critical points, in ad-dition to the design strength of the member. Code limitations on the service load stresses are sum-marized as follows (see Code Sects. 18.4 and 18.5):

Concrete:

1. At release (transfer) of prestress, before time-dependent losses:

   a. Compression $0.60\, f'_{ci}$

   b. Tension (except at ends) $3\sqrt{f'_{ci}}$

   c. Tension at ends* of simply
      supported members $6\sqrt{f'_{ci}}$

2. Under service loads:

   a. Compression $0.45\, f'_c$

   b. Tension in precompressed
      tensile zone when deflections
      are calculated based on gross
      section $6\sqrt{f'_c}$

   c. Tension in precompressed
      tensile zone when deflections
      are calculated based on bilinear
      relationships (see Sect. 4.6.3.) $12\sqrt{f'_c}$

Prestressing steel:

a. Tension due to tendon jacking
   force: $0.85\, f_{pu}$ or $0.94\, f_{py}$

b. Tension immediately after prestress transfer:

   Stress-relieved strand: $0.7\, f_{pu}$

   Low-relaxation strand: $0.74\, f_{pu}$

---

*May be considered at 50 strand diameters from end (see Sect. 4.2.3).

It is common practice in the precast, pre-stressed concrete industry to follow the above recommendations with the following clarifica-tions:

Tension in precompressed tensile zone at ser-vice loads:

Hollow-core and solid flat slabs: $6\sqrt{f'_c}$

Stemmed deck members and beams: $12\sqrt{f'_c}$

Initial stress in steel due to jacking force:

Stress-relieved strand: $0.7\, f_{pu}$

Low-relaxation strand: $0.75\, f_{pu}$

These values should not be exceeded without consulting the product manufacturer.

Calculations of stresses at critical points follow classical straight line theory as illustrated in Fig. 4.2.6.

Composite members:

It is usually more economical to place cast-in-place composite topping without shoring the member, especially for deck members. This means that the weight of the topping and simultaneous construction live load must be carried by the pre-cast member alone. Additional superimposed dead and live loads are carried by the composite sec-tion.

The following examples illustrate a tabular form of superimposing the stresses caused by the pres-tress force and the dead and live load moments.

Sign Convention:

The customary sign convention used in the de-sign of precast, prestressed concrete members for service load stresses is positive (+) for compres-sion and negative (−) for tension. This conven-tion is used throughout this Handbook.

Example 4.2.8 Calculation of critical stresses – straight strands

Given:

Span = 36 ft

Select 4HC12 + 2

Fig. 4.2.6 Calculation of service load stresses

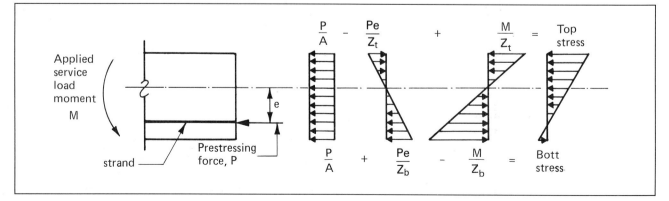

Load	Transfer Pt. at Release $P = P_o$		Midspan at Release $P = P_o$		Midspan at Service load $P = P$	
	f_b	f_t	f_b	f_t	f_b	f_t
P/A	+ 525	+525	+ 525	+ 525	+479	+ 479
Pe/Z	+ 931	− 744	+ 931	− 744	+850	− 680
M_d/Z	− 164	+ 131	− 751	+ 600	−751	+ 600
M_{top}/Z					− 271	+ 217
M_{sd}/Z^c					− 175	+ 128
M_ℓ/Z^c					− 437	+ 318
Stresses	+ 1292	− 88	+ 705	+ 381	− 305	+1062
Allowable	$0.6\,f'_{cl}$	$6\sqrt{f'_{cl}}$	$0.6f'_{cl}$	$0.6f'_{cl}$	$6\sqrt{f'_c}$	$0.45f'_c$
Stresses	2400	− 379	2400	2400	− 465	2700
	OK	OK	OK	OK	OK	OK

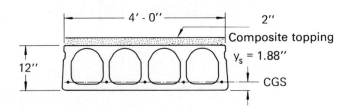

Section properties:

	Non-composite	Composite
A	= 265 sq in.	361 sq in.
I	= 4771 in.⁴	7209 in.⁴
y_b	= 6.67 in.	8.10 in.
y_t	= 5.33 in.	5.90 in.
Z_b	= 715.3 in.³	890.0 in.³
Z_t	= 895.1 in.³	1221.9 in.³
wt	= 276 plf	376 plf
e	= 4.79 in.	

Superimposed dead load = 20 psf = 80 plf
Superimposed live load = 50 psf = 200 plf

Precast concrete:
f'_c = 6000 psi
f'_{ci} = 4000 psi
E_c = 4700 ksi
Normal weight

Topping concrete:
f'_c = 4000 psi
E_c = 3800 ksi
Normal weight

Prestressing steel:
5 – 1/2 in. dia. 270K low-relaxation strand
A_{ps} = 5 × 0.153 = 0.765 sq in.
Straight strands

Problem:

Find critical service load stresses.

Solution:

Prestress force:
P_i = 0.765 (0.75 × 270) = 155 kips
P_o (assume 10% initial loss)
   = 0.90 (155) = 139 kips
P (assume 18% total loss)
   = 0.82 (155) = 127 kips

Midspan service load moments:
M_d = 0.276 (36)² (12/8) = 537 in. kips
M_{top} = 0.100 (36)² (12/8) = 194 in. kips
M_{sd} = 0.080 (36)² (12/8) = 156 in. kips
M_ℓ = 0.200 (36)² (12/8) = 389 in. kips

Allow $6\sqrt{f'_c}$ tension at service load

See table above for stresses

**Example 4.2.9 Calculation of critical stresses —
single point depressed strand**

Given:

Span = 70 ft
Superimposed dead load = 10 psf = 80 plf

Superimposed live load = 35 psf = 280 plf

Select 8DT24 as shown

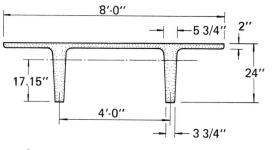

Concrete:

f'_c = 5000 psi

f'_{ci} = 3500 psi

Normal weight

Prestressing steel:

12–½″ dia. 270K low-relaxation strand

A_{ps} = 12 × 0.153 = 1.836 sq in.

Section properties:

A = 401 sq in.
I = 20,985 in.⁴
y_b = 17.15 in.
y_t = 6.85 in.
Z_b = 1224 in.³
Z_t = 3063 in.³
wt = 418 plf = 52 psf

Eccentricities, single point depression:

e_e = 5.48 in., e_c = 13.90 in.

e @ 0.4ℓ = 12.22 in.

e′ = 13.90 − 5.48 = 8.42 in.

e @ 0.4ℓ = 12.22 in.

e′ = 13.90 − 5.48 = 8.42 in.

Problem:

Find critical service load stresses.

Solution:

Prestress force:

P_i = 1.836 (0.75 × 270) = 372 kips

P_o (assume 10% initial loss)

= 0.90 (372) = 335 kips

P (assume 20% total loss)

= 0.80(372) = 298 kips

Service load moments:

at midspan:

M_d = 0.418(70)² (12/8) = 3072 in.-kips

M_{sd} = 0.080(70)² (12/8) = 588 in.-kips

M_ℓ = 0.280(70)² (12/8) = 2058 in.-kips

at 0.4ℓ:

M_d = 3072 (0.96) = 2949 in.-kips

M_{sd} = 588 (0.96) = 564 in.-kips

M_ℓ = 2058 (0.96) = 1976 in.-kips

Allow 12 $\sqrt{f'_c}$ final tension

See table below for stresses.

In this example, a release strength of f'_{ci} = $\frac{2129}{0.6}$ = 3548 psi should be provided. Also deflection should be checked.

Load	Transfer Pt. at Release P = P_o		Midspan at Release P = P_o		0.4ℓ at Service load P = P	
	f_b	f_t	f_b	f_t	f_b	f_t
P/A	+ 835	+ 835	+ 835	+ 835	+ 743	+ 743
Pe/Z	+1500	− 599	+3804	− 1520	+2985	−1189
M_d/Z	− 290	+ 116	−2510	+1003	−2409	+ 962
M_{sd}/Z^c					− 461	+ 184
M_ℓ/Z^c					−1614	+ 645
Stresses	+2045	+ 352	+2129	+ 318	− 756	+1345
Allowable	0.60 f'_{cl}	0.60 f'_{cl}	0.60 f'_{cl}	0.60 f'_{cl}	12$\sqrt{f'_c}$	0.45 f'_c
Stresses	+2100	+2100	2100	2100	− 848	2250
	OK	OK	HIGH	OK	OK	OK

Example 4.2.10 Tensile force to be resisted by top reinforcement

Given:

Span = 24 ft

24IT26 as shown

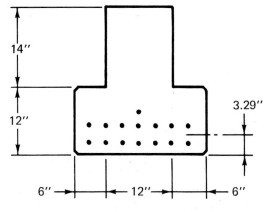

Concrete:

f'_c = 6000 psi

f'_{ci} = 4000 psi

Prestressing steel:

15–1/2″ dia. 270K low-relaxation strand

A_{ps} = 15 (0.153) = 2.295 sq in.

Section properties:

A = 456 sq in.

I = 24,132 in.⁴

y_b = 10.79 in.

y_t = 15.21 in.

Z_b = 2237 in.³

Z_t = 1587 in.³

wt = 475 plf

e = 7.5 in.

Problem:

Find critical service load stresses.

Solution:

Prestress force:

$$P_i = 2.295(0.75)(270) = 465 \text{ kips}$$

P_o (assume 10% initial loss)

$$= 0.90 (465) = 418 \text{ kips}$$

Moment due to member weight:

at midspan:

$$M_d = 0.475(24)^2(12/8) = 410 \text{ in.-kips}$$

at 50 strand diameters (2.08 ft) transfer point:

$$M_d = \frac{wx}{2} (\ell - x) = \frac{0.475(2.08)}{2} (24 - 2.08)(12)$$

$$= 130 \text{ in.-kips}$$

See table below for stresses.

Since the tensile stress exceeds allowable limits, reinforcement is required to resist the total tensile force, as follows:

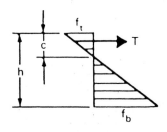

$$c = \frac{f_t}{f_t - f_b} (h) = \frac{-976}{-976 -2260} (26)$$

$$= 7.84 \text{ in.}$$

Load	Transfer Point at Release $P = P_o$		Midspan at Release $P = P_o$	
	f_b	f_t	f_b	f_t
P/A	+ 917	+917	+ 917	+917
Pe/Z	+ 1401	− 1975	+ 1401	− 1975
M_d/Z	− 58	+ 82	− 183	+ 258
Stresses	+2260	− 976	+2135	− 800
Allowable Stresses	0.6 f'_{ci}	6$\sqrt{f'_{ci}}$	0.6 f'_{ci}	3$\sqrt{f'_{ci}}$
	2400	− 379	2400	− 190
	OK	HIGH	OK	HIGH

$$T = \frac{c\, f_t\, b}{2} = \frac{7.84(976)(12)}{2} = 45{,}921 \text{ lb}$$

Similarly, the tension at midspan can be found as 34,000 lb.

The Commentary to the Code,[4] Sect. 18.4.1 (b) and (c), recommends that reinforcement be proportioned to resist this tensile force at a stress of $0.6f_y$, but not more than 30 ksi. Using reinforcement with $f_y = 60$ ksi:

0.6 (60) = 36 ksi, use 30 ksi

$$A_s \text{ (end)} = \frac{45.9}{30} = 1.53 \text{ sq in.}$$

$$A_s \text{ (midspan)} = \frac{34.0}{30} = 1.13 \text{ sq in.}$$

Top strands used as stirrup supports may also be used to resist this tensile force.

4.2.3 Prestress Transfer and Strand Development

In a pretensioned member, the prestress force is transferred to the concrete by bond. The length required to accomplish this transfer is called the "transfer length," and is assumed by the Code to be approximately 50 times the nominal diameter of the strand.

However, the length required to develop the full design strength of the strand is much longer, and is specified in the Code in Sect. 12.9.1 by the equation:

$$\ell_d = (f_{ps} - 2/3 f_{se})d_b \qquad \text{(Eq. 4.2.11)}$$

In the Commentary to the Code, the variation of strand stress along the development length is shown as in Fig. 4.2.7

For convenience, this curve may be approximated by straight lines. Also, to be consistent with the 50 diameter transfer length specified in other Code sections, the value of f_{se}, the stress which must be transferred is assumed to be 150 ksi.

Fig. 4.10.4 is a curve plotted according to the above assumptions, and can be used as a design aid as illustrated in Example 4.2.11.

Example 4.2.11 Use of Fig. 4.10.4 – Design stress for underdeveloped strand

Given:
Span = 12 ft
4HC8 as shown

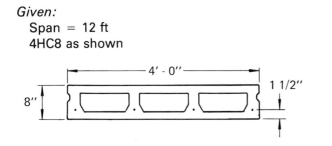

Fig. 4.2.7 Variation of steel stress with development length

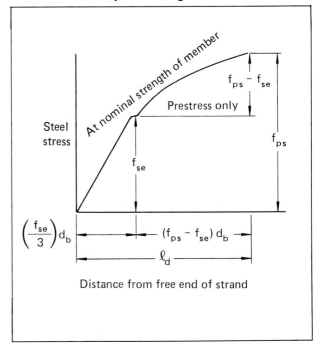

Distance from free end of strand

Concrete:
$f'_c = 5000$ psi
Normal weight

Prestressing steel:
4–1/2″ diameter 270K strands
$A_{ps} = 4 \times 0.153 = 0.612$ sq in.

If the strand is fully developed (see Fig. 4.10.3):

$$C\bar{\omega}_p = C\,\frac{A_{ps}f_{pu}}{b d_p f'_c} = \frac{1.06(0.612)(270)}{48(6.5)(5)}$$

$$= 0.11$$

From Fig. 4.10.3 with $f_{se} = 150$ ksi
$f_{ps}/f_{pu} = 0.97$
$f_{ps} = 0.97(270) = 262$ ksi

The maximum development length available is:
$\ell/2 = 12 \times 12/2 = 72$ in.

From Fig. 4.10.4 the maximum $f_{ps} = 244$ ksi

This value should be used to calculate the design strength (ϕM_n) of the member at midspan.

4.3 Shear

The shear design of precast concrete members is covered in Chapter 11 of ACI 318-83. The shear resistance of precast concrete elements must meet the requirement:

$$V_u \leq \phi V_n$$

where $V_n = V_c + V_s$, and:

PCI Design Handbook

V_c = nominal shear strength of concrete

V_s = nominal shear strength
of shear reinforcement

ϕ = 0.85

For flat deck members (hollow-core and solid slabs), and others proven by test, no shear reinforcement is required if the factored shear force, V_u, does not exceed the design shear strength of the concrete, ϕV_c. For other members, the minimum shear reinforcement is usually adequate. (Note: If V_u is less than $\phi V_c/2$, no shear reinforcement is required.)

The critical section for shear and torsion is indicated in the Code to be at a distance "d" from the face of the support for non-prestressed members and "h/2" for prestressed members. However, precast concrete members on which the load is not applied at the top of the member, such as L-shaped beams, the distance "d" or "h" should be measured from the point of load application to the bottom, or, conservatively, the critical section taken at the face of the support. By Sect. 11.1.2 of the Code, no concentrated load shall occur between the face of the support and the critical section.

Although the Code expresses the equations for shear in terms of forces, it is generally more convenient to work shear design problems using stresses. The shear stress can be expressed by:

$$v_u = V_u/\phi b_w d \qquad \text{(Eq. 4.3.1)}$$

4.3.1 Shear Resistance of Non-Prestressed Concrete

In the absence of torsion and axial forces, the nominal shear resistance of concrete is given by:

$$v_c = 2\sqrt{f'_c} \qquad \text{(Eq. 4.3.2)}$$

or if one performs a more detailed analysis:

$$v_c = \left(1.9\sqrt{f'_c} + 2500\rho_w \frac{V_u d}{M_u}\right) \le 3.5\sqrt{f'_c}$$
$$\text{(Eq. 4.3.3)}$$

A more detailed analysis of elements subjected to significant axial or torsional forces is contained in the Code.

Example 4.3.1 Design of shear reinforcement–non-prestressed member

Given:

A spandrel beam as shown:

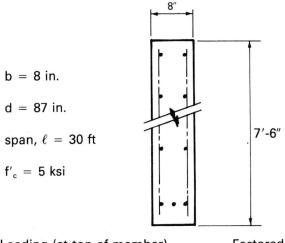

b = 8 in.

d = 87 in.

span, ℓ = 30 ft

f'_c = 5 ksi

Loading (at top of member)	Factored
Self wt = 0.75 × 1.4 =	1.05
Dead load = 3.21 × 1.4 =	4.49
Live load = 1.31 × 1.7 =	2.23
	w_u = 7.77 k/ft

Problem:

Determine what size welded wire fabric will satisfy the shear requirements.

Solution:

Determine V_u at distance d from support:

V_u = 7.77(30/2 − 87/12) = 60.22 kips

v_u = $V_u / \phi b_w d$ = 60,220/(0.85)(8)(87) = 102 psi

$v_c = 2\sqrt{f'_c} = 2\sqrt{5000}$ = 141 psi > 102 > 1/2 v_c therefore minimum shear reinforcement is required.

ACI 318-83, Eq. 11-14 requires a minimum amount of reinforcement be provided as follows:

A_v = 50b_ws/f_y = 50(8) (12)/60,000
= 0.080 sq in./ft

From Table 11.2.10, select a WWF that has vertical wires:

W2 @ 6 in. A_v = 0.08 sq in./ft

4.3.2 Shear Resistance of Prestressed Concrete Members

Shear design of prestressed concrete members is covered in ACI 318-83 by Eqs. 11-10 through 11-13, repeated below in terms of stresses.

Either Eq. 4.3.4 or the lesser of Eqs. 4.3.5 or 4.3.7 may be used, however Eq. 4.3.4. is valid only if the effective prestress force is at least equal to 40% of the tensile strength of the prestressing strand. The Code places certain upper and lower limits on the use of these equations, which are shown in Fig. 4.3.1.

$$v_c = 0.6\sqrt{f'_c} + 700\frac{V_u d}{M_u} \qquad \text{(Eq. 4.3.4)}$$

$$\frac{V_u d}{M_u} \le 1$$

$$v_{ci} = 0.6\sqrt{f'_c} + \frac{V_d + \dfrac{V_i M_{cr}}{M_{max}}}{b_w d} \qquad \text{(Eq. 4.3.5)}$$

$$M_{cr} = \left(\frac{I}{y_b}\right)(6\sqrt{f'_c} + f_{pe} - f_d) \qquad \text{(Eq. 4.3.6)}$$

$$v_{cw} = 3.5\sqrt{f'_c} + 0.3f_{pc} + \frac{V_p}{b_w d} \qquad \text{(Eq. 4.3.7)}$$

The value of d in Eq. 4.3.4 is the distance from the extreme compression fiber to the centroid of the reinforcement. In all other equations, d need not be less than 0.8h.

In unusual cases, such as members which carry heavy concentrated loads, or short spans with high superimposed loads, it may be necessary to construct a shear resistance diagram (v_c) and superimpose upon that a unit shear (v_u) diagram. The procedure is illustrated in Fig. 4.3.1.

The steps for constructing the shear resistance diagram are as follows:

1. Draw a horizontal line at a value of $2\sqrt{f'_c}$ (Note: The Code requires that this minimum be reduced to $1.7\sqrt{f'_c}$ when the stress in the strand after all losses is less than $0.4\,f_{pu}$. For precast, prestressed members the value will always be above $0.4\,f_{pu}$.)
2. Construct the curved portion of the diagram. For this, either Eq. 4.3.5 or, more conservatively, Eq. 4.3.4 may be used. Usually it is ad-

Fig. 4.3.1 Shear design

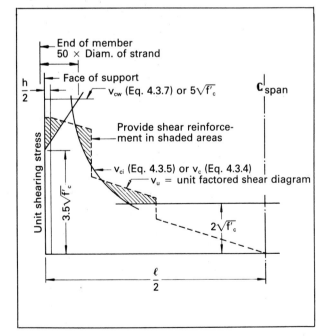

equate to find 3 points on the curve. For uniformly loaded simple span members, Fig. 4.10.5 may be used to quickly determine the points on the curve.

3. Draw the upper limit line, v_{cw} from Eq. 4.3.7 if Eq. 4.3.5 has been used in step 2, or $5\sqrt{f'_c}$ if Eq. 4.3.4 has been used.
4. The diagonal line at the upper left of Fig. 4.3.1 delineates the upper limit of the shear resistance diagram in the prestress transfer zone. This line starts at a value of $3.5\sqrt{f'_c}$ at the end of the member, and intersects the v_{cw} line or $5\sqrt{f'_c}$ line at 50 strand diameters from the end of the member.

Example 4.3.2 Construction of applied and resisting design shear diagrams

Given:

2HC8 with span and loadings shown

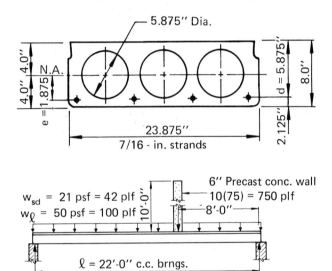

Section properties:

A = 110 sq in.
I = 843 in.⁴
y_b = 4.0 in.
b_w = 6.25 in.
wt = 57 psf = 114 plf

Concrete:

f'_c = 5000 psi

Normal weight

Solution:

1. Determine factored loads

Uniform dead = 1.4 (42 + 114)
= 218 plf

Uniform live = 1.7 (100)
= 170 plf

Concentrated dead = 1.4 (2 × 750)
= 2100 lb

2. Construct shear diagram as shown in Fig. 4.3.2
3. Construct the shear resistance diagram as described in previous section.

   a. Construct line at $2\sqrt{f'_c} = 141$ psi

   b. Construct v_c line by Eq. 4.3.4

$$v_c = 0.6\sqrt{f'_c} + 700\frac{V_u d}{M_u}$$

$$= 42.4 + 4112.5\frac{V_u}{M_u}$$

Fig. 4.3.2 Diagrams for Example 4.3.2

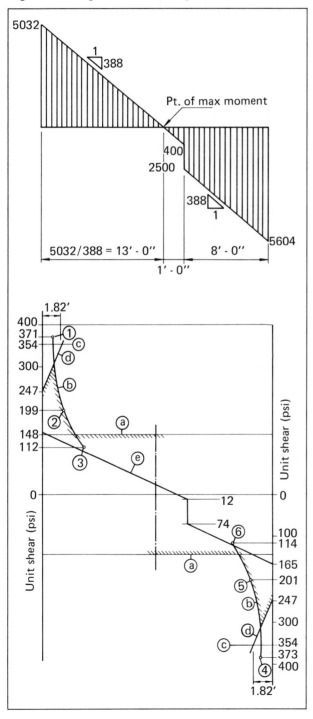

At 1, 2, and 4 ft from each end:

V_u (left) $= 5032 - 388x$

M_u (left) $= \left(5032x - \dfrac{388x^2}{2}\right)12$

V_u (right) $= 5604 - 388x$

M_u (right) $= \left(5604x - \dfrac{388x^2}{2}\right)12$

Point	x	V_u	M_u	$4112.5\dfrac{V_u}{M_u}$	v_c
1	1	4644	58056	329.0	371.4
2	2	4256	111456	157.0	199.4
3	4	3480	204288	70.1	112.5
4	1	5216	64920	330.4	372.8
5	2	4828	125184	158.6	201.0
6	4	4052	231744	71.9	114.3

   c. Construct upper limit line at $5\sqrt{f'_c}$ = 354 psi

   d. Construct diagonal line at transfer zone from $3.5\sqrt{f'_c} = 247$ psi at end of member to 354 psi at $50d_b = 50(7/16) = 21.9$ in. $= 1.82$ ft

   e. Construct v_u diagram by dividing V_u points by $\phi b_w d = 0.85(6.25)(6.4) = 34$
(Note: The Code permits the use of $0.8h$ in lieu of d in this equation.)

   5032/34 = 148

   400/34 = 12

   2500/34 = 74

   5604/34 = 165

It is apparent from construction of these diagrams that no shear reinforcement is required.

4.3.3 Design Using Design Aids

   Figs. 4.10.5 through 4.10.10 are design aids to assist in determing the shear strength of precast, prestressed members.

Lightweight concrete:

   Figs. 4.10.5, 4.10.9, and 4.10.10 employ a coefficient, $\lambda = \dfrac{f_{ct}/6.7}{\sqrt{f'_c}}$ for use with lightweight concrete. For normal weight concrete, λ is equal to 1.0. If the value of f_{ct} is not known, $\lambda = 0.85$ for sand-lightweight concrete, and 0.75 for all-lightweight concrete. Figs. 4.10.6 to 4.10.8 provide separate charts for normal weight and lightweight concrete. In these charts, it is assumed that f_{ct} is not known and the material is sand-lightweight.

Fig. 4.10.5 is useful for finding the points on the v_c curve, when the procedure illustrated in Example 4.3.2 is used.

Example 4.3.3 Use of Fig. 4.10.5—Concrete shear strength by Eq. 11-10 (ACI 318-83)

Given:

PCI standard single tee 10LST48

Simple span, uniformly loaded

2-1/2 in. topping

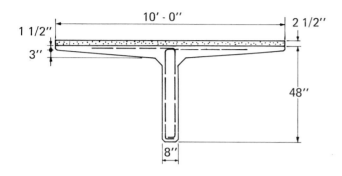

Concrete:

Precast: f'_c = 5000 psi
Sand-lightweight
f_{ct} = 470 psi

Topping: f'_c = 4000 psi
Normal weight

Span = 80 ft

Problem:

(Determine v_c at x = 14 ft from the support face.)

Solution:

x/ℓ = 14/80 = 0.175

d (at that section) = 38 in.

ℓ/d = 80 × 12/38 = 25

To use Fig. 4.10.5 determine the parameters:

$$\lambda = (f_{ct}/6.7)/\sqrt{f'_c}$$
$$= (470/6.7)/\sqrt{5000} = 0.99$$
$$\lambda^2 f'_c = 0.98 \times 5000 = 4900 \text{ psi}$$

Enter Fig. 4.10.5 as the dashed arrows show, at x/ℓ = 0.175 and proceed to the right to ℓ/d = 25, and then downward to $\lambda^2 f'_c$ = 4900. The value, v_c, can be read off the right margin as:

v_c = 170 psi

Figs 4.10.6 through 4.10.8 allow the graphical solution of Eq. 4.3.4 (Eq. 11-10 of ACI 318-83) for simple spans with uniform loads. They may also be used for other loadings with small error, provided the majority of the load is uniform. The charts

are shown for f'_c = 5000 psi. However, Eq. 4.3.4 is relatively insensitive to f'_c so they can be used for strengths of 4000 to 6000 psi with an error of less than 10%.

Example 4.3.4 Use of Figs. 4.10.6 through 4.10.8—Graphical solution of Eq. 11-10 (ACI 318-83)

Given:

PCI standard single tee 8LST36

2-1/2 in. topping

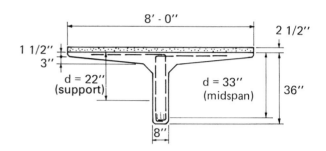

Concrete:

Precast: f'_c = 5000 psi
Sand-lightweight

Topping: f'_c = 4000 psi
Normal weight

Prestress:

13–1/2″ diameter 270K strands
A_{ps} = 13 (0.153) = 1.99 sq in.

Single depression at midspan

Span = 70 ft

Dead load, w_d = 723 lb per ft

Live load, w_ℓ = 600 lb per ft

Shear reinforcement f_y = 40,000 psi

Problem:

Find the maximum value of excess shear stress, $v_u - v_c$, along the span in compliance with Eq. 11-10 (ACI 318-83).

Solution (see Fig. 4.3.3):

To use Figs. 4.10.6 through 4.10.8, first determine the drape pattern.

The strands in this example are draped 33 − 22 = 11 in., which is equal to d/3. This is defined as a shallow drape and Fig. 4.10.7 applies.

The parameters needed for use of Fig. 4.10.7 are determined as follows:

$$\ell/d = 70 \times 12/(33 + 2.5) = 23.7$$

for determining v_u, d = 0.8h

$$= 0.8 (38.5) = 30.8 \text{ in.}$$

$$v_u \text{ at support} = \frac{(1.4w_d + 1.7w_\ell)(\ell/2)}{\phi b_w d}$$

$$= \frac{(1.4 \times 723 + 1.7 \times 600) \times 70/2}{0.85 \times 8 \times 30.8}$$
$$= 340 \text{ psi}$$

The graphical solution (Fig. 4.3.4) follows these steps:

    a. Draw a diagonal line from v_u = 340 psi at support to v_u = 0 at midspan

    b. Draw diagonal line from $v_c = 3.5\lambda\sqrt{f'_c}$ at support to $5\lambda\sqrt{f'_c}$ at 50 strand diameters to meet requirements of Sect. 11.4.3 of the Code.
       50 (0.5) = 25 in. = 0.03 ℓ

    c. Draw a curved line at ℓ/d = 23.7

    d. Draw a vertical line at d/2 from support

    e. Design shear reinforcement for shaded area (see example 4.3.8)

Fig. 4.10.9 is used to obtain the design shear strength at h/2 from the support by Code Eq. 11-13. It is conservative to assume that the excess shear ($v_u - v_c$) varies linearly from a maximum at the support to zero at midspan for a uniformly loaded member.

Example 4.3.5 Use of Fig. 4.10.9 Concrete shear strength at support by Eq. 11-13 (ACI 318-83)

Given:

    14 in. deep double tee as shown

    Span = 38 ft

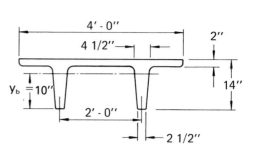

Concrete:

    Precast: f'_c = 5000 psi

             Normal weight

    A = 180 sq in.

Prestressing steel:

    6–7/16'' diameter 270 K strands

    Prestress force after losses, P = 108 kips

    Single depression at midspan

    e at support = 4.4 in. (d = 8.4 in.)

    e at midspan = 7.5 in.

Problem:

Determine the nominal shear stress, v_{cw}, at the support by Eq. 11-13 (ACI 318-83).

Solution:

(Shear stresses will be used for convenience.)

The parameters required to use Fig. 4.10.9 are determined as follows:

    d = 0.8h = 0.8 × 14 = 11.2 in.

Fig. 4.3.3 Solution of Example 4.3.4

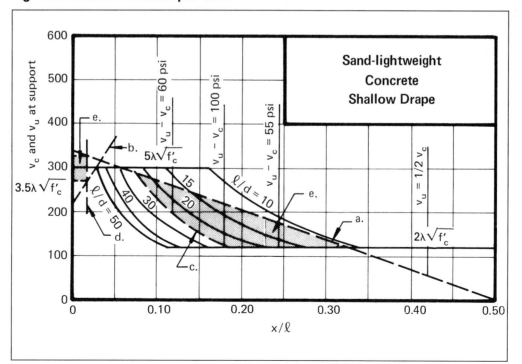

(Note: d in Eq. 11-13 is the distance from the extreme compression fiber to the centroid of the prestressing tendons or 0.8h, whichever is greater.)

$$e' = 7.5 - 4.4 = 3.1 \text{ in.}$$

$$b_w = 2(2.5 + 4.5)/2 = 7.0 \text{ in.}$$

(average width of stems)

$$\frac{e'A}{\ell b_w d} = \frac{3.1(180)}{38.0(12)(7.0)(11.2)} = 0.016$$

$$\text{P/A at h/2(7 in.)} = \frac{108(1000)}{180} \left(\frac{7}{7/16(50)} \right)$$

$$= 192 \text{ psi say } 200 \text{ psi}$$

Enter Fig. 4.10.9, as indicated by the arrow, at $e'A/\ell b_w d = 0.016$. Proceed horizontally to P/A = 200 and vertically to the $\lambda^2 f'_c = 5000$ line.

Then read at the right margin:

$$v_{cw} = 310 \text{ psi}$$

If $w_d = 72$ psf and $w_\ell = 50$ psf, then

$$V_u = [1.4(72) + 1.7(50)](4)(38/2) = 14{,}100 \text{ lb}$$

$$v_u = 14{,}100/(0.85)(7.0)(11.2) = 212 \text{ psi}$$

Since v_u is more than one-half of v_c (in this case v_{cw}) stirrups are required by Eq. 11-13.

Fig. 4.10.10 provides an alternate solution for v_{cw} in Eq. 11-13 (ACI 318-83) on the basis of computed principal tensile stress of $4\lambda\sqrt{f'_c}$ as provided in Sect. 11.4.2.2 (ACI 318-83).

Example 4.3.6 Use of Fig. 4.10.10—v_{cw} corresponding to principal tensile stress = $4\lambda\sqrt{f'_c}$

Given:

Same section as Example 4.3.5

Problem:

Is it advantageous to use the alternate value of V_{cw}, as stipulated in Section 11.4.2.2 (ACI 318-83), which results in a principal tensile stress of $4\lambda\sqrt{f'_c}$?

Solution:

(Shear stresses will be used for convenience.)

To find the shear stress, v_{cw}, which results in a principal tensile stress of $4\lambda\sqrt{f'_c} = 4 \times \sqrt{5000} = 283$ psi, compute the normal stress at the center of gravity of the section or at the junction of the web and the flange when the centroidal axis is in the flange.

In this example, the center of gravity is below the intersection of the web and the flange, hence the normal stress amounts to:

$$f_c = P/A = 200 \text{ psi as above}$$

'(Note that the general form of the equation for normal stress is $f_c = P/A \pm Pe/Z \pm M/Z$. At the center of gravity, the last two terms of the equation drop out since c = 0 in the calculation $Z = I_g/c$). Enter Fig. 4.10.10 with the above value of $4\lambda\sqrt{f'_c} = 283$ psi and proceed horizontally to the curve of $f_c = 200$ psi. Proceeding vertically down from the intersection, yields a value of:

$$v_{cw} = 370 \text{ psi}$$

This value is larger than the 310 psi obtained by Eq. 11-13. It is therefore, advantageous to use $v_{cw} = 370$ psi in $v_u - v_{cw}$ when computing the required shear reinforcement.

4.3.4 Shear Reinforcement

Shear reinforcement is required in all concrete members, except as noted in Section 11.5.5 (ACI 318-83). The minimum area required by the ACI Code is determined using Eq. 11-14:

$$A_v = 50 \ b_w s/f_y \qquad \text{(Eq. 4.3.8)}$$

or, alternatively for prestressed members only, using Eq. 11-15:

$$A_v = \frac{A_{ps}}{80} \frac{f_{pu}}{f_y} \frac{s}{d} \sqrt{\frac{d}{b_w}} \qquad \text{(Eq. 4.3.9)}$$

Fig. 4.10.11 is a graphical solution for the minimum shear reinforcement by Eq. 4.3.9.

Example 4.3.7 Use of Fig. 4.10.11 — Minimum shear reinforcement by Eq. 11-15 (ACI 318-83)

Given:

Single tee of Example 4.3.4

Stirrups

#3, 2-leg ($f_y = 40$ ksi)

A_v per stirrup = 2(0.11) = 0.22 sq in.

Problem:

Determine the minimum amount of shear reinforcement required by Eq. 11-15 of the Code.

Solution:

Determine $b_w d = 8(0.8 \times 38.5) = 246$ sq in.

Enter Fig. 4.10.11 at left with $A_{ps} = 1.99$ sq in., proceed right to the line f_y grade 40 and $f_{pu} = 270$ ksi, then up to $b_w d = 246$ and right to read $A_v = 0.13$ sq. in. per ft.

$$\text{Required stirrup spacing} = \frac{12(0.22)}{0.13}$$

$$= 20.3 \text{ in., say 20 in.}$$

Maximum spacing (Sect. 11.5.4 of the Code) = 0.75h or 24 in.

$$0.75(38.5) = 29 \text{ in.}$$

Shear reinforcement requirements are defined in ACI 318-83 by Eq. 11-17. This equation can be rewritten in terms of shear stresses as:

$$A_v = \frac{(v_u - v_c)b_w s}{f_y} \qquad \text{(Eq.4.3.10)}$$

Fig. 4.10.12 is used to design shear reinforcement by Eq. 4.3.10 for a given excess shear. Stirrup size, strength or spacing can be varied. Welded wire fabric may also be used for shear reinforcement in accordance with Sects. 11.5.1 and 12.13.2 (ACI 318-83).

Example 4.3.8 Use of Fig. 4.10.12—Shear reinforcement

Given:

Single tee of Examples 4.3.4 and 4.3.7.

Problem:

Determine the required spacing for #3, two-leg stirrups ($f_y = 40$ ksi).

Solution:

From Fig. 4.10.12, the minimum stirrup spacing of 20 in. will resist $(v_u - v_c)b_w$ of 440 lb/in., or

$$v_u - v_c = 440/8 = 55 \text{ psi}$$

By scaling from the plot in Fig. 4.3.3, 55 psi is at about $0.25\ell = 17.5$ ft from the end. The maximum value of $v_u - v_c$ can be scaled at about 100 psi. Therefore, $(v_u - v_c) b_w = 800$. From Fig. 4.10.12, this requires a spacing of 11 in. Also note that from about 0.11ℓ (about 8 ft) to the support, minimum spacing is adequate. From about 0.42ℓ to midspan, $v_u < 1/2 \, v_c$ and no stirrups are required.

4.3.5 Horizontal Shear Transfer in Composite Members

In order for a precast, prestressed member with topping to behave compositely, full transfer of the horizontal shear forces must be assured at the interface of the precast member and the cast-in-place topping. The procedure recommended in this section is based on Sect. 17.5.3 of ACI 318-83. An alternate method is given in Sect. 17.5.2.

The horizontal shear force, F_h, which must be resisted is the total force in the topping; compression in positive moment regions, and tension in negative moment regions, as shown in Fig. 4.3.4.

In a composite member which has an interface surface that is intentionally roughened, but does not have horizontal shear ties, F_h should not exceed $80\phi b_v \ell_{vh}$, where b_v is the width of the interface surface and ℓ_{vh} is the horizontal shear length as defined in Fig. 4.3.5.

The area of horizontal shear ties required in length ℓ_{vh} (see Fig. 4.3.5) may be calculated by:

$$A_{cs} = \frac{F_h}{\phi \, \mu_e \, f_y} \qquad \text{(Eq. 4.3.11)}$$

where:

A_{cs} = area of horizontal shear ties, sq in.

F_h = horizontal shear force, lb

f_y = yield strength of horizontal shear ties, psi

μ_e = effective shear-friction coefficient as defined in Sect. 6.7

$$= \frac{1000\lambda A_{cr}\mu}{V_u}$$

$\phi = 0.85$

For composite members, $\mu = 1.0\lambda$ and $A_{cr} = b_v \ell_{vh}$, thus:

$$\mu_e = \frac{1000\lambda^2 b_v \ell_{vh}}{F_h} \le 2.9 \qquad \text{(Eq.4.3.12)}$$

(See Table 6.7.1)

The value of F_h is limited to:

$$F_h \,(\text{max}) = 0.25 f'_c b_v \ell_{vh} \le 1000 b_v \ell_{vh} \qquad \text{(Eq.4.3.13)}$$

where f'_c is the lesser compressive strength of the precast member or the composite topping.

Sect. 17.6.1 of ACI 318-83 also requires that ties, when required, be spaced no more than four times the least dimension of the supported element, nor 24 in., and meet the minimum shear reinforcement requirements of Sect. 11.5.5.3.

$$A_{cs}(\text{min}) = \frac{50 b_v \ell_{vh}}{f_y} \qquad \text{(Eq.4.3.14)}$$

Example 4.3.9 Horizontal shear design for composite beam

Given:

Inverted tee beam with 2 in. composite topping (See Example 4.2.6)
Beam length = 20'-0"

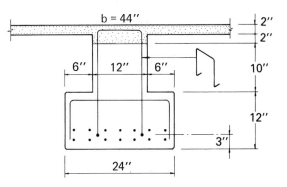

Fig. 4.3.4 Horizontal shear in composite section

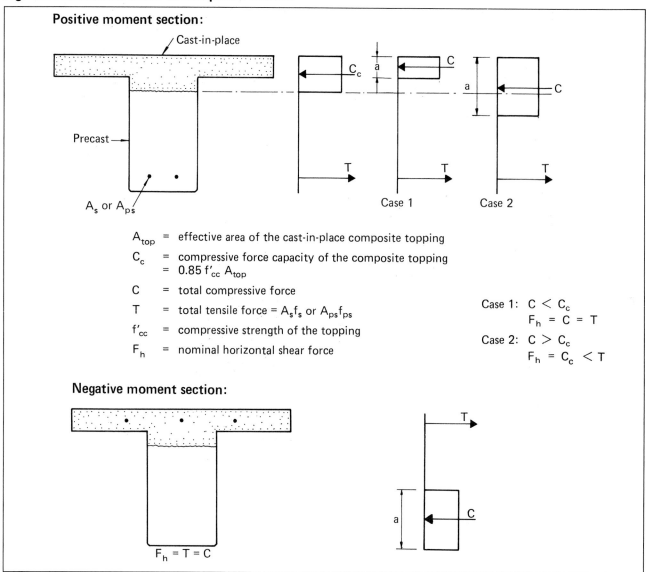

Positive moment section:

A_{top} = effective area of the cast-in-place composite topping

C_c = compressive force capacity of the composite topping
= $0.85 f'_{cc} A_{top}$

C = total compressive force

T = total tensile force = $A_s f_s$ or $A_{ps} f_{ps}$

f'_{cc} = compressive strength of the topping

F_h = nominal horizontal shear force

Case 1: $C < C_c$
$F_h = C = T$

Case 2: $C > C_c$
$F_h = C_c < T$

Negative moment section:

$F_h = T = C$

Fig. 4.3.5 Horizontal shear length

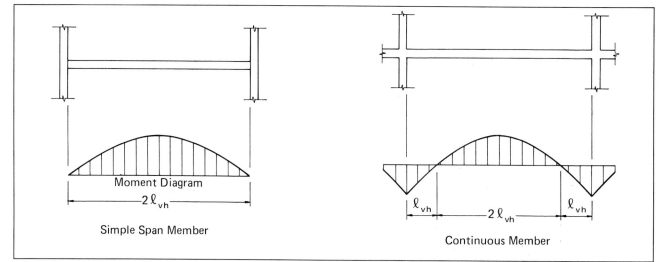

Moment Diagram

$2\ell_{vh}$

Simple Span Member

ℓ_{vh} $2\ell_{vh}$ ℓ_{vh}

Continuous Member

Concrete:
$$f'_c \text{ (precast)} = 5000 \text{ psi}$$
$$f'_{cc} \text{ (topping)} = 3000 \text{ psi}$$

Prestressing steel:
14–1/2" diameter 270K strands
$$A_{ps} = 14 \times 0.153 = 2.142 \text{ sq in.}$$
Tie steel: $f_y = 60,000$ psi

Problem:

Determine the tie requirements to transfer horizontal shear force.

Solution:
$$b_v = 12 \text{ in.}$$
$$\ell_{vh} = \frac{20(12)}{2} = 120 \text{ in.}$$
$$A_{top} = 2(44) + 2(12) = 112 \text{ sq in.}$$
$$C_c = 0.85 f'_{cc} A_{top} = 0.85(3)(112)$$
$$= 285.6 \text{ kips}$$
$$f_{ps} = 243 \text{ ksi (see Ex.4.2.6)}$$
$$A_{ps} f_{ps} = 2.142(243) = 520.5 \text{ kips} > 285.6$$

Therefore, $F_h = 285.6$ kips
$$80\phi b_v \ell_{vh} = 80(0.85)(12)(120)/1000$$
$$= 97.9 < 285.6$$

Therefore, ties are required.
$$\lambda = 1.0 \text{ (normal weight concrete)}$$
$$\mu_e = \frac{1000\lambda^2 b_v \ell_{vh}}{F_h} = \frac{1000(1.0)(12)(120)}{285,600}$$
$$= 5.04 > 2.9 \text{ use } 2.9$$
$$A_{cs} = \frac{F_h}{\phi\mu_e f_y} = \frac{285.6}{0.85(2.9)(60)}$$
$$= 1.93 \text{ sq in.}$$

Check minimum requirements:
$$A_{cs}(min) = \frac{50b_v \ell_{vh}}{f_y} = \frac{50(12)(120)}{60,000}$$
$$= 1.20 \text{ sq in.}$$

Use No. 3 ties, area per tie $= 2(0.11)$
$$= 0.22 \text{ sq in.}$$
Maximum tie spacing $= 4(4) = 16$ in. < 24 in.
$$s = \frac{120(0.22)}{1.93} = 13.7 \text{ in.}$$

Use No. 3 ties at 13 in. o.c.

4.4 Torsion

ACI 318-83 prescribes the torsion design method to be used for non-prestressed concrete members, but specifically excludes prestressed members. It has been common practice to use the Code method for prestressed members, with modifications proposed by Zia and McGee[5] which include the effects of prestress. This method is still applicable. The Collins-Mitchell "compression field theory"[6] illustrated for a prestressed member in Sect. 4.4.2 is also applicable to non-prestressed members, but is not recognized by the Code.

The critical section for shear and torsion by ACI 318-83 is "d" from the face of the support for non-prestressed members and "h/2" for prestressed members. However, by Sect.11.1.2 of the Code, no concentrated load shall occur between the face at the support and the critical section.

4.4.1 Combined Shear and Torsion Strength–Non-Prestressed Members

The torsion design procedures prescribed by ACI 318-83 are illustrated by the following example:

Example 4.4.1 Torsion design–non-prestressed member

Given:

Precast load bearing spandrel beam shown

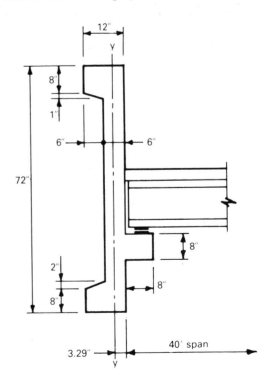

Span of spandrel beam = 30 ft clear
f'_c = 5000 psi, normal weight concrete
Reinforcement f_y = 60,000 psi
d = 69 in.

Loads (kips/ft):

D.L.:
Precast floor   60 psf (20ft) = 1.2 (1.4) = 1.68
Topping   25   (20) = 0.5 (1.4) = 0.70
Superimposed 10   (20) = 0.2 (1.4) = 0.28
Window   = 0.50 (1.4) = 0.07
Spandrel   = 0.63 (1.4) = 0.88

L.L.:   50 psf (20) = 1.00 (1.7) = 1.70
                                    w_u = 5.31

Problem:

Determine torsion reinforcement requirements

Solution:

1. Compute torsion moment (T_u) at critical section, assumed to be 5'-0" from face of support:
   V_u = w_u(15 − 5) = 5.31(10) = 53.1 kips
   w_u for torsion = 1.68 + 0.70 + 0.28 + 1.70
                   = 4.36 kips/ft
   Eccentricity = 2/3 (8) + 3.29 = 8.62 in.
   T_u = w_u (e) (ℓ/2 − 5) = 4.36 (8.62) (10)
                   = 376 in.-kips

2. Determine if torsion effects must be considered.
   If $T_u \geq \phi$ (0.5 $\sqrt{f'_c}$ $\Sigma x^2 y$) must consider torsion.

   $\Sigma x^2 y$ = $6^2(72)$ + $6^2(8)(2)$ + $8^2(8)$ = 3680 in.³

   $$\phi(0.5\sqrt{f'_c}\ \Sigma x^2 y) = \frac{0.85(0.5)\sqrt{5000}\ (3680)}{1000}$$
   $$= 110.6 \text{ in.-kips}$$

   376 > 110.6, consider torsion

3. Determine the torsion moment strength provided by concrete

   $$T_c = \frac{0.8\sqrt{f'_c}\ \Sigma x^2 y}{\sqrt{1 + \left(\frac{0.4\ V_u}{C_t T_u}\right)^2}}$$

   where:

   $$C_t = \frac{b_w d}{\Sigma x^2 y} = \frac{6(69)}{3680} = 0.1125$$

   $$T_c = \frac{0.8\sqrt{5000}\ (3680)}{\sqrt{1 + \frac{0.4\ (53.1)}{0.1125\ (376)}}} = 170 \text{ in.-kips}$$

4. Determine torsion reinforcement requirements:
   $$T_u = \phi T_n = \phi(T_c + T_s)$$
   or
   $$T_s = \frac{T_u}{\phi} - T_c = \frac{376}{0.85} - 170$$
   $$= 272 \text{ in.-kips}$$

By Sect. 11.6.9.4 (ACI 318-83)
$T_s \leq 4T_c$ = 4(170) = 680 OK

By Eq. 11-23 of ACI 318-83

$$T_s = \frac{A_t \alpha_t x_1 y_1 f_y}{s}$$

or

$$A_t = \frac{T_s(s)}{\alpha_t x_1 y_1 f_y}$$

Assume x_1 = 4 in.; y_1 = 70 in.
   α_t = 0.66 + 0.33(y_1/x_1) ≤ 1.5
        = 0.66 + 0.33(70/4) = 6.44, use α_t = 1.5
$$A_t = \frac{272(12)}{1.5(4)(70)(60)} = 0.13 \text{ sq in./ft}$$
$$= 0.011 \text{ sq in./in.}$$

This is the required area of steel in each leg of the closed stirrup for torsion only. The shear steel requirement must be added to A_t. The minimum area of closed stirrups is:

   $A_v + 2A_t$ = $50b_w s/f_y$

Placement of closed ties in a 6-in. web is difficult. Consider re-design with greater web thickness, or arrange reinforcement as follows:

5. Determine the longitudinal bars A_ℓ distributed around the closed stirrups by one of three expressions:

$$A_\ell = 2A_t \left(\frac{x_1 + y_1}{s}\right) = 2(0.011)\left(\frac{4+70}{1}\right)$$

$$= 1.63 \text{ sq in.}$$

or

$$A_\ell = \left[\frac{400xs}{f_y}\left(\frac{T_u}{T_u + \dfrac{V_u}{3C_t}}\right) - 2A_t\right]\left(\frac{x_1 + y_1}{s}\right)$$

but $2A_t$ (in this eq.) $\geq \dfrac{50b_w s}{f_y}$

$$= \frac{50(6)(1)}{60,000} = 0.005 < 2A_t$$

$$A_\ell = \left[\frac{400(6)(1)}{60,000}\left(\frac{376}{376 + \dfrac{53.1}{3(0.1125)}}\right)\right.$$

$$\left. - 2(0.011)\right]\left(\frac{4+70}{1}\right) = 0.46$$

Use $A_\ell = 1.63$ sq in. distributed around perimeter.

4.4.2 Torsion with Shear and Flexure—Prestressed Members

This section describes the procedure developed by Collins and Mitchell.[6] The approach utilizes the compression field theory in which it is assumed that, after cracking, the concrete element can carry no tension and that the shear and torsion are carried by a field of diagonal compression.

A design procedure is presented below as a step-by-step method; Ref. 6 is recommended for a comprehensive presentation of the subject.

Step 1: Determine M_u, V_u, T_u, M_s, V_s, and T_s (factored and unfactored load effects) along the length of the member.

Step 2: Determine the minimum reinforcement requirements. A minimum amount of reinforcement should be provided so that a reserve of strength will exist after any initial cracking occurs.

The minimum reinforcement requirements may be waived if the member is designed for factored loads one-third greater than those determined by analysis.

For members which are not subject to moving loads, minimum reinforcement requirements are satisfied if either:

a. At locations of maximum flexural moment: $M_n \geq 1.2 M_{cr}$, or

b. All the loads are multiplied by $1.2\, M_{cr}/M_n$.

For elements subjected to moving loads or having variable loading ratios, this requirement will be satisfied if:

$$M_n \geq 1.2M_{cr}, \quad V_n \geq 1.2V_{cr}, \quad T_n \geq 1.2T_{cr}$$

In the case of combined flexure, shear, and torsion, the cracking loads may be determined by:

$$\left(\frac{M_{cr}}{M_{ocr}}\right)^2 + \left(\frac{V_{cr}}{V_{ocr}}\right)^2 + \left(\frac{T_{cr}}{T_{ocr}}\right)^2 = 1 \quad \text{(Eq. 4.4.1)}$$

where the pure flexural, shear and torsional loads are:

$$M_{ocr} = Z_{ten}(7.5\lambda\sqrt{f_c'} + f_{pe}) \quad \text{(Eq. 4.4.2)}$$

$$V_{ocr} = b_w d(4\lambda\sqrt{f_c'})\sqrt{1 + f_{pc}/(4\lambda\sqrt{f_c'})} + V_p$$

$$\text{(Eq. 4.4.3)}$$

$$T_{ocr} = (A_c^2/P_c)(4\lambda\sqrt{f_c'})\sqrt{1 + f_{pc}/(4\lambda\sqrt{f_c'})}$$

$$\text{(Eq. 4.4.4)}$$

where:

A_c = area enclosed by outside perimeter of cross section
= gross area of concrete cross section plus area of enclosed voids

P_c = outside perimeter of cross section

The equation for T_{ocr} may be used for hollow sections provided the least wall thickness is not less than $0.75\, A_c/P_c$.

In calculating cracking loads, it may be assumed that:

$$\frac{T_{cr}}{V_{cr}} = \frac{T_u}{V_u} \quad \text{(Eq. 4.4.5a)}$$

and:

$$\frac{M_{cr}}{V_{cr}} = \frac{M_u}{V_u} \geq d \quad \text{(Eq. 4.4.5b)}$$

The influence of axial loads on the magnitude of the cracking loads can be accounted for by substituting:

$$f_{pc}' = f_{pc} \pm N/A \quad \text{(Eq. 4.4.6)}$$

and:

$$f_{pe}' = f_{pe} \pm N/A \quad \text{(Eq. 4.4.7)}$$

for f_{pc} and f_{pe}, respectively, in the cracking load expressions, where N is the axial load.

Fig. 4.4.1 Effective shear area

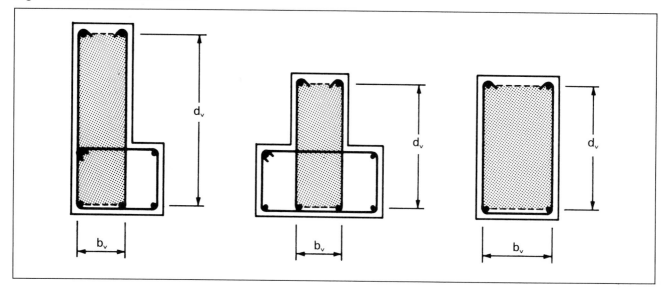

Fig. 4.4.2 Area enclosed by the shear flow

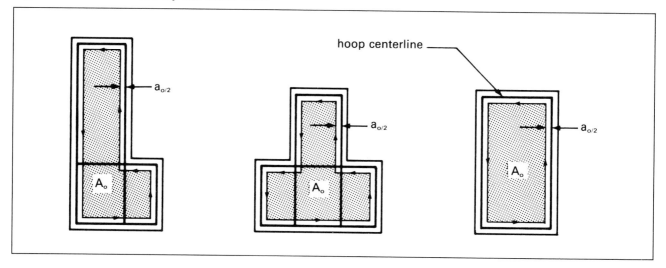

Step 3: Determine if there is a need for torsional reinforcement. Torsion may be neglected if:

$$T_u < \phi(0.25\,T_{ocr})$$

Step 4: Check diagonal crushing. Determine nominal shear stress:

$$\tau_n = \frac{(V_n - V_p)}{b_v d_v} + \frac{T_n p_h}{A_{oh}^2} \quad \text{(Eq. 4.4.8)}$$

where:

b_v = minimum effective web width within depth d_v assuming spalling of unrestrained concrete cover as shown in Fig. 4.4.1

d_v = effective shear depth equal to vertical distance between centers of longitudinal bars in corners of stirrups

A_{oh} = area enclosed by stirrup centerline

p_h = perimeter of stirrup centerline

PCI Design Handbook

To ensure that the transverse reinforcement will yield before diagonal crushing of the concrete occurs, calculate:

For $f_y = 40$ ksi

$$\theta \min = 10 + 99.71\ (\tau_n/f'_c) \qquad \text{(Eq. 4.4.9)}$$

$$\theta \max = 80 - 105.95\ (\tau_n/f'_c) \qquad \text{(Eq. 4.4.10)}$$

For $f_y = 60$ ksi

$$\theta \min = 10 + 83.75\ (\tau_n/f'_c) \qquad \text{(Eq. 4.4.11)}$$

$$\theta \max = 80 - 122.58\ (\tau_n/f'_c) \qquad \text{(Eq. 4.4.12)}$$

If $\theta \min > \theta \max$, then diagonal crushing of the concrete indicates that the cross section size is inadequate to carry the applied torsion. When this occurs, increase the size of the cross section and recalculate τ_n.

If $\theta \min \leq \theta \max$, select a value θ such that:

$$\theta \min \leq \theta \leq \theta \max$$

Selecting a lower value of θ will result in less stirrup reinforcement and more longitudinal reinforcement.

Step 5: Ensure adequate control of diagonal cracking at service loads.

This condition will be satisfied if the cracking load determined in Step 2 exceeds the service load, i.e., $M_{cr} \geq M_s$, $V_{cr} \geq V_s$, $T_{cr} \geq T_s$, where M_s, V_s, and T_s are at service loads.

If the cracking load does not exceed the service load, adequate control of diagonal cracking at service load will be provided if the following three conditions are met:

a. At service loads, the strain in transverse reinforcement is not excessive, which may be considered satisfied if either:

$f_y \leq 40$ ksi, or

$$\tan\theta \geq \left[\frac{f_y}{29}\frac{V_s}{V_n}\right]^2 \left[1 - \frac{f_y}{29}\frac{f_{pc}}{f'_c}\right]\left[1 - \left(\frac{V_{cr}}{V_s}\right)^3\right]^2$$

where f_y is in ksi.

b. Spacing of transverse reinforcement ≤ 12 in.

c. Spacing of longitudinal non-prestressed or prestressed reinforcement at the cracked faces of the member ≤ 12 in.

Step 6: Design transverse reinforcement. At regions near supports, transverse reinforcement may be calculated using V_u and T_u which occur at a distance $d_v/(2\tan\theta)$

from the support, providing τ_n was calculated with Eq. 4.4.8 using values of V_n and T_n at the face of the support.

Calculate the equivalent depth of compression in torsion:

$$a_o = \frac{A_{oh}}{p_h}\left[1 - \sqrt{1 - \frac{T_n p_h}{(0.85)(f'_c)(A_{oh}{}^2)}\left(\tan\theta + \frac{1}{\tan\theta}\right)}\right]$$

(Eq. 4.4.14)

Calculate area enclosed by shear flow path, as illustrated in Fig. 4.4.2:

$$A_o = A_{oh} - a_o p_h/2 \qquad \text{(Eq. 4.4.15)}$$

Calculate transverse reinforcement for torsion:

$$A_t/s = T_n \tan\theta/(2A_o f_y) \qquad \text{(Eq. 4.4.16)}$$

Calculate transverse reinforcement for shear:

$$A_v/s = (V_n - V_p)\tan\theta/(f_y d_v) \qquad \text{(Eq. 4.4.17)}$$

Calculate the total required transverse reinforcement per unit length:

$$A_{tt}/s = A_v/s + 2A_t/s \qquad \text{(Eq. 4.4.18)}$$

Step 7: Design longitudinal reinforcement. The longitudinal reinforcement is designed to resist the applied moment, M_u, axial load, N_u, and an axial tension, $\triangle N_u$ caused by the shear and torsion.

$$\triangle N_u = \frac{1}{\tan\theta}\sqrt{(V_u - \phi V_p)^2 + \left(\frac{T_u p_o}{2A_o}\right)^2}$$

(Eq. 4.4.19)

where $p_o = p_h - 4a_o$

Thus, the longitudinal reinforcement requirements will be satisfied if the member has a positive flexural resistance of:

$$M_n^+ \geq (M_u/\phi_f) + (d_v \triangle N_u/2\phi_v) \qquad \text{(Eq. 4.4.20)}$$

and a negative flexural resistance of:

$$M_n^- \geq (d_v \triangle N_u/2\phi_v) - (M_u/\phi_f) \qquad \text{(Eq. 4.4.21)}$$

where:

$\phi_v = 0.85$

$\phi_f = 0.9$

However, if a section under consideration is closer than $d_v/\tan\theta$ from the inner edge of the bearing area, the negative flexural resistance need not exceed:

$$M_n^- \le \frac{d_v T_u p_o}{2 \tan \theta \phi_v 2A_o} - M_u/\phi_f \qquad \text{(Eq. 4.4.22)}$$

The negative flexural resistance at interior continuous supports need not exceed:

$$M_n^- \le \frac{d_v T_u p_o}{2 \tan \theta \phi_v 2A_o} + M_u/\phi_f \qquad \text{(Eq. 4.4.23)}$$

In order that crushing of the web does not occur at the bearing points, the cross sectional area and bearing length should be such that the effective shear depth, d_{ve}, at the end of the member is:

$$d_{ve} \ge \frac{V_n/b_v}{0.012 \ \theta' \ f_c' - T_n p_h/A_{oh}^2} \qquad \text{(Eq. 4.4.24)}$$

where:

$\theta' = \theta - 10$ if $\theta \le 45°$
$\theta' = 80 - \theta$ if $\theta \ge 45°$
$d_{ve} = \ell_b/\tan\theta$

If d_{ve} is less than shown in Eq. 4.4.24, then confinement reinforcement needs to be at the end of the member.

Step 8: After all the transverse and longitudinal reinforcement have been determined, it is necessary to make sure this reinforcement is adequately anchored. The following guidelines are suggested:

a. Transverse reinforcement should be anchored as required by Sect. 12.13 of ACI 318-83.

b. Transverse reinforcement should have a longitudinal reinforcing bar or prestressing strand at each interior corner. The nominal diameter of this bar or strand should be equal to the diameter of the transverse reinforcement when shear controls and should be not less than $s(\tan \theta/16)$ when torsion reinforcement is required. In either case, the minimum diameter should be 1/2 in.

c. The spacing of the transverse reinforcement should not exceed $d_v/(3 \tan\theta)$ for shear and $p_h/(8 \tan\theta)$ for torsion.

d. Except at supports of simple spans and at free ends of cantilevers, the longitudinal reinforcement must extend beyond the point at which it is no longer required for a distance of 12 bar diameters. The additional requirements of ACI 318-83 to extend this reinforcement a distance equal to the effective depth of the member may be waived.

Example 4.4.2 Torsion of a prestressed concrete member

Given:

Typical precast, prestressed concrete spandrel beam shown in Fig. 4.4.3.

Dead load of deck = 89.5 psf
Live load = 50 psf

Beam properties:

A = 696 sq in.
wt = 725 plf
I = 364,520 in.⁴
y_b = 33.2 in.
y_t = 41.8 in.
Z_b = 10,990 in.³
Z_t = 8,720 in.³
f_c' = 5000 psi normal weight concrete
f_y = 60 ksi

Prestressing:

6 – 1/2 in. dia. 270K stress-relieved strand
A = 6(0.153) = 0.918 sq in.
f_{se} = 150 ksi
d_p = 69 in.
e = 69 – 41.8 = 27.2 in.
C_c = 1-1/4 in.

Problem:

Determine shear and torsion reinforcement required for the spandrel beam.

Solution:

1. Calculate V_s, M_s, T_s, V_u, M_u, T_u

   a. Determine loads on beam

   Service loads:

   Beam = 0.725 kips/ft
   Deck = 0.0895(60/2)(4) = 10.74 kips/stem
   Live load = 0.050(60/2)(4) = 6.0 kips/stem

   Factored loads:

   Beam = 1.4(0.725) = 1.02 kips/ft
   Deck = 1.4(10.74) = 15.04 kips/stem
   Live = 1.7(6.0) = 10.20 kips/stem

   b. Values of V_s, M_s, T_s, V_u, M_u, T_u along the length of the beam are shown in Fig. 4.4.4. Service loads are in parentheses.

2. Determine minimum reinforcement requirements. (Note: Only enough calculations to illustrate the design concepts will be presented here.)

Calculate cracking moment at midspan. Prestress force is:

$P = A_{ps}f_{se} = 0.918(150) = 138$ kips
$f_{pc} = P/A = 138/696 = 0.198$ ksi
$f_{pe} = P/A + Pe/Z_b = 0.198 + 138(27.2)/10,990$
   $= 0.540$ ksi

Fig. 4.4.3 Structure of Example 4.4.2

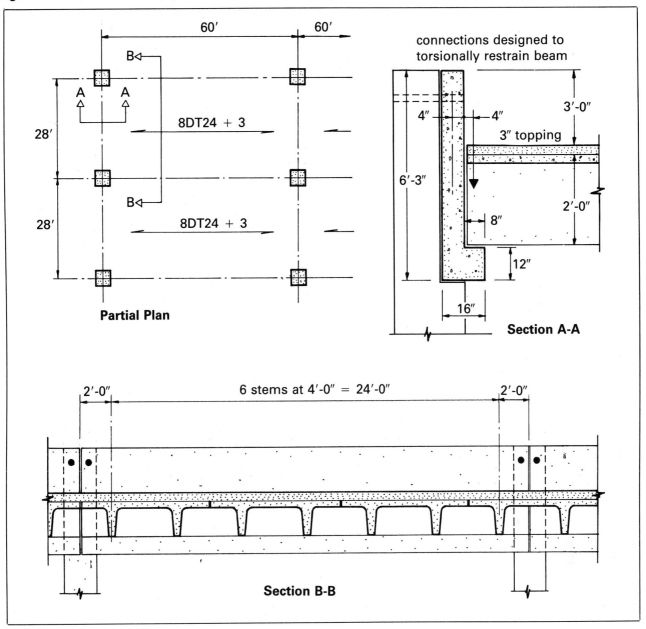

M_{ocr} = Z_b $(7.5\sqrt{f'_c} + f_{pe})$
= $(10,990/12,000)[7.5(\sqrt{5000}) + 540]$
= 980 ft-kips at midspan

Calculate M_n. Determine f_{ps} from Fig. 4.10.3.

$C\omega_{pu}$ = $CA_{ps}f_{pu}/bd_pf'_c$
= $1.06(0.918)(270)/[(8)(69)(5)]$
= 0.095

f_{ps} = $0.97(270)$ = 261.9 ksi

a = $A_{ps}f_{ps}/(0.85f'_cb)$
= $0.918(261.9)/[0.85(5)(8)]$ = 7.07 in.

M_n = $A_{ps}f_{ps} (d_p - a/2)$ = 0.918(261.9)
(69 − 7.07/2)/12 = 1312 ft-kips
> $1.2M_{cr}$ = 1.2(980) = 1176 ft-kips

Therefore minimum reinforcement requirements are satisfied.

3. Determine if torsional reinforcement is required:

By Eq. 4.4.4:

P_c = $75 + 8 + 63 + 8 + 12 + 16$ = 182 in.

T_{ocr} = $(696^2/182)(4\sqrt{5000})$
$\times \sqrt{1 + 198/(4\sqrt{5000})}/12,000$
= 81.8 ft-kips

Max T_u = 58.9 ft-kips
> $0.25T_{ocr}$ = 0.25(81.8) = 20.45 ft-kips

Therefore torsion reinforcement is required.

Fig. 4.4.4 Shears, moments and torsion on beam of Example 4.4.2

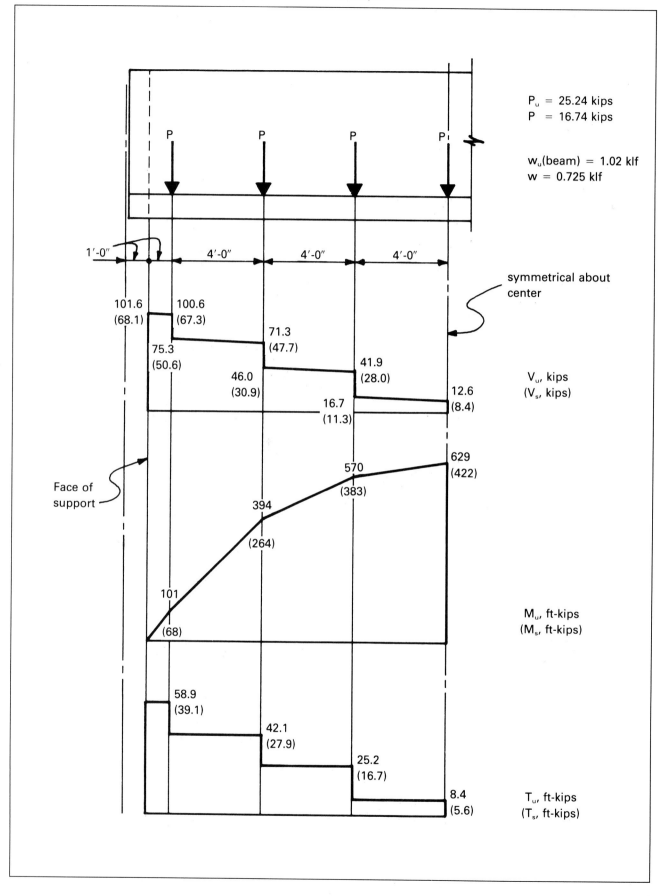

4. Check cross-section size and choose angle θ:

It is necessary to determine if the member has adequate cross-section to prevent crushing of the web concrete at the face of the support. Assume that the load applied by the double tee stem nearest the column will be transferred to the column, such that no shear stresses are introduced into the spandrel web. The minimum values of V_n and T_n are:

$$V_n = V_u/\phi = 75.3/0.85 = 88.6 \text{ kips}$$
$$T_n = T_u/\phi = 42.1/0.85 = 49.5 \text{ ft-kips}$$
$$= 594 \text{ in.-kips}$$

Assume $b_v = 8 - [2(1.5)] = 5 \text{ in.}$

and $d_v = 75 - [2(2)] = 71 \text{ in.}$

$A_{oh} = 5(71) + 8(9) = 427 \text{ sq in.}$

$p_h = 5+71+13+9+8+63 = 169 \text{ in.}$

By Eq. 4.4.8:
$$\tau_n = \frac{88.6}{5(71)} + \frac{594(169)}{427^2} = 0.800 \text{ ksi}$$

$\tau_n/f_c' = 0.800/5 = 0.160$

By Eq. 4.4.11: θ min $= 10 + 83.75(0.160)$
$$= 23.4°$$

By Eq. 4.4.12: θ max $= 80 - 122.58(0.160)$
$$= 60.4°$$

Since θ min $< \theta$ max, diagonal crushing will not occur and the section size is adequate.

Choose $\theta = 35°$

5. Check diagonal crack control requirements: Determine V_{cr}:

By Eq. 4.4.3:
$$V_{ocr} =$$
$$8(69)(4\sqrt{5000}) \sqrt{1 + 198/(4\sqrt{5000})} + 0$$
$$= 203,570 \text{ lb} = 203.6 \text{ kips}$$

By Eq. 4.4.5, at face of support:
$$T_{cr} = (T_u/V_u)V_{cr} = (58.9/101.6)V_{cr} = 0.58V_{cr}$$
$$M_{cr} = (M_u/V_u)V_{cr} = 0$$

By Eq. 4.4.1:
$$(V_{cr}/203.6)^2 + (0.58V_{cr}/81.8)^2 = 1$$
$$V_{cr} = 116.0 \text{ kips} > V_u = 101.6 \text{ kips}$$

Therefore, diagonal cracking requirements are satisfied.

6. Design transverse reinforcement. Consider the region near the support. Calculate V_u and T_u at a distance $d_v/2\tan\theta$ from the face of the support.

$d_v/2\tan\theta = 71/2(\tan 35) = 50.7 \text{ in.}$

From Fig. 4.4.4:
$$V_u = 75.3 - [(50.7 - 12)/48](75.3 - 71.3)$$
$$= 72.1 \text{ kips}$$
$$V_n = V_u/\phi = 72.1/0.85 = 84.8 \text{ kips}$$

$$T_u = 42.1(12) = 505 \text{ in.-kips}$$
$$T_n = T_u/\phi = 505/0.85 = 594 \text{ in.-kips}$$

Calculate torsional depth of compression. By Eq. 4.4.14:

$$a_o = \frac{427}{169}\left[1 - \sqrt{1 - \frac{594(169)}{0.85(5)(427)^2}\left(\tan 35 + \frac{1}{\tan 35}\right)}\right]$$
$$= 0.38 \text{ in.}$$

By Eq. 4.4.15: $A_o = 427 - [(0.38)(169)/2]$
$$= 395 \text{ sq in.}$$

By Eq. 4.4.16:
$$A_t/s = 594(\tan 35)/[2(395)(60)]$$
$$= 0.0088 \text{ sq in./in.}$$

By Eq 4.4.17:
$$A_v/s = (84.8 - 0)\tan 35/[60(71)]$$
$$= 0.0139 \text{ sq in./in.}$$

By Eq 4.4.18:
$$A_{tt}/s = 0.0139 + 2(0.0088) = 0.031 \text{ sq in./in.}$$

Use No. 4 closed stirrups at 12 in. centers.
$$A_{tt} = 0.40/12 = 0.033 \text{ sq in./in.}$$

7. Design longitudinal reinforcement.
   (a) Consider section at midspan:

By Eq. 4.4.19:
$$p_o = p_h - 4a_o = 169 - 4(0.38) = 167.5 \text{ in.}$$

From Fig. 4.4.4:
$$T_u = 8.4(12) = 100.8 \text{ in.-kips}$$
$$V_u = 12.6 \text{ kips}$$

$$\Delta N_u = \frac{1}{\tan 35} \sqrt{(12.6 - 0)^2 + \left[\frac{100.8(167.5)}{2(395)}\right]^2}$$
$$= 35.4 \text{ kips}$$

By Eq. 4.4.20:
Min. $M_n^+ = 629/0.9 + (71/12)(35.4)/[2(0.85)]$
$$= 822 \text{ ft-kips}$$

M_n supplied by strand
$$= 1312 \text{ ft-kips (see step 2) OK}$$

(Note: if required, the additional capacity supplied by the longitudinal reinforcing bars could also be counted.)

   (b) Consider section at face of support.
      $M_u = 0$

By Eq. 4.4.19:

From Fig. 4.4.4:
$$T_u = 58.9(12) = 706.8 \text{ in-kips}$$
$$V_u = 101.6 \text{ kips}$$

$$\Delta N_u = \frac{1}{\tan 35} \sqrt{(101.6)^2 + \left[\frac{(706.8)(167.5)}{2(395)}\right]^2}$$

Fig. 4.4.5 Required moment capacities along length of spandrel beam

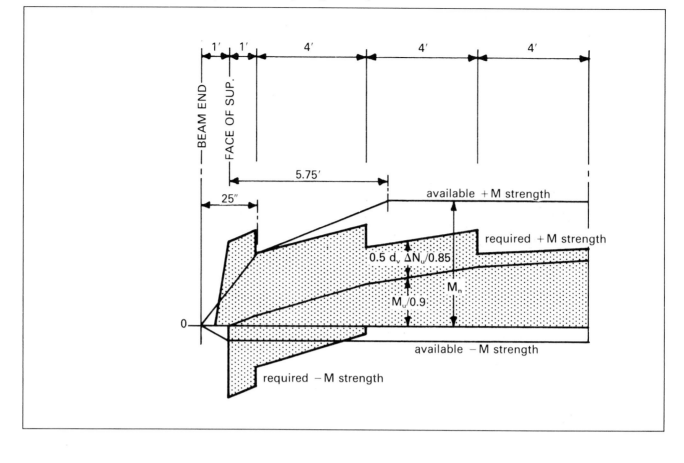

Fig. 4.4.6 Table of values used in Example 4.4.2.

Distance from end of beam, ft	1	2	2	6	6	10	10	14	units
M_u/ϕ	0	112	112	438	438	633	633	699	ft-kips
V_u	102	101	75	71	46	42	17	13	kips
T_u	59	59	42	42	25	25	8	8	ft-kips
ΔN_u (Eq.4.4.19)	259	256	186	183	112	109	39	35	kips
$\dfrac{0.5\, d_v \Delta N_u}{0.85}$	901	892	647	636	390	378	134	123	ft-kips
req'd $+M$ (Eq. 4.4.20)	901	1004	759	1073	828	1012	767	822	ft-kips
req'd $-M$ (Eq. 4.4.21)	901	780	535	198	—	—	—	—	ft-kips
or, to $d_v/\tan\theta = 101.4$ in. (Eq. 4.4.22)	745	626	415	90	—	—	—	—	ft-kips
$-M_n$ (top corner bars)	146 (fully developed @ F.S.)								ft-kips
$+M_n$ (prestressed strand)		751 @ 50 d_b = 25 in.			1312 @ ℓ_d = 81 in.				ft-kips

= 259.0 kips

Use lesser of Eqs. 4.4.21 or 4.4.22:

$$\text{Min. } M_n^- = \frac{71}{12}\left(\frac{259}{2(0.85)}\right) = 901 \text{ ft-kips}$$

$$\text{Min. } M_n^- = \frac{71}{2\tan 35}\left(\frac{706.8(167.5)}{2(395)(0.85)}\right)$$

$$= 8939 \text{ in.-kips} = 745 \text{ ft-kips}$$

Moment capacity supplied by the 2-No. 4 top corner bars:

Since spandrel beams such as this are usually cast with the long dimension horizontal, the bars do not need to be considered "top bars", and will be fully developed at face of support.

d = 73 in.; b = 16 in.

a = $A_s f_y / 0.85 b d f_c'$
    = 0.40(60)/(0.85)(16)(73)(5) = 0.0048 in.

M_n = $A_s f_y (d - a/2)$ = 0.40(60)(73 − 0.0024)
     = 1752 in.-kips = 146 ft-kips

Provide additional reinforcement, properly anchored, to resist 745 − 146 = 599 ft-kips at the support. Also provide bars at other locations required as indicated in Fig. 4.4.5.

8. Check reinforcement details.

Minimum corner bar diameter:
   $(s)\tan\theta/16$ = $(12)(\tan 35)/16$
   = 0.525 in., use No. 5 bars

Maximum transverse reinforcement spacing
   = $d_s/3\tan\theta$ = 71/(3 $\tan 35$) = 33.8 in. or
   = $p_h/8\tan\theta$ = 169/(8 $\tan 35$) = 30.2 in. > 12 in.

OK

In addition to shear and torsion reinforcement, design ledge and hanger reinforcement as described in Sect. 6.14.

4.5 Loss of Prestress

Loss of prestress is the reduction of tensile stress in prestressing tendons due to shortening of the concrete around the tendons, relaxation of stress within the tendons and external factors which reduce the total initial force before it is applied to the concrete. ACI 318-83 identifies the following sources of loss of prestress:

(a) Anchorage seating loss

(b) Elastic shortening of concrete

(c) Creep of concrete

(d) Shrinkage of concrete

(e) Relaxation of tendon stress

(f) Friction loss due to intended or unintended curvature in post-tensioning tendons.

Accurate determination of losses is more important in some prestressed concrete members than in others. Losses have no effect on the ultimate strength of a flexural member unless the tendons are unbonded or if the final stress after losses is less than 0.50 f_{pu}. Underestimation or overestimation of losses can affect service conditions such as camber, deflection and cracking.

4.5.1 Sources of Stress Loss

Anchorage seating loss and friction

These two sources of loss are mechanical. They represent the difference between the tension applied to the tendon by the jacking unit and the initial tension available for application to the concrete by the tendon. Their magnitude can be determined with reasonable accuracy and, in many cases, they are fully or partially compensated for by overjacking.

Elastic shortening of concrete

The concrete around the tendons shortens as the prestressing force is applied to it. Those tendons which are already bonded to the concrete shorten with it.

Shrinkage of concrete

Loss of stress in the tendon due to shrinkage of the concrete surrounding it is proportional to that part of the shrinkage that takes place after the transfer of prestress force to the concrete.

Creep of concrete and relaxation of tendons

Losses due to creep of concrete and relaxation of tendons complicate stress loss calculations. The rate of loss due to each of these factors changes when the stress level changes and the stress level is changing constantly throughout the life of the structure. Therefore, the rates of loss due to creep and relaxation are constantly changing.

4.5.2 Range of Values for Total Losses

Total loss of prestress in typical members will range from about 25,000 to 50,000 psi for normal weight concrete members, and from about 30,000 to 55,000 psi for sand-lightweight members.

Maximum and minimum loss

The total amount of prestress loss due to elastic shortening, creep, shrinkage, and relaxation need not be more than the values given below if the tendon stress immediately after anchoring does not exceed 0.70 f_{pu} for stress-relieved and 0.75 f_{pu} for low-relaxation strand:

Type of strand	Max. Prestress Loss, psi	
	Normal Weight Concrete	Sand-Light-weight Concrete
Stress-relieved strand	50,000	55,000
Low-relaxation strand	40,000	45,000

The load tables in Part 2 have a lower limit on loss of 30,000 psi. When low-relaxation strand is used, actual losses may be less.

4.5.3 Estimating Prestress Losses

This section is based on the report of a task group sponsored by ACI-ASCE Committee 423, Prestressed Concrete[7]. That report gives simple equations for estimating losses of prestress which would enable the designer to estimate the various types of prestress loss rather than using a lump sum value. It is believed that these equations, intended for practical design applications, would provide fairly realistic values for normal design conditions. For unusual design situations and special structures, more detailed and complex analyses may be warranted.

$$\text{T.L.} = ES + CR + SH + RE \quad \text{(Eq. 4.5.1)}$$

where: T.L. = total loss (psi), and the other terms are losses due to:
  ES = elastic shortening
  CR = creep of concrete
  SH = shrinkage of concrete
  RE = relaxation of tendons

$$ES = K_{es} E_s f_{cir}/E_{ci} \quad \text{(Eq. 4.5.2)}$$

where: K_{es} = 1.0 for prestressed members
  E_s = modulus of elasticity of prestressing tendons (usually 27.5 to 28×10^6 psi)
  E_{ci} = modulus of elasticity of concrete at time prestress is applied
  f_{cir} = net compressive stress in concrete at center of gravity of tendons immediately after the prestress has been applied to the concrete (see Eq. 4.5.3)

$$f_{cir} = K_{cir}\left(\frac{P_i}{A} + \frac{P_i e^2}{I_g}\right) - \frac{M_g e}{I_g} \quad \text{(Eq. 4.5.3)}$$

where: K_{cir} = 0.9 for pretensioned members
  P_i = initial prestress force (after anchorage seating loss)

e = eccentricity of center of gravity of tendons with respect to center of gravity of concrete at the cross section considered
A = area of gross concrete section at the cross section considered
I_g = moment of inertia of gross concrete section at the cross section considered
M_g = bending moment due to dead weight of member being prestressed and to any other permanent loads in place at time of prestressing

$$CR = K_{cr} (E_s/E_c)(f_{cir} - f_{cds}) \quad \text{(Eq. 4.5.4)}$$

where: K_{cr} = 2.0 normal weight concrete
  = 1.6 sand-lightweight concrete
  f_{cds} = stress in concrete at center of gravity of tendons due to all superimposed permanent dead loads that are applied to the member after it has been prestressed (see Eq. 4.5.5)
  E_c = modulus of elasticity of concrete at 28 days

$$f_{cds} = M_{sd}(e)/I_g \quad \text{(Eq. 4.5.5)}$$

where: M_{sd} = moment due to all superimposed permanent dead loads applied after prestressing

$$SH = (8.2 \times 10^{-6}) K_{sh} E_s \times (1-0.06V/S)(100-RH) \text{(Eq. 4.5.6)}$$

where: K_{sh} = 1.0 for pretensioned members
  V/S = volume to surface ratio
  RH = average ambient relative humidity (see Fig 3.3.2)

$$RE = [K_{re} - J(SH + CR + ES)]C \quad \text{(Eq. 4.5.7)}$$

where values of K_{re}, J and C are taken from Tables 4.5.1 and 4.5.2.

where: f_{pi} = P_i/A_{ps}
  f_{pu} = ultimate strength of prestressing tendons
  A_{ps} = area of prestressing tendons

4.5.4 Critical Locations

Computations for stress losses due to elastic shortening and creep of concrete are based on the compressive stress in the concrete at the center of gravity (cgs) of the tendons.

Table 4.5.1 Values of K_{re} and J

Type of tendon	K_{re}	J
270 Grade stress-relieved strand or wire	20,000	0.15
250 Grade stress-relieved strand or wire	18,500	0.14
240 or 235 Grade stress-relieved wire	17,600	0.13
270 Grade low-relaxation strand	5,000	0.040
250 Grade low-relaxation wire	4,630	0.037
240 or 235 Grade low-relaxation wire	4,400	0.035
145 or 160 Grade stress-relieved bar	6,000	0.05

Table 4.5.2 Values of C

f_{pi}/f_{pu}	Stress-relieved strand or wire	Stress-relieved bar or low-relaxation strand or wire
0.80		1.28
0.79		1.22
0.78		1.16
0.77		1.11
0.76		1.05
0.75	1.45	1.00
0.74	1.36	0.95
0.73	1.27	0.90
0.72	1.18	0.85
0.71	1.09	0.80
0.70	1.00	0.75
0.69	0.94	0.70
0.68	0.89	0.66
0.67	0.83	0.61
0.66	0.78	0.57
0.65	0.73	0.53
0.64	0.68	0.49
0.63	0.63	0.45
0.62	0.58	0.41
0.61	0.53	0.37
0.60	0.49	0.33

For bonded tendons, stress losses are computed at that point on the span where flexural tensile stresses are most critical. In members with straight, parabolic or approximately parabolic tendons this is usually mid-span. In members with tendons deflected at mid-span only the critical point is generally near the 0.4 point of the span. Since the tendons are bonded, only the stresses at the critical point need to be considered. Stresses or stress changes at other points along the member do not affect the stresses or stress losses at the critical point.

Example 4.5.1 Loss of prestress

Given:

10LDT 32 + 2 as shown

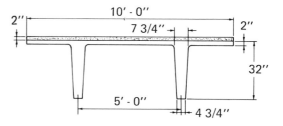

Span = 70 ft

No superimposed dead load except topping

RH = 75%

Section properties (untopped):

A = 615 sq in.

I = 59,720 in.^{4}

Z_b = 2717 in.^{3}

V/S = 615/364 = 1.69 in.

wt = 491 plf

wt of topping = 250 plf

Concrete:

Precast: f'_c = 5000 psi E_c = 3.0 × 10^6 psi

         f'_{ci} = 3500 psi E_{ci} = 2.5 × 10^6 psi

         Sand-lightweight

Topping: Normal weight

Prestressing steel:

12–1/2″ dia. 270K low-relaxation strands

A_{ps} = 12(0.153) = 1.836 sq in.

E_s = 28 × 10^6 psi

Depressed at mid-span

e_e = 12.81 in.

e_c = 18.73 in.

Problem:

Determine total loss of prestress.

Solution:

For depressed strand, critical section is at 0.4ℓ. Determine moments, eccentricity, and prestress force.

$$M @ 0.4\ell = \frac{wx}{2}(\ell - x) = \frac{w(0.4\ell)}{2}(\ell - 0.4\ell)$$

$$= 0.12\, w\ell^2$$

$$M_g = 0.12(0.491)(70)^2 = 289 \text{ ft-kips}$$

$$M_{sd} = 0.12(0.250)(70)^2 = 147 \text{ ft-kips}$$

$$e \text{ at } 0.4\ell = 12.81 + 0.8(18.73 - 12.81)$$
$$= 17.55 \text{ in.}$$

Assume compensation for anchorage seating loss during prestressing.

$$P_i = 0.75 \, A_{ps}f_{pu} = 0.75(1.836)(270)$$
$$= 371.8 \text{ kips}$$

Determine f_{cir} and f_{cds}:

$$f_{cir} = K_{cir}\left(\frac{P_i}{A} + \frac{P_i e^2}{I_g}\right) - \frac{M_g e}{I_g}$$

$$= 0.9\left(\frac{371.8}{615} + \frac{371.8(17.55)^2}{59,720}\right) - \frac{289(12)(17.55)}{59,720}$$

$$= 1.252 \text{ ksi} = 1252 \text{ psi}$$

$$f_{cds} = M_{sd}(e)/I_g$$
$$= 147(12)(17.55)/59,720$$
$$= 0.52 \text{ ksi} = 520 \text{ psi}$$

$$ES = K_{es} \, E_s \, f_{cir}/E_{ci}$$
$$= (1)(28 \times 10^6)(1252)/(2.5 \times 10^6) = 14,022 \text{ psi}$$

$$CR = K_{cr}(E_s/E_c)(f_{cir} - f_{cds})$$
$$CR = (1.6)(28 \times 10^6/3.0 \times 10^6)(1252 - 520)$$
$$= 10,931 \text{ psi}$$

$$SH = 8.2 \times 10^{-6} \, K_{sh} \, E_s(1 - 0.06 \, V/S)(100 - RH)$$
$$= 8.2 \times 10^{-6}(1)(28 \times 10^6)[1 - 0.06(1.69)](100-75)$$
$$= 5157 \text{ psi}$$

$$RE = [K_{re} - J(SH + CR + ES)]C$$

From Table 4.5.1
$$K_{re} = 5000$$
$$J = 0.04$$
$$f_{pi}/f_{pu} = 0.75$$

From Table 4.5.2
$$C = 1.0$$

$$RE = [5000 - 0.04(5157 + 10,931 + 14,022)](1)$$
$$= 3796 \text{ psi}$$

$$T.L. = ES + CR + SH + RE$$
$$= 14,022 + 10,931 + 5157 + 3796$$
$$= 33,906 \text{ psi} = 33.9 \text{ ksi}$$

Final prestress force $= 371.8 - 33.9(1.836)$
$$= 309.6 \text{ kips}$$

4.6 Camber and Deflection

Most precast, prestressed concrete flexural members will have a net positive (upward) camber at the time of transfer of prestress, caused by the eccentricity of the prestressing force. This camber may increase or decrease with time, de-

Table 4.6.1 Maximum permissible computed deflections

Type of member	Deflection to be considered	Deflection limitation
Flat roofs not supporting or attached to nonstructural elements likely to be damaged by large deflections	Immediate deflection due to live load	$\dfrac{\ell*}{180}$
Floors not supporting or attached to nonstructural elements likely to be damaged by large deflections	Immediate deflection due to live load	$\dfrac{\ell}{360}$
Roof or floor construction supporting or attached to nonstructural elements likely to be damaged by large deflections	That part of the total deflection occurring after attachment of nonstructural elements (sum of the long-time deflection due to all sustained loads and the immediate deflection due to any additional live load)‡	$\dfrac{\ell†}{480}$
Roof or floor construction supporting or attached to nonstructural elements not likely to be damaged by large deflections		$\dfrac{\ell§}{240}$

* Limit not intended to safeguard against ponding. Ponding should be checked by suitable calculations of deflection, including added deflections due to ponded water, and considering long-time effects of all sustained loads, camber, construction tolerances, and reliability of provisions for drainage.

† Limit may be exceeded if adequate measures are taken to prevent damage to supported or attached elements.

‡ Long-time deflection shall be determined in accordance with Section 9.5.2.5 or 9.5.4.2 but may be reduced by amount of deflection calculated to occur before attachment of nonstructural elements. This amount shall be determined on basis of accepted engineering data relating to time-deflection characteristics of members similar to those being considered.

§ But not greater than tolerance provided for nonstructural elements. Limit may be exceeded if camber is provided so that total deflection minus camber does not exceed limit.

pending on the stress distribution across the member under sustained loads. Camber tolerances are suggested in Part 8 of this Handbook.

Limitations on instantaneous deflections and time-dependent cambers and deflections are specified in the ACI Code. Table 9.5(b) of the Code is reprinted for reference (see Table 4.6.1).

The following sections contain suggested methods for computing cambers and deflections. There are many inherent variables that affect camber and deflection, such as concrete mix, storage method, time of release of prestress, time of erection and placement of superimposed loads, relative humidity, etc. *Because of this, calculated long-time values should never be considered any better than estimates*. Non-structural components attached to members which could be affected by camber variations, such as partitions or folding doors, should be placed with adequate allowance for error. Calculation of topping quantities should also recognize the imprecision of camber calculations.

It should also be recognized that camber of precast, prestressed members is a result of the placement of the strands needed to resist the design moments and service load stresses. It is not practical to alter the forms of the members to produce a desired camber. Therefore, cambers should not be specified, but their inherent existence should be recognized.

4.6.1 Initial Camber

Initial camber can be calculated using conventional moment-area equations. Fig. 4.10.13 has equations for the camber caused by prestress force for the most common strand patterns used in precast, prestressed members. Figs. 11.1.3 and 11.1.4 provide deflection equations for typical loading conditions and more general camber equations.

Example 4.6.1 Calculation of initial camber

Given:

8DT24 of Example 4.2.9.

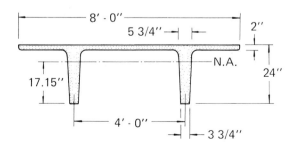

Section Properties

$$
\begin{aligned}
A &= 401 \text{ in.}^2 \\
I &= 20{,}985 \text{ in.}^4 \\
y_b &= 17.15 \text{ in.} \\
y_t &= 6.85 \text{ in.} \\
Z_b &= 1224 \text{ in.}^3 \\
Z_t &= 3063 \text{ in.}^3 \\
wt &= 418 \text{ plf} \\
&\quad\; 52 \text{ psf}
\end{aligned}
$$

Concrete:

$f'_c = 5000$ psi

Normal weight (150 pcf)

$E_c = 33w\sqrt{wf'_c} = 33(150)\sqrt{(150)(5000)}$

$\quad = 4287$ ksi

$f'_{ci} = 3500$ psi

$E_{ci} = 33(150)\sqrt{(150)(3500)} = 3587$ ksi

(Note: The values of E_c and E_{ci} could also be read from Fig. 11.2.2)

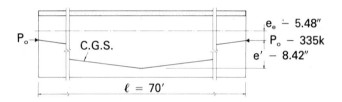

Problem:

Find the initial camber at time of transfer of prestress.

Solution:

The prestress force at transfer and strand eccentricities are calculated in Example 4.2.9 and are shown in the illustration above.

Calculate the upward component using equations given in Fig. 4.10.13.

$$
\begin{aligned}
\Delta \uparrow &= \frac{P_o e_e \ell^2}{8 E_{ci} I} + \frac{P_o e' \ell^2}{12 E_{ci} I} \\[6pt]
&= \frac{335 (5.48) (70 \times 12)^2}{8 (3587) (20{,}985)} \\[6pt]
&\quad + \frac{(335) (8.42) (70 \times 12)^2}{12 (3587) (20{,}985)} \\[6pt]
&= 2.15 + 2.20 = 4.35 \text{ in.} \uparrow
\end{aligned}
$$

Deduct deflection caused by weight of member:

$$
\Delta \downarrow = \frac{5 \, w\ell^4}{384 \, E_{ci} I}
$$

$$= \frac{5\left(\dfrac{0.418}{12}\right)(70 \times 12)^4}{384\,(3587)\,(20{,}985)} = 3.00 \text{ in.} \downarrow$$

Net camber at release $= 4.35 \uparrow - 3.00 \downarrow$
$= 1.35 \text{ in.} \uparrow$

4.6.2 Elastic Deflections

Calculation of instantaneous deflections of both prestressed and non-prestressed members caused by superimposed service loads follow classical methods of mechanics. Design equations for various load conditions are given in Part 11 of this Handbook. If the bottom tension in a simple span member does not exceed the modulus of rupture, the deflection is calculated using the uncracked moment of inertia of the section. The modulus of rupture of concrete is defined in Chapter 9 of the Code as:

$$f_r = 7.5\,\lambda\,\sqrt{f'_c} \qquad \text{(Eq. 4.6.1)}$$

(See Sect. 4.3.3 for definition of λ)

4.6.3 Bilinear Behavior

Sect. 18.4.2 of the Code requires that "bilinear moment-deflection relationships" be used to calculate instantaneous deflections when the bottom tension exceeds $6\sqrt{f'_c}$. This means that the deflection before the member has cracked is calculated using the gross (uncracked) moment of inertia, I_g, and the *additional* deflection after cracking is calculated using the moment of inertia of the cracked section. This is illustrated graphically in Fig. 4.6.1.

In lieu of a more exact analysis, the empirical relationship:

$$I_{cr} = nA_{ps}d^2\,(1 - 1.67\sqrt{n\rho_p}) \qquad \text{(Eq. 4.6.2)}$$

may be used to determine the cracked moment of inertia. Table 4.10.14 gives coefficients for use in solving this equation.

Example 4.6.2 Deflection calculation using bilinear moment-deflection relationships

Given:

8DT24 of Examples 4.2.9 and 4.6.1.

Problem:

Determine the total instantaneous deflection caused by the specified uniform live load.

Fig. 4.6.1 Bilinear moment-deflection relationship

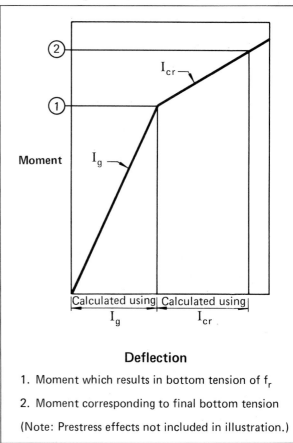

Deflection

1. Moment which results in bottom tension of f_r

2. Moment corresponding to final bottom tension

(Note: Prestress effects not included in illustration.)

Solution:

Determine $f_r = 7.5\sqrt{f'_c} = 530$ psi

From Example 4.2.9 the final tensile stress is 756 psi, which is more than 530 psi, so the bilinear behavior must be considered.

Determine I_{cr} from Table 4.10.14:

A_{ps} = 1.836 sq in. (See Ex. 4.2.9)

d at midspan $= e_c + y_t = 13.90 + 6.85$
$= 20.75$ in.*

$\rho_p = \dfrac{A_{ps}}{bd} = \dfrac{1.836}{(96)(20.75)} = 0.00092$

C $= 0.0052$

I_{cr} $= Cbd^3 = 0.0052(96)(20.75)^3$
$= 4460$ in.⁴

Determine the portion of the live load that would result in a bottom tension of 530 psi.

$756 - 530 = 226$ psi

---

* It is within the precision of the calculation method and observed behavior to use midspan d and to calculate the deflection at midspan, although the maximum tensile stress in this case is assumed at 0.4ℓ.

The tension caused by live load alone is 1614 psi, therefore, the portion of the live load that would result in a bottom tension of 530 psi is:

$$\frac{1614 - 226}{1614}(0.280) = 0.241 \text{ kips/ft}$$

and

$$\Delta_g = \frac{5w\ell^4}{384\,E_c I_g} = \frac{5\left(\dfrac{0.241}{12}\right)(70\times12)^4}{384\,(4287)(20,985)}$$
$$= 1.45 \text{ in.}$$

$$\Delta_{cr} = \frac{5\left(\dfrac{0.039}{12}\right)(70\times12)^4}{384\,(4287)(4460)} = 1.10 \text{ in.}$$

Total deflection $= 1.45 + 1.10 = 2.55$ in.

4.6.4 Effective Moment of Inertia

The Code allows an alternative to the method of calculation described in the previous section. An effective moment of inertia, I_e, can be determined and the deflection then calculated by substituting I_e for I_g in the deflection calculation.

The equation for effective moment of inertia is:

$$I_e = \left(\frac{M_{cr}}{M_a}\right)^3 I_g + \left[1 - \left(\frac{M_{cr}}{M_a}\right)^3\right]I_{cr} \qquad \text{(Eq.4.6.3)}$$

Fig. 4.6.2 Effective moment of inertia

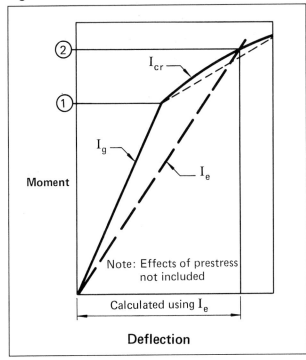

Moment

② I_{cr}

① I_g

I_e

Note: Effects of prestress not included

Calculated using I_e

Deflection

The difference between the bilinear method and the I_e method is illustrated in Fig.4.6.2.

The use of I_e with prestressed concrete members is described in a paper by Branson.[8] The value of M_{cr}/M_a for use in determining live load deflections can be expressed as:

$$\frac{M_{cr}}{M_a} = 1 - \left(\frac{f_{t\ell} - f_r}{f_\ell}\right) \qquad \text{(Eq.4.6.4)}$$

where $f_{t\ell}$ = final calculated total stress in the member

f_ℓ = calculated stress due to live load

Example 4.6.3 Deflection calculation using effective moment of inertia

Given:

Same section and loading conditions of Examples 4.2.9, 4.6.1 and 4.6.2.

Problem:

Determine the deflection caused by live load using the I_e method.

Solution:

From the table of stresses in Example 4.2.9:

$f_{t\ell}$ = 756 psi (tension)

f_ℓ = 1614 psi (tension)

f_r = $7.5\sqrt{f'_c}$ = 530 psi (tension)

$$\frac{M_{cr}}{M_a} = 1 - \left(\frac{756 - 530}{1614}\right) = 0.860$$

$$\left(\frac{M_{cr}}{M_a}\right)^3 = (0.860)^3 = 0.636$$

$$1 - \left(\frac{M_{cr}}{M_a}\right)^3 = 1 - 0.636 = 0.364$$

I_e = 0.636 (20,985) + 0.364(4460)
= 14,970 in.⁴

I_e can also be found using Fig. 4.10.15:

f_e = 756 − 530 = 226 psi

$$\frac{f_e}{f_\ell} = \frac{226}{1614} = 0.14$$

$$\frac{I_{cr}}{I_g} = \frac{4460}{20,985} = 0.213$$

Follow arrows on the chart:

$$\frac{I_e}{I_g} = 0.71$$

I_e = 0.71 (20,985) = 14,899 in.⁴

$$\Delta\ell = \frac{5w\ell^4}{384\ E_c\ I_e} = \frac{5\left(\frac{0.280}{12}\right)(70 \times 12)^4}{384\ (4287)\ (14{,}970)}$$

$$= 2.36 \text{ in.}$$

4.6.5 Long-Time Camber/Deflection

ACI 318-83 provides an equation for estimating the long-time deflection of non-prestressed reinforced concrete members (Sect. 9.5.2.5):

$$\lambda = \frac{\xi}{1+50\rho'} \qquad \text{(Eq. 4.6.5)}$$

where ξ is a factor related to length of time, and ρ' is the ratio of compressive reinforcement. No such guide is given for prestressed concrete.

The determination of long-time cambers and deflections in precast, prestressed members is somewhat more complex because of (1) the effect of prestress and the loss of prestress over time, (2) the strength gain of concrete after release of prestress, and because (3) the camber or deflection is important not only at the "initial" and "final" stages, but also at erection, which occurs at some intermediate stage, usually from 30 to 60 days after casting.

It has been customary in the design of precast, prestressed concrete to estimate the camber of a member after a period of time by multiplying the initial calculated camber by some factor, usually based on the experience of the designer. To properly use these "multipliers," the upward and downward components of the initial calculated camber should be separated in order to take into account the effects of loss of prestress, which only affect the upward component.

Fig. 4.6.3 provides suggested multipliers which can be used as a guide in estimating long-time cambers and deflections for typical members, i.e., those members which are within the span-depth ratios recommended in this Handbook (see Sect. 3.2.2). Derivation of these multipliers is contained in a paper by Martin.[9]

Long-time effects can be substantially reduced by adding non-prestressed reinforcement in prestressed concrete members. The reduction effects proposed by Shaikh and Branson[10] can be applied to the approximate multipliers of Fig. 4.6.3 as follows:

$$C_2 = \frac{C_1 + A_s/A_{ps}}{1 + A_s/A_{ps}} \qquad \text{(Eq. 4.6.6)}$$

where
C_1 = multiplier from table
C_2 = revised multiplier
A_s = area of non-prestressed reinforcement
A_{ps} = area of prestressed steel

Fig. 4.6.3 Suggested multipliers to be used as a guide in estimating long-time cambers and deflections for typical members

		Without Composite Topping	With Composite Topping
	At erection:		
(1)	Deflection (downward) component — apply to the elastic deflection due to the member weight at release of prestress	1.85	1.85
(2)	Camber (upward) component — apply to the elastic camber due to prestress at the time of release of prestress	1.80	1.80
	Final:		
(3)	Deflection (downward) component — apply to the elastic deflection due to the member weight at release of prestress	2.70	2.40
(4)	Camber (upward) component — apply to the elastic camber due to prestress at the time of release of prestress	2.45	2.20
(5)	Deflection (downward) — apply to elastic deflection due to superimposed dead load only	3.00	3.00
(6)	Deflection (downward) — apply to elastic deflection caused by the composite topping	—	2.30

	(1) Release	Multiplier	(2) Erection	Multiplier	(3) Final
Prestress	4.41 ↑	1.80 × (1)	7.94 ↑	2.45 × (1)	10.80 ↑
w_d	3.00 ↓	1.85 × (1)	5.55 ↓	2.7 × (1)	8.10 ↓
	1.41 ↑		2.39 ↑		2.70 ↑
w_{sd}			0.48 ↓	3.0 × (2)	1.44 ↓
			1.91 ↑		1.26 ↑
w_ℓ					2.38 ↓
					1.12 ↓

Example 4.6.4 Use of multipliers for determining long-time cambers and deflections

Given:

8DT24 of Examples 4.2.9, 4.6.1, 4.6.2 and 4.6.3.
Non-structural elements are attached, but not likely to be damaged by deflections (light fixtures, etc.).

Problem:

Estimate the camber and deflection and determine if it meets the requirements of Table 9.5(b) of the Code (see Table 4.6.1).

Solution:

Calculate the instantaneous deflections caused by the superimposed dead and live loads.

$$\Delta_d = \frac{5w\ell^4}{384\, E_c I} = \frac{5\left(\dfrac{0.080}{12}\right)(70 \times 12)^4}{384\,(4287)(20{,}985)}$$
$$= 0.48 \text{ in. } \downarrow$$

$\Delta_\ell = 2.36$ in. ↓ (see Example 4.6.3)

For convenience, a tabular format is used (above).

The estimated critical cambers and deflections would then be:

At erection of the member after

w_{sd} is applied	= 1.80 in.
"Final" long-time camber	= 1.12 in.

The deflection limitation of Table 9.5(b) for the above condition is $\ell/240$.

$$(70 \times 12)/240 = 3.50 \text{ in.}$$

Total deflection occurring after attachment of non-structural elements:

$$\Delta_\ell = (1.80 - 1.12) + 2.36$$
$$= 3.04 \text{ in. } < 3.50 \text{ in. OK}$$

4.7 Compression Members

Precast and prestressed concrete columns and load bearing wall panels are usually proportioned on the basis of strength design. Stresses under service conditions, particularly during handling and erection (especially wall panels) must also be considered. The procedures in this section are based on Chapter 10 of the Code and on the recommendations of the PCI Committee on Prestressed Concrete Columns[11] (referred to in this section as "the Recommended Practice").

4.7.1 Strength Design of Precast Concrete Compression Members

The capacity of a reinforced concrete compression member with eccentric loads is most easily determined by constructing a capacity interaction curve. Points on this curve are calculated using the compatibility of strains and solving the equations of equilibrium as prescribed in Chapter 10 of the Code. Solution of these equations is illustrated in Fig. 4.7.1.

ACI 318-83 waives the minimum vertical reinforcement requirements for compression members if the concrete is prestressed to at least an average of 225 psi after all losses. In addition, the Recommended Practice permits the elimination of column ties, if the nominal capacity is multiplied by 0.85. Interaction curves for typical prestressed square columns and wall panels are provided in Part 2.

Construction of an interaction curve usually follows these steps:

Step 1: Determine P_o for $M_n = 0$. (See Fig. 4.7.1(c))

Step 2: Determine M_o for $P_n = 0$. This is normally done by neglecting the reinforcement above the neutral axis and determining the moment capacity by one of the methods described in Sect. 4.2.1.

Step 3: For non-prestressed columns, P_{nb} and M_{nb} at the balance point may be determined (see Fig. 4.7.1(d)). For prestressed columns, the yield point of the prestressed reinforcement is not well defined and the stress-strain relationship is non-linear over a broad range (see Fig. 11.2.5).

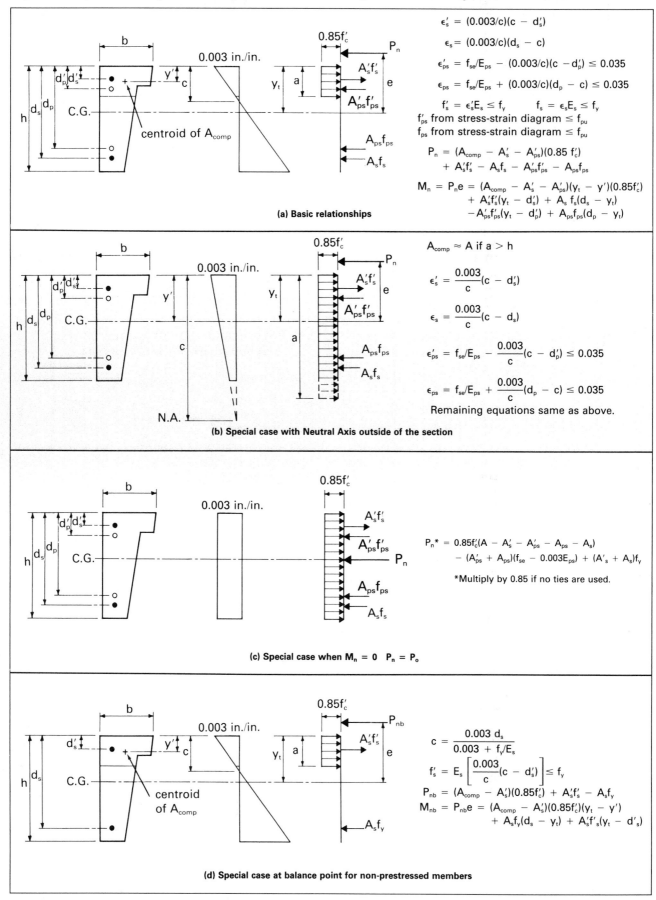

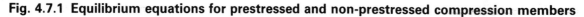

$\epsilon'_s = (0.003/c)(c - d'_s)$

$\epsilon_s = (0.003/c)(d_s - c)$

$\epsilon'_{ps} = f_{se}/E_{ps} - (0.003/c)(c - d'_p) \leq 0.035$

$\epsilon_{ps} = f_{se}/E_{ps} + (0.003/c)(d_p - c) \leq 0.035$

$f'_s = \epsilon'_s E_s \leq f_y \qquad f_s = \epsilon_s E_s \leq f_y$

f'_{ps} from stress-strain diagram $\leq f_{pu}$
f_{ps} from stress-strain diagram $\leq f_{pu}$

$P_n = (A_{comp} - A'_s - A'_{ps})(0.85 f'_c)$
$\quad + A'_s f'_s - A_s f_s - A'_{ps} f'_{ps} - A_{ps} f_{ps}$

$M_n = P_n e = (A_{comp} - A'_s - A'_{ps})(y_t - y')(0.85 f'_c)$
$\quad + A'_s f'_s (y_t - d'_s) + A_s f_s (d_s - y_t)$
$\quad - A'_{ps} f'_{ps} (y_t - d'_p) + A_{ps} f_{ps} (d_p - y_t)$

(a) Basic relationships

$A_{comp} \approx A$ if $a > h$

$\epsilon'_s = \dfrac{0.003}{c}(c - d'_s)$

$\epsilon_s = \dfrac{0.003}{c}(c - d_s)$

$\epsilon'_{ps} = f_{se}/E_{ps} - \dfrac{0.003}{c}(c - d'_p) \leq 0.035$

$\epsilon_{ps} = f_{se}/E_{ps} + \dfrac{0.003}{c}(d_p - c) \leq 0.035$

Remaining equations same as above.

(b) Special case with Neutral Axis outside of the section

$P_n^* = 0.85 f'_c (A - A'_s - A'_{ps} - A_{ps} - A_s)$
$\quad - (A'_{ps} + A_{ps})(f_{se} - 0.003 E_{ps}) + (A'_s + A_s) f_y$

*Multiply by 0.85 if no ties are used.

(c) Special case when $M_n = 0$ $P_n = P_o$

$c = \dfrac{0.003 d_s}{0.003 + f_y/E_s}$

$f'_s = E_s \left[\dfrac{0.003}{c}(c - d'_s) \right] \leq f_y$

$P_{nb} = (A_{comp} - A'_s)(0.85 f'_c) + A'_s f'_s - A_s f_y$

$M_{nb} = P_{nb} e = (A_{comp} - A'_s)(0.85 f'_c)(y_t - y')$
$\quad + A_s f_y (d_s - y_t) + A'_s f'_s (y_t - d'_s)$

(d) Special case at balance point for non-prestressed members

Step 4: For each additional point on the interaction curve, proceed as follows:

    a. Select a value of "c" and calculate a = $\beta_1 c$

    b. Determine the value of A_{comp} from the geometry of the section (shaded portion in Fig. 4.7.1(a)).

    c. Determine the strain in the reinforcement assuming that $\epsilon = 0.003$ at the compression face of the column. For prestressed reinforcement, add the strain due to the effective prestress $\epsilon_{se} = f_{se}/E_{ps}$.

    d. Determine the stress in the reinforcement. For non-prestressed reinforcement, $f_s = \epsilon_s E_s \leq f_y$. For prestressed reinforcement, the stress is determined from stress-strain relationship (see Fig. 11.2.5). If the maximum factored moment occurs near the end of a prestressed element, where the strand is not fully developed, an appropriate reduction in the value of f_{ps} should be made as described in Sect. 4.2.3.

    e. Calculate P_n and M_n by statics.

    f. Calculate ϕP_n and ϕM_n. The Code prescribes that the ϕ factor for compression elements is 0.7, except that it can vary from 0.7 at a point where $\phi P_n = 0.10 f'_c A$ to 0.9 where $\phi P_n = 0$. This ϕ variation is accomplished on the interaction curve by first constructing the curve with $\phi = 0.7$. A straight line is then drawn from the point on that curve where $\phi P_n = 0.10 f'_c A$ to a point on the $\phi P_n = 0$ line corresponding to ϕM_o calculated with $\phi = 0.9$.

Step 5: Calculate the maximum factored axial resistance specified by the Code as
$0.80 \phi P_o$ for tied columns
$0.85 \phi P_o$ for spiral columns.

For cross sections which are not rectangular, it is necessary to determine separate curves for each direction of the applied moment. Further, since most architectural column cross-sections are not rectangular, the "a" distance only defines the depth of the rectangular concrete stress distribution. Instead of using a/2, as for a rectangular cross-section, it is necessary to calculate the actual centroid of the compression area which is indicated as y'.

As noted in Step 4d above, the flexural resistance is reduced for prestressed elements at locations within a distance equal to the strand development length from each end. The flexural resistance of the prestressed reinforcement in this zone can be supplemented by non-prestressed reinforcement that is anchored to end plates, or otherwise developed.

The interaction curves in Part 2 are based on a maximum value of $f_{ps} = f_{se}$, which is equivalent to a development length equal to the assumed transfer length. The required area of end reinforcement can be determined by matching interaction curves, or can be approximated by the following equation if the bar locations approximately match the strand locations:

$$A_s = \frac{A_{ps} f_{se}}{f_y}$$

where
    A_s = required area of bars
    f_{se} = strand stress after losses
    f_y = yield strength of bars

The effects of adding end reinforcement to a 24 × 24 in. prestressed concrete column, thus improving moment capacity in the end 2 ft, are shown in Fig. 4.7.2.

Example 4.7.1 Construction of interaction curve for a precast, reinforced concrete column

Given:

Column cross-section shown

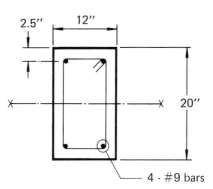

Concrete:
f'_c = 5000 psi

Reinforcement:
Grade 60
f_y = 60,000 psi
E_s = 29,000 ksi

Problem:

Construct interaction curve for bending about x-x axis.

Fig. 4.7.2 End reinforcement in a precast, prestressed concrete column

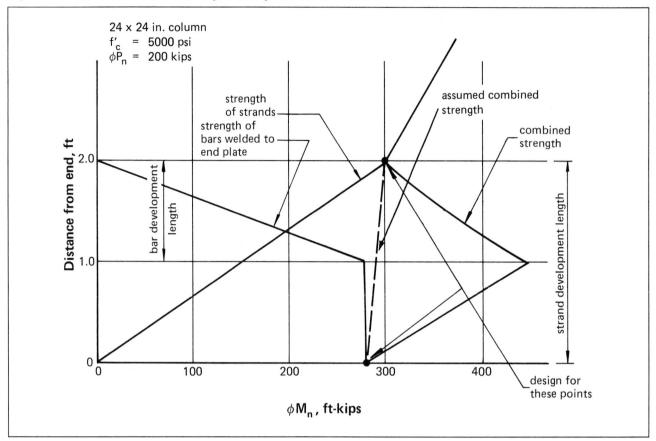

Solution:

Determine following parameters:

$\beta_1 = 0.85 - 0.05 = 0.80$

$d = 20 - 2.5 = 17.5$ in.

$d' = 2.5$ in.

$y_t = 10$ in.

$0.85 f'_c = 0.85(5) = 4.25$ ksi

$A_g = 12 \times 20 = 240$ sq in.

$A_s = A'_s = 2.00$ sq in.

Step 1 — Determine P_o from Fig. 4.7.1(c):

$P_o = 0.85 f'_c (A_g - A'_s - A_s) + (A'_s + A_s) f_y$

$\phi P_o = 0.70 [4.25(240 - 4) + (4)(60)]$

$\quad = 870$ kips

Step 2 — Determine P_{nb} and M_{nb} from Fig. 4.7.1(d):

$c = \dfrac{0.003d}{0.003 + f_y/E_s} = \dfrac{0.003(17.5)}{0.003 + 60/29,000}$

$\quad = 10.36$ in.

$f'_s = 29,000 \left[\dfrac{0.003}{10.36} (10.36 - 2.5) \right]$

$\quad = 66.0 > 60$

therefore $f'_s = f_y = 60$ ksi

$A_{comp} = ab = \beta_1 c\, b = 0.80(10.36)(12)$

$\quad = 99.5$ sq in.

$y' = \dfrac{a}{2} = \dfrac{0.80(10.36)}{2} = 4.14$ in.

$P_{nb} = (99.5 - 2)4.25 + 2(60) - 2(60)$

$\quad = 414.4$ kips

$\phi P_{nb} = 0.70(414.4) = 290$ kips

$M_{nb} = (97.5)(10 - 4.14)(4.25)$

$\quad + 2.0(60)(17.5 - 10)$

$\quad + 2.0(60)(10 - 2.5)$

$\phi M_{nb} = 0.70(2428 + 900 + 900)$

$\quad = 2960$ in-kips $= 247$ ft-kips

Step 3 — Determine M_o — Use conservative solution neglecting compressive reinforcement:

PCI Design Handbook

$$a = \frac{A_s f_y}{0.85 \, f'_c b} = \frac{2.0(60)}{4.25(12)} = 2.35 \text{ in.}$$

$$M_o = A_s f_y \left(d - \frac{a}{2}\right) = (2.0)(60)(17.5 - 2.35/2)$$
$$= 1959 \text{ in.-kips}$$

For $\phi = 0.7$, $\phi M_o = 1371$ in.-kips = 114 ft-kips
(This point is found for curve projection)

For $\phi = 0.9$, $\phi M_o = 1763$ in.-kips = 147 ft kips

To determine intermediate points on the curve:

Step 4a — Set $a = 6$ in., $c = \dfrac{6}{0.80} = 7.5$ in.

Step 4b — $A_{comp} = 6(12) = 72$ sq in.

Step 4d — Use Fig. 4.7.1(a):

$$f'_s = 29,000 \left[\frac{0.003}{7.5}(7.5 - 2.5)\right]$$
$$= 58.0 \text{ ksi} < f_y$$

$$f_s = 29,000 \left[\frac{0.003}{7.5}(17.5 - 7.5)\right]$$
$$= 116 \text{ ksi} > f_y$$

Use $f_s = f_y = 60$ ksi

Steps 4e and 4f —

$$P_n = (72 - 2)4.25 + 2.0(58) - 2.0(60)$$
$$= 293.5 \text{ kips}$$

Fig. 4.7.3 Interaction curve for Example 4.7.1

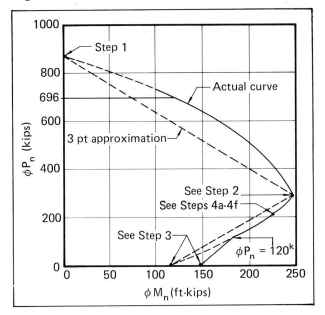

$$\phi P_n = 0.7(293.5) = 205 \text{ kips}$$
$$\phi M_n = 0.70 \, [(72 - 2)(10 - 3) \, 4.25$$
$$+ \; 2.0(60)(17.5 - 10)$$
$$+ \; 2.0(58)(10 - 2.5)]$$
$$= 0.70 \, (2082.5 + 900 + 870)$$
$$= 2697 \text{ in.-kips} = 225 \text{ ft-kips}$$

(Note: Steps 4a to 4f can be repeated for as many points as desired.)

A plot of these points is shown as Fig. 4.7.3.

Step 5 — Also calculate:

Maximum design load $= 0.80 \, \phi P_o$
$$= 0.80(870) = 696 \text{ kips}$$

Transition point for $\phi = 0.7$ to $\phi = 0.9$
$$= 0.10 \, f'_c A_g = 0.10(5)(240) = 120 \text{ kips}$$

Example 4.7.2 Calculation of interaction points for prestressed concrete compression members

Given:

Hollow-core wall panel shown

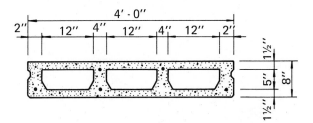

Concrete:

$\quad f'_c = 6000$ psi

$\quad A = 204$ sq in.

Prestressing steel:

$\quad f_{pu} = 270$ ksi

$\quad E_s = 28,000$ ksi

$\quad f_{se} = 150$ ksi

$\quad$ 5–3/8″ dia., 270K strands

$\quad A_{ps}$ (bott) $= 3(0.085) = 0.255$ sq in.

$\quad A'_{ps}$ (top) $= 2(0.085) = 0.170$ sq in.

Problem:

Calculate a point on the design interaction curve for $a = 2$ in.

Solution:

Step 1: $\beta_1 = 0.85 - 2(.05) = 0.75$

$\quad\quad a = 2$ in., $c = \dfrac{2}{0.75} = 2.67$ in.

Step 2: $A_{comp} = 48(1.5) + 12(2 - 1.5)$
$= 78$ sq in.

$$y' = \frac{48(1.5)(1.5/2) + 12(0.5)(1.5 + 0.5/2)}{78}$$

$= 0.83$ in.

Step 3: $\dfrac{f_{se}}{E_s} = \dfrac{150}{28,000} = 0.00536$ in./in.

Step 4: From Fig. 4.7.1(a)

$$\epsilon'_{ps} = 0.00536 - \frac{0.003}{2.67}(2.67 - 1.5)$$

$= 0.00536 - 0.00131 = 0.00405$ in./in.

From Fig. 11.2.5, this strain is on the linear part of the curve:

$f'_{ps} = \epsilon'_{ps}E_s = 0.00405(28,000) = 113$ ksi

$\epsilon_{ps} = 0.00536 + \dfrac{0.003}{2.67}(6.5 - 2.67)$

$= 0.00536 + 0.00430 = 0.00966$

From Fig. 11.2.5, $f_{ps} = 245$ ksi

From Fig. 4.7.1(a):

$P_n = (A_{comp})\,0.85f'_c - A'_{ps}f'_{ps} - A_{ps}f_{ps}$

$= 78\,(0.85)(6) - 0.170(113) - 0.255(245)$

$= 397.8 - 19.2 - 62.5 = 316.1$ kips

$\phi P_n = 0.7(316.1) = 221.3$ kips

$M_n = 397.8(4 - 0.83) - 19.2(4 - 1.5)$
$\qquad + 62.5(6.5 - 4)$

$= 1261.0 - 48.0 + 156.3 = 1369$ in.-kips

$= 114$ ft-kips

$\phi M_n = 0.7(1369) = 958$ in.-kips

$= 79.8$ ft-kips

Since no lateral ties are used in this member, the values should be multiplied by 0.85.

$\phi P_n = 0.85(221.3) = 188.1$ kips

$\phi M_n = 0.85\,(79.8) = 67.8$ ft-kips

Note that this is for *fully developed* strand. If the capacity at a point near the end of the transfer zone is desired, then $f_{ps} \leq f_{se} = 150$ ksi.

$\phi P_n = 0.85\,(0.7)\,[397.8 - 19.2 - 0.255\,(150)]$

$= 202.5$ kips

$\phi M_n = 0.85\,(0.7)\,[1261.0 - 48.0$
$\qquad + 0.255(150)(6.5 - 4)]$

$= 778.6$ in.-kips $= 64.9$ ft-kips

(Note: In prestressed wall panels, the effects of unsymmetrical prestress should also be investigated.)

4.7.2 Slenderness Effects

Sects. 10.10 and 10.11 of ACI 318-83 contain provisions for evaluating slenderness effects (buckling) of columns. Use of these provisions is described in Part 3 of this Handbook. Additional recommendations are given in the Recommended Practice.

4.7.3 Service Load Stresses

There are no limitations in ACI 318-83 on service load stresses in compression members subject to bending. The Recommended Practice suggests that, for prestressed members, the limitations of Sect. 18.4 of the Code be applied. For non-prestressed members, stresses and crack control are discussed in Sects. 4.2.2.1 and 5.2.4. Handling stresses are nearly always more critical than service load stresses.

4.7.4 Effective Width of Wall Panels

The Recommended Practice specifies that the portion of a wall considered as effective for supporting concentrated loads or for determining the effects of slenderness shall be the least of the following:

a. The center-to-center distance between loads.

b. The length of the loaded portion plus six times the wall thickness on either side (Fig. 4.7.4).

c. The width of the rib (in ribbed wall panels) plus six times the thickness of the wall between ribs on either side of the rib (Fig. 4.7.4).

d. 0.4 times the actual height of the wall.

4.7.5 Varying Section Properties of Compression Members

Architectural wall panels will frequently be of a configuration that varies over the unsupported height of the panel. While there are precise methods of determining the effects of slenderness for such members, the approximate nature of the analysis procedures used do not warrant such precision. Example 4.7.3 illustrates approximate

Fig. 4.7.4 Effective width of wall panels

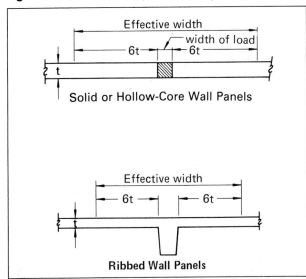

Solid or Hollow-Core Wall Panels

Effective width

6t — 6t

Ribbed Wall Panels

methods for determining section properties used in evaluating slenderness effects.

Example 4.7.3 Approximate section properties of an architectural mullion panel

Given:

The load bearing architectural wall panel shown below.

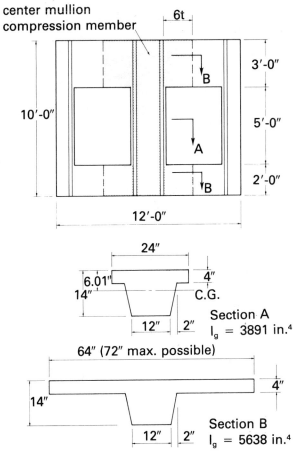

center mullion
compression member

10'-0"

3'-0"

B

5'-0"

A

2'-0"

B

12'-0"

24"

6.01"

14"

4"

C.G.

12" 2"

Section A
$I_g = 3891$ in.[4]

64" (72" max. possible)

14"

4"

12" 2"

Section B
$I_g = 5638$ in.[4]

$f'_c = 5000$ psi; $E_c = 4300$ ksi

Problem:

Determine an approximate moment of inertia for slenderness analysis.

Solution:

One method is to determine a simple span deflection with a uniform load as follows:

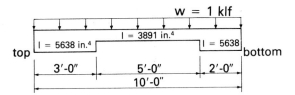

w = 1 klf

top $I = 5638$ in.[4] $I = 3891$ in.[4] $I = 5638$ bottom

3'-0"   5'-0"   2'-0"
10'-0"

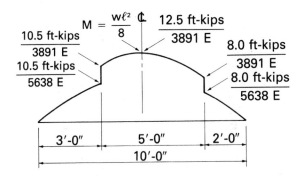

$M = \dfrac{w\ell^2}{8}$ ℄

10.5 ft-kips / 3891 E   12.5 ft-kips / 3891 E   8.0 ft-kips / 3891 E

10.5 ft-kips / 5638 E   8.0 ft-kips / 5638 E

3'-0"   5'-0"   2'-0"
10'-0"

Using the moment area method, the center or mid-height deflection is:

$\Delta_o = 0.013$ in.

$$\Delta_o = \frac{5wL^4}{384\,E\,I_{equiv}}$$

$$I_{equiv} = \frac{5wL^4}{\Delta_o\,384E}$$

$$= \frac{5(1/12)(10 \times 12)^4}{0.013(384)(4300)}$$

$$= 4025 \text{ in.}^4$$

A second, more approximate method would be to use a "weighted average" of the moments of inertia of the two sections. In this case:

$$I_{equiv} = \frac{I_1 h_1 + I_2 h_2}{h_1 + h_2}$$

$$= \frac{5638(3 + 2) + 3891(5)}{10} = 4764 \text{ in}^4$$

Once an equivalent moment of inertia is determined, slenderness effects are evaluated by one of the methods described in Sect. 3.5.

4.7.6 Piles

4.7.6.1 General

The pile designs considered here are based upon structural capacity alone. The ability of the soil to carry these loads must be established by load tests or evaluated by soils engineers.

In the following design procedure for pretensioned concrete piles, load capacity is limited by the service load stresses. An overall factor of safety based on the nominal strength ($\phi = 1.0$) of the section is computed, and limits suggested for various loading conditions. Stresses caused by transporting, handling and driving should also be considered. Experience has shown that the frictional and bearing resistance of the soil will control the design more often than the strength and service load stresses on the pile.

The values used in sample calculations are based on a concrete strength of 6,000 psi. In many areas, higher concrete strengths have been effectively used in prestressed pile design. Engineers should check with local prestressed concrete pile producers to determine what concrete strengths are available in their areas as well as sizes and types, i.e., square, octagonal, or round.

The values in the pile load table (Table 2.7.1) may be modified to fit the actual service conditions and to maintain reasonable safety factors consistent with the character of the applied loads and how and where the piles are to be used.

4.7.6.2 Strengths under Direct Loads

(a) Nominal strength

It is assumed that the concrete stress at failure of a concrete pile will be $0.85 f'_c$. The amount of prestress remaining in the tendons must be deducted. Assuming an ultimate concrete strain of 0.003 it can be shown that only about 60% of the effective prestress, f_{pc}, is left in the member when it reaches its nominal strength. Thus, if a 6000 psi concrete pile is prestressed to an effective prestress of 700 psi, the nominal strength can be computed as:

$$P_n = (0.85 f'_c - 0.60 f_{pc}) A$$

$$P_n = (0.85 \times 6000 - 0.60 \times 700) A = 4680A$$

Where A = gross sectional area of the concrete in square inches, and P_n is the concentric nominal strength in pounds.

(b) Service load

For a concentric load on a short column pile, a factor of safety of between 2.0 and 3.0 is usually adequate.

Based upon a recent study by the Portland Cement Association, which has been accepted by most current building codes, the formula for service loads on concentrically loaded short column prestressed piles is:

$$N = (0.33 f'_c - 0.27 f_{pc}) A$$

Thus, for 6,000 psi concrete and 700 psi prestress

$$N = (0.33 \times 6,000 - 0.27 \times 700) A = 1790 A$$

If the service load is 1790 A, the overall safety factor for the pile as given in Section (a) will be $\frac{4,680}{1,790} = 2.61$ which is considered quite adequate.

For a pile with $f'_c = 6,000$ psi and an effective prestress of 1,200 psi, the corresponding overall safety factor will be approximately the same, 2.65. The value of $0.2 f'_c$ is considered to be about the desirable upper limit for the prestressing force, and a value of 700 psi is recommended as the desirable lower limit for all piles over about 40 ft in length. The overall safety factors for this range of prestress fall within the acceptable limits as indicated above.

(c) Unsupported length

It is suggested that service loads based on the short column value of $(0.33 f'_c - 0.27 f_{pc}) A$ be limited to values of h/r up to 60

where

h = unsupported length of the pile
r = radius of gyration

For piles considered fully fixed at one end and hinged at the other end, it is suggested that h be taken as 0.7 of the length between hinge and assumed point of fixity. For piles fully fixed at both ends, h may be taken as 0.5 of the length between the assumed points of fixity.

4.7.6.3 Moment Resisting Capacities

(a) Service load stresses

(1) Allowing no tension. Under this criterion, if the effective prestress is $f_{pc} = P/A$, the moment capacity is $M = f_{pc} I/c$, where c = distance from extreme fiber to neutral axis. This criterion of "no tension" is much too conservative for prestressed piles under normal conditions. Zero tension will result in an overall safety factor of about 3 or more, which is greater than required for bending and greater than the safety factors used in steel or reinforced concrete design.

(2) The modulus of rupture for 6,000 psi concrete is generally taken to be $7.5 \sqrt{f'_c}$ or ap-

proximately 600 psi. Allowing tension up to about 50% of the modulus of rupture, the allowable tension for normal bending can be taken as $4\sqrt{f_c'}$ or 300 psi. If the prestress is 700 psi, the total stress available for bending is 1,000 psi and the moment capacity is M = 1000 l/c.

This usually gives a safety factor of 2.5 or more. For earthquake and other transient loads this is conservative and the allowable tension may be increased to 600 psi. This gives a moment capacity of M = 1300 l/c.

(b) Strength design

The nominal moment capacity of a pretensioned pile can be computed by the ultimate strength of the tendons multiplied by a proper lever arm. As an approximation, the total ultimate strength in all the tendons can be used. The lever arm is approximately 0.37 t for solid square piles and 0.32 t for solid circular and octagonal piles. For hollow piles, the lever arm will be a little longer, approximately 0.38 t for square piles and 0.34 t for circular and octagonal piles.

Thus, the approximate nominal moment strength is given by:

M_n = 0.37 t $A_{ps} f_{pu}$ for solid square piles
M_n = 0.32 t $A_{ps} f_{pu}$ for solid circular and octagonal piles
M_n = 0.38 t $A_{ps} f_{pu}$ for hollow square piles
M_n = 0.34 t $A_{ps} f_{pu}$ for hollow circular and octagonal piles

where A_{ps} = total steel area of all the tendons in square inches, f_{pu} = ultimate strength of the tendons in psi, and t = diameter or thickness of the pile in inches.

(c) Allowable service load moment

In general, the allowable moment should be based on an allowable tensile stress based on the modulus of rupture as in (a) and checked by the nominal strength moment as in (b) to insure a factor of safety of 2 for normal loading. For wind, earthquake, or other short-time loads a safety factor of 1.5 is considered adequate. In corrosive conditions the engineer should make an evaluation to determine whether to reduce or eliminate the allowable tension.

4.7.6.4 Combined Moment and Direct Load

(a) Service load stresses

By the elastic theory, the existence of direct load delays the cracking of the concrete piles and thereby increases the moment carrying capacity. For example, if the prestress in the concrete is 700 psi, an external load induces 400 psi, and the allowable tensile stress is 300 psi, the total fiber stress available for bending moment is 1400 psi, and the moment capacity is:

M = 1400 l/c

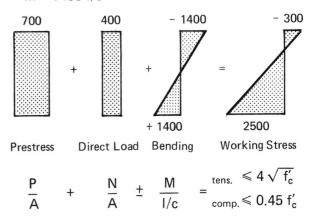

Prestress   Direct Load   Bending   Working Stress

$$\frac{P}{A} + \frac{N}{A} \pm \frac{M}{l/c} = \begin{array}{l} \text{tens.} \leqslant 4\sqrt{f_c'} \\ \text{comp.} \leqslant 0.45\, f_c' \end{array}$$

Piles controlled by compression should also be checked against the interaction formula:

$$\frac{N}{N'} + \frac{M}{M'} \leqslant 1$$

where N′ and M′ are the allowable axial loads and moments, respectively.

(b) Strength design

The nominal moment capacity of a prestressed pile is reduced by the presence of external direct load, because the area of the compression zone is increased and the available lever arm for the resisting steel is correspondingly reduced.

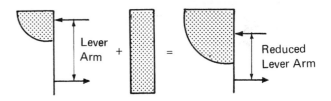

It can be roughly estimated that, for an external load producing 700 psi in the concrete, the nominal moment capacity is given by:

M_n = 0.29 t $A_{ps} f_{pu}$ for solid square piles
M_n = 0.25 t $A_{ps} f_{pu}$ for solid round and octagonal piles

For hollow piles the lever arm is a little longer, 0.30 t and 0.26 t respectively.

At the pile head, where combined moment and direct load may be critical, the transfer in which the prestress increases from zero to full value (about 50 tendon diameters) must be considered.

(c) Allowable service load moment

With the presence of external direct load, the

moment capacity of the piles should be first computed by the elastic theory as in (a), and the nominal strength checked as in (b) to insure a safety factor of 2. For wind, earthquake, or other short-time loads, a safety factor of 1.5 is considered adequate. In corrosive conditions the engineer should make an evaluation to determine whether to reduce or eliminate the allowable tension.

Example 4.7.4 Bearing pile

For a 12″ square solid pile, compute the allowable load and moments.

Given:

A = 144 sq in.
I = 1,728 in.^{4}
I/c = 288 in.^{3}

Prestress with 6 - 7/16″ diameter 270K strands.

A_{ps} = 0.69 sq in.
f_{pu} = 270,000 psi or 31,000 lb. per strand
f'_c = 6000 psi

If the strands are stressed initially to 0.7 f_{pu} and total losses are assumed to be 22%, effective prestress = 0.7 x 0.78 x 31,000 = 16,900 lb/strand

$$f_{pc} = \frac{6 \times 16,900}{144} = 705 \text{ psi}$$

(a) Direct Load

Using the formula N = $(0.33 f'_c - 0.27 f_{pc})$ A
N = 1,790 A

Allowable N = 1,790 x 144 = 258 kips or 129 tons.

Nominal strength is given by:

P_n = $(0.85 f'_c - 0.60 f_{pc})$ A
= $(0.85 \times 6,000 - 0.60 \times 705)$ 144 = 673 kips.

Thus, factor of safety $\frac{673}{258}$ = 2.61

(b) Moment Capacity

For an allowable tension of 300 psi:
M = f_c I/c
= (300 + 705) 288/1000 = 289 in.-kips

Nominal moment strength:

M_n = 0.37 t A_{ps} f_{pu} = 0.37 x 12 x 0.69 x 270
= 827 in.-kips

Thus, factor of safety = $\frac{827}{289}$ = 2.86, which is higher than necessary. Thus, for transient loads, the allowable tension could be increased beyond 300 psi.

(c) Allowable Unsupported Length

This will vary with the load. The maximum h/r for which the short column formula should be used is 60. From pile properties in Table 2.7.1, r = 3.46 in. for a 12-in. square pile. Thus, for a sustained load of 258 kips, and assuming pile fully fixed at both ends, the allowable unsupported length

$$\ell_u = \frac{60r}{0.5} = 415 \text{ in.} = 34.6 \text{ ft}$$

Example 4.7.5 Sheet pile

For a 12″ x 36″ sheet pile, compute allowable moment.

Given:

A = 432 sq in.
I = 5184 in.^{4}
I/c = 864 in.^{3} (per pile)

Prestress with 20 - 1/2″ 270 K strands

A_{ps} = 3.06 sq in.
f_{pu} = 270,000 psi or 41,300 lb per strand
f'_c = 6000 psi

Effective prestress = 0.7 x 0.78 x 41,300
= 22,560 lb/strand

$$f_{pc} = \frac{20 \times 22,560}{432} = 1045 \text{ psi}$$

Moment Capacity

(a) Service loads

For allowable tension of 300 psi:

Allowable M = (300 + 1045) x 864/1000
= 1160 in.-kips

or $\frac{1160}{3}$ = 386 in.-kips per foot of wall.

(b) Strength design

Nominal moment strength, M_n = 0.37 t A_{ps} f_{pu}
= (0.37 x 12 x 3.06 x 270) = 3670 in.-kips

indicating a factor of safety of $\frac{3670}{1160}$ = 3.1,

which is more than sufficient.

Thus, the above allowable tension of 300 psi may be increased for transient loads, except in corrosive conditions.

4.8 Special Considerations

This section outlines solutions of special situations which may arise in the design of a precast floor or roof system. Since production methods of products vary, local producers should be consulted. Also, test data may indicate that the conservative guidelines presented here may be exceeded for a specific application.

4.8.1 Load Distribution

Frequently, floors and roofs are subjected to line loads, for example from walls, and concentrated loads. The ability of hollow-core systems to transfer or distribute loads laterally through grouted shear keys has been demonstrated in several published tests,[12-15] and many unpublished tests. Research is continuing, and the recommendations here may be refined in the future. Based on tests,

analysis and experience, the PCI Hollow-Core Slab Producers Committee recommends that line and concentrated loads be resisted by an effective section as described in Fig. 4.8.1. Exception: If the total deck width, perpendicular to the span, is less than the span, modification may be required. Contact local producers for recommendations.

Load distribution of stemmed members may not necessarily follow the same pattern, because of different torsional resistance properties. Research on this matter[15] was being conducted at the time of publication.

Example 4.8.1 Load distribution

Given:

An untopped hollow-core floor with 4 ft wide slabs, and supporting a load bearing wall and concentrated loads as shown in Fig. 4.8.2.

Problem:

Determine the design loads for the slab sup-

Fig. 4.8.1 Assumed load distribution

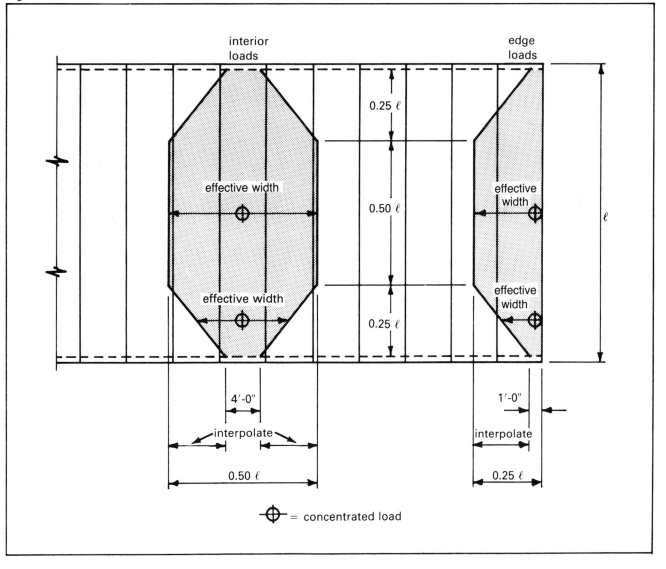

porting the wall and concentrated loads.

Solution:

(Note: Each step corresponds to a line number in table in Fig. 4.8.2)

1. Calculate the shears and moments for the non-distributable (uniform) loads:

   $w_u = 1.4(56 + 10) + 1.7(40) = 160$ psf

2. Calculate the shears and moments for the distributable (concentrated and line) loads:

   $w_u = 1.4(650) + 1.7(1040) = 2678$ lb/ft
   $P_{1u} = 1.4(500) + 1.7(1000) = 2400$ lb
   $P_{2u} = 1.4(1000) + 1.7(3000) = 6500$ lb

3. Calculate effective width along the span:

   At the support, width = 4.0 ft
   At 0.25ℓ (6.25 ft), width = 0.5ℓ = 12.5 ft
   Between x = 0 and x = 6.25 ft:
   width = 4 + (x/6.25)(12.5 − 4) = 4 + 1.36x

4. Divide distributable shears and moments from step 2 by the effective widths from step 3.

5. Add the distributed shears and moments to the non-distributable shears and moments from step 1.

Once the moments and shears are determined, the slabs are designed as described in Sects. 4.2, 4.3, and 4.6.

This method is suitable for computer solution. For manual calculations, the procedure can be simplified by investigating only critical sections. For example, shear may be determined by dividing all distributable loads by 4 ft, and flexure at midspan can be checked by dividing the distributable loads by 0.5ℓ.

4.8.2 Effect of Openings

Openings may be provided in precast decks by: (1) saw cutting after the deck is installed and grouted, (2) forming (blocking out) or sawing in the plant, or (3) using short units with steel headers or other connections. In hollow-core or solid slabs, structural capacity is least affected by orienting the longest dimension of an opening parallel to a span, aligning several openings parallel to a span, or by coring small holes to cut the fewest strands. Openings in stemmed members must not cut through the stem, and should be narrower than the distance between stems less the top stem width less 4 in.

Following are some conservative guidelines regarding design of hollow-core slabs around openings. Some producers may have data to support a different procedure:

1. An opening located near the end of the span and extending into the span less than the lesser of 0.125ℓ or 4 ft may be neglected when designing for flexure in the midspan region.

2. Strand development must be considered on each side of an opening which cuts strand. (See Sect. 4.2.3.)

3. Slabs which are adjacent to openings which are long ($\ell/4$ or more), or occur near midspan, may be considered to have a free edge for flexural design.

4. Slabs which are adjacent to openings closer to the end than $3\ell/8$ may be considered to have a free edge for shear design.

4.8.3 Continuity

Precast deck members are normally used as part of a simple span system. However, when reinforcement is required at supports for structural integrity ties or diaphragm connections, limited continuity can be achieved. The amount of reinforcement is usually too low to develop significant moment capacity, but may be considered for reducing service load deflections. It is recommended that full simple span positive moment capacity be provided for strength design in all deck members because of the uncertain moment-curvature conditions existing at the supports at ultimate loads.

When top steel is required at supports, it may be placed in composite topping (if available), or, in slabs, in the grout keys or concreted into cores.

Advantage may be taken of limited continuity when using rational design procedures for fire resistance—see Sect. 9.3.

4.8.4 Cantilevers

The method by which precast, prestressed members resist cantilever moments depends on (1) the method of production, (2) the length and loading requirements of the cantilever, and (3) the size of the project (amount of repetition).

If a cantilever is not long enough to fully develop top strands, a reduced value of f_{ps} must be used, as discussed in Sect. 4.2.3. As with reinforcing bars, top strands often do not bond as well as the ACI development equation indicates, especially in dry cast systems. In many cases it is necessary, or at least desirable, to debond the top strands in positive moment regions, and the bottom strands in the cantilever.

In some cases it is preferable to design cantilevers as reinforced concrete members, using deformed reinforcing bars to provide the negative

Fig. 4.8.2 Example 4.8.1

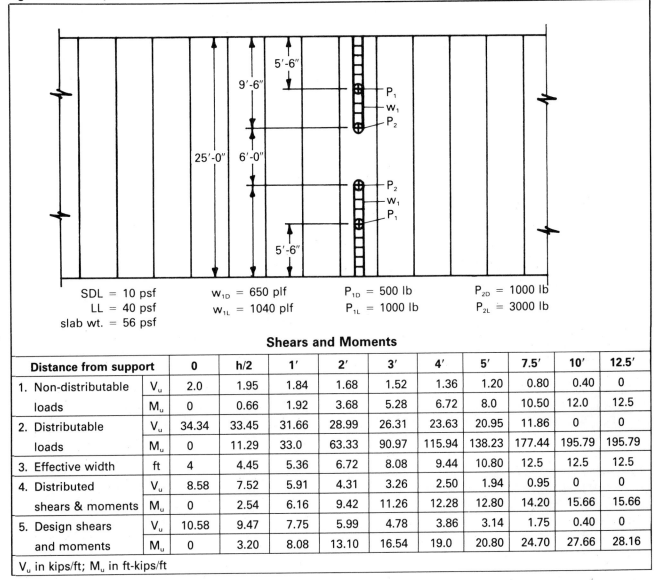

		SDL = 10 psf	w_{1D} = 650 plf	P_{1D} = 500 lb	P_{2D} = 1000 lb
LL = 40 psf	w_{1L} = 1040 plf	P_{1L} = 1000 lb	P_{2L} = 3000 lb
slab wt. = 56 psf			

Shears and Moments

Distance from support		0	h/2	1'	2'	3'	4'	5'	7.5'	10'	12.5'
1. Non-distributable	V_u	2.0	1.95	1.84	1.68	1.52	1.36	1.20	0.80	0.40	0
loads	M_u	0	0.66	1.92	3.68	5.28	6.72	8.0	10.50	12.0	12.5
2. Distributable	V_u	34.34	33.45	31.66	28.99	26.31	23.63	20.95	11.86	0	0
loads	M_u	0	11.29	33.0	63.33	90.97	115.94	138.23	177.44	195.79	195.79
3. Effective width	ft	4	4.45	5.36	6.72	8.08	9.44	10.80	12.5	12.5	12.5
4. Distributed	V_u	8.58	7.52	5.91	4.31	3.26	2.50	1.94	0.95	0	0
shears & moments	M_u	0	2.54	6.16	9.42	11.26	12.28	12.80	14.20	15.66	15.66
5. Design shears	V_u	10.58	9.47	7.75	5.99	4.78	3.86	3.14	1.75	0.40	0
and moments	M_u	0	3.20	8.08	13.10	16.54	19.0	20.80	24.70	27.66	28.16

V_u in kips/ft; M_u in ft-kips/ft

moment resistance. In machine-made products, the steel can be placed in grout keys, composite topping, or concreted into cores.

Top tension under service loads should be limited to $6\sqrt{f'_c}$ or less so that the section remains uncracked, allowing better prediction of service load deflections.

Consultation with local producers is recommended before choosing a method of reinforcement for cantilevers. Some have developed standard methods that work best with their particular system, and have proven them with tests or experience.

4.8.5 Composite Topping with Hollow-Core Slabs

Most precast floor and roof systems are untopped, but a composite, cast-in-place concrete topping is sometimes used in floor construction. The composite action adds stiffness and strength for gravity loads, and may be required as a diaphragm to transfer lateral loads.

Tests have shown that the normal finished surface of hollow-core and stemmed deck members will develop the 80 psi on the interface surface specified in Chapter 17 of ACI 318-83, if the precast surface is thoroughly cleaned before topping is placed. Sect. 4.3.5 of this Handbook describes design procedures for horizontal shear transfer. In nearly all cases, the allowable 80 psi is enough to develop the full strength of the topping on precast decks.

The strength of the topping may be determined from the design requirements for the deck. Producer's load tables usually indicate either 3000 or 4000 psi. The load tables in Part 2 are based on a topping strength of 3000 psi.

4.9 References

1. "Building Code Requirements for Reinforced Concrete (ACI 318-83)", American Concrete Institute, Detroit, MI.

2. Naaman, A. E., "Ultimate Analysis of Prestressed and Partially Prestressed Sections by Strain Compatibility", *PCI Journal*, V.22, No. 1, Jan-Feb, 1977.

3. Gergely, P. and Lutz, L. A., "Maximum Crack Width in Reinforced Concrete Flexural Members", *Causes, Mechanisms, and Control of Cracking in Concrete*, SP-20, American Concrete Institute, Detroit, MI.

4. "Commentary on Building Code Requirements for Reinforced Concrete (ACI 318-83)", American Concrete Institute, Detroit, MI.

5. Zia, Paul and McGee, W. D., "Torsion Design of Prestressed Concrete", *PCI Journal*, V. 19, No. 2, March-April, 1974.

6. Collins, M. P. and Mitchell, D., "Shear and Torsion Design of Prestressed and Non-Prestressed Concrete Beams", *PCI Journal*, V. 25, No. 5, Sept-Oct, 1980.

7. Zia, Paul, Preston, H. K., Scott, N. L., and Workman, E. B., "Estimating Prestress Losses", *Concrete International*, V. 1, No. 6, June, 1979.

8. Branson, D. E., "The Deformation of Noncomposite and Composite Prestressed Concrete Members", *Deflections of Concrete Structures*, SP-43, American Concrete Institute, Detroit, MI.

9. Martin, L. D., "A Rational Method of Estimating Camber and Deflections of Precast, Prestressed Concrete Members", *PCI Journal*, V. 22, No. 1, Jan-Feb, 1977.

10. Shaikh, A. F., and Branson, D. E., "Non-tensioned Steel in Prestressed Concrete Beams", *PCI Journal*, V. 15, No. 1, Feb, 1970.

11. PCI Committee on Prestressed Columns, "Recommended Practice for the Design of Prestressed Columns and Walls", *PCI Journal*, Scheduled for publication 1985.

12. LaGue, David J., "Load Distribution Tests on Precast Prestressed Hollow-Core Slab Construction", *PCI Journal*, Vol. 16, No. 6, Nov-Dec, 1971.

13. Johnson, Ted and Ghadiali, Zohair, "Load Distribution Test on Precast Hollow-Core Slabs with Openings", *PCI Journal*, Vol. 17, No. 5, Sept-Oct, 1972.

14. Pfeifer, Donald W. and Nelson, Theodore A., "Tests to Determine the Lateral Distribution of Vertical Loads in a Long-Span Hollow-Core Floor Assembly", *PCI Journal*, Vol. 28, No. 6, Nov-Dec, 1983.

15. "Lateral Distribution of Loads on Prestressed Concrete Decks" PCI sponsored research project at University of Washington. Scheduled for completion 1985.

FLEXURE

Fig. 4.10.1 Flexural resistance coefficients for elements with non-prestressed, partially prestressed and prestressed reinforcement

Procedure:

Design:

1. Determine $K_u = \dfrac{M_u/\phi - A_s'f_y'(d - d_s')}{f_c'bd^2}$

2. Find $\overline{\omega}$ from table

3. For prestressed reinforcement, estimate f_{ps}

4. Select A_{ps}, A_s and A_s' from:
$A_{ps}f_{ps} + A_sf_y - A_s'f'_y = \overline{\omega}bdf_c'$

5. Check assumed value of f_{ps}

Analysis:

1. Determine $\overline{\omega} = \dfrac{A_{ps}f_{ps} + A_sf_y - A_s'f'_y}{bdf_c'}$

2. Find K_u from table

3. Determine $\phi M_n = \phi [K_uf_c'bd^2 + A_s'f_y'(d - d_s')]$

Basis:

$K_u = \overline{\omega}(1 - 0.59\overline{\omega})$

M_u in units of in.-lb

	f_c', psi	$\omega_{max} = \overline{\omega}_{max}$ for non-prestressed elements					
f_y, ksi		3000	4000	5000	6000	7000	8000
40		0.371	0.371	0.349	0.328	0.306	0.284
50		0.344	0.344	0.324	0.304	0.283	0.263
60		0.321	0.321	0.302	0.283	0.264	0.245

	Values of K_u									
$\overline{\omega}$	0.000	0.001	0.002	0.003	0.004	0.005	0.006	0.007	0.008	0.009
0.00	0.0000	0.0010	0.0020	0.0030	0.0040	0.0050	0.0060	0.0070	0.0080	0.0090
0.01	0.0099	0.0109	0.0119	0.0129	0.0139	0.0149	0.0158	0.0168	0.0178	0.0188
0.02	0.0198	0.0207	0.0217	0.0227	0.0237	0.0246	0.0256	0.0266	0.0275	0.0285
0.03	0.0295	0.0304	0.0314	0.0324	0.0333	0.0343	0.0352	0.0362	0.0371	0.0381
0.04	0.0391	0.0400	0.0410	0.0419	0.0429	0.0438	0.0448	0.0457	0.0466	0.0476
0.05	0.0485	0.0495	0.0504	0.0513	0.0523	0.0532	0.0541	0.0551	0.0560	0.0569
0.06	0.0579	0.0588	0.0597	0.0607	0.0616	0.0625	0.0634	0.0644	0.0653	0.0662
0.07	0.0671	0.0680	0.0689	0.0699	0.0708	0.0717	0.0726	0.0735	0.0744	0.0753
0.08	0.0762	0.0771	0.0780	0.0789	0.0798	0.0807	0.0816	0.0825	0.0834	0.0843
0.09	0.0852	0.0861	0.0870	0.0879	0.0888	0.0897	0.0906	0.0914	0.0923	0.0932
0.10	0.0941	0.0950	0.0959	0.0967	0.0976	0.0985	0.0994	0.1002	0.1011	0.1020
0.11	0.1029	0.1037	0.1046	0.1055	0.1063	0.1072	0.1081	0.1089	0.1098	0.1106
0.12	0.1115	0.1124	0.1132	0.1141	0.1149	0.1158	0.1166	0.1175	0.1183	0.1192
0.13	0.1200	0.1209	0.1217	0.1226	0.1234	0.1242	0.1251	0.1259	0.1268	0.1276
0.14	0.1284	0.1293	0.1301	0.1309	0.1318	0.1326	0.1334	0.1343	0.1351	0.1359
0.15	0.1367	0.1375	0.1384	0.1392	0.1400	0.1408	0.1416	0.1425	0.1433	0.1441
0.16	0.1449	0.1457	0.1465	0.1473	0.1481	0.1489	0.1497	0.1505	0.1513	0.1521
0.17	0.1529	0.1537	0.1545	0.1553	0.1561	0.1569	0.1577	0.1585	0.1593	0.1601
0.18	0.1609	0.1617	0.1625	0.1632	0.1640	0.1648	0.1656	0.1664	0.1671	0.1679
0.19	0.1687	0.1695	0.1703	0.1710	0.1718	0.1726	0.1733	0.1741	0.1749	0.1756
0.20	0.1764	0.1772	0.1779	0.1787	0.1794	0.1802	0.1810	0.1817	0.1825	0.1832
0.21	0.1840	0.1847	0.1855	0.1862	0.1870	0.1877	0.1885	0.1892	0.1900	0.1907
0.22	0.1914	0.1922	0.1929	0.1937	0.1944	0.1951	0.1959	0.1966	0.1973	0.1981
0.23	0.1988	0.1995	0.2002	0.2010	0.2017	0.2024	0.2031	0.2039	0.2046	0.2053
0.24	0.2060	0.2067	0.2074	0.2082	0.2089	0.2096	0.2103	0.2110	0.2117	0.2124
0.25	0.2131	0.2138	0.2145	0.2152	0.2159	0.2166	0.2173	0.2180	0.2187	0.2194
0.26	0.2201	0.2208	0.2215	0.2222	0.2229	0.2236	0.2243	0.2249	0.2256	0.2263
0.27	0.2270	0.2277	0.2283	0.2290	0.2297	0.2304	0.2311	0.2317	0.2324	0.2331
0.28	0.2337	0.2344	0.2351	0.2357	0.2364	0.2371	0.2377	0.2384	0.2391	0.2397
0.29	0.2404	0.2410	0.2417	0.2423	0.2430	0.2437	0.2443	0.2450	0.2456	0.2463
0.30	0.2469									

FLEXURE

Fig. 4.10.2 Coefficients, K_u', for determining flexural design strength — bonded prestressing steel

Procedure:

1. Determine $\omega_{pu} = \dfrac{A_{ps}}{bd_p}\dfrac{f_{pu}}{f_c'}$

2. Find K_u' from table

3. Determine $\phi M_n = K_u' \dfrac{bd_p^2}{12,000}$ (ft–kips)

Basis:

$$K_u' = \frac{\phi\, f_{ps} f_c'}{f_{pu}}(\omega_{pu})\left[1 - (0.59\omega_{pu})\left(\frac{f_{ps}}{f_{pu}}\right)\right]$$

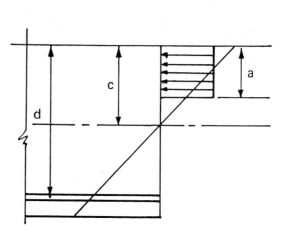

Note: K_u' from this table is approximately equivalent to $\phi K_u f_c'$ from Table 4.10.1.

Table values are based on a strain compatibility analysis, using a stress-strain curve for prestressing strand similar to that shown in Fig. 11.2.5. Asterisk(*) indicates $\omega_{pu} > 0.366\beta_1$, and $\phi M_n = \phi[f_c' b d_p^2 (0.36\,\beta_1 - 0.08\,\beta_1^2)]$

Values of K_u'

f_u'	ω_{pu}	.00	.01	.02	.03	.04	.05	.06	.07	.08	.09
3000 psi	0.0	0	26	52	78	103	128	153	178	202	225
	0.1	248	271	293	315	337	379	379	399	419	439
	0.2	458	477	495	513	529	562	562	577	592	607
	0.3	621	634	646	658	670	670*	670*	670*	670*	670*
4000 psi	0.0	0	35	70	104	138	171	204	237	269	300
	0.1	331	361	391	420	449	477	505	532	559	585
	0.2	610	636	660	683	706	728	750	770	790	809
	0.3	827	845	861	878	893	894*	894*	894*	894*	894*
5000 psi	0.0	0	47	93	138	183	227	271	314	356	397
	0.1	438	477	516	554	592	629	665	700	735	769
	0.2	802	834	865	895	924	951	978	1004	1029	1052
	0.3	1074	1066*	1066*	1066*	1066*	1066*	1066*	1066*	1066*	1066*
6000 psi	0.0	0	60	118	176	234	290	346	401	454	506
	0.1	557	607	656	704	752	798	843	887	931	973
	0.2	1014	1052	1090	1127	1161	1194	1215*	1215*	1215*	1215*
	0.3	1215*	1215*	1215*	1215*	1215*	1215*	1215*	1215*	1215*	1215*
7000 psi	0.0	0	74	148	220	291	362	431	499	564	629
	0.1	692	753	814	873	931	988	1043	1097	1149	1199
	0.2	1247	1293	1337	1341*	1341*	1341*	1341*	1341*	1341*	1341*
	0.3	1341*	1341*	1341*	1341*	1341*	1341*	1341*	1341*	1341*	1341*
8000 psi	0.0	0	92	182	271	358	444	529	610	690	768
	0.1	845	920	993	1064	1133	1201	1267	1331	1391	1449
	0.2	1441*	1441*	1441*	1441*	1441*	1441*	1441*	1441*	1441*	1441*
	0.3	1441*	1441*	1441*	1441*	1441*	1441*	1441*	1441*	1441*	1441*

FLEXURE

Fig. 4.10.3 Values of f_{ps} by stress-strain relationship — bonded strand

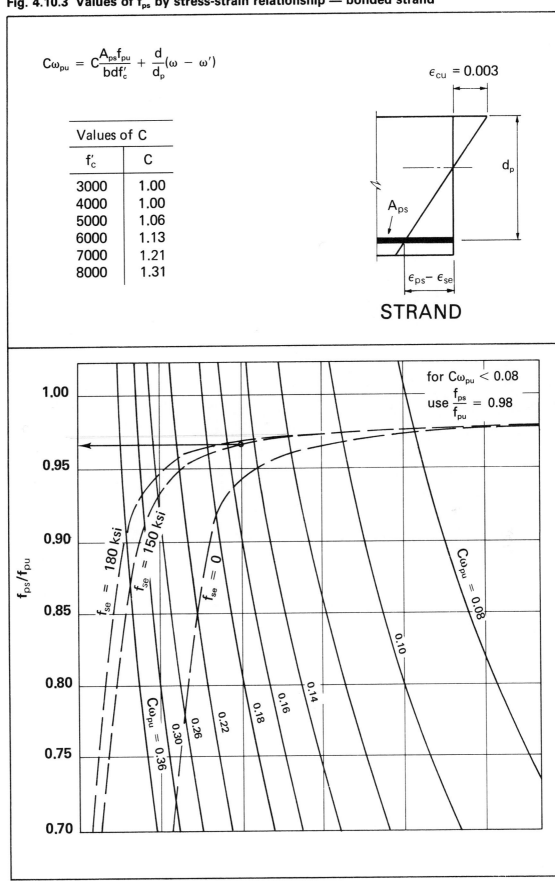

$$C\omega_{pu} = C\frac{A_{ps}f_{pu}}{bdf'_c} + \frac{d}{d_p}(\omega - \omega')$$

Values of C

f'_c	C
3000	1.00
4000	1.00
5000	1.06
6000	1.13
7000	1.21
8000	1.31

$\epsilon_{cu} = 0.003$

STRAND

for $C\omega_{pu} < 0.08$

use $\dfrac{f_{ps}}{f_{pu}} = 0.98$

Fig. 4.10.4 Design stress for underdeveloped strand

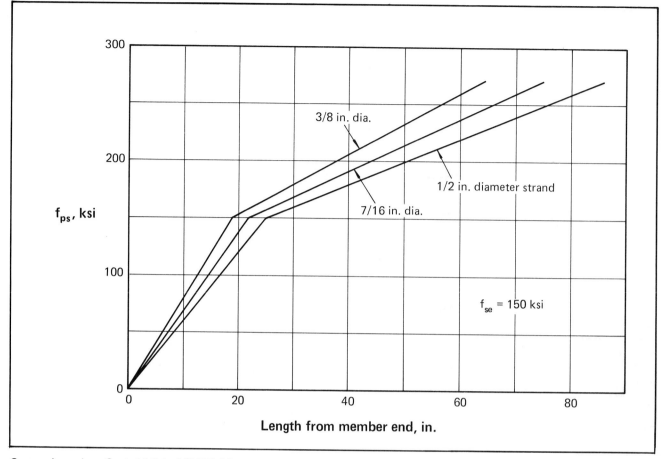

Curves based on Sect. 12.9.1, ACI 318-83

SHEAR

Fig. 4.10.5 Concrete shear strength by Eq. 11-10 (ACI 318-83)

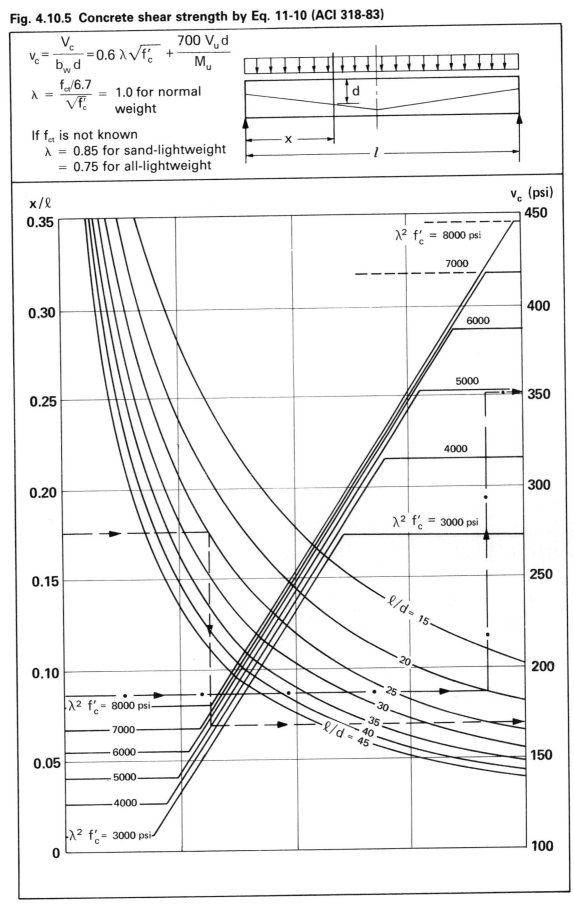

$$v_c = \frac{V_c}{b_w d} = 0.6\,\lambda\,\sqrt{f'_c} + \frac{700\,V_u d}{M_u}$$

$$\lambda = \frac{f_{ct}/6.7}{\sqrt{f'_c}} = 1.0 \text{ for normal weight}$$

If f_{ct} is not known
$\lambda = 0.85$ for sand-lightweight
$= 0.75$ for all-lightweight

SHEAR

Fig. 4.10.6 Shear design by Eq. 11-10 (ACI 318-83)   **Straight strands**

Notes:

1. Applicable to simple span, uniformly loaded members only
2. f'_c = 5000 psi — Error less than 10% for 4000 psi to 6000 psi

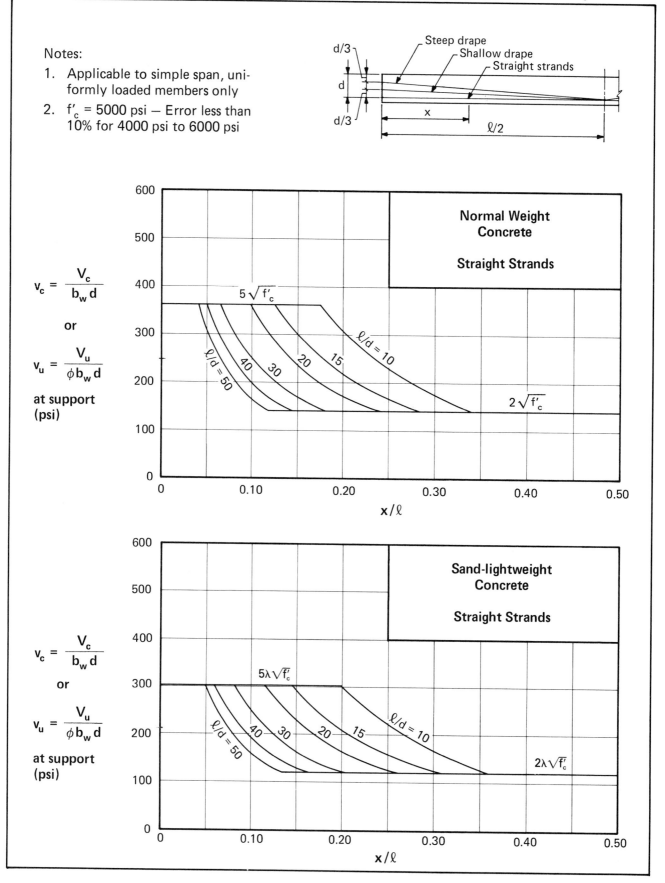

$$v_c = \frac{V_c}{b_w d}$$

or

$$v_u = \frac{V_u}{\phi b_w d}$$

at support (psi)

SHEAR

Fig. 4.10.7 Shear design by Eq. 11-10 (ACI 318-83)   Shallow drape

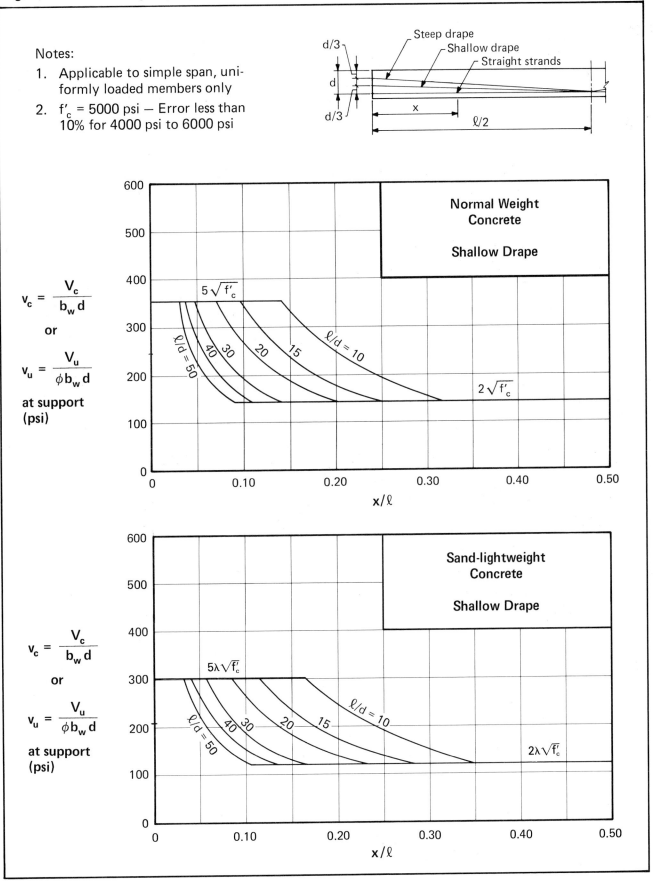

Notes:

1. Applicable to simple span, uniformly loaded members only
2. f'_c = 5000 psi — Error less than 10% for 4000 psi to 6000 psi

$$v_c = \frac{V_c}{b_w d}$$

or

$$v_u = \frac{V_u}{\phi b_w d}$$

at support (psi)

Normal Weight Concrete

Shallow Drape

$5\sqrt{f'_c}$

$\ell/d = 50$ 40 30 20 15 $\ell/d = 10$

$2\sqrt{f'_c}$

x/ℓ

$$v_c = \frac{V_c}{b_w d}$$

or

$$v_u = \frac{V_u}{\phi b_w d}$$

at support (psi)

Sand-lightweight Concrete

Shallow Drape

$5\lambda\sqrt{f'_c}$

$\ell/d = 50$ 40 30 20 15 $\ell/d = 10$

$2\lambda\sqrt{f'_c}$

x/ℓ

SHEAR

Fig. 4.10.8 Shear design by Eq. 11-10 (ACI 318-83)   Steep drape

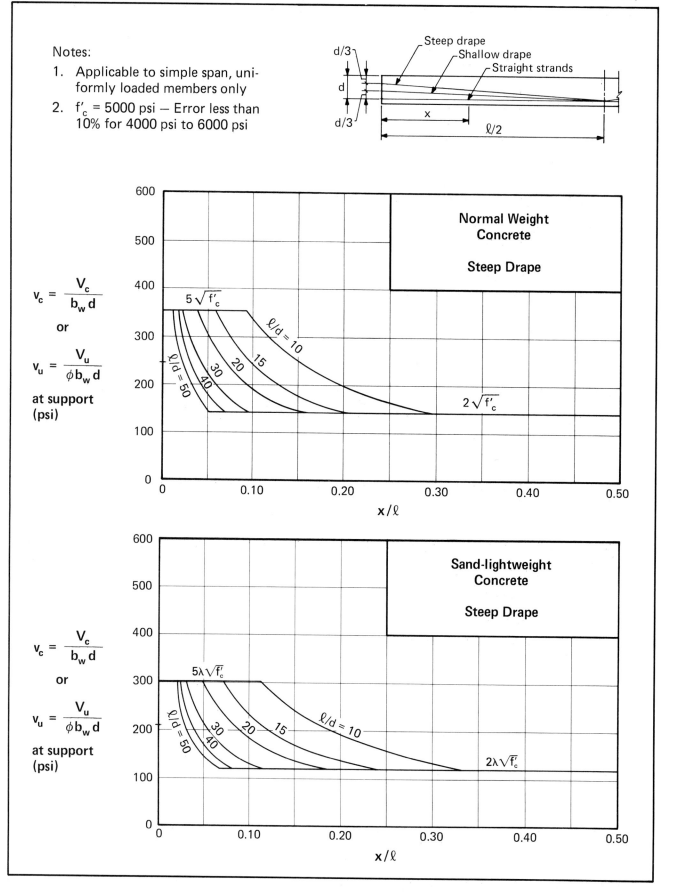

Notes:

1. Applicable to simple span, uniformly loaded members only
2. f'_c = 5000 psi — Error less than 10% for 4000 psi to 6000 psi

$$v_c = \frac{V_c}{b_w d}$$

or

$$v_u = \frac{V_u}{\phi b_w d}$$

at support (psi)

Normal Weight Concrete

Steep Drape

$5\sqrt{f'_c}$

$\ell/d = 10$

$\ell/d = 50$ 40 30 20 15

$2\sqrt{f'_c}$

x/ℓ

$$v_c = \frac{V_c}{b_w d}$$

or

$$v_u = \frac{V_u}{\phi b_w d}$$

at support (psi)

Sand-lightweight Concrete

Steep Drape

$5\lambda\sqrt{f'_c}$

$\ell/d = 10$

$\ell/d = 50$ 40 30 20 15

$2\lambda\sqrt{f'_c}$

x/ℓ

SHEAR

Fig. 4.10.9 Concrete shear strength at support by Eq. 11-13 (ACI 318-83)

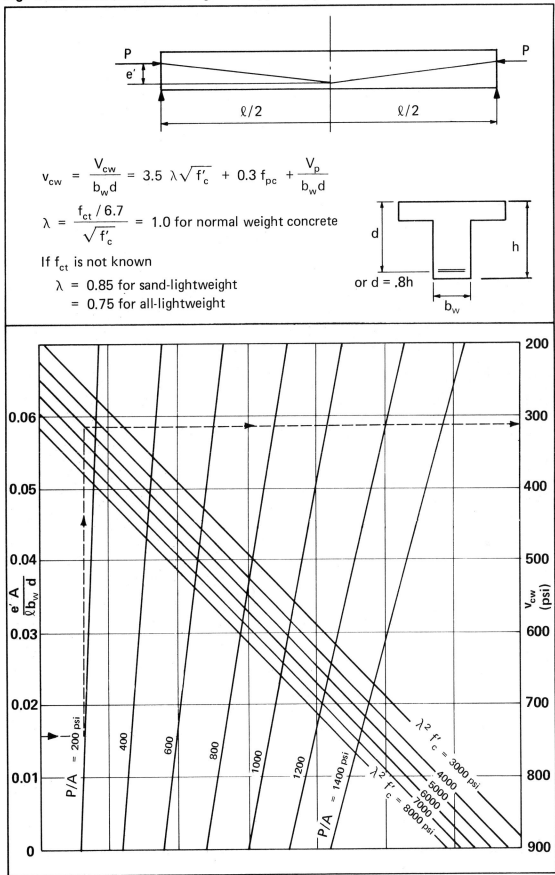

$$v_{cw} = \frac{V_{cw}}{b_w d} = 3.5\,\lambda\sqrt{f'_c} + 0.3\,f_{pc} + \frac{V_p}{b_w d}$$

$$\lambda = \frac{f_{ct}/6.7}{\sqrt{f'_c}} = 1.0 \text{ for normal weight concrete}$$

If f_{ct} is not known

λ = 0.85 for sand-lightweight

   = 0.75 for all-lightweight

or d = .8h

SHEAR

Fig. 4.10.10 v_{cw} corresponding to principal tensile stress = $4\lambda\sqrt{f'_c}$

$$4\lambda\sqrt{f'_c} = \sqrt{v_{cw}^2 + \left(\frac{f_c}{2}\right)^2} - \frac{f_c}{2}$$

$$v_{cw} = \frac{V_{cw}}{b_w d} = \sqrt{\left(4\lambda\sqrt{f'_c} + \frac{f'_c}{2}\right)^2 - \left(\frac{f_c}{2}\right)^2}$$

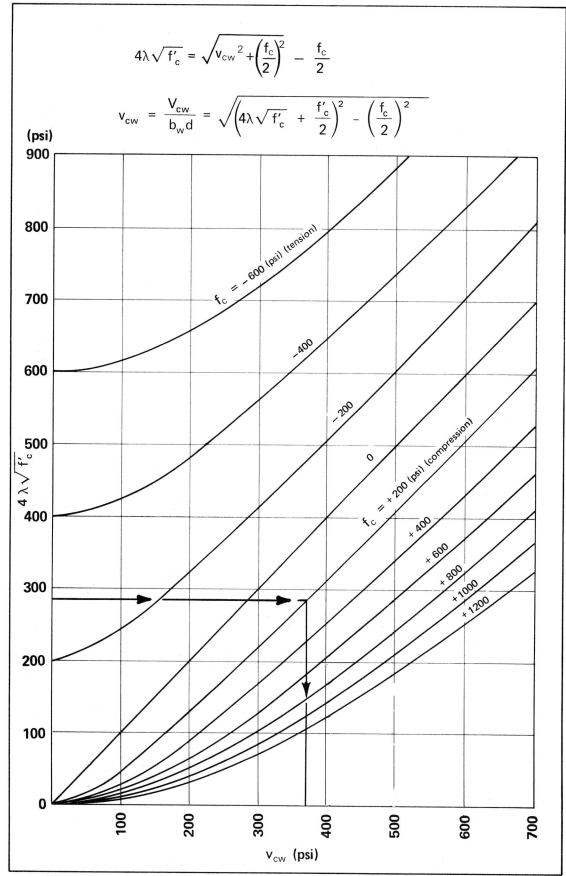

PCI Design Handbook

SHEAR

Fig. 4.10.11 Minimum shear reinforcement by Eq. 11-15 (ACI 318-83)

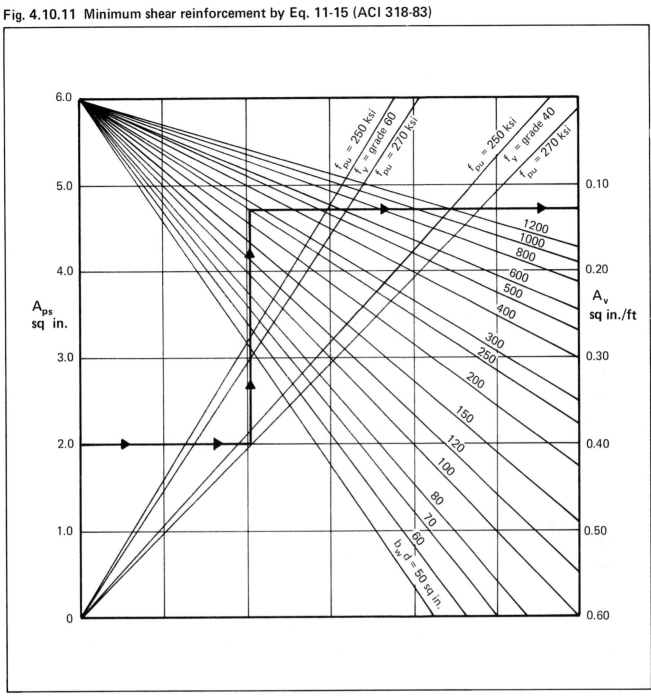

SHEAR

Fig. 4.10.12 Shear reinforcement

$$A_v = \frac{(v_u - v_c)\, b_w\, s}{f_y}$$

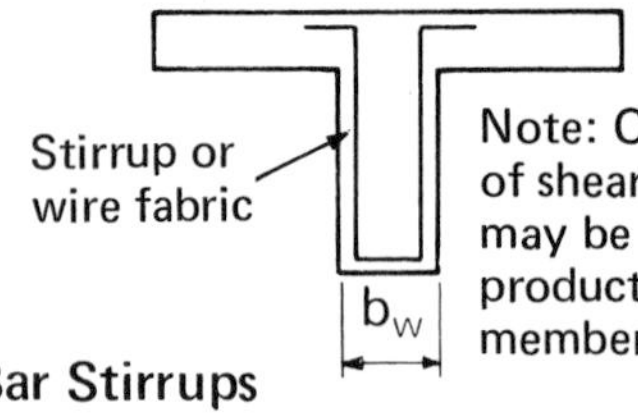

Stirrup or wire fabric

b_w

Note: Other configurations of shear reinforcement may be used to simplify production of precast members.

Vertical Deformed Bar Stirrups

Stirrup spacing (in.)	Maximum values of $(v_u - v_c)\, b_w$ (lb./in.)						Stirrup spacing (in.)
	f_y = 40,000 psi			f_y = 60,000 psi			
	No. 3 A_v = 0.22	No. 4 A_v = 0.40	No. 5 A_v = 0.62	No. 3 A_v = 0.22	No. 4 A_v = 0.40	No. 5 A_v = 0.62	
2.0	4400	8000	12400	6600	12000	18600	2.0
2.5	3520	6400	9920	5280	9600	14880	2.5
3.0	2933	5333	8267	4400	8000	12400	3.0
3.5	2514	4571	7086	3771	6857	10629	3.5
4.0	2200	4000	6200	3300	6000	9300	4.0
4.5	1956	3556	5511	2933	5333	8267	4.5
5.0	1760	3200	4960	2640	4800	7440	5.0
5.5	1600	2909	4509	2400	4364	6764	5.5
6.0	1467	2667	4133	2200	4000	6200	6.0
7.0	1257	2286	3543	1886	3429	5314	7.0
8.0	1100	2000	3100	1650	3000	4650	8.0
9.0	978	1778	2756	1467	2667	4133	9.0
10.0	880	1600	2480	1320	2400	3720	10.0
11.0	800	1455	2255	1200	2182	3382	11.0
12.0	733	1333	2067	1100	2000	3100	12.0
13.0	677	1231	1908	1015	1846	2862	13.0
14.0	629	1143	1771	943	1714	2657	14.0
15.0	587	1067	1653	880	1600	2480	15.0
16.0	550	1000	1550	825	1500	2325	16.0
17.0	518	941	1459	776	1412	2188	17.0
18.0	489	889	1378	733	1333	2067	18.0
20.0	440	800	1240	660	1200	1860	20.0
22.0	400	727	1127	600	1091	1691	22.0
24.0	367	667	1033	550	1000	1550	24.0

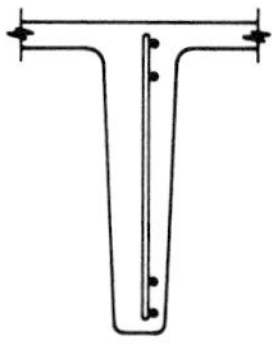

Welded Wire Fabric as Shear Reinforcement (f_y = 60,000 psi)

Spacing of vertical wire (in.)	Maximum values of $(v_u - v_c)\, b_w$ (lb./in.)								Spacing of vertical wire (in.)
	One row				Two rows				
	Vertical wire				Vertical wire				
	W7.5 A_v = 0.074	W5.5 A_v = 0.054	W4 A_v = 0.040	W2.9 A_v = 0.029	W7.5 A_v = 0.148	W5.5 A_v = 0.108	W4 A_v = 0.080	W2.9 A_v = 0.058	
2	2220	1620	1200	870	4440	3240	2400	1740	2
3	1480	1080	800	580	2960	2160	1600	1160	3
4	1110	810	600	435	2220	1620	1200	870	4
6	740	540	400	290	1480	1080	800	580	6

CAMBER AND DEFLECTION

Fig. 4.10.13 Camber equations for typical strand profiles

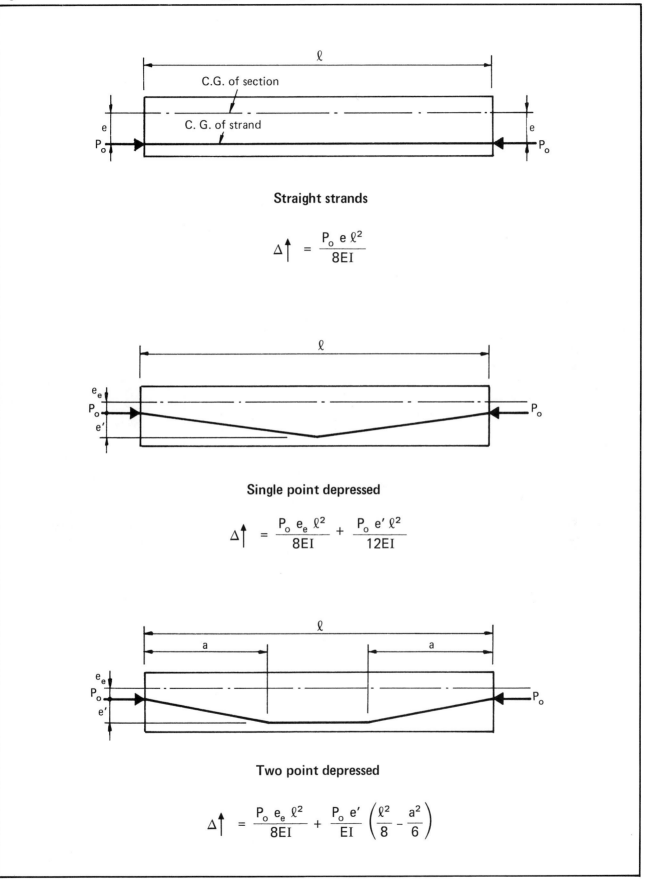

Straight strands

$$\Delta\uparrow = \frac{P_o\, e\, \ell^2}{8EI}$$

Single point depressed

$$\Delta\uparrow = \frac{P_o\, e_e\, \ell^2}{8EI} + \frac{P_o\, e'\, \ell^2}{12EI}$$

Two point depressed

$$\Delta\uparrow = \frac{P_o\, e_e\, \ell^2}{8EI} + \frac{P_o\, e'}{EI}\left(\frac{\ell^2}{8} - \frac{a^2}{6}\right)$$

CAMBER AND DEFLECTION

Fig. 4.10.14 Moment of inertia of transformed section—prestressed members

$$I_{cr} = nA_{ps}\,d^2\,(1 - 1.67\sqrt{n\rho_p})$$

$$= C\ \text{(from table)}\ \times\ bd^3$$

$$\rho_p = \frac{A_{ps}}{bd} \qquad n = \frac{E_s}{E_c}$$

Values of Coefficient, C

| | ρ_p | f'_c, psi | | | | | |
		3000	4000	5000	6000	7000	8000
Normal Weight Concrete	.0005	.0037	.0032	.0029	.0027	.0025	.0023
	.0010	.0071	.0062	.0056	.0051	.0048	.0045
	.0015	.0102	.0090	.0081	.0075	.0070	.0065
	.0020	.0132	.0116	.0105	.0097	.0090	.0085
	.0025	.0159	.0141	.0128	.0118	.0110	.0104
	.0030	.0186	.0165	.0150	.0138	.0129	.0122
	.0035	.0211	.0187	.0171	.0158	.0148	.0140
	.0040	.0235	.0209	.0191	.0177	.0166	.0157
	.0045	.0258	.0230	.0210	.0195	.0183	.0173
	.0050	.0279	.0250	.0229	.0213	.0200	.0189
	.0055	.0300	.0269	.0247	.0230	.0216	.0204
	.0060	.0320	.0287	.0264	.0246	.0231	.0219
	.0065	.0338	.0305	.0281	.0262	.0247	.0234
	.0070	.0356	.0322	.0297	.0277	.0261	.0248
	.0075	.0373	.0338	.0312	.0292	.0275	.0262
	.0080	.0390	.0354	.0327	.0306	.0289	.0275
	.0085	.0405	.0369	.0342	.0320	.0303	.0288
	.0090	.0420	.0383	.0355	.0333	.0315	.0300
	.0095	.0434	.0397	.0369	.0346	.0328	.0313
	.0100	.0447	.0410	.0382	.0359	.0340	.0324
Sand-Lightweight Concrete	.0005	.0054	.0047	.0042	.0039	.0036	.0034
	.0010	.0101	.0089	.0081	.0074	.0069	.0065
	.0015	.0145	.0128	.0116	.0107	.0100	.0094
	.0020	.0185	.0164	.0149	.0138	.0129	.0121
	.0025	.0222	.0197	.0180	.0167	.0156	.0147
	.0030	.0256	.0229	.0209	.0194	.0182	.0172
	.0035	.0288	.0258	.0237	.0220	.0207	.0196
	.0040	.0318	.0286	.0263	.0245	.0230	.0218
	.0045	.0346	.0312	.0287	.0268	.0253	.0240
	.0050	.0372	.0337	.0311	.0290	.0274	.0260
	.0055	.0396	.0360	.0333	.0312	.0294	0.280
	.0060	.0418	.0381	.0354	.0332	.0314	.0299
	.0065	.0439	.0402	.0373	.0351	.0333	.0317
	.0070	.0458	.0421	.0392	.0369	.0350	.0334
	.0075	.0475	.0438	.0410	.0387	.0367	.0351
	.0080	.0491	.0455	.0426	.0403	.0383	.0367
	.0085	.0505	.0470	.0442	.0418	.0399	.0382
	.0090	.0518	.0484	.0456	.0433	.0413	.0396
	.0095	.0530	.0498	.0470	.0447	.0427	.0410
	.0100	.0540	.0510	.0483	.0460	.0440	.0423

Fig. 4.10.15 Effective moment of inertia by Eq. 9-7 (ACI 318-83)

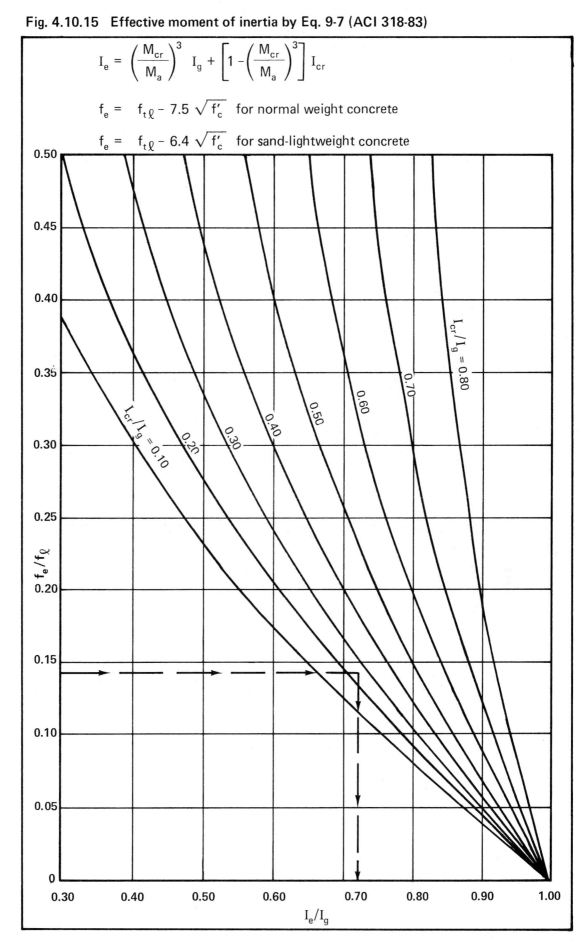

$$I_e = \left(\frac{M_{cr}}{M_a}\right)^3 I_g + \left[1 - \left(\frac{M_{cr}}{M_a}\right)^3\right] I_{cr}$$

$$f_e = f_{t\ell} - 7.5\sqrt{f'_c} \quad \text{for normal weight concrete}$$

$$f_e = f_{t\ell} - 6.4\sqrt{f'_c} \quad \text{for sand-lightweight concrete}$$

PART 5
PRODUCT HANDLING AND ERECTION BRACING

			Page No.
5.1	Notation		5–2
5.2	Product Handling		5–2
	5.2.1	Introduction	5–2
	5.2.2	Structural Design Criteria	5–3
	5.2.3	Form Suction and Impact Factors	5–3
	5.2.4	Stress Limitations	5–3
	5.2.5	Safety Factors	5–5
	5.2.6	Prestressed Wall Panels	5–5
	5.2.7	Handling Considerations	5–6
	5.2.8	Handling Devices	5–10
	5.2.9	Lateral Stability	5–14
	5.2.10	Storage	5–16
	5.2.11	Transportation	5–17
	5.2.12	Erection	5–19
	5.2.13	Design Example - Flat Panel	5–19
5.3	Erection Bracing		5–25
	5.3.1	Introduction	5–25
	5.3.2	Loads	5–27
	5.3.3	Factors of Safety	5–27
	5.3.4	Bracing Equipment and Materials	5–27
	5.3.5	Examples	5–29
5.4	References		5–36

PRODUCT HANDLING AND ERECTION BRACING

5.1 Notation

A = area; average effective area around one reinforcing bar

A = load amplification factor

A_s = area of reinforcement

a = dimension defined in section used

b = dimension defined in section used

d_c = concrete cover to the center of the reinforcement

E_c = modulus of elasticity of concrete

F = multiplication factor (Fig. 5.2.7)

f = fundamental frquency of vibration

f_b, f_t = stress in bottom and top fiber, respectively

f'_c = concrete compressive strength

f'_{ci} = concrete compressive strength at the time considered

f'_r = allowable flexural tensile stress computed using the gross concrete section

f_s = stress in steel

f_y = yield strength of steel

g = acceleration due to gravity

h_1 = distance from centroid of tensile reinforcement to neutral axis

h_2 = distance from extreme tension fiber to neutral axis

I = moment of inertia

M_x, M_y, M_z = see Figs. 5.2.4, 5.2.5, and 5.2.9

P, P_H, P_V = see Figs. 5.2.8 and 5.2.9

R = reaction (with subscripts)

T = tension on cable

t = thickness

W = total load

w = weight per unit length or area

w_b = wind load on beam

w_c = wind load on column

$w_{d\ell}$ = weight of member

w_{dy} = equivalent static load to approximate dynamic effects

y_b, y_c, y_t = see Figs. 5.2.8 and 5.2.9

y_{max} = instantaneous maximum displacement

y_t = time dependent displacement

Z_b, Z_t = section modulus with respect to bottom and top respectively

β_y = stability factor (Sect. 5.2.9)

λ = deflection amplification factor

λ = see Sect. 5.2.4

μ = coefficient of angular friction

ρ' = reinforcement ratio for non-prestressed compression reinforcement

ϕ = strength reduction factor

5.2 Product Handling

5.2.1 Introduction

The loads and forces on precast and prestressed members, especially wall panels, during production, transportation or erection will frequently be more critical than in-place loads. This is because concrete strengths are lower and support points and orientation are usually different. The production cycle, equipment used to handle the pieces, and site restrictions will also influence the design.

Most structural products are manufactured in standard steel molds with fixed dimensions. Standard product dimensions are shown in Part 2 and in manufacturer's catalogs. Architectural wall panels are formed in a variety of sizes and shapes, usually designed by the project architect.

The most economical element for a project is usually the largest, considering:

1. Stability and stresses on the element during handling.

2. Transportation size and weight regulations and equipment restrictions.

3. Available crane capacity at both the plant and the project site. Position of the crane must be considered, since capacity is a function of reach.

4. Storage space, truck turning radius, and other site restrictions.

The shape of the member must be such that concrete thickness is structurally adequate and that placement of reinforcement and concrete can be done easily.

To remove a member from a form without partially dismantling the form, and to avoid trapping

Fig. 5.2.1 Draft on sides of forms

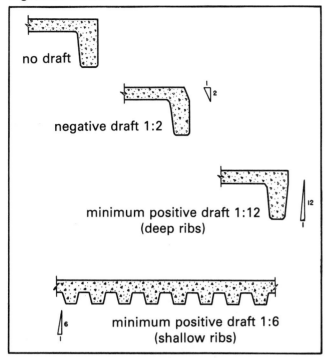

air bubbles, the sides must have adequate slope or draft (see Fig. 5.2.1).

5.2.2 Structural Design Criteria

Precast products must be designed for the loadings which occur during each phase of their existence, as shown in Fig. 5.2.2. The items which affect the forces imposed during each phase are listed below:

1. *Stripping.*
   a. Orientation of member — horizontal, vertical or some angle between.
   b. Form suction and impact — see Table 5.2.1.
   c. Number and location of handling devices.
   d. Member weight and weight of any additional items which must be lifted, such as forms which remain with the member during stripping.
2. *Yard Handling and Storage.*
   a. Orientation of the member.
   b. Location of temporary support points.
   c. Location with respect to other stored members.
   d. Orientation with respect to the sun.
3. *Transportation to the Job Site.*
   a. Orientation of the member.
   b. Location of horizontal and vertical supports.
   c. Condition of the transporting vehicle, roads and site.
   d. Dynamic considerations during movement.

4. *Erection.*
   a. Lifting point locations.
   b. Orientation and tripping (rotating).
   c. Location of temporary supports.
   d. Temporary loadings.
5. *In-place.*
   See Parts 3 and 4.

5.2.3. Form Suction and Impact Factors

To account for the forces on the member caused by form suction and impact, it is common practice to apply a multiplier to the member weight and treat the resulting force as an equivalent static service load. The multipliers cannot be quantitatively derived, so are based on experience. Table 5.2.1 provides typical values.

5.2.4. Stress Limitations

Stress limits for prestressed members during production are specified in Chapter 18 of ACI 318-83, and are discussed in Part 4. However, codes do not restrict stresses on conventionally reinforced members. When exposed to view these products may be designed for handling to (a) limit stresses so that cracks are not visible, or (b) control cracking in accordance with Sect. 4.2.2.1, considering the lower concrete strengths and actual flexural conditions. When not exposed to view, design is in accordance with the strength design requirements for reinforced concrete in ACI 318-83.

Handling without Cracking

Under this criterion, surfaces remain free of discernible cracks by limiting the flexural tension to the modulus of rupture modified by a safety factor. If the modulus of rupture is $7.5 \lambda \sqrt{f'_{ci}}$, and a safety factor of 1.5 is used:

$$f'_r = 5 \lambda \sqrt{f'_{ci}} \qquad \text{(Eq. 5.2.1)}$$

where:

f'_r = allowable flexural tensile stress computed using the gross concrete section

f'_{ci} = concrete compressive strength at the time considered

For normal weight concrete:

λ = 1.0

For lightweight concrete, if the splitting tensile strength, f_{ct}, is known:

Fig. 5.2.2. Typical handling methods

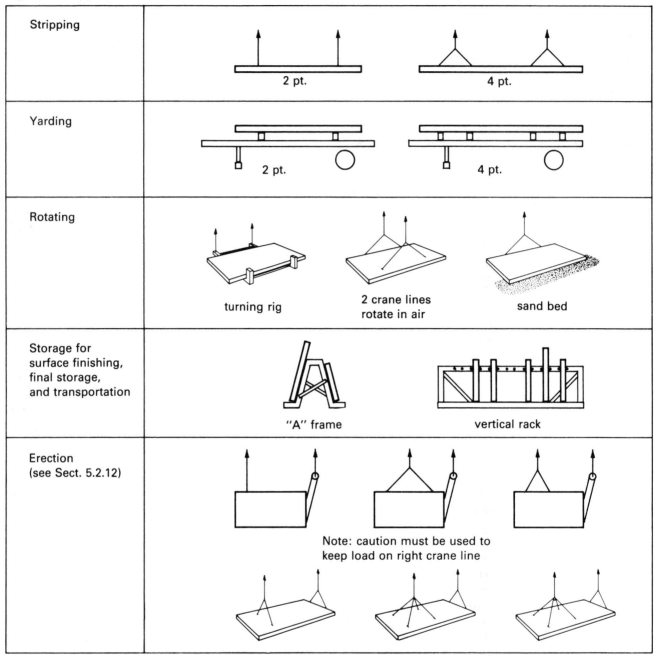

Stripping	
Yarding	
Rotating	turning rig / 2 crane lines rotate in air / sand bed
Storage for surface finishing, final storage, and transportation	"A" frame / vertical rack
Erection (see Sect. 5.2.12)	Note: caution must be used to keep load on right crane line

$$\lambda = f_{ct}/6.7$$

If f_{ct} is not known:

λ = 0.75 for all-lightweight concrete

= 0.85 for sand-lightweight concrete

Handling with Controlled Cracking

The amount and location of reinforcing steel has a negligible effect on performance until a crack develops. As flexural tension increases above the modulus of rupture, hairline cracks will develop and extend a distance into the element. If cracks are narrow and closely spaced, the structural adequacy of the element will remain unimpaired because reinforcement will be fully protected.

For members exposed to view, it is recommended that crack widths be limited to the following:

Exposed to weather: 0.005 in.
Not exposed to weather: 0.010 in.

Design procedures are illustrated in Examples 4.2.7 and 5.2.2.

Whether or not the product is designed to prevent or to control cracking, a minimum reinforcement ratio of 0.001 should be provided.

Table 5.2.1 Equivalent static load multipliers[1] to account for stripping and dynamic forces

Product Type	Finish	
	Exposed aggregate with retarder	Smooth mold (form oil only)
Stripping		
Flat, with removable side forms, no false joints or reveals	1.2	1.3
Flat, with false joints and/or reveals	1.3	1.4
Fluted, with proper draft[4]	1.4	1.6
Sculptured	1.5	1.7
Yard handling[2] and erection[3]		
All products	1.2	
Travel[2]		
All products	1.5	

1. These factors are used in flexural design of panels and are not to be applied to required safety factors on lifting devices. At stripping, suction between product and form introduces forces, which are treated here by introducing a multiplier on product weight. It would be more accurate to establish these multipliers based on the actual contact area and a suction factor independent of product weight.

2. Certain unfavorable conditions in road surface, equipment, etc., may require use of higher values.

3. Under certain circumstances may be higher.

4. For example, tees, channels and fluted panels.

5.2.5 Safety Factors

When designing for stripping and handling, the following safety factors are recommended:

1. For handling and erection devices attached to or embedded in the concrete, use the greater of 4 times the actual weight supported or 2.5 times the modified load (Table 5.2.1). For example, a device used for stripping a sculptured panel with exposed aggregate should have a safety factor of 2.5 × 1.5 = 3.75 or 4. Use 4.

2. For flexural tension in unreinforced members, or members designed "without cracking", use a factor of 1.5 applied to the modulus of rupture as discussed in Sect. 5.2.4.

3. When strength design procedures are used, the appropriate load factors and ϕ-factors from ACI 318-83 should be used with the equivalent loads calculated using Table 5.2.1.

5.2.6 Prestressed Wall Panels

When the handling procedures for wall panels cause the limitations of Sect. 5.2.4 to be exceeded,

the panel can be prestressed, using either pretensioning or post-tensioning. Design is based on Chapter 18 of ACI 318-83, as described in Part 4, with the further restriction that tensile stresses during handling should not exceed f_t' given by Eq. 5.2.1.

It is recommended that the average stress due to prestressing, after losses, be limited to a range of 150-800 psi. The prestressing force should be concentric with the effective cross section in order to minimize camber, although some manufacturers prefer to have a slight inward bow in the in-place position to counteract thermal bow (see Sect. 3.3.2). It should be noted that, since concentrically prestressed members do not camber, the form adhesion may be significant compared with other prestressed members.

In order to minimize the possibility of splitting cracks in thin pretensioned members, the strand diameter should not exceed those shown in Table 5.2.2.

When pretensioning facilities are not available, wall panels may be post-tensioned. In this case, care must be taken to assure proper transfer of

Table 5.2.2 Suggested maximum strand diameter

Panel thickness, in.	Strand diameter, in.
2½	⅜
2½ to 3½	⁷⁄₁₆
3½ and thicker	½

force at the anchorage and protection of anchors and tendons against corrosion. Straight strands or bars may be used, or, to reduce the number of anchors, the method shown in Fig. 5.2.3. Plastic coated tendons with a low coefficient of angular friction (μ = 0.03 to 0.05) are looped within the panel, and anchors installed at only one end. The tendons remain unbonded.

It should be noted that if an unbonded tendon is cut, the prestress is lost. This can sometimes happen if an unplanned opening is put in at a later date.

5.2.7 Handling Considerations

The number and location of lifting devices are chosen to keep stresses within the allowable limits, which depends on whether the "no cracking" or controlled cracking criteria are used. It is desirable to use the same lifting devices for both stripping and erection; however, additional devices may be required to rotate the member to its final position.

Panels that are stripped by rotating about one edge with lifting devices at the opposite edge will develop moments as shown in Fig. 5.2.4. When panels are stripped this way, care should be taken to prevent spalling of the edge along which rotation occurs. A compressible material or sand bed will help protect this edge.

Members that are stripped flat from the mold will develop the moments shown in Fig. 5.2.5.

To determine stresses in flat panels for either rotation or flat stripping, the calculated moment may be assumed to be resisted by the effective widths shown.

In some plants, tilt tables or turning rigs are used to reduce stripping stresses, as shown in Fig. 5.2.6.

When a panel is ribbed or is of a configuration or size such that stripping by rotation or tilting is not practical, vertical pick-up points on the top surface can be used. These lift points should be located to minimize the tensile stresses on the exposed face.

Since the section modulus with respect to the top and bottom faces may not be the same, the designer must select the controlling design limitation:

1. Tensile stresses on both faces to be less than that which would cause cracking.
2. Tensile stress on one face to be less than that which would cause cracking, with controlled cracking permitted on the other face.
3. Controlled cracking permitted on both faces.

If only one of the faces is exposed to view, it will generally control the design.

Fig. 5.2.3 Example of post-tensioned wall panel

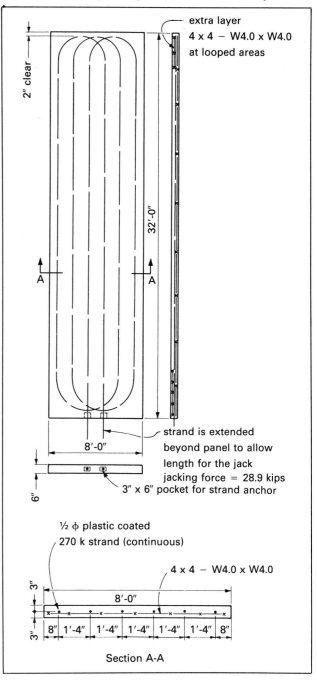

extra layer
4 x 4 — W4.0 x W4.0
at looped areas

2" clear

32'-0"

A   A

8'-0"

6"

strand is extended beyond panel to allow length for the jack
jacking force = 28.9 kips
3" x 6" pocket for strand anchor

½ φ plastic coated
270 k strand (continuous)

4 x 4 — W4.0 x W4.0

3"

3"

8'-0"

8"   1'-4"   1'-4"   1'-4"   1'-4"   1'-4"   8"

Section A-A

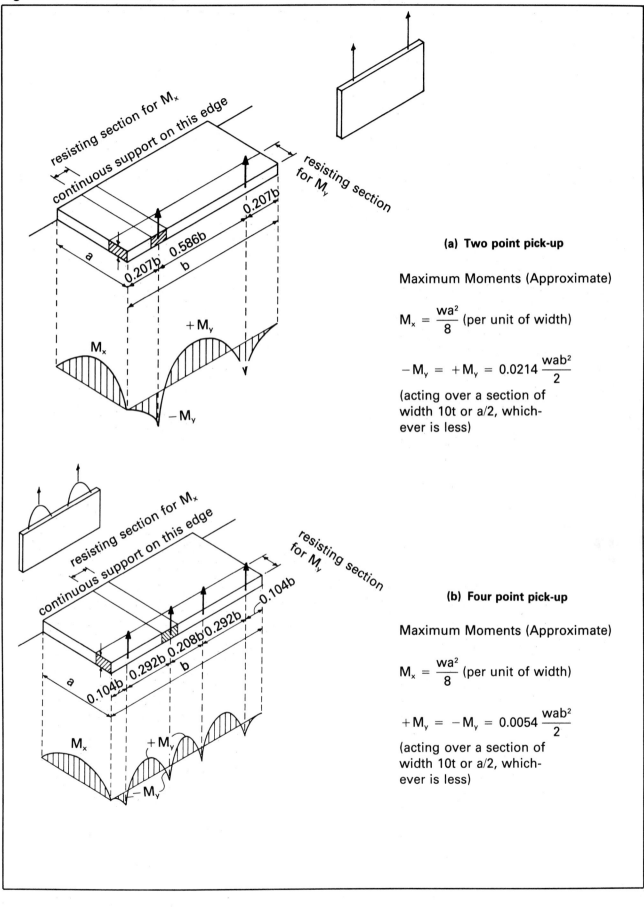

(a) Two point pick-up

Maximum Moments (Approximate)

$$M_x = \frac{wa^2}{8} \text{ (per unit of width)}$$

$$-M_y = +M_y = 0.0214 \frac{wab^2}{2}$$

(acting over a section of width 10t or a/2, whichever is less)

(b) Four point pick-up

Maximum Moments (Approximate)

$$M_x = \frac{wa^2}{8} \text{ (per unit of width)}$$

$$+M_y = -M_y = 0.0054 \frac{wab^2}{2}$$

(acting over a section of width 10t or a/2, whichever is less)

Fig. 5.2.5 Moments developed in panels stripped flat

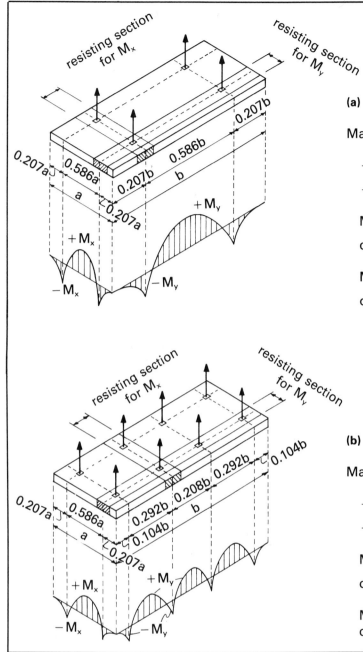

(a) Two point pick-up

Maximum Moments (Approximate)

$$+M_x = -M_x = 0.0107 \, wba^2$$

$$+M_y = -M_y = 0.0107 \, wab^2$$

M_x acting over a section of width 0.29a, 10t, or b/2, whichever is less

M_y acting over a section of width 0.29b, 10t, or a/2, whichever is less

(b) Four point pick-up

Maximum Moments (Approximate)

$$-M_x = +M_x = 0.0054 \, wa^2b$$

$$-M_y = +M_y = 0.0027 \, wab^2$$

M_x acting over a section of width 0.29a, 10t, or b/2, whichever is less

M_y acting over a section of width 0.29b, 10t, or a/2, whichever is less

Lift line forces for a two point lift, using inclined lines, is shown in Fig. 5.2.7. When the angle of lift line to the horizontal is small, the component of force parallel to the longitudinal axis may generate a significant moment. While this effect can and should be accounted for, it is not recommended that it be allowed to dominate design moments. Rather, consideration should be given to using spreader beams, two cranes or other mechanisms to reduce the angle of lift. Any such special handling requirements should be clearly shown on the shop drawings.

In addition to longitudinal bending moments, a transverse bending moment may be caused by the orientation of the pick-up points with respect to the transverse dimension (Fig. 5.2.8). For the section shown, a critical moment could occur between the ribs because of the thin cross-section.

The design guidelines listed above apply to elements of constant cross-section. For elements of varying cross-section, the location of lift points is usually determined by trial and error. Rolling blocks can be used on long elements of varying section (Fig. 5.2.10), which makes the forces in the lifting

PCI Design Handbook

Fig. 5.2.6 Stripping from a tilt table

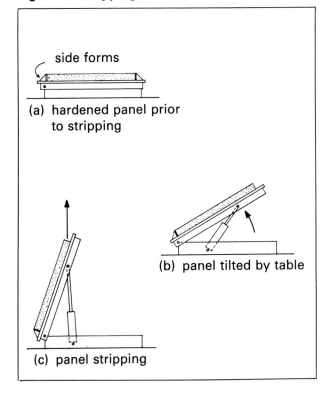

(a) hardened panel prior to stripping

side forms

(b) panel tilted by table

(c) panel stripping

Fig. 5.2.7 Determination of force in inclined lift lines

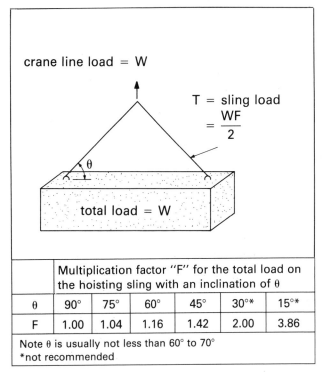

crane line load = W

T = sling load $= \dfrac{WF}{2}$

total load = W

	Multiplication factor "F" for the total load on the hoisting sling with an inclination of θ					
θ	90°	75°	60°	45°	30°*	15°*
F	1.00	1.04	1.16	1.42	2.00	3.86

Note θ is usually not less than 60° to 70°
*not recommended

Fig. 5.2.8 Pick-up points for equal stresses of a ribbed member

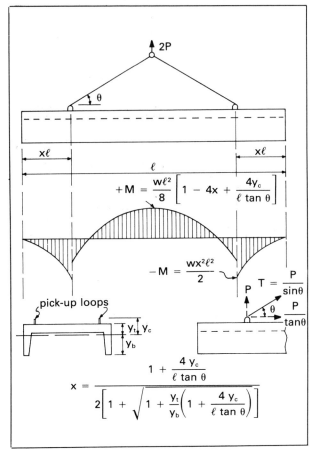

$$+M = \frac{w\ell^2}{8}\left[1 - 4x + \frac{4y_c}{\ell \tan \theta}\right]$$

$$-M = \frac{wx^2\ell^2}{2}$$

$$T = \frac{P}{\sin\theta}$$

pick-up loops

$$x = \frac{1 + \dfrac{4y_c}{\ell \tan \theta}}{2\left[1 + \sqrt{1 + \dfrac{y_t}{y_b}\left(1 + \dfrac{4y_c}{\ell \tan \theta}\right)}\right]}$$

Fig. 5.2.9 Moments caused by eccentric lifting

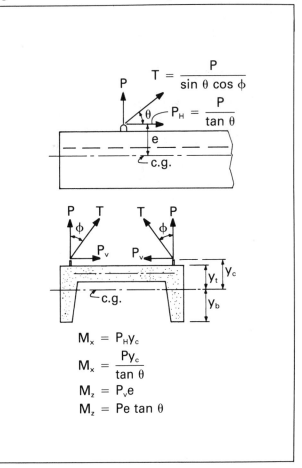

$$T = \frac{P}{\sin\theta \cos\phi}$$

$$P_H = \frac{P}{\tan\theta}$$

$$M_x = P_H y_c$$

$$M_x = \frac{Py_c}{\tan\theta}$$

$$M_z = P_v e$$

$$M_z = Pe \tan\theta$$

Fig. 5.2.10 Arrangement for equalizing lifting loads

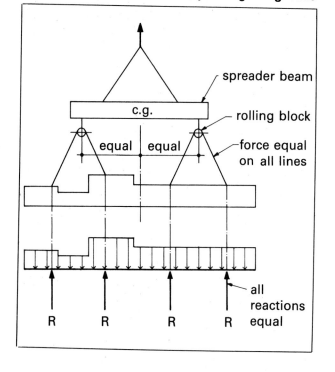

lines equal. The member can then be analyzed as a beam with varying load supported by equal reactions.

The force in inclined lift lines can be determined from Fig. 5.2.7.

5.2.8 Handling devices

The most common lifting devices are prestressing strand or cable loops projecting from the concrete, threaded inserts, or special proprietary devices.

Since lifting devices are subject to dynamic loads, ductility of the material is part of the design requirement. Deformed reinforcing bars should not be used since the deformations result in stress concentrations from the shackle pin. Also, reinforcing bars are often hard-grade or re-rolled rail steel with little ductility and low impact strength at cold temperatures. Smooth bars of a known steel grade may be used if adequate embedment or mechanical anchorage is provided. The diameter must be such that localized failure will not occur by bearing on the shackle pin.

Prestressing strand is often used for lifting loops. The variables involved make it almost impossible to calculate a capacity which can be used for all situations. Generally, producers will establish standard criteria for use in handling the standard products manufactured by that plant. Table 5.2.3 is an example which has been used successfully.

Reduced capacities for shorter embedment lengths may be suitable. In shallow products, providing a 90° bend can reduce the required embedment length significantly. Lightly rusted strand has better bond than bright strand.

The diameter of the bend of the loop should be at least 4 in. For smaller diameters, the loop capacities in Table 5.2.3 should be reduced to:

1 in. dia. — 70 %
2 in. dia. — 85 %
3 in. dia. — 90 %

The angle of incline of lifting has little effect on the strand lifting loop capacity if the angle from the horizontal is more than about 20°. Typical handling methods are usually such that this angle is no less than 60°.

Table 5.2.3 Capacity of ½ in. diameter, 270 ksi strands used as lifting loops

Lifting angle		Embedment length (in.)	Single loop (kips)	Double loop (kips)	Triple loop (kips)
45 degrees		16	5	8.5	11.5
		22	8	13	17.5
		28	10	18	23
		34	11	23	29
Vertical		16	7.5	12.5	16.5
		22	11.5	19	24.5
		28	15.5	25.5	33
		34	16	32.5	41

1. These values are limited by slippage rather than strand strength, with a factor of safety of 4. For other strand diameters, multiply table values by 0.75 for ⅜ in. diameter, 0.85 for ⁷⁄₁₆ in. diameter, and 1.1 for 0.6 in. diameter.

2. Minimum $f_c' = 3000$ psi.

3. Multiple strand loops must be fabricated to ensure equal force on each strand.

Table 5.2.4 Capacity of 7 × 19 aircraft cable used as lifting loops[1]

Diameter (in.)	Capacity (kips)[2]
3/8	3.6
7/16	4.4
1/2	5.7

1. 7 strands with 19 wires each.
2. Based on a single strand with a factor of safety of 4 applied to the minimum breaking strength of galvanized cable. The user should consider embedment, loop diameter and other factors discussed for strand loops. Aircraft cables are usually wrapped around reinforcement in the precast product.

Increased capacity can be obtained by the use of multiple strand loops. To avoid localized overstress, care should be taken when fabricating the loop to ensure that all strands are bent the same. This may be accomplished by using a bent thin-wall conduit over the strand in the area of the bend.

Some precast producers also use aircraft cable and standard wire rope for lifting devices (see Table 5.2.4).

Lifting heavy members with threaded inserts should be carefully assessed. When properly designed for both insert and concrete capacities, threaded inserts have many advantages. However, correct usage is sometimes difficult to inspect during handling operations. In order to ensure that an embedded insert acts primarily in tension, a swivel plate as indicated in Fig. 5.2.11 should be used.

A common problem is the failure of the bolt to engage enough threads in the insert to develop its strength. To guard against this, some manufacturers use a long bolt with a nut between the bolt head and the swivel plate. After the bolt has "bottomed out", the nut is turned against the swivel plate.

Connection hardware should not be used for lifting or handling of any but the lightest units, unless approved by the designer.

Example 5.2.1 Stripping forces

Given:

A ribbed wall panel as shown. Span ℓ = 40'-0". f'_c at stripping = 3000 psi (normal weight).

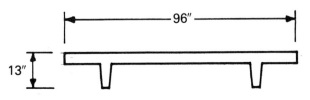

Fig. 5.2.11 Swivel plate

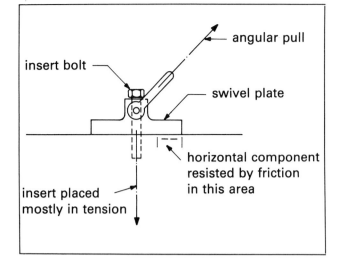

Problem:

Determine stripping forces and corresponding stresses.

Solution:

Half section properties:

$$y_b = 10.18 \text{ in.}$$
$$y_t = 2.81 \text{ in.}$$
$$Z_b = 156 \text{ in.}^3$$
$$Z_t = 565 \text{ in.}^3$$
$$\text{wt.} = 192 \text{ plf}$$

$$\frac{y_t}{y_b} = 0.276$$

Stripping load:

assume a load multiplier of 1.4 (Table 5.2.1)
w = 1.4 (192) = 269 plf

Case No. 1 — Neglecting the moment due to eccentric pick-up points.

Pick-up location for equal tension on each face Fig. 5.2.8, with y_c = 0:

$$x = \cfrac{1}{2\left[1 + \sqrt{1 + \cfrac{y_t}{y_b}}\,\right]} = \frac{1}{2(1 + \sqrt{1.276})}$$

$$= 0.235$$

$$-M = \frac{wx^2\ell^2}{2} = \frac{0.269}{2}(0.235 \times 40)^2$$

$$= 11.88 \text{ ft-kips}$$

$$f_t = \frac{-M}{Z_t} = \frac{11,880\ (12)}{565} = 252\ \text{psi}$$

$$< 5\sqrt{3000} = 274\ \text{psi} \quad \text{OK}$$

Case No. 2 — Accounting for moments due to eccentric pick-up points

For $\theta = 45$ deg

Assume: $y_c = y_t + 3'' = 5.81$ in. (Fig. 5.2.8)

$$\frac{4\ y_c}{\ell \tan\theta} = \frac{4(5.81)}{12(40)\tan 45} = 0.048$$

$$x = \frac{1 + \dfrac{4\ y_c}{\ell \tan\theta}}{2\left[1 + \sqrt{1 + \dfrac{y_t}{y_b}\left(1 + \dfrac{4\ y_c}{\ell \tan\theta}\right)}\right]}$$

$$= \frac{1.048}{2[1 + \sqrt{1 + 0.276\ (1.048)}]} = 0.245$$

$$-M = \frac{0.269}{2}\ (0.245 \times 40)^2 = 12.92\ \text{ft-kips}$$

$$f_t = \frac{12,920\ (12)}{565} = 274\ \text{psi} = 5\sqrt{3000} \quad \text{OK}$$

Example 5.2.2 Locating pick-up points

Given:

The window unit shown is to be cast face down and stripped vertically.

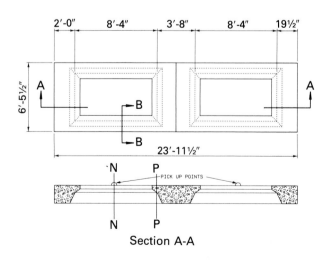

Section A-A

Problem:

Locate pick-up points to minimize tensile stresses in the concrete.

Solution:

Dead load of member—assume 1.6 multiplier (Table 5.2.1):

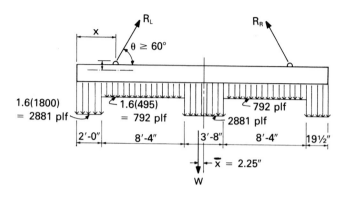

$$W = 16.67\ (792) + 7.292\ (2881) = 34,211\ \text{lb}$$

Lifting loops should be placed symmetrically about the center of gravity of the member.

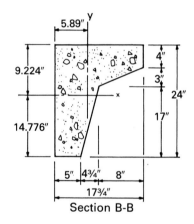

Section B-B

Assume critical cracking stress will occur in the narrow sections of the unit (Section B-B):

Section Properties

$$A = 237.6\ \text{in.}^2$$
$$I = 10,969\ \text{in.}^4$$
$$y_t = 9.224\ \text{in.}$$
$$y_b = 14.776\ \text{in.}$$

For equal stresses on each face:

$$\frac{-My_t}{I} = \frac{+My_b}{I}$$

$$-M = \frac{y_b}{y_t}\ (+M) = \frac{14.776}{9.224}\ (+M) = 1.60(+M)$$

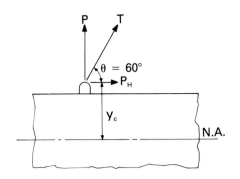

$$P_H = \frac{P}{\tan\theta} = \frac{34,211/2}{\tan60°} = 9876 \text{ lb}$$

See Fig. 5.2.9.

$$y_c = y_t + 3 = 12.22 \text{ in.}$$

$$M_x = \frac{12.22\ (9876)}{12} = 10,060 \text{ ft-lb}$$

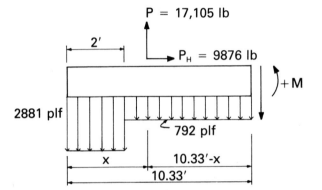

$$+M = 17,105\ (10.33 - x) - 792\ \frac{(8.33)^2}{2}$$
$$- 2881\ (2)\ 9.33 + 10,060$$
$$= -17,105\ x + 105,520$$

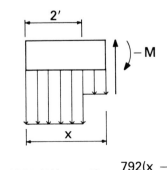

$$-M = 2881\ (2)(x - 1) + \frac{792(x - 2)^2}{2}$$
$$= 396\ x^2 + 4178x - 4178$$

$$396x^2 + 4178x - 4178$$
$$= 1.60\ (-17,105\ x + 105,520)$$

$$x^2 + 91.14x = 499$$

$$x = 5.15 \text{ ft}$$

Use: x = 5 ft

$$\begin{aligned}+M &= 105,520 - 17,105\ (5)\\ &= 19,996 \text{ ft-lb}\\ &= 239.9 \text{ in.-kips}\end{aligned}$$

$$\begin{aligned}-M &= 396\ (5)^2 + 4178\ (5) - 4178\\ &= 26,614 \text{ ft-lb} = 319.3 \text{ in.-kips}\end{aligned}$$

$$\begin{aligned}f_t &= \frac{(-M)y_t}{I} = \frac{319,300\ (9.224)}{2\ (10,969)}\\ &= 134 \text{ psi}\end{aligned}$$

$$\begin{aligned}f_b &= \frac{(+M)y_b}{I} = \frac{239,900\ (14.776)}{2\ (10,969)}\\ &= 162 \text{ psi}\end{aligned}$$

Using an allowable stress of $5\sqrt{f'_{ci}}$, a stripping strength as low as 1050 psi would theoretically be permitted. Thus, reinforcement would not be required.

To illustrate a "controlled cracking" design allowing a crack width of 0.005 in., assume the moments on each narrow section are:

$$-M = 270 \text{ in.-kips}$$
$$+M = 165 \text{ in.-kips}$$

(This would require a stripping strength greater than 2000 psi.)

Referring to Sect. 4.2.2.1:

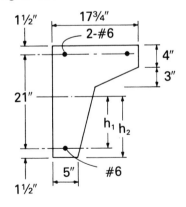

d = 22.5 in.

For +M:

$$h_1 = 14.776 - 1.5 = 13.27 \text{ in.}$$
$$h_2 = 14.78 \text{ in.}$$
$$h_2/h_1 = 1.04$$

$A \approx 5(2)(1.5) = 15 \text{ in.}^2$

$d_c = 1.5 \text{ in.}$

From Eq. 4.2.9 with w = 0.005 in.

$f_s = 22.4 \text{ ksi}$

From Eq. 4.2.10:

$$A_s = \frac{+M}{0.9f_s d} = \frac{165}{0.9(22.4)(22.5)} = 0.36 \text{ in.}$$

$$\rho_{s\ min} = \frac{200}{f_y} = \frac{200}{60,000} = 0.00333$$

$A_{s\ min} = 0.00333 (5) 22.5 = 0.375 \text{ in.}^2$

Use 1-#6 bar $A_s = 0.44 \text{ in.}^2$

For $-M$:

$h_2 = 9.22 \text{ in.}; h_1 = 9.22 - 1.5 = 7.72 \text{ in.}$

$h_2/h_1 = 9.22/7.72 = 1.19$

For 2 bars,

$A = 17.75(2)(1.5)/2 = 26.6 \text{ in.}^2$

$f_s = 16.2 \text{ ksi}$

$$A_s = \frac{-M}{0.9f_s d} = \frac{270}{0.9(16.2)(22.5)} = 0.82 \text{ in.}^2$$

Use 2-#6 bars, $A_s = 0.88 \text{ in.}^2$

5.2.9 Lateral Stability

Attention should be given to temporary stresses and stability of long members with narrow compression flanges during handling, transportation and erection.[2]

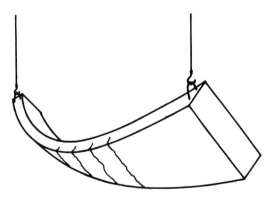

Consider a member of the general shape as follows:

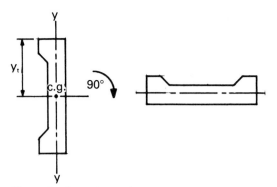

First, the location of the centroid of the member and the moment of inertia of the member about the weak axis is calculated. Next, the member is assumed to be rotated 90 deg., freely supported at its ends, and allowed to bend about the y-axis.

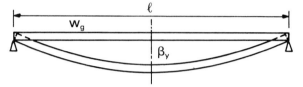

The factor β_y due to the member's self weight, w, over a span, ℓ, is calculated. If the member is prismatic and is of uniform weight per foot and has a constant moment of inertia I_y:

$$\beta_y = \frac{5}{384} \frac{w\ell^4}{E_c I_y} \qquad \text{(Eq. 5.2.2)}$$

The distance from the top face to the centroid of the member is y_t.

When the member is lifted at the crane hooks, the factor of safety against lateral buckling is:

$$\text{F.S.} = \frac{y_t}{\beta_y} \qquad \text{(Eq. 5.2.3)}$$

Allowing for dynamic effects, a factor of safety of 2 should be required. Thus, for safe hanging, $y_t \geqslant 2\beta_y$.

For safe handling of long members, resistance to lateral buckling can be improved by several methods:

1. Design adequate moment of inertia I_y.
2. Specify very high-strength concrete (increase modulus of elasticity).
3. Keep weight, w, low, if possible.
4. Reduce β_y by moving lifting loops away from ends of the members.
5. Attach temporary lateral bracing to compression flange or provide strongbacks, stiffening trusses or pipe frames. Sometimes two or more units can be transported together, side by side, and tied together to provide the necessary lateral strength.

Fig. 5.2.12 Moments, shears, and deflection of a beam with overhangs

Loading and support	Reactions, constraining moments, and vertical shear	Bending moment M and maximum bending moment	Deflection y, maximum deflection, and end slope θ
Equal overhangs, uniform load $W = w\ell$ down is (−)	$R_1 = R_2 = \dfrac{W}{2}$ (A to B) $V = -\dfrac{W(c-x)}{l}$ (B to C) $V = W\left(\dfrac{1}{2} - \dfrac{x+c}{l}\right)$ (C to D) $V = \dfrac{W(c+d-x)}{l}$	(A to B) $M = -\dfrac{W}{2l}(c-x)^2$ (B to C) $M = -\dfrac{W}{2l}[c^2 - x(d-x)]$ $M = -\dfrac{Wc^2}{2l}$ at B and D $M = -\dfrac{W}{2l}\left(c^2 - \dfrac{d^2}{4}\right)$ at $x = \dfrac{d}{2}$ If $d > 2c$, $M = 0$ at $x = \dfrac{d}{2} \pm \left[\dfrac{d^2}{4} - c^2\right]^{\frac12}$ If $c = 0.207l$, $M = -\dfrac{Wl}{46.62}$ at $x = 0 = d$ and $M = +\dfrac{Wl}{46.62}$ at $x = \dfrac{d}{2}$ x is considered positive on both sides of the origin.	(A to B) $y = -\dfrac{Wx}{24EIl}[6c^2(d+x) - x^2(4c-x) - d^3]$ (B to C) $y = -\dfrac{Wx(d-x)}{24EIl}[x(d-x) + d^2 - 6c^2]$ $y = -\dfrac{Wc}{24EIl}[3c^2(c+2d) - d^3]$ at A and D $y = -\dfrac{Wd^2}{384EIl}(5d^2 - 24c^2)$ at $x = \dfrac{d}{2}$ If $2c < d < 2.449c$, the maximum deflection between supports is $y = +\dfrac{W}{96EIl}(6c^2 - d^2)^2$ at $x = \dfrac{d}{2} \pm \left[3\left(\dfrac{d^2}{4} - c^2\right)\right]^{\frac12}$ $\theta = +\dfrac{W}{24EIl}(6c^2d + 4c^3 - d^3)$ at A $\theta = -\dfrac{W}{24EIl}(6c^2d + 4c^3 - d^3)$ at D
Unequal overhangs, uniform load $W = w\ell$ down is (−)	$R_1 = \dfrac{W}{2d}(c+d-e)$ $R_2 = \dfrac{W}{2d}(d+e-c)$ (A to B) $V = -\dfrac{W}{l}(c-x)$ (B to C) $V = R_1 - \dfrac{W}{l}(c+x)$ (C to D) $V = -\dfrac{W}{l}(d+e-x)$	(A to B) $M = -\dfrac{W}{2l}(c-x)^2$ (B to C) $M = -\dfrac{W}{2l}(c+x)^2 + R_1 x$ (C to D) $M = -\dfrac{W}{2l}(e+d-x)^2$ $M = -\dfrac{Wc^2}{2l}$ at B $M = -\dfrac{We^2}{2l}$ at C M_{max} between supports $= \dfrac{W}{2l}(c^2 - x_1^2)$ at $x = x_1$ $= \dfrac{c^2 + d^2 - e^2}{2d}$ If $x_1 > c$, $M = 0$ at x $= x_1 \pm [x_1^2 - c^2]^{\frac12}$ x is considered positive on both sides of the origin.	(A to B) $y = -\dfrac{Wx}{24EIl}\big[2d(e^2 + 2c^2) + 6c^2 x - x^2(4c-x) - d^3\big]$ (B to C) $y = -\dfrac{Wx(d-x)}{24EIl}\left\{x(d-x) + d^2 - 2(c^2+e^2) - \dfrac{2}{d}[e^2 x + c^2(d-x)]\right\}$ (C to D) $y = -\dfrac{W(x-d)}{24EIl}\big[2d(c^2 + 2e^2) + 6e^2(x-d) - (x-d)^2(4e+d-x) - d^3\big]$ $y = -\dfrac{Wc}{24EIl}[2d(e^2 + 2c^2) + 3c^3 - d^3]$ at A $y = -\dfrac{We}{24EIl}[2d(c^2 + 2e^2) + 3e^3 - d^3]$ at D This case is too complicated to obtain a general expression for critical deflections between the supports. $\theta = +\dfrac{W}{24EIl}(4c^3 + 4c^2d - d^3 + 2de^2)$ at A $\theta = -\dfrac{W}{24EIl}(2c^2d + 4de^2 - d^3 + 4e^3)$ at D

Example 5.2.3 Lateral buckling

Given:

The 3 in. thick by 8 ft high panel shown. (Note: This example is for illustration only. Panels 3 in. thick by 40 ft long are *not* recommended.)
f'_c = 5000 psi, wt = 300 plf

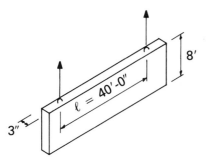

Problem:

Determine if lateral buckling is critical.

Solution:

$$\beta_y = \frac{5}{384} \frac{w\ell^4}{E_c I_y}$$

$$I_y = \frac{1}{12} \times 96 \times 3^3 = 216 \text{ in.}^4$$

$$E_c = 4.3 \times 10^6 \text{ psi}$$

$$\beta_y = \frac{5}{384} \times \frac{300}{12} \times \frac{(40 \times 12)^4}{4.3 \times 10^6 \times 216} = 18.6 \text{ in.}$$

$$y_t = 96/2 = 48 \text{ in.} > 2\beta_y \quad \text{OK}$$

5.2.10 Storage

Wherever possible, an element should be stored on only two points of support located at or near those used for stripping and handling. Thus, the design for stripping and handling will usually control. Where points other than those used for stripping or handling are used for storage, the storage condition must be checked.

If support is provided at more than two points, and the design based on more than two supports, precautions must be taken so that the element does not bridge over one of the supports due to differential support settlement. Particular care must be taken for prestressed elements, with consideration made for the effect of prestressing. Designing for equal stresses on both faces will help to minimize deformations in storage. Moments, shears and deformations (not including creep) for uniformly loaded members with different support arrangements are shown in Fig. 5.2.12.

Fig. 5.2.13 Panel warpage in storage

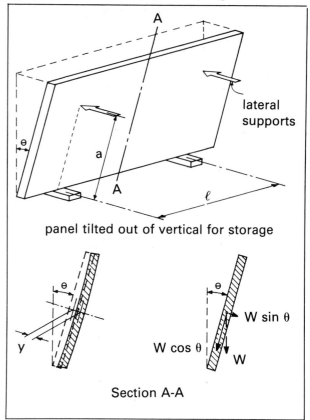

panel tilted out of vertical for storage

Section A-A

Warpage in storage may be caused by temperature or shrinkage differential between surfaces, creep and storage conditions. Warpage cannot be totally eliminated, although it can be minimized by providing blocking so that the panel remains plane. Where feasible, the member should be oriented in the yard so that the sun does not overheat one side. (See Sect. 3.3.2 for a discussion of thermal bowing.) Storing members so that flexure is resisted about the strong axis will minimize stresses and deformations.

For the support conditions shown in Fig. 5.2.13, warping can occur in both directions. By superposition, the total instantaneous deflection, y_{max}, at the maximum point can be estimated by:

$$y_{max} = \frac{5 \text{ w sin}\theta}{384 \text{ E}_c} \left[\frac{a^4}{I_c} + \frac{\ell^4}{I_b} \right] \qquad \text{(Eq. 5.2.4)}$$

where:

w	= panel weight, lb/in.
E_c	= modulus of elasticity of concrete, psi
a	= panel support height, in.
ℓ	= horizontal distance between supports, in.
I_c, I_b	= moment of inertia of uncracked section in the respective directions for 1 in. width of panel, in.[4]

PCI Design Handbook

Fig. 5.2.14 Effect of compression reinforcement on creep

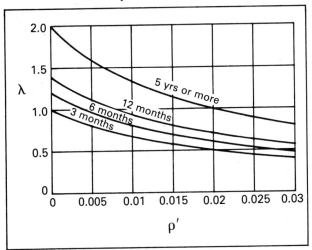

This instantaneous deflection should be modified by a factor to account for the time-dependent effects of creep and shrinkage. ACI 318-83 suggests the total deformation, y_t, at any time can be estimated as:

$$y_t = y_{max} (1 + \lambda) \qquad \text{(Eq. 5.2.5)}$$

y_t = time dependent displacement

y_{max} = instantaneous displacement

λ = amplification due to creep and shrinkage (Fig. 5.2.14)

ρ' = reinforcement ratio for nonprestressed compression reinforcement, A'_s/bt

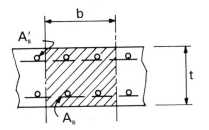

5.2.11 Transportation

The method used for transporting precast products to the job site can affect the structural design because of size and weight limitations and dynamic effects.

Except for long prestressed deck members, most products are transported on either flatbed or low-boy trailers. These trailers deform during hauling. Thus, support at more than two points can be achieved only after considerable modification of the trailer, and even then results may be doubtful.

Size and weight limitations vary from one state to another, so a check of local regulations is necessary when large units are moved. Loads are further restricted on some secondary roads during spring thaws.

The common payload for standard trailers without special permits is 20 tons with width and height restricted to 8 ft and length to 40 ft. Low-boy trailers permit the height to be increased to about 10 to 12 ft. However, low-boys cost more to operate and have a shorter bed length. In some states, a total height (roadbed to top of load) of 13 ft 6 in. is allowed without special permit. This height may require special routing to avoid low overpasses and overhead wires.

Maximum width with permit varies among states — and even among cities — from 10 to 14 ft. Some states allow lengths over 70 ft with only a simple permit, while others require, for any load over 55 ft, a special permit, escorts front and rear and travel limited to certain times of the day. In some states, weights of up to 100 tons are allowed with permit, while in other states there are very severe restrictions on loads over 25 tons.

These restrictions add to the cost of precast units, and should be compared with savings realized by combining smaller units into one large unit. When possible, a precast unit, or several units combined, should approximate the usual payload of 20 tons. For example, an 11 ton unit may not be economical, because only one can be shipped per load, while two 10 ton units could be shipped on one load.

Erection is facilitated when members are transported in the same orientation they will have in the structure. For example, single-story wall panels can be transported on A-frames with the panels upright (Fig. 5.2.15). A-frames also provide good lateral support and the desired two points of vertical support. Longer units can be transported on their sides to take advantage of the increased stiffness compared with flat shipment (Fig. 5.2.16). In all cases, the panel support locations should be consistent with the panel design. Panels with large openings frequently require strong-backs, braces or ties to keep stresses within the design values (Fig. 5.2.17).

During transportation, units are usually supported with one or both ends cantilevered. For members not symmetrical with respect to the bending axis, the expressions given in Fig. 5.2.18 can be used for determining the location of supports to give equal tensile stresses.

Dynamic Stresses

During transportation, members may be sub-

Fig. 5.2.15 Transportation of single-story panels

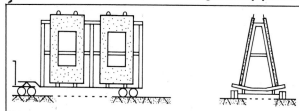

Fig. 5.2.16 Transportation of multi-story panels

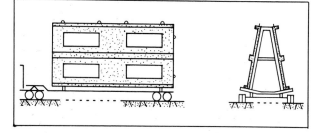

Fig. 5.2.17 Methods of temporary strengthening of panels with significant openings

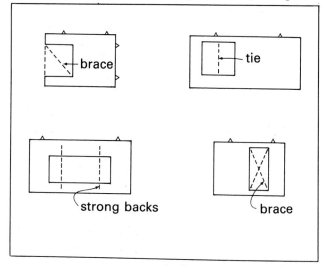

jected to dynamic forces. Significant stresses may result, particularly if the member has a natural frequency close to that of the support system.

When a member is supported near the ends, an equivalent static load, w_{dy}, can be substituted for the member dead load to approximate the dynamic effect:

$$w_{dy} = Aw_{d\ell} \qquad \text{(Eq. 5.2.6)}$$

where

A = the amplification factor from Fig. 5.2.19

$w_{d\ell}$ = the weight of the member

f = fundamental frequency of vibration, hz (cps), from Fig. 5.2.20

E_c = modulus of elasticity of the member

I = moment of inertia of the section about the bending axis normal to the displacement

g = acceleration due to gravity (386 in/sec²)

ℓ = length of the member

The curves in Fig. 5.2.19 are based on a single degree of freedom, of exciting truck frequencies

Fig. 5.2.18 Equations for equal tensile stresses top and bottom — unsymmetrical members

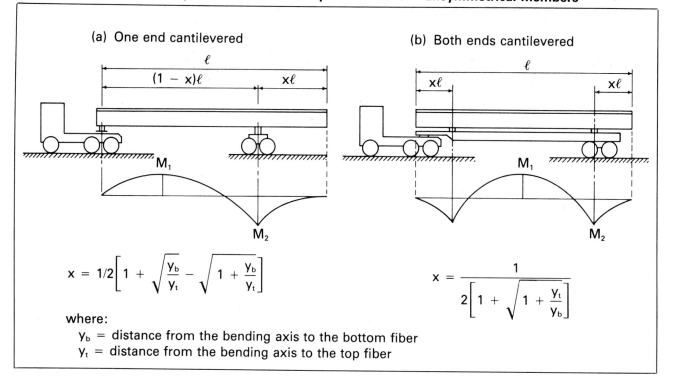

(a) One end cantilevered

$(1 - x)\ell$ $x\ell$

M_1

M_2

$$x = \frac{1}{2}\left[1 + \sqrt{\frac{y_b}{y_t}} - \sqrt{1 + \frac{y_b}{y_t}}\right]$$

where:

y_b = distance from the bending axis to the bottom fiber

y_t = distance from the bending axis to the top fiber

(b) Both ends cantilevered

$x\ell$ $x\ell$

M_1

M_2

$$x = \frac{1}{2\left[1 + \sqrt{1 + \frac{y_t}{y_b}}\right]}$$

Fig 5.2.19 Load amplification factors

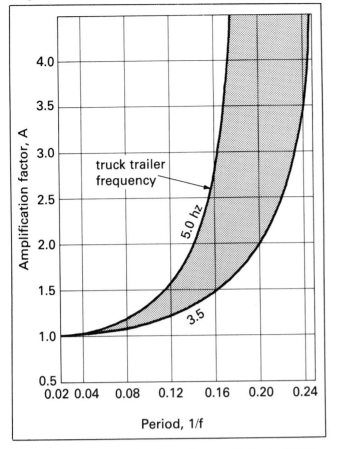

Fig. 5.2.20 Fundamental frequency of simple beams with cantilevers at each end

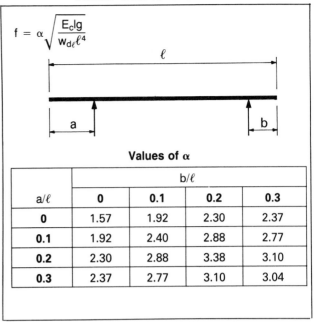

$$f = \alpha \sqrt{\frac{E_c I g}{w_{d\ell} \ell^4}}$$

Values of α

a/ℓ	b/ℓ			
	0	0.1	0.2	0.3
0	1.57	1.92	2.30	2.37
0.1	1.92	2.40	2.88	2.77
0.2	2.30	2.88	3.38	3.10
0.3	2.37	2.77	3.10	3.04

$I = 216$ in.^{4}

$w_{d\ell} = 75$ plf $= 6.25$ lb/in.

$f = 2.40 \sqrt{\dfrac{(4.03 \times 10^6)(216)(386)}{6.25(20 \times 12)^4}} = 9.66$

$1/f = 1/9.66 = 0.104; \quad A = 1.3$

$w_{dy} = 1.3(75) = 97$ lb/ft

from 3.5 to 5 hz, and a damping coefficient of 0.02. An exciting frequency of 5 hz is in the higher range of those reported for highway trucks.

The curve given is very approximate, since it is based on averages, thus it should be used conservatively. For members with a period, $1/f$, less than about 0.16, it may be advisable to revise the hauling procedure.

This procedure is not an equivalent of impact, but an approximation of the effects of vibration, or "bounce". Thus, it can occur in either direction, in effect reducing the dead load on the member, which could be critical. For example, in a prestressed member, the dead load is frequently depended upon to keep top tension within limits.

Example 5.2.4 Dynamic stresses on a solid slab

Given:

A 20 ft long, 6 in. thick solid slab is supported 2 ft from its ends. $E_c = 4.03 \times 10^6$ psi. Normal weight concrete.

Problem:

Find the equivalent static load, w_{dy}.

Solution:

From Fig. 5.2.20 for $a/\ell = b/\ell = 0.1$, $\alpha = 2.40$

5.2.12 Erection

Precast members frequently must be re-oriented from the position used to transport to that which it will be in the final construction. The analysis for this "tripping" (rotating) operation is similar to that used during other handling stages. Fig. 5.2.21 shows maximum moments for several commonly used tripping techniques.

When using two crane lines the center of gravity must be between them in order to prevent a sudden shifting of the load while it is being rotated. To insure that this is avoided, the stability condition shown in Fig. 5.2.22 must be met. The capacities of lifting devices must be checked for the forces imposed during the tripping operation, since the directions vary.

5.2.13 Design Example — Flat Panel

This example illustrates the use of many of the recommendations in this section. It is intended to be illustrative and general only. Each plant will have its own preferred methods of manufacture.

Fig. 5.2.21 Typical tripping (rotating) positions for erection of wall panels

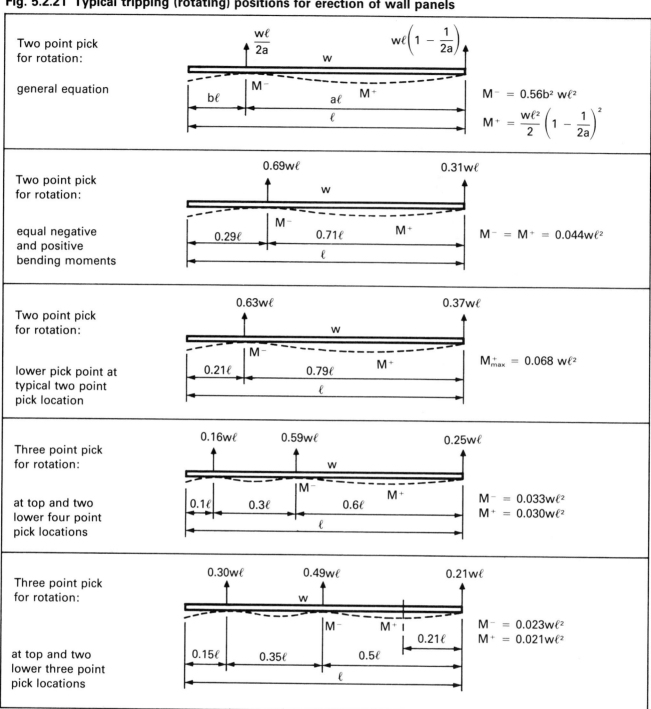

Given:

A flat panel used as a non-load bearing facade on a two-story structure, as shown in Fig. 5.2.23.

Wind Load – 20 psf pressure or suction

f'_c = 5000 psi @ 28 days

f_{ci} = 2000 psi @ stripping

Cracks in rear face permitted without width re-

striction. Crack width in exposed face limited to 0.005 in.

Solution:

Establish Handling Procedures

Casting: Face down. Use same mix for exposed aggregate surface (retarded) and smooth white side bands. Use gray concrete backup.

Fig. 5.2.22 Stability during erection

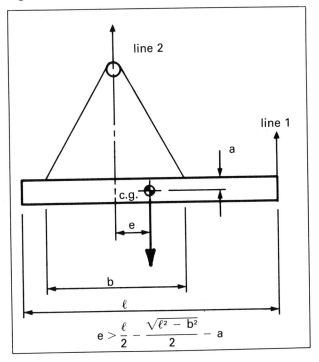

$$e > \frac{\ell}{2} - \frac{\sqrt{\ell^2 - b^2}}{2} - a$$

Fig. 5.2.23 Example of Sect. 5.2.13

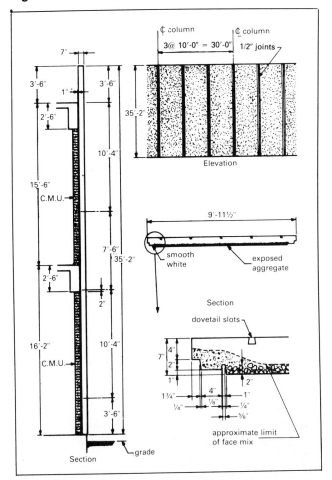

Stripping: Due to edge detail and inside crane headroom, panel cannot be turned on edge directly in mold, therefore strip flat and move to sand bed (or turning equipment) for turning.

Storage: Since panels will be stored for several months and storage yard is subject to settlement, thus negating possible four-point support, and since bowing must be avoided, store on edge.

Determine Handling Multipliers (Table 5.2.1)

Stripping: Exposed flat surface has deep exposure (heavy retarder); side rails removed prior to stripping; drafts on edge detail are good: use 1.2.

Use strand loops in back of panel (plant practice).

Yard handling: Turning: use 1.2; transport to storage: use 1.2.

Shipping: Distance traveled is 150 miles and jobsite roadways are bumpy: use 1.7 (exceeds the value 1.5 recommended in Table 5.2.1 since travel conditions are considered rather severe and cracking is limited).

Erection: Use 1.5 (based on engineer's judgment instead of value from Table 5.2.1).

Handling devices: Use 4 (from Sect. 5.2.5).

Section Properties (use 7″ thick × 9′ – 11½″ wide)

$A = 836$ in.^{2}

$Z_b = Z_t = 976$ in.^{3}

$I = 3416$ in.^{4}

Unit weight @ 150 pcf = 870 plf
                      or 87 psf

Total weight = 30.6 kips

Establish Allowable Tensile Stresses

@ stripping, yard handling and storage

$$f_r' = \frac{5\sqrt{2000}}{1000} = 0.224 \text{ ksi (from Sect. 5.2.4)}$$

@ shipping and erection

$$f_r' = \frac{5\sqrt{5000}}{1000} = 0.354 \text{ ksi}$$

Check Handling Stresses — Stripping:

a. Longitudinal bending

Two point pick-up, Fig. 5.2.5(a)

a = 10.0 ft   b = 35.2 ft

0.29 b = 112 in.; 10t = 70 in.; a/2 = 60 in.

$$Z = \frac{60(7)^2}{6} = 490 \text{ in.}^3$$

$$M_y = 0.0107 \text{ wab}^2$$

$$= 0.0107(0.087)(10)(1.2)(35.2)^2(12)$$

$$= 166 \text{ in.-kips}$$

$$f_t = f_b = \frac{166}{490} = 0.339 > 0.224 \text{ ksi}$$

Therefore, 2-point stripping no good

Four point pick-up, Fig. 5.2.5(b)

$$M_y = 0.0027 \text{ wab}^2$$
$$= 0.0027(0.087)(10)(1.2)(35.2)^2(12)$$
$$= 41.9 \text{ in.-kips}$$

Additional moment due to lifting angle (Fig. 5.2.9)

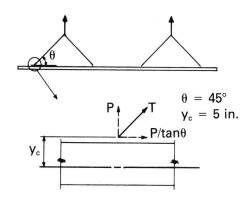

θ = 45°
y_c = 5 in.

$$M_y = \frac{P y_c}{\tan\theta} = \left[\frac{0.87(1.2)(35.2)}{4(2)}\right]\frac{5}{\tan 45°}$$

$$M_y = 23 \text{ in.-kips}$$

$$M_{total} = 41.9 + 23 = 64.9 \text{ in.-kips}$$

$$f_t = f_b = 0.132 < 0.224 \text{ ksi} \qquad \text{OK}$$

Use 4-point pick-up for stripping

Note: By inspection of support structure and 4-point pick-up stripping locations, it is determined that shifting lifting loops toward ends slightly will avoid interference between loops and edge beams thus avoiding a delay in erection while loops are burned off. Effect on stresses is minor.

b. Transverse bending — beam strip properties

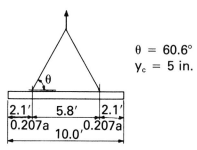

θ = 60.6°
y_c = 5 in.

0.29a = 35 in., 10t = 70 in., b/2 = 211 in.

Z = 35 (7)²/6 = 286 in.³

$$M_x = 0.0054(0.087)(1.2)(10)^2(12)(35.2)$$
$$= 23.7 \text{ in.-kips}$$

$$M_x \text{ (lifting angle)} = \left[\frac{1.2(0.087)(10)(35.2)}{(4)(2)}\right]$$

$$\times \frac{5}{\tan 60.6°} = 12.9 \text{ in.-kips}$$

Total M_x = 23.7 + 12.9 = 36.6 in.-kips

$f_t = f_b$ = 36.6/286 = 0.128 ksi < 0.224 ksi   OK

Final stripping loop locations:

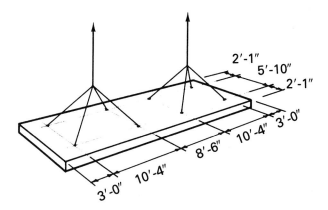

Check Handling Stresses — Turning:

Stresses for turning from edge to edge are excessive. Note also that edge detail may restrict this type of turning, due to excessive shear loading on insert cast into edge.

Therefore turn as follows:

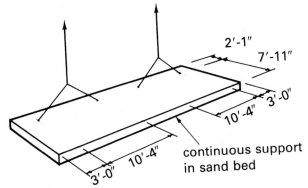

continuous support in sand bed

Transverse moments:

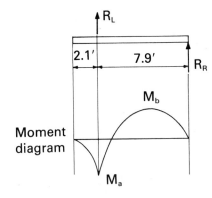

Moment diagram

$w = 0.087(1.2)(35.2/4) = 0.92$ k/ft

$R_L = \dfrac{0.92(10^2/2)}{7.9} = 5.81$ kips

$R_R = 0.92(10) - 5.81 = 3.39$ kips

$M_a = (0.92)(2.1)^2(12)/2 = 24.3$ in.-kips

M_b maximum at $3.39/0.92 = 3.68$ ft.

$M_b = [3.39(3.68) - 0.92(3.68)^2/2]12$
$= 74.9$ in.-kips

Using same resisting section as for stripping:

$f_t = 24.3/286 = 0.085$ ksi < 0.224 OK

$f_b = 74.9/286 = 0.262$ ksi > 0.224

Therefore, concrete should have strength of $(262/5)^2 = 2746$–say 2750 psi

Longitudinal bending similar to stripping.

Use 4-point turning with one edge in sand bed.

Check Handling Stresses — Shipping:

The following factors were considered in determining shipping method:

Alternatives

(1) Ship flat: • strength with 2-point support
               • permits required in 3 states for each load (since panel weight of

30,600 lb restricts loads to one panel each)

(2) Ship vertical: • requires low-boy trailer with 35 ft well with maximum height of 3 ft which would restrict total height to 13.5 ft (6 in. support material)

(3) Special frames: • fabricate special frame so panel could be set at about a 45° angle and be within non-permit restrictions of 13.5 ft high by 8.0 ft wide

Other factors:

• Total number of pieces - 132

• Erection rate - 10 pieces per day

• Drivers not permitted to drop loads on job after working hours (union regulations at job site)

• Permit loads on bridge restricted to traveling between 9:30 A.M. and 2:30 P.M.

To avoid disrupting normal production in plant, loading must be done in P.M. (same cranes used for stripping (A.M.), yarding and loading).

These considerations led to the conclusion that 30 trailers would be necessary to properly supply the job. This can be demonstrated as follows with each group of trailers being 10 (A, B, C):

	Mon	Tues	Wed	Thurs	Fri
Load	A	B	C	A	B
Ship		A	B	C	A
Erect			A	B	C
Return Trailer			A	B	C

3 groups of 10 required = 30 trailers

Reconsider alternatives

(2) Vertical - 30 low-boys not available

(3) Special frames @ 45° - with 2 panels per trailer

Check panel bending.
Longitudinal:

$w = 0.87(1.7)(\sin 45°) = 1.05$ kips/ft

From Fig. 5.2.5 for full panel width:
$M_y = 2(0.0107)(1.05)(35.2)^2(12)$
$= 334$ in.-kips

$f = 334/976 = 0.342$ ksi < 0.354 OK

Since stress is high and a long panel traveling 150 miles is subject to possible dynamic forces, provide a 3rd frame for lateral support only, to avoid possible harmonic motion.

$$30(3) = 90 \text{ frames required}$$

Cost estimated @ $250 each or $22,500

(1) Flat - cost of permits $65 × 150 loads
$$= \$\ 9,750$$

possible saving = $12,750

To achieve this saving, we must provide proper support. Two point support is desirable since flexing of trailer normally will lower or raise any additional supports thus causing bridging and unanticipated stresses.

4-Point Support

* Too far apart to consider for shipping since the trailer will deflect and cause 2 or 3 point support at times during transportation.

Try adjusting support points so that 10'-4" is reduced to 5'-0" which is a practical limit to insure 4 points will have support at all times during transportation.

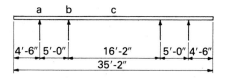

A moment distribution of this condition results in moments as follows:

M_a = 0.0082$w\ell^2$ = 0.0082(0.87)(1.7) ×
(35.2)²(12) = 180.3 in.-kips
tension in back face

M_b = 0.014$w\ell^2$ = 307.8 in.-kips
tension in back face

M_c = 0.0125$w\ell^2$ = 274.8 in.-kips
tension in front face

Stresses:

$f_a = \dfrac{180.3}{976}$ = 0.185 ksi < 0.354 ksi   OK

$f_b = \dfrac{307.8}{976}$ = 0.315 ksi   OK

$f_c = \dfrac{274.8}{976}$ = 0.281 ksi   OK

Ship with supports as shown

Note: Had stress been excessive, supports could be shifted toward center of panel until cantilever condition is such that it produces a stress of 354 ksi.

An alternate solution provides 4-point support for the precast piece with 2 supports on the truck bed as shown:

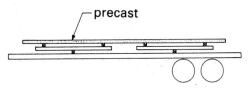

Check Handling Stresses — Erection:

Try 3-point pick as follows:

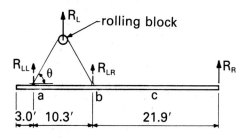

Longitudinal bending: with rolling block, reactions at stripping loops are equal.

w   = 0.87 (1.5) = 1.31 kips/ft

Distance from R_L to R_R = $35.2 - 3.0 - \dfrac{10.3}{2}$
$$= 27.0 \text{ ft}$$

$R_L = \dfrac{1.31(35.2)^2/2}{27.0}$ = 30.0 kips

$R_{LL} = R_{LR}$ = 30.0/2 = 15.0 kips

R_R = 35.2(1.31) − 30.0 = 16.1 kips

M_c maximum at 16.1/1.31 = 12.3 ft.

M_c = [16.1(12.3 − 1.31(12.3)²/2]12

$$= 1187 \text{ in.-kips}$$

f   = $\dfrac{1187}{976}$ = 1.2 ksi—too high

Since this stress is much too high, it is apparent that it cannot be brought within limits by adjusting pick points. Therefore, erect as follows (stresses less critical than stripping):

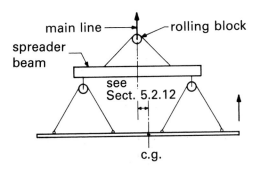

Lift from truck in horizontal position with one line and rotate in air to vertical position with 2nd crane line.

Note: Must use tag line opposing 2nd crane line to avoid rapid movement, or a more suitable solution is to shift lifting points toward bottom of piece to insure that the main line is below the center of gravity of the piece. This then insures that a vertical force is required at the top to rotate the piece. See Sect. 5.2.12.

5.3 Erection Bracing

5.3.1 Introduction

This section deals with the temporary bracing which may be necessary to maintain structural stability of a precast structure during construction. When possible, the final connections should be used to provide at least part of the erection bracing, but additional bracing apparatus is frequently required to resist all of the temporary loads. These temporary loads include wind, seismic, eccentric dead loads including construction loads, unbalanced conditions due to erection sequence and incomplete connections.

Proper planning of the construction process is essential for efficient and safe erection. Sequence of erection must be established early and the effects accounted for in the bracing analysis and the preparation of shop drawings.

The responsibility for the erection of precast concrete may vary as follows:

1. The precast concrete manufacturer supplies the product erected, either with his own forces, or by an independent erector.

2. The manufacturer is responsible only for supplying the product, F.O.B. plant or jobsite. Erection is done either by the general contractor, or an independent erector under a separate agreement.

3. The products are purchased by an independent erector who has a contract to furnish the complete precast concrete package.

Regardless of the contractual arrangement, responsibility for stability during erection must be clearly understood. Design for erection conditions must be in accordance with all local, state and federal regulations. It is desirable that this design be directed or approved by a professional engineer.

Erection drawings instruct the erector on how to assemble the components into the final structure, and should be prepared for all projects. When temporary bracing is required, additional bracing drawings are recommended. These may show such items as sequence of erection, bracing hardware and procedures, and instructions on removal. For large and/or complex projects, a pre-job conference prior to the preparation of erection drawings may be desirable. The purpose of such a meeting is to discuss erection methods and sequence and coordination with other trades.

Handling Equipment

The type of jobsite handling equipment selected may influence the erection sequence, and hence affect the temporary bracing requirements. Several types of erection equipment are available, including truck-mounted and crawler mobile cranes, hydraulic cranes, tower cranes, monorail systems, derricks and others. The PCI Recommended Practice for Erection of Precast Concrete[3] provides more information on the uses of each.

Surveying and Layout

Before products are shipped to the jobsite, a field check of the project should be made to ensure that prior construction is suitable to accept the precast units. This check should include location, line and grade of bearing surfaces, notches, blockouts, anchor bolts, cast-in hardware, and dimensional deviations. Site conditions such as access ramps, overhead electrical lines, truck access, etc., should also be checked. Any discrepancies between actual conditions and those shown on drawings should be corrected before erection is started.

Surveys should be required before, during and after erection:

1. Before, so that the starting point is clearly established and any potential difficulties with the support structure are determined early.

Fig. 5.3.1 Temporary loading conditions that affect stability

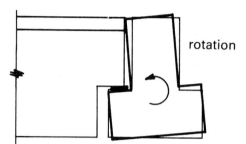

(a) Columns with eccentric loads from other framing components tend to make the column lean out of plumb. Cable bracing can be used to keep the column plumb.

(b) Unbalanced loads due to partially complete erection may result in beam rotation. The erection drawings should note these conditions.

deflection of supporting slab results in rotation of panel

deflected position of supporting member due to weight of panels

(c) Loading one entire elevation of a structural frame with cladding panels tends to make the building lean out of plumb.

(d) Examples of rotations and deflections of framing members caused by cladding panels. May result in alignment problems and require temporary connections.

2. During, to maintain alignment.

3. After, to ensure that the products have been erected within tolerances.

5.3.2 Loads

Wind: Wind loads used for the design during erection should be based on local codes, tempered by judgment. For example, in hurricane regions, the maximum code load is usually not used, since there is sufficient warning of a hurricane so that additional bracing can be installed. Note that it is possible that more surface area is exposed to wind pressure than when the structure is totally enclosed.

Earthquake: In seismic regions, the degree to which earthquake loads are considered for construction design is a decision which should be made by the engineer designing the temporary bracing system, unless the question is covered by local codes or project specifications. Often, seismic loading is neglected unless the project is expected to be shut down in a temporary condition for an extended period of time.

Construction loads: This includes materials stored on floor members such as masonry, drywall or other finishing materials, and construction equipment such as buggies used for concrete placement. A value of 25 psf has been used without creating an excessive burden on the bracing requirements.

Others: Fig. 5.3.1 shows other temporary loading conditions which affect stability and bracing design.

5.3.3 Factors of Safety

Safety factors used for temporary loading conditions are a matter of judgment, and should consider failure mode (brittle or ductile), predictability of loads, consequences of possible failure, quality control of products and construction, opportunity for human error, and economics. The total factor of safety also depends on load factors and capacity reduction factors used in the design of the entire bracing system. These must be consistent with code requirements and other sections of this Handbook.

The values shown in Table 5.3.1 will usually provide safe margins for the temporary conditions considered here.

5.3.4 Bracing Equipment and Materials

For most one- and two-story high components that require bracing, steel pipe braces similar to

Table 5.3.1 Suggested factors of safety for construction loads

Bracing for wind loads	2
Bracing inserts cast into precast members	3
Reusable hardware	4
Lifting inserts	4

those shown in Fig. 5.3.2 are used. A wide range of bracing types are available from a number of suppliers, who should be consulted for dimensions and capacities. Pipe braces resist both tension and compression. When long braces are used in compression, it may be necessary to provide lateral restraint to the brace to prevent buckling.

Cable guys with turnbuckles are normally used for higher structures. Since wire rope used in cable guys can resist only tension, they are usually used in combination with other cable guys in an opposite direction. Compression struts, which are usually the precast components, are needed to complete truss action of the bracing system.

A large number of types of wire rope is also available.[4] Typically, wire rope is constructed of three basic components: (1) wires that form the strands, (2) multi-wire strands laid helically around a core and (3) the core (see Fig. 5.3.3).

The wire may be iron, stainless steel, monel, or bronze, but for construction uses is nearly always high carbon steel. The core, which is the foundation for the wire rope, is made of either fiber or steel. The most commonly used cores are: fiber core (FC), independent wire rope core (IWRC), and wire strand core (WSC).

Rope is classified by the construction type. For example, a 6 × 7 FC consists of 6 strands of 7 wires each, wrapped around a fiber core. A 6 × 19 IWRC has 6–19-wire strands wrapped around an independent wire rope core. Strength of wire rope is dependent on the component materials. Grades include: traction steel (TS), mild plow steel (MPS), plow steel (PS), improved plow steel (IPS), and extra improved plow steel (EIPS).

Table 5.3.2 shows the properties of several commonly used wire rope sizes. It is recommended that the minimum size rope used for bracing be ½ in. diameter.

The elongation or "stretch" of wire ropes must be considered in designing bracing. Elongation comes from two sources: constructional stretch and elastic stretch. Constructional stretch is dependent on the classification and results primarily from a reduction in diameter as load is applied and the strands compact against each other. Approximate ranges of constructional stretch are shown in Table 5.3.2. Wire ropes may be pre-

Fig. 5.3.2 Typical pipe braces

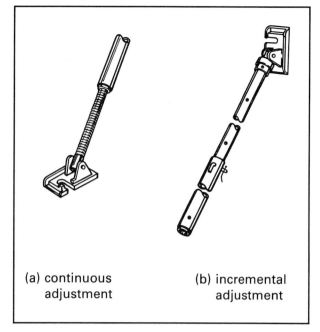

(a) continuous   (b) incremental
    adjustment   adjustment

Fig. 5.3.3 Wire rope

stretched to remove some of the constructional stretch.

Elastic stretch is caused by the deformation of the metal itself when load is applied. As with constructional stretch, a precise value is difficult to establish, but the following equation gives adequate results:

$$\text{Elastic stretch} = \frac{PL}{AE}$$

where: P = change in load
       L = length
       A = area of wire rope
       E = modulus of elasticity

Example 5.3.1 Stretch of wire rope

Given:
A 70 ft long, ¾ in. diameter, 6 × 7 FC wire rope resisting a tension force of 12 kips.

Problem:
Determine the total stretch.

Solution:
Constructional stretch (use 0.75%) = 0.0075(70)(12) = 6.3 in.

$$\text{Elastic stretch} = \frac{12(70)(12)}{0.288(10,000)} = 3.5 \text{ in.}$$

Total = 6.3 + 3.5 = 9.8 in.

Table 5.3.2 Properties of wire rope

Nominal diameter in.	Fiber core (FC) E = 10,000 ksi Constructional stretch:0.5-0.75%		Wire core (IWRC) E = 13,000 ksi Constructional stretch:0.25-0.5%	
	Area, in.²	Nominal strength (kips)	Area, in.²	Nominal strength (kips)
½	0.096	20.6	0.113	22.2
⅝	0.150	31.8	0.176	34.2
¾	0.288	45.4	0.254	48.8
⅞	0.294	61.4	0.345	66.0
1	0.384	79.4	0.451	85.4
1⅛	0.486	99.6	0.571	107.0
1¼	0.600	122.0	0.705	131.6
1⅜	0.726	146.2	0.853	157.2
1½	0.864	172.4	1.015	185.4

Properties based on 6 × 7 classification. 6 × 19 classification approximately 4% higher.
Based on "Improved Plow Steel". "Extra Improved Plow Steel" approximately 15% higher.

PCI Design Handbook

Fig. 5.3.4 Schematic plan of typical floor — Example 5.3.2

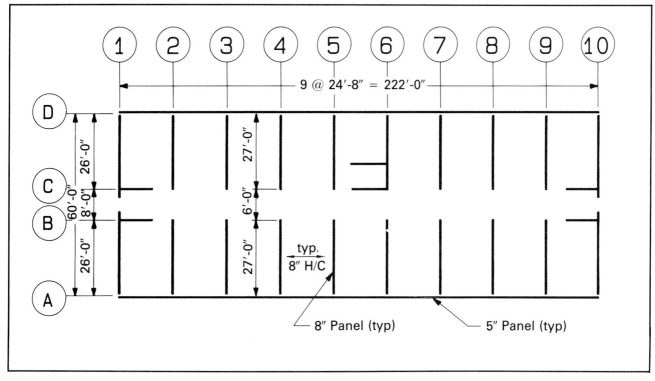

5.3.5 Examples

The following examples demonstrate a suggested procedure to insure structural stability and safety during various stages of construction. Actual loads, factors of safety, equipment used, etc., must be evaluated for each project.

Example 5.3.2 — Load bearing wall panel structure

Given:

18-story hotel with plan as shown in Fig. 5.3.4

Floor to floor — 8'-8"

Floor system — 8 in. hollow-core planks

Wind load — 15 psf. The code requires higher loads at upper levels, but the engineer has judged that for temporary conditions 15 psf is adequate.

No expansion joint

Final stability in the east-west direction depends on the stair and elevator walls at the ends, plus the exterior precast panels on the north and south walls (lines A and D). Diaphragm action of the floor distributes lateral loads to these shear walls.

Problem:

Determine sequence of erection and temporary bracing system.

Solution:

After consultation between precaster and erector, it has been decided to erect the floors and load bearing walls through the sixth floor one floor at a time. Then a second crew and lighter crane will be used to erect the precast wall panels on lines A and D. Erection of structural precast floors and walls will never be more than six levels ahead of the shear walls, lines A and D. Design tasks include:

1. Design single panel with two braces at upper level of any erection phase.

2. Determine diaphragm loads and check if elevator walls can resist these for six levels. If not, reduce the number of levels the floor erection can be ahead of the shear wall placement.

Design single panel:

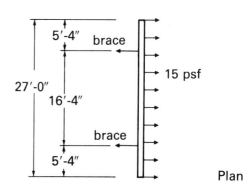

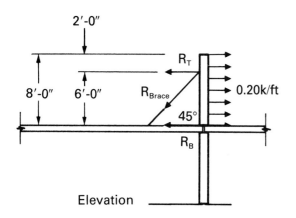

Elevation

Wind load per brace = 0.015(27)/2
= 0.20 kips/ft

$$R_T = \frac{0.20(8^2/2)}{6} = 1.07 \text{ kips}$$

$R_B = 0.20(8) - 1.07 = 0.53$ kips

$R_{brace} = 1.07\sqrt{2} = 1.51$ kips

These loads and reactions can act in either direction. Typically, critical directions would be suction for the inserts in the panel and floor, and pressure for the brace. Forces on top connection:

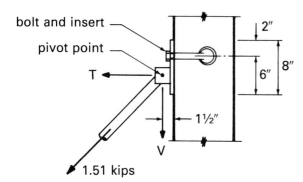

V = 1.07 kips

$$T \approx \frac{1.51}{\sqrt{2}} + \frac{1.07(1.5)}{6-1} = 1.39 \text{ kips}$$

Design of inserts and bolts is discussed in Part 6. Design of the connection at the bottom of the brace is similar. Expansion bolts in hollow-core plank must be placed to avoid the cores. Vertical tie connections between wall panels is usually adequate to take the wind shear at the base of the panel.

Check panel section:
Consider the horizontal span:

w = 0.015(8) = 0.12 kips/ft

$-M = wa^2/2 = 0.12(5.33)^2 (12)/2 = 20.5$ in.-kips

$+M = w\ell^2/8 - (-M)$
$= 0.12(16.33)^2(12)/8 - 20.5 = 27.6$ in.-kips

$Z = bd^2/6 = 96(8)^2/6 = 1024$ in.^{3}

$f = 27.6/1024 = 0.027$ ksi $< 5\sqrt{f'_c}$ OK

Determine diaphragm loads:
Shielding of the wind from adjacent structures is not considered in building design, because of the possibility that they may eventually be razed. However wind shielding by adjacent panels during erection is more predictable, and may be used with judgment realizing that erection sequences can change.

Various wind directions must also be considered. For this structure, temporary loading from north or south wind is less critical than the final condition, so it can be neglected. Two possible wind directions will be considered:

1. Wind from east or west. Apply full wind to end wall, with the other walls assumed to be 50% shielded (some wind can flow over the tops of the walls), except at the ends. Assume full wind applied to 25% of the length of the interior panels as wind flows along the sides of the building. See Fig. 5.3.5.

At 18th level:
0.015(8)(60) + 0.015(4)(60)(9) = 39.6 kips

At other levels:
0.015(8)(60) + 0.015(6.75)(2)(8)(9)
= 21.7 kips

2. Wind from any 45° angle. Apply full wind component to end wall with 50% shielding at upper level. Shielding of lower interior panels is determined from the geometry (see Fig. 5.3.6).

At 18th level:
0.015(0.707)(8)(60) + 0.015(0.707)(4)(60)(9)
= 28.0 kips

At other levels:
0.015(0.707)(8)(60) + 0.015(0.707)(24)(8)(9)
= 23.4 kips

The above shielding effects are based on empirical judgment, and will vary among structures and designers. Note that the resultant of loads for condition 1 acts at the center of the structure, while for condition 2 it acts eccentrically.

The design of a diaphragm and shear walls is discussed in Part 3, and connections in Part 6.

Example 5.3.3 — Single-story industrial building
Given:
A single-story building with the plan shown schematically in Fig. 5.3.7.
Totally precast structure — columns, inverted

Fig. 5.3.5 Distribution of east-west wind load — Example 5.3.2

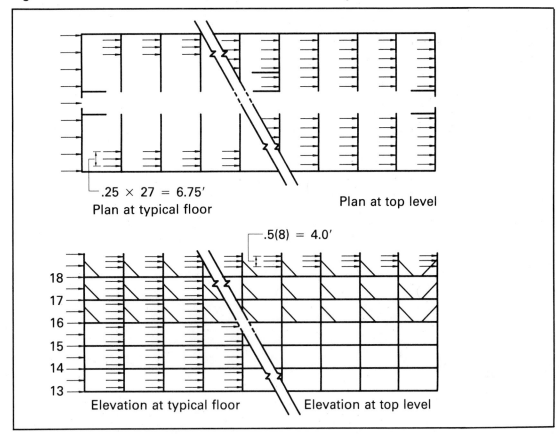

.25 × 27 = 6.75'

Plan at typical floor

Plan at top level

.5(8) = 4.0'

Elevation at typical floor

Elevation at top level

Fig. 5.3.6 Distribution of wind from 45° angle —Example 5.3.2

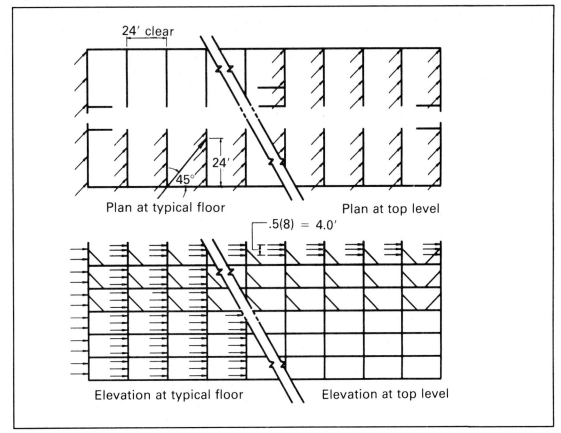

24' clear

45°

24'

Plan at typical floor

Plan at top level

.5(8) = 4.0'

Elevation at typical floor

Elevation at top level

Fig. 5.3.7 Schematic plan of structure of Example 5.3.3

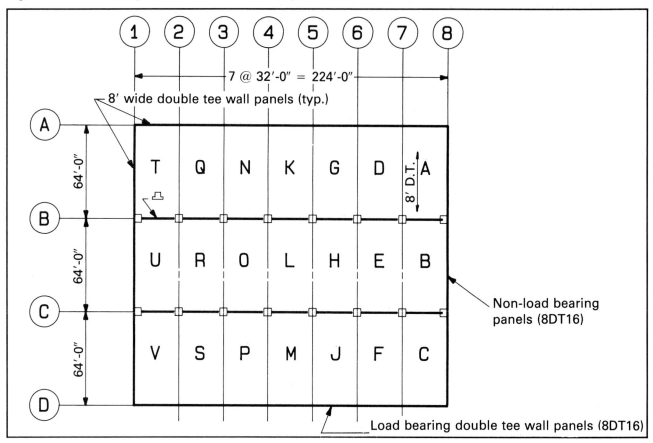

tee beams, double tee roof, load bearing and non-load bearing wall panels.

Wind load = 10 psf

Final stability by exterior panels acting as shear walls.

No expansion joint.

Erection sequence: With a crane in the center bay, start at column line 8 and move toward line 1, erecting all three bays progressively.

Problem:

Determine temporary erection bracing systems.

Solution:

To demonstrate the thought process the designer should go through, the following temporary conditions will be considered:

1. Erect columns B8 and B7 with inverted tee beam between them and check as free standing on the base plate.

w_b = 0.01(2.33) = 0.023 kips/ft

P = 0.023(32.5)/2 = 0.37 kips

w_c = 0.01(1) = 0.01 kips/ft

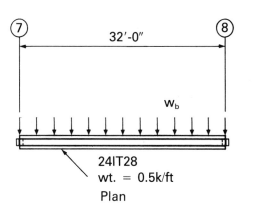

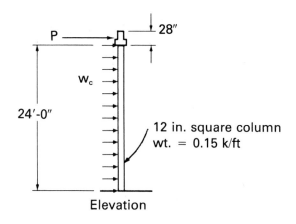

Moment at column base = 0.37(24) + 0.01(24)²/2
                      = 11.8 ft-kips

Dead load at column base = 0.5(32.5)/2 + 0.15(24)
                         = 11.7 kips

Design base plate and anchor bolts as described in Sect. 6.10.

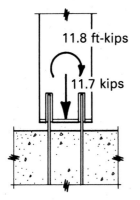

2. Erect wall panels on line A from 7 to 8. This will result in two possibilities:

a. If the base of the wall has some moment resisting capacity, the base should be designed for:

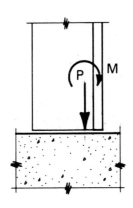

M = 0.01(8)(0.5)(28²/2) = 15.7 ft-kips each stem

P = 0.34(28)/2 = 4.75 kips each stem

b. If the wall panel has no moment resisting

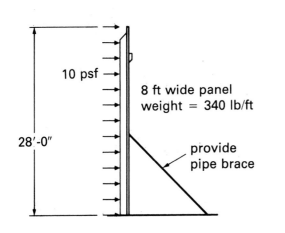

capacity, a brace must be designed as shown in Ex. 5.3.2. Be sure to check the unsupported length of the brace to prevent buckling under compressive loads.

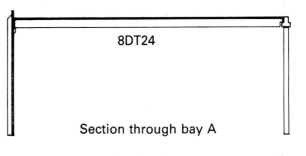

Section through bay A

3. Erect double tee roof members in bay A. Assuming the wall panel has moment resisting capacity at the base, two loading conditions must be checked:

a. Dead loads only:

Load from roof tee = 0.052(4)(64/2)
                   = 6.66 kips/stem

M = 6.66(9/12) = 5.0 ft-kips

P = 6.66 + 4.75 = 11.4 kips

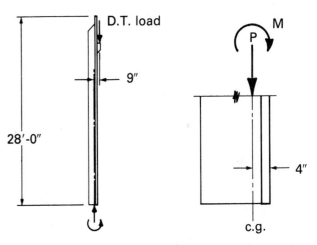

Check beam/column connection. This will be critical when all four roof tees in bay A are in place:

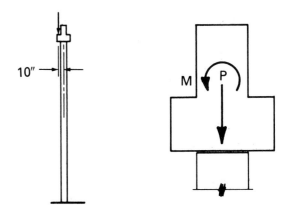

Load from double tees
= 6.66 kips/stem times 4 stems = 26.6 kips

M = 26.6(10/12) = 22.2 ft-kips

P = 26.6 + 0.5(32/2) = 34.6 kips

Check column base connection:

M = 22.2 ft-kips (from above)

P = 34.6 + 0.15(24) = 38.2 kips

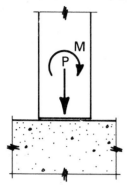

Column base connection

b. Dead load plus wind load. In this example, it is apparent that the loading for 1 and 2 above will be more critical than this condition.

The designer should always try to use the permanent connections for erection stability. In this example, the final connections between the wall and roof deck and the beam and roof deck will provide some degree of moment resistant capacity. This will reduce the moments on some of the connections considered above.

4. Erect wall panels on line 8 from A to B. (Note: If the crane reach is limited, these may be erected immediately after the first roof tee is placed.) These will be connected to the roof with the permanent diaphragm transfer connections. We now have a rigid "box" which will provide stability to the remainder of the structure.

5. Continue erection working from the rigid box in bay A. It is unlikely that other temporary conditions will be more critical than those encountered in bay A. However, each should be considered and analyzed if appropriate.

Example 5.3.4 Multilevel parking structure

Given:

A typical 7-story parking structure shown schematically in Fig. 5.3.8.

Structural system — untopped double tees on L-shaped and inverted tee interior beams and load bearing spandrel panels. Multilevel columns are spliced at level 4. No expansion joint.

Loads: Wind — 10 psf. Construction loads — 5 psf. This is lower than recommended in Sect. 5.3.2 because there are virtually no interior finishing materials, such as masonry or drywall in this construction.

Final stability will be provided by:

Long direction: The ramped floors acting as a truss.

Short direction: Shear walls as shown.

Problem:

Outline critical design conditions during erection, and show a detailed erection sequence.

Solution:

Outline of critical erection design conditions:

1. Free standing columns — design columns and base plates in accordance with Parts 3, 4 and 6.

2. Determine bracing forces for wind loads from either direction.

3. Select wire rope sizes.

4. Determine forces on inserts used for bracing and select anchorages in accordance with Part 6.

5. Check column designs with temporary loading and bracing.

6. Check diaphragm design and determine which permanent connections must be made during erection. Determine need for temporary connections.

7. Check inverted tee beams and their connections for loading on one side only.

Sequence of erection: Start at column line 1 and, with crane in north bay, erect vertically and back out of structure at line 9 in the following sequence:

1. Erect columns A1, B1, A2 and B2 — (lower tier). Check as free standing columns with 10 psf wind on surface.

2. Erect 3 levels in bay A.

3. Install X-bracing between A1 and B1. Check capacity required with full wind load on exposed surfaces.

4. Erect column A3 and shearwall.

5. Erect 3 levels in bay B.

6. Erect 4th level in bay A.

7. Install X-bracing at line B between B2 and shear wall. Check capacity required with full wind load on exposed surfaces.

8. Splice 2nd tier columns at A1, A2, A3, B1 and B2. Check capacity of welded splice with upper tier as free standing columns with 10 psf wind on surfaces.

Fig. 5.3.8 Parking structure of Example 5.3.4

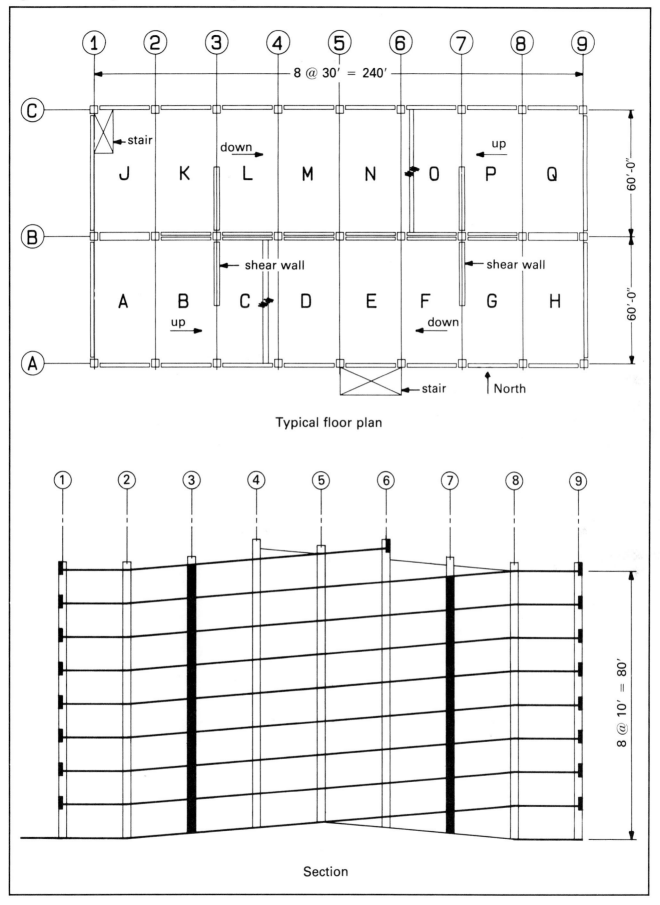

Typical floor plan

Section

9. Erect columns C1 and C2.
10. Erect 4 levels in bay J.
11. Install X-bracing from columns B1 to C1. Check capacity required with full load on exposed surfaces.
12. Erect columns A4 and B4.
13. Erect to 3rd level in bay C.
14. Install X-bracing between B3 and B4.
15. Erect bay A to 6th level.
16. Erect bay J to 6th level.
17. Erect bay A to roof.
18. Erect bay B to 6th level.
19. Erect column C3.
20. Erect bay K to 4th level.
21. Erect bay J to roof.
22. Erect columns A5 and B5.
23. Erect bay D to 3rd level.
24. Install X-bracing from A5 to B5. Check capacity required.
25. Erect bay C to 5th level.
26. Erect column C4.
27. Erect bay L to 4th level.
28. Erect bay K to 6th level.
29. Erect bay B to roof.
30. Continue in same sequence.

Additional notes:

1. After each level is erected weld coil rod from spandrel to reinforcement in double tee. (See Fig. 5.3.9).
2. Install additional X-bracing on line B from 6 to 7 and 7 to 8 after framing is erected.
3. Install X-bracing on line 9 from A to B and B to C when framing is erected.
4. Install cables to upper tier columns as required to keep columns plumb.
5. Check upper tier columns as free standing with full wind load on exposed surfaces.

5.4 References

1. Tanner, John, "Architectural Panel Design and Production Using Post-Tensioning" *PCI Journal,* v. 22, no. 3, May-June, 1977.

2. Anderson, A. R., "Lateral Stability of Long Prestressed Concrete Beams", *PCI Journal,* v. 16, no. 3, May-June 1971.

3. PCI Erectors Committee, "Recommended Practic for Erection of Precast Concrete," Prestressed Concrete Institute, Chicago, IL (To be published in 1985).

4. *Wire Rope Users Manual,* Committee of Wire Rope Producers, American Iron and Steel Institute, Washington, D. C., 1979.

Fig. 5.3.9 Typical load-bearing spandrel — Example 5.3.4

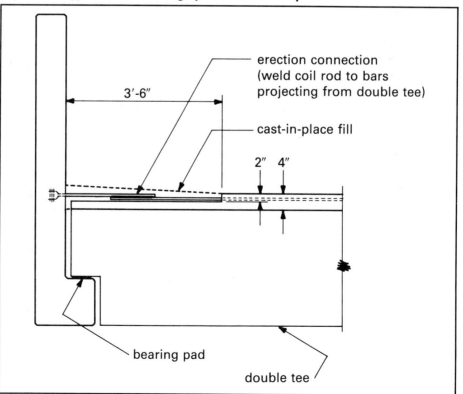

PART 6
DESIGN OF CONNECTIONS

Page No.

6.1 Notation 6–2
6.2 General 6–3
6.3 Loads and Load Factors 6–3
6.4 Connection Design Criteria 6–3
6.5 Connection Hardware and Load Transfer Devices 6–4
    6.5.1 Reinforcing Bars 6–4
    6.5.2 Welded Headed Studs 6–6
    6.5.3 Deformed Bar Anchors 6–10
    6.5.4 Bolts and Threaded Connectors 6–10
    6.5.5 Inserts Cast in Concrete 6–10
    6.5.6 Structural Steel 6–11
    6.5.7 Post-Tensioning Steel 6–14
    6.5.8 Bearing Pads 6–14
    6.5.9 Connection Angles 6–17
6.6 Friction 6–18
6.7 Shear-Friction 6–18
6.8 Bearing on Plain Concrete 6–19
6.9 Reinforced Concrete Bearing 6–20
6.10 Column Base Plates 6–21
6.11 Concrete Brackets or Corbels 6–23
6.12 Structural Steel Haunches 6–25
6.13 Dapped-End Connections 6–28
    6.13.1 Flexure and Axial Tension in the Extended End 6–28
    6.13.2 Direct Shear 6–28
    6.13.3 Diagonal Tension at Reentrant Corner 6–30
    6.13.4 Diagonal Tension in the Extended End 6–30
    6.13.5 Anchorage of Reinforcement 6–30
    6.13.6 Detailing Considerations 6–30
    6.13.7 Alternate Placement of Reinforcement 6–30
6.14 Beam Ledges 6–33
6.15 Hanger Connections 6–34
    6.15.1 Cazaly Hanger 6–34
    6.15.2 Loov Hanger 6–36
6.16 Moment Connections 6–38
6.17 Connection of Non-Load Bearing Wall Panels 6–38
6.18 Connection of Load Bearing Wall Panels 6–39
    6.18.1 Vertical Joints 6–39
    6.18.2 Horizontal Joints 6–41
    6.18.3 Structural Integrity 6–43
    6.18.4 Typical Details 6–43
6.19 References 6–43
6.20 Design Aids 6–48

DESIGN OF CONNECTIONS

6.1 Notation

a $\quad$ = shear span

A_b $\quad$ = area of bar or stud

A_{cr} $\quad$ = area of crack face

A_e $\quad$ = effective slab bearing area

A_f $\quad$ = area of flexural reinforcement

A_h $\quad$ = area of shear reinforcement parallel to flexural tension reinforcement

A_ℓ $\quad$ = area of longitudinal reinforcement in beam ledge

A_n $\quad$ = area of reinforcement required to resist axial tension

A_o $\quad$ = lateral surface area of failure surface

A_s $\quad$ = area of reinforcement

A'_s $\quad$ = area of vertical reinforcement near end of steel haunch

A_{sh} $\quad$ = area of reinforcement for horizontal or diagonal cracks

A_v $\quad$ = diagonal tension reinforcement in dapped end

A_{vf} $\quad$ = area of shear-friction reinforcement

A_w $\quad$ = area of weld

A_1 $\quad$ = loaded area

A_2 $\quad$ = maximum area of the portion of the supporting surface that is geometrically similar to and concentric with the loaded area

b $\quad$ = width

C.E. $\quad$ = carbon equivalent

C $\quad$ = compressive force

C $\quad$ = confinement factor

C_r $\quad$ = reduction coefficient (see Eq. 6.8.1)

C_{es} $\quad$ = reduction coefficient for edge distance (see Eq. 6.5.3)

d $\quad$ = depth to centroid of reinforcement

d_b $\quad$ = bar or stud diameter

d_e $\quad$ = distance from center of load to beam end

d_e $\quad$ = edge distance in direction of load

d_h $\quad$ = head diameter of stud

D $\quad$ = durometer

e_i $\quad$ = center of bolt to horizontal reaction

e_v $\quad$ = eccentricity of vertical load

e_x, e_y $\quad$ = eccentricity of load in x,y directions

f $\quad$ = unit stress

f_{bu} $\quad$ = factored bearing stress

f_{ct} $\quad$ = splitting tensile strength of concrete

f'_c $\quad$ = compressive strength of concrete

f_r $\quad$ = resultant stress on weld

f_s $\quad$ = design strength of steel

f_{ue} $\quad$ = see Sect. 6.18.2

f_v $\quad$ = yield strength of A_v

f_w $\quad$ = design strength of weld (see Table 6.20.1)

f_x $\quad$ = combined shear and torsion stress in horizontal direction

f_y $\quad$ = combined shear and torsion stress in vertical direction

f_y $\quad$ = yield strength of reinforcement or structural steel

f_{ys} $\quad$ = yield strength of A_{sh}

F $\quad$ = connection forces

F_s $\quad$ = factored friction force

ΣF $\quad$ = greatest sum of factored anchor bolt forces on one side of the column

g $\quad$ = gage of angle

h $\quad$ = total depth

I_p $\quad$ = polar moment of inertia

I_{xx}, I_{yy} $\quad$ = moment of inertia of weld segment with respect to its own axes

j_u $\quad$ = for resisting lever arm, used in $j_u d$

ℓ $\quad$ = length of joint

ℓ_b $\quad$ = bearing length

ℓ_d $\quad$ = development length

ℓ_e $\quad$ = embedment length

ℓ_ℓ $\quad$ = angle leg length

ℓ_p $\quad$ = projection of corbel, beam ledge or dapped end

ℓ_w $\quad$ = length of weld

M_t $\quad$ = torsional moment

M_u $\quad$ = factored moment

n $\quad$ = number of studs in a group

N $\quad$ = unfactored horizontal or axial force

N_u $\quad$ = factored horizontal or axial force

P $\quad$ = applied load

P_c = nominal tensile strength of concrete element

P_n = nominal strength of joint

P_s = nominal tensile strength of steel element

P_u = applied factored load

P_u = factored tension load

P_x, P_y = applied force in x or y direction

R_e = reduction factor for load eccentricity

s = distance from free edge to center of bearing

s = spacing of concentrated loads

S = shape factor

t = thickness

t_g = grout thickness

t_w = effective throat thickness of weld

T = tensile force

V = unfactored vertical or shear force

V_c = nominal shear strength of concrete element

V_d = unfactored vertical dead load

V_n = nominal bearing or shear strength of an element

V_r = nominal strength provided by reinforcement

V_s = nominal shear strength of steel element

V_u = factored shear force

V_{ud} = factored dead load force normal to the friction face

w = dimension (see specific applications)

w = uniform load

x,y = surface dimensions of assumed failure plane around stud group

x_c = distance from centerline of bolt to face of column

x_o = base plate projection

x_t = distance from centerline of bolt to centerline of reinforcement

Z_s = plastic section modulus of structural steel section

Δ = horizontal deformation of bearing pad

α = angle of reinforcement placement

θ = angle of assumed crack plane

λ = coefficient for use with lightweight concrete (see Sect. 5.2.4)

μ = shear-friction coefficient

μ_e = effective shear-friction coefficient

μ_s = static coefficient of friction

ϕ = strength reduction factor

6.2 General

The design of connections is one of the most critical engineering phases in the design of a precast concrete structure. There are many successful solutions to each connection problem, and the design methods and examples included in this part are not the only proper ones. Information is included on the design of common precast concrete connections, intended for use by those with an understanding of engineering mechanics and structural design, and in no case should it replace good structural engineering judgment.

The purpose of a connection is to transfer load and/or provide stability. Within any one connection, there may be several load transfers, and each one must be considered by the designer. In the sections that follow, different methods of transferring load will be examined separately, then it will be shown how some of these are combined in typical connection situations. More complete information on connections can be found in References 1 and 2 in Sect. 6.19.

6.3 Loads and Load Factors

With noted exceptions, such as bearing pads, the design methods in this part are based on strength design relationships, incorporating the load factors and strength reduction factors (ϕ-factors) specified in ACI 318-83.

In addition to gravity loads, wind and seismic forces, forces from restraint of volume change strains, forces induced into wall panels by restrained differential movements between the panel and the structure, and the forces required for stability and equilibrium should be considered. Determination of these forces is covered in Part 3. It is recommended that connections of flexural members be designed for a minimum tensile force of 0.20 times the vertical dead load, acting parallel with the member span, unless properly designed bearing pads are used (see Sect. 6.5.8).

It is undesirable for the connection to be the weak link in a precast framing system. Therefore, it is recommended that most connections be designed with an additional load factor of at least 1.3. Insensitive connections, such as column bases, do not need the additional factor.

6.4 Connection Design Criteria

Precast concrete connections must meet a variety of design and performance criteria, and not all connections are required to meet the same criteria. These criteria include:

1. *Strength:* A connection must have the strength to resist the forces to which it will be subjected during its lifetime, including those caused by volume change restraint and those required to maintain stability.

2. *Ductility:* This is the ability to accommodate relatively large deformations without failure. In connections, ductility is achieved by designing so that steel devices yield prior to concrete failure.

3. *Volume change accommodation:* Restraint of creep, shrinkage and temperature change strains can cause severe stresses on precast concrete members and their supports. These stresses must be considered in the design, but it is usually far better if the connection will allow some movement to take place, thus relieving the stresses.

4. *Durability:* When exposed to weather, or used in a corrosive atmosphere, steel elements should be adequately covered by concrete, or be painted or galvanized. Stainless steel is sometimes used, especially in architectural concrete.

5. *Fire resistance:* Connections in which weakening by fire would jeopardize the structure's stability should be protected to the same degree as that required for the members that are connected.

6. *Constructability:* The following items should be kept in mind when designing connections:
   a. Standardize products
   b. Avoid reinforcement and hardware congestion
   c. Avoid penetration of forms
   d. Reduce post-stripping work
   e. Be aware of material sizes and limitations
   f. Consider clearances and tolerances
   g. Avoid non-standard production and erection tolerances
   h. Use standard hardware items and as few sizes as possible
   i. Use repetitious details
   j. Plan for the shortest possible hoist hook-up time
   k. Provide for field adjustment
   l. Provide accessibility
   m. Use connections that are not susceptible to damage in handling.

6.5 Connection Hardware and Load Transfer Devices

A wide variety of hardware, including reinforcing bars, studs, coil inserts, structural steel shapes, bolts, threaded rods and others, are used in connections. These devices provide for load transfer through anchorage in concrete by bond or a shear cone type failure mechanism. When possible, it is preferable to have steel material failure govern the connection strength, because such failures are more predictable and ductile. Load transfer should be as direct as possible to reduce the complexity and increase the efficiency of the connection.

6.5.1 Reinforcing Bars

Reinforcing bars (and prestressing strand) are usually anchored by bonding to the concrete. Very often, there is insufficient length available to anchor the bars by bond alone, and supplemental mechanical anchorage is required. This can be accomplished by hooks or welded cross-bars as shown in Fig. 6.11.2. Load transfer between bars may be by welding, lap splices or various types of mechanical couplers. Required development lengths and standard hook dimensions are given in Part 11.

Reinforcing bar welding

Welding of reinforcing bars is covered by AWS D1.4-79, "Structural Welding Code–Reinforcing Steel",[3] by the American Welding Society. Weldability is defined in that publication as a function of the chemical composition, as shown in the mill report by the following formula:

$$C.E. = \% \, C + \frac{\% \, Mn}{6} + \frac{\% \, Cu}{40} + \frac{\% \, Ni}{20}$$
$$+ \frac{\% \, Cr}{10} - \frac{\% \, Mo}{50} - \frac{\% \, V}{10} \qquad \text{(Eq. 6.5.1)}$$

where C.E. = carbon equivalent

The last three elements usually only appear as trace elements, so are often not included in the mill report. For bars that are to be welded, the carbon equivalent should be requested with the order from the mill.

AWS D1.4-79 indicates that most reinforcing bars can be welded. However, the preheat and other quality control measures that are required for bars with high carbon equivalents are very difficult to achieve. Except for welding shops with proven quality control procedures that meet AWS D1.4-79, it is recommended that carbon equivalents be limited to 0.45% for No. 7 and larger bars, and 0.55% for No. 6 and smaller bars.

Most reinforcing bars which meet ASTM A615, Grade 60, will not meet the above chemistry spec-

Fig. 6.5.1 Typical reinforcing bar welds

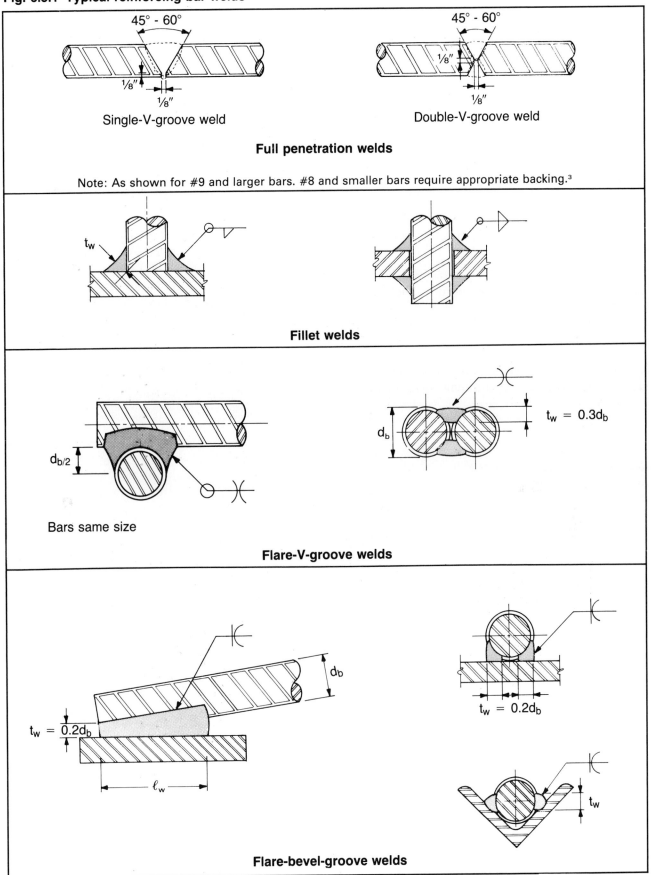

45° - 60°

1/8"

1/8"

Single-V-groove weld

45° - 60°

1/8"

1/8"

Double-V-groove weld

Full penetration welds

Note: As shown for #9 and larger bars. #8 and smaller bars require appropriate backing.[3]

t_w

Fillet welds

$d_{b/2}$

Bars same size

d_b

$t_w = 0.3d_b$

Flare-V-groove welds

$t_w = 0.2d_b$

ℓ_w

d_b

$t_w = 0.2d_b$

t_w

Flare-bevel-groove welds

ifications. A615, Grade 40 bars may or may not meet the above specifications. Bars which meet ASTM A706 are specially formulated to be weldable, and are now available in most parts of North America.

Fig. 6.5.1 shows the most common welds used with reinforcing bars. Full penetration groove welds can be considered the same as the nominal strength of the bar. The design strength of the other weld types can be calculated using the values from Table 6.20.1.[1] The total design strength of the weld is $f_w \ell_w t_w$

where:

f_w = unit design strength from Table 6.20.1

ℓ_w = length of weld

t_w = effective throat thickness of weld (Fig. 6.5.1)

Tables 6.20.3 through 6.20.5 show welding required to develop the full strength of reinforcing bars.

The welded cross-bar detail shown in Fig. 6.5.1 is not included in AWS D1.4. However, it has been used in numerous structures and tests[7,8], and when the diameter of the cross bar is at least the same size, the full strength of the main bar has been shown to be developed.

Reinforcing bars should not be welded within 2 bar diameters of a bend.

AWS D1.4 requires that tack welds be made using the same preheat and quality control requirements as permanent welds, and prohibits them unless authorized by the engineer.

Reinforcing bar couplers

Proprietary bar coupling devices are available as an alternative to lap splices or welding. Manufacturers of these devices furnish design information and test data. Refs. 5 and 6 contain more detailed information.

6.5.2 Welded Headed Studs*

Welded headed studs are designed to resist direct tension, shear or a combination of the two. Either the strength of the concrete or of the steel may be critical, and both must be checked. The design equations given below are applicable to studs which are previously welded to steel plates or members, and embedded in unconfined concrete. Confinement of the concrete, either from applied compressive loads or from reinforcement,

*The material in this section is based on a report by A.F. Shaikh for the PCI Committee on Connection Details (see Ref. 22).

Fig. 6.5.2 Shear cone development for welded headed studs

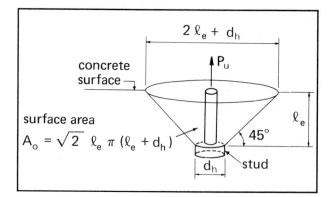

is known to increase the capacity, however due to limited research, acceptable design equations which include confinement are not available.

Tension

The design tensile strength governed by concrete failure is:

$$\phi P_c = \phi A_o (2.8 \lambda \sqrt{f'_c})* \qquad \text{(Eq. 6.5.2)}$$

where:

$\phi = 0.85$

A_o = area of the assumed failure surface which, for a single stud not located near a free edge, is taken to be that of a 45° truncated cone as shown in Fig. 6.5.2.

Using the 45° cone area and $\phi = 0.85$, Eq 6.5.2 may be written as:

$$\phi P_c = 10.7 \ell_e (\ell_e + d_h) \lambda \sqrt{f'_c} \qquad \text{(Eq. 6.5.2a)}$$

Or, for simplicity, conservatively neglecting the diameter of the head:

$$\phi P_c = 10.7 \ell_e^2 \lambda \sqrt{f'_c} \qquad \text{(Eq. 6.5.2b)}$$

For a stud located closer to a free edge than the embedment length, ℓ_e, the design tensile strength given by Eqs. 6.5.2, 6.5.2a and 6.5.2b should be reduced by multiplying it by C_{es}:

$$C_{es} = \frac{d_e}{\ell_e} \leq 1.0 \qquad \text{(Eq. 6.5.3)}$$

*In this Handbook, the subscript "u", e.g. M_u, V_u, P_u, etc., denotes only the applied factored forces. The subscript "n", and in this section, "c" and "s", denote the nominal strength. The design strength, formerly called "ultimate strength" is the nominal strength multiplied by the strength reduction factor, ϕ, for example ϕM_n, ϕV_n, ϕP_n. Thus, the design of a member or component requires that $M_u \leq \phi M_n$, $V_u \leq \phi V_n$, $P_u \leq \phi P_n$, etc.

In presenting equations for strength design, this leads to the apparent algebraic redundancy of having the term ϕ on both sides of the equation. However, to avoid the inadvertent neglect of the ϕ-factor, the equations are given in terms of "design strength" rather than "nominal strength."

Fig. 6.5.3 Pullout surface areas for stud groups

Note: For each case, use lesser of P_{c1} or P_{c2}

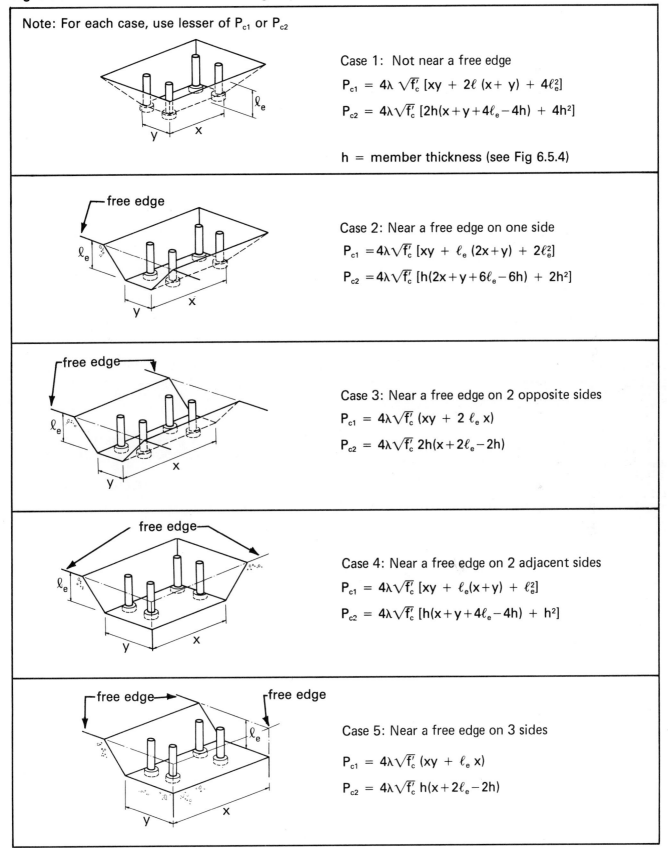

Case 1: Not near a free edge

$$P_{c1} = 4\lambda \sqrt{f_c'}\,[xy + 2\ell(x+y) + 4\ell_e^2]$$

$$P_{c2} = 4\lambda\sqrt{f_c'}\,[2h(x+y+4\ell_e-4h) + 4h^2]$$

h = member thickness (see Fig 6.5.4)

Case 2: Near a free edge on one side

$$P_{c1} = 4\lambda\sqrt{f_c'}\,[xy + \ell_e(2x+y) + 2\ell_e^2]$$

$$P_{c2} = 4\lambda\sqrt{f_c'}\,[h(2x+y+6\ell_e-6h) + 2h^2]$$

Case 3: Near a free edge on 2 opposite sides

$$P_{c1} = 4\lambda\sqrt{f_c'}\,(xy + 2\,\ell_e x)$$

$$P_{c2} = 4\lambda\sqrt{f_c'}\,2h(x+2\ell_e-2h)$$

Case 4: Near a free edge on 2 adjacent sides

$$P_{c1} = 4\lambda\sqrt{f_c'}\,[xy + \ell_e(x+y) + \ell_e^2]$$

$$P_{c2} = 4\lambda\sqrt{f_c'}\,[h(x+y+4\ell_e-4h) + h^2]$$

Case 5: Near a free edge on 3 sides

$$P_{c1} = 4\lambda\sqrt{f_c'}\,(xy + \ell_e x)$$

$$P_{c2} = 4\lambda\sqrt{f_c'}\,h(x+2\ell_e-2h)$$

where d_e is the distance measured from the stud axis to the free edge. If a stud is located in the corner of a concrete member, Eq. 6.5.3 should be applied twice, once for each edge distance.

For a group of studs, the concrete failure surface may be along a truncated pyramid rather than separate shear cones, as shown in Fig. 6.5.3.

For this case, the design tensile strength is:

$$\phi P_c = \phi\lambda\sqrt{f'_c}\,(2.8\,A_{slope} + 4\,A_{flat}) \qquad \text{(Eq. 6.5.4)}$$

where:

A_{slope} = area of the sloping sides
A_{flat} = area of the flat bottom of the truncated pyramid

For stud groups in thin members, the failure surface may penetrate the thickness of the member as shown in Fig. 6.5.4. The strengths based on this type of failure are P_{c2} values given in Fig. 6.5.3. For design, select the least of P_{c1}, P_{c2} or the sum of the individual capacities. Tables 6.20.8 through 6.20.12 are provided to calculate these values.

The design tensile strength per stud as governed by steel failure is:

$$\phi P_s = 0.9 A_b f_y = 54,000\,A_b \qquad \text{(Eq. 6.5.5)}$$

where $\phi = 1.0$ and $f_y = 60,000$ psi. Table 6.20.6 tabulates the maximum design strengths from the above equations.

Shear

The design shear strength governed by concrete failure should be taken as the least of the values given by the following equations:

$$\phi V_c = \phi 800\,A_b\lambda\sqrt{f'_c} \qquad \text{(Eq. 6.5.6)}$$
$$\phi V_c = \phi 2\pi d_e^2\lambda\sqrt{f'_c} \qquad \text{(Eq. 6.5.7)}$$

where $\phi = 0.85$

For groups of studs, the design shear strength, based on concrete strength, should be taken as the least of:

1. Strength of the weakest stud, based on the above equations, times the number of studs,
2. Strength based on the d_e of the weakest row of studs times the number of rows, or
3. Strength based on the d_e of the row of studs farthest from the free edge.

Note: These are based on "normal" arrangement of studs. For arrangements which are very unsymmetrical or unusual, a separate analysis, which considers the "zipper" effect, should be made.

Example 6.5.1 Shear strength of stud groups

Given:
A stud group in a column subject to the shear force shown.
$f'_c = 5000$ psi (normal weight)

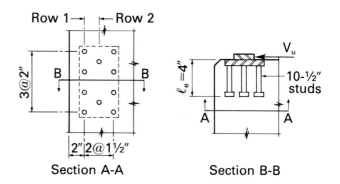

Section A-A   Section B-B

Fig. 6.5.4 Pullout surface areas for stud groups in thin sections

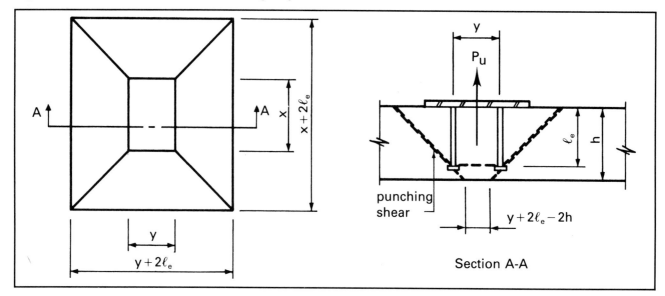

Problem:

Find the maximum shear strength.

Solution:

A_b = 0.20 sq in.

(a) ϕV_c = 0.85(800)(0.20)(1) $\sqrt{5000}$ = 9617 lb/stud

(b) For d_e = 2 in.

ϕV_c = 0.85(2)(3.14)(2)² (1)$\sqrt{5000}$ = 1510 lb

(c) For d_e = 3.5 in.

ϕV_c = 0.85(2)(3.14)(3.5)² (1)$\sqrt{5000}$ = 4624 lb

(d) For d_e = 5 in.

ϕV_c = 0.85(2)(3.14)(5)² (1)$\sqrt{5000}$ = 9436 lb

Maximum capacity of the group:

1. 10(1510) = 15,000 lb

2. 4(1510)(3) = 18,120 lb
    or 2(4624)(3) = 27,744 lb

3. 4(9436) = 37,744 lb

Thus condition 1 controls.

The design shear strength as governed by steel strength is:

$$\phi V_s = 0.75\ A_b\ f_s = 45,000\ A_b \qquad \text{(Eq. 6.5.8)}$$

where ϕ = 1.0

Table 6.20.7 tabulates the maximum capacities from the above equations.

Combined shear and tension

The design strength of studs under combined tension and shear should satisfy the following interaction equations:

$$\text{Concrete:}\ \frac{1}{\phi}\left[\left(\frac{P_u}{P_c}\right)^2 + \left(\frac{V_u}{V_c}\right)^2\right] \leq 1.0$$
$$\text{(Eq. 6.5.9)}$$

where ϕ = 0.85

$$\text{Steel:}\ \frac{1}{\phi}\left[\left(\frac{P_u}{P_s}\right)^2 + \left(\frac{V_u}{V_s}\right)^2\right] \leq 1.0$$
$$\text{(Eq. 6.5.10)}$$

where ϕ = 1.0

P_u and V_u are the factored tension and shear loads.

Plate thickness

Thickness of plates to which studs are attached should be at least ⅔ of the diameter of the stud.

Example 6.5.2 Capacity of welded headed studs

Given:

Bracket on column as shown.

f'_c = 5000 psi (normal weight)
Factored load on bracket = 75 kips

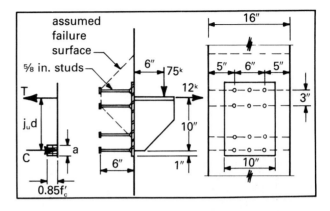

Problem:

Determine if studs are adequate to resist the loads shown.

Solution:

(a) Check concrete strength:

Tension (top group of studs) from Table 6.20.6:

d_e = 5 in., ℓ_e = 6 in., ⅝ in. studs

ϕP_c = 6(27.4) = 164.4 kips

This is the cumulative capacity of six individual cones, reduced for edge distance. It can also be determined from Eqs. 6.5.2a and 6.5.3.

Or ϕP_c from Table 6.20.10 (Case 3):

y   = 3 in., x = 16 in. ℓ_e = 6 in.
ϕP_c = 67.5 kips
P_c = 67.5/0.85 = 79.4 kips

This is the capacity of a truncated pyramid accounting for the stud spacing and controls the design.

A moment-resisting couple is formed:

C = T = 0.85 f'_c ba = 67.5 kips

comp. block, a $\approx \dfrac{67.5}{0.85(5)(10)}$ = 1.59 in.

$j_u d$ = 11 − 1.59/2 = 10.2 in.

P_u = T + N_u = M_u/j_ud + N_u
= 75(6)/10.2 + 12
= 56.1 kips

Check shear (all studs):

From Table 6.20.7:

f'_c = 5000 psi, $d_e > 9$ in., 5/8 in. studs

$\phi V_c = 12(14.7) = 176.4$ kips

To satisfy Eq. 6.5.7 ($\phi V_c \leq \phi P_c$):

From Table 6.20.10:

x = 16 in. y = 11.5 in., ℓ_e = 6 in.

$\phi V_c = \phi P_c = 106.8$ kips (controls)

V_c = 106.8/0.85 = 125.6

Combined capacity:

From Eq. 6.5.8:

$$\frac{1}{\phi}\left[\left(\frac{P_u}{P_c}\right)^2 + \left(\frac{V_u}{V_c}\right)^2\right] \leq 1.0$$

$$\frac{1}{0.85}\left[\left(\frac{56.1}{79.4}\right)^2 + \left(\frac{75}{125.6}\right)^2\right]$$

$= 1.01 \approx 1.0$ OK

(b) Check steel strength:

Tension in top group of studs:

From Table 6.20.6 for 5/8 in. studs:

$P_s = 6 (16.6) = 99.6$ kips

(Could also be determined from Eq. 6.5.5)

$C = T = 0.85 f'_c ba = 99.6$ kips

comp. block, a $= \dfrac{99.6}{0.85(5)(10)} = 2.34$ in.

$j_u d = 11 - 2.34/2 = 9.83$ in.

$P_u = M_u/j_u d + N_u = 75(6)/9.83 + 12$
$= 57.8$ kips

Shear in studs:

From Table 6.20.7 for 5/8 in. studs:

$V_s = 12 (13.8) = 165.6$ kips

(Could also be determined from Eq. 6.5.8)

Combined capacity:

From Eq. 6.5.10:

$$\left(\frac{57.8}{99.6}\right)^2 + \left(\frac{75}{165.6}\right)^2$$

$= 0.34 + 0.21 = 0.55 < 1.0$ OK

6.5.3 Deformed Bar Anchors

Deformed bar anchors are automatically welded to steel plates, similar to headed studs. They are anchored to the concrete by bond, and the development length can be taken the same as Grade 60 reinforcing bars (see Table 11.2.7).

6.5.4 Bolts and Threaded Connectors

In most connections, bolts are shipped loose and threaded into inserts. Occasionally a precast concrete member will be cast with a threaded connector projecting from the face. This is usually undesirable because of possible damage during handling. When embedded in such a manner, design for concrete strength is similar to that for studs.

High strength bolts are used infrequently in precast concrete connections because it is questionable as to whether the tension can be held when tightened against concrete. When used, AISC recommendations should be followed.

Table 6.20.13 gives allowable working and design strengths for most commonly used threaded fasteners.

High strength threaded rods

Rods with threads and specially designed nuts and couplers are available with properties similar to Grade 60 reinforcing bars and post-tensioning bars. Design information is give in Part 11.

6.5.5 Inserts Cast in Concrete

Loop inserts of the type shown in Fig. 6.5.5 can be investigated in a manner similar to that for welded studs, using Eqs. 6.5.2, 6.5.6 and 6.5.7 for the concrete tensile and shear strengths. The strength as controlled by steel can be taken from manufacturers' catalogs, or calculated based on wire strengths shown in Table 6.20.15 or the strength of the bolt or threaded rod shown in Tables 6.20.13 and 6.20.14.

An evaluation of published test results leads to certain characteristics common to most of the available inserts:[7]

1. Controlling strength conditions of various types of inserts are similar.

2. Pullout strength decreases with decreasing unit weight of concrete.

3. For inserts located in zones of potential flexural cracking, the pullout strength should be reduced by about 10%.

Fig. 6.5.5 Shear cone development for loop inserts

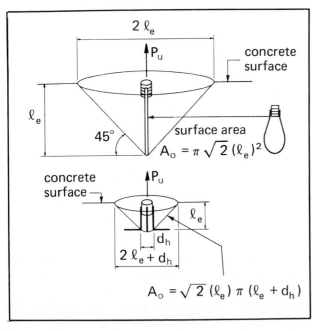

$$A_o = \pi \sqrt{2} \, (\ell_e)^2$$

$$A_o = \sqrt{2} \, (\ell_e) \, \pi \, (\ell_e + d_h)$$

Fig. 6.5.6 Expansion inserts

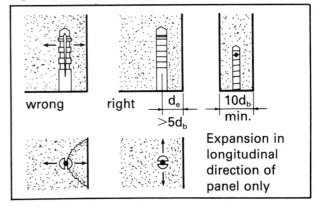

wrong   right $>5d_b$ $10d_b$ min.

Expansion in longitudinal direction of panel only

4. Sustained vibrations decrease strength up to 30%.

Expansion inserts

All expansion inserts are proprietary, and design values can be taken from manufacturers' catalogs, although tensile strength can be estimated from Eq. 6.5.2b where ℓ_e = minimum recommended length. Because of slip of the anchor in the hole, deeper embedment does not proportionally increase the capacity. Edge distances for expansion inserts are more critical than for cast-in inserts. Expanding the insert in the direction of the edge should be avoided (Fig. 6.5.6).

The performance of expansion inserts when subjected to stress reversals, vibrations or earthquake loading is not sufficiently known. Their use for these load conditions should be carefully considered by the designer.

Anchorage strength depends entirely on the lateral force (wedge action) on the concrete. Therefore, it is advisable to limit their use to connections with more shear than tension applied to them.

6.5.6 Structural Steel

Structural steel plates, angles, wide-flange beams, channels, tubes, etc., are often used in connections (see Sect. 6.12). When designed using factored loads, it is appropriate to use plastic section properties and yield strengths. Plastic sec-

tion moduli and shape factors are given in Table 6.20.16. The shear yield strength of structural steel has been commonly taken as $0.55f_y$.

Welding of structural steel

Nearly all structural steel used in precast concrete connections is ASTMA A-36 steel. Thus, it is readily weldable with standard equipment.[4]

Stainless steel plates are weldable to other stainless steel or to low carbon steel. The general procedure for welding low carbon steels should be followed, taking into account the stainless steel characteristics that differ, such as higher thermal expansion and lower thermal conductivity. Ref. 8 contains information on weldability of stainless steel.

The design strength of welds for use with factored loads can be determined by multiplying the allowable working stress from the AISC Manual by 1.67.[1] This is the basis of Table 6.20.1. The most commonly used welds in connections are full penetration welds or fillet welds. Table 6.20.2 gives the strengths of 45° fillet welds.

Minimization of cracking in concrete around welded connections

When welding is performed on components that are embedded in concrete, thermal expansion and distortion of the steel may destroy bond between the steel and concrete or induce cracking or spalling in the surrounding concrete.

The extent of cracking and distortion is dependent on the amount of heat generated during welding and the stiffness of the steel member. Heat may be reduced by: (1) use of low-heat welding rods of small size; (2) use of intermittent rather than continuous welds; or (3) using smaller welds in multiple passes.

Distortion can be minimized by using thicker steel sections—a minimum of ⅜ in. is recommended for plates. Providing a space around the metal on the surface with sealing foam or weather stripping may also reduce potential for damage.

Weld groups

Weld groups are more efficient than linear welds to resist bending or torsion moments created by eccentric loads. Examples of this type of loading are shown in Fig. 6.5.7. Design procedure is illustrated by Fig. 6.5.8, where:

f_x = combined shear and torsion stress in horizontal direction

    = $P_x/A_w + M_t y/I_p$

f_y = combined shear and torsion stress in vertical direction

    = $P_y/A_w + M_t x/I_p$

f_r = resultant stress on the weld

    = $\sqrt{(f_x)^2 + (f_y)^2}$

P_x = applied force in x direction

P_y = applied force in y direction

y = vertical distance from c.g. of weld group to point under investigation

x = horizontal distance from c.g. of weld group to point under investigation

I_p = polar moment of inertia

    = $I_x + I_y = \Sigma I_{xx} + \Sigma A_w \bar{y}^2 + \Sigma I_{yy} + \Sigma A_w \bar{x}^2$

M_t = torsional moment = $P_x e_y + P_y e_x$

t_w = effective throat thickness of weld

A_w = area of weld = weld length × t_w

I_{xx}, I_{yy} = moment of inertia of weld segment with spect to its own axes

Fig. 6.5.7 Typical eccentric loadings and weld groups

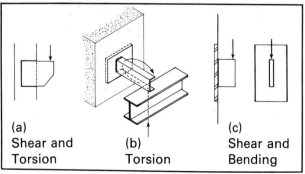

| (a) | (b) | (c) |
| Shear and Torsion | Torsion | Shear and Bending |

Fig. 6.5.8 Eccentric bracket connection

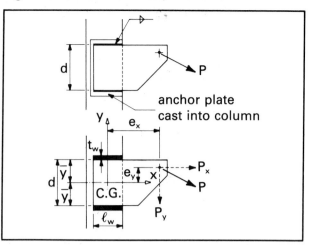

For computing nominal stresses the locations of the lines of weld are defined by edges along which the fillets are placed, rather than to the center of the effective throat. This makes little difference, since the throat dimension is usually small. By treating the welds as lines with $t_w = 1$, the physical properties of weld groups are simplified. The most commonly occurring weld groups are listed in Fig. 6.5.9.

The equations given above and in Fig. 6.5.9 are for elastic section properties, which is inconsistent, but conservative, when used with factored loads and design strength of weld material. The plastic section properties of the weld group could be calculated, or the properties divided by the appropriate "shape factor" from Table 6.20.16, if a less conservative solution is desired. Methods shown in Ref. 12 may also be used.

Example 6.5.3 Design of a weld group

Given:

Corner angle connection as shown

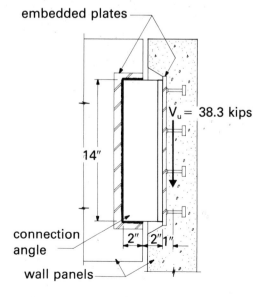

Fig. 6.5.9 Properties of weld groups treated as lines ($t_w = 1$)

Section b = width; d = depth		Section Modulus $I_x/\bar{y}$	Polar Moment of Inertia, I_p about Center of Gravity
1.	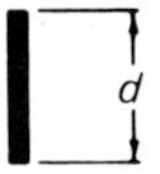	$S = \dfrac{d^2}{6}$	$I_p = \dfrac{d^3}{12}$
2.	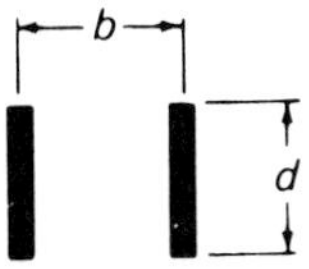	$S = \dfrac{d^2}{3}$	$I_p = \dfrac{d(3b^2 + d^2)}{6}$
3.	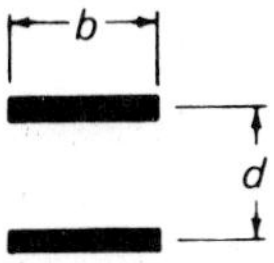	$S = bd$	$I_p = \dfrac{b(3d^2 + b^2)}{6}$
4.	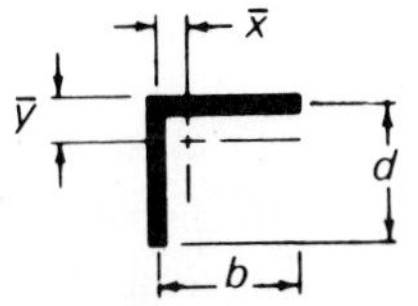 $\bar{y} = \dfrac{d^2}{2(b+d)}$ $\bar{x} = \dfrac{b^2}{2(b-d)}$	$S = \dfrac{4bd + d^2}{6}$	$I_p = \dfrac{(b+d)^4 - 6b^2d^2}{12(b+d)}$
5.	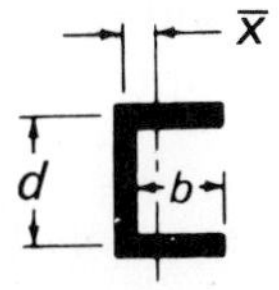 $\bar{x} = \dfrac{b^2}{2b+d}$	$S = bd + \dfrac{d^2}{6}$	$I_p = \dfrac{8b^3 + 6bd^2 + d^3}{12}$ $- \dfrac{b^4}{2b+d}$
6.	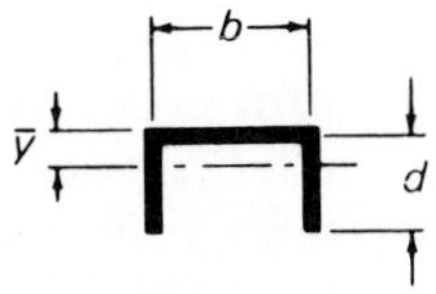 $\bar{y} = \dfrac{d^2}{b+2d}$	$S = \dfrac{2bd + d^2}{3}$	$I_p = \dfrac{b^3 + 6b^2d + 8d^3}{12}$ $- \dfrac{d^4}{2d+b}$
7.	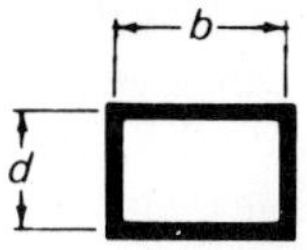	$S = bd + \dfrac{d^2}{3}$	$I_p = \dfrac{(b+d)^3}{6}$
8.	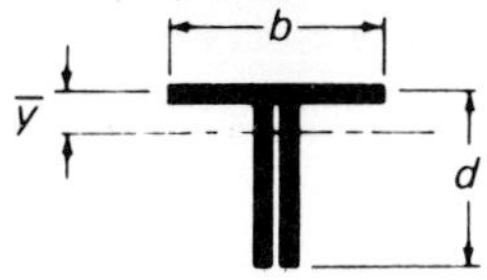 $\bar{y} = \dfrac{d^2}{b+2d}$	$S = \dfrac{2bd + d^2}{3}$	$I_p = \dfrac{b^3 + 8d^3}{12} - \dfrac{d^4}{b+2d}$
9.	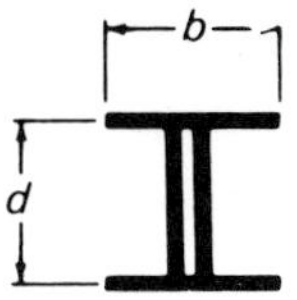	$S = bd + \dfrac{d^2}{3}$	$I_p = \dfrac{b^3 + 3b^2 + d^3}{6}$
10.	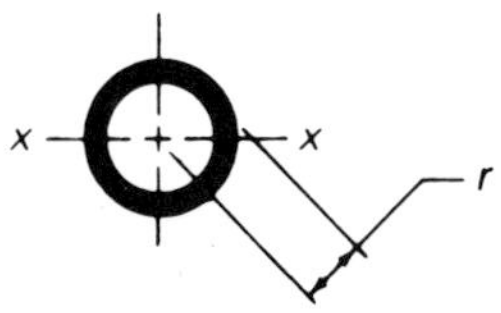	$S = \pi r^2$	$I_p = 2\pi r^3$

Angle size = 4 × 4 × ½ × 1'-2"

f_y = 36 ksi, E70 electrodes

Applied factored load = 38.3 kips

Problem:
Determine required weld size.

Solution:

Apply additional factor for precast connections
$V_u = P_y = 1.3(38.3) = 49.8$ kips

Find c.g. of weld group:

$$\bar{x} = \frac{2(2) + 14(0)}{14 + 2(2)} = 0.22 \text{ in.}$$

by symmetry, $\bar{y}$ = 7 in.

$A_w = [2(2) + 14]t_w = 18t_w$

$$I_p = 2[t_w(2)^3/12 + 2t_w(0.78)^2 + 2(t_w)^3/12$$
$$+ 2(t_w)(7^2)] + t_w(14)^3/12 + 14(t_w)^3/12$$
$$+ 14(t_w)(0.22)^2$$
$$= 429.1t_w + 1.50(t_w)^3$$

Since second term is very small, it may be neglected.

Alternatively, using Fig. 6.5.9, case 5:

b = 2 in., d = 14 in.

$\bar{x} = b^2/(2b + d) = 4(4 + 14) = 0.22$ in.

$$I_p = \frac{8b^3 + 6bd^2 + d^3}{12} - \frac{b^4}{2b + d}$$

$$= \frac{8(2)^3 + 6(2)(14)^2 + 14^3}{12} - \frac{2^4}{2(2) + 14}$$

$$= 429.1 \text{ in.}^4$$

$e_x = 1 + 4 - 0.22 = 4.78$ in.

$M_t = P_y e_x = 49.8(4.78) = 238$ in.-kips

$x = 2 - 0.22 = 1.78$

$f_y = 49.8/18t_w + 238(1.78)/429.1t_w$

$\quad = 2.77/t_w + 0.99/t_w = 3.76/t_w$

$f_x = 0 + 238(7)/429.1t_w = 3.88/t_w$

$f_r = \sqrt{(3.88/t_w)^2 + (3.76/t_w)^2}$

$\quad = 5.40/t_w$

or, $t_w = 5.40/f_r$

From Table 6.20.1, design strength of E70 weld = 35 ksi

$t_w = 5.40/35 = 0.154$ in.

For a 45° fillet weld, leg size = 0.154/0.707
$$= 0.218 \text{ in.}$$

Use ¼ in. fillet weld.

6.5.7 Post-Tensioning Steel

Either 7-wire strand or bar tendons are used in connections; bars being more common. In order to reliably measure the prestressing force, a strand tendon must be at least 15 or 20 ft long. The primary reason for using prestressing steel in connections is its high ultimate strength and ductility. Post-tensioning is sometimes used in long continuous columns, walls and beams (see Sect. 9.10, Precast Segmental Construction).

6.5.8 Bearing Pads

Bearing pads are used to distribute vertical loads and reactions over the bearing area and to allow limited horizontal or rotational movement. Their use has proven beneficial and often may be necessary for satisfactory performance of precast concrete structures.

Several materials are commonly used for bearing pads:

1. AASHTO-grade chloroprene pads are made with 100 percent chloroprene (neoprene) as the only elastomer and conform to the requirements of the AASHTO Standard Specifications for Highway Bridges (1977), Sect. 25. Inert fillers are used with the chloroprene and the resulting pad is black in color and of a smooth uniform texture. While allowable compressive stresses are somewhat lower than other pad types, these pads allow the greatest freedom in movement at the bearing. *Note:* chloroprene pads which do not meet the AASHTO Specifications are not recommended for use in precast concrete structures.

2. Pads reinforced with randomly oriented fibers have been used successfully in recent years. These pads are usually black, and the short reinforcing fibers are clearly visible. Vertical load capacity is increased by the reinforcement, but tolerance of rotations and horizontal movement is somewhat less than chloroprene pads. No national standard specifications are available for this material.

3. Cotton-duck fabric reinforced pads are generally used where a higher compressive strength is desired. These pads are often yellow-orange in color and are reinforced with closely spaced, horizontal layers of fabric, bonded in the elas-

tomer. The horizontal reinforcement layers are easily observed at the edge of the pad. Section 2.10.3(L) of the AASHTO Standard Specifications for Bridges and Military Specification MIL-C-882C discuss this material.

4. Chloroprene pads laminated with alternate layers of bonded steel or fiberglass are often used in bridges, but seldom in building construction. The above mentioned AASHTO specification covers these pads.

5. A multimonomer plastic bearing strip is manufactured expressly for bearing purposes. It is a commonly used material for the bearing support of hollow-core slabs, and is highly suitable for this application. The material has a compressive strength higher than the typical design range of concrete used in precast construction.

6. Tempered hardboard strips are also used with hollow-core slabs to prevent concrete-to-concrete bearing. Use with caution when moisture conditions exist.

7. TFE (trade name Teflon) coated materials are often used in bearing areas when large horizontal movements are anticipated, for example at "slip" joints or expansion joints. The TFE is normally reinforced by bonding to an appropriate backing material, such as steel. Fig. 6.5.10 shows a typical bearing detail using TFE, and Fig. 6.5.11 shows the range of friction coefficients that may be used for design.

Design recommendations

Recent research[9] has shown that most of the stress-relieving characteristics of elastomeric bearing pads (1 through 4 above) are due to slippage rather than pad deformation. Following are recommendations which, along with Figs. 6.5.12 and 6.5.13, can be used to select bearing pads.

1. Use unfactored service loads for design.

Fig. 6.5.10 Typical TFE bearing pad detail

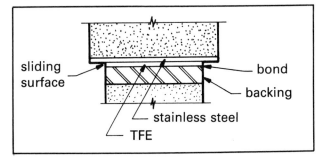

Fig. 6.5.11 TFE friction coefficients

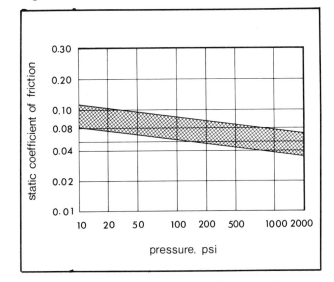

2. At the suggested maximum uniform compressive stress, instantaneous vertical strains of 10 to 20% can be expected. This number may double if the bearing surfaces are not parallel. In addition, time-dependent creep strains may add another 25 to 100%, depending on the magnitude of sustained dead load.

3. The length and width of unreinforced pads should be at least five times the thickness for stability.

4. A minimum thickness of ¼ in. for joists and double tee stems, and ⅜ in. for beams is recommended.

5. Shear stress has been shown to be a function of slip in the chloroprene and random fiber reinforced pads, Fig. 6.5.13.

6. The portion of pad outside of the covered bearing surface should be ignored in calculating pad stresses, stability and movements.

7. Shape factors, S, for unreinforced pads should be greater than 2 when used under tee stems, and greater than 3 under beams.

8. The sustained dead load compressive stress on unreinforced chloroprene pads should be less than 500 psi.

9. The volume change strains shown in Sect. 3.3.1 may be reduced by one half when calculating horizontal movement, $\triangle$, because of compensating creep and slip in the bearing pad.

Fig. 6.5.12 Single layer bearing pads free to slip

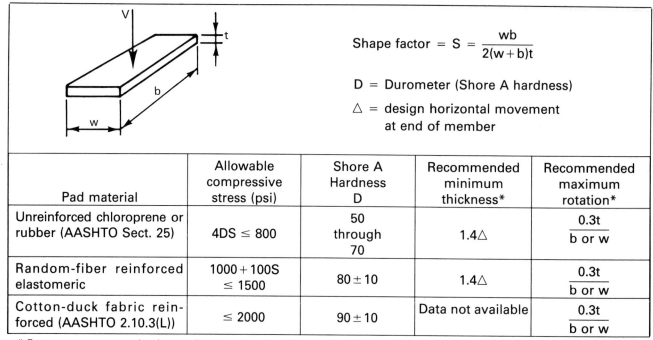

Shape factor $= S = \dfrac{wb}{2(w+b)t}$

D = Durometer (Shore A hardness)

$\triangle$ = design horizontal movement at end of member

Pad material	Allowable compressive stress (psi)	Shore A Hardness D	Recommended minimum thickness*	Recommended maximum rotation*
Unreinforced chloroprene or rubber (AASHTO Sect. 25)	$4DS \le 800$	50 through 70	$1.4\triangle$	$\dfrac{0.3t}{b \text{ or } w}$
Random-fiber reinforced elastomeric	$1000+100S \le 1500$	80 ± 10	$1.4\triangle$	$\dfrac{0.3t}{b \text{ or } w}$
Cotton-duck fabric reinforced (AASHTO 2.10.3(L))	≤ 2000	90 ± 10	Data not available	$\dfrac{0.3t}{b \text{ or } w}$

* For movement or rotation in two directions, use the higher value. The values in the table are based on sliding criteria. If sliding is not critical or testing indicates more advantageous conditions, thinner pads may be used.

Fig. 6.5.13 Shear resistance of bearing pads

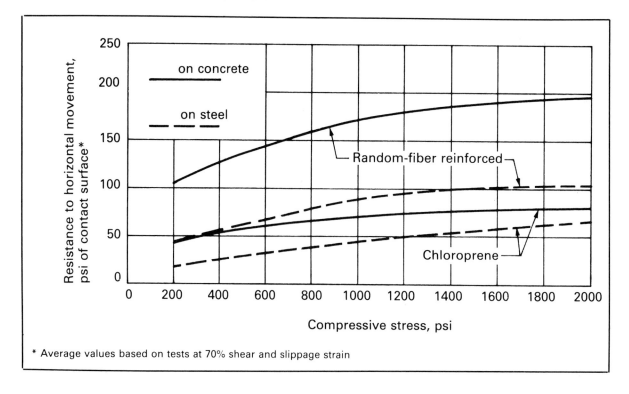

* Average values based on tests at 70% shear and slippage strain

PCI Design Handbook

Fig. 6.5.14 Design relationships for connection angles

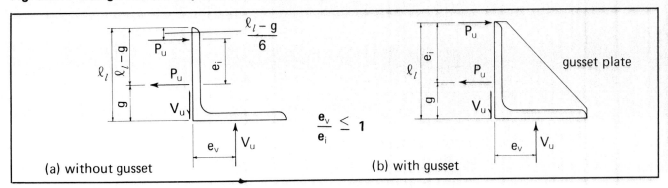

(a) without gusset   (b) with gusset

6.5.9 Connection Angles

Angles used to support precast members can be designed by statics as shown in Fig. 6.5.14. In addition to the applied vertical and horizontal loads, the design should include all loads induced by restraint of relative movement between the precast member and the supporting member. The minimum thickness of non-gusseted angles loaded in shear as shown in Fig. 6.5.15 can be determined by:

$$t = \sqrt{\frac{4\, V_u\, e_v}{\phi\, f_y b}} \qquad \text{(Eq. 6.5.11)}$$

where

    $\phi = 0.90$

    b = length of the angle

    design e_v = specified e_v + ½ in.

The tension on the bolt can be calculated by:

$$P_u = V_u \frac{e_v}{e_i} \qquad \text{(Eq. 6.5.12)}$$

For angles loaded axially, Fig. 6.5.16, either in tension or compression, the minimum thickness of non-gusseted angles can be calculated by:

$$t = \sqrt{\frac{4\, N_u g}{\phi\, f_y b}} \qquad \text{(Eq. 6.5.13)}$$

where

    $\phi = 0.90$

    g = gage of the angle (see Fig. 6.5.16)

    b = length of the angle

Fig. 6.5.15 Vertical loads on connection angle

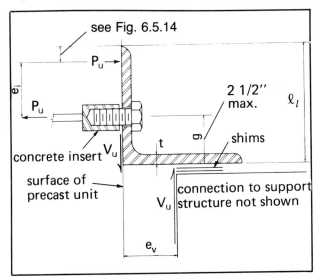

Fig. 6.5.16 Horizontal loads on connection angle

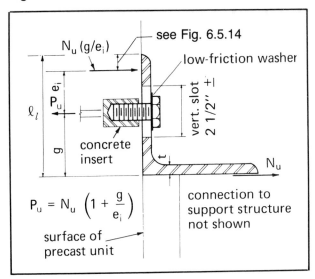

Table 6.5.1 Minimum edge distance for punched, reamed or drilled holes, measured from center of hole, in.***

Bolt Diameter (in.)	At Sheared Edges	At Rolled Edges of Plates, Shapes or Bars or Gas Cut Edges**
1/2	7/8	3/4
5/8	1-1/8	7/8
3/4	1-1/4	1
7/8	1-1/2*	1-1/8
1	1-3/4*	1-1/4
1-1/4	2-1/4	1-5/8

* These may be 1-1/4 in. at the ends of beam connection angles.

** All edge distances in this column may be reduced 1/8 in. when the hole is at a point where stress does not exceed 25% of the maximum allowable stress in the element.

*** When oversized or slotted holes are used, edge distances should be increased to maintain the clear distance from edge of hole to free edge provided by distances tabulated.

The minimum edge distance for bolt holes is shown in Table 6.5.1.

Tables 6.20.17 and 6.20.18 may be used for the design of connection angles.

Connections may be made by welding instead of bolting and the welds designed in accordance with Sect. 6.5.6.

Table 6.6.1 Static coefficients of friction of dry materials*

Material	μ_s
Elastomeric to steel or concrete	Fig. 6.5.13
Concrete to concrete	0.8
Concrete to steel	0.4
Steel to steel (not rusted)	0.25
TFE to stainless steel	Fig. 6.5.11
Hardboard to concrete	0.5
Multimonomer plastic (non-skid) to concrete	1.2
Multimonomer plastic (smooth) to concrete	0.4

*Reduce by 10-25% for wet conditions

6.6 Friction

The static coefficients of friction shown in Table 6.6.1 are conservative values for use in determining the *upper limit* of volume change forces for members without "hard" connections. Thus, the maximum force resulting from the frictional restraint of axial movements can be determined by:

$$F_s = \mu_s V_{ud} \qquad \text{(Eq. 6.6.1)}$$

where

F_s = factored friction force

μ_s = static coefficient of friction as given in Table 6.6.1

V_{ud} = factored dead load force normal to the friction face

The coefficients in Table 6.6.1 should be divided by 5 if friction is to be depended upon for support of temporary construction loads.

6.7 Shear-Friction

Shear-friction is an extremely useful tool in connection design, and other applications in precast and prestressed concrete structures.

Use of the shear-friction theory is recognized by Sect. 11.7 of ACI 318-83, which states that shear friction is "to be applied where it is appropriate to consider shear transfer across a given plane, such as an existing or potential crack, an interface between dissimilar materials, or an interface between two concretes cast at different times."

A basic assumption used in applying the shear-friction concept is that concrete within the direct shear area of the connection will crack in the most undesirable manner. Ductility is achieved by placing reinforcement across this anticipated crack so that the tension developed by the reinforcing bars will provide a force normal to the crack. This normal force in combination with "friction" at the crack interface provides the shear resistance. The shear-friction analogy can be adapted to designs for reinforced concrete bearing, corbels, daps, composite sections, and other connection devices.

An "effective shear-friction coefficient," μ_e, may be used when the concept is applied to precast concrete connections. The shear-friction reinforcement nominally perpendicular to the assumed crack plane can be determined by:

$$A_{vf} = \frac{V_u}{\phi f_y \mu_e} \qquad \text{(Eq. 6.7.1)}$$

Table 6.7.1 Shear-friction coefficients

Crack interface condition	Recommended μ	Maximum μ_e	Maximum V_u, lb
1. Concrete to concrete, cast monolithically	1.4λ	3.4	$0.30\,\lambda^2\,f'_c\,A_{cr} \leq 1000\,\lambda^2\,A_{cr}$
2. Concrete to hardened concrete with roughened surface	1.0λ	2.9	$0.25\lambda^2\,f'_c\,A_{cr} \leq 1000\,\lambda^2\,A_{cr}$
3. Concrete to concrete	0.6λ	2.2	$0.20\,\lambda^2\,f'_c\,A_{cr} \leq 800\,\lambda^2\,A_{cr}$
4. Concrete to steel	0.7λ	2.4	$0.20\,\lambda^2\,f'_c\,A_{cr} \leq 800\,\lambda^2\,A_{cr}$

where

ϕ = 0.85

A_{vf} = area of reinforcement nominally perpendicular to the assumed crack plane, sq in.

f_y = yield strength of A_{vf}, psi (equal to or less than 60,000 psi)

V_u = applied factored shear force, parallel to the assumed crack plane, lb (limited by the values given in Table 6.7.1)

$$\mu_e = \frac{1000\,\lambda A_{cr}\mu}{V_u} \qquad \text{(Eq. 6.7.2)}$$

λ = 1.0 for normal weight concrete

= $(f_{ct}/6.7)\sqrt{f'_c}$ for sand-lightweight or all-lightweight concrete. If f_{ct} is unknown: 0.85 for sand-lightweight concrete; 0.75 for all-lightweight concrete

f_{ct} = splitting tensile strength of concrete, psi

μ = value from Table 6.7.1

A_{cr} = area of the crack interface, sq in. (varies depending on the type of connection)

When axial tension is present, additional reinforcement area should be provided:

$$A_n = \frac{N_u}{\phi f_y} \qquad \text{(Eq. 6.7.3)}$$

where

A_n = area of reinforcement required to resist axial tension, sq in.

N_u = applied factored horizontal tensile force nominally perpendicular to the assumed crack plane, lb

ϕ = 0.85

All reinforcement, either side of the assumed crack plane, should be properly anchored by de-velopment length, welding to angles or plates, or hooks.

6.8 Bearing on Plain Concrete

For uniform bearing on plain concrete the design bearing strength (Fig. 6.8.1) is:

$$\phi V_n = \phi\, C_r\, (0.85 f'_c\, A_1)\sqrt{A_2/A_1} \leq 1.2 f'_c\, A_1 \qquad \text{(Eq. 6.8.1)}$$

Fig. 6.8.1 Bearing on plain concrete

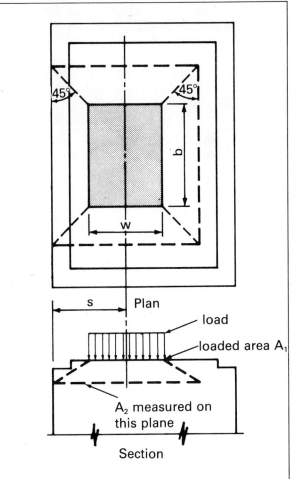

where

ϕV_n = design bearing strength, lb

ϕ = 0.70

$C_r = \left(\dfrac{sw}{200}\right)^{N_u/V_u}$

= 1.0 when reinforcement is provided in direction of N_u, in accordance with Sect. 6.9, or when N_u is zero. sw should not be taken greater than 9.0 sq in.

A_1 = loaded area

A_2 = maximum area of the portion of the supporting surface that is geometrically similar to and concentric with the loaded area

To avoid accidental spalling or cracking at the ends of thin stemmed members, minimum reinforcement equal to $N_u/\phi f_y$, but not less than one No. 3 bar, is recommended when the bearing area is less than 20 sq in.

6.9 Reinforced Concrete Bearing

If the applied load, V_u, exceeds the design bearing strength, ϕV_n, as calculated by Eq. 6.8.1, reinforcement is required in the bearing area. (Note: The PCI Committee on Connection Details recommends that *all* precast members be designed for reinforced bearing, except solid and hollow-core slabs.) This reinforcement can be designed by shear-friction as discussed in Sect. 6.7. Referring to Fig. 6.9.1, the reinforcement $A_{vf} + A_n$ nominally parallel to the direction of the axial load, N_u, is determined by Eqs. 6.7.1 through 6.7.3 with A_{cr} the lesser of bw/sinθ or bh/cosθ.

Vertical reinforcement across potential horizontal cracks can be calculated by:

$$A_{sh} = \frac{(A_{vf} + A_n) f_y}{\mu_e f_{ys}} \qquad \text{(Eq. 6.9.1)}$$

where

$\mu_e = \dfrac{1000\lambda A_{cr}\mu}{(A_{vf} + A_n)f_y}$ (Eq.6.9.2)

f_{ys} = yield strength of A_{sh}, psi

A_{cr} = 1.7ℓ_db, sq in.

b = average member width, in.

ℓ_d = development length of A_{vf} bars, in.

Stirrups or mesh used for diagonal tension reinforcement can be considered to act as A_{sh} reinforcement.

When members are subjected to bearing stresses in excess of the limits indicated in Sect. 10.15 of ACI 318-83, confinement reinforcement in all directions may be required. Design of this steel is beyond the scope of this Handbook.

Fig. 6.9.1 Reinforced concrete bearing

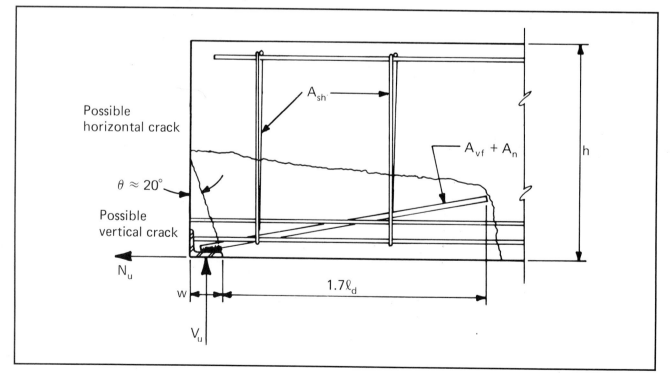

PCI Design Handbook

Example 6.9.1 Reinforced bearing for a rectangular beam

Given:

PCI standard rectangular beam 16RB28

V_u = 94 kips (includes all load factors)

N_u = 20 kips

bearing pad = 4 in. × 14 in.

f_y for all reinforcement = 60,000 psi

f'_c = 5000 psi (normal weight)

Problem:
Determine reinforcement requirements for the end of the member.

Solution:

$$A_{cr} = \text{lesser of } \frac{bw}{\sin\theta} \text{ or } \frac{bh}{\cos\theta}$$

$$\frac{bw}{\sin\theta} = 16(4)/0.342 = 187 \text{ sq in.}$$

$$\frac{bh}{\cos\theta} = 16(28)/0.940 = 477 \text{ sq in.}$$

Use A_{cr} = 187 sq in.

By Eq. 6.7.2:

$$\mu_e = \frac{1000\lambda\, A_{cr}\mu}{V_u} = \frac{1000(1)(187)(1.4)}{94,000}$$
$$= 2.79 < 3.4$$

By Eq. 6.7.1:

$$A_{vf} = \frac{V_u}{\phi\, f_y\, \mu_e} = \frac{94,000}{0.85(60,000)(2.79)}$$
$$= 0.66 \text{ sq in.}$$

By Eq. 6.7.3:

$$A_n = \frac{N_u}{\phi\, f_y} = \frac{20,000}{0.85(60,000)}$$
$$= 0.39 \text{ sq in.}$$

$A_{vf} + A_n$ = 0.66 + 0.39 = 1.05 sq in.

Use 3– #6 = 1.32 sq in.

By Eq. 6.9.2:

$$\mu_e = \frac{1000\lambda\, A_{cr}\mu}{(A_{vf} + A_n)\, f_y}$$

$1.7\ell_d$ from Table 11.2.7 = 30.6 in.

$A_{cr} = 1.7\ell_d b = (30.6)(16) = 489.6$ sq in.

$$\mu_e = \frac{1000(1)(489.6)(1.4)}{1.05(60,000)} = 10.88 > 3.4$$

Use μ_e = 3.4

$$A_{sh} = \frac{(A_{vf} + A_n)f_y}{\mu_e\, f_{ys}} = \frac{1.05(60,000)}{3.4(60,000)}$$
$$= 0.31 \text{ sq in.}$$

Use 3– #3 stirrups = 0.33 sq in.

6.10 Column Base Plates

Column bases must be designed for both erection loads and loads which occur in service, the former often being more critical. Two commonly used base plate details are shown in Fig. 6.10.1 although other details are also frequently used.

If, in the analysis for erection loads or temporary construction loads, before grout is placed under the plate, all the anchor bolts are in compression, the base plate thickness required to satisfy bending is determined from:

$$t = \sqrt{\frac{(\Sigma F)\, 4x_c}{\phi\, bf_y}} \qquad \text{(Eq. 6.10.1)}$$

where

ϕ = 0.90

x_c, b from Fig. 6.10.1, in.

f_y = yield strength of the base plate, psi

ΣF = greatest sum of anchor bolt factored forces on one side of the column, lb

If the analysis indicates the anchor bolts on one or both sides of the column are in tension, the base plate thickness is determined by:

$$t = \sqrt{\frac{(\Sigma F)\, 4x_t}{\phi\, bf_y}} \qquad \text{(Eq. 6.10.2)}$$

x_t from Fig. 6.10.1

Under loads which occur at service, the base plate thickness may be controlled by bearing on the concrete or grout. In this case, the base plate thickness is determined by:

$$t = x_o \sqrt{\frac{2\, f_{bu}}{\phi\, f_y}} \qquad \text{(Eq. 6.10.3)}$$

Fig. 6.10.1 Column base connections

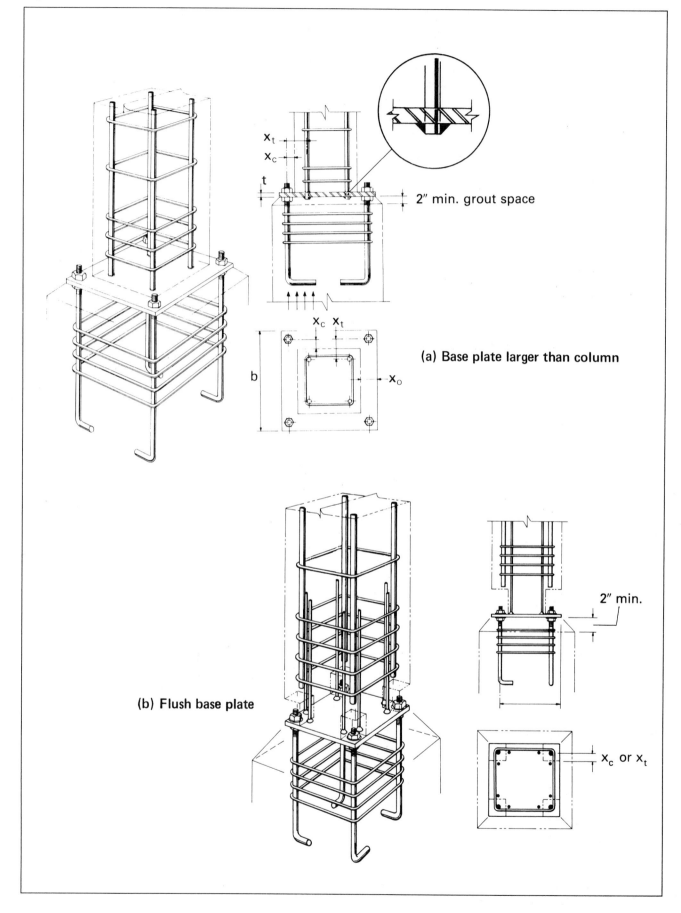

x_t
x_c
t
2″ min. grout space

x_c x_t

b

x_o

(a) Base plate larger than column

(b) Flush base plate

2″ min.

x_c or x_t

where

x_o from Fig. 6.10.1

$$f_{bu} = 0.70(0.85) f'_c = 0.595 f'_c$$

Table 6.20.19 may be used for base plate design.

Nominal base plate shearing stresses should not exceed $0.55 f_y$.

The anchor bolt diameter is determined by the tension or compression on the stress area of the threaded portion of the bolt. Anchor bolts may be ASTM A307 bolts or, more frequently, threaded rods of ASTM A36 steel.

In most cases, both base plate and anchor bolt stresses can be significantly reduced by using properly placed shims during erection.

When the bolts are near a free edge, as in a pier or wall, the buckling of the bolts before grouting may be a consideration. Confinement reinforcement, as shown in Fig. 6.10.1, should be provided in such cases. A minimum of 4 No. 3 ties at about 3 in. centers is recommended for confinement.

The strength of the concrete when the bolt is in tension may be critical and can be determined by assuming a shear cone pull-out failure as described for headed studs (Sect. 6.5.2).

The length of the anchor bolt should be such that the concrete will develop the desired strength of the bolt in bond and bearing on the hook projection or bolt head. Bearing area of bolt heads can be increased by welding a washer or steel plate to the bolt head. Nominal bond stress on smooth anchor bolts should not exceed 250 psi. The nominal confined bearing stress on the hook or bolt head should not exceed $\phi f'_c$. The bottom of the bolt should be a minimum of 4 in. above the bottom of a footing, and above the footing reinforcement.

6.11 Concrete Brackets or Corbels

ACI 318-83 prescribes the design method for corbels, based on Refs. 10 and 11. The equations in this section follow those recommendations, and are subject to the following limitations (see Figs. 6.11.1 and 6.11.2):

1. $a/d \leq 1$

2. $N_u \leq V_u$

3. $\phi = 0.85$ for all calculations

4. Anchorage at the front face must be provided by welding or other positive means.

Fig. 6.11.1 Corbel force diagrams and typical reinforcement–wall panels

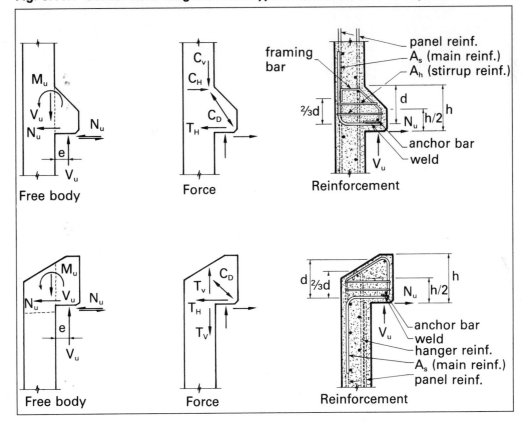

Fig. 6.11.2 Design of concrete corbels

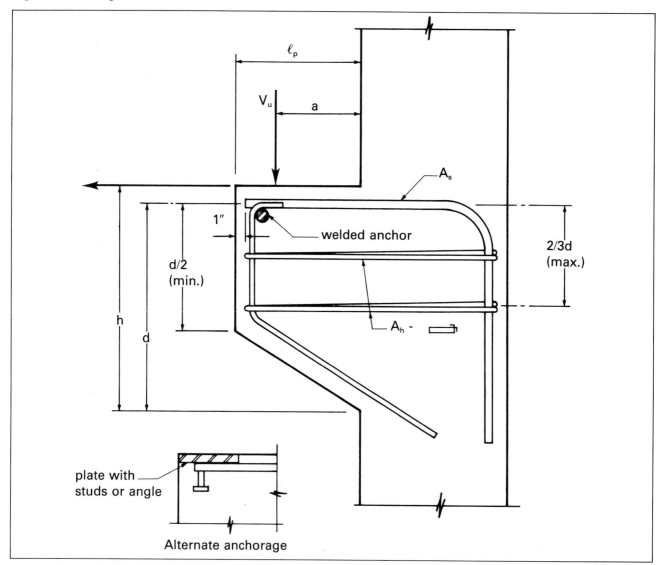

5. Concentrated loads on continuous corbels may be distributed as for beam ledges, Sect. 6.14.

The area of primary tension reinforcement, A_s, is the greater of $A_f + A_n$ as calculated below, or $2A_{vf}/3 + A_n$ as calculated in Sect. 6.7:

$$A_f = \frac{V_u a + N_u (h-d)}{\phi f_y d} \qquad \text{(Eq. 6.11.1)}$$

$$A_n = \frac{N_u}{\phi f_y} \qquad \text{(Eq. 6.11.2)}$$

For convenience, the equations can be rewritten so that A_s shall be the greater of Eq. 6.11.3 or 6.11.4:

$$A_s = \frac{1}{\phi f_y}\left[V_u \left(\frac{a}{d}\right) + N_u \left(\frac{h}{d}\right) \right] \qquad \text{(Eq. 6.11.3)}$$

$$A_s = \frac{1}{\phi f_y}\left[\frac{2}{3}\frac{V_u}{\mu_e} + N_u \right] \qquad \text{(Eq. 6.11.4)}$$

$$A_s \text{ (min.)} = 0.04(f'_c/f_y)bd \qquad \text{(Eq. 6.11.5)}$$

$$A_h \geq 0.5(A_s - A_n) \qquad \text{(Eq. 6.11.6)}$$

A_h to be distributed within 2d/3 of A_s.

The shear strength of a corbel is limited by:

Normal weight concrete:

$$V_n \leq 0.2f'_c bd \leq 800 bd \qquad \text{(Eq. 6.11.7)}$$

Lightweight or sand-lightweight concrete:

$$V_n \leq (0.2 - 0.07a/d)f'_c bd$$
$$\leq (800 - 280a/d)bd \qquad \text{(Eq. 6.11.8)}$$

Example 6.11.1 Reinforced concrete corbel

Given:

A concrete corbel similar to that shown in Fig. 6.11.2

$V_u = 80$ kips (includes all load factors)

$N_u = 15$ kips

f_y = Grade 60 (weldable)

$f'_c = 5000$ psi (normal weight)

Bearing pad – 14 in. $\times$ 6 in.

$b = 14$ in.

$\ell_p = 8$ in.

Problem:

Find corbel depth and reinforcement.

Solution:

Try $h = 14$ in., $d = 13$ in.

$a = 3/4 \ell_p = 6$ in.

By Eq. 6.11.3:

$$A_s = \frac{1}{\phi f_y} \left[V_u \left(\frac{a}{d} \right) + N_u \left(\frac{h}{d} \right) \right]$$

$$= \frac{1}{0.85(60)} \left[80 \left(\frac{6}{13} \right) + 15 \left(\frac{14}{13} \right) \right]$$

$$= 1.04 \text{ sq in.}$$

By Eq. 6.11.4:

$$\mu_e = \frac{1000 \, bh \, \mu}{V_u} = \frac{1000(14)(14)(1.4)}{80,000}$$

$$= 3.43 > 3.4 \quad \text{Use } 3.4$$

$$A_s = \frac{1}{\phi f_y} \left[\frac{2V_u}{3\mu_e} + N_u \right]$$

$$= \frac{1}{0.85(60)} \left[\frac{2(80)}{3(3.4)} + 15 \right]$$

$$= 0.60 < 1.04$$

By Eq 6.11.5:

min. $A_s = 0.04 \, bd \, (f'_c/f_y) = 0.04 \, (14)(13)(5/60)$

$$= 0.61 \text{ sq in.} < 1.04$$

Provide $2 - \#7$ bars $= 1.20$ sq in.

The A_s reinforcement could also be estimated from Table 6.20.20:

For $b = 14$ in. and $\ell_p = 8$ in., interpolation shows that for $h = 14$ in., the corbel would have a strength of about 89 kips with $A_s = 2 - \#7$.

By Eq 6.11.6:

$A_h = 0.5 \, (A_s - A_n) = 0.5 \, [1.04 - 15/(0.85)(60)]$

$$= 0.37 \text{ sq in.}$$

Provide $2 - \#3$ closed ties $= 0.44$ sq in.

6.12 Structural Steel Haunches

Structural steel shapes, such as wide flange beams, double channels, tubes or vertical plates, often serve as haunches or brackets as illustrated in Fig. 6.12.1. The capacity of these members can be calculated by statics, using the assumptions shown in Figs. 6.12.2 and 6.12.3.[7]

The design strength of the section is:

$$V_c = \frac{0.85 \, f'_c \, b\ell_e}{1 + 3.6e/\ell_e} \qquad \text{(Eq. 6.12.1)}$$

where

V_c = nominal strength of the section controlled by concrete, lb

$e = a + \ell_e/2$, in.

a = shear span, in.

ℓ_e = embedment depth, in.

b = effective width of the compression block, in.

For the additional contribution of reinforcement welded to the embedded shape, and properly developed in the concrete, and with $A'_s = A_s$:

$$V_r = \frac{2A_s f_y}{1 + \dfrac{6e/\ell_e}{4.8s/\ell_e - 1}} \qquad \text{(Eq. 6.12.2)}$$

then $V_n = (V_c + V_r)$; and $V_u \leq \phi V_n$

Notation for Eqs. 6.12.1 and 6.12.2 are illustrated in Fig. 6.12.3.

The following clarifications and limitations are recommended:[13]

1. In a column with closely spaced ties above and below the haunch, the effective width, b, can be assumed as the width of the confined region, or 2.5 times the width of the steel section, whichever is less.

2. For thin-walled members, such as the tube shown in Fig. 6.12.3, the inside should be filled with concrete to prevent local buckling.

3. When the supplemental reinforcement, A_s and A'_s, is anchored both above and below the

Fig. 6.12.1 Structural steel haunches

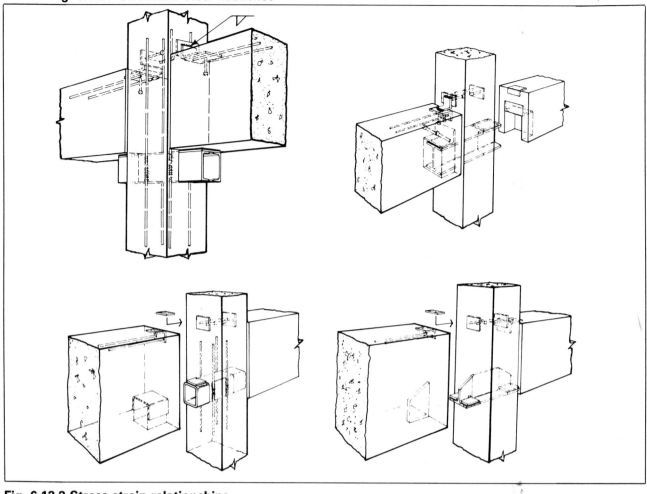

Fig. 6.12.2 Stress-strain relationships

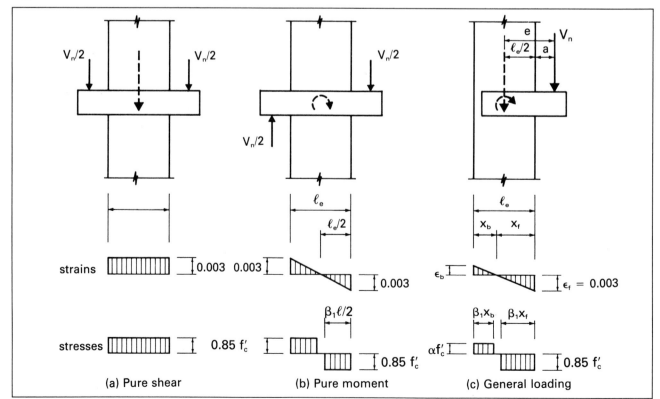

(a) Pure shear   (b) Pure moment   (c) General loading

PCI Design Handbook

Fig. 6.12.3 Assumptions and notation–steel haunch design

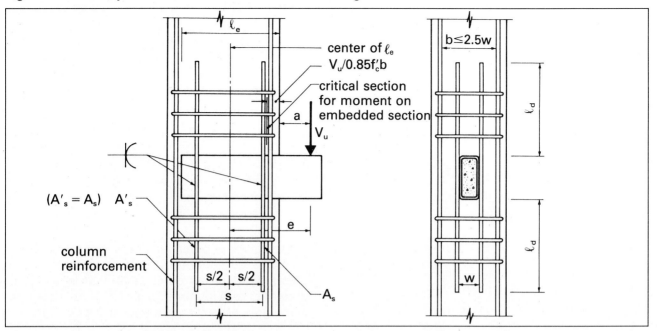

members, as in Fig. 6.12.3, it can be counted twice.

4. The critical section for bending of the steel member is located a distance $V_u/(0.85 f'_c b)$ in from the face of the column.

If the steel section projects from both sides, as in Fig. 6.12.2(a), a minimum eccentricity of $e/\ell_e = 0.5$ is recommended in Eq. 6.12.1.

The design strength of the steel section can be determined by:

Flexural design strength:

$$\phi V_n = \frac{\phi Z_s f_y}{a + V_u/(0.85 f'_c b)}$$ (Eq. 6.12.3)

Shear design strength:

$$\phi V_n = \phi(0.55 f_y h t)$$ (Eq. 6.12.4)

where

Z_s = plastic section modulus of the steel section (see Table 6.20.16)

f_y = yield strength of the steel

h,t = depth and thickness of steel web, respectively

$\phi = 0.90$

(Note: Plastic design criteria for structural steel does not require the use of ϕ-factor. However, the load factors used are 1.7(D + L). Therefore, when using steel plastic design with concrete load fac-

tors (1.4D + 1.7L), the use of $\phi = 0.90$ is recommended in order to provide approximately the same overall factor of safety.)

Horizontal forces, N_u, are resisted by bond on the perimeter of the section. If the bond stress resulting from factored loads exceeds 250 psi, headed studs or reinforcing bars can be welded to the section.

Example 6.12.1 Design of structural steel haunch

Given:

The structural steel haunch shown.

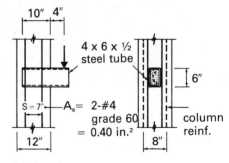

$f'_c = 5000$ psi

f_y (reinforcement) = 60,000 psi (weldable)

f_y (structural steel) = 36,000 psi

Problem:

Find the design strength.

Solution:

Effective width, b = confined area (8 in.) or 2.5w = 2.5(4) = 10 in.

Use b = 8 in.   e = 4 + 10/2 = 9 in.

$$V_c = \frac{0.85f'_c b\ell_e}{1 + 3.6e/\ell_e} = \frac{0.85(5)(8)(10)}{1 + 3.6(9)/(10)}$$

$$= 80.2 \text{ kips}$$

Since the A_s bars are anchored above and below, they can be counted twice.

$$A_s = 2 - \#4 = 2(2)(0.2) = 0.80 \text{ sq in.}$$

$$V_r = \frac{2A_s f_y}{1 + \dfrac{6e/\ell_e}{4.8s/\ell_e - 1}} = \frac{2(0.80)(60)}{1 + \dfrac{6(9)/10}{4.8(7)/(10) - 1}}$$

$$= 29.2 \text{ kips}$$

$$\phi V_n = 0.85(80.2 + 29.2) = 93.0 \text{ kips}$$

Alternate solution using Tables 6.20.21 and 6.20.22:

For b = 8 in., a = 4 in., ℓ_e = 10 in.

Read ϕV_c = 68 kips

For $A_s = 2 - \#4$, anchored 2 sides, ℓ_e = 10 in.

Read V_r = 28 kips

$$\phi V_n = \phi V_c + \phi V_r = 68 + 0.85(28) = 92 \text{ kips}$$

Steel shear capacity:

$$\phi V_n = \phi(0.55f_y ht) = 0.9(0.55)(36)(6)(2)(0.5)$$

$$= 106.9 \text{ kips}$$

Steel flexure capacity:

$$Z_s = \frac{bh^2}{4}\left[1 - \left(1 - \frac{2w}{b}\right)\left(1 - \frac{2t}{h}\right)^2\right]$$

$$= \frac{4(6)^2}{4}\left[1 - \left(1 - \frac{1.0}{4}\right)\left(1 - \frac{1.0}{6}\right)^2\right]$$

$$= 17.25 \text{ in.}^3$$

Assume V_u = 85 kips, $V_u/0.85f'_c b$ = 2.50 in.

$$\phi V_n = \frac{\phi Z_s f_y}{a + V_u/0.85f'_c b} = \frac{0.9(17.25)(36)}{4 + 2.50}$$

$$= 86.0 \text{ kips}$$

Steel flexure controls

6.13 Dapped-End Connections

Design of connections which are recessed, or dapped into the end of the member, requires the investigation of several potential failure modes. These are illustrated in Fig. 6.13.1 and listed below with the reinforcement required for each consideration.

1. Flexure (cantilever bending) and axial tension in the extended end. Provide flexural reinforcement, A_f, plus axial tension reinforcement, A_n.
2. Direct shear at the junction of the dap and the main body of the member. Provide shear-friction reinforcement composed of A_f and A_h, plus axial tension reinforcement, A_n.
3. Diagonal tension emanating from the reentrant corner. Provide shear reinforcement, A_{sh}.
4. Diagonal tension in the extended end. Provide shear reinforcement composed of A_h and A_v.
5. Diagonal tension in the undapped portion. This is resisted by providing a full development length for A_s beyond the potential crack.[14]

Each of these potential failure modes should be investigated separately. The reinforcement requirements are not cumulative, that is, A_s is the greater of that required by 1 or 2, not the sum. A_h is the greater of that required by 2 or 4.

6.13.1 Flexure and Axial Tension in the Extended End

The horizontal reinforcement is determined in a manner similar to that for column corbels, Sect. 6.11. Thus:

$$A_s = A_f + A_n$$

$$= \frac{1}{\phi f_y}\left[V_u\left(\frac{a}{d}\right) + N_u\left(\frac{h}{d}\right)\right] \quad (\text{Eq. 6.13.1})$$

where

ϕ = 0.85*

a = shear span, in., measured from load to center of A_{sh}

h = depth of the member above the dap, in.

d = distance from top to center of the reinforcement, A_s, in.

f_y = yield strength of the flexural reinforcement, psi

6.13.2 Direct Shear

The potential vertical crack shown in Fig. 6.13.1 is resisted by a combination of A_s and A_h. This reinforcement can be calculated by Eqs. 6.13.2 through 6.13.5.

---

*To be theoretically correct, Eq. 6.13.1 should have $j_u d$ in the denominator. The use of ϕ = 0.85 instead of 0.90 (flexure) compensates for this approximation.

Fig. 6.13.1 Required reinforcement in dapped-end connections

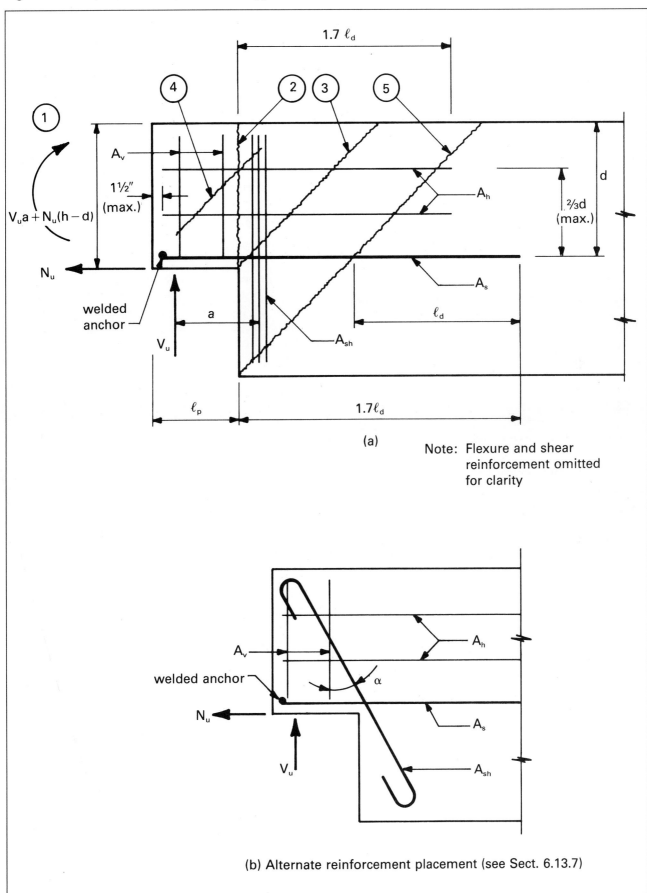

(a)

Note: Flexure and shear reinforcement omitted for clarity

(b) Alternate reinforcement placement (see Sect. 6.13.7)

$$A_s = \frac{2V_u}{3\,\phi f_y \mu_e} + A_n \qquad \text{(Eq. 6.13.2)}$$

$$A_n = \frac{N_u}{\phi f_y} \qquad \text{(Eq. 6.13.3)}$$

$$A_h = 0.5(A_s - A_n) \qquad \text{(Eq. 6.13.4)}$$

where

ϕ = 0.85

f_y = yield strength of A_s, A_n, A_h, psi

$$\mu_e = \frac{1000\lambda\,bh\,\mu}{V_u} \qquad \text{(Eq. 6.13.5)}$$

The shear strength of the extended end is limited by:

Normal weight concrete:

$$V_n \le 0.2 f'_c\,bd \le 800\,bd \qquad \text{(Eq. 6.13.6)}$$

Lightweight or sand-lightweight concrete:

$$\begin{aligned} V_n &\le (0.2 - 0.07a/d)f'_c\,bd \\ &\le (800 - 280a/d)\,bd \end{aligned} \qquad \text{(Eq. 6.13.7)}$$

6.13.3 Diagonal Tension at Reentrant Corner

The reinforcement required to resist diagonal tension cracking starting from the reentrant corner can be calculated from:

$$A_{sh} = \frac{V_u}{\phi f_y} \qquad \text{(Eq. 6.13.8)}$$

where

ϕ = 0.85

V_u = applied factored load, lb

A_{sh} = vertical or diagonal bars across potential diagonal tension crack, sq in.

f_y = yield strength of A_{sh}, psi

6.13.4 Diagonal Tension in the Extended End

Additional reinforcement is required in the extended end, as shown in Fig. 6.13.1, such that:

$$\phi V_n = \phi (A_v f_y + A_h f_y + 2\lambda\,bd\sqrt{f'_c}) \qquad \text{(Eq. 6.13.9)}$$

At least one half of the reinforcement required in this area should be placed vertically. Thus:

$$\text{min. } A_v = \frac{1}{2f_y}\left(\frac{V_u}{\phi} - 2\lambda\,bd\sqrt{f'_c}\right) \qquad \text{(Eq. 6.13.10)}$$

6.13.5 Anchorage of Reinforcement

Horizontal bars A_s should be extended a minimum of $1.7\ell_d$ past the end of the dap, or ℓ_d past crack 5, Fig. 6 13.1, and anchored at the end of the beam by welding to cross bars, angles or plates. Horizontal bars A_h should be extended a minimum of $1.7\ell_d$ past the end of the dap, and anchored at the end of the beam by hooks or other suitable means. Vertical or diagonal bars A_{sh} and A_v should be properly anchored by hooks as required by ACI 318-83. Welded wire fabric may be used for reinforcement, and should be anchored in accordance with ACI 318-83.

6.13.6 Detailing Considerations

Experience has shown that the depth of the extended end should not be less than one-half the depth of the beam, unless the beam is significantly deeper than necessary for other than structural reasons.

Diagonal tension reinforcement, A_{sh}, should be placed as closely as practical to the reentrant corner. This reinforcement requirement is not additive to other shear reinforcement requirements.

Reinforcement requirements may be met with welded headed studs, deformed bar anchors or welded wire fabric.

If the flexural stress, calculated for the full depth of section using factored loads and gross section properties, exceeds $6\sqrt{f'_c}$ immediately beyond the dap, longitudinal reinforcement should be placed in the beam to develop the required flexural strength.

6.13.7 Alternate Placement of Reinforcement

As an alternate to placing reinforcement as shown in Fig. 6.13.1a, diagonal bars can be placed as shown in Fig. 6.13.1b. The requirements for reinforcement placed in this manner can be determined by:

$$A_{sh} = \frac{V_u}{\phi\,f_y\cos\alpha} \approx \frac{V_u\sqrt{a^2 + d^2}}{\phi\,f_y d} \qquad \text{(Eq. 6.13.11)}$$

$$A_s = \frac{N_u h}{\phi\,f_y d} \qquad \text{(Eq. 6.13.12)}$$

but not less than that determined in Sect. 6.13.2.

If the diagonal bars can be adequately anchored into the extended end, they may also be used as at least partial replacement for A_v and A_h requirements shown in Sect. 6.13.4.

Fig. 6.13.2 Dapped-end beam of Example 6.13.1

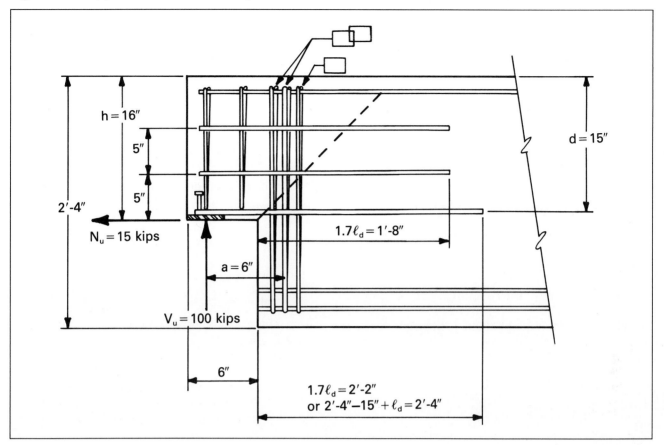

Example 6.13.1 Reinforcement for dapped-end beam

Given:

The 16RB28 beam with a dapped end as shown in Fig. 6.13.2.

V_u = 100 kips (includes all load factors)

N_u = 15 kips

f'_c = 5000 psi (normal weight)

f_y for all reinforcement = 60 ksi (weldable)

Problem:

Determine the requirements for reinforcement A_s, A_h, A_{sh}, and A_v shown in Fig. 6.13.1

Solution:

Assume: Shear span, a = 6 in.

d = 15 in.

1. Flexure in extended end:

By Eq. 6.13.1:

$$A_s = \frac{1}{\phi\, f_y}\left[V_u\left(\frac{a}{d}\right) + N_u\left(\frac{h}{d}\right)\right]$$

$$= \frac{1}{0.85 \times 60}\left[100\left(\frac{6}{15}\right) + 15\left(\frac{16}{15}\right)\right]$$

$$= 1.10 \text{ sq in.}$$

Use 4–#5, A_s = 1.24 sq in.

2. Direct shear:

$$\mu_e = \frac{1000\, \lambda b h\, \mu}{V_u} = \frac{1000(1)(16)(16)(1.4)}{100,000}$$

$$= 3.58 > 3.4 \text{ Use } 3.4$$

By Eqs. 6.13.2 and 6.13.3:

$$A_s = \frac{2V_u}{3\phi f_y \mu_e} + \frac{N_u}{\phi f_y}$$

$$= \frac{2(100)}{3(0.85)(60)(3.4)} + \frac{15}{0.85(60)}$$

$$= 0.38 + 0.29 = 0.67 \text{ sq in.}$$

Use A_s = 1.10 sq in.

By Eq. 6.13.4:

$A_h = 0.5(A_s - A_n) = 0.5 (1.10 - 0.29)$
$\qquad = 0.41$ sq in.

Use 2 – #3 U-bars, A = 0.44 sq in.

Check shear strength, Eq. 6.13.6:

$\phi V_n = \phi(800bd) = 0.85(800)(16)15/1000$
$\qquad = 163.2$ kips > 100 OK

3. Diagonal tension at reentrant corner:

By Eq. 6.13.8:

$A_{sh} = \dfrac{V_u}{\phi\, f_y} = \dfrac{100}{0.85(60)} = 1.96$ sq in.

Use 5 – #4 closed ties = 2.00 sq in.

4. Diagonal tension in the extended end:

Concrete capacity $= 2\lambda\sqrt{f'_c}\, bd$

$\qquad = 2\,(1)\sqrt{5000}\,(16)(15)/1000 = 33.9$ kips

By Eq. 6.13.10:

$A_v = \dfrac{1}{2\,f_y}\left(\dfrac{V_u}{\phi} - 2\lambda\sqrt{f'_c}\,bd\right)$

$\qquad = \dfrac{1}{2\,(60)}\left(\dfrac{100}{0.85} - 33.9\right) = 0.70$ sq in.

Try 2 – #4 = 0.80 sq in.

Check Eq. 6.13.9:

$\phi V_n = \phi\,(A_v\, f_y + A_h\, f_y + 2\lambda\sqrt{f'_c}\,bd)$
$\qquad = 0.85\,[0.80\,(60) + 0.44\,(60) + 33.9]$
$\qquad = 92.1$ kips < 100

Change A_h to 2 – #4

$\phi V_n = 110.4$ kips > 100 OK

Check anchorage requirements:

A_s bars:

From Table 11.2.7:

f_y = 60,000 psi, f'_c = 5000 psi, #5 bars

ℓ_d = 15 in. past 45° diagonal crack from corner

Total = 28 in. − 15 in. + ℓ_d = 28 in., or

$1.7\ell_d$ = 26 in. beyond dap

A_h bars:

From Table 11.2.7, for #4 bars:

$1.7\ell_d$ = 20 in. beyond dap

Fig. 6.14.1 Design of beam ledges

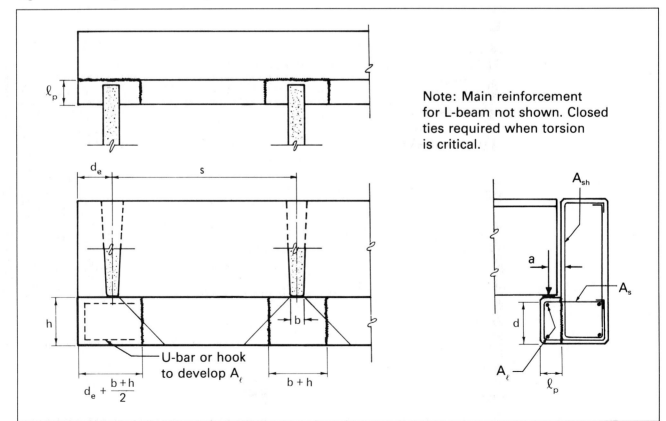

Note: Main reinforcement for L-beam not shown. Closed ties required when torsion is critical.

6.14 Beam Ledges

The design shear strength of continuous beam ledges supporting concentrated loads, as illustrated in Fig. 6.14.1, can be determined by the lesser of Eq. 6.14.1 and 6.14.2:

for $s > b + h$

$$\phi V_n = 3 \phi h \lambda \sqrt{f'_c} (2\ell_p + b + h) \quad \text{(Eq. 6.14.1)}$$

$$\phi V_n = \phi h \lambda \sqrt{f'_c} (2\ell_p + b + h + 2d_e) \quad \text{(Eq.6.14.2)}$$

for $s < b + h$, and equal concentrated loads, use the lesser of Eq. 6.14.1a, 6.14.2a or 6.14.3

$$\phi V_n = 1.5 \phi h \lambda \sqrt{f'_c} (2\ell_p + b + h + s) \quad \text{(Eq. 6.14.1a)}$$

$$\phi V_n = \phi h \lambda \sqrt{f'_c} \left(\ell_p + \frac{b + h}{2} + d_e + s\right) \quad \text{(Eq. 6.14.2a)}$$

where

h = depth of the beam ledge, in.

ℓ_p = ledge projection, in.

b = width of bearing area, in.

s = spacing of concentrated loads, in.

d_e = distance from center of load to the end of the beam, in.

If the ledge supports a continuous load or closely spaced concentrated loads, the design shear strength is:

$$\phi V_n = 24 \phi h \lambda \sqrt{f'_c} \quad \text{(Eq. 6.14.3)}$$

where ϕV_n is the design shear strength in pounds per foot.

If the applied factored load exceeds the strength as determined by Eqs. 6.14.1, 6.14.2 or 6.14.3, the ledge should be designed in accordance with Sect. 6.13.

Flexural reinforcement, A_s, computed by Eq. 6.13.1 should be provided in the beam ledge, and "hanger steel", A_{sh}, computed by Eq. 6.13.8 should be provided in the beam stem. Such reinforcement may be uniformly spaced over a width of 6h on either side of the bearing, but not to exceed ½ the distance to the next load. Bar spacing should not exceed the ledge depth, h, or 18 in. A_{sh} need not be additive to shear and torsion reinforcement designed in accordance with Sects. 4.3 and 4.4.*

Longitudinal reinforcement, calculated by Eq. 6.14.4 should be placed in *both* the top and bottom of the ledge portion of the beam:

*A consensus on the design procedure for hanger reinforcement, A_{sh}, has not been reached as of publication, and future recommendations may differ from that shown here. References 15-17 discuss the issue.

$$A_\ell = 200\ell_p \, d/f_y \quad \text{(Eq. 6.14.4)}$$

Example 6.14.1 Design of a beam ledge

Given:

8 ft wide double tees resting on a standard L-beam similar to that shown in Fig. 6.14.1. Layout of tees is irregular so that a stem can be placed at any point on the ledge.

V_u per stem = 18 kips

N_u per stem = 3 kips

h = 12 in.

d = 11 in.

b = 3 in.

ℓ_p = 6 in.

s = 48 in.

f'_c = 5000 psi (normal weight)

f_y = 40 ksi

Problem:

Investigate shear strength and determine reinforcement for the ledge.

Solution:

Min. $d_e = b/2 = 1.5$ in.

Since $s > b + h$ and $d_e < 2\ell_p + b + h$, use Eq. 6.14.2:

$$\phi V_n = \phi h \lambda \sqrt{f'_c} (2\ell_p + b + h + 2d_e)$$

$$= 0.85(12)(1) \sqrt{5000} \, [2(6) + 3 + 12 + 2(1.5)]/1000 = 21.6 \text{ kips} > 18$$

Shear span, $a = 3\ell_p/4 + 1½ = 6$ in.

By Eq. 6.13.1:

$$A_s = \frac{1}{\phi f_y} \left[V_u\left(\frac{a}{d}\right) + N_u\left(\frac{h}{d}\right)\right]$$

$$= \frac{1}{0.85(40)} [18(6/11) + 3(12/11)]$$

$$= 0.39 \text{ sq in.}$$

$6h = 6$ ft $> s/2$

Therefore distribute reinforcement over s/2 each side of the load.

$(s/2)(2) = 4$ ft

Maximum bar spacing = h = 12 in.

#3 @ 12 in. = 0.44 sq in. in each 4 ft

Place 2 additional bars at the beam end to provide equivalent reinforcement for stem placed near the end.

By Eq. 6.13.8:

$$A_{sh} = \frac{V_u}{\phi f_y} = \frac{18}{0.85(40)} = 0.529 \text{ sq in.}$$

$$\frac{0.529}{4} = 0.132 \text{ sq in./ft}$$

$$A_{sh} = \#3 @ 10 \text{ in.} = 0.132 \text{ sq in./ft}$$

Note: For placing convenience, the fabricator may elect to place A_s at the same spacing as A_{sh}.

By Eq. 6.14.4:

$$A_\ell = \frac{200\ell_p d}{f_y} = \frac{200(6)(11)}{40,000} = 0.33 \text{ sq in.}$$

Use 2 – #4 top and bottom = 0.40 sq in.

Note: Also check shear and torsion requirements, Sects. 4.3 and 4.4.

6.15 Hanger Connections

Hangers are similar to dapped ends, except that the extended, or bearing, end is steel instead of concrete. They are used when it is desired to keep the structural depth very shallow. Examples are shown in Fig. 6.15.1.

6.15.1 Cazaly Hanger[18]

The Cazaly hanger has three basic components (Fig. 6.15.2a). Design assumptions are as follows (Fig. 6.15.2b):

1. The cantilevered bar is usually proportioned so that the interior reaction from the concrete is 0.33 V_u. The hanger strap should then be proportioned to yield under a tension of 1.33 V_u.

$$A_s = \frac{1.33 V_u}{\phi f_y} \qquad \text{(Eq. 6.15.1)}$$

where:

f_y = yield strength of the strap material

ϕ = 0.90

2. V_u may be assumed to be applied 0.5 in. from the face of the seat. The remaining part of the moment arm is the width of the joint, g. It is therefore important that the joint width used in analysis is not exceeded in the field.

3. The moment in the cantilevered bar is then given by:

$$\begin{aligned} M_u &= V_u (0.5 + g + 0.375s) \\ &= \phi f_y bd^2/6 \end{aligned} \qquad \text{(Eq. 6.15.2)}$$

Fig. 6.15.1 Hanger connections

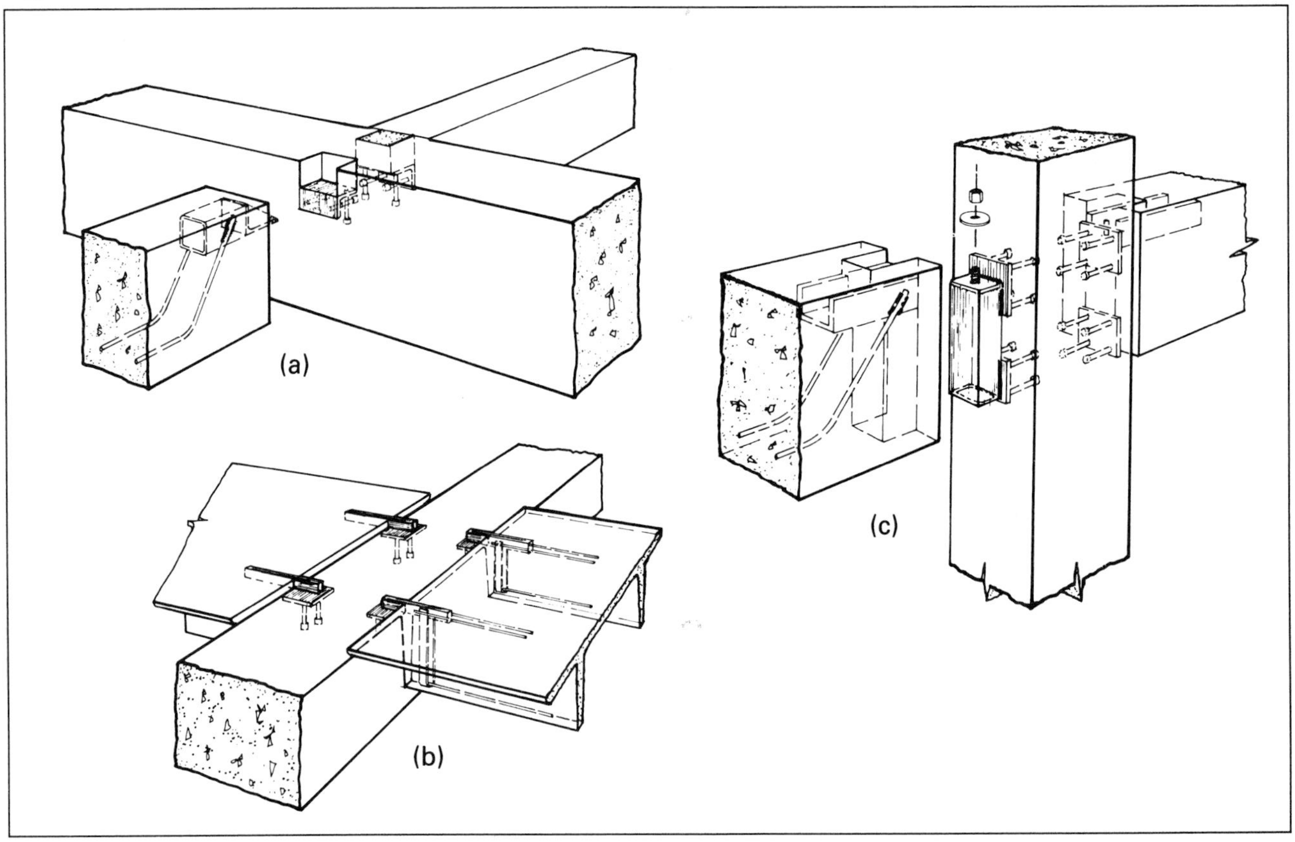

(a)

(b)

(c)

Fig. 6.15.2 Cazaly hanger

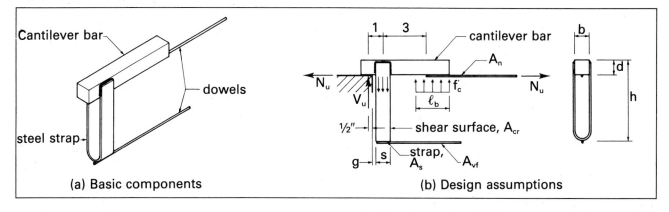

(a) Basic components   (b) Design assumptions

where

f_y = yield strength of the bar material

ϕ = 0.90

Other notation is shown in Fig. 6.15.2b

If the bar is proportioned to take this moment at the yield stress, but using elastic section properties, the shear and tensile forces can usually be neglected.

4. The bearing pressure creating the interior reaction may be calculated as in Sect. 6.8. If the width of the member in which the hanger is cast = b_1, then:

$$f_{bu} = 0.85\phi f'_c \sqrt{b_1/b} \qquad \text{(Eq. 6.15.3)}$$

where ϕ = 0.7

The bearing length, ℓ_b, is then given by:

$$\ell_b = \frac{V_u/3}{b\, f_{bu}} \qquad \text{(Eq. 6.15.4)}$$

The exterior cantilever should have a minimum length of $(g + 1)$ in. Most hangers in practice have cantilever lengths of 2½ to 3½ in.

5. To maintain the conditions of equilibrium assumed, the interior cantilever must have a length:

$(1.5 + 3g + s + 0.5\ell_b)$ in.

6. The minimum total length of bar is then:

$(2.5 + 4g + 2s + 0.5\ell_b)$ in.   (Eq. 6.15.5)

7. Longitudinal dowels, A_n, are welded to the cantilevered bar to transmit the axial force, N_u:

$$A_n = \frac{N_u}{\phi f_y} \qquad \text{(Eq. 6.15.6)}$$

where

f_y = yield strength of the dowel

ϕ = 0.90

8. The lower dowel, A_{vf}, and the area confined within the strap can be conservatively proportioned using effective shear-friction described in Sect. 6.7.

$$A_{vf} = \frac{V_u}{\phi f_y \mu_e} \qquad \text{(Eq. 6.15.7)}$$

where

ϕ = 0.85

f_y = yield strength of lower dowels, psi

$$\mu_e = \frac{1400 \lambda^2 bh}{V_u} \le 3.4 \qquad \text{(Eq. 6.15.8)}$$

V_u (max.) = $0.30 \lambda^2 f'_c\, bh \le 1000 \lambda^2 bh$

Example 6.15.1 Design of a Cazaly hanger

Given:

Hanger similar to that shown in Fig 6.15.1b

f'_c = 5000 psi (both member and support)

f_y (reinforcing bars) = 60 ksi

f_y (structural steel) = 36 ksi

V_u = 24 kips

N_u = 4 kips

Width of member in which hanger is cast

b_1 = 6 in.

g = 1 in.

Problem:

Size the hanger components.

Solution:

By Eq. 6.15.1:

$$A_s \text{ (strap)} = \frac{1.33 V_u}{\phi f_y}$$

$$= \frac{1.33(24)}{0.9(36)} = 0.99 \text{ in.}^2$$

Use $\frac{1}{4} \times 2$ in. strap; $A_s = 0.25(2)(2) = 1.00$ in.^{2}

Use $\frac{3}{16}$ in. fillet weld; Table 6.20.2, E70 electrodes:

$$\ell_w = \frac{1.33(24)}{2(4.64)} = 3.44 \text{ in.}$$

Weld 2 in. across top, $\frac{3}{4}$ in. down sides
$= 3.50$ in.

By Eq. 6.15.2:

$$M_u = V_u(0.5 + g + 0.375s) = 24(2.25)$$
$$= 54 \text{ in.-kips}$$

$$Z_{req'd} = \frac{M_u}{\phi f_y} = \frac{54}{0.9(36)} = 1.67 \text{ in.}^3 = \frac{bd^2}{6}$$

Try 2 in. wide bar:

$$d = \sqrt{\frac{6(1.67)}{2}} = 2.24 \text{ in. min.}$$

Use $2 \times 2\frac{1}{4}$ in. bar

By Eqs. 6.15.3 and 6.15.4:

$$f_{bu} = 0.85\phi f'_c \sqrt{b_1/b} = 0.85(0.7)(5)\sqrt{6/2}$$
$$= 5.15 \text{ ksi}$$

$$\ell_b = \frac{V_u/3}{f_{bu}(b)} = \frac{24/3}{5.15(2)} = 0.78 \text{ in.}$$

Min. interior cantilever
$= 1.5 + 3g + s + 0.5\ell_b$
$= 1.5 + 3 + 2 + 0.78/2 = 6.89$ in.

Min. total length (Eq. 6.15.5)
$= 2.5 + 4g + 2s + 0.5\ell_b = 10.89$ in.
Use bar $2 \times 2\frac{1}{4} \times 12$ in.

By Eq. 6.15.6:

$$A_n = \frac{N_u}{\phi f_y} = \frac{4}{0.9(60)} = 0.07 \text{ sq in.}$$

Use 1–#3 dowel

Try $h = 16$ in.; by Eqs. 6.15.7 and 6.15.8:

$$\mu_e = \frac{1400\lambda^2 bh}{V_u} = \frac{1400(1)(2)(16)}{24,000} = 1.87$$

$$A_{vf} = \frac{V_u}{\phi f_y \mu_e} = \frac{24}{0.85(60)(1.87)} = 0.25 \text{ sq in.}$$

Use 1–#5 dowel

Also check welding requirements.

6.15.2 Loov Hanger[19]

The hanger illustrated in Fig. 6.15.3 is designed using the following equations:

$$A_{sh} = \frac{V_u}{\phi f_y \cos \alpha} \qquad \text{(Eq. 6.15.9)}$$

where

$\phi = 0.85$

f_y = yield strength of A_{sh}

$$A_n = \frac{N_u}{\phi f_y} \left(1 + \frac{h - d}{d - a/2}\right) \qquad \text{(Eq. 6.15.10)}$$

where

$\phi = 0.90$

f_y = yield strength of A_n

The steel bar is proportioned so that the bearing strength of the concrete is not exceeded, and to provide sufficient weld length to develop the diagonal bars. Bearing strength is discussed in Sect. 6.8. However, if the bar is at the top of the member as in Fig. 6.15.3, there is no "geometrically similar" area larger than the edge of the bar, and:

$$f_{bu} = 0.85 \phi f'_c = 0.6f'_c \qquad \text{(Eq. 6.15.11)}$$

The connection should be detailed so that the reaction, the center of compression and the center of the diagonal bars meet at a common point, as shown in Fig. 6.15.3b. The compressive force, C, is assumed to act at a distance $a/2$ from the top of the bearing plate. Thus:

$$a = \frac{C}{b_1 f_{bu}} \qquad \text{(Eq. 6.15.12)}$$

where

$$C = V_u \tan\alpha + \frac{N_u (h - d)}{d - a/2} \qquad \text{(Eq. 6.15.13)}$$

For most designs, the horizontal bars, A_n, are placed very close to the bottom of the steel bar. Thus, the term $(h - d)$ can be assumed equal to zero, simplifying Eqs. 6.15.10 and 6.15.13.

Tests have indicated a weakness in shear in the vicinity of the hangers, so it is recommended that stirrups in the beam end be designed to carry the total shear.

Example 6.15.2 Design of a Loov hanger

Given:
   Hanger similar to that shown in Fig. 6.15.3.
   Design for same data as Example 6.15.1
   $\alpha = 30°$

Fig. 6.15.3 Loov hanger

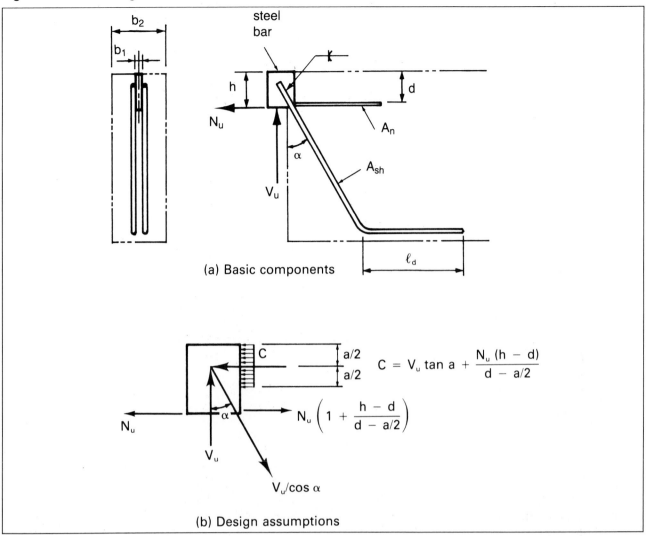

(a) Basic components

$$C = V_u \tan a + \frac{N_u (h - d)}{d - a/2}$$

$$N_u \left(1 + \frac{h - d}{d - a/2} \right)$$

$V_u/\cos \alpha$

(b) Design assumptions

Problem:

Size the hanger components.

Solution:

$$A_{sh} = \frac{V_u}{\phi \, f_y \, \cos\alpha} = \frac{24}{0.85(60)\cos 30} = 0.54 \text{ sq in.}$$

Use 2–#5 bars
Detail A_n so it is near the bottom of the steel
bar
$h - d \approx 0$

$$A_n = \frac{N_u}{\phi \, f_y} = \frac{4}{0.9(60)} = 0.07 \text{ sq in.}$$

Use 1–#3 dowel

By Eq. 6.15.11:

$f_{bu} = 0.85 \, \phi \, f'_c = 0.85 \, (0.7)(5) = 2.98 \text{ ksi}$
$C = V_u \tan\alpha = 24 \tan 30° = 13.9 \text{ kips}$

Assume $b_1 = 1$ in.

$$a = \frac{C}{b_1 \, f_{bu}} = \frac{13.9}{1(2.98)} = 4.65 \text{ in.}$$

$a/2 = 2.33$

Min. weld length, #5 bar, E70 electrode (Table
6.20.3) = 2¼ in.

Provide end bearing plate as follows:

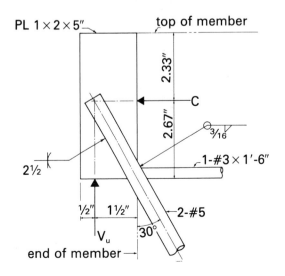

PL 1 × 2 × 5″

top of member

2.33″

2.67″

C

³⁄₁₆

1-#3 × 1'-6″

2½

½″ 1½″

2-#5

V_u

30°

end of member

Fig. 6.16.1 Moment connections

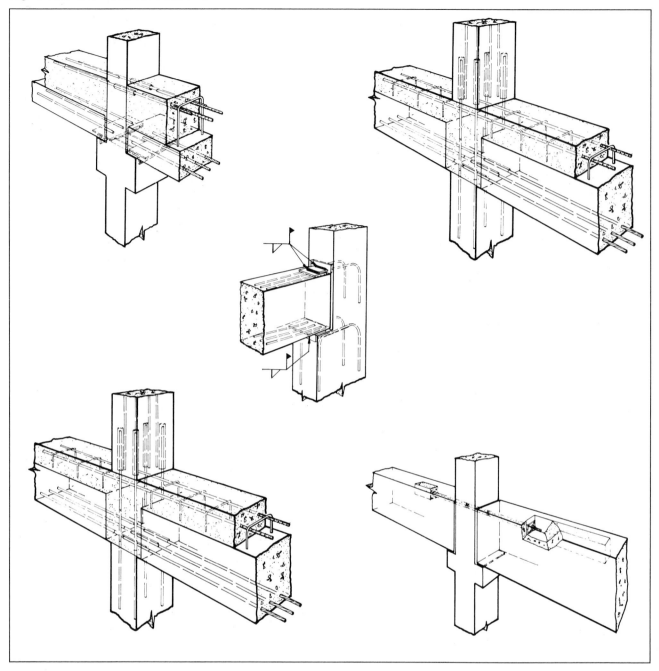

6.16 Moment Connections

Moment connections are sometimes required in building frames, as discussed in Part 3. Moment-resisting capacity is attained by designing the connection so that a force couple can be developed. Tensile capacity through the connection can be attained by studs, deformed bar anchors, inserts, welding, post-tensioning or combinations of these.

The erection process must be considered, since in most cases the connection is not complete when some of the load is first applied.

6.17 Connection of Non-Load Bearing Wall Panels

The design of connections of non-load bearing architectural wall panels follows the same principles as structural connections, except the loads are generally less. Usually, the connection will transfer only the self-weight of the panel and wind loads. The details are most important, not only for the design of the connection itself, but also for the design of the supporting structure. Typical details are shown in Sect. 7.2.

6.18 Connection of Load Bearing Wall Panels

Connections for load bearing wall panels are an essential part of the structural support system and the stability of the structure may depend upon them. In addition to the weight of the panels, the connections must resist and transfer dead, live, wind and earthquake loads, and effects of volume changes.

Erected load bearing walls may have both horizontal and/or vertical joints across which forces must be transferred. Fig. 6.18.1 indicates, for separate cases, the principal exterior forces and the resulting joint force systems. In buildings, superpositions of forces and various combinations of panel and joint assemblies must be considered.

Distribution of lateral forces to shear walls depends largely on adequate connections of floors to walls. In addition to the transfer of vertical shear forces due to lateral loads, vertical joints may also be subject to shear forces induced by differential loads on adjacent panels. Joint and connection details of exterior bearing walls are especially critical since the floor elements are usually connected at this elevation and a waterproofing detail must be incorporated.

6.18.1 Vertical Joints

Vertical joints may be designed so that the wall panels form one structural unit or act independently.

Hinge connection

A hinge connection transfers compression and tension forces but not moments. This is usually done at floor levels through floor diaphragms and tie beams. The joint between floor levels usually is "open" so the panels resist lateral loads independently according to their relative rigidity (Fig. 6.18.2). Sound and waterproofing details may also have to be considered.

Grooved joint connection

Grooved joints are continuous and usually filled with grout. The minimum groove dimension should be 1½ in. deep and 3 in. wide (Fig. 6.18.3). The joint strength can be evaluated by shear-friction even if shrinkage, creep, and temperature movements have caused a crack at the wall-grout interface.

Fig. 6.18.1 Exterior forces and joint force systems

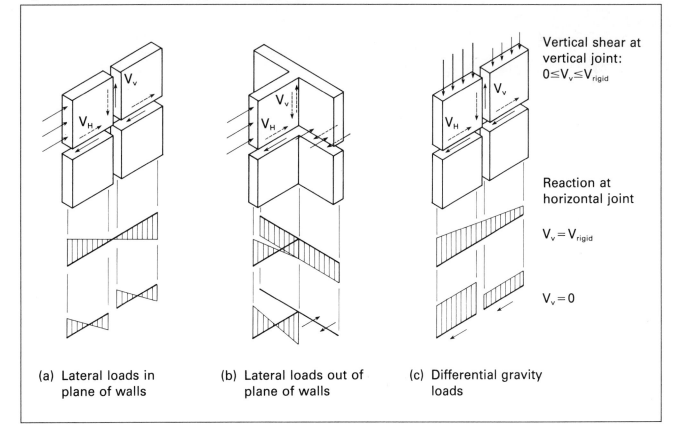

(a) Lateral loads in plane of walls

(b) Lateral loads out of plane of walls

(c) Differential gravity loads

Vertical shear at vertical joint:
$0 \leq V_v \leq V_{rigid}$

Reaction at horizontal joint

$V_v = V_{rigid}$

$V_v = 0$

Fig. 6.18.2 Wall to wall hinge connections at floor levels

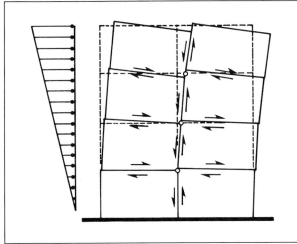

Fig. 6.18.3 Grooved joint connections

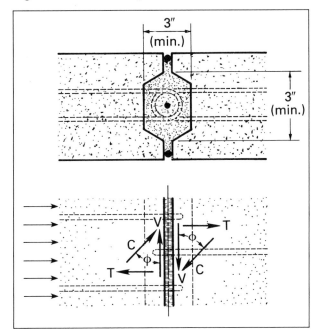

Fig. 6.18.4 Mechanical connections

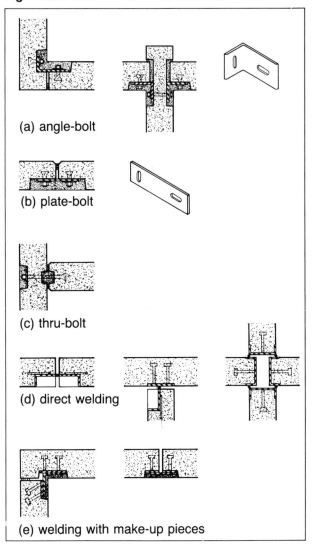

(a) angle-bolt

(b) plate-bolt

(c) thru-bolt

(d) direct welding

(e) welding with make-up pieces

Mechanical connection

Mechanical connections consist of anchorage devices cast into the wall panels and steel sections (plates, angles, bars, etc.) crossing the joint. The strength is usually controlled by the capacity of the cast-in anchorage (Fig. 6.18.4); connection of the steel section to the anchorage device can be made by bolting, welding, or grouting.

Tie beam connections at floor levels may act with the mechanical connections. The relative participation in resisting applied forces depends on their force-deformation characteristics. The ultimate capacity is the sum of the strength of the tie beams and the mechanical connections.

Once the connection forces have been established, evaluation of connection strength is made according to strength of materials and the principles developed in other sections of this Handbook.

Keyed joint connection

Keyed joints can either be reinforced or nonreinforced (Fig. 6.18.5). Test results indicate similar deformation behavior, but also show that reinforced joints are stronger. Reinforcement is required in high seismic zones.

As shown in Fig. 6.18.6, the resistance can be limited by:

(a) cracking of grout concrete parallel to joint,

(b) diagonal cracks across joints,

(c) crushing of key edges or joint concrete at key edges, or

(d) slippage along contact area.

Fig. 6.18.5 Keyed joint connection

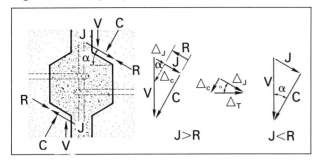

For (a) the shear-friction concept applies (Sect. 6.7). For (b), (c) and (d) the strength of the connection is usually a function of the compressive strength of the grout, the bond strength of the grout to the precast concrete, and the profile of the keys. As shown in Fig. 6.18.5, the vertical shear force can be resolved into tension and compression components with ϕ as the apparent friction coefficient and α the angle of the key.

Depending on the number of keys per floor, the unit forces per key resulting from the vertical shear force V are:

$$J = V \sin\alpha$$
$$C = V \cos\alpha$$

The joint force J is resisted by the shear-friction force R developed in the plane of J, with:

$$R = C \tan \phi$$

Assuming a conservative value of $\tan \phi = 0.60$, a sliding along J will not occur if:

$$R > J$$

which is the case for $\alpha \leqslant 30°$.

For $\alpha > 30°$ and $R < J$, a tension force T develops which must be absorbed by horizontal reinforcing of the joint. According to Fig. 6.18.5:

$$\Delta T = \frac{\Delta J}{\cos\alpha} = \frac{J - R}{\cos\alpha} = \frac{V(\sin\alpha - \cos\alpha \tan\phi)}{\cos\alpha}$$

$$= V(\tan\alpha - \tan \phi) \qquad \text{(Eq. 6.18.1)}$$

The sum of the unit tension forces can be absorbed at each floor level by horizontal ties or by uniformly distributed horizontal reinforcing bars protruding from the wall.

6.18.2 Horizontal Joints

Horizontal joints in load bearing wall construction occur at floor levels and at the transition to foundation or transfer beams. The principal forces

Fig. 6.18.6 Forces in keyed joint

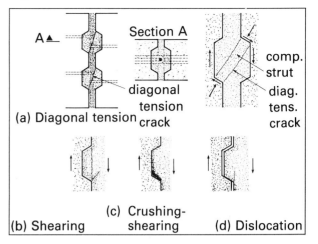

(a) Diagonal tension
(b) Shearing
(c) Crushing-shearing
(d) Dislocation

to be transferred are vertical and horizontal loads from panels above and from the diaphragm action of floor slabs. The resulting forces are:

(a) normal to joint—compression or tension;
(b) horizontal to joint—horizontal shear;
(c) vertical to joint at face—vertical shear; and
(d) perpendicular to joint—compression or tension from floor to diaphragm (Fig. 6.18.7).

Because of the limited frame action that can be developed perpendicular to a wall, moment stresses in the joint are normally only of minor importance.

Fig. 6.18.7 Typical interior horizontal connection

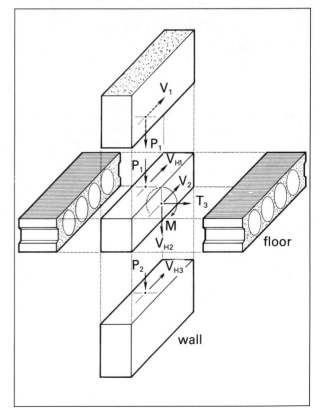

Fig. 6.18.8 Typical joints in a bearing wall building

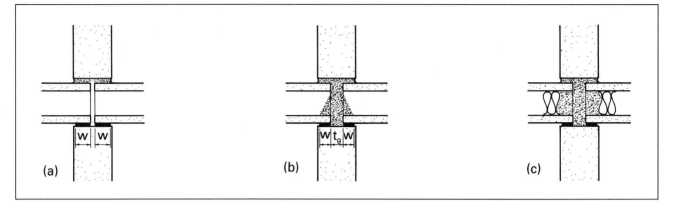

Axial load transfer through horizontal joints[20]

Fig. 6.18.8 shows three joint details used in multistory load bearing buildings with hollow-core slabs used for the floors. For the condition of Fig. 6.18.8 (a), the joint strength is:

$$\phi P_n = \phi 0.85 A_e f'_c R_e \qquad \text{(Eq. 6.18.2)}$$

where:

P_n = nominal strength of the joint
A_e = effective slab bearing area = $2\,w b_w$
w = bearing length (Fig. 6.18.8)
b_w = net web width of slab
f'_c = design concrete strength of slab
R_e = reduction factor for eccentricity of load
  = $1 - (2e/h)$
e = eccentricity of load (occurs when floor spans or loads on either side of wall are unequal, or at end walls)
h = slab thickness
ϕ = 0.7

When the space between slab ends is grouted, load is shared by the slab ends and grout columns according to their stiffnesses. The splitting strength of the wall may also limit the joint capacity. The effect of grout flowing solidly into the slab ends is to confine the grout column. If the space between slab ends is less than about 1-1/2 in., an unconfined grout column will add little strength.

Table 6.18.1 Equivalent bearing strength, f_{ue} (ksi)

	Grout strength, psi		
	3000	4000	5000
Slab cores not filled	4.5	5.9	5.9
Slab cores filled	5.8	6.5	7.1
Valid for f'_c = 5000 psi or higher; slabs supported on multimonomer plastic bearing strips. For other conditions see Ref. 20.			

The strength of the connection can be determined by:

$$\phi P_n = \phi t_g \ell f_{ue} R_e \qquad \text{(Eq. 6.18.3)}$$

where:

t_g = grout thickness
ℓ = length of joint (parallel to wall) being considered
f_{ue} = equivalent bearing strength from Table 6.18.1. Accounts for distribution of load between grout column and slab ends
R_e = $1 - 2e/h$

Example 6.18.1 Design of grouted horizontal joint

Given:

An 18-story bearing wall building with 8 in. precast concrete walls and 8 in. hollow-core floors and roof. Floor slabs span 28 ft and bear on multimonomer plastic bearing strips.

f'_c (precast concrete) = 5000 psi

Loads:
   Roof: DL = 15 psf; LL = 30 psf
   Floors: DL = 10 psf; LL = 40 psf
   Hollow core = 60 psf
   Walls = 800 plf/story
No LL reduction for example

Problem:
   Find grouting requirements for interior joint.
Solution:
   Loads:
      Roof: w_u = 28 [1.4(60 + 15) + 1.7(30)]/1000
        = 4.37 klf

      Floors: w_u = 28 [1.4(60 + 10) + 1.7(40)]/1000
        = 4.65 klf

      Walls: w_u = 1.4(800)/1000
        = 1.12 klf/story

Accumulate loads above floor noted:

Floor	w_u	Σw_u
18	4.37 + 1.12	5.49
17	4.65 + 1.12	11.26
16	5.77	17.03
15	5.77	22.80
14	5.77	28.57
13	5.77	34.34
12	5.77	40.11
11	5.77	45.88
10	5.77	51.65
9	5.77	57.42
8	5.77	63.19
7	5.77	68.96
6	5.77	74.73
5	5.77	80.50
4	5.77	86.27
3	5.77	92.04
2	5.77	97.81

Evaluate capacity of ungrouted joint:
Assume the ratio of web width to total width of the slabs = 0.3

$$\phi P_n = \phi \, 0.85 \, A_e \, f'_{c(slab)} \, R_e$$

$$= 0.7(0.85)(3 + 3)(0.3 \times 12)(5)\left(1 - \frac{2(0)}{8}\right)$$

$$= 64.26 \text{ kips/ft}$$

Adequate floors 8 through roof.

Evaluate capacity of grouted joint:
Try f'_c of grout = 3000 psi
Use t_g = 2 in. R_e = 1.0 as above

From Table 6.18.1:
Slab cores not filled, f_{ue} = 4.5 ksi

From Eq. 6.18.3:

$$\phi P_n = 0.7(2)(12)(4.5)(1) = 75.6 \text{ kips}$$

Adequate for floors 6 and 7.

Slab cores filled, f_{ue} = 5.8 ksi

$$\phi P_n = 0.7(2)(12)(5.8)(1)$$
$$= 97.4 \text{ kips} \approx 97.81 \text{ Say OK (may choose to specify a slightly higher grout strength.)}$$

Typical examples of connections through horizontal joints are shown in Fig. 6.18.9. Since forces are concentrated at a few points, they must be redistributed into the panels above and below. Connections should have more ductility and strength than the vertical tie fastened to it.

6.18.3 Structural Integrity[21]

Minimum tensile ties should be provided at the joints to resist the following forces (Figs. 6.18.10 and 6.18.11):

$T_{1,u} \geq$ 1500 lb/ft × span of floor slabs (ft) (cross tie)

$T_{2,u} \geq$ 16,000 lb (peripheral tie)

$T_{3,u} \geq$ 2½% of service load on wall
$\geq$ 1500 lb/ft × distance between ties (ft) (longitudinal tie)

$T_{4,u} \geq$ 3000 lb/ft × distance between ties (ft) (vertical tie)

6.18.4 Typical Details

Typical wall to floor and wall to foundation details are shown in Figs. 6.18.12–6.18.14.

6.19 References

1. Martin, L. D., and Korkosz, W. J., "Connections for Precast Prestressed Concrete Buildings — Including Earthquake Resistance," Technical Report No. 2, 1982, Prestressed Concrete Institute, Chicago, IL.

2. PCI Committee on Connection Details, "Manual on Design and Detailing of Connections for Precast Prestressed Concrete," scheduled for publication 1985. Prestressed Concrete Institute, Chicago, IL.

3. "Structural Welding Code–Reinforcing Steel," AWS D1.4-79, American Welding Society, Miami, FL, 1979.

4. "Structural Welding Code–Steel", AWS D1.1-79, American Welding Society, Miami, FL, 1979.

5. "Reinforcement Anchorages and Splices", Concrete Reinforcing Steel Institute, Schaumburg, IL, 1980.

6. ACI Committee 439, "Mechanical Connections of Reinforcing Bars", Concrete International, V.5, No.1, January 1983.

7. Reichard, T. W., Carpenter, E. F., and Leyendecker, E. V., "Design Loads for Inserts Embedded in Concrete," Building Science Series 42.N.B.S. (U.S.)Code:BSSNVB, Superintendent of Documents, U.S. Government Printing Office, Washington, D.C., May, 1972.

8. The Procedure Handbook of Arc Welding, Twelfth Edition, The Lincoln Electric Company, Cleveland, Ohio, June 1973.

9. Iverson, J. K., and Pfeiffer, D. W., "Criteria for Design of Bearing Pads", Research Pro-

Fig. 6.18.9 Connections through horizontal joints

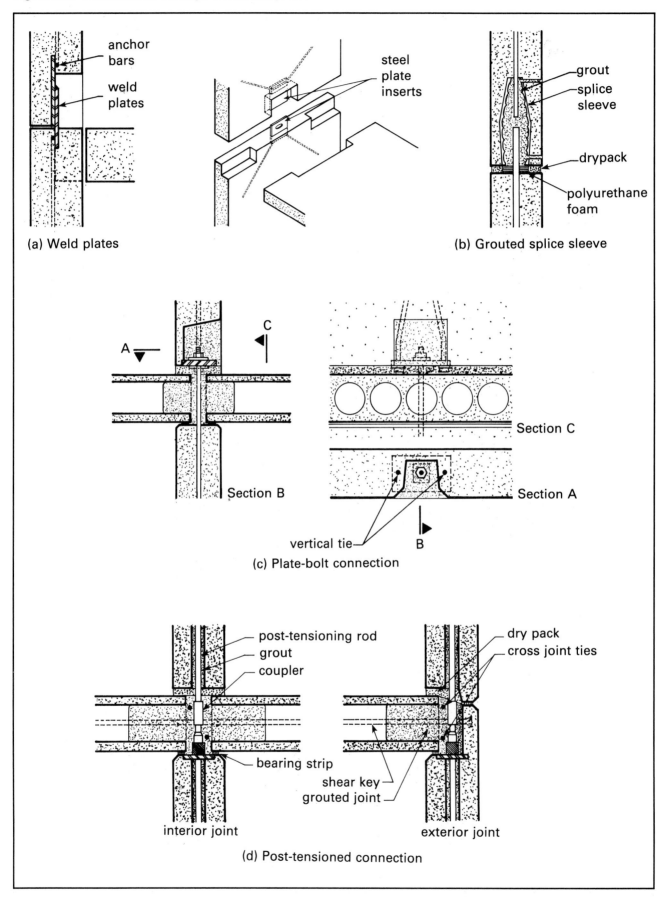

(a) Weld plates

(b) Grouted splice sleeve

(c) Plate-bolt connection

(d) Post-tensioned connection

Fig. 6.18.10 Recommended tie forces in precast concrete bearing wall buildings

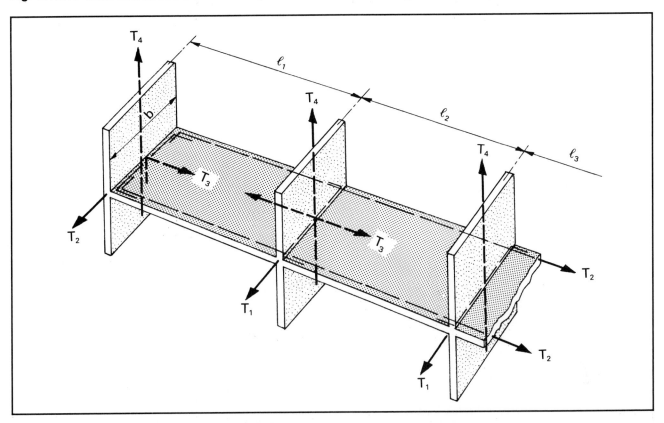

ject for PCI, scheduled for publication 1985, Prestressed Concrete Institute, Chicago, IL.

10. Kriz, L. B. and Raths, C. H., "Connections in Precast Concrete Structures — Strength of Corbels," *PCI Journal*, V.10, No.1, February, 1965.

11. Mattock, A. H., "Design Proposals for Reinforced Concrete Corbels", *PCI Journal*, V.21, No.2, May-June, 1976.

12. *Manual of Steel Construction*, Eighth Edition, American Institute of Steel Construction, Chicago, IL.

13. Marcakis, K. and Mitchell D., "Precast Concrete Connections with Embedded Steel Members," *PCI Journal*, V.25, No.4, July-August, 1980.

14. Mattock, A. H. and Chan, T. C., "Design and Behavior of Dapped-End Beams," *PCI Journal*, V.24, No.6, November-December, 1979.

Fig. 6.18.11 Typical tie arrangement

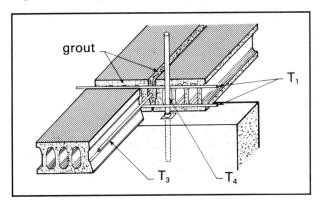

Fig. 6.18.12 Wall to foundation connections

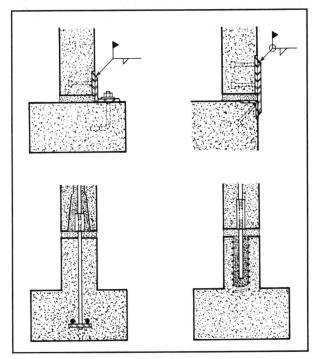

Fig. 6.18.13 Floor to bearing wall connections

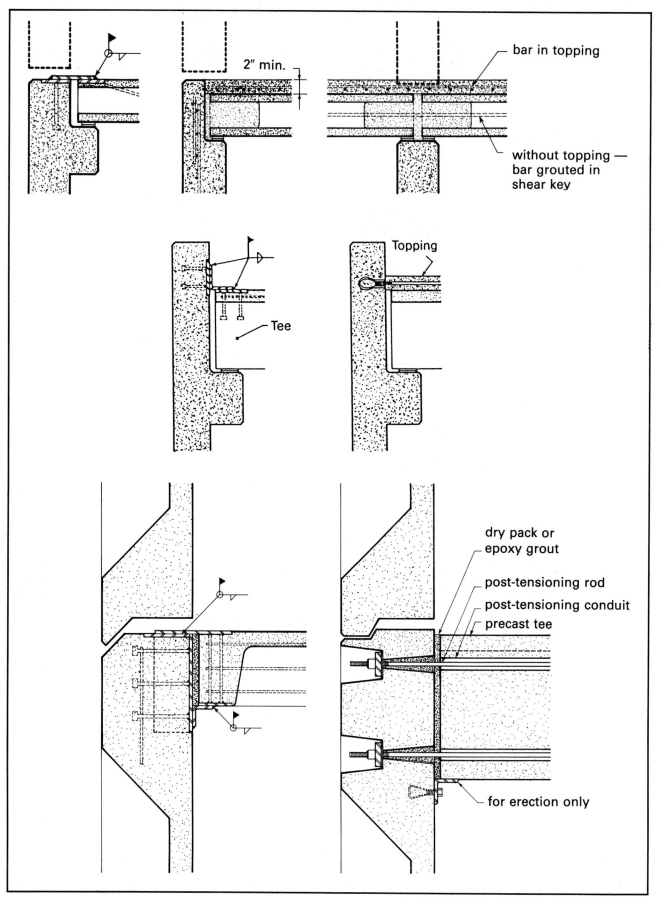

PCI Design Handbook

Fig. 6.18.14 Floor to shear wall connections

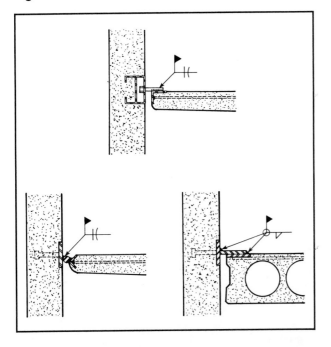

15a. Sher Ali Mizza and Richard W. Furlong, "Serviceability Behavior and Failure Mechanisms of Concrete Inverted T-Beam Bridge Bent Caps", *Journal of the American Concrete Institute*, V.80, No.4, July-August, 1983.

15b. Sher Ali Mizza and Richard W. Furlong, "Strength Criteria for Concrete Inverted T-Girders", *ASCE Structural Journal*, August, 1983.

16. Collins, Michael P. and Mitchell, Denis, "Shear and Torsion Design of Prestressed and Non-Prestressed Concrete Beams," *PCI Journal*, V.25, No.5, September/October, 1980.

17. Raths, Charles H., "Spandrel Beam Behavior and Design", *PCI Journal*, V.29, No.2, March-April, 1984.

18. Cazaly, L. and Huggins, M., "Canadian Prestressed Concrete Institute Handbook," Canadian Prestressed Concrete Institute, Ottawa, Ontario, 1964.

19. Loov, Robert, "A Precast Beam Connection Designed for Shear and Axial Load," *PCI Journal*, V.13, No.3, June 1968.

20. Johal L. and Hanson, N., "Design for Vertical Loads on Horizontal Connections in Large Panel Structures," *PCI Journal*, V.27, No.1, January-February, 1982.

21. Speyer, Irwin J., "Considerations for the Design of Precast Concrete Bearing Wall Buildings to Withstand Abnormal Loads," *PCI Journal*, V.21, No.2, March-April, 1976.

22. Shaikh, A. Fattah and Yi, Whayong, "In-Place Strength of Welded Headed Studs," scheduled for publication 1985, *PCI Journal*.

CONNECTIONS

Table 6.20.1 Allowable working stress and design strength of welds*

Electrode	Allowable Working Stress (ksi)	Design[†] Strength (ksi)
E60	18	30
E70	21	35
E80	24	40
E90	27	45
E100	30	50

*Based on AISC Spec. for buildings. For bridges, use 90% of values.
[†]Use factored loads and $\phi = 1.0$ with these values.

Table 6.20.2 Strength of fillet welds for building construction*

Filet Weld Size	E60 Electrodes		E70 Electrodes	
	Working Stress (k/in.)	Design[†] Strength (k/in.)	Working Stress (k/in.)	Design[†] Strength (k/in.)
1/8	1.59	2.65	1.86	3.09
3/16	2.39	3.98	2.78	4.64
1/4	3.18	5.30	3.71	6.19
5/16	3.98	6.63	4.64	7.73
3/8	4.77	7.95	5.57	9.28
7/16	5.57	9.28	6.50	10.83
1/2	6.36	10.61	7.42	12.37
9/16	7.16	11.93	8.35	13.92
5/8	7.95	13.26	9.28	15.47

*Use 90% of values for bridges. Assumes 45° fillet.
[†]Use factored loads and $\phi = 1.0$ with these values.

Table 6.20.3 Minimum length of weld to develop full strength of bar. Weld parallel to bar length.

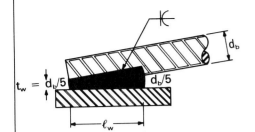 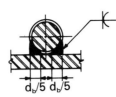 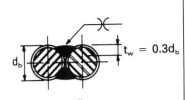

splice

Electrode	Bar size	Min plate thickness for weld length, in.							Min. splice length, in.	Bar size
		¼	5/16	3/8	7/16	½	9/16	5/8		
E70	3	1¼	1¼	1¼					1¼	3
	4	1¾	1¾	1¾					1½	4
	5	2¼	2¼	2¼					2	5
	6	2¾	2½	2½	2½				2¼	6
	7		3	3	3	3			2½	7
	8			3½	3½	3½			3	8
	9			4	4	4	4	4	3¼	9
	10				4¼	4¼	4¼	4¼	3¾	10
	11				5¼	4¾	4¾	4¾	4	11
E80	3	1¼	1¼	1¼					1	3
	4	1½	1½	1½					1¼	4
	5	2	2	2					1¾	5
	6	2¾	2¼	2¼	2¼				2	6
	7		3	2¾	2¾	2¾			2¼	7
	8			3¼	3	3			2½	8
	9			4	3½	3½	3½	3½	3	9
	10				4¼	3¾	3¾	3¾	3¼	10
	11				5¼	4½	4¼	4¼	3½	11
E90	3	1	1	1					1	3
	4	1½	1½	1½					1¼	4
	5	2	1¾	1¾					1½	5
	6	2¾	2¼	2	2				1¾	6
	7		3	2½	2½	2½			2	7
	8			3¼	2¾	2¾			2¼	8
	9			4	3½	3	3	3	2½	9
	10				4¼	3¾	3½	3½	3	10
	11				5¼	4½	4	3¼	3¼	11

Table is based on reinforcing bar f_y = 60 ksi.   Plate shear yield = 19.8 ksi.

CONNECTIONS

Table 6.20.4 Size of fillet weld required to develop full strength of bar

Bar perpendicular to plate,

welded one side

Plate f_y = 36 ksi

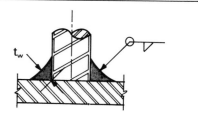

	Grade 40 bar					
	E70 Electrode		**E80 Electrode**		**E90 Electrode**	
Bar size	**Nominal weld size (in.)**	**Min. plate thickness (in.)**	**Nominal weld size (in.)**	**Min. plate thickness (in.)**	**Nominal weld size (in.)**	**Min. plate thickness (in.)**
3	1/8	1/4	1/8	1/4	1/8	1/4
4	3/16	1/4	3/16	1/4	3/16	1/4
5	1/4	5/16	3/16	5/16	3/16	5/16
6	1/4	5/16	1/4	5/16	1/4	3/8
7	5/16	3/8	5/16	3/8	1/4	3/8
8	3/8	7/16	5/16	7/16	5/16	7/16
9	7/16	1/2	3/8	1/2	5/16	1/2
10	7/16	9/16	3/8	9/16	3/8	9/16
11	1/2	5/8	7/16	5/8	3/8	5/8
	Grade 60 bar					
3	3/16	1/4	3/16	1/4	3/16	1/4
4	1/4	5/16	1/4	5/16	1/4	5/16
5	5/16	3/8	5/16	3/8	1/4	3/8
6	3/8	7/16	5/16	1/2	5/16	1/2
7	7/16	9/16	3/8	9/16	3/8	9/16
8	1/2	5/8	7/16	5/8	3/8	5/8
9	9/16	11/16	1/2	11/16	7/16	11/16
10	5/8	3/4	9/16	3/4	1/2	13/16
11	11/16	13/16	5/8	13/16	9/16	7/8

CONNECTIONS

Table 6.20.5 Size of fillet weld required to develop full strength of bar

Bar perpendicular to plate,

welded both sides

Plate $f_y = 36$ ksi

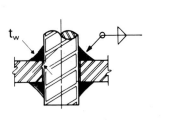

	Grade 40 bar					
	E70 Electrode		E80 Electrode		E90 Electrode	
Bar size	Nominal weld size (in.)	Min. plate thickness (in.)	Nominal weld size (in.)	Min. plate thickness (in.)	Nominal weld size (in.)	Min. plate thickness (in.)
3	1/8	3/16	1/8	3/16	1/8	3/16
4	1/8	1/4	1/8	1/4	1/8	1/4
5	1/8	5/16	1/8	5/16	1/8	5/16
6	3/16	3/8	1/8	3/8	1/8	3/8
7	3/16	7/16	3/16	7/16	1/8	7/16
8	3/16	1/2	3/16	1/2	3/16	1/2
9	1/4	1/2	3/16	9/16	3/16	9/16
10	1/4	9/16	1/4	5/8	3/16	5/8
11	1/4	5/8	1/4	5/8	1/4	11/16
	Grade 60 bar					
3	1/8	1/4	1/8	1/4	1/8	1/4
4	3/16	3/8	1/8	3/8	1/8	3/8
5	3/16	7/16	3/16	7/16	3/16	7/16
6	1/4	1/2	3/16	1/2	3/16	1/2
7	1/4	5/8	1/4	5/8	3/16	5/8
8	5/16	11/16	1/4	11/16	1/4	11/16
9	5/16	3/4	5/16	3/4	1/4	3/4
10	3/8	13/16	5/16	7/8	5/16	7/8
11	3/8	15/16	3/8	15/16	5/16	15/16

CONNECTIONS

Table 6.20.6 Tensile strength of welded headed studs and bolts (see Fig. 6.5.2)

Edge dist., d_e	Stud length, ℓ_e	Maximum Design Tensile Strength, ϕP_c, Limited by Concrete Strength (kips)											
		Normal weight concrete[1]						Sand-lightweight concrete[1]					
		Diameter, d_b, in.						Diameter, d_b, in.					
		¼	⅜	½	⅝	¾	⅞	¼	⅜	½	⅝	¾	⅞
2 in.	2.5	4.5	4.9	5.3	5.7	5.7	5.9	3.9	4.2	4.5	4.8	4.8	5.0
	4.0	6.8	7.2	7.6	7.9	7.9	8.1	5.8	6.1	6.4	6.8	6.8	6.9
	5.0	8.3	8.7	9.1	9.5	9.5	9.6	7.1	7.4	7.7	8.0	8.0	8.2
	6.0	9.8	10.2	10.6	11.0	11.0	11.2	8.4	8.7	9.0	9.3	9.3	9.5
	7.0	11.3	11.7	12.1	12.5	12.5	12.7	9.6	10.0	10.3	10.6	10.6	10.8
	8.0	12.9	13.2	13.6	14.0	14.0	14.2	10.9	11.3	11.6	11.9	11.9	12.1
3 in.	2.5	5.7	6.1	6.6	7.1	7.1	7.3	4.8	5.2	5.6	6.0	6.0	6.2
	4.0	10.2	10.8	11.3	11.9	11.9	12.2	8.7	9.2	9.6	10.1	10.1	10.4
	5.0	12.5	13.1	13.6	14.2	14.2	14.5	10.6	11.1	11.6	12.1	12.1	12.3
	6.0	14.8	15.3	15.9	16.5	16.5	16.7	12.5	13.0	13.5	14.0	14.0	14.2
	7.0	17.0	17.6	18.2	18.7	18.7	19.0	14.5	15.0	15.4	15.9	15.9	16.2
	8.0	19.3	19.9	20.4	21.0	21.0	21.3	16.4	16.9	17.4	17.8	17.8	18.1
4 in.	2.5	5.7	6.1	6.6	7.1	7.1	7.3	4.8	5.2	5.6	6.0	6.0	6.2
	4.0	13.6	14.4	15.1	15.9	15.9	16.3	11.6	12.2	12.9	13.5	13.5	13.8
	5.0	16.6	17.4	18.2	18.9	18.9	19.3	14.1	14.8	15.4	16.1	16.1	16.4
	6.0	19.7	20.4	21.2	21.9	21.9	22.3	16.7	17.4	18.0	18.7	18.7	19.0
	7.0	22.7	23.5	24.2	25.0	25.0	25.3	19.3	19.9	20.6	21.2	21.2	21.5
	8.0	25.7	26.5	27.2	28.0	28.0	28.4	21.9	22.5	23.2	23.8	23.8	24.1
5 in.	2.5	5.7	6.1	6.6	7.1	7.1	7.3	4.8	5.2	5.6	6.0	6.0	6.2
	4.0	13.6	14.4	15.1	15.9	15.9	16.3	11.6	12.2	12.9	13.5	13.5	13.8
	5.0	20.8	21.8	22.7	23.6	23.6	24.1	17.7	18.5	19.3	20.1	20.1	20.5
	6.0	24.6	25.5	26.5	27.4	27.4	27.9	20.9	21.7	22.5	23.3	23.3	23.7
	7.0	28.4	29.3	30.3	31.2	31.2	31.7	24.1	24.9	25.7	26.5	26.5	26.9
	8.0	32.2	33.1	34.0	35.0	35.0	35.5	27.3	28.1	28.9	29.7	29.7	30.1
6 in.	2.5	5.7	6.1	6.6	7.1	7.1	7.3	4.8	5.2	5.6	6.0	6.0	6.2
	4.0	13.6	14.4	15.1	15.9	15.9	16.3	11.6	12.2	12.9	13.5	13.5	13.8
	5.0	20.8	21.8	22.7	23.6	23.6	24.1	17.7	18.5	19.3	20.1	20.1	20.5
	6.0	29.5	30.6	31.8	32.9	32.9	33.5	25.1	26.0	27.0	28.0	28.0	28.5
	7.0	34.0	35.2	36.3	37.5	37.5	38.0	28.9	29.9	30.9	31.8	31.8	32.3
	8.0	38.6	39.7	40.9	42.0	42.0	42.6	32.8	33.8	34.7	35.7	35.7	36.2
7 in.	2.5	5.7	6.1	6.6	7.1	7.1	7.3	4.8	5.2	5.6	6.0	6.0	6.2
	4.0	13.6	14.4	15.1	15.9	15.9	16.3	11.6	12.2	12.9	13.5	13.5	13.8
	5.0	20.8	21.8	22.7	23.6	23.6	24.1	17.7	18.5	19.3	20.1	20.1	20.5
	6.0	29.5	30.6	31.8	32.9	32.9	33.5	25.1	26.0	27.0	28.0	28.0	28.5
	7.0	39.7	41.0	42.4	43.7	43.7	44.4	33.8	34.9	36.0	37.1	37.1	37.7
	8.0	45.0	46.3	47.7	49.0	49.0	49.7	38.3	39.4	40.5	41.6	41.6	42.2
8 in.	2.5	5.7	6.1	6.6	7.1	7.1	7.3	4.8	5.2	5.6	6.0	6.0	6.2
	4.0	13.6	14.4	15.1	15.9	15.9	16.3	11.6	12.2	12.9	13.5	13.5	13.8
	5.0	20.8	21.8	22.7	23.6	23.6	24.1	17.7	18.5	19.3	20.1	20.1	20.5
	6.0	29.5	30.6	31.8	32.9	32.9	33.5	25.1	26.0	27.0	28.0	28.0	28.5
	7.0	39.7	41.0	42.4	43.7	43.7	44.4	33.8	34.9	36.0	37.1	37.1	37.7
	8.0	51.4	53.0	54.5	56.0	56.0	56.7	43.7	45.0	46.3	47.6	47.6	48.2

Maximum Design Tensile Strength, P_s of Studs, Limited by Steel Strength (kips)[3]	Diameter	¼	⅜	½	⅝	¾	⅞
	P_s	2.7	6.0	10.6	16.6	23.9	32.5

(1) $f'_c = 5000$ psi, for other strengths multiply by $\sqrt{f'_c/5000}$
(2) For stud groups, also check Tables 6.20.8 - 6.20.12
(3) See Table 6.20.13 for bolts.

CONNECTIONS

Table 6.20.7 Shear strength of welded headed studs and bolts

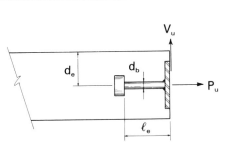

Table 6.20.7 Shear strength of welded headed studs and bolts

		Maximum Design Shear Strength, ϕV_c, Limited by Concrete Strength (kips)											
		Normal weight concrete ($\lambda = 1.0$)						Sand-lightweight concrete ($\lambda = 0.85$)					
f'_c	Edge dist., d_e	Diameter, d_b, in.						Diameter, d_b, in.					
		1/4	3/8	1/2	5/8	3/4	7/8	1/4	3/8	1/2	5/8	3/4	7/8
4000 psi	2	1.4	1.4	1.4	1.4	1.4	1.4	1.1	1.1	1.1	1.1	1.1	1.1
	3	2.1	3.0	3.0	3.0	3.0	3.0	1.8	2.6	2.6	2.6	2.6	2.6
	4	2.1	4.7	5.4	5.4	5.4	5.4	1.8	4.0	4.6	4.6	4.6	4.6
	5	2.1	4.7	8.4	8.4	8.4	8.4	1.8	4.0	7.2	7.2	7.2	7.2
	6	2.1	4.7	8.4	12.2	12.2	12.2	1.8	4.0	7.2	10.3	10.3	10.3
	7	2.1	4.7	8.4	13.2	16.5	16.5	1.8	4.0	7.2	11.2	14.1	14.1
	8	2.1	4.7	8.4	13.2	19.0	21.6	1.8	4.0	7.2	11.2	16.1	18.4
	9	2.1	4.7	8.4	13.2	19.0	25.8	1.8	4.0	7.2	11.2	16.1	22.0
	or more												
5000 psi	2	1.5	1.5	1.5	1.5	1.5	1.5	1.3	1.3	1.3	1.3	1.3	1.3
	3	2.4	3.4	3.4	3.4	3.4	3.4	2.0	2.9	2.9	2.9	2.9	2.9
	4	2.4	5.3	6.0	6.0	6.0	6.0	2.0	4.5	5.1	5.1	5.1	5.1
	5	2.4	5.3	9.4	9.4	9.4	9.4	2.0	4.5	8.0	8.0	8.0	8.0
	6	2.4	5.3	9.4	13.6	13.6	13.6	2.0	4.5	8.0	11.6	11.6	11.6
	7	2.4	5.3	9.4	14.7	18.5	18.5	2.0	4.5	8.0	12.5	15.7	15.7
	8	2.4	5.3	9.4	14.7	21.2	24.2	2.0	4.5	8.0	12.5	18.0	20.5
	9	2.4	5.3	9.4	14.7	21.2	28.9	2.0	4.5	8.0	12.5	18.0	24.6
	or more												
6000 psi	2	1.7	1.7	1.7	1.7	1.7	1.7	1.4	1.4	1.4	1.4	1.4	1.4
	3	2.6	3.7	3.7	3.7	3.7	3.7	2.2	3.2	3.2	3.2	3.2	3.2
	4	2.6	5.8	6.6	6.6	6.6	6.6	2.2	4.9	5.6	5.6	5.6	5.6
	5	2.6	5.8	10.3	10.3	10.3	10.3	2.2	4.9	8.8	8.8	8.8	8.8
	6	2.6	5.8	10.3	14.9	14.9	14.9	2.2	4.9	8.8	12.7	12.7	12.7
	7	2.6	5.8	10.3	16.2	20.3	20.3	2.2	4.9	8.8	13.7	17.2	17.2
	8	2.6	5.8	10.3	16.2	23.3	26.5	2.2	4.9	8.8	13.7	19.8	22.5
	9	2.6	5.8	10.3	16.2	23.3	31.7	2.2	4.9	8.8	13.7	19.8	26.9
	or more												
Maximum Design Shear Strength, V_s, of Studs Limited by Steel Strength (kips)*		**Diameter**		1/4	3/8	1/2	5/8	3/4	7/8				
		V_s		2.2	5.0	8.8	13.8	19.9	27.1				

*See Table 6.20.13 for bolts

CONNECTIONS

Table 6.20.8 Concrete design strength for stud or insert groups — Case 1

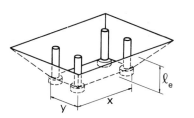

Case 1 — Not near a free edge

$$P_{c1} = 4\lambda\sqrt{f'_c}\,[xy + 2\ell_e(x + y) + 4\ell_e^2]$$

for thin sections, also check P_{c2} (Fig 6.5.3)

$\phi = 0.85$

$\lambda f'_c = 5000$ psi

for other values, multiply by $\lambda\sqrt{f'_c}/5000$

Maximum Tensile Strength, $\phi\,P_c$, of a Stud Group, kips

ℓ_e, in.	Dim. y, in.	2	4	6	8	10	12	14	16	18	20	22	24
	0	10	13	16	18	21	24	27	30	33	35	38	41
	2	14	18	22	26	30	34	38	42	46	49	53	57
	4	18	23	28	33	38	43	48	53	59	64	69	74
2.5	6	22	28	34	40	47	53	59	65	72	78	84	90
	8	26	33	40	48	55	62	70	77	85	92	99	107
	10	30	38	47	55	64	72	81	89	98	106	115	123
	12	34	43	53	62	72	82	91	101	111	120	130	139
	0	23	27	32	36	41	45	50	54	59	63	68	72
	2	28	34	40	45	51	57	62	68	74	79	85	90
	4	34	41	48	54	61	68	75	81	88	95	102	109
4	6	40	48	55	63	71	79	87	95	103	111	119	127
	8	45	54	63	72	81	90	100	109	118	127	136	145
	10	51	61	71	81	92	102	112	122	132	143	153	163
	12	57	68	79	90	102	113	124	136	147	158	170	181
	0	48	54	61	68	75	81	88	95	102	109	115	122
	2	55	63	71	79	87	95	103	111	119	127	135	143
	4	63	72	81	90	100	109	118	127	136	145	154	163
6	6	71	81	92	102	112	122	132	143	153	163	173	183
	8	79	90	102	113	124	136	147	158	170	181	192	204
	10	87	100	112	124	137	149	162	174	187	199	212	224
	12	95	109	122	136	149	163	176	190	204	217	231	244
	0	81	90	100	109	118	127	136	145	154	163	172	181
	2	92	102	112	122	132	143	153	163	173	183	193	204
	4	102	113	124	136	147	158	170	181	192	204	215	226
8	6	112	124	137	149	162	174	187	199	212	224	236	249
	8	122	136	149	163	176	190	204	217	231	244	258	271
	10	132	147	162	176	191	206	221	235	250	265	279	294
	12	143	158	174	190	206	222	238	253	269	285	301	317
	0	124	136	147	158	170	181	192	204	215	226	238	249
	2	137	149	162	174	187	199	212	224	236	249	261	274
	4	149	163	176	190	204	217	231	244	258	271	285	299
10	6	162	176	191	206	221	235	250	265	279	294	309	324
	8	174	190	206	222	238	253	269	285	301	317	333	348
	10	187	204	221	238	255	271	288	305	322	339	356	373
	12	199	217	235	253	271	290	308	326	344	362	380	398
	0	176	190	204	217	231	244	258	271	285	299	312	326
	2	191	206	221	235	250	265	279	294	309	324	338	353
	4	206	222	238	253	269	285	301	317	333	348	364	380
12	6	221	238	255	271	288	305	322	339	356	373	390	407
	8	235	253	271	290	308	326	344	362	380	398	416	434
	10	250	269	288	308	327	346	365	385	404	423	442	462
	12	265	285	305	326	346	367	387	407	428	448	468	489

CONNECTIONS

Table 6.20.9 Concrete design strength for stud or insert groups — Case 2

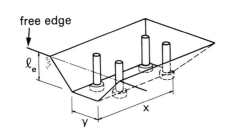

Case 2 — Near a free edge on one side

$$P_{c1} = 4\lambda\sqrt{f'_c}\,[xy + \ell_e(2x + y) + 2\ell_e^2]$$
for thin sections, also check P_{c2} (Fig 6.5.3)

$\phi = 0.85$

$\lambda\, f'_c = 5000$ psi
for other values, multiply by $\lambda\sqrt{f'_c}/5000$

Maximum Tensile Strength, $\phi\, P_{cr}$ of a Stud Group, kips

ℓ_e, in.	Dim. y, in.												
		2	**4**	**6**	**8**	**10**	**12**	**14**	**16**	**18**	**20**	**22**	**24**
2.5	0	6	9	12	15	18	21	23	26	29	32	35	37
	2	9	13	17	21	25	29	33	37	41	45	49	52
	4	11	17	22	27	32	37	42	47	52	57	62	67
	6	14	20	26	33	39	45	51	58	64	70	76	82
	8	17	24	31	39	46	53	61	68	75	83	90	97
	10	19	28	36	45	53	62	70	78	87	95	104	112
	12	22	31	41	50	60	70	79	89	99	108	118	127
4	0	14	18	23	27	32	36	41	45	50	54	59	63
	2	17	23	28	34	40	45	51	57	62	68	74	79
	4	20	27	34	41	48	54	61	68	75	81	88	95
	6	24	32	40	48	55	63	71	79	87	95	103	111
	8	27	36	45	54	63	72	81	90	100	109	118	127
	10	31	41	51	61	71	81	92	102	112	122	132	143
	12	34	45	57	68	79	90	102	113	124	136	147	158
6	0	27	34	41	48	54	61	68	75	81	88	95	102
	2	32	40	48	55	63	71	79	87	95	103	111	119
	4	36	45	54	63	72	81	90	100	109	118	127	136
	6	41	51	61	71	81	92	102	112	122	132	143	153
	8	45	57	68	79	90	102	113	124	136	147	158	170
	10	50	62	75	87	100	112	124	137	149	162	174	187
	12	54	68	81	95	109	122	136	149	163	176	190	204
8	0	45	54	63	72	81	90	100	109	118	127	136	145
	2	51	61	71	81	92	102	112	122	132	143	153	163
	4	57	68	79	90	102	113	124	136	147	158	170	181
	6	62	75	87	100	112	124	137	149	162	174	187	199
	8	68	81	95	109	122	136	149	163	176	190	204	217
	10	74	88	103	118	132	147	162	176	191	206	221	235
	12	79	95	111	127	143	158	174	190	206	222	238	253
10	0	68	79	90	102	113	124	136	147	158	170	181	192
	2	75	87	100	112	124	137	149	162	174	187	199	212
	4	81	95	109	122	136	149	163	176	190	204	217	231
	6	88	103	118	132	147	162	176	191	206	221	235	250
	8	95	111	127	143	158	174	190	206	222	238	253	269
	10	102	119	136	153	170	187	204	221	238	255	271	288
	12	109	127	145	163	181	199	217	235	253	271	290	308
12	0	95	109	122	136	149	163	176	190	204	217	231	244
	2	103	118	132	147	162	176	191	206	221	235	250	265
	4	111	127	143	158	174	190	206	222	238	253	269	285
	6	119	136	153	170	187	204	221	238	255	271	288	305
	8	127	145	163	181	199	217	235	253	271	290	308	326
	10	135	154	173	192	212	231	250	269	288	308	327	346
	12	143	163	183	204	224	244	265	285	305	326	346	367

Table 6.20.10 Concrete design strength for stud or insert groups — Case 3

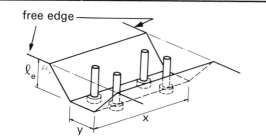

Case 3 — Near a free edge on 2 opposite sides

$$P_{c1} = 4\lambda\sqrt{f'_c}\,(xy + 2\ell_e x)$$

for thin sections, also check P_{c2} (Fig 6.5.3)

$\phi = 0.85$

$\lambda f'_c = 5000$ psi

for other values, multiply by $\lambda\sqrt{f'_c}/5000$

Maximum Tensile Strength, ϕP_{cr} of a Stud Group, kips

ℓ_e, in.	Dim. y, in.	Dimension x, in.											
		2	**4**	**6**	**8**	**10**	**12**	**14**	**16**	**18**	**20**	**22**	**24**
2.5	0	3	6	8	11	14	17	20	23	25	28	31	34
	2	4	8	12	16	20	24	28	32	36	40	44	48
	4	5	10	15	20	25	31	36	41	46	51	56	61
	6	6	12	19	25	31	37	44	50	56	62	68	75
	8	7	15	22	29	37	44	51	59	66	74	81	88
	10	8	17	25	34	42	51	59	68	76	85	93	102
	12	10	19	29	38	48	58	67	77	87	96	106	115
4	0	5	9	14	18	23	27	32	36	41	45	50	54
	2	6	11	17	23	28	34	40	45	51	57	62	68
	4	7	14	20	27	34	41	48	54	61	68	75	81
	6	8	16	24	32	40	48	55	63	71	79	87	95
	8	9	18	27	36	45	54	63	72	81	90	100	109
	10	10	20	31	41	51	61	71	81	92	102	112	122
	12	11	23	34	45	57	68	79	90	102	113	124	136
6	0	7	14	20	27	34	41	48	54	61	68	75	81
	2	8	16	24	32	40	48	55	63	71	79	87	95
	4	9	18	27	36	45	54	63	72	81	90	100	109
	6	10	20	31	41	51	61	71	81	92	102	112	122
	8	11	23	34	45	57	68	79	90	102	113	124	136
	10	12	25	37	50	62	75	87	100	112	124	137	149
	12	14	27	41	54	68	81	95	109	122	136	149	163
8	0	9	18	27	36	45	54	63	72	81	90	100	109
	2	10	20	31	41	51	61	71	81	92	102	112	122
	4	11	23	34	45	57	68	79	90	102	113	124	136
	6	12	25	37	50	62	75	87	100	112	124	137	149
	8	14	27	41	54	68	81	95	109	122	136	149	163
	10	15	29	44	59	74	88	103	118	132	147	162	176
	12	16	32	48	63	79	95	111	127	143	158	174	190
10	0	11	23	34	45	57	68	79	90	102	113	124	136
	2	12	25	37	50	62	75	87	100	112	124	137	149
	4	14	27	41	54	68	81	95	109	122	136	149	163
	6	15	29	44	59	74	88	103	118	132	147	162	176
	8	16	32	48	63	79	95	111	127	143	158	174	190
	10	17	34	51	68	85	102	119	136	153	170	187	204
	12	18	36	54	72	90	109	127	145	163	181	199	217
12	0	14	27	41	54	68	81	95	109	122	136	149	163
	2	15	29	44	59	74	88	103	118	132	147	162	176
	4	16	32	48	63	79	95	111	127	143	158	174	190
	6	17	34	51	68	85	102	119	136	153	170	187	204
	8	18	36	54	72	90	109	127	145	163	181	199	217
	10	19	38	58	77	96	115	135	154	173	192	212	231
	12	20	41	61	81	102	122	143	163	183	204	224	244

CONNECTIONS

Table 6.20.11 Concrete design strength for stud or insert groups — Case 4

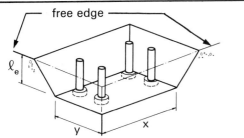

free edge

ℓ_e

y

x

Case 4 — Near a free edge on 2 adjacent sides

$$P_{c1} = 4\lambda\sqrt{f'_c}\,[xy + \ell_e(x + y) + \ell_e^2]$$

for thin sections, also check P_{c2} (Fig 6.5.3)

$\phi = 0.85$

$\lambda f'_c = 5000$ psi

for other values, multiply by $\lambda\sqrt{f'_c}/5000$

Maximum Tensile Strength, $\phi\,P_c$, of a Stud Group, kips

ℓ_e, in.	Dim. y, in.	Dimension x, in.											
		2	4	6	8	10	12	14	16	18	20	22	24
2.5	0	3	5	6	7	9	10	12	13	14	16	17	19
	2	6	8	11	13	16	18	21	24	26	29	31	34
	4	8	12	16	19	23	27	30	34	38	41	45	49
	6	11	16	20	25	30	35	40	44	49	54	59	64
	8	13	19	25	31	37	43	49	55	61	67	73	79
	10	16	23	30	37	44	51	58	65	72	80	87	94
	12	18	27	35	43	51	59	68	76	84	92	100	109
4	0	7	9	11	14	16	18	20	23	25	27	29	32
	2	10	14	17	20	24	27	31	34	37	41	44	48
	4	14	18	23	27	32	36	41	45	50	54	59	63
	6	17	23	28	34	40	45	51	57	62	68	74	79
	8	20	27	34	41	48	54	61	68	75	81	88	95
	10	24	32	40	48	55	63	71	79	87	95	103	111
	12	27	36	45	54	63	72	81	90	100	109	118	127
6	0	14	17	20	24	27	31	34	37	41	44	48	51
	2	18	23	27	32	36	41	45	50	54	59	63	68
	4	23	28	34	40	45	51	57	62	68	74	79	85
	6	27	34	41	48	54	61	68	75	81	88	95	102
	8	32	40	48	55	63	71	79	87	95	103	111	119
	10	36	45	54	63	72	81	90	100	109	118	127	136
	12	41	51	61	71	81	92	102	112	122	132	143	153
8	0	23	27	32	36	41	45	50	54	59	63	68	72
	2	28	34	40	45	51	57	62	68	74	79	85	90
	4	34	41	48	54	61	68	75	81	88	95	102	109
	6	40	48	55	63	71	79	87	95	103	111	119	127
	8	45	54	63	72	81	90	100	109	118	127	136	145
	10	51	61	71	81	92	102	112	122	132	143	153	163
	12	57	68	79	90	102	113	124	136	147	158	170	181
10	0	34	40	45	51	57	62	68	74	79	85	90	96
	2	41	48	54	61	68	75	81	88	95	102	109	115
	4	48	55	63	71	79	87	95	103	111	119	127	135
	6	54	63	72	81	90	100	109	118	127	136	145	154
	8	61	71	81	92	102	112	122	132	143	153	163	173
	10	68	79	90	102	113	124	136	147	158	170	181	192
	12	75	87	100	112	124	137	149	162	174	187	199	212
12	0	48	54	61	68	75	81	88	95	102	109	115	122
	2	55	63	71	79	87	95	103	111	119	127	135	143
	4	63	72	81	90	100	109	118	127	136	145	154	163
	6	71	81	92	102	112	122	132	143	153	163	173	183
	8	79	90	102	113	124	136	147	158	170	181	192	204
	10	87	100	112	124	137	149	162	174	187	199	212	224
	12	95	109	122	136	149	163	176	190	204	217	231	244

CONNECTIONS

Table 6.20.12 Concrete design strength for stud or insert groups — Case 5

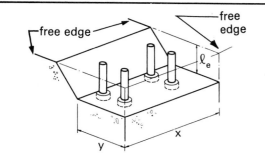

Case 5 — Near a free edge on 3 sides

$P_{c1} = 4\lambda\sqrt{f'_c}\,(xy + \ell_e x)$
  for thin sections, also check P_{c2} (Fig 6.5.3)
$\phi = 0.85$
$\lambda\,f'_c = 5000$ psi
  for other values, multiply by $\lambda\sqrt{f'_c}/5000$

Maximum Tensile Strength, $\phi\,P_c$, of a Stud Group, kips

ℓ_e, in.	Dim. y, in.	Dimension x, in. 2	4	6	8	10	12	14	16	18	20	22	24
	0	1	3	4	6	7	8	10	11	13	14	16	17
	2	3	5	8	10	13	15	18	20	23	25	28	31
	4	4	7	11	15	18	22	26	29	33	37	40	44
2.5	6	5	10	14	19	24	29	34	38	43	48	53	58
	8	6	12	18	24	30	36	42	48	53	59	65	71
	10	7	14	21	28	35	42	49	57	64	71	78	85
	12	8	16	25	33	41	49	57	66	74	82	90	98
	0	2	5	7	9	11	14	16	18	20	23	25	27
	2	3	7	10	14	17	20	24	27	31	34	37	41
	4	5	9	14	18	23	27	32	36	41	45	50	54
4	6	6	11	17	23	28	34	40	45	51	57	62	68
	8	7	14	20	27	34	41	48	54	61	68	75	81
	10	8	16	24	32	40	48	55	63	71	79	87	95
	12	9	18	27	36	45	54	63	72	81	90	100	109
	0	3	7	10	14	17	20	24	27	31	34	37	41
	2	5	9	14	18	23	27	32	36	41	45	50	54
	4	6	11	17	23	28	34	40	45	51	57	62	68
6	6	7	14	20	27	34	41	48	54	61	68	75	81
	8	8	16	24	32	40	48	55	63	71	79	87	95
	10	9	18	27	36	45	54	63	72	81	90	100	109
	12	10	20	31	41	51	61	71	81	92	102	112	122
	0	5	9	14	18	23	27	32	36	41	45	50	54
	2	6	11	17	23	28	34	40	45	51	57	62	68
	4	7	14	20	27	34	41	48	54	61	68	75	81
8	6	8	16	24	32	40	48	55	63	71	79	87	95
	8	9	18	27	36	45	54	63	72	81	90	100	109
	10	10	20	31	41	51	61	71	81	92	102	112	122
	12	11	23	34	45	57	68	79	90	102	113	124	136
	0	6	11	17	23	28	34	40	45	51	57	62	68
	2	7	14	20	27	34	41	48	54	61	68	75	81
	4	8	16	24	32	40	48	55	63	71	79	87	95
10	6	9	18	27	36	45	54	63	72	81	90	100	109
	8	10	20	31	41	51	61	71	81	92	102	112	122
	10	11	23	34	45	57	68	79	90	102	113	124	136
	12	12	25	37	50	62	75	87	100	112	124	137	149
	0	7	14	20	27	34	41	48	54	61	68	75	81
	2	8	16	24	32	40	48	55	63	71	79	87	95
	4	9	18	27	36	45	54	63	72	81	90	100	109
12	6	10	20	31	41	51	61	71	81	92	102	112	122
	8	11	23	34	45	57	68	79	90	102	113	124	136
	10	12	25	37	50	62	75	87	100	112	124	137	149
	12	14	27	41	54	68	81	95	109	122	136	149	163

CONNECTIONS

Table 6.20.13 Allowable loads on bolts

Bolt diameter (in.)	Nominal area (sq in.)	Tensile stress area (sq in.)	Threaded A-36 rods				ASTM A-307 Bolts			
			Tension		Shear		Tension		Shear	
			Design strength (kips)	Service load (kips)	Design strength (kips)	Service load (kips)	Design strength (kips)	Service load (kips)	Design strength (kips)	Service load (kips)
1/2	0.196	0.142	5.11	3.12	3.88	2.12	4.69	2.84	3.56	1.96
5/8	0.307	0.226	8.14	4.97	6.08	3.32	7.46	4.52	5.57	3.07
3/4	0.442	0.334	12.02	7.36	8.75	4.77	11.02	6.69	8.02	4.42
7/8	0.601	0.462	16.63	10.16	11.90	6.49	15.25	9.23	10.91	6.01
1	0.785	0.606	21.82	13.33	15.54	8.48	20.00	12.11	14.25	7.85
1 1/4	1.227	0.969	34.88	21.32	24.29	13.25	31.98	19.38	22.27	12.27

Table 6.20.14 Capacity of coil bolts and threaded coil rods

Bolt diameter (in.)	Minimum Coil Penetration (in.)	Tensile Strength (P_s)	Shear Strength (V_s)
1/2	1 1/2	13,500	8,100
3/4	2	18,470	11,080
1	2 1/2	37,870	22,720
1 1/4	2 1/2	54,960	32,980
1 1/2	3	83,340	50,000

Table 6.20.15 Capacity of round wire used in concrete inserts

Leg Wire diameter (in.)	Wire Grade	Yield Strength (lb)
0.218	C1008	2000
0.223	C1038	3900
0.225	C1038	3700
0.240	C1008	2900
0.260	C1008	3550
0.281	C1035	6000
0.306	C1035	6900
0.340	C1035	7500
0.375	C1008	7450
0.440	C1035	12,000

Table 6.20.16 Plastic section moduli and shape factors

Section	Plastic section modulus, Z_s, in.^{3}	Shape factor
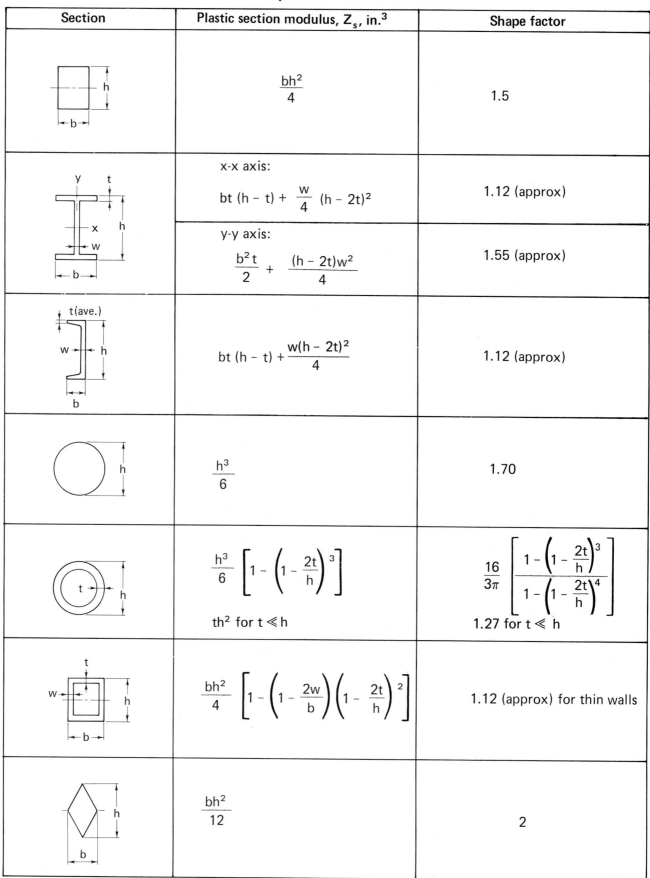	$\dfrac{bh^2}{4}$	1.5
	x-x axis: $bt(h-t) + \dfrac{w}{4}(h-2t)^2$	1.12 (approx)
	y-y axis: $\dfrac{b^2t}{2} + \dfrac{(h-2t)w^2}{4}$	1.55 (approx)
	$bt(h-t) + \dfrac{w(h-2t)^2}{4}$	1.12 (approx)
	$\dfrac{h^3}{6}$	1.70
	$\dfrac{h^3}{6}\left[1-\left(1-\dfrac{2t}{h}\right)^3\right]$ th^2 for $t \ll h$	$\dfrac{16}{3\pi}\left[\dfrac{1-\left(1-\dfrac{2t}{h}\right)^3}{1-\left(1-\dfrac{2t}{h}\right)^4}\right]$ 1.27 for $t \ll h$
	$\dfrac{bh^2}{4}\left[1-\left(1-\dfrac{2w}{b}\right)\left(1-\dfrac{2t}{h}\right)^2\right]$	1.12 (approx) for thin walls
	$\dfrac{bh^2}{12}$	2

CONNECTIONS

Table 6.20.17 Shear strength of connection angles

$$t = \sqrt{\frac{4 V_u e_v}{\phi f_y b}} \text{ , in.}$$

ϕ = 0.90

b = width of angle, in.

f_y = yield strength of angle steel = 36,000 psi

ϕV_n, lb. per inch of width

Angle thickness t	$e_v = 3/4''$	$e_v = 1''$	$e_v = 1\text{-}1/2''$	$e_v = 2''$	$e_v = 2\text{-}1/2''$
5/16"	1055	791	527	396	316
3/8"	1519	1139	759	570	456
7/16"	2067	1550	1034	775	620
1/2"	2700	2025	1350	1013	810
9/16"	3417	2563	1709	1281	1025
5/8"	4219	3164	2109	1582	1266

Table 6.20.18 Axial strength of connection angles

$$t = \sqrt{\frac{4 N_u g}{\phi f_y b}}$$

ϕ = 0.90

b = width of angle, in.

f_y = yield strength of angle steel = 36,000 psi

ϕN_n, lb. per inch of width

Angle thickness t	$l_l = 5''$ g = 3''	$l_l = 6''$ g = 4''	$l_l = 7''$ g = 5''	$l_l = 8''$ g = 6''
5/16"	264	198		
3/8"	380	285	228	
7/16"	517	388	310	258
1/2"	675	506	405	338
9/16"	854	641	513	427
5/8"	1055	791	633	527

CONNECTIONS

Table 6.20.19 Column base plate thickness requirements

Thickness Required For Concrete Bearing

f_{bu} (psi)	$x_o = 3''$	$x_o = 4''$	$x_o = 5''$
500	5/8	3/4	1
1000	3/4	1	1 3/8
1500	1	1 3/8	1 5/8
2000	1 1/8	1 1/2	1 7/8
2500	1 1/4	1 5/8	2
3000	1 3/8	1 7/8	2 1/4
3500	1 1/2	2	2 1/2
4000	1 5/8	2 1/8	2 5/8

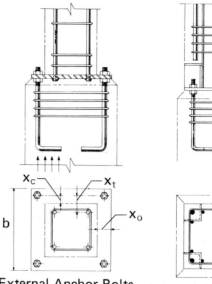

External Anchor Bolts   Internal Anchor Bolts

Thickness Required for Bolt Loading

Tension On External Anchor Bolts

b (in.)	No. & Diameter Of A 36 Or A 307 Anchor Bolts Per Side							
	$2 - \frac{3}{4}''$ $x_t = 3.75''$	$2 - \frac{3}{4}''$ $x_t = 4.25''$	$2 - 1''$ $x_t = 3.75''$	$2 - 1''$ $x_t = 4.25''$	$2 - 1\frac{1}{4}''$ $x_t = 3.75''$	$2 - 1\frac{1}{4}''$ $x_t = 4.25''$	$2 - 1\frac{1}{2}''$ $x_t = 3.75''$	$2 - 1\frac{1}{2}''$ $x_t = 4.25''$
	1	1 1/8	1 3/8	1 1/2	1 3/4	1 7/8	2 1/8	2 1/4
	1	1	1 3/8	1 3/8	1 5/8	1 3/4	2	2 1/8
	7/8	1	1 1/4	1 3/8	1 1/2	1 5/8	1 7/8	2
	7/8	1	1 1/8	1 1/4	1 1/2	1 1/2	1 3/4	1 7/8
	7/8	7/8	1 1/8	1 1/8	1 3/8	1 1/2	1 5/8	1 3/4
	3/4	7/8	1	1 1/8	1 3/8	1 3/8	1 5/8	1 5/8
	3/4	3/4	1	1 1/8	1 1/4	1 3/8	1 1/2	1 5/8
	3/4	3/4	1	1	1 1/4	1 1/4	1 1/2	1 1/2
	3/4	3/4	1	1	1 1/8	1 1/4	1 3/8	1 1/2

Compression On Anchor Bolts Or Tension On Internal Anchor Bolts

b (in.)	No. & Diameter Of A 36 Or A 307 Anchor Bolts Per Side							
	$2 - \frac{3}{4}''$ $x_c = 1.5''$	$2 - \frac{3}{4}''$ $x_c = 2.0''$	$2 - 1''$ $x_c = 1.5''$	$2 - 1''$ $x_c = 2.0''$	$2 - 1\frac{1}{4}''$ $x_c = 1.5''$	$2 - 1\frac{1}{4}''$ $x_c = 2.0''$	$2 - 1\frac{1}{2}''$ $x_c = 1.5''$	$2 - 1\frac{1}{2}''$ $x_c = 2.0''$
12	3/4	3/4	7/8	1	1 1/4	1 3/8	1 3/8	1 5/8
14	3/4	3/4	7/8	1	1	1 1/4	1 1/4	1 1/2
16	3/4	3/4	3/4	7/8	1	1 1/8	1 1/4	1 3/8
18	3/4	3/4	3/4	7/8	1	1 1/8	1 1/8	1 1/4
20	3/4	3/4	3/4	7/8	7/8	1	1 1/8	1 1/4
22	3/4	3/4	3/4	3/4	7/8	1	1	1 1/8
24	3/4	3/4	3/4	3/4	7/8	1	1	1 1/8
26	3/4	3/4	3/4	3/4	3/4	7/8	1	1 1/8
28	3/4	3/4	3/4	3/4	3/4	7/8	7/8	1

CONNECTIONS

Table 6.20.20 Design strength of concrete brackets, corbels, or haunches

Design strength by Eqs. 6.11.3
or 6.11.4 for following criteria:

f_y = 60,000 psi

N_u = 0.2 V_u

b = width of bearing

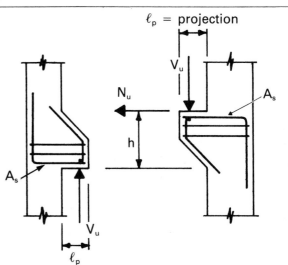

Values of ϕV_n, kips

b = 6"

A_s / h	4" Projection								6" Projection								8" Projection							
	4	6	8	10	12	14	16	18	6	8	10	12	14	16	18	20	8	10	12	14	16	18	20	22
2-#4	14	23	29	35	40	44	47	48	17	22	27	31	35	38	41	0	18	22	26	29	32	35	0	0
2-#5	23	35	46	51	55	59	62	65	26	35	42	49	55	60	65	67	28	34	40	45	50	55	59	62
2-#6	33	51	58	64	69	73	77	81	38	50	61	69	73	77	81	84	40	49	58	65	72	79	84	87
2-#7	44	61	69	76	82	88	93	97	51	68	76	82	88	93	97	102	54	67	78	88	93	97	102	106
2-#8	58	71	80	89	96	103	109	114	67	80	89	96	103	109	114	119	71	88	96	103	109	114	119	124
2-#9	67	80	92	101	110	117	124	131	80	92	101	110	117	124	131	137	90	101	110	117	124	131	137	142

b = 8"

A_s / h	4" Projection								6" Projection								8" Projection							
	4	6	8	10	12	14	16	18	6	8	10	12	14	16	18	20	8	10	12	14	16	18	20	22
2-#4	14	23	29	35	40	44	0	0	17	22	27	31	35	0	0	0	18	22	26	29	0	0	0	0
2-#5	23	35	46	55	62	65	69	72	26	35	42	49	55	60	65	69	28	34	40	45	50	55	59	62
2-#6	33	51	66	72	77	82	86	90	38	50	61	70	79	86	90	94	40	49	58	65	72	79	84	90
2-#7	44	69	78	86	93	99	104	109	51	68	83	93	99	104	109	114	54	67	78	89	98	107	114	118
2-#8	58	80	91	100	109	116	122	128	67	89	100	109	116	122	128	134	71	88	103	116	122	128	134	139
2-#9	73	92	104	115	124	133	140	147	84	104	115	124	133	140	147	154	90	111	124	133	140	147	154	160

b = 10"

A_s / h	4" Projection								6" Projection								8" Projection							
	4	6	8	10	12	14	16	18	6	8	10	12	14	16	18	20	8	10	12	14	16	18	20	22
2-#4	14	23	29	35	0	0	0	0	17	22	27	0	0	0	0	0	18	22	0	0	0	0	0	0
2-#5	23	35	46	55	62	69	74	77	26	35	42	49	55	60	65	0	28	34	40	45	50	55	0	0
2-#6	33	51	66	79	84	89	94	98	38	50	61	70	79	86	93	99	40	49	58	65	72	79	84	90
2-#7	44	69	86	94	102	108	114	119	51	68	83	96	107	114	119	124	54	67	78	89	98	107	115	122
2-#8	58	89	100	110	119	127	134	140	67	89	108	119	127	134	140	146	71	88	103	116	128	140	146	152
2-#9	73	101	115	126	137	146	154	162	84	112	126	137	146	154	162	169	90	111	130	146	154	162	169	175

b = 12"

A_s / h	4" Projection								6" Projection								8" Projection							
	4	6	8	10	12	14	16	18	6	8	10	12	14	16	18	20	8	10	12	14	16	18	20	22
2-#4	14	23	29	0	0	0	0	0	17	22	0	0	0	0	0	0	18	0	0	0	0	0	0	0
2-#5	23	35	46	55	62	69	0	0	26	35	42	49	55	0	0	0	28	34	40	45	0	0	0	0
2-#6	33	51	66	79	90	96	100	105	38	50	61	70	79	86	93	99	40	49	58	65	72	79	84	90
2-#7	44	69	90	102	109	116	122	128	51	68	83	96	107	117	127	133	54	67	78	89	98	107	115	122
2-#8	58	91	109	119	128	137	144	151	67	89	108	125	137	144	151	157	71	88	103	116	128	140	150	160
2-#9	73	110	124	137	147	157	166	174	84	112	136	147	157	166	174	181	90	111	130	147	163	174	181	188
3-#4	22	34	44	53	60	66	0	0	25	33	40	47	52	0	0	0	27	33	38	44	0	0	0	0
3-#5	34	53	69	82	93	98	103	107	39	52	63	73	82	90	97	104	42	51	60	68	75	82	88	94
3-#6	49	76	98	108	116	123	130	136	56	75	91	105	118	129	136	141	60	74	87	98	108	118	127	135
3-#7	67	104	118	129	139	148	156	164	77	102	124	139	148	156	164	171	82	101	118	133	148	161	171	177
3-#8	87	121	137	151	163	174	183	193	100	133	151	163	174	183	193	201	107	131	154	174	183	193	201	209
3-#9	110	137	156	172	186	199	211	221	127	156	172	186	199	211	221	231	135	166	186	199	211	221	231	240

Table 6.20.20 (continued) Design strength of concrete brackets, corbels, or haunches

b	A_s	10″ Projection								12″ Projection								14″ Projection							
	h →	10	12	14	16	18	20	22	24	12	14	16	18	20	22	24	26	14	16	18	20	22	24	26	28
b = 6″	2–#4	18	22	25	28	30	0	0	0	19	22	24	27	0	0	0	0	19	22	24	0	0	0	0	0
	2–#5	29	34	39	43	47	51	55	58	30	34	38	42	45	48	52	55	30	34	37	40	44	47	49	52
	2–#6	42	49	56	62	68	73	79	83	42	49	54	60	65	70	74	79	43	49	54	58	63	67	71	75
	2–#7	56	67	76	85	93	100	106	109	58	66	74	82	88	95	101	107	59	66	73	79	85	91	97	102
	2–#8	74	87	99	109	114	119	124	128	76	87	97	106	116	124	128	133	77	86	95	104	112	119	126	133
	2–#9	93	110	117	124	131	137	142	147	96	110	123	131	137	142	147	152	97	109	120	131	141	147	152	157
b = 8″	2–#4	18	22	25	0	0	0	0	0	19	22	0	0	0	0	0	0	19	0	0	0	0	0	0	0
	2–#5	29	34	39	43	47	51	55	0	30	34	38	42	45	48	0	0	30	34	37	40	44	0	0	0
	2–#6	42	49	56	62	68	73	79	83	42	49	54	60	65	70	74	79	43	49	54	58	63	67	71	75
	2–#7	56	67	76	85	93	100	107	113	58	66	74	82	88	95	101	107	59	66	73	79	85	91	97	102
	2–#8	74	87	99	110	121	131	139	144	76	87	97	106	116	124	132	140	77	86	95	104	112	119	126	133
	2–#9	93	110	126	140	147	154	160	166	96	110	123	135	146	157	166	171	97	109	120	131	141	151	160	168
b = 10″	2–#4	18	0	0	0	0	0	0	0	0	0	0	0	0	0	0	0	0	0	0	0	0	0	0	0
	2–#5	29	34	39	43	47	0	0	0	30	34	38	42	0	0	0	0	30	34	37	0	0	0	0	0
	2–#6	42	49	56	62	68	73	79	83	42	49	54	60	65	70	74	79	43	49	54	58	63	67	71	0
	2–#7	56	67	76	85	93	100	107	113	58	66	74	82	88	95	101	107	59	66	73	79	85	91	97	102
	2–#8	74	87	99	110	121	131	140	148	76	87	97	106	116	124	132	140	77	86	95	104	112	119	126	133
	2–#9	93	110	126	140	153	165	175	181	96	110	123	135	146	157	167	177	97	109	120	131	141	151	160	168
b = 12″	2–#4	0	0	0	0	0	0	0	0	0	0	0	0	0	0	0	0	0	0	0	0	0	0	0	0
	2–#5	29	34	39	0	0	0	0	0	30	34	0	0	0	0	0	0	30	0	0	0	0	0	0	0
	2–#6	42	49	56	62	68	73	79	0	42	49	54	60	65	70	0	0	43	49	54	58	63	0	0	0
	2–#7	56	67	76	85	93	100	107	113	58	66	74	82	88	95	101	107	59	66	73	79	85	91	97	102
	2–#8	74	87	99	110	121	131	140	148	76	87	97	106	116	124	132	140	77	86	95	104	112	119	126	133
	2–#9	93	110	126	140	153	165	177	188	96	110	123	135	146	157	167	177	97	109	120	131	141	151	160	168
	3–#4	28	33	37	0	0	0	0	0	28	32	0	0	0	0	0	0	29	0	0	0	0	0	0	0
	3–#5	43	51	58	65	71	77	82	0	44	51	57	62	68	73	0	0	45	51	56	61	65	0	0	0
	3–#6	62	73	84	93	102	110	118	125	64	73	82	90	97	105	111	118	65	73	80	87	94	101	107	112
	3–#7	85	100	114	127	139	150	160	170	87	99	111	122	133	142	152	160	88	99	109	119	128	137	145	153
	3–#8	111	130	149	166	181	196	209	216	113	130	145	160	173	186	198	209	115	129	143	155	167	179	189	200
	3–#9	140	165	188	210	221	231	240	249	143	164	184	202	219	236	249	257	146	164	181	197	212	226	240	253

CONNECTIONS

Table 6.20.20 (continued) Design strength of concrete brackets, corbels, or haunches

b = 14"

	4″ Projection								6″ Projection								8″ Projection							
h → / A_s	4	6	8	10	12	14	16	18	6	8	10	12	14	16	18	20	8	10	12	14	16	18	20	22
2-#4	14	23	29	0	0	0	0	0	17	22	0	0	0	0	0	0	18	0	0	0	0	0	0	0
2-#5	23	35	46	55	62	0	0	0	26	35	42	49	0	0	0	0	28	34	40	0	0	0	0	0
2-#6	33	51	66	79	90	99	106	110	38	50	61	70	79	86	93	0	40	49	58	65	72	79	0	0
2-#7	44	69	90	107	116	123	129	135	51	68	83	96	107	117	127	135	54	67	78	89	98	107	115	122
2-#8	58	91	116	127	137	145	153	160	67	89	108	125	140	153	160	166	71	88	103	116	128	140	150	160
2-#9	73	115	133	146	157	167	176	185	84	112	136	157	167	176	185	192	90	111	130	147	163	177	190	199
3-#4	22	34	44	53	60	0	0	0	25	33	40	47	0	0	0	0	27	33	38	0	0	0	0	0
3-#5	34	53	69	82	93	103	109	113	39	52	63	73	82	90	97	0	42	51	60	68	75	82	0	0
3-#6	49	76	99	115	123	131	137	143	56	75	91	105	118	129	140	149	60	74	87	98	108	118	127	135
3-#7	67	104	125	138	148	158	166	174	77	102	124	143	158	166	174	181	82	101	118	133	148	161	173	184
3-#8	87	129	146	161	174	185	195	205	100	133	161	174	185	195	205	213	107	131	154	174	193	205	213	222
3-#9	110	147	167	184	199	212	225	236	127	167	184	199	212	225	236	246	135	166	195	212	225	236	246	256

b = 16"

	4″ Projection								6″ Projection								8″ Projection							
h → / A_s	4	6	8	10	12	14	16	18	6	8	10	12	14	16	18	20	8	10	12	14	16	18	20	22
2-#6	33	51	66	79	90	99	107	0	38	50	61	70	79	86	0	0	40	49	58	65	72	0	0	0
2-#7	44	69	90	107	122	129	136	141	51	68	83	96	107	117	127	135	54	67	78	89	98	107	115	122
2-#8	58	91	118	134	144	153	161	168	67	89	108	125	140	153	166	174	71	88	103	116	128	140	150	160
2-#9	73	115	140	154	166	176	186	194	84	112	136	158	176	186	194	202	90	111	130	147	163	177	190	202
3-#6	49	76	99	118	130	137	144	151	56	75	91	105	118	129	140	149	60	74	87	98	108	118	127	135
3-#7	67	104	133	145	156	166	175	183	77	102	124	143	161	175	183	190	82	101	118	133	148	161	173	184
3-#8	87	136	155	170	183	195	206	216	100	133	162	183	195	206	216	225	107	131	154	174	193	210	225	233
3-#9	110	156	177	195	211	225	237	249	127	168	195	211	225	237	249	260	135	166	195	220	237	249	260	269
4-#6	65	102	131	144	155	164	173	181	75	100	121	140	157	173	181	188	80	99	115	131	144	157	169	180
4-#7	89	138	157	172	186	198	209	219	102	136	165	186	198	209	219	228	109	134	157	178	197	214	228	236
4-#8	116	161	182	201	217	232	245	257	134	177	201	217	232	245	257	268	142	175	205	232	245	257	268	278
4-#9	147	183	208	230	248	265	281	295	169	208	230	248	265	281	295	308	180	222	248	265	281	295	308	320

b = 18"

	4″ Projection								6″ Projection								8″ Projection							
h → / A_s	4	6	8	10	12	14	16	18	6	8	10	12	14	16	18	20	8	10	12	14	16	18	20	22
2-#6	33	51	66	79	90	99	0	0	38	50	61	70	79	0	0	0	40	49	58	65	0	0	0	0
2-#7	44	69	90	107	122	135	141	147	51	68	83	96	107	117	127	135	54	67	78	89	98	107	115	0
2-#8	58	91	118	140	151	160	168	175	67	89	108	125	140	153	166	177	71	88	103	116	128	140	150	160
2-#9	73	115	147	162	174	185	194	203	84	112	136	158	177	194	203	211	90	111	130	147	163	177	190	202
3-#6	49	76	99	118	135	143	151	157	56	75	91	105	118	129	140	149	60	74	87	98	108	118	127	135
3-#7	67	104	135	153	164	174	183	191	77	102	124	143	161	176	190	199	82	101	118	133	148	161	173	184
3-#8	87	136	163	179	193	205	216	226	100	133	162	187	205	216	226	235	107	131	154	174	193	210	225	240
3-#9	110	164	186	205	221	236	249	261	127	168	205	221	236	249	261	272	135	166	195	220	244	261	272	282
4-#6	65	102	132	151	162	172	181	189	75	100	121	140	157	173	186	196	80	99	115	131	144	157	169	180
4-#7	89	139	165	181	195	207	219	229	102	136	165	191	207	219	229	238	109	134	157	178	197	214	230	245
4-#8	116	169	192	211	228	243	257	269	134	177	211	228	243	257	269	281	142	175	205	232	257	269	281	291
4-#9	147	193	219	242	261	279	295	310	169	219	242	261	279	295	310	323	180	222	260	279	295	310	323	336

b = 20"

	4″ Projection								6″ Projection								8″ Projection							
h → / A_s	4	6	8	10	12	14	16	18	6	8	10	12	14	16	18	20	8	10	12	14	16	18	20	22
2-#6	33	51	66	79	90	0	0	0	38	50	61	70	0	0	0	0	40	49	58	0	0	0	0	0
2-#7	44	69	90	107	122	135	146	153	51	68	83	96	107	117	127	0	54	67	78	89	98	107	0	0
2-#8	58	91	118	140	157	166	174	182	67	89	108	125	140	153	166	177	71	88	103	116	128	140	150	160
2-#9	73	115	149	169	181	192	202	211	84	112	136	158	177	194	210	220	90	111	130	147	163	177	190	202
3-#6	49	76	99	118	135	149	156	163	56	75	91	105	118	129	140	0	60	74	87	98	108	118	0	0
3-#7	67	104	135	159	171	181	190	199	77	102	124	143	161	176	190	203	82	101	118	133	148	161	173	184
3-#8	87	136	170	187	201	213	225	235	100	133	162	187	210	225	235	245	107	131	154	174	193	210	225	240
3-#9	110	172	195	214	231	246	260	272	127	168	205	231	246	260	272	283	135	166	195	220	244	265	283	294
4-#6	65	102	132	157	169	179	188	196	75	100	121	140	157	173	186	199	80	99	115	131	144	157	169	180
4-#7	89	139	172	189	203	216	228	238	102	136	165	191	214	228	238	248	109	134	157	178	197	214	230	245
4-#8	116	177	201	221	238	254	268	281	134	177	216	238	254	268	281	292	142	175	205	232	257	280	292	303
4-#9	147	202	230	253	273	292	308	323	169	224	253	273	292	308	323	337	180	222	260	292	308	323	337	350

CONNECTIONS

Table 6.20.20 (continued) Design strength of concrete brackets, corbels, or haunches

b = 14"

A_s \ h	| 10" Proj. 10	12	14	16	18	20	22	24	| 12" Proj. 12	14	16	18	20	22	24	26	| 14" Proj. 14	16	18	20	22	24	26	28
2-#4	0	0	0	0	0	0	0	0	0	0	0	0	0	0	0	0	0	0	0	0	0	0	0	0
2-#5	29	34	0	0	0	0	0	0	30	0	0	0	0	0	0	0	0	0	0	0	0	0	0	0
2-#6	42	49	56	62	68	0	0	0	42	49	54	60	0	0	0	0	43	49	54	0	0	0	0	0
2-#7	56	67	76	85	93	100	107	113	58	66	74	82	88	95	101	0	59	66	73	79	85	91	0	0
2-#8	74	87	99	110	121	131	140	148	76	87	97	106	116	124	132	140	77	86	95	104	112	119	126	133
2-#9	93	110	126	140	153	165	177	188	96	110	123	135	146	157	167	177	97	109	120	131	141	151	160	168
3-#4	28	33	0	0	0	0	0	0	28	0	0	0	0	0	0	0	0	0	0	0	0	0	0	0
3-#5	43	51	58	65	71	0	0	0	44	51	57	62	0	0	0	0	45	51	56	0	0	0	0	0
3-#6	62	73	84	93	102	110	118	125	64	73	82	90	97	105	111	118	65	73	80	87	94	101	107	112
3-#7	85	100	114	127	139	150	160	170	87	99	111	122	133	142	152	160	88	99	109	119	128	137	145	153
3-#8	111	130	149	166	181	196	210	222	113	130	145	160	173	186	198	209	115	129	143	155	167	179	189	200
3-#9	140	165	188	210	229	246	256	265	143	164	184	202	219	236	251	265	146	164	181	197	212	226	240	253

b = 16"

A_s \ h	| 10" Proj. 10	12	14	16	18	20	22	24	| 12" Proj. 12	14	16	18	20	22	24	26	| 14" Proj. 14	16	18	20	22	24	26	28
2-#6	42	49	56	62	0	0	0	0	42	49	54	0	0	0	0	0	43	49	0	0	0	0	0	0
2-#7	56	67	76	85	93	100	107	0	58	66	74	82	88	95	0	0	59	66	73	79	85	0	0	0
2-#8	74	87	99	110	121	131	140	148	76	87	97	106	116	124	132	140	77	86	95	104	112	119	126	133
2-#9	93	110	126	140	153	165	177	188	96	110	123	135	146	157	167	177	97	109	120	131	141	151	160	168
3-#6	62	73	84	93	102	110	118	125	64	73	82	90	97	105	111	0	65	73	80	87	94	101	0	0
3-#7	85	100	114	127	139	150	160	170	87	99	111	122	133	142	152	160	88	99	109	119	128	137	145	153
3-#8	111	130	149	166	181	196	210	222	113	130	145	160	173	186	198	209	115	129	143	155	167	179	189	200
3-#9	140	165	188	210	229	248	265	279	143	164	184	202	219	236	251	265	146	164	181	197	212	226	240	253
4-#6	83	98	112	124	136	147	157	167	85	97	109	120	130	140	149	157	86	97	107	117	126	134	142	150
4-#7	113	133	152	169	185	200	214	227	116	133	148	163	177	190	202	214	118	132	146	159	171	182	193	204
4-#8	148	174	198	221	242	261	278	288	151	173	194	213	231	248	264	279	154	173	190	207	223	238	253	266
4-#9	187	220	251	280	295	308	320	332	191	219	245	270	292	314	332	343	194	218	241	262	282	302	320	337

b = 18"

A_s \ h	| 10" Proj. 10	12	14	16	18	20	22	24	| 12" Proj. 12	14	16	18	20	22	24	26	| 14" Proj. 14	16	18	20	22	24	26	28
2-#6	42	49	56	0	0	0	0	0	42	49	0	0	0	0	0	0	43	0	0	0	0	0	0	0
2-#7	56	67	76	85	93	100	0	0	58	66	74	82	88	0	0	0	59	66	73	79	0	0	0	0
2-#8	74	87	99	110	121	131	140	148	76	87	97	106	116	124	132	140	77	86	95	104	112	119	126	0
2-#9	93	110	126	140	153	165	177	188	96	110	123	135	146	157	167	177	97	109	120	131	141	151	160	168
3-#6	62	73	84	93	102	110	118	0	64	73	82	90	97	105	0	0	65	73	80	87	94	0	0	0
3-#7	85	100	114	127	139	150	160	170	87	99	111	122	133	142	152	160	88	99	109	119	128	137	145	153
3-#8	111	130	149	166	181	196	210	222	113	130	145	160	173	186	198	209	115	129	143	155	167	179	189	200
3-#9	140	165	188	210	229	248	265	281	143	164	184	202	219	236	251	265	146	164	181	197	212	226	240	253
4-#6	83	98	112	124	136	147	157	167	85	97	109	120	130	140	149	157	86	97	107	117	126	134	142	150
4-#7	113	133	152	169	185	200	214	227	116	133	148	163	177	190	202	214	118	132	146	159	171	182	193	204
4-#8	148	174	198	221	242	261	279	296	151	173	194	213	231	248	264	279	154	173	190	207	223	238	253	266
4-#9	187	220	251	280	306	323	336	348	191	219	245	270	292	314	334	353	194	218	241	262	282	302	320	337

b = 20"

A_s \ h	| 10" Proj. 10	12	14	16	18	20	22	24	| 12" Proj. 12	14	16	18	20	22	24	26	| 14" Proj. 14	16	18	20	22	24	26	28
2-#6	42	49	0	0	0	0	0	0	42	0	0	0	0	0	0	0	0	0	0	0	0	0	0	0
2-#7	56	67	76	85	93	0	0	0	58	66	74	82	0	0	0	0	59	66	73	0	0	0	0	0
2-#8	74	87	99	110	121	131	140	0	76	87	97	106	116	124	0	0	77	86	95	104	112	0	0	0
2-#9	93	110	126	140	153	165	177	188	96	110	123	135	146	157	167	177	97	109	120	131	141	151	160	168
3-#6	62	73	84	93	102	0	0	0	64	73	82	90	0	0	0	0	65	73	80	0	0	0	0	0
3-#7	85	100	114	127	139	150	160	170	87	99	111	122	133	142	152	160	88	99	109	119	128	137	145	0
3-#8	111	130	149	166	181	196	210	222	113	130	145	160	173	186	198	209	115	129	143	155	167	179	189	200
3-#9	140	165	188	210	229	248	265	281	143	164	184	202	219	236	251	265	146	164	181	197	212	226	240	253
4-#6	83	98	112	124	136	147	157	167	85	97	109	120	130	140	149	157	86	97	107	117	126	134	142	0
4-#7	113	133	152	169	185	200	214	227	116	133	148	163	177	190	202	214	118	132	146	159	171	182	193	204
4-#8	148	174	198	221	242	261	279	296	151	173	194	213	231	248	264	279	154	173	190	207	223	238	253	266
4-#9	187	220	251	280	306	331	350	362	191	219	245	270	292	314	334	353	194	218	241	262	282	302	320	337

Table 6.20.20 (continued) Design strength of concrete brackets, corbels, or haunches

4", 6", 8" Projection

b	A$_s$	4" Projection h=4	6	8	10	12	14	16	18	6" Projection h=6	8	10	12	14	16	18	20	8" Projection h=8	10	12	14	16	18	20	22
b = 22"	2-#6	33	51	66	79	90	0	0	0	38	50	61	70	0	0	0	0	40	49	58	0	0	0	0	0
	2-#7	44	69	90	107	122	135	146	0	51	68	83	96	107	117	0	0	54	67	78	89	98	0	0	0
	2-#8	58	91	118	140	159	172	180	188	67	89	108	125	140	153	166	177	71	88	103	116	128	140	150	0
	2-#9	73	115	149	175	188	199	210	219	84	112	136	158	177	194	210	224	90	111	130	147	163	177	190	202
	3-#6	49	76	99	118	135	149	161	168	56	75	91	105	118	129	140	0	60	74	87	98	108	118	0	0
	3-#7	67	104	135	161	177	188	197	206	77	102	124	143	161	176	190	203	82	101	118	133	148	161	173	184
	3-#8	87	136	176	194	209	222	233	244	100	133	162	187	210	230	244	253	107	131	154	174	193	210	225	240
	3-#9	110	172	203	223	240	256	269	282	127	168	205	237	256	269	282	294	135	166	195	220	244	265	285	303
	4-#6	65	102	132	158	175	185	195	203	75	100	121	140	157	173	186	199	80	99	115	131	144	157	169	180
	4-#7	89	139	179	196	211	224	236	247	102	136	165	191	214	235	247	257	109	134	157	178	197	214	230	245
	4-#8	116	181	209	230	248	264	278	291	134	177	216	248	264	278	291	303	142	175	205	232	257	280	300	315
	4-#9	147	211	239	263	284	303	320	336	169	224	263	284	303	320	336	350	180	222	260	294	320	336	350	364
b = 24"	2-#6	33	51	66	79	0	0	0	0	38	50	61	0	0	0	0	0	40	49	0	0	0	0	0	0
	2-#7	44	69	90	107	122	135	0	0	51	68	83	96	107	0	0	0	54	67	78	89	0	0	0	0
	2-#8	58	91	118	140	159	176	186	194	67	89	108	125	140	153	166	0	71	88	103	116	128	140	0	0
	2-#9	73	115	149	177	194	206	216	226	84	112	136	158	177	194	210	224	90	111	130	147	163	177	190	202
	3-#6	49	76	99	118	135	149	161	0	56	75	91	105	118	129	0	0	60	74	87	98	108	0	0	0
	3-#7	67	104	135	161	183	194	203	212	77	102	124	143	161	176	190	203	82	101	118	133	148	161	173	184
	3-#8	87	136	176	201	216	229	241	252	100	133	162	187	210	230	248	262	107	131	154	174	193	210	225	240
	3-#9	110	172	211	231	249	265	279	292	127	168	205	237	265	279	292	304	135	166	195	220	244	265	285	303
	4-#6	65	102	132	158	179	191	201	209	75	100	121	140	157	173	186	199	80	99	115	131	144	157	169	180
	4-#7	89	139	180	203	219	232	244	255	102	136	165	191	214	235	254	265	109	134	157	178	197	214	230	245
	4-#8	116	181	217	238	257	273	288	301	134	177	216	250	273	288	301	314	142	175	205	232	257	280	300	320
	4-#9	147	219	248	273	295	314	332	348	169	224	273	295	314	332	348	362	180	222	260	294	325	348	362	376

10", 12", 14" Projection

b	A$_s$	10" Projection h=10	12	14	16	18	20	22	24	12" Projection h=12	14	16	18	20	22	24	26	14" Projection h=14	16	18	20	22	24	26	28
b = 22"	2-#6	42	49	0	0	0	0	0	0	42	0	0	0	0	0	0	0	0	0	0	0	0	0	0	0
	2-#7	56	67	76	85	0	0	0	0	58	66	74	0	0	0	0	0	59	66	0	0	0	0	0	0
	2-#8	74	87	99	110	121	131	0	0	76	87	97	106	116	0	0	0	77	86	95	104	0	0	0	0
	2-#9	93	110	126	140	153	165	177	188	96	110	123	135	146	157	167	177	97	109	120	131	141	151	160	0
	3-#6	62	73	84	93	102	0	0	0	64	73	82	90	0	0	0	0	65	73	80	0	0	0	0	0
	3-#7	85	100	114	127	139	150	160	170	87	99	111	122	133	142	152	0	88	99	109	119	128	137	0	0
	3-#8	111	130	149	166	181	196	210	222	113	130	145	160	173	186	198	209	115	129	143	155	167	179	189	200
	3-#9	140	165	188	210	229	248	265	281	143	164	184	202	219	236	251	265	146	164	181	197	212	226	240	253
	4-#6	83	98	112	124	136	147	157	167	85	97	109	120	130	140	149	0	86	97	107	117	126	134	0	0
	4-#7	113	133	152	169	185	200	214	227	116	133	148	163	177	190	202	214	118	132	146	159	171	182	193	204
	4-#8	148	174	198	221	242	261	279	296	151	173	194	213	231	248	264	279	154	173	190	207	223	238	253	266
	4-#9	187	220	251	280	306	331	354	375	191	219	245	270	292	314	334	353	194	218	241	262	282	302	320	337
b = 24"	2-#6	42	0	0	0	0	0	0	0	0	0	0	0	0	0	0	0	0	0	0	0	0	0	0	0
	2-#7	56	67	76	0	0	0	0	0	58	66	0	0	0	0	0	0	59	0	0	0	0	0	0	0
	2-#8	74	87	99	110	121	0	0	0	76	87	97	106	0	0	0	0	77	86	95	0	0	0		
	2-#9	93	110	126	140	153	165	177	188	96	110	123	135	146	157	167	0	97	109	120	131	141	151	0	0
	3-#6	62	73	84	93	0	0	0	0	64	73	82	0	0	0	0	0	65	73	0	0	0	0	0	0
	3-#7	85	100	114	127	139	150	160	0	87	99	111	122	133	142	0	0	88	99	109	119	128	0	0	0
	3-#8	111	130	149	166	181	196	210	222	113	130	145	160	173	186	198	209	115	129	143	155	167	179	189	200
	3-#9	140	165	188	210	229	248	265	281	143	164	184	202	219	236	251	265	146	164	181	197	212	226	240	253
	4-#6	83	98	112	124	136	147	157	0	85	97	109	120	130	140	0	0	86	97	107	117	126	0	0	0
	4-#7	113	133	152	169	185	200	214	227	116	133	148	163	177	190	202	214	118	132	146	159	171	182	193	204
	4-#8	148	174	198	221	242	261	279	296	151	173	194	213	231	248	264	279	154	173	190	207	223	238	253	266
	4-#9	187	220	251	280	306	331	354	375	191	219	245	270	292	314	334	353	194	218	241	262	282	302	320	337

CONNECTIONS

Table 6.20.21 Design of structural steel haunches — concrete

Values are for design strength of concrete by Eq. 6.12.1 for following criteria:

$f'_c = 5000$ psi; for other concrete strengths multiply values by $f'_c/5000$.

Adequacy of structural steel section should be checked.

Additional design strength, ϕV_r, can be obtained with reinforcing bars — see Table 6.20.22.

$V_u \leq \phi (V_c + V_r)$
$\phi = 0.85$

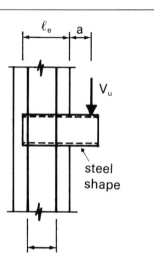

Values of ϕV_c (kips)

Shear span, a (in.)	Embedment, ℓ_e (in.)	Effective Width of Section (in.)											
		6	7	8	9	10	11	12	13	14	15	16	17
	6	33	38	43	49	54	60	65	70	76	81	87	92
	8	47	55	62	70	78	86	94	102	109	117	125	133
	10	62	72	82	92	103	113	123	133	144	154	164	174
	12	76	89	102	115	128	140	153	166	179	191	204	217
2	14	92	107	122	137	153	168	183	198	214	229	244	259
	16	107	124	142	160	178	196	213	231	249	267	285	302
	18	122	142	163	183	203	224	244	264	284	305	325	345
	20	137	160	183	206	229	252	274	297	320	343	366	389
	22	152	178	203	229	254	280	305	330	356	381	407	432
	6	25	29	33	38	42	46	50	54	58	63	67	71
	8	38	44	50	57	63	69	75	82	88	94	101	107
	10	51	60	68	77	85	94	102	111	119	128	136	145
	12	65	76	87	98	108	119	130	141	152	163	173	184
4	14	79	92	106	119	132	145	159	172	185	198	211	225
	16	94	109	125	141	156	172	187	203	219	234	250	266
	18	108	126	145	163	181	199	217	235	253	271	289	307
	20	123	144	164	185	205	226	246	267	287	308	328	349
	22	138	161	184	207	230	253	276	299	322	345	368	391
	6	20	24	27	30	34	37	41	44	47	51	54	58
	8	32	37	42	47	53	58	63	68	74	79	84	89
	10	44	51	58	66	73	80	87	95	102	109	117	124
	12	57	66	75	85	94	104	113	123	132	141	151	160
6	14	70	82	93	105	116	128	140	151	163	175	186	198
	16	84	97	111	125	139	153	167	181	195	209	223	237
	18	98	114	130	146	163	179	195	211	228	244	260	276
	20	112	130	149	168	186	205	223	242	261	279	298	317
	22	126	147	168	189	210	231	252	273	294	315	336	357
	6	17	20	23	26	29	31	34	37	40	43	46	48
	8	27	32	36	41	45	50	54	59	63	68	72	77
	10	38	45	51	57	64	70	76	83	89	95	102	108
	12	50	58	67	75	83	92	100	108	117	125	133	142
8	14	62	73	83	94	104	115	125	135	146	156	167	177
	16	75	88	101	113	126	138	151	163	176	188	201	214
	18	89	103	118	133	148	163	177	192	207	222	236	251
	20	102	119	136	153	170	187	204	222	239	256	273	290
	22	116	135	155	174	193	213	232	251	271	290	309	329

PCI Design Handbook

CONNECTIONS

Table 6.20.22 Design of structural steel haunches — reinforcement

Values are for additional design strength
obtained from reinforcement by Eq. 6.12.2
for following criteria:

A_s = 2 bars welded to steel shape
$A'_s = A_s$
Reinforcement anchored in only one direction.

When reinforcement, A_s and A'_s, is anchored
both above and below steel shape, it can
be counted twice (values may be doubled).

For design strength of concrete, ϕV_c —
see Table 6.20.21.

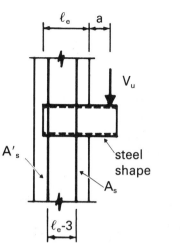

Values of V_r (kips)

Shear span, a (in.)	Embedment, ℓ_e (in.)	Reinforcing bar size f_y = 40 ksi						Reinforcing bar size f_y = 60 ksi					
		#4	#5	#6	#7	#8	#9	#4	#5	#6	#7	#8	#9
2	6	7	11	15	21	27	35	10	16	23	32	41	52
	8	10	15	22	30	39	49	14	23	33	44	58	73
	10	11	18	25	35	45	57	17	26	38	52	68	86
	12	12	19	28	38	50	63	19	29	42	57	74	94
	14	13	21	30	40	53	66	20	31	44	60	79	100
	16	14	21	31	42	55	69	21	32	46	63	82	104
	18	14	22	32	43	57	72	21	33	48	65	85	107
	20	14	23	33	44	58	73	22	34	49	67	87	110
	22	15	23	33	45	59	75	22	35	50	68	89	112
4	6	5	8	12	16	21	27	8	12	18	24	31	40
	8	8	12	18	24	31	40	12	18	27	36	47	60
	10	10	15	21	29	38	48	14	22	32	44	57	73
	12	11	17	24	33	43	54	16	25	36	49	64	82
	14	12	18	26	36	47	59	17	27	39	53	70	88
	16	12	19	28	38	49	62	18	29	42	57	74	93
	18	13	20	29	39	51	65	19	30	43	59	77	98
	20	13	21	30	41	53	67	20	31	45	61	80	101
	22	14	21	31	42	55	69	20	32	46	63	82	104
6	6	4	7	10	13	17	21	6	10	14	19	25	32
	8	7	10	15	20	26	33	10	16	22	30	40	50
	10	8	13	19	25	33	42	12	19	28	38	50	63
	12	9	15	21	29	38	48	14	22	32	44	57	72
	14	10	16	23	32	42	53	16	24	35	48	63	79
	16	11	17	25	34	45	57	17	26	38	51	67	85
	18	12	18	27	36	47	60	18	28	40	54	71	89
	20	12	19	28	38	49	62	18	29	41	56	74	93
	22	13	20	29	39	51	64	19	30	43	58	76	96
8	6	4	6	8	11	14	18	5	8	12	16	21	27
	8	6	9	13	17	23	29	9	13	19	26	34	43
	10	7	11	16	22	29	37	11	17	25	34	44	55
	12	9	13	19	26	34	43	13	20	29	39	51	65
	14	9	15	21	29	38	48	14	22	32	43	57	72
	16	10	16	23	31	41	52	15	24	35	47	61	78
	18	11	17	24	33	43	55	16	25	37	50	65	83
	20	11	18	26	35	46	58	17	27	39	52	68	87
	22	12	19	27	36	47	60	18	28	40	55	71	90

PART 7
SPECIAL TOPICS FOR
ARCHITECTURAL PRECAST CONCRETE

Page No.

7.1 Introduction 7–2
    7.1.1 Structural Design and Analysis 7–2
    7.1.2 Load Transfer—General Methods 7–3
    7.1.3 Application 7–3
    7.1.4 Design Analysis 7–3
    7.1.5 Design Objectives 7–4

7.2 Connections of Non-Load Bearing Panels 7–4
    7.2.1 Design Fundamentals 7–4
    7.2.2 Connection Details 7–6

7.3 Precast Concrete Used as Forms 7–14
    7.3.1 General 7–14
    7.3.2 Design Considerations 7–14
    7.3.3 Construction Considerations 7–15

7.4 Column Covers and Mullions 7–18

7.5 Veneered Panels 7–19
    7.5.1 General 7–19
    7.5.2 Structural Clay Products 7–21
    7.5.3 Natural Stone 7–22

7.6 Design Example—Window Wall Panel 7–26

7.7 References 7–30

7.1 Introduction

Architectural precast concrete products are those that, through application, finish, shape, color or texture, contribute to the appearance of the structure.

The rapid growth in the use of architectural precast concrete has created many innovations in architecture. Architectural precast concrete is available in complex shapes which serve not only as non-load bearing (cladding) walls, but may also combine an attractive appearance with the ability to serve as main structural members. The successful and economical use of architectural precast concrete depends on an understanding of the production and erection limitations, as well as the structural behavior of the element.

7.1.1 Structural Design and Analysis

Architectural precast concrete construction can be considered in three parts:

1. The precast elements individually.
2. The support of the precast element, i.e., the beam, wall, column, foundation or any other part of the structure which provides vertical and horizontal support for the element.
3. The connection that joins the precast element to its support system.

The other parts of this Handbook give detailed information and procedures for analyzing and designing all types of precast concrete elements and structures—architectural as well as structural. The following sections are also applicable to architectural precast concrete:

Part or Section	Title
1	Materials
2.6	Load Bearing Wall Panels
3.3	Volume Changes
3.4	Component Analysis
3.7	Shear Wall Buildings
3.10	Earthquake Analysis
4.2	Flexure
4.3	Shear
4.4	Torsion
4.7	Compression Members
5	Product Handling and Erection Bracing
6.4	Connection Design Criteria
6.5	Connection Hardware and Load Transfer Devices
6.8	Bearing on Plain Concrete
6.11	Concrete Brackets or Corbels
6.12	Structural Steel Haunches
6.18	Connection of Load Bearing Wall Panels
8	Tolerances for Precast and Prestressed Concrete
9	Thermal, Acoustical, Fire and Other Considerations
10.2	Guide Specifications for Architectural Precast Concrete
10.3	Code of Standard Practice for Precast Concrete

Following is a brief summary of structural engineering principles discussed in detail in the above listed sections:

The potential for element volume change requires consideration. Volume changes may be a result of:

1. Elastic and inelastic (creep) deformations.
2. Shrinkage.
3. Expansion and contraction resulting from temperature change.

Potential movement in the support system must also be considered. This can result from:

1. Elastic and inelastic deformation from gravity loads.
2. Horizontal displacement resulting from wind and earthquake.
3. Foundation movement.
4. Temperature change.
5. Shrinkage of concrete.

In most cases, potential movements may be estimated by analysis, and provisions made to accommodate these movements.

The structural design of architectural precast concrete requires that, after all loads are determined, the following be considered:

1. Forces and strains caused by handling, transportation and erection.
2. Acceptable crack width and location.
3. Strain gradients and restraint forces from thermal and moisture differentials through the panel.
4. Localized wind forces and the response of a precast element to these transient loads.
5. Forces transferred to the element and connections due to distortion of the structural frame.
6. Differential deflections between the precast element and the supporting structure.
7. Accommodation of tolerances allowed in the supporting structure.
8. Experience with various types of connections.

The designer must recognize that loads and behavior cannot be established precisely, particularly with respect to members exposed to the environment. This imprecision will generally not affect the safety of the member, provided that reasonable values have been established and the above factors considered.

7.1.2 Load Transfer—General Methods

The forces which must be considered in the design of architectural precast concrete structures are:

1. Those caused by the precast member itself, e.g., the self weight and earthquake forces.

2. Those caused by loads, such as wind, snow, floor live loads, soil or fluid pressure, or construction loads, that are externally applied to the element or transferred to the element by the behavior of the supporting structure.

3. Those that develop as a result of restraint of volume changes or support system movement. These forces are generally concentrated at the connections.

All non-load bearing elements should be designed to accommodate movement freely and, whenever possible, with no redundant supports, except where necessary to restrain bowing. Relatively simple analytical procedures provide the forces required for connection design. The calculations required for movement accommodation are more complex. The designer can use the simplified methods discussed in Part 3, or computer analysis programs.

When redundant supports are necessary or when movement is to be resisted, the load-deformation characteristics of the element, connections and support system should be taken into account. Design of connections that restrain support system elements may need to consider the load-deformation characteristics of the connection.

7.1.3 Application

The most common applications of architectural precast concrete in building construction are those in which the precast elements function as walls. Concrete is a nearly ideal wall material, since it has excellent sound transmission characteristics, is fire resistant, and is durable. In addition, architectural precast concrete provides almost unlimited potential for economically achieving esthetic design objectives. Architectural precast elements are capable of functioning as a wall without backing. Large elements provide the complete wall, and often are made with integral window openings.

It is apparent that precast walls which are of sufficient mass for requirements related to manufacturing, sound transmission, and fire resistance are also capable of supporting significant loads. Thus, using precast concrete walls, which separate or enclose space, to carry loads is an economical alternative to the use of separate structural systems and wall materials.

In this Handbook, a distinction is frequently made between load bearing and non-load bearing elements:

1. A non-load bearing (cladding) element is one which could theoretically be removed from the structure without significantly affecting the integrity of the building. Although non-load bearing walls are normally provided only to enclose space, they are subjected to loads such as wind and earthquake.

2. A load bearing element resists and transfers loads applied from other elements. Therefore, a load bearing member cannot be removed without affecting the strength or stability of the building.

The use of architectural precast concrete walls as shear walls to resist and transfer horizontal loads is another efficient application. In this case, the connections between panels and other structural elements are the primary design consideration. Precast panels can also be attached to existing frames to improve lateral load resistance, such as in upgrading structures in seismic zones.

Architectural precast concrete elements also often function as beams to transfer gravity loads to supports. The beams may also participate in the transfer of lateral loads. For example, the beams may serve as struts between lateral load resisting elements, such as columns or shear walls, transferring the load as an axial force. In other cases, precast beams interact with columns to form a moment frame as a lateral load resisting system.

In composite construction, precast concrete elements may be used as forms for cast-in-place concrete (Sect. 7.3). This is especially suitable for combining architectural and structural functions in load bearing facades, or for improving ductility in locations of high seismic risk.

7.1.4 Design Analysis

In some cases, the structural design of a single precast element can be completed with very little consideration of other materials and elements in the structure. The weight of the element and the superimposed loads are simply transferred to supports, and the element can be considered independently of the structure. Occasionally, however, it is necessary to consider the characteristics of other materials and elements within the structure. For example, neglecting the relative movement between two support points may lead to inaccurate estimates of connection forces.

In other cases, architectural precast elements may interact with other parts of the structure in the transfer of loads. The design of the interacting system and the individual precast elements must be based on analysis of the whole system.

The designer of a precast building can choose to transfer loads through architectural precast concrete elements or intentionally avoid significant load transfers through the precast elements. In the preliminary design phase, the structural engineer should therefore recognize that he is able to choose the structural characteristics rather than simply analyze a predetermined set of criteria.

7.1.5 Design Objectives

Structural integrity of the completed structure is the primary objective. Deflections must be limited to acceptable levels, and distress that could result in instability of an individual element or of the complete structure must be prevented. The inherent stiffness of architectural precast concrete panels will significantly stiffen a structure, thereby reducing deflections and improving stability.

Economy is an important design objective. The total cost of a completed structure is the determining factor. What seems to be a relatively expensive precast element may result in the most economical building because of the reduction or elimination of other costs. In some cases, precast elements are not economical due to improper application or the necessity for complex on-site construction procedures. The designer should attempt to optimize the structure by using precast panels to serve several functions, and take advantage of the economy gained by standardization.

Standardization reduces costs because fewer molds are required. Also, productivity in all phases of manufacture and erection is improved through repetition of familiar tasks. There is also less chance of error.

A very strict discipline is required of the designer to avoid a large number of non-standard units. Preliminary planning and budgeting should recognize the probability that the number of different units will increase as the design progresses. If non-standard units are unavoidable, costs can be minimized if they can be cast from a "master mold" with simple modifications for the special pieces, rather than requiring special molds.[1]

The esthetic design objectives for the structure should be a matter of concern to the structural engineer. A precast element or system may achieve all other design objectives but fall short of esthetic objectives through treatment of structural features only.

7.2 Connections of Non-Load Bearing Panels

7.2.1 Design Fundamentals

To assure the satisfactory performance of architectural precast concrete elements, it is important that connections be properly conceived and designed. In general, the design of connections of non-load bearing elements follows the procedures given in Part 6.

The designer should attempt to provide simple load paths through the connections and ductility within the connections. This will reduce the sensitivity of the connection and the necessity to precisely calculate loads and forces from, for example, volume changes and frame distortions. The number of load transfer points should be kept to a practical minimum. It is preferable that no more than two connections per panel be used to transfer gravity loads. Load transfer should be as direct as possible. Fig. 7.2.1 shows examples of several load transfer mechanisms.

Connections and assemblies should develop sufficient ductility to preclude brittle failures. For example, inserts in concrete should be attached to and/or hooked around reinforcing steel, provided with confinement reinforcement or otherwise be terminated to effectively transfer forces to the concrete and/or reinforcement. It is desirable that the pullout strength of concrete and strength of welds be greater than the tension or bending strength of the steel.

The design of the panel connection and the supporting frame are interdependent. For example, if a supporting spandrel beam lacks sufficient rotational stiffness, the precast connections should be designed to transmit the vertical loads to the centerline of the beam, or other provisions made to avoid torsional rotation. This may induce additional bending in the panel and connection because of the increased eccentricity.

Support systems which result in statically determinate force systems are recommended, because the forces are more predictable. It is usually better to support the entire weight of the panel at one level.

Connection hardware must be permanently protected from the elements, or corrosion resistant materials used. Frequently, special fire protection is required.

The connection details should be standardized as much as possible. This not only facilitates production and erection, but also reduces the chance for error. Adequate tolerances and clearances, as discussed in Part 8, are required.

Fig. 7.2.1 Connections illustrating number of force transfers

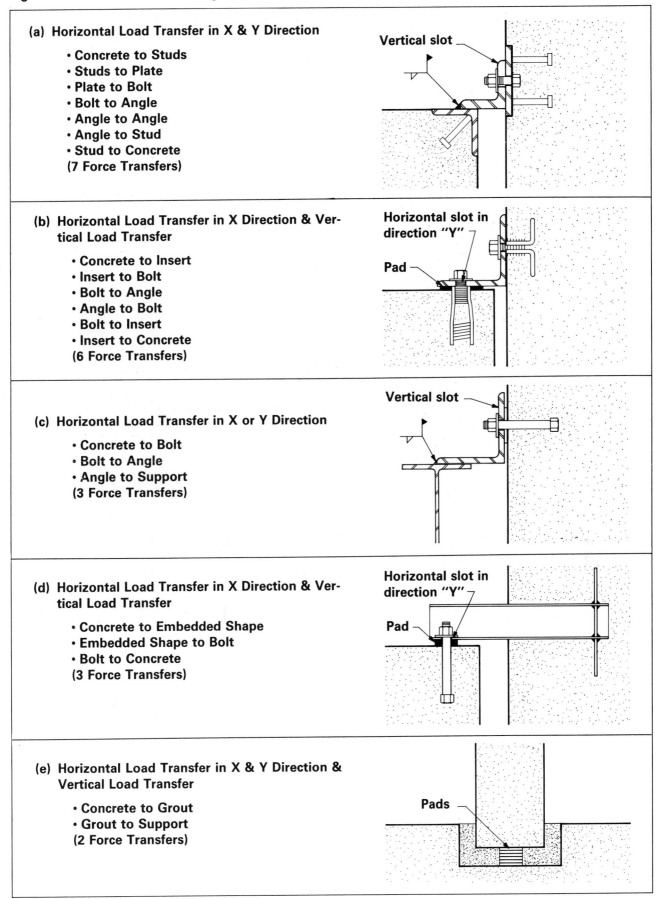

(a) Horizontal Load Transfer in X & Y Direction

- Concrete to Studs
- Studs to Plate
- Plate to Bolt
- Bolt to Angle
- Angle to Angle
- Angle to Stud
- Stud to Concrete

(7 Force Transfers)

Vertical slot

(b) Horizontal Load Transfer in X Direction & Vertical Load Transfer

- Concrete to Insert
- Insert to Bolt
- Bolt to Angle
- Angle to Bolt
- Bolt to Insert
- Insert to Concrete

(6 Force Transfers)

Horizontal slot in direction "Y"

Pad

(c) Horizontal Load Transfer in X or Y Direction

- Concrete to Bolt
- Bolt to Angle
- Angle to Support

(3 Force Transfers)

Vertical slot

(d) Horizontal Load Transfer in X Direction & Vertical Load Transfer

- Concrete to Embedded Shape
- Embedded Shape to Bolt
- Bolt to Concrete

(3 Force Transfers)

Horizontal slot in direction "Y"

Pad

(e) Horizontal Load Transfer in X & Y Direction & Vertical Load Transfer

- Concrete to Grout
- Grout to Support

(2 Force Transfers)

Pads

Production considerations

Connections should allow economical fabrication of the precast elements. The hardware should not interfere with concrete placement or cause finishing problems, nor require penetration through molds. The size of reinforcing bars should be limited, because large bars require anchorage lengths and hook sizes that may be impractical. It is often better to use welded cross bars or other types of mechanical anchorage than to rely on bond. Connection details with reinforcing bars crossing each other require careful dimensional checking to insure sufficient cover.

Plates, angles or other steel shapes embedded in precast concrete must be securely attached to the forms to prevent their becoming misaligned or skewed.

Inserts must be placed accurately because their capacity depends on the depth of embedment, spacing and distance from free edges. They should be kept free of dirt and protected so that concrete does not enter during casting.

When possible, connections should be dimensioned to the nearest 1/2 in. The minimum clearance between the various items within a connection should not be less than 1/2 in., with 3/4 in. preferred. It should be remembered that the real dimensions of reinforcing bars are approximately 1/8 in. greater than the nominal, because of the deformations.

Erection considerations

Ease and speed of erection are heavily dependent on simplicity and ruggedness of assembly details. Field patching and finishing should be kept to a minimum. If possible, the details should allow erection to proceed in nearly all kinds of weather. The panels should be capable of being erected without temporary shoring. Ideally, it should be possible to complete the connection by working downhand from the top of the erected member or from a stable deck.

The maximum feasible adjustability should be provided in all directions, preferably at least 1 in. For example, the supporting beam may rotate or deflect under the weight of the precast panels, making it necessary to adjust the connections during erection. Significant support rotation and deflection are more common in structural steel than in concrete frames. The supporting beam should be designed to minimize rotation and deflection.

Connections should be detailed so that hoisting equipment can be quickly released. It may be necessary to provide temporary connections that are released after final adjustments are made. Loading conditions during erection may be more critical than other phases, due to eccentricities, construction loads or impact.

When cast-in-place concrete, grout or drypack is required to complete a connection, the detail should provide for self-forming if possible. When not practical, the connection should allow for easy placement and removal of formwork.

7.2.2 Connection Details

Figs. 7.2.2 through 7.2.11 show a number of typical connection arrangements for cladding panels.[2,10] There are obviously many possible combinations of anchors, plates, bolts, angles, etc., to form various connection assemblies. In this section, the connections have been broadly categorized as bearing, tie-back or alignment connections. The details shown are not to be considered "standard", but are presented as ideas on which to build.

Direct bearing connections (Fig. 7.2.2)

The actual load transfer is usually through shims as shown in (a). The joint then may be caulked or dry-packed. The remainder of the details illustrate various methods of attachment to maintain proper horizontal position: (b) and (c) use rods or bolts in inserts in the upper panel, grouted into holes in the lower panel or support; (d) through (g) show various welding arrangements; (h) uses an anchor bolt projecting from the supporting member; (i) and (j) have reinforcing bars projecting from the panel which are grouted into sleeves. If there is tension on the connector, as when the panel is used as a shear wall, the sleeve may be formed by post-tensioning conduit. Detail (k) uses a drilled-in expansion anchor. Caution: details (b), (c), (i) and (j) should be limited to warm weather applications, or provisions made to prevent water from entering the grout hole and freezing prior to grouting.

Steel or concrete haunch bearing connections (Fig. 7.2.3)

A typical concrete corbel cast with a cladding panel to support it is shown in (a). Design procedures are given in Sect. 6.11. Various types of steel haunches are shown in (b) through (f). The projecting haunch can be a rolled steel section such as a tube (b), angle (c), wide-flange (d), solid bar (e), or structural tee (f). It can be embedded in the panel as in (c), (d) and (e), or welded on after stripping as in (b) and (f). Steel haunch design is given in Sect. 6.12, and anchor design for the types shown in (b) and (f) through (h) is given in Sect. 6.5.2. "Back-up" bars welded to the haunch to maintain alignment are shown in (b), (c) and (f). The side location of the back-up bar shown allows small longitudinal movement caused by volume changes.

Variations in which the architectural precast

Fig. 7.2.2 Direct bearing connections

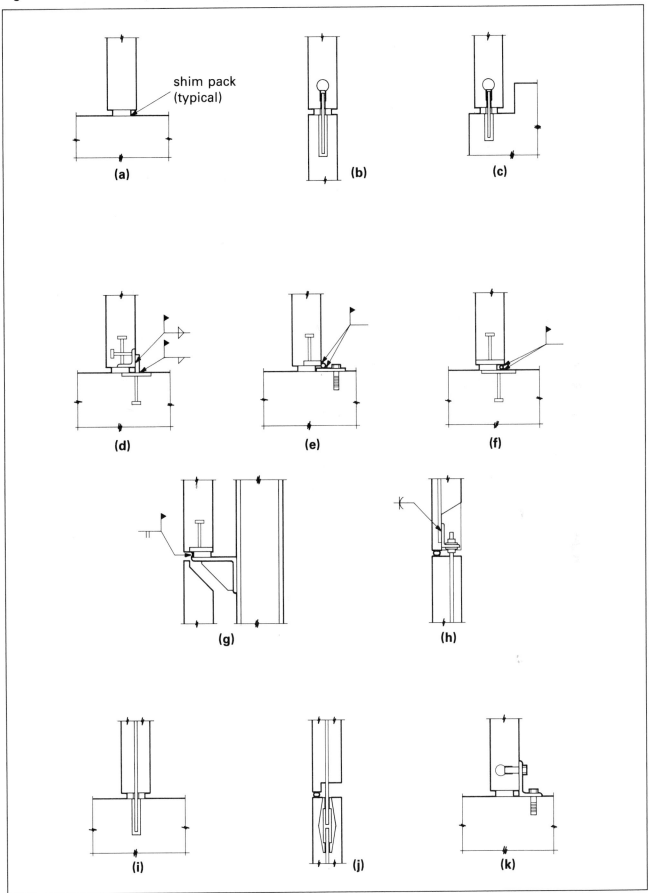

Fig. 7.2.3 Steel or concrete haunch bearing connections

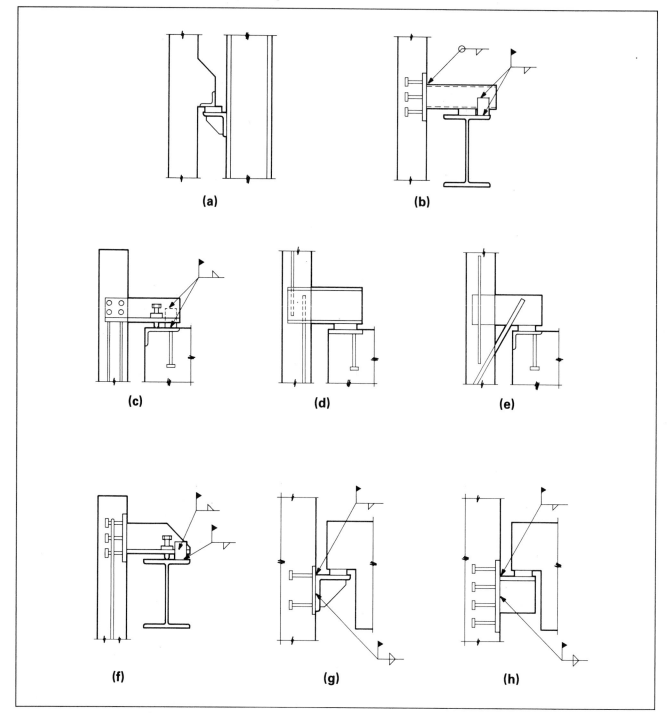

Note: Use of shims vs. levelling bolts depends on local practice or preference.

Fig. 7.2.4 Angle seat bearing connections

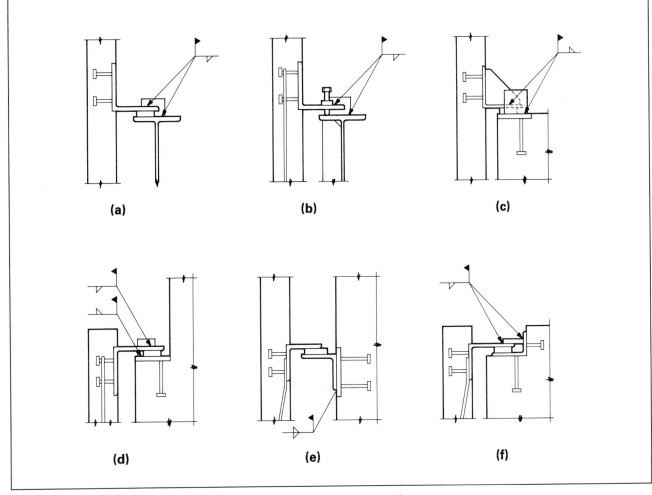

(a)   (b)   (c)

(d)   (e)   (f)

Note: Use of shims vs. levelling bolts depends on local practice or preference.

concrete element may serve as a horizontal sun screen are shown in (g) and (h). The member should be reinforced similar to a dapped-end beam (see Sect. 6.13).

Angle seat bearing connections (Fig. 7.2.4)

The most commonly used hardware items used to support cladding panels are steel angles. Fig. 7.2.4 shows several of an almost infinite variety of combinations. Depending on the load to be supported, the angle may need to be stiffened as shown in (c). Design of connection angles is given in Sect. 6.5.16. Design of headed studs or bolts attached to the angles in (a) through (f) is given in Sect. 6.5.2. Note that in (b), (d), (e) and (f) confinement reinforcement is shown around the embedded studs. This adds ductility to the connection and is highly recommended, especially in earthquake zones.

Fig. 7.2.5 Welded plate tie-back connections

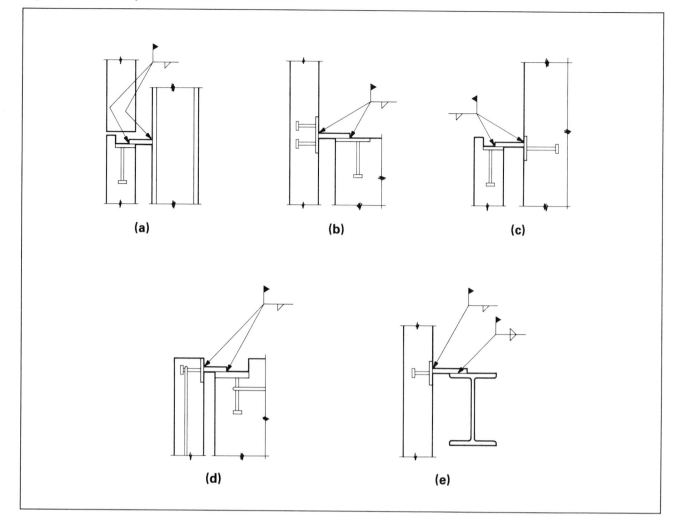

(a)　　　　　　　　(b)　　　　　　　　(c)

(d)　　　　　　　　(e)

Tie-back connections (Figs. 7.2.5 through 7.2.9)

Nearly every precast panel requires tie-back connections in addition to the bearing (support) connections. The simplest use a welded plate or flat bar as shown in Fig. 7.2.5. Some special conditions may require angles instead of plates. Fig. 7.2.6 (a) and (b) show plan views of panels connected to steel columns. Fig. 7.2.6 (c) and (d) show situations where the desired tie-back point in the

panel is some distance from the structural element to which the panel is tied.

Fig. 7.2.7 shows several examples of bolts into inserts; Fig. 7.2.8 details are similar, except threaded rods are used instead of bolts, usually because of location or the amount of adjustment required. Fig. 7.2.9 illustrates how both bolting and welding are frequently used. When bolts are used in connections, slots or oversize holes should be provided to permit adjustments.

Fig. 7.2.6 Welded angle tie-back connections

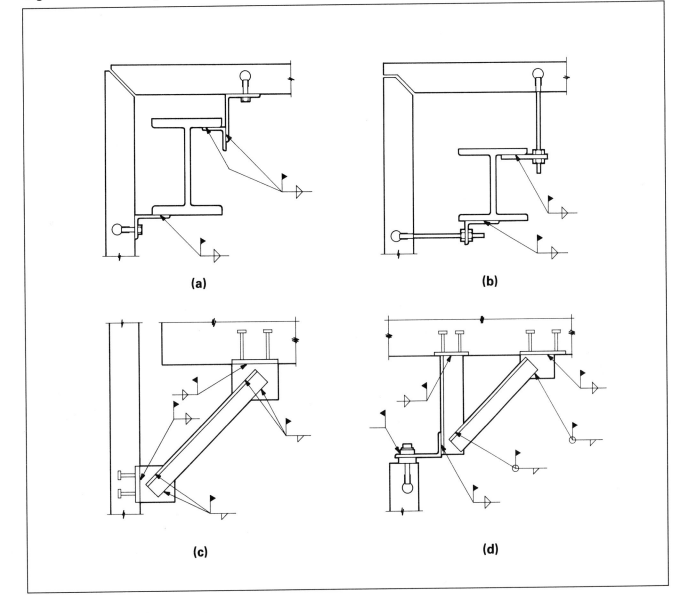

(a)

(b)

(c)

(d)

Fig. 7.2.7 Bolted tie-back connections

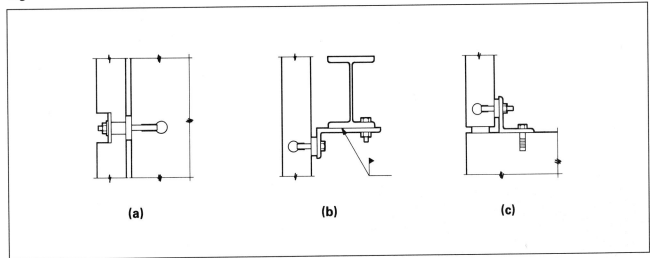

(a)

(b)

(c)

Fig. 7.2.8 Threaded rod tie-back connections

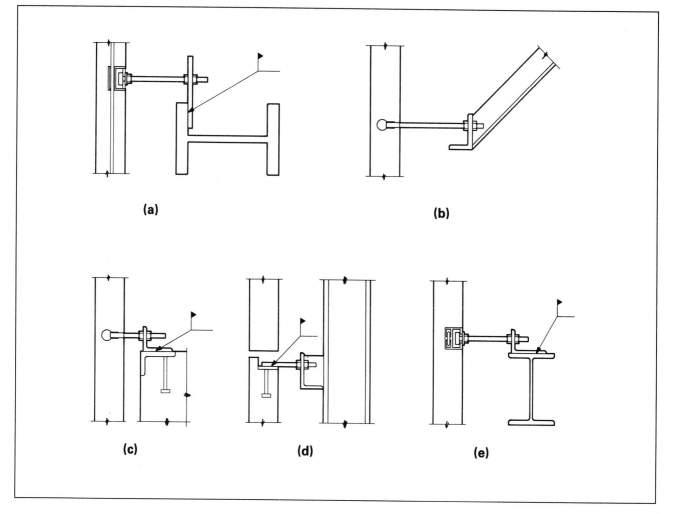

(a)

(b)

(c)

(d)

(e)

Fig. 7.2.9 Bolted and welded tie-back connections

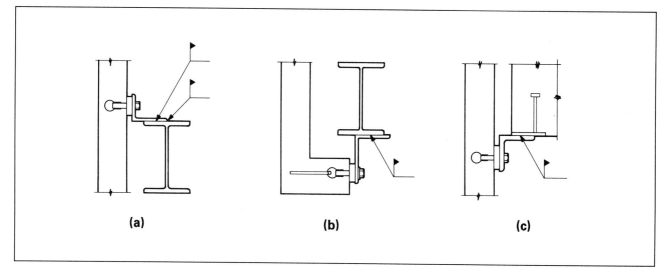

(a)

(b)

(c)

PCI Design Handbook

Alignment connections (Figs. 7.2.10 and 7.2.11)

The third major category of connections are those used to align two panels. They may be either welded (Fig. 7.2.10), or bolted (Fig. 7.2.11). With bolted connections, slotted holes, as shown in (b), should be used to permit adjustment during erection and panel movement in service.

Fig. 7.2.10 Welded alignment connections

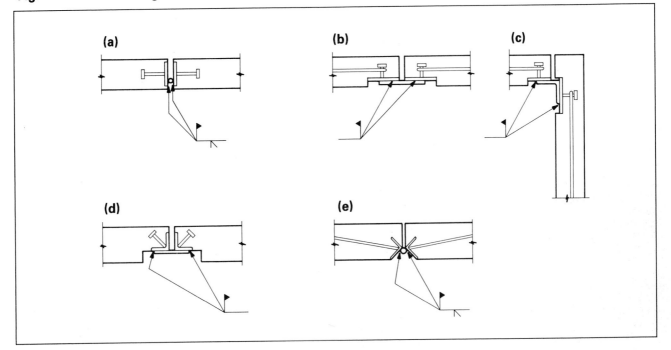

Fig. 7.2.11 Bolted alignment connections

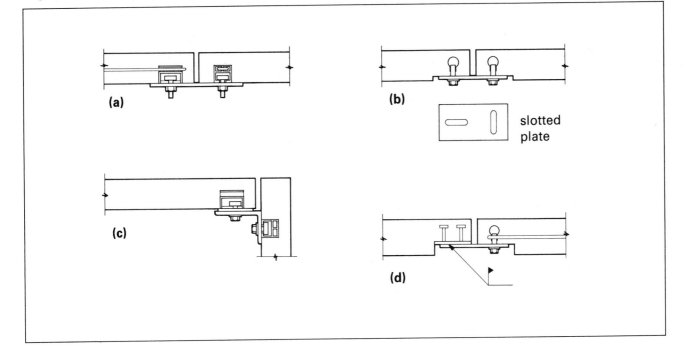

7.3 Precast Concrete Used as Forms

7.3.1 General

This section provides design and detailing information for structures in which architectural precast concrete is used as the formwork for cast-in-place structural concrete.[3]

In most cases, the architectural precast concrete is considered solely as a form, serving only decorative purposes after the cast-in-place concrete has achieved design strength. This is accomplished by providing open or compressible joints between abutting precast panels, and neglecting or eliminating bond at the interface of the precast and cast-in-place concrete. When the architectural precast concrete unit is non-composite with the cast-in-place concrete, the reinforcing steel extending from the precast into the cast-in-place concrete need only be of sufficient strength to support the formwork unit.

In other cases, it may be desirable to detail the structure so that the precast and cast-in-place concrete act compositely, thus combining the strength of both. It is then necessary to provide shear transfer as for other composite assemblies. Sect. 4.3.5 describes design procedures.

7.3.2 Design Considerations

Following are recommendations for design of precast panels used as forms for cast-in-place concrete:

Lateral pressure of fresh concrete. With the following limitations, the formulas below may be used to determine the forces to be resisted by forms, ties, and bracing:[4]

1. Material is normal weight concrete, with a unit weight of 145 to 155 pcf.
2. Vibration is by internal means only; form vibrators, if used, will require a modification of the formulas.
3. The depth vibrated, or revibrated, does not exceed 4 ft below the concrete surface.
4. Retarding admixtures, if used, will require a modification of the formulas.
5. For heights greater than 18 ft, an interval of 2 hr should elapse after each 18 ft lift prior to continuation.
6. Slump not greater than 4 in. at point of placement.

Columns, placement rate up to 25 ft/hr:

$p = 150 + 9000R/T$, maximum 3000 psf or 150h, whichever is least

Walls, placement rate less than 7 ft/hr:

$p = 150 + 9000R/T$, maximum 2000 psf or 150h, whichever is least

Walls, placement rate 7 to 10 ft/hr:

$p = 150 + 43{,}400/T + 2800R/T$, maximum 2000 psf or 150h, whichever is least

Walls, placement rate greater than 10 ft/hr:

$p = 150h$

where:

p = lateral pressure, psf
R = placement rate, ft/hr
T = temperature of concrete in forms, °F
h = height of fresh concrete above point considered, ft

Lateral forces on precast concrete forms. In addition to pressure induced by fresh concrete, the applied lateral forces on forms and bracing should be in accordance with the local building code, but not less than:

1. Wind—15 psf applied to all exposed surfaces, unless local codes specifically permit less.
2. Columns—2% of the total dead load supported by the form, applied as a horizontal load to the top of the form.
3. Walls—100 lb per linear ft of wall, applied to the top of the wall form.

Vertical loads on precast concrete forms. The vertical load supported by precast concrete forms should consist of all superimposed dead loads, including an allowance for storage of construction materials, and a live load of not less than 25 psf.

Formwork accessories. In the design of formwork accessories such as form ties and form anchors, a minimum safety factor of 2 based on the ultimate strength of the accessory is recommended except that yield point must not be exceeded.

Design assumptions.

1. Forms are generally supported in a manner which will permit them to act as continuous beams. It is generally sufficiently accurate to assume that the form acts as a fixed beam between adjacent lateral supports. The engineer must judge whether to include deformation of the form supports in the analysis.
2. Partially open concrete shapes are often used to form columns. These U-shapes will develop internal stresses which are a function of the stiffness of the sides. Design assumptions are shown in Fig. 7.3.1.
3. The internal stresses, produced by the temperature rise of the form as the cast-in-place concrete is placed, are usually neglected. Moments

developed due to uniform temperature rise for U-shaped forms are shown in Fig. 7.3.1.

4. The cross-sectional area of column form panels is usually neglected when determining either the immediate (elastic) or long term (creep) column shortening. The thickness of panels is usually included when investigating the effect of various temperature changes across the column.

Deflection. Form deflection should generally be limited to 1/360 of the unsupported height or length. Deflection of beam forms, and warping of wall forms, may result from differential shrinkage of precast and cast-in-place concrete, as well as the dead load or lateral pressure of the cast-in-place concrete.

Fig. 7.3.1 Forces due to internal pressure

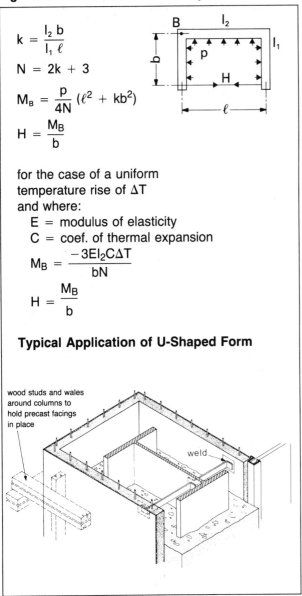

$$k = \frac{l_2 \, b}{l_1 \, \ell}$$

$$N = 2k + 3$$

$$M_B = \frac{p}{4N} \,(\ell^2 + kb^2)$$

$$H = \frac{M_B}{b}$$

for the case of a uniform temperature rise of ΔT
and where:

    E = modulus of elasticity
    C = coef. of thermal expansion

$$M_B = \frac{-3El_2C\Delta T}{bN}$$

$$H = \frac{M_B}{b}$$

Typical Application of U-Shaped Form

wood studs and wales around columns to hold precast facings in place

weld

Cambering of architectural precast forms to compensate for deflections is expensive and should be avoided.

Crack control. In order to minimize form cracking due to the pressures induced by fresh concrete, the design procedures for crack control in Parts 4 and 5 should be followed. Where the member is long enough to develop bonded strand, pretensioning may be used in the precast form units.

Composite design. Interaction between precast forms and cast-in-place concrete can be achieved by providing for the transfer of shear forces at the interface. Effective composite behavior will significantly increase the strength of members and reduce deflections. Design methods and shear transfer requirements are discussed in Part 4.

In zones of seismic activity, present codes treat precast walls subjected to lateral loads as non-ductile, with an appropriately high multiplier to establish design base shear. Continuity and ductility may be achieved by casting in place spandrel beams and columns using the wall panels as forms. The ductility of walls partially depends on the location of reinforcement. Ductile behavior is significantly improved if the reinforcement is located at the ends of the walls. Thus, a usually inactive curtain wall can become a major lateral load (seismic) resisting element.

When the precast and cast-in-place concrete are designed to act compositely, the form joints must be located away from points of high moment.

In compression members, axial loads will tend to be distributed initially to the two components in accordance with their individual axial stiffnesses. Over time, there will be some redistribution of load from the more highly stressed components to the other due to creep.

7.3.3 Construction Considerations

Realistic assumptions with regard to construction techiques are required. It must be determined (or specified) how the precast panels will be supported during concreting in order to design them.

Concrete form panels should be erected and temporarily braced to proper grade and alignment in such a way that the tolerances specified for the finished structure can be met. Temporary bracing for the panels generally consists of adjustable pipe bracing from panel to floor slab. Supports, braces, and form ties must be stiff enough so that their elastic deformation will not significantly affect the assumed load distribution. Form ties may be attached as shown in Fig. 7.3.2, or welded to plates cast in the panels. Column forms may use column clamps or be wrapped with steel bands to aid in resisting hydrostatic pressure.

Attachments between the precast form and other elements, such as steel columns, must be detailed to provide the necessary field adjustments. Contact area between precast concrete and external braces, clamps or bands should be protected from staining and chipping.

Fig. 7.3.2 Typical form ties

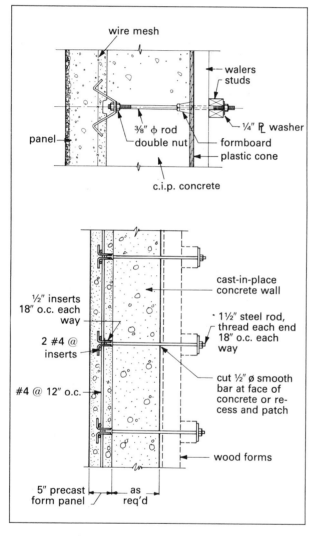

Joints. Architectural precast concrete surfaces require tight joints so that concrete does not leak and mar decorative facings. Methods used to close joints include buttering on the inside with mortar (units may also be bedded on mortar when designed to accept load transfer through the joint), and gasketing with low density, closed cell neoprene rubber or other durable, permanent, resilient materials.

Where form panels are non-composite, the joint material should prevent load transfer. Where form panels are intended to act compositely with the cast-in-place concrete, the joint material must be mortar or other non-staining material of sufficient strength to transfer the intended loads. For joints exposed to the environment, mortar is usually raked back from the face, and the joint caulked.

Some precaution and special details may be required to take care of differential shrinkage or creep between cast-in-place concrete and precast concrete used as forms for columns or load-bearing walls.[3,6] Stress relief may be simply handled in the design of the joints in the precast forms at the top and bottom of vertical structural components carrying axial loads.

Horizontal construction joints in the cast-in-place concrete are generally 3 in. below the top edge of the panels used as permanent forms rather than in line with horizontal form joints. This reduces the possibility of water leakage through the construction joints.

Example 7.3.1 Precast panel used as a wall form

Given:

The wall panel section shown below.
Panel concrete strength: f'_c = 5000 psi
C.i.p. concrete placement rate: 4 ft per hr
Maximum height of c.i.p. concrete: 4 ft
C.i.p. concrete temperature: 90°F

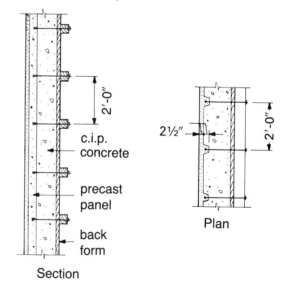

Problem:

Determine panel reinforcement requirements for pressure of wet concrete.

Solution:

Assume a 5 in. panel thickness is necessary to provide anchorage for the form ties. Either a 5 in. flat panel could be used or, as in this case, a ribbed section with a 5 in. rib depth. The rib spacing is set by the strength of the ties. Since the placement rate is less than 7 ft per hr:

$$p = 150 + 9000R/T = 150 + 9000(4)/90$$
$$= 550 \text{ psf}$$

or a maximum of:

$$p = 150(4) = 600 \text{ psf. Use 550 psf}$$

Check flexure:

Neglecting effect of ribs, assume the 2½ in. panel acts as a fixed beam of 24 in. span:

$$M = p\ell^2/12 = 550(2)^2/12 = 183 \text{ ft-lb}$$

$$f = \frac{6M}{bd^2} = \frac{6(183)(12)}{12(2.5)^2} = 176 \text{ psi}$$

Maximum allowable tension to prevent cracking

$$= 5\sqrt{5000}$$
$$= 353 \text{ psi} > 176 \text{ OK}$$

Provide reinforcement to develop an ultimate moment capacity $>1.4M$

$$= 1.4(183) = 256 \text{ ft-lb}$$

Try $6 \times 6 - W4.0 \times W4.0$ in center of panel.
$A_s = 0.08 \text{ in.}^2/\text{ft}$

$$a = \frac{A_s f_y}{0.85 f'_c b} = \frac{0.08(60)}{0.85(5)(12)} = 0.09 \text{ in.}$$

$$d = 2.5/2 = 1.12 \text{ in.}$$

$$\phi M_n = \phi A_s f_y (d - a/2)$$
$$= 0.90(0.08)(60,000)(1.12 - 0.09/2)/12$$
$$= 385 \text{ ft-lb} > 256 \text{ OK}$$

Check shear:

$$V_u = 1.4(550) \, 2/2 = 770 \text{ lb}$$

$$\phi V_n = \phi 2\sqrt{f'_c} \, b_w d = 0.85(2)\sqrt{5000}(12)(1.12)$$
$$= 1616 \text{ lb} > 770 \text{ OK}$$

Also check panel for stripping and handling—see Part 5.

Example 7.3.2 Column cover acting as a form

Given:

The column shown below.
Precast concrete strength: $f'_c = 5000$ psi
C.i.p. concrete placement rate: 8 ft per hr
Maximum height of c.i.p. concrete: 6 ft
C.i.p. concrete temperature: 90 °F

Problem:

Determine panel reinforcement and tie requirements.

Solution:

Design for pressure of wet concrete:

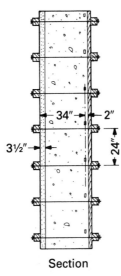

Section

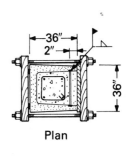

Plan

$$p = 150 + 9000R/T = 150 + 9000(8)/90$$
$$= 950 \text{ psf}$$

or a maximum of:

$$p = 150(6) = 900 \text{ psf} < 950. \text{ Use 900 psf}$$

Check flexure:

$$b = 36 - 2 - 3.50/2$$
$$= 32.25 \text{ in.} = 2.69 \text{ ft}$$

$$\ell = 36 - 3.50$$
$$= 32.50 \text{ in.} = 2.71 \text{ ft}$$

$$I_2 = I_1; \quad k = I_2/I_1(b/\ell) = 0.99$$

$$N = 2k + 3 = 4.98$$

$$M_B = \frac{p}{4N}(\ell^2 + kb^2)$$

$$= \frac{900}{4(4.98)}[(2.71)^2 + (0.99)(2.69)^2]$$

$$= 653 \text{ ft-lb}$$

$$f = \frac{6M}{bt^2} = \frac{6(653)(12)}{12(3.5)^2}$$

$$= 321 \text{ psi} < 5\sqrt{5000} = 353 \text{ psi OK}$$

Provide reinforcement to develop design strength $> 1.4M$:

$$M_u = 1.4(653) = 914 \text{ ft-lb}$$

Try W4 wire at 4 in. o.c.

$$A_s = 0.12 \text{ in.}^2/\text{ft}$$

$$a = \frac{A_s f_y}{0.85 f'_c b} = \frac{0.12(60)}{0.85(5)(12)} = 0.14 \text{ in.}$$

$$d = 3.50/2 = 1.75 \text{ in.}$$

$$\phi M_n = \phi A_s f_y (d - a/2)$$

$$= 0.90(0.12)(60,000)(1.75 - 0.14/2)/12$$

$$= 907 \text{ ft-lb} \approx 914 \text{ ft-lb say OK}$$

Tie force:

$$H = M/b = 653/2.69 = 243 \text{ lb/ft}$$

If ties are spaced 2 ft o.c.,

tie force = 2(243) = 486 lb

Use 3/16 × 1 in. A-36 steel strap. Weld to embedded plates.

Check shear:

$$V_u = 1.4(900) (2.71/2) = 1707 \text{ lb}$$

$$\phi V_n = \phi 2 \sqrt{f_c'}\, b_w d = 0.85(2)\sqrt{5000}(12)(1.75)$$

$$= 2524 \text{ lb} > 1707 \text{ OK}$$

Also check form for stripping and handling.

Example 7.3.3 Precast fascia as form for beam

Given:

Precast concrete fascia forming a beam supporting 24-ft span hollow-core floor as shown below. Beam is supported on 2 ft square columns at 20 ft on center.

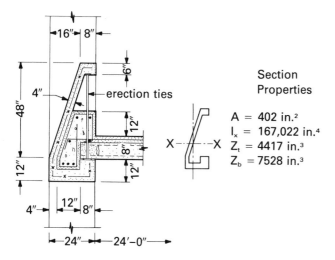

Section Properties

$A = 402 \text{ in.}^2$
$I_x = 167,022 \text{ in.}^4$
$Z_t = 4417 \text{ in.}^3$
$Z_b = 7528 \text{ in.}^3$

Problem:

Check stresses of fascia panel during erection.

Solution:

Loads:

Hollow-core slabs:
0.055(24/2 + 8/12)   = 0.70 k/ft
3 in. lightweight topping: 0.03 (24.2) = 0.36
C.i.p. lightweight concrete beam: = 0.30
Precast concrete fascia:
area = 420 sq in.   = 0.44
Total dead load   = 1.80 k/ft

Construction live load: 0.025 (24/2) = 0.30 k/ft

Clear span = 20 − 2 = 18 ft

$$M_{d\ell} = 1.80 (18)^2/8 = 72.9 \text{ ft-kip}$$

$$M_{\ell\ell} = 0.30 (18)^2/8 = \underline{12.2}$$

Total   = 85.1

Stresses:

$$f_t = 85,100 (12)/4417 = 231 \text{ psi compression}$$

$$f_b = 85,100 (12)7528$$

$$= 136 \text{ psi} < 5\sqrt{5000} = 353 \text{ psi OK}$$

$$V_u = [1.4 (1.80) + 1.7 (0.3)] (18/2) = 27.3 \text{ kips}$$

$b_w = 4 \text{ in.}, d = 57 \text{ in.}$

$$\phi V_n = \phi 2 \sqrt{f_c'}\, b_w d$$

$$= 0.85 (2) \sqrt{5000} (4) (57)/1000$$

$$= 27.4 \text{ kips} > 27.3 \text{ kips OK}$$

Also check for stripping and handling.

7.4 Column Covers and Mullions

The use of precast concrete panels as covers over steel or cast-in-place concrete columns and beams, and as mullions, is a common method of achieving architectural expression, special shapes, or fire rating.[10]

Column covers are usually supported by the structural column or the floor, and are themselves designed to transfer no vertical load other than their own weight. The vertical load of each length of column cover section is usually supported at one elevation, and tied back top and bottom to the floors for lateral load transfer and stability. In order to minimize erection costs and horizontal joints, it is desirable to make the covers as long as possible, subject to limitations imposed by weight and handling.

With adequate shear connectors between the cover and the column, the cover can add stiffness to the column. In tall buildings, this combined stiffness may significantly reduce the calculated drift.

Mullions are vertical elements serving to separate glass areas. They generally resist only wind loads applied from the adjacent glass, and must be stiff enough to maintain deflections within the limitations imposed by the window manufacturer. Since mullions are often thin, they are sometimes prestressed to prevent cracking.

Column covers and mullions are usually major focal points in a structure, and esthetic success requires that careful thought be given to all facets of design and erection. Following are some items which should be considered:

1. Since column covers and mullions are often isolated elements forming a long vertical line, any variation from a vertical plane is readily observable. This variation is usually the result

of the tolerances allowed in the structural frames. To some degree these variations can be handled by precast connections with adjustability. The designer should plan a clearance of at least 1½ in. between the panel and structure. For steel columns, the designer should consider the clearances around splice plates and projecting bolts.

2. Provide support at only one elevation for vertical loads, and at additional locations for lateral loads and stability. When access is available, consider providing an intermediate connection for lateral support and restraint of bowing.

3. Column covers and mullions which project from the facade will be subjected to shearing wind loads. Connections must be provided to resist these forces.

4. Members which are exposed to the environment will be subjected to temperature and humidity change. Horizontal joints between abutting precast column covers and mullions should be wide enough to permit length changes and rotation from temperature gradients. The behavior of thin flexible members will be improved by prestressing.

5. Due to vertical loads and the effects of creep and shrinkage in cast-in-place concrete columns, structural columns will tend to shorten. The width of the horizontal joint between abutting precast covers should be sufficient to permit this shortening to occur freely.

6. The designer must clearly envision the erection process. Column cover connections are often difficult to reach and, once made, difficult to adjust. This problem of access is compounded when all four sides of a column are covered for a height exceeding the length of a precast cover. Sometimes this problem can be solved by welding the lower piece to the column and anchoring the upper piece to the lower with dowels set in front, or by a mechanical device that does not require access.

7. Use of insulation on the interior face of the column cover reduces heat loss at these locations. Such insulation will also minimize temperature differentials between exterior columns and the interior of the structure.

Typical connection details are illustrated in Fig. 7.4.1. A connection detail for a single-story column cover is shown in (b), while (c) illustrates a detail for a multi-story column cover. Generally, the precast elements forming column covers and mullions are supported near the bottom; however they may also be hung as illustrated by (d). Where welds are shown, bolts can be substituted. The column covers are detailed to provide 1½ in. minimum clearance. In some cases, a larger clearance may be required. All joints are caulked after completion of the connection.

7.5 Veneered Panels

7.5.1 General

Precast concrete panels faced with brick, tile or natural stone combine the rich beauty of traditional materials with the strength, versatility, and economy of precast concrete. Some examples are shown in Fig. 7.5.1.

Structural design of veneered precast concrete units is the same as for other precast concrete wall panels, except that special consideration must be given to the veneer material and its attachment to the concrete. The physical properties of the facing material must be compared with the properties of the concrete back-up. These properties include:

1. Tensile, compressive and shear strength
2. Modulus of elasticity
3. Coefficient of thermal expansion
4. Volume change with moisture absorption.

Some natural stone veneers exhibit different properties depending on the orientation of the applied force with respect to the natural planes of the material.

Cover depth of reinforcing steel should be a minimum of ½ in. at the veneer surface, maintained by non-corrosive spacers. Galvanized reinforcement is recommended at cover depths of less than ¾ in. Precautions should be taken to avoid any materials that could corrode and cause discoloration of the veneer.

Because of the difference in material properties, veneered panels are somewhat more susceptible to bowing than all-concrete units. This is a consideration in the reinforcement design. In some cases, reinforcing trusses are used to add stiffness; in others, concrete ribs are formed on the back of the panel.

The method of attachment of facing materials, through bond and/or mechanical anchorage, may depend on stresses to which the panel will be subjected, such as:

1. Shrinkage of concrete during curing
2. Stresses imposed during handling and erection
3. Thermal response caused by different coefficients of expansion and by thermal gradients (see Part 3)
4. Service loads.

Fig. 7.4.1 Typical connections for column covers

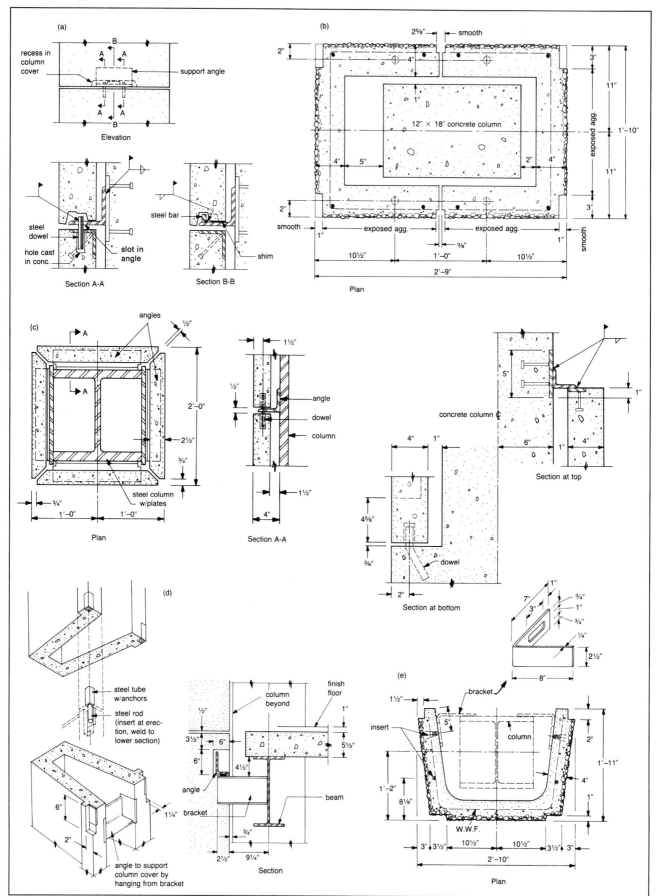

PCI Design Handbook

If the veneer is bonded to the backup, the differential shrinkage of the concrete and veneer will cause outward bowing in a simple span panel. The flat surfaces of some veneers, such as cut stone, reveal bowing more prominently than other finishes. Therefore bowing may be a critical consideration even though the rigidity of the cut stone helps to resist bowing.

Cracking in the veneer may occur if the bonding or anchoring details force the veneer pieces to follow the bowing. This is particularly critical where the face materials are large (cut stone) and the concrete shrinkage is significant. A good mix design, quality control, and moist curing will help reduce shrinkage.

Even with concrete shrinkage kept to the lowest possible level, there will still be some interaction with the facing material either through bond or the mechanical anchors of the facing units. This interaction will be minimized if a bond-breaker is used between the facing material and the concrete. Connections of natural stone to the concrete should be made with mechanical anchors which can accommodate some relative in-plane movement, a necessity if bond-breakers are used. One exception is the limestone industry which uses rigid, rather than flexible connectors.

7.5.2 Structural Clay Products

Clay products which are bonded directly to concrete include brick, structural facing tile, and architectural terra cotta (ceramic veneer). Ceramic glaze units may craze from freeze-thaw cycles or the bond of the facing may fail on exposure. Manufacturers of the clay products should be consulted to determine products suitable for exterior exposure.

Since ceramic-glazed units have very low permeance to water vapor, it is recommended that a vapor barrier be installed on the warm side of walls enclosing areas in which the average relative humidity may be expected to exceed 50%.

Clay product faced precast panels may be designed as concrete members, neglecting the effect of the face veneer. Differential shrinkage and thermal expansion are also usually neglected. However, if the panel is to be prestressed, the effect of composite behavior and the resulting prestress eccentricity should be considered.

Sizes. Brick manufacturers should be consulted early in the design stage to determine available shapes, sizes and size deviations. Brick may be available from some suppliers in veneer thickness, usually 1/3 to 1/4 standard brick thickness.

Architectural terra cotta is a custom product and, within limitations, is produced in sizes for specific jobs. Two thicknesses and sizes of units are usually manufactured: 1¼ in. thick units, including dovetails spaced 5 in. on centers, may be 20 × 30 in.; 2¼ in. thick units including dovetails spaced 7 in. on centers, may be 32 × 48 in.

Bond. The nature of the surfaces of the brick are important for bond to the backup concrete. Smooth, dense, heavily sanded or glazed surfaces are not usually satisfactory where high bond is required. Textures which give good bond include: scored finish, in which the surface is grooved as it comes from the die; combed finish where the surface is altered by parallel scratches; and roughened finish produced by wire cutting or wire brushing to remove the smooth surface or die skin from the extrusion process.

With most clay products used to face precast concrete panels, metal ties are not required to attach them to the concrete. Occasionally, a die skin or heavily sanded brick is specified, and metal ties, such as those used in cavity wall construction (Fig. 7.5.2) may be required.

Fig. 7.5.1 Applications of veneer faced precast concrete

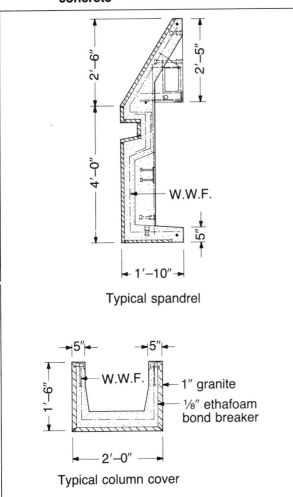

Typical spandrel

Typical column cover

Fig. 7.5.2 Wall ties

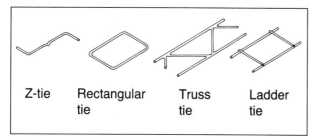

Z-tie   Rectangular   Truss   Ladder
          tie   tie   tie

Absorption. Brick with an initial rate of absorption less than 20 g per min. per 30 sq in. when tested in accordance with ASTM C67 is not required to be wetted. However, brick with a higher rate of absorption should be wetted prior to placement of the concrete. Terra cotta units must be soaked in water for at least 1 hr, and should be damp at the time of concrete placement.

Properties. Physical properties of brick vary considerably, depending on the source and grade of brick. Table 7.5.1 shows typical physical properties of brick from four different regions of the United States.[7]

The modulus of elasticity of brick ranges from 1.4 to 5.0 $\times$ 10[6] psi and Poisson's ratio from 0.04 to 0.11. The modulus of elasticity varies with the compressive strength to a compressive value of approximately 5000 psi. After this, there is little change. The average coefficient of thermal expansion of brick is 3.6 $\times$ 10[-6] in./in./°F. The thermal expansion of clay units is not the same as the thermal expansion of brick-faced precast panels due to joints. A moisture expansion of 0.0002 in./in. is recommended in the design of brick and tile panels.

The compressive strength of structural clay tile varies from 2000 to 8000 psi, the modulus of elasticity from 1.8 to 6.2 $\times$ 10[6] psi and Poisson's ratio from 0.05 to 0.10. Tile has an average coefficient of thermal expansion of 3.3 $\times$ 10[-6] in./in./°F.

The compressive strength of terra cotta units usually ranges from 8000 to 11,000 psi and the average coefficient of thermal expansion is 4.0 $\times$ 10[-6] in./in./°F.

Properties of clay products vary widely, and consultation with local suppliers is recommended.

7.5.3 Natural Stone[8,9]

Natural stone facings are used in various sizes, shapes and colors to provide an infinite number of pattern and color possibilities.

Sizes. Stone veneers used for precast facing are usually thinner than those used for conventionally set stone. Marble may be fabricated with a minimum 7/8 in. thickness. Marble pieces at this thickness will be limited to an area of approximately 20 sq ft with a maximum dimension of about 8 ft. Larger pieces tend to crack from differential shrinkage or thermal movement. Limestone may have a minimum thickness of 1½ in. At this thickness, the pieces will have an area of 4 to 12 sq ft. Thicker pieces may be as large as 18 ft high by 8 ft wide. Minimum thickness of granite depends on the sawings, handling and finishing, and type of anchor. At the present time, U.S. fabricators will supply granite with a minimum thickness of 1½ to 2 in., while foreign suppliers may supply granite as thin as 7/8 in. Depending on the color range, the size of granite pieces varies from 3 $\times$ 3 ft to 5 $\times$ 7 ft. The length and width of veneer materials should be sized to a tolerance of plus zero, minus 1/8 in. Flatness tolerances for finished surfaces depend on the type of finish. For example, the granite industry tolerances vary from 3/64 in. for a polished surface to 3/16 in. for flame (thermal) finish when measured with a 4 ft straightedge. Thickness variations are not important since concrete will provide a uniform back face, except at corner butt joints. In such cases, the finished edges should be within 1/16 in. of specified thickness.

Table 7.5.1 Average physical properties of brick[7]

Source of Brick	Compressive Strength*		Modulus of Rupture		Tensile Strength,	Shearing Strength,
	Flat-wise, psi	Edge-wise, psi	Flat-wise, psi	Edge-wise, psi	psi	psi
Chicago	3280	3350	1225	1340	417	1100
Detroit	3540	3270	670	680	222	1165
Mississippi	3410	3625	820	760	317	1590
New England	8600	11470	1550	1640	601	3550

*The ratio of compressive strength of clay and shale brick tested on edge to the compressive strength tested flatwise ranges from 0.74 to 2.3.

PCI Design Handbook

Anchorage of stone facing. It is recommended that there be no concrete bonding between stone veneer and concrete backup in order to minimize bowing, cracking and staining of the veneer. Mechanical anchors should be used to secure the veneer.

The following methods have been used to break the bond between the veneer and concrete: (1) a liquid bondbreaker of sufficient thickness to provide a low shear modulus, applied to the veneer back surface prior to placing the concrete; (2) a 6 mil polyethylene sheet; and (3) a 1/8 in. polyethylene foam pad.

Stone veneer is supplied with holes predrilled in the back surface for the attachment of mechanical anchors. Preformed anchors fabricated from stainless steel, Type 304, are usually used. The number and location of anchors are scheduled (Figs. 7.5.3 through 7.5.5) based on the height, width, thickness and strength of the stone veneer, and on the desired mechanical bond. Anchor size and spacing in veneers of questionable strengths or with natural planes of weakness may require special analysis. Depth of holes should be approximately one-half the thickness of the veneer (minimum depth of 1/2 in.) and for granite and marble, the holes should be approximately 50% oversize to allow for differential movement between the stone and the concrete.

Typical anchor details and test data for veneers are shown in Figs. 7.5.3 through 7.5.5.

It is advisable to conduct tests on the stone anchoring system proposed for use. Test samples should be a typical panel section of about 1 sq ft and approximate as closely as possible actual panel anchoring conditions. A bond breaker is placed between stone and concrete during sample manufacture to eliminate any bond between veneer and concrete surface. Each test sample contains one or more anchors connecting stone to concrete back-up and a minimum of 10 tests are needed to determine pull-out and shear strength of each anchor.

Veneer jointing. Joints between veneer pieces on a precast element should be a minimum of 1/4 in. The veneer pieces may be spaced with a chemically neutral, resilient, non-removable gasket which will not stain the veneer nor adversely affect the sealant to be applied later. The gaskets are of a size and configuration that will provide a pocket to receive the sealant and also prevent any backing concrete from entering the joint between

Fig. 7.5.3 Typical anchor details and test data for marble veneers

Anchor Test Data

	Shear, lb	Tension, lb
Avg. Failure[1]	2500[2]	425[2]
Safety Factor	30 to 1[3]	5 to 1[3]

(1) Tests performed with 100% bond failure built into panel
(2) Per anchor
(3) based on anchor schedule

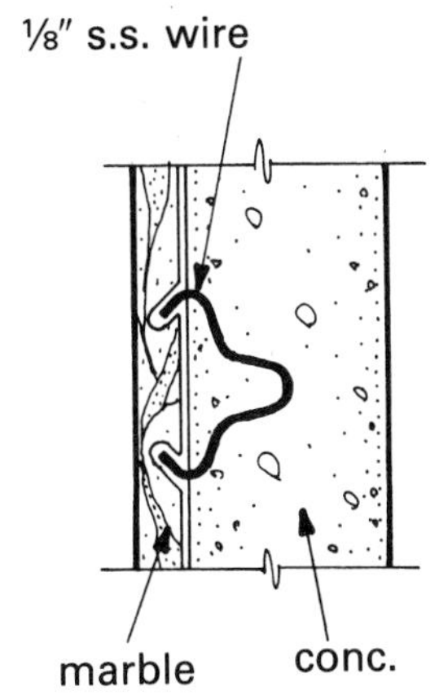

1/8" s.s. wire

marble   conc.

Anchor Schedule

		Width of Marble Facing Units			
		0'-6" to 1'-0"	1'-0" to 3'-6"	3'-6" to 4'-0"	Over 4'-0"
Height of Marble Facing Unit	Up To 4'-6"	One Anchor Top, One Anchor Bottom	Two Anchors Top, Two Anchors Bottom	Two Anchors Top, Two Anchors Bottom	Three Anchors Top, Three Anchors Bottom
	4'-6" And Over	Add One Center Anchor	Add One Center Anchor	Add Two Center Anchors	Add Two Center Anchors

Note: Top and bottom anchor holes of each marble facing unit should be drilled near the top and bottom edge and should be 3/4 in. deep at 35° angles in the back of the marble facing units and opposing each other in the vertical position to receive the preformed anchors. Anchor holes centered in the backs of the marble facing units, when necessary, should be drilled in the same manner as the top and bottom anchors.

the veneer units. Shore A hardness of the gasket should be less than 20.

Caulking should be of a type that will not stain the veneer material. In some projects caulking may be installed more economically and satisfactorily at the same time as the caulking between precast elements.

Properties. The strength of natural stone depends on several factors: the rift and cleavage of crystals, the degree of cohesion, the interlocking of crystals, and the nature of cementing materials present. The properties will vary with the locality from which it is quarried. Sedimentary and metamorphic rocks will exhibit different strengths when measured parallel and perpendicular to their original bedding planes.

The range of properties for some building stones are given in Table 7.5.2. These may be used as guides for preliminary design, but specific data should be obtained from suppliers where the design properties are critical. In most cases, the use of lower-bound data is satisfactory.

Samples. For cut stone finishes, samples and mock-up units are particularly important, and tests on the behavior of the unit for anticipated temperature changes may be required. To the degree possible, durability of the combined unit should be investigated.

Full scale mock-up units with cut stone in actual production sizes, along with casting and curing of the units under realistic production conditions, are recommended for each new or major application or configuration of the cut stones. Bowing should be measured over several weeks in the normal storage area and final details of stone sizes and fastening determined to suit observed behavior.

Acceptance criteria for the cut stone should be established prior to fabrication.

Fig. 7.5.4 Typical anchor details and test data for granite veneers

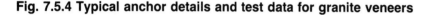

Anchor A

Tensile strength of one wire: 2346 lb
Shear strength of one wire: 1759 lb

Anchor B

Tensile strength of anchor: 4162 lb
Shear strength of anchor: 3121 lb

Capacity depends on depth of holes and grout material.

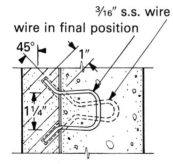

³⁄₁₆″ s.s. wire
wire in final position
45°
1″
1¼″

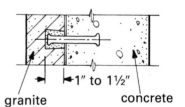

granite   concrete
1″ to 1½″

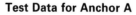

⁷⁄₁₆″ dia.   ¼″ dia.

Test Data for Anchor A

Type of Hole	Shear Parallel to Anchor, lb	Shear Perpendicular to Anchor, lb	Tension, lb
Diamond Cored Drilled	2400 to 2650	3200 to 3500	795-1115
Tungsten Carbide Drilled	2200 (approx.)	3000 (approx.)	425

Note:
(1) Factor of safety of anchor shear and tensile capability should be a minimum of 8.
(2) Toed-out, the anchor has approximately 50% more tensile capability than a toed-in anchor.
(3) As a general rule, one anchor should be used for every 2 sq ft.

PCI Design Handbook

Fig. 7.5.5 Typical anchor details for limestone veneers

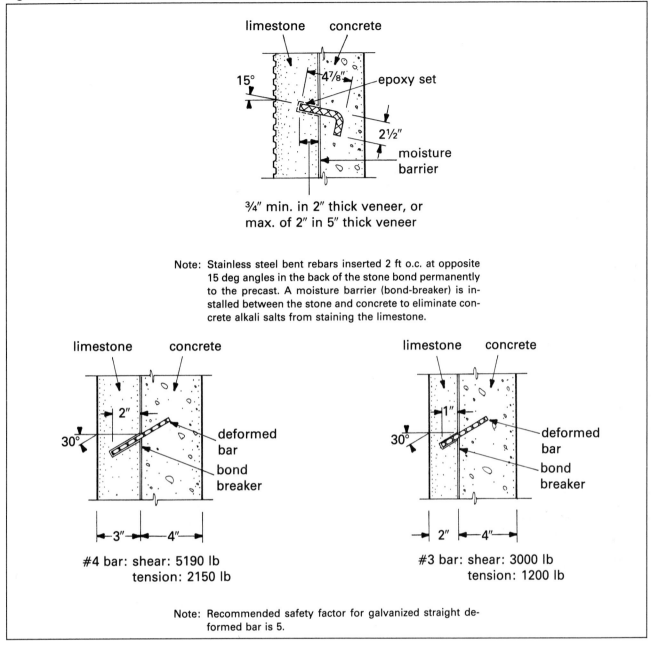

¾" min. in 2" thick veneer, or
max. of 2" in 5" thick veneer

Note: Stainless steel bent rebars inserted 2 ft o.c. at opposite
15 deg angles in the back of the stone bond permanently
to the precast. A moisture barrier (bond-breaker) is installed between the stone and concrete to eliminate concrete alkali salts from staining the limestone.

#4 bar: shear: 5190 lb
tension: 2150 lb

#3 bar: shear: 3000 lb
tension: 1200 lb

Note: Recommended safety factor for galvanized straight deformed bar is 5.

Table 7.5.2 Range of natural stone properties

Kind	Compression, psi	Tension, psi	Modulus of Rupture, psi	Shear, psi	Modulus of Elasticity, psi x 10⁶	Coef. of Thermal Expansion, in./in./°F x 10⁻⁶
Granite	10,000-40,000	600-1000	1100-3000	2000-4300	5.7-9.6	4.5 (avg)
Limestone	4000-10,000	300-375	700-1200	900-1800	3.0-5.4	2.4-3.0
Marble	6000-15,300[1] 7500-16,750[2]	400-2300	1100-2600 1100-2700	1650-4800 2350-4300	1.9-13.0 4.3-14.8	3.7-12.3
Serpentine	11,000-28,000	800-1600		2600-5000	4.8-9.6	
Sandstone	5000-20,000	280-500	500-1000	300-3000	1.9-7.7	
Slate	7000-31,000	3000-4300	4000-9000	2000-3600	9.0-15.0	

[1]Parallel   [2]Perpendicular

7.6 Design Example—Window Wall Panel

This example illustrates the design of a 2-story high non-load bearing window wall with deep reveals. The project is a 4-story building with a structural steel frame and cast-in-place concrete floor slabs. Precast concrete window wall panels are supported at their lowest points, and tied back to the structure at the tops. They transfer only horizontal loads to the steel frame. The vertical loads are transferred directly to the foundation (see Fig. 7.6.1).

Given:

Wind load = 15 psf, both during erection and after
f'_c = 5000 psi at 28 days
f'_c = 2000 psi at stripping

Problem:

Exposed face to be designed "crack free"
Cracks in rear face permitted without width restriction.

Solution:

Step 1:

Design panel for handling—see Part 5.

Each piece will be cast, stripped, stored, and shipped in the flat, face-down position.

Determine handling multipliers (Table 5.2.1)

Stripping: surface is not retarded, use 1.7
Yard handling: 1.2
Shipping: 1.5
Erection: 1.5

Fig. 7.6.1 Window wall panel example

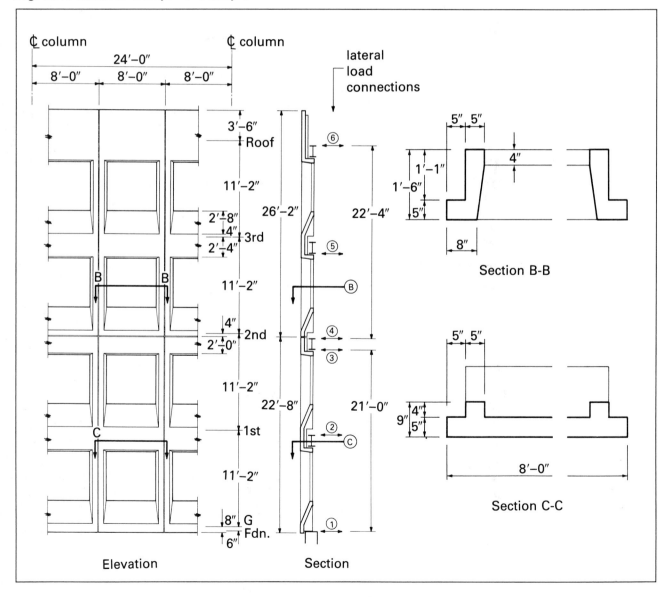

Determine allowable tensile stress in exposed face (Sect. 5.2.4.1)

At stripping, yard handling and storage:
$$f_r' = 5\sqrt{2000} = 224 \text{ psi}$$

At shipping and erection:
$$f_r' = 5\sqrt{5000} = 354 \text{ psi}$$

Check handling stresses

Since all handling except erection is to be in the flat position, using the same lift points, it is apparent that stripping is the critical design condition. After several trials and adjustment of lift point locations, it is determined that the tensile stresses at the critical section (Section C-C, Fig. 7.6.1) are too sensitive to slight relocations. Therefore, select locations so that there is compression in the face at this section. The loading and moment diagrams are shown in Fig. 7.6.2.

Fig. 7.6.2. Loads and moments on panel at stripping

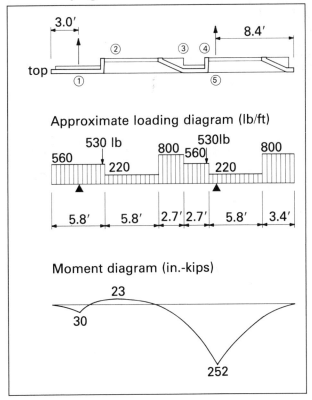

Approximate loading diagram (lb/ft)

Moment diagram (in.-kips)

Section properties at Section B-B:

A = 202 in.²	I = 6044 in.⁴
y_b = 8.00 in.	Z_b = 755 in.³
y_t = 10.00 in.	Z_t = 604 in.³

Section properties at Section C-C:

A = 520 in.²	I = 1801 in.⁴
y_b = 2.85 in.	Z_b = 632 in.³
y_t = 6.15 in.	Z_t = 293 in.³

Stress Summary:

Location	Moments (in-kips) (1)	(2)	Tensile stress (psi)	
5	−252	−428	709	high
4	−176	−299	1020	high
3	− 72	−122	416	high
2	+ 23	+ 39	52	>0
1	− 30	− 51	174	<224

(1) No multiplier
(2) 1.7 multiplier

Design panel reinforcement for stripping

Stresses at points 3, 4 and 5 exceed $5\sqrt{f_{ci}'}$. Therefore more than minimum reinforcement is required. Since these points are on the rear face of the panel, where crack control is not required, select reinforcement using the "z" factor, Sect. 4.2.2.1, as well as to satisfy the strength requirements, Sect. 4.2.1. From Fig. 4.2.5, the maximum allowable z = 175.

Reinforcement required at point 5, Fig. 7.6.2:

M_u per window jamb = 428/2 = 214 in-kips
d = 18 − 2 = 16 in.; b = 8 in.
Try 1−#5, grade 60, A_s = 0.31 in.

$$a = \frac{A_s f_y}{0.85 b f_c'} = \frac{0.31(60)}{0.85(8)(2)}$$
$$= 1.37 \text{ in.}$$

$$\phi M_n = \phi A_s f_y (d - a/2) = 0.9(0.31)(60)(16 - 0.68)$$
$$= 256 \text{ in.-kips} > 214 \text{ OK}$$

$$z = f_s \sqrt[3]{d_c A}$$

$f_s = 0.6 f_y = 36$ ksi; $d_c = 2$ in.
A = 4(5) = 20 in.²
$$z = 36\sqrt[3]{2(20)} = 123 \text{ kips/in.} < 175 \text{ OK}$$

Reinforcement required at point 4, Fig. 7.6.2 (use same for point 3):

M_u = 299 in.-kips
d = 9 − 1.5 = 7.5 in.; b = 96 in.

Try 2−#6 (one each rib), grade 60, A_s = 0.88 in.²

$$a = \frac{0.88(60)}{0.85(96)(2)} = 0.32 \text{ in.}$$

$$\phi M_n = 0.9(0.88)(60)(7.5 - 0.16)$$
$$= 349 \text{ in.-kips} > 299 \text{ OK}$$

d_c = 1.5, A = (3)(5) = 15 in.
$$z = 36\sqrt[3]{1.5(15)} = 102 \text{ kips/in.} < 175 \text{ OK}$$

Note: Although the section at points 3 and 4 is adequate without a strongback, extend the No. 5 bars in the ribs through the opening to provide stiffness to Section C-C.

Reinforcement required at points 1 and 2:

Since the tensile stress is below $5\sqrt{2000}$, only minimum reinforcement is required.

At point 1:
minimum A_s = 0.001bt = 0.001(area)
= 0.001(202) = 0.202 in.^{2}

At point 2:
minimum A_s = 0.001 bt = 0.001(96)(5)
= 0.48 in.^{2}

Transverse Bending

By inspection, the section is very stiff in the transverse direction. Therefore, use minimum reinforcement as shown in the summary, Fig. 7.6.3.

Step 2:

Design for condition during erection:

Overall construction sequence:
1. Erect structural steel frame.
2. Cast floor slabs.
3. Set precast concrete wall panels.

Precast concrete erection sequence:
1. Set bottom panel with lateral connections at points 1 and 3 only (Fig. 7.6.1).
2. Make lateral connection at point 2.
3. Set upper panel with lateral connection at points 4 and 6 only.
4. Make lateral connection at point 5.

Wind load on panel:
w = 15(8) = 120 lb/ft = 0.12 kips/ft

Determine loads on connections (see Fig. 7.6.4)

Before connection at point 2 is made:

a. Due to eccentric panel weight (Detail A):
Weight = 14 kips; e = 5 in.
$$R_1 = (-R_3) = \frac{14(5)}{21(12)} = 0.28 \text{ kips}$$

b. Due to wind:
$$R_1 = R_3 = \frac{0.015(8)(22.67)}{2} = 1.36 \text{ kips}$$

c. Load from upper panel on lower panel:
With connection 2 made before setting the upper

Fig. 7.6.3 Reinforcement summary

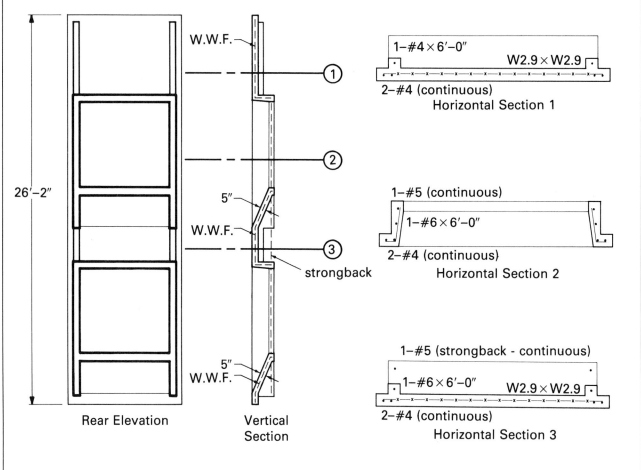

Rear Elevation   Vertical Section

26'-2"

W.W.F.

5"

W.W.F.

strongback

5"
W.W.F.

1
2
3

1–#4 × 6'–0"
W2.9 × W2.9
2–#4 (continuous)
Horizontal Section 1

1–#5 (continuous)
1–#6 × 6'–0"
2–#4 (continuous)
Horizontal Section 2

1–#5 (strongback - continuous)
1–#6 × 6'–0"
W2.9 × W2.9
2–#4 (continuous)
Horizontal Section 3

panel, the lateral loads at 1, 2 and 3 are indeterminate. To simplify, conservatively assume connection 2 has not been made.

Weight of upper panel = 13.3 kips
e (detail A) = 13 − 5 = 8 in.

$$R_1 = (-R_3) = \frac{13.3(8)}{21(12)} = 0.42 \text{ kips}$$

   d. Design loads (ACI 318-83, Sect. 9.2):
      Two connections per panel

        $U = 1.4D + 1.7L$
          $= 1.4(0.28 + 0.42)/2 = 0.49$ kips
        $U = 0.75(1.4D + 1.7L + 1.7W)$
          $= 0.75[1.4(0.70) + 1.7(1.36)]/2 = 1.23$ kips

   e. Loads on upper panel connections:

      An analysis similar to a through d on the upper panels shows that the maximum design load on the connection is 0.92 kips. For simplicity, design all connections for a lateral load of 1.23 kips. A connection load factor (see Sect. 6.3) of 1.3 should be applied to this.
        1.3(1.23) = 1.60 kips

Check panel bending

   Before connection at point 2 is made:

   $M = w\ell^2/8 = 0.12(21.0)^2/8$
          $= 6.62$ ft-kips = 79.4 in.-kips

   $M_u = 0.75(1.7W) = 0.75(1.7)(79.4) = 101$ in.-kips

   This is less critical than condition at stripping.

Connection design (Fig. 7.6.4)

   Assume the panel weight is transferred through shims at the second floor and foundation (Details A and B). The shims must be large enough to transfer load in plain concrete bearing (Sect. 6.8).

   Weight of upper panel   = 13.3 kips
   Weight of lower panel   = <u>14.0</u>
                                  27.3 kips

   Use 1.3 connection load factor.
   V_u (each connection) = 27.3(1.4)(1.3)/2
                      = 24.8 kips

   For more efficient erection, use same size shims throughout.

   Foundation concrete f'_c = 3000 psi.

   Rearrange Eq. 6.8.1. Conservatively assume A_2

Fig. 7.6.4 Connection design—example problem

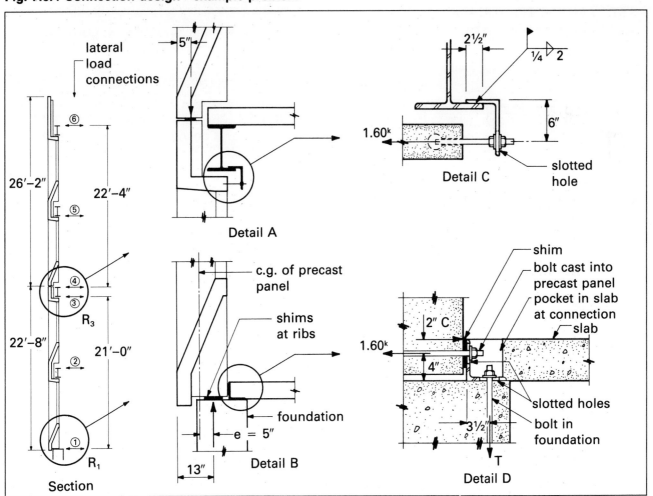

$= A_1$, and $C_r = 1$, since connections will be designed to take lateral loads.

$$A_1 = \frac{V_u}{\phi(0.85)f'_c} = \frac{24.8}{0.7(0.85)(3)} = 13.9 \text{ sq. in.}$$

Use standard size plastic or steel shims with a minimum area of 13.9 sq. in.

Connection at point 1:

Angle and bolt design: Choose a 6 in. long angle and minimum ⅝ in. diameter bolts. Note: Smaller bolts are likely to be damaged during handling.

From Eq. 6.5.13, with g = 4 in.; b = 6 in.; $f_y = 36$ ksi

$$t = \sqrt{\frac{4N_u g}{\phi f_y b}} = \sqrt{\frac{4(1.60)(4)}{0.9(36)(6)}} = 0.36 \text{ in.}$$

Use 6 × 6 × ⅜ angle 6 in. long

From Table 6.20.13, the minimum ⅝ in. bolt is adequate. The bolt in the foundation resists both tension and shear:

Taking moments about the upper toe of the angle:

$3.5T_u = 2(1.60)$; $T_u = 0.91$ kips

The interaction equation, Eq. 6.5.10, is applicable:

From Table 6.20.13, $P_s = 7.46$ kips; $V_s = 5.57$ kips

$$\left(\frac{P_u}{P_s}\right)^2 + \left(\frac{V_u}{V_s}\right)^2 = \left(\frac{0.91}{7.46}\right)^2 + \left(\frac{1.60}{5.57}\right)^2$$
$$= 0.015 + 0.083 = 0.098 < 1.0 \text{ OK}$$

Connections 2, 3 and 5:

Determine angle thickness:

Try 5 in. long angle.

From Eq. 6.5.13 with g = 6 in.; b = 5 in.

$$t = \sqrt{\frac{4N_u g}{\phi f_y b}} = \sqrt{\frac{4(1.60)(6)}{0.9(36)(5)}} = 0.49 \text{ in.}$$

Use angle 8 × 4 × ½ × 5 in. long.

Design weld to beam flange:

Refer to Sect. 6.5.6.

Try ¼ in. weld by 2 in. long each side of angle
$t_w = 0.707(0.25) = 0.177$ in.

Conservatively use elastic section properties. From Fig. 6.5.9, Case 2:

$$S = d^2/3 = (2.0)^2/3 = 1.33 \text{ in.}^3$$

From Table 6.20.1:

Use E70 electrodes, allowable strength = 35 ksi

$$f = \frac{P}{A} + \frac{M}{S} = \frac{1.60}{0.177(2.0)(2)} + \frac{1.60(6)}{1.33}$$

$$= 2.26 + 7.22 = 9.48 \text{ ksi} < 35 \text{ OK}$$

Connection at point 4:

By inspection, the minimum ⅝ in. diameter bolt is OK

Weld to beam:

Try ¼ in. weld by 2 in. long (one side only).

Table 6.5.9, Case 1:
$S = d^2/6 = (2.0)^2/6 = 0.67$

$$f = \frac{1.60}{0.177(2)} + \frac{1.60(5)}{0.67}$$

$$= 4.52 + 11.94 = 16.46 \text{ ksi} < 35 \text{ OK}$$

7.7 References

1. *Architectural Precast Concrete*, Prestressed Concrete Institute, Chicago, Illinois, 1973.

2. Phillips, W. R. and Sheppard, D. A., "Plant Cast Precast and Prestressed Concrete—A Design Guide," Prestressed Concrete Institute, Chicago, Illinois, 1977.

3. ACI Committee 347, "Precast Concrete Units Used as Forms for Cast-in-Place Concrete," *ACI Journal*, Proceedings V. 66, No. 10, October, 1969, pp. 798–813.

4. ACI Committees 347, "Recommended Practice for Concrete Formwork (ACI 347-78)," American Concrete Institute, Detroit, Michigan, 1978, 37 pp.

5. Birkeland, Halvard W., "Differential Shrinkage in Composite Beams," *ACI Journal*, Proceedings V. 56, No. 11, May, 1960, pp. 1123–1136.

6. Kulka, Felix, Lin, T. Y., and Yang, Y.C., "Prestressed Concrete Building Construction Using Precast Wall Panels," *PCI Journal*, V. 20, No. 1, January-February, 1975, pp. 62–72.

7. Plummer, Harry C., "Brick and Tile Engineering," Brick Institute of America, McLean, Virginia, 1962.

8. McDaniel, W. Bryant, "Marble-Faced Precast Panels," *PCI Journal*, V. 12, No. 4, August, 1967, pp. 29–37.

9. "Marble-Faced Precast Panels," National Association of Marble Producers, 1966.

10. *PCI Manual for Structural Design of Architectural Precast Concrete*, Prestressed Concrete Institute, Chicago, Illinois, 1977.

PART 8
TOLERANCES FOR PRECAST AND PRESTRESSED CONCRETE

		Page No.
8.1 General		8–2
	8.1.1 Definitions	8–2
	8.1.2 Purpose	8–2
	8.1.3 Responsibility	8–2
	8.1.4 Tolerance Acceptability Range	8–3
	8.1.5 Relationships between Different Tolerances	8–3
8.2 Product Tolerances		8–3
	8.2.1 General	8–3
	8.2.2 Overall Dimensions	8–3
	8.2.3 Sweep or Horizontal Alignment	8–3
	8.2.4 Position of Tendons	8–4
	8.2.5 Camber and Differential Camber	8–4
	8.2.6 Weld Plates	8–4
	8.2.7 Haunches of Columns and Wall Panels	8–5
	8.2.8 Warping and Bowing	8–5
	8.2.9 Smoothness	8–6
	8.2.10 Architectural vs. Structural Walls	8–6
8.3 Erection Tolerances		8–7
	8.3.1 General	8–7
	8.3.2 Recommended Erection Tolerances	8–7
	8.3.3 Mixed Building Systems	8–7
	8.3.4 Connections and Bearing	8–7
8.4 Clearances		8–7
	8.4.1 General	8–7
	8.4.2 Joint Clearance	8–13
	8.4.3 Procedure for Determining Clearance	8–13
	8.4.4 Clearance Examples	8–13
8.5 Interfacing Tolerances		8–17
	8.5.1 General	8–17
	8.5.2 Interface Design Approach	8–17
	8.5.3 Characteristics of the Interface	8–17
8.6 References		8–19

TOLERANCES FOR PRECAST AND PRESTRESSED CONCRETE

8.1 General

8.1.1 Definitions

Tolerance—
- The permitted variation from a basic dimension or quantity, as in the length or width of a member
- The range of variation permitted in maintaining a basic dimension, as in an alignment tolerance
- A permitted variation from location or alignment.

Product Tolerances—Those relating to individual precast concrete members.

Erection Tolerances—Those required for acceptable matching of precast members after they are erected.

Interfacing Tolerances—Those associated with other materials in contact with or in close proximity to precast concrete.

Variation—The difference between the actual and the basic dimension. Variations may be either negative (less) or positive (greater).

Basic Dimension—Those shown on contract drawings or in specifications. Basic dimensions apply to size and location. May also be called "nominal" dimensions.

Working Dimension—The planned dimension of a member which considers both its basic dimension and joints or clearances. For example, a member with a basic width of 8'-0" may have a working width of 7'-11¼". Product tolerances are applied to working dimensions.

Actual Dimension—The measured dimension of the member after casting. This may differ from the working dimension due to construction- and material-induced variation.

Alignment Face—The face of a precast element which is to be set in alignment with the face of adjacent elements or features.

Primary Control Surface—A surface on a precast element which is dimensionally controlled during erection. Clearance is generally allowed to vary so the primary control surface can be set within tolerance.

Secondary Control Surfaces—A surface on a precast element, the location of which is dependent on the locational tolerance of the primary control surface plus the product tolerances.

Feature Tolerance—The locational or dimensional tolerance of a feature, such as a corbel or a blockout, with respect to the overall member dimensions.

8.1.2 Purpose

Tolerances are normally established by economical and practical production, erection and interfacing considerations. Once established, they should be shown in the project documents, and used in detailing components. Architectural and structural concepts should be developed with the practical limitations of dimensional control in mind, as the tolerances will affect the dimensions of the completed structure.

Tolerances are required for the following reasons:

Structural. To ensure that structural design accounts for factors sensitive to variations in dimension. Examples include eccentric loadings, bearing areas, and locations of reinforcement and embedded items.

Feasibility. To ensure acceptable performance of joints and interfacing materials in the finished structure.

Visual. To ensure that the variations will be controllable and result in an acceptable looking structure.

Economic. To ensure ease and speed of production and erection.

Legal. To avoid encroaching on property lines and to establish a standard against which the work can be compared.

Contractual. To establish an acceptability range and also to establish responsibility for developing and maintaining specified tolerances.

8.1.3 Responsibility

While the responsibility for specifying and maintaining tolerances of the various elements may vary among projects, it is important that these responsibilities be clearly assigned. The conceptual design of a precast project is the place to begin consideration of dimensional control. The established tolerances or required performance should fall within generally accepted limits and should not be made more rigid than necessary.

Once the tolerances have been specified, and connections which consider those tolerances have been designed, the production of the elements must be organized to assure tolerance compliance. An organized quality control program which emphasizes dimensional control is necessary. Likewise, an erection quality assurance effort which includes a clear definition of responsibilities will aid in assuring that the products are assembled in accordance with the specified erection tolerances. Responsibility should include dimension verification and adjustment, when necessary, of both precast components and any interfacing structure.

8.1.4 Tolerance Acceptability Range

Tolerances must be used as guidelines for acceptablity and not limits for rejection. If specified tolerances are met, the member should be accepted. If not, the member may be accepted if it meets any of the following criteria:

1. Exceeding the tolerance does not affect the structural integrity or architectural performance of the member.
2. The member can be brought within tolerance by structurally and architecturally satisfactory means.
3. The total erected assembly can be modified to meet all structural and architectural requirements.

8.1.5 Relationships between Different Tolerances

A precast member is erected so that its primary control surface is in conformance with the established erection and interfacing tolerances. The secondary control surfaces are generally not directly positioned during erection, but are controlled by the product tolerances. Thus, if the primary control surfaces are within erection and interfacing tolerances, and the secondary surfaces are within product tolerances, the member is erected within tolerance. The result is that the tolerance limit for the secondary surface may be the sum of the product and erection tolerances.

Since tolerances for some features of a precast member may be additive, it must be clear to the erector which are the primary control surfaces. If both primary and secondary control surfaces must be controlled, provisions for adjustment should be included. The accumulated tolerance limits may have to be accommodated in the interface clearance. Surface and feature control requirements should be clearly outlined in the plans and specifications.

On occasion, the structure may not perform properly if the tolerances are allowed to accumulate. Which tolerance takes precedence is a question of economics. The costs associated with each of the three tolerances must be evaluated, recognizing unusual situations. This may include difficult erection requirements, connections which are tolerance sensitive, or production requirements which are set by the available equipment. Any special tolerance requirements should be clearly noted in the contract documents.

8.2 Product Tolerances

8.2.1 General

Detailed product tolerances are given in the full report of the PCI Committee on Tolerances[1] as well as in the *Manual for Quality Control for Plants and Production of Precast and Prestressed Concrete Products*[2] and *Manual for Quality Control for Plants and Production of Architectural Precast Concrete Products*.[3] The products included are listed in Table 8.2.1. Discussion of the more critical tolerances are given in the following sections. The values shown have become the consensus standards of the precast concrete industry. More rigid tolerances may significantly increase costs, so should not be specified unless absolutely necessary.

8.2.2 Overall Dimensions

Tolerances for the overall dimensions of most common products are given in Table 8.2.1. Architectural precast concrete panels have plan dimension tolerances that vary with panel size from ± 1/8 in. for a dimension under 10 ft to ± 1/4 in. for a dimension of 20 to 40 ft. Overall thickness of wall panels are listed in Table 8.2.1 under depth tolerances.

Top and bottom slabs (flanges) of box beams and hollow-core slabs are dependent on position of cores. Flange thickness tolerances are not given for hollow-core slabs. Instead, measured flange areas cannot be less than 85% of nominal calculated area.

The committee report emphasizes that the recommended tolerances are only guidelines. Different values may be applicable in some cases, and each project should be considered individually.

8.2.3 Sweep or Horizontal Alignment

Horizontal misalignment, or sweep, usually occurs as a result of form and member width tolerances. It can also result from prestressing with lateral eccentricity, which should be considered in the design. Joints should be dimensioned to accommodate such variations.

Tolerances generally vary with length of unit, for example, ± 1/8 in. per 10 ft. Top limits vary from ± 3/8 in. for wall panels and hollow-core slabs to ± 3/4 in. for joists usually used in composite construction.

Table 8.2.1 Typical tolerances for precast, prestressed concrete products*

Product Tolerances	Products
Length ** —	
± 1/2 in.	6,7,8,9,13
± 3/4 in.	3,5
± 1 in.	1,2,4,11,12
Width ** —	
± 1/4 in.	1,2,3,5,6,7,8,9,12
+ 3/8 in., − 1/4 in.	4
± 3/8 in.	11,13
Depth —	
+ 1/4 in., − 1/8 in.	10
± 1/4 in.	1,2,3,5,6,7,8,9,12
+ 1/2 in., − 1/4 in.	4
± 3/8 in.	11
± 1/2 in.	13
Flange Thickness —	
+ 1/4 in., − 1/8 in.	1,2,8,10,12
± 1/4 in.	3,4,13
Web Thickness —	
± 1/8 in.	1,8,10,12
± 1/4 in.	2,3
+ 3/8 in., − 1/4 in.	4
± 3/8 in.	5
Position of Tendons —	
± 1/4 in.	1,2,3,4,5,6,8,9,11,12
± 1/8 in.	10
Camber, variation from design —	
± 1/4 in. per 10 ft., ± 3/4 in. max.	1,2,12
± 1/8 in. per 10 ft., ± 1 in. max.	4
± 3/4 in. max.	3
± 1/2 in. max.	5
Camber, differential —	
1/4 in. per 10 ft., 3/4 in max.	1,2,5
Bearing Plates, position —	
± 1/2 in.	1,2,3,12
± 5/8 in.	4
Bearing Plates, tipping and flushness —	
± 1/8 in.	1,2,3,4,12

*For more details such as graphic descriptions of features to which tolerances apply and tolerances for sleeves, blockouts, inserts, plates, end squareness, surface smoothness, etc., see committee report.[1]

**See Sect. 8.2.2 for dimensional tolerances for architectural wall panels.

Key:
1 = double tee
2 = single tee
3 = building beam (rect. and ledger)
4 = I-beam
5 = box beam
6 = column
7 = hollow-core slab
8 = ribbed wall panel
9 = insulated wall panel
10 = architectural wall panel
11 = pile
12 = joist
13 = step unit

8.2.4 Position of Tendons

It is a common practice to use 5/8 in. diameter holes in end dividers (bulkheads or headers) for 3/8 to 1/2 in. diameter strands, since it is costly to switch end dividers for different strand diameters. Thus, better accuracy is achieved when using larger diameter strands.

Generally, individual tendons must be positioned within ± 1/4 in. of design position and bundled tendons within ± 1/2 in. Hollow-core slabs have greater individual strand tolerances as long as the center of gravity of the strand group is within ± 1/4 in. and a minimum cover of 3/4 in. is maintained.

8.2.5 Camber and Differential Camber

Design camber is generally based on camber at release of prestress; thus, camber measurements on products should be made as soon after stripping as possible. Differential camber refers to the final in-place condition of adjacent products.

It is important that cambers are measured at the same time of day, preferably in the early hours before the sun has begun to warm the members. Cambers for all units used in the same assembly should be checked at the same age.

If a significant variation in camber from calculated values is observed, the cause should be determined or the effect of the variation on the performance of the member evaluated. If differential cambers exceed recommended tolerances, additional effort is often required to erect the members in a manner which is satisfactory for the intended use.

The final installed differential between two adjacent cambered members erected in the field may be the combined result of member differential cambers, variations in support elevations, and any adjustments made to the members during erection.

For most flexural members, maximum camber variation from design camber is ± 3/4 in., and maximum differential camber between adjacent units of the same design is 3/4 in. This may be increased for joists used in composite construction. Recommendations for hollow-core slabs are not definitive, because of production variations between systems, and require discussion with the producer.

8.2.6 Weld Plates

In general, it is easier to hold plates to closer tolerances at the bottom of the member (or against the side form) than with plates cast on top of the member. Bottom and side plates can be fastened to the form and hence are less susceptible to movement caused by vibration. This applies to position of weld plates as well as tipping and flushness.

The tolerance on weld plates is less restrictive than for bearing plates. The position tolerance is $\pm$ 1 in. for all products except for hollow-core slabs with a tolerance of $\pm$ 2 in.; tipping and flushness tolerance is $\pm$ 1/4 in.

8.2.7 Haunches of Columns and Wall Panels

The importance of haunch location tolerances depends on the connection at the base of the member. Since base connections usually allow some flexibility, it is more important to control dimensions from haunch to haunch in multilevel columns or walls than from haunch to end of member.

The haunch to haunch tolerance is $\pm$ 1/8 in. Bearing surface squareness tolerance is $\pm$ 1/8 in. per 18 in. with a maximum of $\pm$ 1/4 in., except for architectural precast concrete panels, with a tolerance of $\pm$ 1/8 in., and columns, with a maximum of $\pm$ 1/8 in. in short direction and $\pm$ 3/8 in. in long direction.

8.2.8 Warping and Bowing

Warping and bowing tolerances affect panel edge matchup during erection, and the appearance of the erected members. They are especially critical with architectural panels.

Warping is a variation from plane in which the corners of the panel do not fall within the same plane. Warping tolerances are given in terms of corner variation, as shown in Fig. 8.2.1. The allowable variation from the nearest adjacent corner is 1/16 in. per ft.

Bowing differs from warping in that two opposite edges of a panel may fall in the same plane, but the portion between is out of plane (Fig. 8.2.2). Bowing tolerance is L/360, where L is the length of bow. Maximum tolerance on differential bowing between panels of the same design is 1/2 in.

The effects of differential temperature and moisture absorption between the inside and outside of a panel, and prestress eccentricity, should be considered in design of the panel and its connections to minimize bowing and warping. Pre-erection storage conditions may also affect warping and bowing (see Sects. 3.3.2 and 5.2.10).

Thin panels are more likely to bow, and the tolerances should be more liberal. Table 8.2.2 gives thicknesses, related to panel dimensions, below which the warping and bowing tolerances given above may not apply. (Note: Table 8.2.2 is not intended to limit panel thickness.) For example, a panel that is 16 × 8 ft, and less than 6 in. thick, may require greater warping and bowing toler-

Table 8.2.2 Minimum thickness for use of normal bowing and warping tolerances*, in.

Panel dimensions, ft	8	10	12	16	20	24	28	32
4	3	4	4	5	5	6	6	7
6	3	4	4	5	6	6	6	7
8	4	5	5	6	6	7	7	8
10	5	5	6	6	7	7	8	8

*Thinner panels may require more liberal tolerances

Fig. 8.2.1 Definition of panel warping

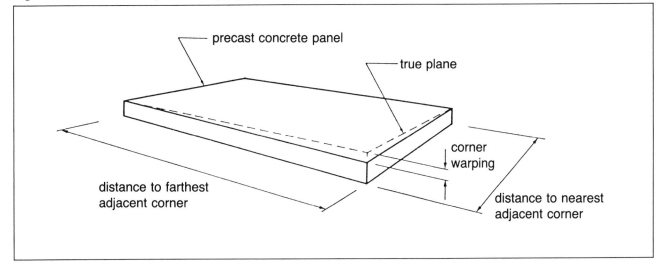

precast concrete panel

true plane

corner warping

distance to farthest adjacent corner

distance to nearest adjacent corner

ance than indicated above.

Similarly, panels made from concrete with over 3/4 in. aggregate, panels using two significantly different concrete mixes, and veneered and insulated panels may require special consideration. In all cases, the local precaster should be consulted regarding overall economic and construction feasibility.

8.2.9 Smoothness

Local smoothness describes the condition where small areas of the surface may be out of plane, as shown in Fig. 8.2.3. The tolerance for this type of variation is 1/4 in. per 10 ft for all products. The tolerance is usually checked with a 10 ft straight edge or the equivalent, as explained in Fig. 8.2.3.

8.2.10 Architectural vs. Structural Walls

When discussing tolerances, "architectural panel" refers to a class of tolerances specified, and not necessarily to the use of the member in the final structure. Architectural panels usually require more rigid tolerances than structural members for esthetic reasons.

Double tees, hollow-core slabs and solid slabs are often used for exterior facades, but are not classed as "architectural panels". The same degree of accuracy should not be expected, since the manufacturing method is the same as for the structural product for which these elements are more commonly used. If more rigid tolerances are required, they should be clearly indicated in the contract documents, and subsequent higher costs anticipated.

Fig. 8.2.2 Definition of panel bowing

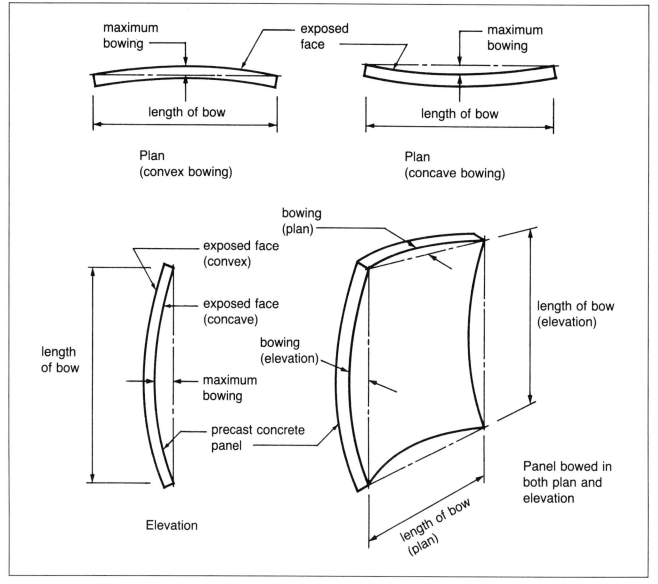

PCI Design Handbook

Fig. 8.2.3 Local smoothness variation

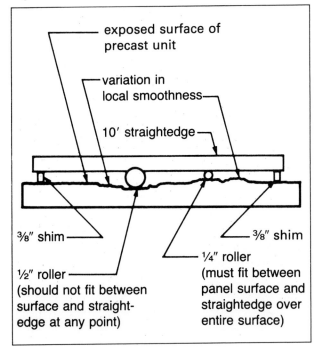

exposed surface of
precast unit

variation in
local smoothness

10' straightedge

⅜" shim

¼" roller
(must fit between
panel surface and
straightedge over
entire surface)

½" roller
(should not fit between
surface and straight-
edge at any point)

⅜" shim

8.3 Erection Tolerances

8.3.1 General

Erection tolerance values are those to which the primary control surfaces of the member should be set. The final location of other features and surfaces will be the result of the combination of the erection tolerances and the product tolerances given in Sect. 8.2.

Because erection is equipment- and site-dependent, there may be good reason to vary some of the recommended tolerances to account for unique project conditions. Combining liberal product tolerances with restrictive erection tolerances may place an unrealistic burden on the erector. Thus, the designer should review proposed tolerances with manufacturers and erectors prior to deciding on the final project tolerances.

To minimize erection problems, the dimensions of the in-place structure should be checked prior to starting precast erection. After erection, and before other trades interface with the precast concrete members, it should be verified that the precast elements are erected within tolerances.

8.3.2 Recommended Erection Tolerances

Figs. 8.3.1 to 8.3.5 show erection tolerances for:
- Precast element to precast element
- Precast element to cast-in-place concrete
- Precast element to masonry
- Precast element to steel construction

8.3.3 Mixed Building Systems

Mixed building systems combine precast and prestressed concrete with other materials, usually cast-in-place concrete, masonry or steel. Each industry has its own recommended erection tolerances which apply when its products are used exclusively. The compatibility of those tolerances with the precast tolerances should be checked and adjusted when necessary.

Example 8.4.2 shows one problem that can occur when erection tolerances are chosen for each system without considering the project as a whole.

8.3.4 Connections and Bearing

The details of connections must be considered when specifying erection tolerances. Space must be provided to make the connection under the most adverse combination of tolerances.

Bearing length is measured in the direction of the span, and bearing width is measured perpendicular to the span. Bearing length is often not the same as the length of the end of a member over the support, as shown in Fig. 8.3.6. When they differ, it should be noted on erection drawings.

8.4 Clearances

8.4.1 General

Clearance is the space between adjacent members and provides a buffer area where erection and production tolerance variations can be absorbed. The following items should be addressed when determining the appropriate clearance to provide in the design:
- Product tolerance
- Type of member
- Size of member
- Location of member
- Member movement
- Function of member
- Erection tolerance
- Fireproofing of steel
- Thickness of plates, bolt heads, and other projecting elements.

Of these factors, product tolerances and member movement are the most significant. As shown in the examples, it may not always be practical to account for all possible factors in the clearance provided.

Fig. 8.3.1 Erection tolerances - beams and spandrels

a = Plan location from building grid datum... ±1 in.
a₁ = Plan location from centerline of steel*.. ±1 in.

b = Bearing elevation** from nominal elevation at support
      Maximum low .... ½ in.
      Maximum high .... ¼ in.

c = Maximum plumb variation over height of element
      Per 12 in. height .... ⅛ in.
      Maximum .... ½ in.

d = Maximum jog in alignment of matching edges
      Architectural exposed edges.. ¼ in.
      Visually noncritical edges.. ½ in.

e = Joint width
      Architectural exposed joints .... ± ¼ in.
      Hidden joints .... ± ¾ in.
      Exposed structural joint *not* visually critical................................. ± ½ in.

f = Bearing length*** (span direction) .... ± ¾ in.

g = Bearing width*** .... ± ½ in.

   *For precast elements erected on a steel frame, this tolerance takes precedence over tolerance dimension "a".

  **Or member top elevation where member is part of a frame without bearings.

 ***This is a setting tolerance and should not be confused with structural performance requirements set by the architect/engineer.

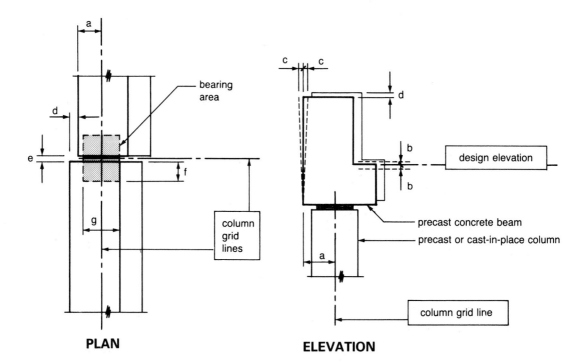

PLAN   **ELEVATION**

Precast element to precast concrete, cast-in-place concrete, masonry, or structural steel

Fig. 8.3.2 Erection tolerances — floor and roof members

a = Plan location from building grid datum .... ± 1 in.
a₁ = Plan location from centerline of steel* .... ± 1 in.

b = Top elevation from nominal top elevation at member ends
    Covered with topping .... ± ¾ in.
    Untopped floor .... ± ¼ in.
    Untopped roof .... ± ¾ in.

c = Maximum jog in alignment of matching edges
    (both topped and untopped construction) .... 1 in.

d = Joint width
    0 to 40 ft member length .... ± ½ in.
    41 to 60 ft member length .... ± ¾ in.
    61 ft plus .... ± 1 in.

e = Differential top elevation as erected
    Covered with topping .... ¾ in.
    Untopped floor .... ¼ in.
    Untopped roof .... ¾ in.

f = Bearing length** (span direction) .... ± ¾ in.
g = Bearing width** .... ± ½ in.
h = Differential bottom elevation of exposed hollow-core slabs*** .... ¼ in.

*For precast concrete erected on a steel frame building, this tolerance takes precedence over tolerance
on dimension "a".
**This is a setting tolerance and should not be confused with structural performance requirements set
by the architect/engineer.
***Untopped installations will require a larger tolerance.

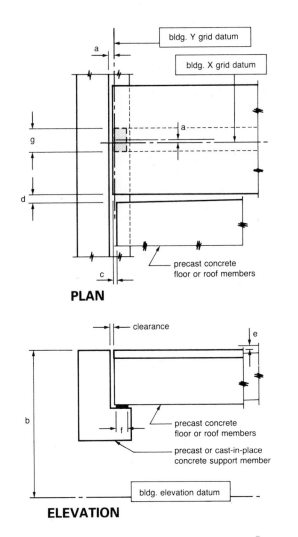

PLAN

ELEVATION

**Precast element to precast or
cast-in-place concrete or masonry**

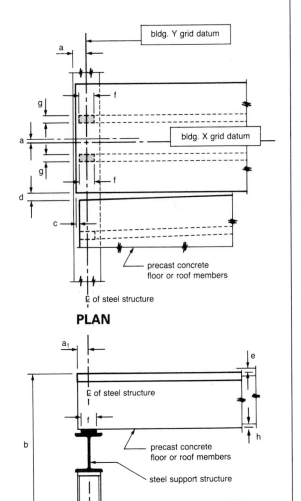

PLAN

ELEVATION

Precast element to structural steel

Fig. 8.3.3 Erection tolerances — columns

a = Plan location from building grid datum
    Structural applications .... ± ½ in.
    Architectural applications .... ± ⅜ in.

b = Top elevation from nominal top elevation
    Maximum low .... ½ in.
    Maximum high .... ¼ in.

c = Bearing haunch elevation from nominal elevation
    Maximum low .... ½ in.
    Maximum high .... ¼ in.

d = Maximum plumb variation over height of element
    (element in structure of maximum height of 100 ft) .... 1 in.

e = Plumb in any 10 ft of element height .... ¼ in.

f = Maximum jog in alignment of matching edges
    Architectural exposed edges .... ¼ in.
    Visually non-critical edges .... ½ in.

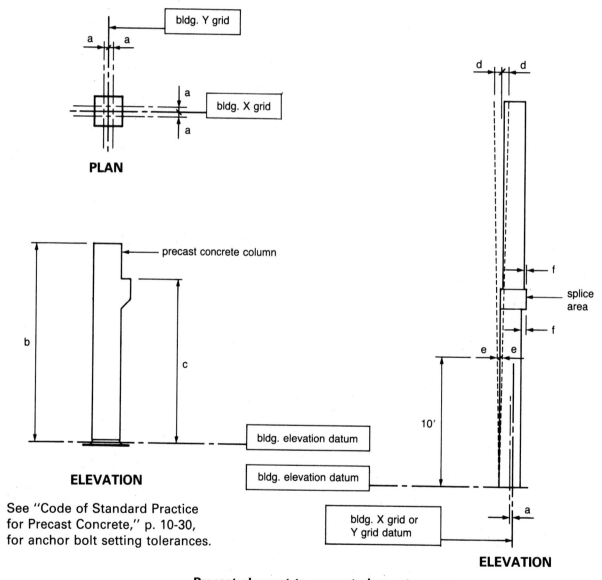

PLAN

precast concrete column

ELEVATION

bldg. elevation datum

See "Code of Standard Practice for Precast Concrete," p. 10-30, for anchor bolt setting tolerances.

splice area

bldg. elevation datum

10'

bldg. X grid or Y grid datum

ELEVATION

Precast element to precast element

Fig. 8.3.4 Erection tolerances — structural wall panels

a = Plan location from building grid datum* .... ± ½ in.
a_1 = Plan location from centerline of steel** .... ± ½ in.
b = Top elevation from nominal top elevation
    Exposed individual panel .... ± ½ in.
    Nonexposed individual panel .... ± ¾ in.
    Exposed relative to adjacent panel .... ½ in.
    Nonexposed relative to adjacent panel .... ¾ in.
c = Bearing elevation from nominal elevation
    Maximum low .... ½ in.
    Maximum high .... ¼ in.
d = Maximum plumb variation over height of structure or 100 ft. whichever is less* ... 1 in.
e = Plumb in any 10 ft of element height .... ¼ in.
f = Maximum jog in alignment of matching edges .... ½ in.
g = Joint width (governs over joint taper) .... ± ⅜ in.
h = Joint taper over length of panel .... ½ in.
h_{10} = Joint taper over 10 ft length .... ⅜ in.
i = Maximum jog in alignment of matching faces
    Exposed .... ⅜ in.
    Nonexposed .... ¾ in.
j = Differential bowing, as erected, between adjacent members of the same design*** ½ in.

*For precast buildings in excess of 100 ft. tall, tolerances "a" and "d" can increase at the rate of ⅛ in.
per story to a maximum of 2 in.
**For precast concrete erected on a steel frame building, this tolerance takes precedence over tolerance
on dimension "a".
***Refer to Sect. 8.2.8 for description of bowing tolerance.

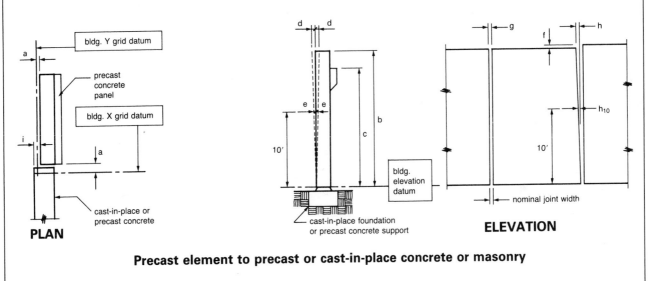

Precast element to precast or cast-in-place concrete or masonry

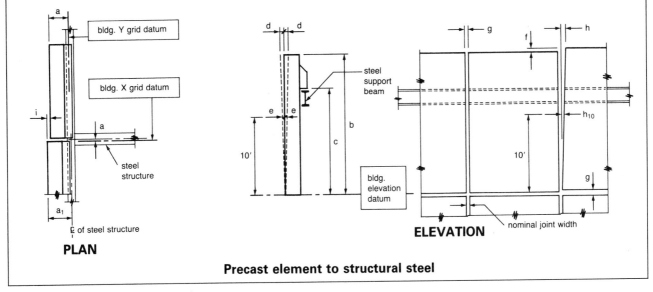

Precast element to structural steel

Fig. 8.3.5 Erection tolerances — architectural wall panels

a	=	Plan location from building grid datum*	± ½ in.
a₁	=	Plan location from centerline of steel**	± ½ in.
b	=	Top elevation from nominal top elevation	
		Exposed individual panel	± ¼ in.
		Nonexposed individual panel	± ½ in.
		Exposed relative to adjacent panel	¼ in.
		Nonexposed relative to adjacent panel	½ in.
c	=	Support elevation from nominal elevation	
		Maximum low	½ in.
		Maximum high	¼ in.
d	=	Maximum plumb variation over height of structure or 100 ft. whichever is less*	1 in.
e	=	Plumb in any 10 ft of element height	¼ in.
f	=	Maximum jog in alignment of matching edges	¼ in.
g	=	Joint width (governs over joint taper)	± ¼ in.
h	=	Joint taper maximum	⅜ in.
h₁₀	=	Joint taper over 10 ft. length	¼ in.
i	=	Maximum jog in alignment of matching faces	¼ in.
j	=	Differential bowing or camber as erected between adjacent members of the same design***	¼ in.

*For precast buildings in excess of 100 ft. tall, tolerances "a" and "d" can increase at the rate of ⅛ in. per story to a maximum of 2 in.

**For precast concrete erected on a steel frame building, this tolerance takes precedence over tolerance on dimension "a".

***Refer to Sect. 8.2.8 for description of bowing tolerance.

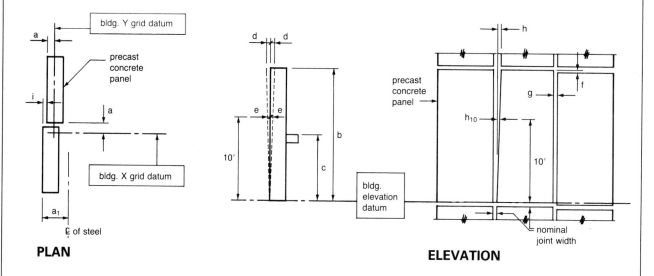

PLAN

ELEVATION

Precast element to precast or cast-in-place concrete, masonry, or structural steel

PCI Design Handbook

Fig. 8.3.6 Relationship between bearing length and length over support

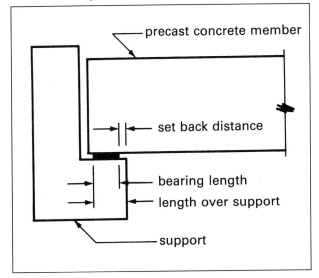

precast concrete member

set back distance

bearing length

length over support

support

8.4.2 Joint Clearance

Joints between architectural panels must accommodate variations in the panel dimensions and the erection tolerances for the panels. They must also provide a good visual line and sufficient width to allow for effective sealing. Generally, the larger the panel, the wider the joints should be. For most situations, architectural panel joints should be designed to be not less than 3/4 in. wide. Tolerances in overall building width and length are normally accommodated in panel joints.

8.4.3 Procedure for Determining Clearance

Following is a systematic approach to selecting an appropriate clearance:

Step 1:
Determine the maximum size of the members involved (basic or nominal dimension plus additive tolerances). This should include not only the precast and prestressed members, but also other materials.

Step 2:
Add the minimum space required for member movement.

Step 3:
Check if this clearance allows the member to be erected within the erection and interfacing tolerances, such as plumbness, face alignment, etc. Adjust the clearance as required to meet all the needs.

Step 4:
Check if the member can physically be erected

with this clearance. Consider the size and location of members in the structure and how connections will be made. Adjust the clearance as required.

Step 5:
Review the clearance to see whether increasing its dimensions will allow easier, more economical erection without adversely affecting esthetics. Adjust the clearance as required.

Step 6:
Review structural considerations such as types of connections involved, sizes required, bearing area requirements, and other structural issues.

Step 7:
Check design to assure adequacy in the event that minimum member size should occur. Adjust clearance as required for minimum bearing and other structural considerations.

Step 8:
Select final clearance.

8.4.4 Clearance Examples

The following examples are given to show the thought process, and may not be the only correct solutions for the situations described.

Example 8.4.1 Clearance determination—single story industrial building

Given:
A double tee roof member 60 ft long, ± 1 in. length tolerance, bearing on ribbed wall members 25 ft high, maximum plan variance ± 1/2 in., variation from plumb 1/4 in. per 10 ft, haunch depth 6 in. beyond face of panel, long-term roof movement – 1/4 in. Refer to Fig. 8.4.1.

Problem:
Find the minimum acceptable clearance.

Solution:
Step 1: Determine Maximum Member Sizes (refer to product tolerances)

Maximum tee length	+ 1 in.
Wall thickness	+ 1/4 in.
Initial clearance chosen	3/4 in. per end

Step 2: Member Movement

Long-term shrinkage and creep will increase the clearance so this movement can be neglected in the initial clearance determination, although it must be considered structurally.

Required clearance adjustment as a result of member movement	0
Clearance chosen	3/4 in.

Fig. 8.4.1 Clearance Example 8.4.1

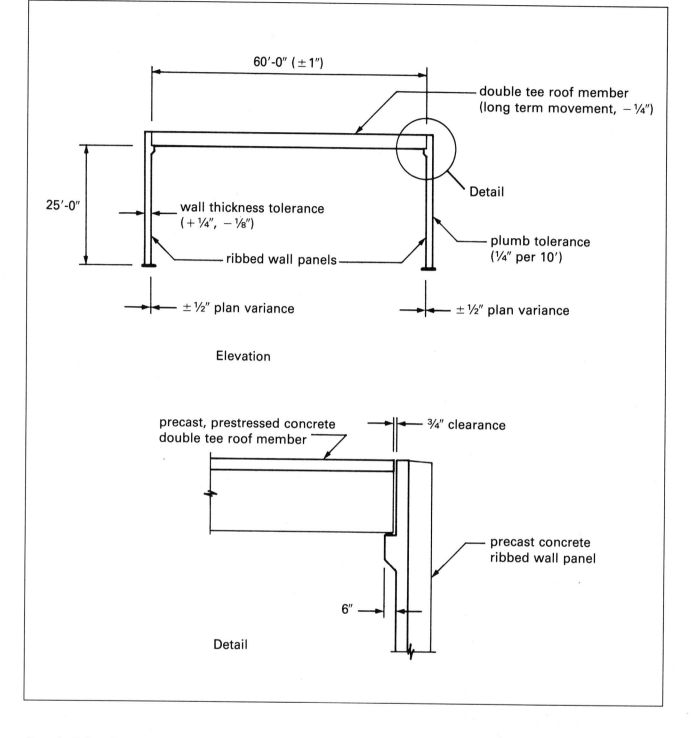

Step 3: Other Erection Tolerances

If the wall panel is set inward toward the building interior 1/2 in. and erected plumb, the clearance should be increased by 1/2 in. If the panel is erected out of plumb outward 1/2 in., no clearance adjustment is needed.

Clearance adjustment required
    for erection tolerances   0
Clearance chosen   3/4 in.

Step 4: Erection Considerations

If all members are fabricated perfectly, then the joint clearance is 3/4 in. at either end (1 1/2 in. total). This is ample space for erection. If all members are at maximum size variance, maximum inward plan variance, and maximum inward variance from plumb, then the total clearance is zero. This situation is undesirable as it would require some rework during erection. A judgment should be made as to the likelihood of maximum product tolerances all occurring in one location. If the likelihood is low, the 3/4 in. clearance needs no adjustment but, if the likelihood is high, the engineer might increase the clearance to 1 in. In this instance, the likelihood has been judged low; therefore, no adjustment has been made.

Clearance chosen   3/4 in.

Step 5: Economy

In single-story construction, increasing the clearance beyond 3/4 in. is not likely to speed up erection as long as product tolerances remain within allowables. No adjustment is required for economic considerations.

Step 6: Review Structural Considerations

Allowing a setback from the edge of the corbel, assumed in this instance to have been set by the engineer at 1¼ in., plus the clearance, the bearing is 4 in. and there should be space to allow member movement. The engineer judges this to be acceptable from a structural and architectural point of view and no adjustment is required for structural considerations.

Step 7: Check for Minimum Member Sizes
(refer to product tolerances)

Tee length	− 1 in. (1/2 in. each end)
Wall thickness	− 1/8 in.
Bearing haunch	No change
Clearance chosen	3/4 in.
Minimum bearing without setback	4⅝ in.
OK in this instance.	

Wall plumbness would also be considered in an actual application.

Step 8: Final Solution

Minimum clearance used   3/4 in. per end
Satisfies all conditions considered.

Note: For simplicity in this example, end rotation, flange skew, and global skew tolerances have not been considered. In an actual situation, these issues should also be considered.

Example 8.4.2 Clearance determination—high rise frame structure

Given:

A 36-story steel frame structure, precast concrete cladding, steel tolerances per AISC, member movement negligible. In this example, precast tolerance for variation in plan is ± 1/4 in. Refer to Fig. 8.4.2.

Problem:

Determine whether or not the panels can be erected plumb and determine the minimum acceptable clearance at the 36th story.

Solution:

Step 1: Product Tolerances
(refer to product tolerances)

Precast cladding thickness	+ 1/4 in., − 1/8 in.
Steel width	+ 1/4 in., − 3/16 in.
Steel sweep (varies)	1/4 in. assumption
Initial clearance chosen	3/4 in.

Step 2: Member Movement

For simplification, assume this can be neglected in this example.

Step 3: Other Erection Tolerances

Steel variance in plan, maximum	2 in.
Initial clearance	3/4 in.
Clearance chosen	2¾ in.

Step 4: Erection Considerations

No adjustment required for erection considerations.

Step 5: Economy

Clearance chosen   2¾ in.

Increasing clearance will not increase economy. No adjustment for economic considerations.

Step 6: Structural Considerations

Clearance chosen   2¾ in.

Expensive connection but possible. No adjustment.

Step 7: Check Minimum Member Sizes at 36th Story
(refer to product tolerances)

Initial clearance	2¾ in.
Precast thickness	− 1/8 in.
Steel width	− 3/16 in.
Steel sweep	− 1/4 in.
Steel variance in plan minimum	− 3 in.
Clearance calculated	6⁵⁄₁₆ in.

Step 8: Final Solution

A clearance of over 6 in. would require an extremely expensive connection for the precast panel, and would produce high torsional stresses in the steel supporting beams. The 6 in. clearance is not practical, although the 2¾ in. minimum initial clearance is still needed. Either the precast panels should be allowed to follow the steel frame or the tolerances for the exterior columns need to be

Fig. 8.4.2. Clearance Example 8.4.2

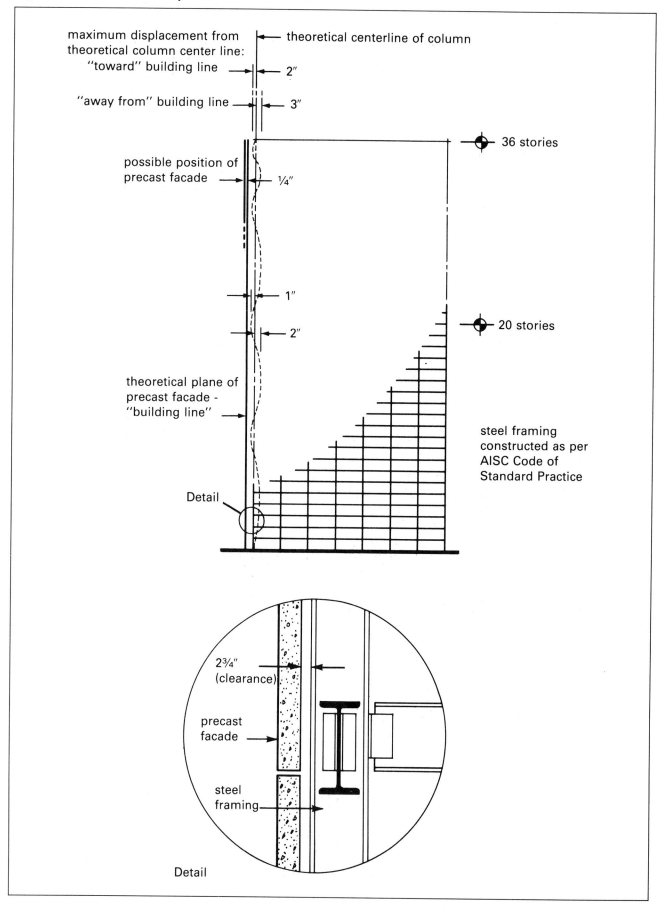

maximum displacement from theoretical column center line:

theoretical centerline of column

"toward" building line — 2"

"away from" building line — 3"

possible position of precast facade — ¼"

36 stories

1"

2"

20 stories

theoretical plane of precast facade - "building line"

steel framing constructed as per AISC Code of Standard Practice

Detail

2¾" (clearance)

precast facade

steel framing

Detail

made more stringent, such as the AISC requirements for elevator columns. The most economical will likely be for the panels to follow the steel frame.

Minimum clearance used   2¾ in.
Allow panels to follow the steel frame.

8.5 Interfacing Tolerances

8.5.1 General

In interfacing with other materials, the tolerances may be very system-dependent. For example, different brands of windows may have different tolerance requirements. If substitutions are made after the initial design is complete, the interface details must be reviewed for the new system.

Following is a partial checklist for consideration in determining interfacing requirements:
- Structural requirements
- Volume change
- Weathering and corrosion
- Waterproofing
- Drainage
- Architectural requirements
- Dimensional considerations
- Vibration considerations
- Fire-rating requirements
- Acoustical considerations
- Economics
- Manufacturing/erection considerations

8.5.2 Interface Design Approach

The following approach is one method of organizing the task of designing the interface between two systems:

Step 1:
Define the interface between the two systems, show shape and location, and determine contractual responsibility. For example, the precast panel furnished by the precaster, the window furnished and installed by the general contractor, and the sealant between the window and the precast concrete furnished and installed by the general contractor.

Step 2:
Define the functional requirements of each interfacing system. For example, the building drain line must have a flow line which allows adequate drainage. This will place limits on where the line must penetrate precast units. Note whether this creates special problems such as conflicting with prestressing strands.

Step 3:
List the tolerances of each interfacing system. For example, determine from manufacturer's specifications what are the external tolerances on the door jamb. Determine from the precast concrete product tolerances what is the tolerance on a large panel door blockout. For the door installation, determine what is the floor surface tolerance in the area of the door.

Step 4:
Select the operational clearances required. For example, determine the magnitude of operational clearances which are needed to align the door to function properly. Then choose dimensions which include necessary clearances.

Step 5:
Check compatibility of the interface tolerances. Starting with the least precise system, check the tolerance requirements and compare against the minimum and maximum dimensions of the interfacing system. If interferences result, alter the nominal dimension of the appropriate system. For example, it is usually more economical to make a larger opening than to specify a non-standard window size.

Step 6:
Establish assembly and installation procedures for the interfacing systems to assure compatibility. Show the preferred adjustments to accommodate the tolerances of the systems. Specify such things as minimum bearing areas, minimum and maximum joint gaps, and other dimensions which will vary as a result of interface tolerances. Consider economic trade-offs such as in-plant work vs. field work, and minor fit-up rework vs. tighter tolerances.

Step 7:
Check the final project specifications as they relate to interfacing. Be aware of subsystem substitutions which might be made during the final bidding and procurement.

8.5.3 Characteristics of the Interface

The following list of questions will help to define the nature of the interface:

1. What specifically is to be interfaced?
2. How does the interface function?
3. Is there provision for adjustment upon installation?

4. How much adjustment can occur without re-work?
5. What are the consequences of an interface tolerance mismatch?
   - Rework requirements (labor and material)
   - Rejection limits
6. What are the high material cost elements of the interface?
7. What are the high labor cost elements of the interface?
8. What are the normal tolerances associated with the system to be interfaced?
9. Are the system interface tolerances simple planar tolerances or are they more complex and three dimensional?
10. Do all of the different products of this type have the same interface tolerance requirements?
11. Do you as the designer of the precast system have control over all the aspects of the interfaces involved? If not, what actions do you need to take to accommodate this fact?

Listed below are common characteristics of most systems of the type listed:

1. Windows and Doors
   - No load transfer through window element
   - Compatible with air and moisture sealant system
   - Open/close characteristics (swing or slide)
   - Compatibility with door locking mechanisms

2. Mechanical Equipment
   - Duct clearances for complex prefabricated ductwork
   - Large-diameter prefabricated pipe clearance requirements
   - Deflections from forces associated with large-diameter piping and valves
   - Expansion/contraction allowances for hot/cold piping
   - Vibration isolation/transfer considerations
   - Acoustical shielding considerations
   - Hazardous gas/fluid containment requirements

3. Electrical Equipment
   - Multiple mating conduit runs
   - Prefabricated cable trays
   - Embedded conduits and outlet boxes
   - Corrosion related to DC power
   - Special insert placement requirements for electrical isolation
   - Location requirements for embedded grounding cables
   - Shielding clearance requirements for special "clean" electrical lines

4. Elevators and Escalators
   - Elevator guide location requirements
   - Electrical conduit location requirements
   - Elevator door mechanism clearances
   - Special insert placement requirements

5. Architectural Cladding
   - Joint tolerance for caulking system
   - Flashing and reglet fit-up. (Lining up reglets from panel to panel is very difficult and often costly. Surface-mounted flashing should be considered.)
   - Expansion and contraction provisions for dissimilar materials
   - Effects of differential thermal gradients

6. Structural Steel and Miscellaneous Steel
   - Details to prevent rust staining of concrete
   - Details to minimize potential for corrosion at field connections between steel and precast concrete
   - Coordination of structural steel expansion/contraction provisions with those of the precast system
   - Special provisions for weld plates or other attachment features for steel structures

7. Masonry
   - Coordination of masonry expansion/contraction provisions with those of the precast system
   - Detailing to assure desired contact bearing between masonry and precast units
   - Detailing to assure desired transfer of load between masonry shear wall and precast frame

8. Roofing
   - Roof camber, both upon erection and long term, as it relates to roof drain placement
   - Fit-up of prefabricated flashing
   - Dimensional effects of added material during reroofing
   - Coordination of structural control joint locations with roofing system expansion/contraction provisions
   - Location of embedded HVAC unit supports

9. Waterproofing
   - Location and dimensions of flashing reglets
   - Location and shape of window gasket grooves
   - Coordination of waterproofing system requirements with structural system expansion provisions
   - Special details around special penetrations

10. Interior Finishes—Floors, Walls, and Ceilings
    - Joints between plank members for direct carpet overlay
    - Visual appearance of joints for exposed ceilings

- Fit-up details to assure good appearance of interior corners
- Appearance of cast-in-place to precast concrete interfaces

11. Interior Walls and Partitions
- Clearance for prefabricated cabinetry
- Interfacing of mating embedded conduit runs

8.6 References

1. PCI Committee on Tolerances, "Tolerances for Precast and Prestressed Concrete," *PCI Journal*, January-February, 1985, V. 30. No. 1.

2. "Manual for Quality Control for Plants and Production of Precast and Prestressed Concrete Products," Third Edition, MNL-116-85, Prestressed Concrete Institute, Chicago, Illinois.

3. "Manual for Quality Control for Plants and Production of Architectural Precast Concrete Products," MNL-117-77, Prestressed Concrete Institute, Chicago, Illinois.

PART 9
THERMAL, ACOUSTICAL, FIRE, AND OTHER CONSIDERATIONS

		Page No.
9.1	Thermal Properties of Precast Concrete	9–2
9.2	Acoustical Properties of Precast Concrete	9–23
9.3	Fire Resistance	9–36
9.4	Sandwich Panels	9–55
9.5	Corrosive Environments	9–62
9.6	Quality Control	9–63
9.7	Concrete Coatings and Joint Sealants	9–64
9.8	Vibration in Concrete Structures	9–65
9.9	Cracking, Repair and Maintenance	9–67
9.10	Precast Segmental Construction	9–68
9.11	Coordination with Mechanical, Electrical and Other Sub-Systems	9–70

THERMAL PROPERTIES OF PRECAST CONCRETE

9.1.1 Notation

a	=	thermal conductance of an air space
A	=	area (all areas are in square feet)
A_d	=	area of doors
A_f	=	area of glass (fenestration)
A_{f_1}	=	area of glass (Wall 1)
A_{f_2}	=	area of glass (Wall 2)
$A_{f\ell}$	=	area of floor, excluding openings
A_o	=	gross area
$A_{of\ell}$	=	gross area of floor
A_{or}	=	gross area of roof
A_{ow}	=	gross area of walls
A_p	=	area of openings in floors
A_r	=	opaque area of roof
A_s	=	area of skylights
A_w	=	opaque area of walls
Btu	=	british thermal unit
C	=	thermal conductance for the specified thickness, Btu/(hr)(ft^2)(deg F)
D	=	heating degree day (65°F base)
f	=	film or surface conductance
k	=	thermal conductivity, (Btu-in.)/(hr)(ft^2)(deg F)
ℓ	=	thickness of a material, in.
M	=	permeance, perms
OTTV	=	overall thermal transfer value
OTTV$_r$	=	overall thermal transfer value for roof, Btu/(hr)(ft^2)
OTTV$_w$	=	overall thermal transfer value for walls, Btu/(hr)(ft^2)
P	=	vapor pressure (all pressures are in inches of Hg)
Q	=	rate of heat flow, Btu/hr
R	=	thermal resistance, (hr)(ft^2)(deg F)/Btu
R_a	=	thermal resistance of air space
R_{fi}, R_{fo}	=	thermal resistances of inside and outside surfaces
R_o	=	initial R-value
R_t	=	total thermal resistance
R_u	=	upgraded R-value
$R_{1,2,...n}$	=	thermal resistance of material layer 1, 2,...n
$R_{materials}$	=	summation of thermal resistance of opaque material layers
RH	=	relative humidity, %
SC	=	shading coefficient of fenestration

SC$_s$	=	shading coefficient of skylight
SF	=	solar factor, Btu/(hr)(ft^2)
t_i	=	indoor temperature, deg F
t_o	=	outdoor temperature, deg F
t_s	=	dew-point temperature, room air, at design maximum relative humidity, deg F
T	=	temperature, deg F
TC	=	specific heat × density × thickness, (Btu)/(ft^2)(deg F)
TD$_{eq}$	=	equivalent temperature difference, indoor-outdoor, deg F
TD$_{eqr}$	=	equivalent temperature difference, roof, deg F
TD$_{eqw}$	=	equivalent temperature difference, walls, deg F
TDR	=	temperature difference ratio
U	=	heat transmittance value [all heat and average transmittance values are Btu/(hr)(ft^2)(deg F)]
U_d	=	heat transmittance value of door
U_f	=	heat transmittance value of glass (fenestration)
$U_{f\ell}$	=	heat transmittance value of floor
U_o	=	average heat transmittance value
$U_{of\ell}$	=	average heat transmittance value of floor
U_{or}, U_{ow}	=	average heat transmittance value of roof, walls
$U_{ow1,2...n}$	=	average heat transmittance value of wall sections 1, 2,...n.
U_p	=	heat transmittance value of openings in floor
U_r	=	heat transmittance value of opaque roof
U_s	=	heat transmittance value of skylight
U_w	=	heat transmittance value of opaque walls
$\triangle T$	=	temperature difference, deg F
$\triangle T_f$	=	temperature difference of fenestration, indoor-outdoor, deg F, cooling design
$\triangle T_s$	=	temperature difference of skylights, indoor-outdoor, deg F, cooling design
μ	=	permeability, permeance of a unit thickness, given material, perm-in.
ϕ	=	wall area/roof area, for use in Fig. 9.1.9

9.1.2 Glossary

British thermal units (Btu) — Approximately the amount of heat to raise one pound of water from 59°F to 60°F.

Degree day (D) — A unit, based on temperature difference and time, used in estimating fuel consumption and specifyng nominal heating load of a building in winter. For any one day, when the mean temperature is less than 65°F, there are as many degree days as there are F degrees difference in temperature between the mean temperature for the day and 65°F.

Dew-point temperature (t_s) — The temperature at which condensation of water vapor begins for a given humidity and pressure as the vapor temperature is reduced. The temperature corresponding to saturation (100 percent relative humidity) for a given absolute humidity at constant pressure.

Film or surface conductance (f) — The time rate of heat exchange by radiation, conduction, and convection of a unit area of a surface with its surroundings. Its value is usually expressed in Btu per (hr) (sq ft of surface area) (deg F temperature difference). Subscripts "i" and "o" are usually used to denote inside and outside surface conductances, respectively.

Heat transmittance (U) — Overall coefficient of heat transmission of thermal transmittance (air-to-air); the time rate of heat flow usually expressed in Btu per (hr) (sq ft of surface area) (deg F temperature difference between air on the inside and air on the outside of a wall, floor, roof, or ceiling). The term is applied to the usual combinations of materials and also single materials such as window glass, and includes the surface conductance on both sides. This term is frequently called the U-value.

Perm — A unit of permeance. A perm is 1 grain per (sq ft of area) (hr) (in. of mercury vapor pressure difference).

Permeability, water vapor (μ) —The property of a substance which permits the passage of water vapor. It is equal to the permeance of 1 in. of a substance. Permeability is measured in perm inches. The permeability of a material varies with barometric pressure, temperature and relative humidity conditions.

Permeance (M) — The water vapor permeance of any sheet or assembly is the ratio of the water vapor flow per unit area per hour to the vapor pressure difference between the two surfaces. Permeance is measured in perms.

Two commonly used test methods are the Wet Cup and Dry Cup Tests. Specimens are sealed over the tops of cups containing either water or desiccant, placed in a controlled atmosphere usually at 50 percent relative humidity, and weight changes measured.

Relative humidity (RH) — The ratio of water vapor present in air to the water vapor present in saturated air at the same temperature and pressure.

Thermal conductance (C) — The time rate of heat flow expressed in Btu per (hr) (sq ft of area) (deg F average temperature difference between two surfaces). The term is applied to specific materials as used, either homogenous or heterogeneous for the thickness or construction stated, not per in. of thickness.

Thermal conductance of an air space (a) — The time rate of heat flow through a unit area of an air space per unit temperature difference between the boundary surfaces. Its value is usually expressed in Btu per (hr) (sq ft of area) (deg F).

Thermal conductivity (k) — The time rate of heat flow by conduction only through a unit thickness of a homogeneous material under steady-state conditions, through unit area, per unit temperature gradient in the direction perpendicular to the isothermal surface. Its unit is (Btu-in.) per (hr) (sq ft of area) (deg F).

Thermal resistance (R) — The reciprocal of a heat transmission coefficient, as expressed by U, C, f, or a. Its unit is (deg F) (hr) (sq ft of area) per Btu. For example, a wall with a U-value of 0.25 would have a resistance value of:

$$R = 1/U = 1/0.25 = 4.0$$

9.1.3 General

Thermal codes and standards specify the heat transmission requirements for buildings in many different ways. Prescriptive standards specify U or R values for each building component, whereas with performance standards, two buildings are equivalent if they use the same amount of energy, regardless of the U or R values of the components. This allows the designer to choose conservation strategies that provide the required performance at the least cost.

Precast and prestressed concrete construction, with their thermal inertia and thermal storage properties, have an advantage over lightweight materials. Procedures to account for the benefits of heavier materials are presented in Sect. 9.1.8.

The trend is toward more insulation with little regard given to its total impact or energy used. Mass effects, glass area, air infiltration, ventilation, building orientation, exterior color, shading or reflections from adjacent structures, surrounding surfaces or vegetation, building aspect ratio, number of stories, wind direction and speed, all have an effect on insulation requirements.

This section is condensed from a more complete treatment given in Reference 4. Except where noted, the information and design criteria are taken or derived from the ASHRAE Handbook[1], and from the ASHRAE Standard 90A-1980.[2] All design criteria are not given in this section, and the criteria may change as the ASHRAE Standard and Handbook are revised. Local codes and latest references must be used for specified values and procedures.

9.1.4 Thermal Properties of Materials, Surfaces, and Air Spaces

The thermal properties of materials and air spaces are based on steady state tests, which measure the heat that passes from the warm side to the cool side of the test specimen. The tests determine the conductivity, k, or, for non-homogeneous sections, compound sections and air spaces, the conductance, C, for the total thickness. The values of k and C do not include surface conductances, f_i and f_o.

The overall thermal resistance of wall, floor, and roof sections is the sum of the resistances, R (reciprocal of k, C, f_i and f_o). The R-values of construction materials are not influenced by the direction of heat flow, but the R-values of surfaces and air spaces differ depending on whether they are vertical, sloping, or horizontal. Also, the R-values of surfaces are affected by the velocity of air at the surfaces and by their reflective properties.

Tables 9.1.1 and 9.1.2 give the thermal resistances of surfaces and 3½ in. air spaces. Table 9.1.3 gives the thermal properties of most commonly used building materials. Only U-values are given for glass because the surface resistances and air space between panes account for nearly all of the U-value. Table 9.1.4 gives the thermal properties of various weight concretes and some standard precast, prestressed concrete products in the "normally dry" condition. Normally dry is the condition of concrete containing an equilibrium amount of free water after extended exposure to warm air at 35 to 50 percent relative humidity.

Thermal conductances and resistances of other building materials are usually reported for oven dry conditions. Normally dry concrete in combination with insulation generally provides about the same R-value as equally insulated oven dry concrete, but because of the moisture content, has the ability to store a greater amount of heat than oven dry concrete. However, higher moisture content in concrete causes higher thermal conductance.

9.1.5 Computation of Thermal Transmittance Values

The heat transmittance (U-values) of a building wall, floor or roof is computed by adding together the R-values of the materials in the section, the surfaces (R_{fi} and R_{fo}) and air spaces (R_a) within the section. The reciprocal of the R's is the U-value:

$$U = \frac{1}{R_{fi} + R_{materials} + R_a + R_{fo}} \qquad \text{(Eq. 9.1.1)}$$

where $R_{materials}$ is the sum of all opaque materials in the wall. A number of typical wall and roof U-values are given in Tables 9.1.5, 9.1.6, and 9.1.7.

The U-values for heating design may be modified to account for the effects of mass, given in Sect. 9.1.8. The U-values for cooling design are used without modification; mass effects are explained in Sect. 9.1.7.

The following examples show use of Tables 9.1.1 through 9.1.4 in calculating R- and U-values for wall and roof assemblies:

Example — Wall

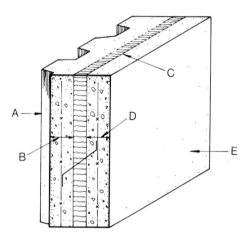

	R Winter	R Summer	Table
A. Surface, outside	0.17	0.25	9.1.1
B. Concrete, 2 in., (110 pcf)	0.38	0.38	9.1.4
C. Polystyrene, 1½ in.	6.00	6.00	9.1.3
D. Concrete, 2½ in., (110 pcf)	0.48	0.48	9.1.4
E. Surface, inside	0.68	0.68	9.1.1
Total R =	7.71	7.79	
U = 1/R =	0.13	0.13	

PCI Design Handbook

Table 9.1.1 Thermal resistances, R_f, of surfaces

| Position of surface | Direction of heat flow | Still air, R_{fi} | | | Moving air, R_{fo} | |
| | | Non reflective surface | Reflective surface | | Non reflective surface | |
			Aluminum painted paper	Bright aluminum foil	15 mph Winter design	7½ mph Summer design
Vertical	Horizontal	0.68	1.35	1.70	0.17	0.25
Horizontal	Up	0.61	1.10	1.32	0.17	0.25
	Down	0.92	2.70	4.55	0.17	0.25

Table 9.1.2 Thermal resistances, R_a, of air spaces[1]

| Position of air space | Direction of heat flow | Air space | | Non reflective surfaces | Reflective surfaces | | |
		Mean temp. °F	Temp. diff. °F		One side[2]	One side[3]	Both sides[3]
Vertical	Horizontal (walls)	Winter 50	10	1.01	2.32	3.40	3.63
		50	30	0.91	1.89	2.55	2.67
	Horizontal (walls)	Summer 90	10	0.85	2.15	3.40	3.69
Horizontal	Up (roofs)	Winter 50	10	0.93	1.95	2.66	2.80
		50	30	0.84	1.58	2.01	2.09
	Down (floors)	50	30	1.23	3.86	8.17	9.60
	Down (roofs)	Summer 90	10	1.00	3.41	8.19	10.07

1. For 3½ in. air space thickness. The values with the exception of those for reflective surfaces, heat flow down, will differ about 10% for air space thickness of ¾ in. to 16 in. Refer to Table 2, Chapter 23 of the ASHRAE Handbook for values of other thicknesses, reflective surfaces, heat flow down.
2. Aluminum painted paper
3. Bright aluminum foil

Example — Roof

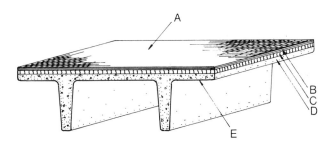

	R Winter	R Summer	Table
A. Surface, outside	0.17	0.25	9.1.1
B. Roofing, built-up	0.33	0.33	9.1.3
C. Polystyrene, 2 in.	8.00	8.00	9.1.3
D. Concrete, 2 in. (145 pcf)	0.15	0.15	9.1.4
E. Surface, inside	0.61	0.92	9.1.1
Total R =	9.26	9.65	
U = 1/R =	0.11	0.10	

Table 9.1.3 Thermal properties of various building materials[1]

| Material | Unit weight, pcf | Resistance, R | | Trans-mittance, U | Specific heat, Btu/(lb.)(°F) |
		Per inch of thick-ness, 1/k	For thick-ness shown, 1/C		
Insulation, rigid					
Cellular glass	8.5	2.86			0.18
Glass fiber, organic bonded	4 - 9	4.00			0.23
Mineral fiber, resin binder	15	3.45			0.17
Mineral fiberboard, wet felted, roof insulation	16 - 17	2.94			—
Cement fiber slabs (shredded wood with magnesia oxysulfide binder)	22	1.75			0.31
Expanded polystyrene extruded cut cell surface	1.8	4.00			0.29
Expanded polystyrene extruded smooth skin surface	1.8 - 3.5	5.00			0.29
Expanded polystyrene molded bead	1.0	5.00			—
Cellular polyurethane	1.5	6.25			0.38
Miscellaneous					
Acoustical tile (mineral fiberboard wet felted)	18	2.86			0.19
Carpet, fibrous pad			2.08		0.34
Carpet, rubber pad			1.23		0.33
Floor tile, asphalt, rubber, vinyl			0.05		0.30
Gypsum board	50	0.88[2]			0.26
Particle board	50	1.06			0.31
Plaster					
cement, sand agg.	116	0.20			0.20
gyp., L.W. agg.	45	0.63[2]			—
gyp., sand agg.	105	0.18			0.20
Roofing, ⅜ in. built-up	70		0.33		0.35
Wood, hard	45	0.91			0.30
Wood, soft	32	1.25			0.33
Plywood	34	1.25			0.29
Glass doors & windows[3]					
Single, winter				1.10	
Single, summer				1.04	
Double, winter[4]				0.59	
Double, summer[4]				0.61	
Doors, metal[5]					
Insulated, winter				0.47	
Insulated, summer				0.46	

1. See Table 9.1.4 for all concretes, including insulating concrete for roof fill.
2. Average value.
3. Does not include correction for sash resistance. Refer to Chapter 23 of the ASHRAE Handbook for sash correction.
4. ¼ in. air space; coating on either glass surface facing air space.
5. Solid polystyrene core with thermal break.

Table 9.1.4 Thermal properties of concrete[1]

Description	Concrete weight, pcf	Thickness, in.	Resistance, R		Specific heat,[3] Btu/(lb.)(°F)
			Per inch of thickness, 1/k	For thickness shown, 1/C	
Concretes including normal weight, lightweight and lightweight insulating concretes	145		0.075		0.19
	140		0.083		
	130		0.11		
	120		0.14		
	110		0.19		
	100		0.24		
	90		0.30		
	80		0.37		
	70		0.45		
	60		0.52		
	50		0.67		
	40		0.83		
	30		1.00		
	20		1.43		
Normal weight tees[2] and solid slabs	145	2		0.15	0.19
		3		0.23	
		4		0.30	
		5		0.38	
		6		0.45	
		8		0.60	
Normal weight hollow-core slabs	145	6		1.07	0.19
		8		1.34	
		10		1.73	
		12		1.91	
Structural lightweight tees[2] and solid slabs	110	2		0.38	0.19
		3		0.57	
		4		0.76	
		5		0.95	
		6		1.14	
		8		1.52	
Structural lightweight hollow-core slabs	110	8		2.00	0.19
		12		2.59	

1. Based on normally dry concrete (see Chapter 4 of Reference 3).
2. Thickness for tees is thickness of slab portion including topping, if used. The effect of the stems generally is not significant, therefore, their thickness and surface area may be disregarded.
3. The specific heat shown is the mean value from test data compiled in Reference 4 and is sufficiently accurate for calculating TC of Fig. 9.1.3. For a more exact value, tests should be performed on the concrete mix being used.

Table 9.1.5 Wall U-values: prestressed tees, hollow-core slabs, solid and sandwich panels; winter and summer conditions[1]

Concrete weight, pcf	Type of wall panel	Thickness, t, and resistance, R, of concrete		Winter $R_{fo}=0.17$, $R_{fi}=0.68$					Summer $R_{fo}=0.25$, $R_{fi}=0.68$				
				Insulation resistance, R									
		t[2]	R	None	4	6	8	10	None	4	6	8	10
145	Solid walls, tees, and sandwich panels	2	0.15	1.00	.20	.14	.11	.09	.93	.20	.14	.11	.09
		3	0.23	.93	.20	.14	.11	.09	.86	.19	.14	.11	.09
		4	0.30	.87	.19	.14	.11	.09	.81	.19	.14	.11	.09
		5	0.38	.81	.19	.14	.11	.09	.76	.19	.14	.11	.09
		6	0.45	.77	.19	.14	.11	.09	.72	.19	.14	.11	.09
		8	0.60	.69	.18	.13	.11	.09	.65	.18	.13	.10	.09
	Hollow core slabs[3]	6(o)	1.07	.52	.17	.13	.10	.08	.50	.17	.13	.10	.08
		(f)	1.86	.37	.15	.11	.09	.08	.36	.15	.11	.09	.08
		8(o)	1.34	.46	.16	.12	.10	.08	.44	.16	.12	.10	.08
		(f)	3.14	.25	.13	.10	.08	.07	.25	.12	.10	.08	.07
		10(o)	1.73	.39	.15	.12	.09	.08	.38	.15	.12	.09	.08
		(f)	4.05	.20	.11	.09	.08	.07	.20	.11	.09	.08	.07
		12(o)	1.91	.36	.15	.11	.09	.08	.35	.15	.11	.09	.08
		(f)	5.01	.17	.10	.08	.07	.06	.17	.10	.08	.07	.06
110	Solid walls, tees, and sandwich panels	2	0.38	.81	.19	.14	.11	.09	.76	.19	.14	.11	.09
		3	0.57	.70	.18	.13	.11	.09	.67	.18	.13	.11	.09
		4	0.76	.62	.18	.13	.10	.09	.59	.18	.13	.10	.09
		5	0.95	.56	.17	.13	.10	.09	.53	.17	.13	.10	.08
		6	1.14	.50	.17	.13	.10	.08	.48	.16	.12	.10	.08
		8	1.52	.42	.16	.12	.10	.08	.41	.16	.12	.10	.08
	Hollow core slabs[3]	8(o)	2.00	.35	.15	.11	.09	.08	.34	.14	.11	.09	.08
		(f)	4.41	.19	.11	.09	.08	.07	.19	.11	.09	.07	.07
		12(o)	2.59	.29	.13	.11	.09	.07	.28	.13	.11	.09	.07
		(f)	6.85	.13	.09	.07	.06	.06	.13	.08	.07	.06	.06

1. When insulations having other R-values are used, U-values can be interpolated with adequate accuracy, or U can be calculated as shown in Sect. 9.1.5. When a finish, air space or any other material layer is added, the new U-value is:

$$\frac{1}{\dfrac{1}{U \text{ from table}} + R \text{ of added finish, air space, or material}}$$

2. Thickness for tees is thickness of slab portion. For sandwich panels, t is the sum of the thicknesses of the wythes.
3. For hollow panels (o) and (f) after thickness designates cores open or cores filled with insulation.

PCI Design Handbook

Table 9.1.6 Roof U-values: concrete units with built-up roofing, winter conditions, heat flow upward[1]

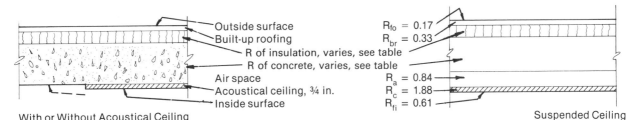

Concrete weight, pcf	Prestressed concrete member	Thickness, t and resistance, R of concrete		Without ceiling				With ceiling							
								Applied direct				Suspended			
				Top insulation resistance, R											
		t^2	R	None	4	10	16	None	4	10	16	None	4	10	16
145	Solid slabs and tees	2	0.15	.79	.19	.09	.06	.29	.13	.07	.05	.24	.12	.07	.05
		3	0.23	.75	.19	.09	.06	.29	.13	.07	.05	.23	.12	.07	.05
		4	0.30	.71	.18	.09	.06	.28	.13	.07	.05	.23	.12	.07	.05
		5	0.38	.67	.18	.09	.06	.27	.13	.07	.05	.22	.12	.07	.05
		6	0.45	.64	.18	.09	.06	.27	.13	.07	.05	.22	.12	.07	.05
		8	0.60	.58	.18	.09	.06	.26	.13	.07	.05	.21	.11	.07	.05
	Hollow core slabs[3]	6 (o)	1.07	.46	.16	.08	.06	.23	.12	.07	.05	.19	.11	.07	.05
		(f)	1.86	.34	.14	.08	.05	.20	.11	.07	.05	.17	.10	.06	.05
		8 (o)	1.34	.41	.16	.08	.05	.22	.12	.07	.05	.18	.11	.06	.05
		(f)	3.14	.24	.12	.07	.05	.16	.10	.06	.04	.14	.09	.06	.04
		10 (o)	1.73	.35	.15	.08	.05	.20	.11	.07	.05	.17	.10	.06	.05
		(f)	4.05	.19	.11	.07	.05	.14	.09	.06	.04	.12	.08	.06	.04
		12 (o)	1.91	.33	.14	.08	.05	.19	.11	.07	.05	.17	.10	.06	.05
		(f)	5.01	.16	.10	.06	.05	.12	.08	.05	.04	.11	.08	.05	.04
110	Solid slabs and tees	2	0.38	.67	.18	.09	.06	.27	.13	.07	.05	.22	.12	.07	.05
		3	0.57	.60	.18	.09	.06	.26	.13	.07	.05	.21	.12	.07	.05
		4	0.76	.53	.17	.08	.06	.25	.12	.07	.05	.21	.11	.07	.05
		5	0.95	.49	.17	.08	.06	.24	.12	.07	.05	.20	.11	.07	.05
		6	1.14	.44	.16	.08	.05	.23	.12	.07	.05	.19	.11	.07	.05
		8	1.52	.38	.15	.08	.05	.21	.11	.07	.05	.18	.10	.06	.05
	Hollow core slabs[3]	8 (o)	2.00	.32	.14	.08	.05	.19	.11	.07	.05	.16	.10	.06	.05
		(f)	4.41	.18	.11	.06	.05	.13	.09	.06	.04	.12	.08	.05	.04
		12 (o)	2.59	.27	.13	.07	.05	.17	.10	.06	.05	.15	.09	.06	.04
		(f)	6.85	.13	.08	.06	.04	.10	.07	.05	.04	.09	.07	.05	.04

1. U-values listed are rounded to two decimal places. When insulations having other R-values are used, U-values can be interpolated with adequate accuracy, or U can be calculated as shown in Sect. 9.1.5.
   When a finish, air space or any material layer is added, the new U-value is:

$$\frac{1}{\dfrac{1}{U \text{ from table}} + R \text{ of added finish, air space, or material}}$$

2. Thickness for tees is thickness of slab portion.
3. For hollow panels (o) and (f) after thickness designates cores open on cores filled with insulation.

Table 9.1.7 Roof U-values: concrete units with built-up roofing, summer conditions, heat flow downward[1]

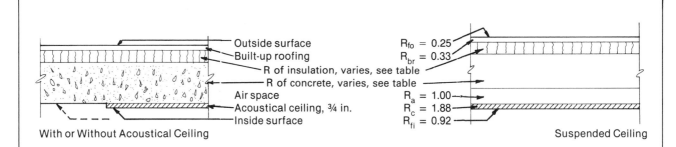

With or Without Acoustical Ceiling

$R_{fo} = 0.25$
$R_{br} = 0.33$
$R_a = 1.00$
$R_c = 1.88$
$R_{fi} = 0.92$

Suspended Ceiling

Concrete weight, pcf	Prestressed concrete member	Thickness, t and resistance, R of concrete		Without ceiling				With ceiling							
								Applied direct				Suspended			
				Top insulation resistance, R											
		t^2	R	None	4	10	16	None	4	10	16	None	4	10	16
145	Solid slabs and tees	2	0.15	.61	.18	.09	.06	.26	.13	.07	.05	.21	.11	.07	.05
		3	0.23	.58	.17	.09	.06	.26	.13	.07	.05	.20	.11	.07	.05
		4	0.30	.56	.17	.08	.06	.25	.13	.07	.05	.20	.11	.07	.05
		5	0.38	.53	.17	.08	.06	.25	.12	.07	.05	.20	.11	.07	.05
		6	0.45	.51	.17	.08	.06	.24	.12	.07	.05	.20	.11	.07	.05
		8	0.60	.48	.16	.08	.06	.24	.12	.07	.05	.19	.11	.07	.05
	Hollow core slabs[3]	6 (o)	1.07	.39	.15	.08	.05	.21	.11	.07	.05	.17	.10	.06	.05
		(f)	1.86	.30	.14	.07	.05	.18	.11	.06	.05	.15	.10	.06	.04
		8 (o)	1.34	.35	.15	.08	.05	.20	.11	.07	.05	.17	.10	.06	.05
		(f)	3.14	.22	.12	.07	.05	.15	.09	.06	.04	.13	.08	.06	.04
		10 (o)	1.73	.31	.14	.08	.05	.19	.11	.07	.05	.16	.10	.06	.04
		(f)	4.05	.18	.10	.06	.05	.13	.09	.06	.04	.11	.08	.05	.04
		12 (o)	1.91	.29	.13	.07	.05	.18	.10	.06	.05	.15	.09	.06	.04
		(f)	5.01	.15	.10	.06	.04	.12	.08	.05	.04	.10	.07	.05	.04
110	Solid slabs and tees	2	0.38	.53	.17	.08	.06	.25	.12	.07	.05	.20	.11	.07	.05
		3	0.57	.48	.16	.08	.06	.24	.12	.07	.05	.19	.11	.07	.05
		4	0.76	.44	.16	.08	.05	.23	.12	.07	.05	.18	.11	.06	.05
		5	0.95	.41	.16	.08	.05	.22	.12	.07	.05	.18	.10	.06	.05
		6	1.14	.38	.15	.08	.05	.21	.11	.07	.05	.17	.10	.06	.05
		8	1.52	.33	.14	.08	.05	.19	.11	.07	.05	.16	.10	.06	.05
	Hollow core slabs[3]	8 (o)	2.00	.29	.13	.07	.05	.18	.10	.06	.05	.15	.09	.06	.04
		(f)	4.41	.17	.10	.06	.05	.12	.08	.06	.04	.11	.08	.05	.04
		12 (o)	2.59	.24	.12	.07	.05	.16	.10	.06	.04	.14	.09	.06	.04
		(f)	6.85	.12	.08	.05	.04	.10	.07	.05	.04	.09	.06	.05	.04

1. When insulations having other R-values are used, U-values can be interpolated with adequate accuracy, or U can be calculated as shown in Section 9.1.5.
   When a finish, air space or any material layer is added, the new U-value is:

$$\frac{1}{(1/U \text{ from table}) + R \text{ of added finish, air space, or material}}$$

2. Thickness for tees is thickness of slab portion.
3. For hollow panels, (o) and (f) after thickness designates cores open or cores filled with insulation.

9.1.6 Heating Design

To determine heat loss through opaque and glass areas, a temperature difference, $\triangle T$ is multiplied by the applicable U-values. For heated buildings, the maximum indoor temperature, t_i, is usually assumed as 72°F. Outdoor design temperatures for major cities are listed in Table 9.1.8. The ASHRAE Standard permits the 97½% values to be used for buildings. The $\triangle T$ for below grade walls is the difference between inside and ground surface temperature, because of the heat capacity of the soil. Fig. 9.1.1 may be used to analyze below grade walls.

Many codes and standards specify weighted average thermal transmittance values, U_o, for walls, floors and roofs. The U_o-value may include more than one element; for example, opaque wall areas are combined with glazed areas to arrive at the U_o for the entire wall. The U_o limits for buildings vary with location, occupancy and height.

For heating designs the U_o-values of walls, roofs, and floors are calculated as follows:

$$U_{ow} = \frac{U_wA_w + U_fA_f + U_dA_d}{A_{ow}} \qquad \text{(Eq. 9.1.2)}$$

$$U_{or} = \frac{U_rA_r + U_sA_s}{A_{or}} \qquad \text{(Eq. 9.1.3)}$$

$$U_{of\ell} = \frac{U_{f\ell}A_{f\ell} + U_pA_p}{A_{of\ell}} \qquad \text{(Eq. 9.1.4)}$$

Note that U_wA_w, U_rA_r, etc., may be a combination of two or more areas.

Fig. 9.1.1 for walls below grade is based on an assumed heat flow path and thermal and physical properties in accordance with the ASHRAE Handbook. As an example of the use of the aid, consider a building located in an area having a TDR = 0.60, with 4 ft of wall below grade. Enter the graph at 4 ft, move vertically to TDR = 0.60, move horizontally and read the U-value = 0.145. By taking the area weighted average of this value and the U-value of the above grade portion of the wall, U_w for use in Eq. 9.1.2 can be determined.

Once U_{ow}, U_{or} and $U_{of\ell}$ are established, the designer can proportion opaque areas and glazed areas so the total heat transmission does not exceed desired values. Fig. 9.1.2 may be used for that purpose.

For example, assume that a local code limits U_{ow} to 0.30. If single glass is used, and the U_w of the opaque area is 0.17, from Fig. 9.1.2(a), the area of glass is limited to 13½ percent of the gross area A_{ow}. Or, from Fig. 9.1.2(b), the area of double glass is limited to 27½ percent of the gross wall area.

9.1.7 Cooling Design

For buildings that are cooled, an indoor temperature, t_i, of 78°F for habitable space is required to comply with the ASHRAE Standard. The outdoor design temperature, t_o, is taken from the last column in Table 9.1.8.

Heat gain through opaque areas is calculated in one step using an equivalent temperature difference, TD_{eq}, that takes into account solar radiant temperature, air temperature, mass, insulation, and color. Heat gain through glass is a function of both conduction caused by the temperature difference and solar gain.

The ASHRAE Standard requires that glass and opaque areas be selected to limit the overall thermal transfer value, OTTV. The overall thermal transfer value for roofs, $OTTV_r$, is limited to a maximum of 8.5 Btu/(hr) (sq ft). The limiting $OTTV_w$ value depends on latitude and is specified by the governing code. Below grade walls are not considered in the calculation of $OTTV_w$.

OTTV values are calculated using the following equations:

$$OTTV_w = \frac{(U_wA_wTD_{eqw}) + (U_fA_f\triangle T_f) + (SF\ SC\ A_f)}{A_{ow}}$$

$$\text{(Eq. 9.1.5)}$$

$$OTTV_r = \frac{(U_rA_rTD_{eqr}) + (U_sA_s\triangle T_s) + (138\ SC_sA_s)}{A_{or}}$$

$$\text{(Eq. 9.1.6)}$$

The first term in these equations reflects heat gain through opaque areas. The TD_{eqw} and TD_{eqr} values are given in Fig. 9.1.3. The second term is for conduction heat gain through glass, and the third term gives the average solar heat gain through glass. Solar factors (SF) are given in Fig. 9.1.4. Shading coefficients (SC) are determined using guidelines in the ASHRAE Handbook or from manufacturer's literature and range from 1.00 for ⅛-in. clear glass to 0.53 for ½-in. heat absorbing glass.

Once the limiting value for OTTV is determined, opaque and glazed areas are proportioned so that the heat gain does not exceed the limit. For example, consider a roof with $\triangle T_s$ = 15°F and U_s = 0.61, SC = 1.0, and U_r = 0.10. The roof structure is precast double tees with 2 in. flanges (see example, Sect. 9.1.5). Concrete is normal weight (150 pcf) with a specific heat of 0.19. From the equation in Fig. 9.1.3(b), TC = (0.19)(150)(2/12) = 4.75, and U/TC = 0.10/4.75 = 0.021. Enter with this value and read TD_{eqr} = 59°F. Determine the percentage of skylights allowed by solving Eq. 9.1.6 for %A_s = (A_s/A_{or}) (100) and substituting appropriate values:

Table 9.1.8 Outdoor temperatures, latitudes, and degree days

City[2]	Latitude deg.	Latitude min.	Winter Temperatures[1] °F 99%	Winter Temperatures[1] °F 97-1/2%	Winter Degree Days[3]	Summer Design Dry Bulb Temperatures 2-1/2%, °F
UNITED STATES						
Albuquerque, NM (AP)	35	00	12	16	4,350	94
Anchorage AK (AP)	61	10	−23	−18	10,860	68
Atlanta, GA (AP)	33	40	17	22	2,960	92
Baltimore, MD (CO)	39	20	14	17	4,110	89
Baltimore, MD (AP)	39	10	10	13	4,650	91
Birmingham, AL (AP)	33	30	17	21	2,550	94
Bismarck, ND (AP)	46	50	−23	−19	8,850	91
Boise, ID (AP)	43	30	3	10	5,810	94
Boston, MA (AP)	42	20	6	9	5,630	88
Burlington, VT (AP)	44	30	−12	−7	8,270	85
Charleston, WV (AP)	38	20	7	11	4,480	90
Charlotte, NC (AP)	35	00	18	22	3,190	93
Casper, WY (AP)	42	50	−11	−5	7,410	90
Chicago, IL (CO)	41	50	−3	2	5,880	91
Chicago, IL (Midway AP)	41	50	−5	0	6,160	91
Chicago, IL (O'Hare AP)	42	00	−8	−4	6,640	89
Cincinatti, OH (CO)	39	10	1	6	4,410	90
Cleveland, OH (AP)	41	20	1	5	6,350	88
Columbia, SC (AP)	34	00	20	24	2,480	95
Concord, NH (AP)	43	10	−8	−3	7,380	87
Dallas, TX (AP)	32	50	18	22	2,360	100
Denver, CO (AP)	39	50	−5	−1	6,280	91
Des Moines, IA (AP)	41	30	−10	−5	6,590	91
Detroit, MI (AP)	42	20	3	6	6,230	88
Fairbanks AK (AP)	64	50	−51	−47	14,280	78
Great Falls, MT (AP)	47	30	−21	−15	7,750	88
Hartford, CT (AP)	41	50	3	7	6,240	88
Houston, TX (CO)	29	50	28	33	1,280	95
Houston, TX (AP)	29	40	27	32	1,400	94
Indianapolis, IN (AP)	39	40	−2	2	5,700	90
Jackson, MS (AP)	32	20	21	25	2,240	95
Kansas City, MO (AP)	39	10	2	6	4,710	96
Las Vegas, NV (AP)	36	10	25	28	2,710	106
Lexington, KY (AP)	38	00	3	8	4,680	91
Little Rock, AR (AP)	34	40	15	20	3,220	96
Los Angeles, CA (AP)	34	00	41	43	2,060	80
Los Angeles, CA (CO)	34	00	37	40	1,350	89
Memphis, TN (AP)	35	00	13	18	3,230	95
Miami, FL (AP)	25	50	44	47	210	90
Milwaukee, WI (AP)	43	00	−8	−4	7,640	87
Minneapolis, MN (AP)	44	50	−16	−12	8,380	89
New Orleans, LA (AP)	30	00	29	33	1,380	92
New York, NY (La Guardia AP)	40	50	11	15	4,810	89
New York, NY (Kennedy AP)	40	40	12	15	5,220	87
Norfolk, VA (AP)	36	50	20	22	3,420	91
Oklahoma City, OK (AP)	35	20	9	13	3,720	97
Omaha, NE (AP)	41	20	−8	−3	6,610	91
Philadelphia, PA (AP)	39	50	10	14	5,140	90
Phoenix, AZ (AP)	33	30	31	34	1,760	107
Pittsburgh, PA (CO)	40	30	3	7	5,050	88
Pittsburgh, PA (AP)	40	30	1	5	5,990	86
Portland, ME (AP)	43	40	−6	−1	7,510	84
Portland, OR (AP)	45	40	17	23	4,640	85
Portland, OR (CO)	45	30	18	24	4,110	86
Providence, RI (AP)	41	40	5	9	5,950	86
Rochester, NY (AP)	43	10	1	5	6,750	88
Salt Lake City, UT (AP)	40	50	3	8	6,050	95
San Francisco, CA (CO)	37	50	38	40	3,000	71
San Francisco, CA (AP)	37	40	35	38	3,020	77
Seattle, WA (CO)	47	40	22	27	4,420	82
Seattle, WA (Tacoma AP)	47	30	21	26	5,140	80
Sioux Falls, SD (AP)	43	40	−15	−11	7,840	91
St. Louis, MO (CO)	38	40	3	8	4,480	94
St. Louis, MO (AP)	38	50	2	6	4,900	94
Tampa, FL (AP)	28	00	36	40	680	91
Trenton, NJ (CO)	40	10	11	14	4,980	88
Washington, DC (National AP)	38	50	14	17	4,220	91
Washington, DC (Andrews AFB)	38	50	10	14	4,220	90
Wichita, KS (AP)	37	40	3	7	4,620	98
Wilmington, DE (AP)	39	40	10	14	4,930	89
CANADA						
Edmonton, AB (AP)	53	34	−29	−25	10,270	82
Halifax, NS (AP, CO)	44	39	1	5	7,360	76
Montreal, PQ (AP)	45	28	−16	−10	8,200	85
Saskatoon, SK (AP, CO)	52	10	−35	−31	10,870	86
St. John's, NF (AP)	47	37	3	7	8,990	75
Saint John, NB (AP, CO)	45	19	−12	−8	8,220	77
Toronto, ON (AP, CO)	43	41	−5	−1	6,830	87
Vancouver, BC (AP)	49	11	15	19	5,520	77
Winnipeg, MB (AP)	49	54	−30	−27	10,680	86

(1) ASHRAE Handbook, Chapter 24.
(2) (CO) stands for city and (AP) for airport. (AP, CO) airport data used for temperatures and city data for degree day value.
(3) Rounded to nearest 10.

PCI Design Handbook

Fig. 9.1.1 Average U-value for walls below grade

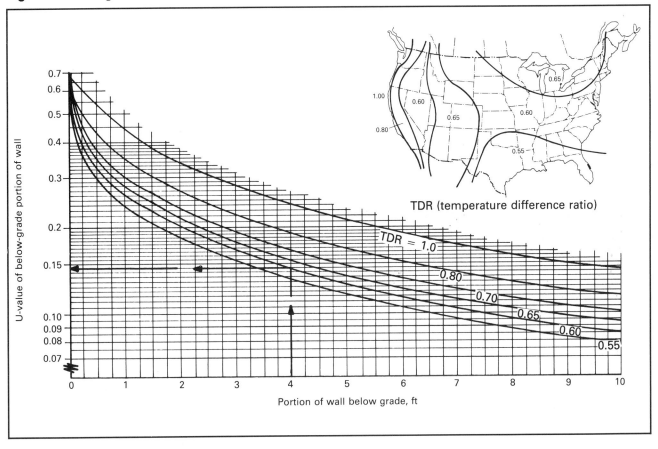

Fig. 9.1.2 Wall-glass heating design charts

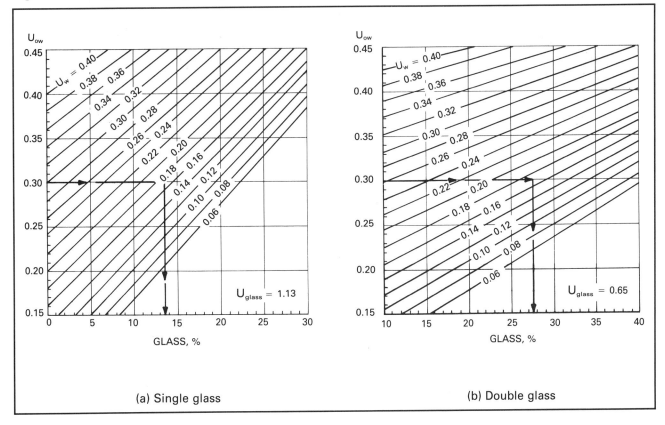

(a) Single glass

(b) Double glass

Fig. 9.1.3 Equivalent temperature differences

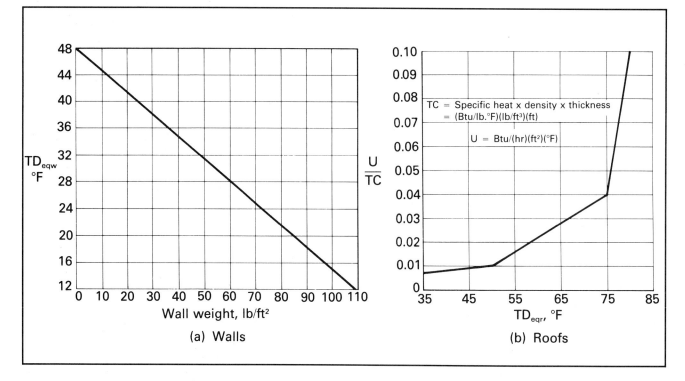

(a) Walls (b) Roofs

Fig. 9.1.4 Solar factor (SF) values

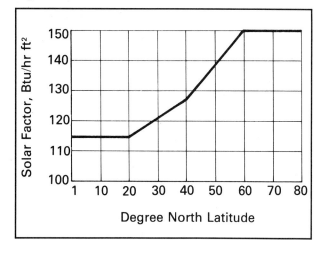

Degree North Latitude

$$\%A_s = \frac{(8.5 - TD_{eqr}U_r)(100)}{(138)\, SC_s + U_s\, \Delta T_s - TD_{eqr}U_r} \quad \text{(Eq. 9.1.7)}$$

$$= \frac{[8.5 - (59)(0.10)](100)}{(138)(1.0) + (0.61)(15) - (59)(0.10)}$$

$$= 1.8\%$$

The ASHRAE Standard requires that the more stringent of heating or cooling criteria be used. In most cases the heating criteria govern.

To simplify selection of the optimum combination of U_w, A_w, TD_{eqw}, U_f, A_f, and SC, design aids have been developed for walls.[3]

9.1.8 Thermal Storage Effects

Analytical and experimental studies have shown that the use of heavy materials in buildings has the effect of reducing heating and cooling loads, compared with lightweight materials, because of greater heat storage capacity.

Unlike cooling design, the ASHRAE Standard does not prescribe a method to account for thermal storage effects for heating. However, it does permit a performance approach to heating design which can be used to take advantage of the benefits of mass.

Energy use differences between light and heavy materials are illustrated in the hour-by-hour computer analyses shown in Fig. 9.1.5.

Fig. 9.1.5(a) compares the heat flow through three walls having the same U-value, but made of different materials. The concrete wall consisted of a layer of insulation sandwiched between inner and outer wythes of 2-in. concrete and weighed 48.3 psf. The metal wall, weighing 3.3 psf, had insulation sandwiched between an exterior metal panel and ½-in. drywall. The wood frame wall weighed 7.0 psf and had wood siding on the outside, insulation between 2×4 studs, and ½-in. drywall on the inside. The walls were exposed to simulated outside temperatures that represented a typical spring day in a moderate climate. The massive concrete wall had lower peak loads by about 13% for heating and 30% for cooling than

Fig. 9.1.5 Heating and cooling load comparisons

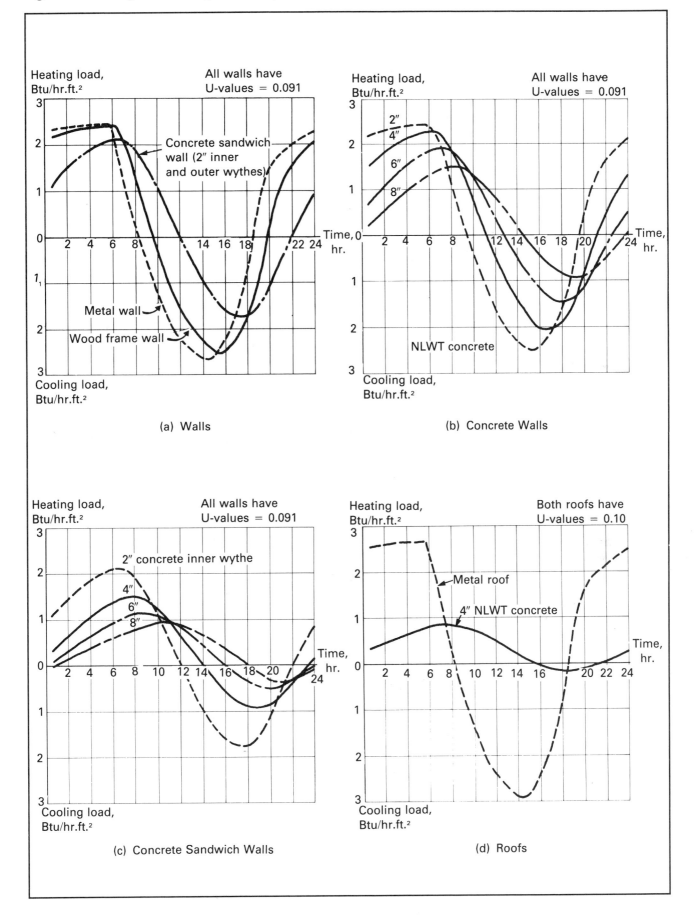

(a) Walls

(b) Concrete Walls

(c) Concrete Sandwich Walls

(d) Roofs

the less massive walls.

Concrete walls of various thicknesses that were exposed to the same simulated outside temperatures, are compared in Fig. 9.1.5(b). The walls had a layer of insulation sandwiched between concrete on the outside and ½-in. drywall on the inside; U-values were the same. The figure shows that the more massive the wall the lower the peak loads and the more the peaks were delayed.

Fig. 9.1.5(c) compares concrete sandwich panels having an outer wythe of 2-in., various thicknesses of insulation, and various thicknesses of inner wythes. All walls had U-values of 0.091 and were exposed to the same simulated outside temperatures. The figure shows that by increasing the thickness of the inner concrete wythe, peak loads were reduced and delayed.

A metal roof is compared to a concrete roof in Fig. 9.1.5(d). Both roof systems had built-up roofing on rigid board insulation on the outside and acoustical tile on the inside. The concrete roof weighed 48.3 psf and the metal roof 1.5 psf. The roofs had identical U-values of 0.10 and were exposed to the same simulated outside temperatures. The figure shows that the concrete roof had lower peak loads by 68% for heating and by 94% for cooling, and peaks were delayed by about 1.8 hours for heating and about 4 hours for cooling.

The rates of many utilities are structured so that lower peaks and delayed peaks can result in significant cost savings.

Other studies have shown that concrete buildings have lower average heating and cooling loads than lightweight buildings for a given insulation level. Thus life-cycle costs will be lower, or less insulation can be used for equivalent performance. The lower R-values that are required for energy consumption equal to lightweight structures are the "equivalent R-values" shown in Figs. 9.1.6 through 9.1.8.

Figs. 9.1.6 through 9.1.8 are based on peak values for occupancy and lighting heat gains of 384 and 205 kBtu/hr, respectively. The hourly average for occupancy and lighting internal gains were 104 and 85 kBtu/hr. No other internal gains were considered. The building parameters chosen were generally conservative so the aids can be used for heating analyses of most commercial buildings with the possible exception of warehouses. They are not applicable to cooling analyses. Fig. 9.1.6 can be used for roofs weighing 52 psf or more and having exposed ceilings.

The following example illustrates the use of Figs. 9.1.6 through 9.1.8:

Consider a single story building in Chicago, having a double tee concrete roof with no sky-lights, and sandwich walls having an inner wythe of 4-in concrete. Other parameters:

Heating degrees days = 6160
U_o maximum permitted for walls = 0.27
U_o maximum for roof = 0.075
A_{ow} = 9000 sq ft
A_w = 7546 sq ft
U_w = 0.125 (R8) for a lightweight wall
A_f = 1334 sq ft, U_f = 1.1
A_d = 120 sq ft, U_d = 0.16
A_r = A_{or} = 20,000 sq ft
U_r = 0.075 (R13.3) for a lightweight roof

The U-values for the lightweight walls and roof satisfy the ASHRAE requirements for heating. This can be shown by substituting appropriate values into equations 9.1.2 and 9.1.3.

Determine equivalent R-values for the concrete walls and roof:

For the roof, enter Fig. 9.1.6 at a heating degree day of 6160. Move vertically to the equivalent R-value line and read 10.8 (U-value = 0.093) on the vertical scale. This compares to an R-value for the low-mass roof of R13.3 (U-value = 0.075).

For the walls, enter Fig. 9.1.8(b) at a heating degree day value of 6160. Move vertically to the R8 curve and read the equivalent R-value of 4.9 (U-value = 0.204) on the vertical scale.

Once the equivalent R-value is determined for the concrete walls and roof, the components should be checked for compliance with the cooling criteria, as discussed in Sect. 9.1.7.

Fig. 9.1.6 Design aid for selecting equivalent R-values for roofs

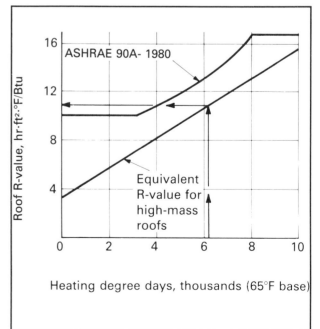

Fig. 9.1.7 Equivalent R-values for concrete walls with interior insulation

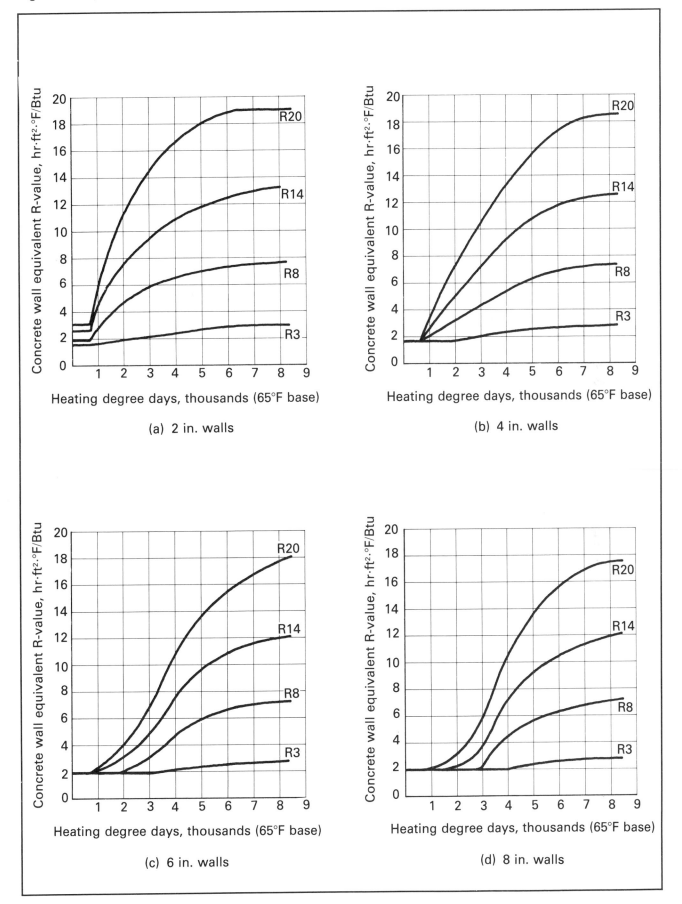

(a) 2 in. walls

(b) 4 in. walls

(c) 6 in. walls

(d) 8 in. walls

Fig. 9.1.8 Equivalent R-values for concrete walls with exterior insulation

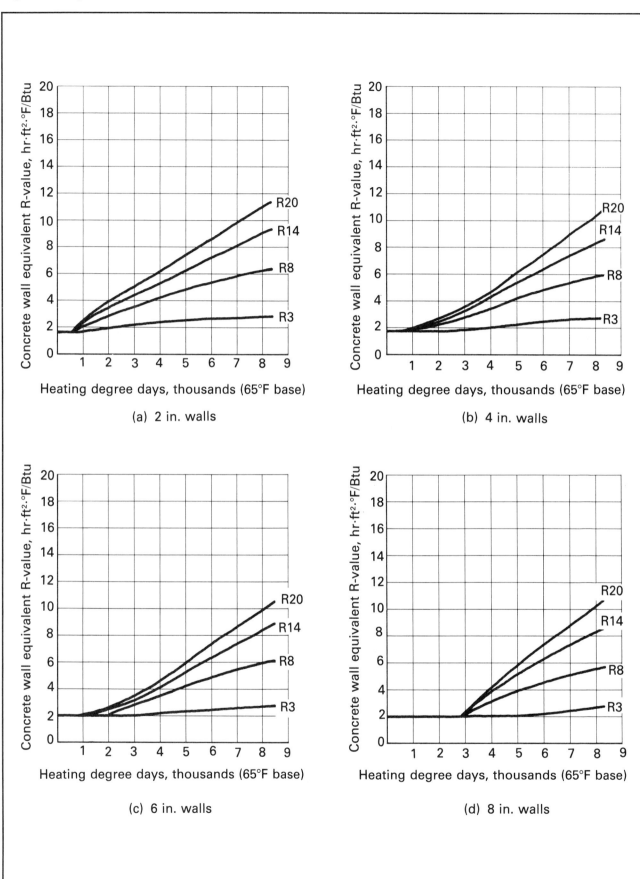

(a) 2 in. walls

(b) 4 in. walls

(c) 6 in. walls

(d) 8 in. walls

PCI Design Handbook

9.1.9 Building Envelope Performance

The ASHRAE Standard and some codes permit the stated U_o-value of any one assembly such as roof/ceiling, wall or floor, to be increased and the U_o-value of other components decreased, provided the overall heat transmission for the entire building envelope does not exceed the total allowed. Fig. 9.1.9 is a nomograph to facilitate these trade-off provisions. The following example illustrates the use of this aid:

Consider a building three stories high (25 ft) having plan dimensions of 80 × 200 ft located in Chicago. From Table 9.1.8 this location has 6160 heating degree days. U_o-values required by the ASHRAE Standard are 0.27 for walls and 0.075 for roofs. The ground floor is on grade with edges insulated to meet requirements. Assume there are windows on only one 200 ft side of the building and that the U_{ow} for that side has been calculated as 0.30. For the other three windowless sides, $U_{ow} = 0.12$.

Calculate the average U_{ow} provided for the walls:

$$U_{ow} = \frac{U_{ow1}A_{ow1} + U_{ow2}A_{ow2}}{A_{ow}}$$

$$= \frac{0.30(200)(25) + 0.12(360)(25)}{(560)(25)}$$

$$= 0.184 \text{ Btu/(hr)(sq ft)°F}$$

To use Fig. 9.1.9, calculate:

$$\phi = \text{wall area/roof area}$$

$$= (560)(25)/(80)(200) = 0.875$$

$$U_{ow} \text{ (provided)} - U_{ow} \text{ (required)}$$

$$= 0.184 - 0.27 = -0.086$$

Enter Fig. 9.1.9 at 0.086, move vertically to the interpolated location of $\phi = 0.875$, and follow the dashed line and read $U_{or} = 0.150$. This revised roof design changes U_{or} from 0.075 to 0.15. The R-value is reduced by $13.33 - 6.67 = 6.66$, representing 1½ to 1¾ in. less insulation.

Fig. 9.1.9 Nomograph for wall-ceiling or wall-roof U-value trade-offs for commercial buildings

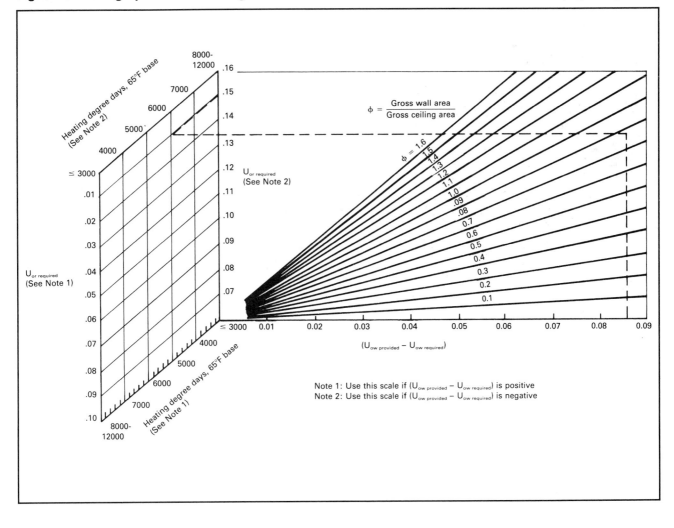

Note 1: Use this scale if ($U_{ow \text{ provided}} - U_{ow \text{ required}}$) is positive
Note 2: Use this scale if ($U_{ow \text{ provided}} - U_{ow \text{ required}}$) is negative

9.1.10 Condensation Control

Moisture which condenses on the interior of a building is unsightly and can cause damage to the building or its contents. Even more undesirable is the condensation of moisture within a building wall or ceiling assembly where it is not readily noticed until damage has occurred. All air in buildings contains water vapor, with warm air carrying more moisture than cold air. In many buildings moisture is added to the air by industrial processes, cooking, laundering, or humidifiers. If the inside surface temperature of a wall, floor or ceiling is too cold, the air contacting this surface will be cooled below its dew-point temperature and leave its excess water on that surfce. Condensation occurs on the surface with the lowest temperature.

Once condensation occurs, the relative humidity of the interior space of a building cannot be increased since any additional water vapor will simply condense on the cold surface. In effect, then, the inside temperature of an assembly limits the relative humidity which may be contained in an interior space.

Prevention of condensation on wall surfaces

The U-value of a wall must be such that the surface temperature will not fall below the dew-point temperature of the room air in order to prevent condensation on the interior surface of a wall.

Fig. 9.1.10 gives U-values for any combination of outside temperatures and inside relative humidities above which condensation will occur on the interior surfaces. For example, if a building were located in an area with an outdoor design temperature of 0°F and it was desired to maintain a relative humidity within the building of 25%, the wall must be designed so that all components have a U-value less than 0.78, otherwise there will be a problem with condensation. In many designs the desire to conserve energy will dictate the use of lower U-values than those required to avoid the condensation problem.

The degree of wall heat transmission resistance that must be provided to avoid condensation may be determined from the following relationship:

$$R_t = R_{fi} \frac{(t_i - t_o)}{(t_i - t_s)} \qquad \text{(Eq. 9.1.8)}$$

Dew-point temperatures to the nearest deg F for various values of t_i and relative humidity are shown in Table 9.1.9.

Determine R_t when the room temperature and relative humidity to be maintained are 70°F and 40%, and t_o during the heating season is −10°F.

From Table 9.1.9 the dew-point temperature t_s is 45°F, and from Table 9.1.1, $R_{fi} = 0.68$.

$$R_t = 0.68 \frac{[70 - (-10)]}{[70 - 45]} = 2.18$$
$$U = 0.46$$

Prevention of condensation within wall construction

Water vapor in air behaves as a gas and will diffuse through building materials at rates which depend on vapor permeabilities of materials to water vapor and vapor pressure differentials. The colder the outside temperature the greater the pressure of the water vapor in the warm inside air to reach the cooler, dried outside air. Also, leakage of moisture laden air into an assembly

Fig. 9.1.10 Relative humidity at which visible condensation occurs on inside surfaces. Inside temperature, 70F°

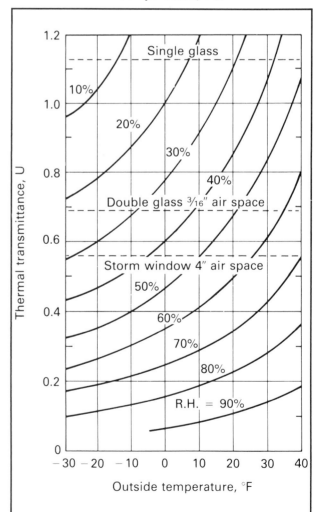

PCI Design Handbook

Table 9.1.9 Dew-point temperatures, °F[1]

Dry bulb or room temperature	Relative humidity (%)									
	10	20	30	40	50	60	70	80	90	100
40	−7	6	14	19	24	28	31	34	37	40
45	−3	9	18	23	28	32	36	39	42	45
50	−1	13	21	27	32	37	41	44	47	50
55	5	17	26	32	37	41	45	49	52	55
60	7	21	30	36	42	46	50	54	57	60
65	11	24	33	40	46	51	55	59	62	65
70	14	27	38	45	51	56	60	63	67	70
75	17	32	42	49	55	60	64	69	72	75
80	21	36	46	54	60	65	69	73	77	80
85	23	40	50	58	64	70	74	78	82	85
90	27	44	55	63	69	74	79	83	85	90

1. Temperatures are based on barometric pressure of 29.92 in. Hg.

through small cracks may be a greater problem than vapor diffusion. The passage of water vapor through material is in itself generally not harmful. It becomes of consequence when, at some point along the vapor flow path, a temperature level is encountered that is below the dew-point temperature and condensation results.

Building materials have water vapor permeances from very low to very high (see Table 9.1.10). When properly used, low permeance materials keep moisture from entering a wall or roof assembly, and materials with higher permeance allow construction moisture and moisture which enters inadvertently or by design to escape.

When a material such as plaster or gypsum board has a permeance which is too high for the intended use, one or two coats of paint is frequently sufficient to lower the permeance to an acceptable level, or a vapor barrier can be used directly behind such products. Polyethylene sheet, aluminum foil and roofing materials are commonly used. Proprietary vapor barriers, usually combinations of foil and polyethylene or asphalt, are frequently used in freezer and cold storage construction.

Concrete is a relatively good vapor barrier. Permeance is a function of the water-cement ratio of the concrete. A low water-cement ratio, such as that used in most precast concrete members, results in concrete with low permeance.

Where climatic conditions demand insulation, a vapor barrier is generally necessary in order to prevent condensation. A closed cell insulation, if properly applied, will serve as its own vapor barrier. For other insulation materials a vapor barrier should be applied to the warm side of the insulation.

For a more complete treatment of the subject of condensation within wall or roof assemblies, see References 1 and 4.

9.1.11 Thermal Bridges

Metal ties through walls or solid concrete paths through sandwich panels as described in Sect. 9.4 may cause localized cold spots. The most significant effect of these cold spots is condensation which may cause annoying or damaging wet streaks.

Table 9.1.10 Typical permeance (M) and permeability (μ) values

Material	M perms	μ perm-in.
Concrete	—	3.2
Wood (sugar pine)	—	0.4 - 5.4
Expanded polystyrene (extruded)	—	1.2
Paint - 2 coats		
Asphalt paint on plywood	0.4	
Enamels on smooth plaster	0.5 - 1.5	
Various primers plus 1 coat flat oil paint on plaster	1.6 - 3.0	
Expanded polystyrene (bead)	—	2.0 - 5.8
Plaster on gypsum lath (with studs)	20.00	
Gypsum wallboard, 0.375 in.	50.00	
Polyethylene, 2 mil	0.16	
Polyethylene, 10 mil	0.03	
Aluminum foil, 0.35 mil	0.05	
Aluminum foil, 1 mil	0.00	
Built-up roofing (hot mopped)	0.00	
Duplex sheet, asphalt laminated aluminum foil one side	0.002[2]	

1. ASHRAE Handbook, Chapter 21, Table 2.
2. Dry-cup.

The effect of these "bridges" on the heat transmittance can be calculated with reasonable accuracy by the zone method described in Chapter 23 of Ref. 1. The net effect of metal ties is to increase the U-value by 10 or 15 percent, depending on type, size and spacing. For example, a wall as shown in Fig. 9.1.11 would have a U-value of 0.13 if the effect of the ties is neglected. If the effect of ¼ in. diameter ties at 16 in. on center is included, U = 0.16; at 24 in. spacing, U = 0.15.

9.1.12 References

1. *ASHRAE Handbook 1981 Fundamentals*, American Society of Heating, Refrigerating, and Air-Conditioning Engineers, Inc., New York, N.Y., 1981.

2. *Energy Conservation in New Building Design*, ASHRAE Standard 90A-1980, American Society of Heating, Refrigerating, and Air-Conditioning Engineers, Inc., New York, N.Y., 1980.

3. "Simplified Thermal Design of Building Envelopes," *Bulletin EB089b*, Portland Cement Association, Skokie, IL, 1981.

4. Balik, J.S., and Barney, G.B., "Thermal Design of Precast Concrete Buildings," *PCI Journal*, V. 29, No. 6, Nov.–Dec., 1984.

Fig. 9.1.11 Example of thermal bridges

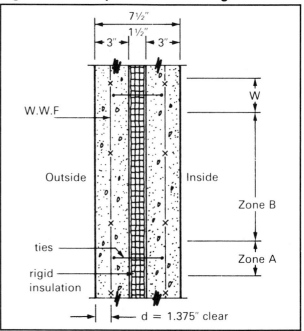

ACOUSTICAL PROPERTIES OF PRECAST CONCRETE

9.2.1 Glossary

Airborne Sound – sound that reaches the point of interest by propagation through air.

Background Level – the ambient sound pressure level existing in a space.

Decibel (dB) – a logarithmic unit of measure of sound pressure or sound power. Zero on the decibel scale corresponds to a standardized reference pressure (20 μPa) or sound power (10^{-12} watt).

Flanking Transmission – transmission of sound by indirect paths other than through the primary barrier.

Frequency (Hz) – the number of complete vibration cycles per second.

Impact Insulation Class (IIC) – a single figure rating of the overall impact sound insulation merits of floor-ceiling assemblies in terms of a reference contour (ASTM E989).

Impact Noise – the sound produced by one object striking another.

Noise – unwanted sound.

Noise Criteria (NC) – a series of curves, used as design goals to specify satisfactory background sound levels as they relate to particular use functions.

Noise Reduction (NR) – the difference in decibels between the space-time average sound pressure levels produced in two enclosed spaces by one or more sound sources in one of them.

Noise Reduction Coefficent (NRC) – the arithmetic average of the sound absorption coefficients at 250, 500, 1000 and 2000 Hz expressed to the nearest multiple of 0.05 (ASTM C423).

RC Curves – a revision of the NC curves based on empirical studies of background sounds.

Reverberation – the persistence of sound in an enclosed or partially enclosed space after the source of sound has stopped.

Sabin -- the unit of measure of sound absorption (ASTM C423).

Sound Absorption Coefficient (α) – the fraction of randomly incident sound energy absorbed or otherwise not reflected off a surface (ASTM C423).

Sound Pressure Level (SPL) – ten times the common logarithm of the ratio of the square of the sound pressure to the square of the standard reference pressure of 20 μPa. Commonly measured with a sound level meter and microphone, this quantity is expressed in decibels.

Sound Transmission Class (STC) – the single number rating system used to give a preliminary estimate of the sound insulation properties of a partition system. This rating is derived from measured values of transmission loss (ASTM E413).

Sound Transmission Loss (TL) – ten times the common logarithm of the ratio, expressed in decibels, of the airborne sound power incident on the partition that is transmitted by the partition and radiated on the other side (ASTM E90).

Structureborne Sound – sound that reaches the point of interest over at least part of its path by vibration of a solid structure.

9.2.2 General

The basic purpose of architectural acoustics is to provide a satisfactory environment in which desired sounds are clearly heard by the intended listeners and unwanted sounds (noise) are isolated or absorbed.

Under most conditions, the architect/engineer can determine the acoustical needs of the space and then design the building to satisfy those needs. Good acoustical design utilizes both absorptive and reflective surfaces, sound barriers and vibration isolators. Some surfaces must reflect sound so that the loudness will be adequate in all areas where listeners are located. Other surfaces absorb sound to avoid echoes, sound distortion and long reverberation times. Sound is isolated from rooms where it is not wanted by selected wall and floor-ceiling constructions. Vibration generated by mechanical equipment must be isolated from the structural frame of the building.

Most acoustical situations can be described in terms of: (1) sound source, (2) sound transmission path, and (3) sound receiver. Sometimes the source strength and path can be controlled and the receiver made more attentive by removing distraction or made more tolerant of disturbance. Acoustical design must include consideration of these three elements.

9.2.3 Approaching the Design Process

Criteria must be established before the acoustical design of a building can begin. Basically a satisfactory acoustical environment is one in which the character and magnitude of all sounds are compatible with the intended space function.

Although a reasonable objective, it is not always easy to express these intentions in quantitative terms. In addition to the amplitude of sound, the properties such as spectral characteristics, continuity, reverberation and intelligibility must be specified.

People are highly adaptable to the sensations of heat, light, odor, sound, etc., with sensitivities varying widely. The human ear can detect a sound intensity of rustling leaves, 10 dB, and can tolerate, if even briefly, the powerful exhaust of a jet engine at 120 dB, 10^{12} times the intensity of the rustling sound.

9.2.4 Dealing with Sound Levels

The problems of sound insulation are usually considerably more complicated than those of sound absorption. The former involves reductions of sound level, which are of greater orders of magnitude than can be achieved by absorption. These large reductions of sound level from space to space can be achieved only by continuous, impervious barriers. If the problem also involves structure borne sound, it may be necessary to introduce resilient layers or discontinuities into the barrier.

Sound absorbing materials and sound insulating materials are used for different purposes. There is not much sound absorption from an 8-in. concrete wall; similarly, high sound insulation is not available from a porous lightweight material that may be applied to room surfaces. It is important to recognize that the basic mechanisms of sound absorption and sound insulation are quite different.

9.2.5 Sound Transmission Loss

Sound transmission loss measurements are made at 16 frequencies at one-third octave intervals covering the range from 125 to 4000 Hz. The testing procedure is ASTM Specification E90, *Laboratory Measurement of Airborne Sound Transmission Loss of Building Partitions*. To simplify specification of desired performance characteristics the single number Sound Transmission Class (STC) was developed.

Airborne sound reaching a wall, floor or ceiling produces vibration in the wall and is radiated with reduced intensity on the other side. Airborne sound transmission loss of walls and floor-ceiling assemblies is a function of its weight, stiffness and vibration damping characteristics.

Weight is concrete's greatest asset when it is used as a sound insulator. For sections of similar design, but different weights, the STC increases approximately 6 units for each doubling of weight as shown in Fig. 9.2.1.

Precast concrete walls, floors and roofs usually do not need additional treatments in order to provide adequate sound insulation. If desired, greater sound insulation can be obtained by using a resiliently attached layer(s) of gypsum board or other building material. The increased transmission loss occurs because the energy flow path is now increased to include a dissipative air column and additional mass.

The acoustical test results of both airborne sound transmission loss and impact insulation of 4, 6 and 8 in. flat panels, a 14 in. double tee, and 6 and 8 in. hollow-core slabs are shown in Figs. 9.2.2, 9.2.3 and 9.2.4.

Table 9.2.1 presents the ratings for various precast concrete walls and floor-ceiling assemblies. The effects of various assembly treatments on sound transmission can also be predicted from results of previous tests as shown in Table 9.2.2. The improvements are additive, but in some cases the total effect may be slightly less than the sum.

9.2.6 Impact Noise Reduction

Footsteps, dragged chairs, dropped objects, slammed doors, and plumbing generate impact noise. Even when airborne sounds are adequately controlled there can be severe impact noise problems.

The test method used to evaluate systems for impact sound insulation is described in ASTM Specification E492, *Laboratory Measurement of Impact Sound Transmission Using the Tapping Machine*. As with the airborne standard, measurements are made at 16 one-third octave intervals but in the range from 100 to 3150 Hz. For performance specification purposes the single number Impact Insulation Class (IIC) is used.

Fig. 9.2.1 Sound transmission class as a function of weight of floor or wall

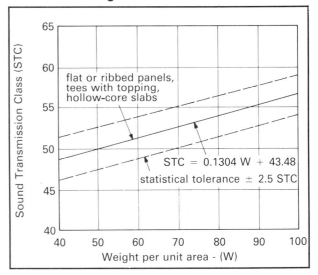

PCI Design Handbook

Fig. 9.2.2 Acoustical test data of solid flat concrete panels — normal weight concrete

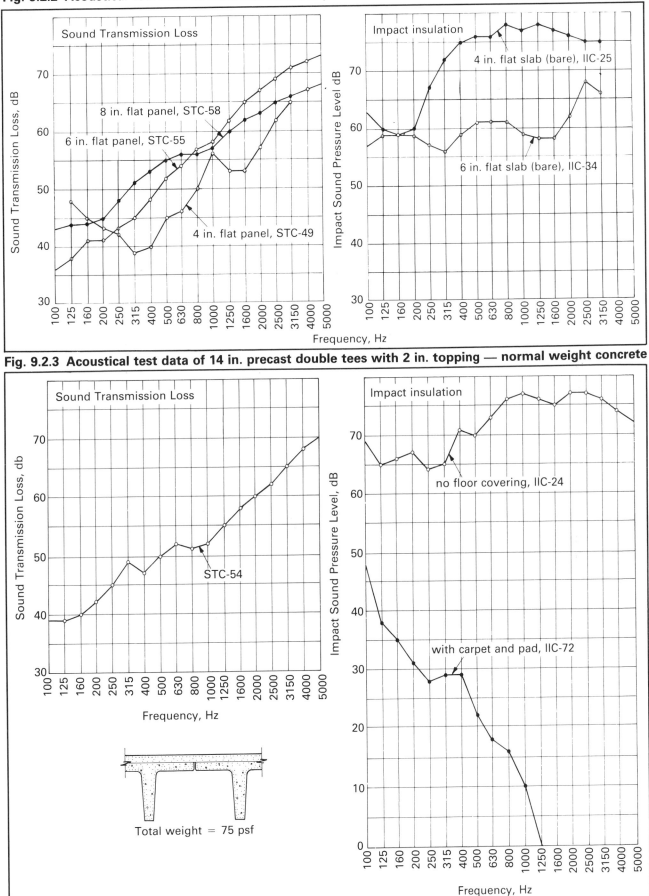

Fig. 9.2.3 Acoustical test data of 14 in. precast double tees with 2 in. topping — normal weight concrete

Fig. 9.2.4 Acoustical test data of hollow-core panels — normal weight concrete

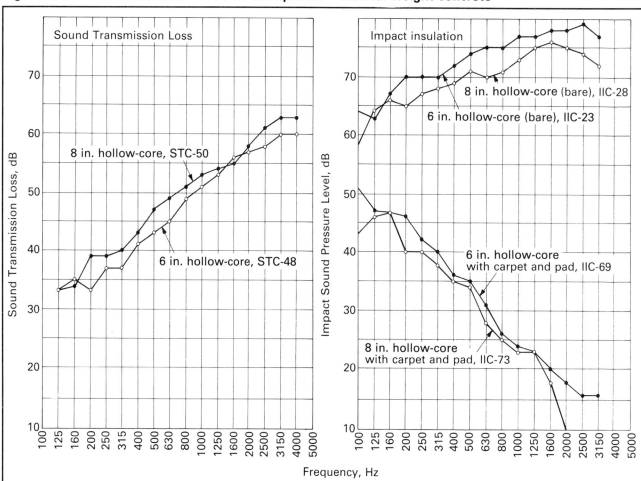

In general, thickness or unit weight of concrete does not greatly affect the transmission of impact sounds as shown in the following table:

Thickness, in.	Unit weight of concrete, pcf	IIC
5	79	23
	114	24
	144	24
10	79	28
	114	30
	144	31

Structural concrete floors in combination with resilient materials effectively control impact sound. One simple solution consists of good carpeting on resilient padding. Table 9.2.2 shows that a carpet and pad over a bare concrete slab will increase the impact noise reduction from 43 to 56 points. The overall efficiency varies according to the characterisitics of the carpeting and padding such as resilience, thickness and weight. So called resilient flooring materials, such as linoleum, rub-

ber, asphalt vinyl, vinyl asbestos, etc., are not entirely satisfactory directly on concrete, nor are parquet or strip wood floors when applied directly. Impact sound also may be controlled by providing a discontinuity in the structure such as would be obtained by adding a resilient-mounted plaster or drywall suspended ceiling or a floating floor consisting of a second layer of concrete cast over resilient pads, insulation boards or mastic. The thickness of floating slabs is usually controlled by structural requirements, however, 8 in. is adequate in most instances.

9.2.7 Absorption of Sound

A sound wave always loses part of its energy as it is reflected by a surface. This loss of energy is termed sound absorption. It appears as a decrease in sound pressure of the reflected wave. The sound absorption coefficient is the fraction of energy incident but not reflected per unit of surface area. Sound absorption can be specified at individual frequencies or as an average of absorption coefficients (NRC).

Table 9.2.1 Airborne sound transmission and impact insulation class ratings from tests of precast concrete assemblies

Assembly No.	Description	STC	IIC
	Wall Systems		
1	4 in. flat panel, 54 psf	49	—
2	6 in. flat panel, 75 psf	55	—
3	Assembly 2 with "Z" furring channels, 1 in. insulation and ½ in. gypsum board, 75.5 psf	62	—
4	Assembly 2 with wood furring, 1½ in. insulation and ½ in. gypsum board, 73 psf	63	—
5	Assembly 2 with ½ in. space, 1⅝ in. metal stud row, 1½ in. insulation and ½ in. gypsum board	63*	—
6	8 in. flat panel, 95 psf	58	—
7	14 in. prestressed tees with 4 in. flange, 75 psf	54	—
	Floor-Ceiling Systems		
8	8 in. hollow-core prestressed units, 57 psf	50	28
9	Assembly 8 with carpet and pad, 58 psf	50	73
10	8 in. hollow-core prestressed units with ½ in. wood block flooring adhered directly, 58 psf	51	47
11	Assembly 10 except ½ in. wood block flooring adhered to ½ in. sound-deadening board underlayment adhered to concrete, 60 psf	52	55
12	Assembly 11 with acoustical ceiling, 62 psf	59	61
13	Assembly 8 with quarry tile, 1¼ in. reinforced mortar bed with 0.4 in. nylon and carbon black spinerette matting, 76 psf	60	54
14	Assembly 13 with suspended ⅝ in. gypsum board ceiling with 3½ in. insulation, 78.8 psf	61	62
15	14 in. prestressed tees with 2 in. concrete topping, 75 psf	54	24
16	Assembly 15 with carpet and pad, 76 psf	54	72
17	Assembly 15 with resiliently suspended acoustical ceiling with 1½ in. mineral fiber blanket above, 77 psf	59	51
18	Assembly 17 with carpet and pad, 78 psf	59	82
19	4 in. flat slabs, 54 psf	49	25
20	5 in. flat slabs, 60 psf	52*	24
21	5 in. flat slab concrete with carpet and pad, 61 psf	52*	68
22	6 in. flat slabs, 75 psf	55	34
23	8 in. flat slabs, 95 psf	58	34*
24	10 in. flat slabs, 120 psf	59*	31
25	10 in. flat slab concrete with carpet and pad, 121 psf	59*	74

*Estimated values

A dense non-porous concrete surface typically absorbs 1 to 2% of incident sound and has an NRC of 0.015. There are specially fabricated units with porous concrete surfaces which provide greater absorption. In the case where additional sound absorption of precast concrete is desired, a coating of acoustical material can be spray applied, acoustical tile can be applied with adhesive, or an acoustical ceiling can be suspended. Most of the spray applied fire retardant materials used to increase the fire resistance of precast concrete and other floor-ceiling systems can also be used to absorb sound. The NRC of the sprayed fiber types range from 0.25 to 0.75. Most cementitious types have an NRC from 0.25 to 0.50.

If an acoustical ceiling were added to Assembly 15 of Table 9.2.1 (as in Assembly No. 17), the sound entry through a floor or roof would be reduced 5

Table 9.2.2 Typical improvements for wall, floor, and ceiling treatments used with precast concrete elements

Treatment	Increase in Ratings	
	Airborne (STC)	Impact (IIC)
Wall furring, ¾ in. insulation and ½ in. gypsum board attached to concrete wall	3	0
Separate metal stud system, 1½ in. insulation in stud cavity and ½ in. gypsum board attached to concrete wall	5 to 10	0
2 in. concrete topping (24 psf)	3	0
Carpet and pad	0	43 to 56
Vinyl tile	0	3
½ in. wood block adhered to concrete	0	20
½ in. wood block and resilient fiber underlayment adhered to concrete	4	26
Floating concrete floor on fiberboard	7	15
Wood floor, sleepers on concrete	5	15
Wood floor on fiberboard	10	20
Acoustical ceiling resiliently mounted	5	27
- if added to floor with carpet	5	10
Plaster or gypsum board ceiling resiliently mounted	10	8
- with insulation in space above ceiling	13	13
Plaster direct to concrete	0	0

dB. In addition, the acoustical ceiling would absorb a portion of the sound after entry and provide a few more decibels of quieting. Use of the following expression can be made to determine the intra-room noise or loudness reduction due to the absorption of sound.

$$NR = 10 \log \frac{A_o + A_a}{A_o} \quad \text{(Eq. 9.2.1)}$$

where
  NR = sound pressure level reduction, dB
  A_o = original absorption, Sabins
  A_a = added absorption, Sabins

Values for A_o and A_a are the products of the absorption coefficients of the various room materials and their surface areas.

A plot of this equation is shown in Fig 9.2.5. For an absorption ratio of 5, the decibel reduction is 7 dB. Note that the decibel reduction is the same, regardless of the original sound pressure level and depends only on the absorption ratio. This is due to the fact that the decibel scale is itself a scale of ratios, rather than difference in sound energy.

While a decibel difference is an engineering quantity which can be physically measured, it is also important to know how the ear judges the change in sound energy due to sound conditioning. Apart from the subjective annoyance factors associated with excessive sound reflection, the ear can make accurate judgments of the relative loud-

ness between sounds. An approximate relation between percentage loudness, reduction of reflected sound and absorption ratio is plotted in Fig. 9.2.6.

The percentage loudness reduction does not depend on the original loudness, but only on the absorption ratio. (The curve is drawn for loudness within the normal range of hearing and does not apply to extremely faint sounds.) Referring again to the absorption ratio of 5, the loudness reduction from Fig. 9.2.6 is approximately 40 percent.

Example 9.2.1 Using Table 9.2.2.

The performance of a 2 in. concrete topping, carpet and pad added to 8 in. hollow-core prestressed floor units is calculated as follows:

Materials	STC	IIC
Bare slab	50	28
2 in. concrete topping	3	0
Carpet and pad	0	50
Totals	53	78

9.2.8 Acceptable Noise Criteria

As a rule, a certain amount of continuous sound can be tolerated before it becomes noise. An "acceptable" level neither disturbs room occupants

PCI Design Handbook

Fig. 9.2.5 Relation of decibel reduction of reflected sound to absorption ratio

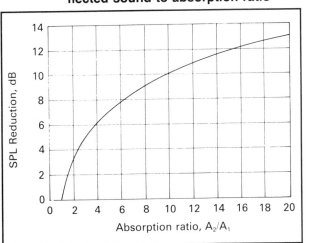

Fig. 9.2.6 Relation of percent loudness reduction of reflected sound to absorption ratio

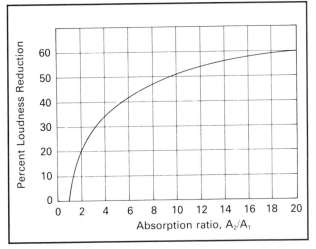

nor interferes with the communication of wanted sound.

The most widely accepted and used noise criteria today are expressed as the Noise Criteria (NC) curves Fig 9.2.7a. The figures in Table 9.2.3 represent general acoustical goals. They can also be compared with anticipated noise levels in specific rooms to assist in evaluating noise reduction problems.

The main criticism of NC curves is that they are too permissive when the control of low or high frequency noise is of concern. For this reason, Room Criteria (RC) Curves were developed (Fig. 9.2.7b).[1,2] RC curves are the result of extensive studies based on the human response to both sound pressure level and frequency and take into account the requirements for speech intelligibility.

A low background level obviously is necessary where listening and speech intelligibility is important. Conversely, higher ambient levels can

Table 9.2.3 Recommended category classification and suggested noise criteria range for steady background noise as heard in various indoor functional activity areas*[1]

Type of space	NC or RC curve
1. Private residences	25 to 30
2. Apartments	30 to 35
3. Hotels/motels	
a. Individual rooms or suites	30 to 35
b. Meeting/banquet rooms	30 to 35
c. Halls, corridors, lobbies	35 to 40
d. Service/support areas	40 to 45
4. Offices	
a. Executive	25 to 30
b. Conference rooms	25 to 30
c. Private	30 to 35
d. Open-plan areas	35 to 40
e. Computer/ business machine areas	40 to 45
f. Public circulation	40 to 45
5. Hospitals and clinics	
a. Private rooms	25 to 30
b. Wards	30 to 35
c. Operating rooms	25 to 30
d. Laboratories	30 to 35
e. Corridors	30 to 35
f. Public areas	35 to 40
6. Churches	25 to 30**
7. Schools	
a. Lecture and classrooms	25 to 30
b. Open-plan classrooms	30 to 35**
8. Libraries	30 to 35
9. Concert Halls	**
10. Legitimate theatres	**
11. Recording studios	**
12. Movie theatres	30 to 35

* Design goals can be increased by 5 dB when dictated by budget constraints or when noise intrusion from other sources represents a limiting condition.
** An acoustical expert should be consulted for guidance on these critical spaces.

Fig. 9.2.7a NC (Noise Criteria) Curves

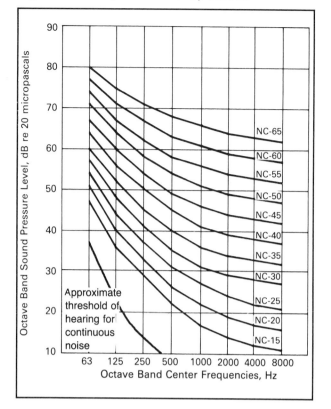

Fig. 9.2.7b RC (Room Criteria) Curves

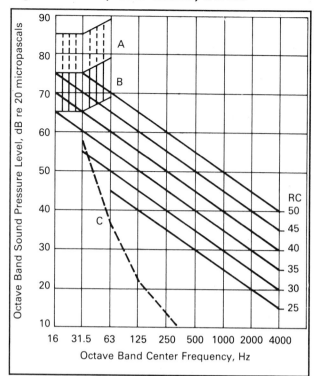

Region A: High probability that noise-induced vibration levels in lightweight wall/ceiling constructions will be clearly feelable; anticipate audible rattles in light fixtures, doors, windows, etc.
Region B: Noise-induced vibration levels in lightweight wall/ceiling constructions may be moderately feelable; slight possibility of rattles in light fixtures, doors, windows, etc.
Region C: Below threshold of hearing for continuous noise.

persist in large business offices or factories where speech communication is limited to short distances. Often it is just as important to be interested in the minimum as in the maximum permissible levels of Table 9.2.3. In an office or residence, it is desirable to have a certain ambient sound level to assure adequate acoustical privacy between spaces, thus minimizing the transmission loss requirements of unwanted sound (noise).

These undesirable sounds may be from an exterior source such as automobiles or aircraft, or they may be generated as speech in an adjacent

classroom or music in an adjacent apartment. They may be direct impact-induced sound such as footfalls on the floor above, rain impact on a lightweight roof construction or vibrating mechanical equipment.

Thus the designer must always be ready to accept the task of analyzing the many potential sources of intruding sound as related to their frequency characteristics and the rates at which they

Fig. 9.2.7c Noise criterion (NC) curves

Noise criterion curves	Octave band center frequency, Hz							
	63	125	250	500	1000	2000	4000	8000
NC-15*	47	36	29	22	17	14	12	11
NC-20*	51	40	33	26	22	19	17	16
NC-25*	54	44	37	31	27	24	22	21
NC-30	57	48	41	35	31	29	28	27
NC-35	60	52	45	40	36	34	33	32
NC-40	64	56	50	45	41	39	38	37
NC-45	67	60	54	49	46	44	43	42
NC-50	71	64	58	54	51	49	48	47
NC-55	74	67	62	58	56	54	53	52
NC-60	77	71	67	63	61	59	58	57
NC-65	80	75	71	68	66	64	63	62

* The applications requiring background levels less than NC-25 are special purpose spaces in which an acoustical consultant should set the criteria.

PCI Design Handbook

occur. The level of toleration that is to be expected by those who will occupy the space must also be established. Figs. 9.2.8 and 9.2.9 are the spectral characteristics of common noise sources.

With these criteria, the problem of sound isolation now must be solved, namely the reduction process between the high unwanted noise source and the desired ambient level. For this solution, two related yet mutually exclusive processes must be incorporated, i.e., sound transmission loss and sound absorption.

9.2.9 Establishment of Noise Insulation Objectives

Often acoustical control is specified as to the minimum insulation values of the dividing partition system. Municipal building codes, lending institutions and the Department of Housing and Urban Development (HUD) list both airborne STC and impact IIC values for different living environments. For example, the HUD minimum property standards[3] are:

Location	STC	IIC
Between living units	45	45
Between living units and public space	50	50

Other community ordinances are more specific, listing the sound insulation criteria with relation to particular ambient environments.[4]

	Grade I Suburban	Grade II Residential Urban & suburban	Grade III Urban
Ambient level	NC or RC 20-25	NC or RC 25-30	NC or RC 35+
Walls Floor-ceiling assemblies	STC 55 STC 55 IIC 55	STC 52 STC 52 IIC 52	STC 48 STC 48 IIC 48

Once the objectives are established, the designer then should refer to available data, e.g., Fig. 9.2.1 or Table 9.2.1, and select the system which best meets these requirements. In this respect, concrete systems have superior properties and can with minimal effort comply with these criteria. When the insulation value has not been specified, selection of the necessary barrier can be determined analytically by (1) identifying exterior and/or interior noise sources, and (2) by establishing acceptable interior noise criteria.

Example 9.2.2 Sound insulation criteria

Assume a precast, prestressed concrete office building is to be erected adjacent to a major highway. Private and semiprivate offices will run along the perimeter of the structure. The first step is to determine the degree of insulation required of the exterior wall system.

Fig. 9.2.8 Sound pressure levels — exterior noise sources

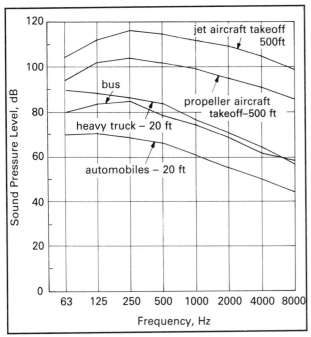

Fig. 9.2.9 Sound pressure levels — interior noise sources

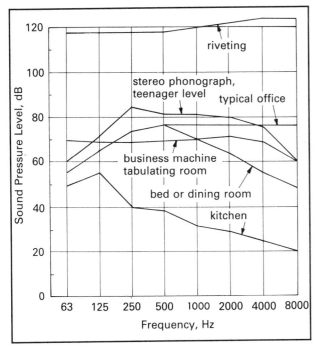

Sound pressure level - (dB)								
Frequency (Hz)	63	125	250	500	1000	2000	4000	8000
Bus traffic source noise (Fig. 9.2.8)	80	83	85	78	74	68	62	58
Private office noise criteria - NC 35 (Fig. 9.2.7)	60	52	45	40	36	34	33	32
Required Insulation	20	31	40	38	38	34	29	26

Sound pressure level - (dB)						
Frequency (Hz)	125	250	500	1000	2000	4000
Required insulation	31	40	38	38	34	29
6 in. Precast solid concrete wall (Fig. 9.2.2)	38	43	52	59	67	72
Deficiencies	—	—	--	—	—	—

The 500 Hz requirement, 38 dB, can be used as the first approximation of the wall STC category. However, if windows are planned for the wall, a system of about 50-55 STC should be selected (see following composite wall discussion). Individual transmission loss performance values of this system are then compared to the calculated need.

The selected wall should meet or exceed the insulation needs at all frequencies. However, to achieve the most efficient design conditions, certain limited deficiencies can be tolerated. Experience has shown that the maximum deficiencies are 3 dB at two frequencies or 5 dB on one frequency point.

9.2.10 Composite Wall Considerations

Doors and windows are often the weak link in an otherwise effective sound barrier. Minimal effects on sound transmission loss will be achieved in most cases by a proper selection of glass (Table 9.2.4).[5] Mounting of the glass in its frame should be done with care to eliminate noise leaks and to reduce the glass plate vibrations.

Sound transmission loss of a door depends upon its material and construction, and the sealing between the door and the frame.[6] There is a mass law dependence of STC on weight (psf) for both wood and steel doors. The approximate relationships are:

For steel doors: STC = 15 + 27 log W
For wood doors: STC = 12 + 32 log W
where W = weight of the door, psf

These relationships are purely empirical and a large deviation can be expected for any given door.

For best results, the distances between adjacent door and/or window openings should be maximized, staggered when posible and held to a minimum area. Minimizing openings retains the acoustical properties of precast concrete. The design characteristics of a door or window system must be analyzed prior to specification. Such qualities as frame design, door construction and glazing thickness are vital performance criteria. Installation procedures must be exact and care given to the frame of each opening. Gaskets, weatherstripping and raised threshold serve as both excellent thermal and acoustical seals and are recommended.

Fig. 9.2.10 can be used to calculate the effective acoustic isolation of a wall system which contains a composite of elements, each with known individual transmission loss data.

Example 9.2.3 Composite wall insulation criteria

To complete the office building wall acoustical design from Sect. 9.2.9, assume the following:

1. The glazing area represents 10% of the exterior wall area.

2. The windows will be double glazed with a 38 STC acoustical insulation rating.

The problem now becomes the task of determining the combined effect of the concrete-glass combination and a redetermination of criteria compliance.

Table 9.2.4 Acoustical properties of glass

Type and overall thickness	Inside light	Construction space	Outside light	STC
Sound transmission class (STC)				
⅛″ Plate or float	—	—	⅛″	23
¼″ Plate or float	—	—	¼″	28
½″ Plate or float	—	—	½″	31
1″ Insulated glass	¼″	½″ Air space	¼″	31
¼″ Laminated	⅛″	0.030 Vinyl	⅛″	34
1½″ Insulated glass	¼″	1″ Air space	¼″	35
¾″ Plate or float	—	—	¾″	36
1″ Insulated glass	¼″	½″ Air space	¼″ Laminated	38
1″ Plate or float	—	—	1″	37
2¾″ Insulated glass	¼″	2″ Air space	½″	39
4¾″ Insulated glass	¼″	4″ Air space	½″	40
6¾″ Insulated glass	¼″	6″ Air space	¼″ Laminated	42

Transmission loss (dB)

Frequency (Hz)

125	160	200	250	315	400	500	630	800	1000	1250	1600	2000	2500	3150	4000
						¼ inch plate glass - 28 STC									
24	22	24	24	21	23	21	23	26	27	33	36	37	39	40	40
						1 inch insulating glass with ½ inch air space - 31 STC									
25	25	22	20	24	27	27	30	32	33	35	34	29	31	33	36
						1 inch insulating glass laminated with ½ inch air space - 38 STC									
30	29	26	28	31	34	35	37	37	38	38	40	41	40	41	44

The maximum deficiency is 3 dB and occurs at only one frequency point. The 6 in. precast concrete wall with double glazed windows will provide the required acoustical insulation.

Floor-ceiling assembly acoustical insulation requirements are determined in the same manner as walls by using Figs 9.2.2 and 9.2.9.

9.2.11 Leaks and Flanking

The performance of a building section with an otherwise adequate STC can be seriously reduced by a relatively small hole or any other path which allows sound to bypass the acoustical barrier. All noise which reaches a space by paths other than

Sound pressure level - (dB)						
Frequency (Hz)	125	250	500	1000	2000	4000
6 in. Precast solid concrete wall (Fig. 9.2.2)	38	43	52	59	67	72
Double glazed windows (Table 9.2.4)	30	28	35	38	41	44
Correction (Fig. 9.2.10)	−2	−6	−7	−11	−15	−19
Combined transmission loss	36	37	45	48	52	53
Insulation requirements	31	40	38	38	34	29
Deficiencies	—	3	—	—	—	—

through the primary barrier is called flanking. Common flanking paths are openings around doors or windows, at electrical outlets, telephone and television connections, and pipe and duct penetrations. Suspended ceilings in rooms where walls do not extend from the ceiling to the roof or floor above allow sound to travel to adjacent rooms.

Anticipation and prevention of leaks begins at the design stage. Flanking paths (gaps) at the perimeters of interior precast walls and floors are generally sealed during construction with grout or drypack. In addition, all openings around penetrations through walls or floors should be as small as possible and must be sealed airtight. The higher the STC of the barrier, the greater the effect of an unsealed opening (see Fig. 9.2.10).

Perimeter leakage more commonly occurs at the intersection between an exterior curtain wall and floor slab. It is of vital importance to seal this gap in order to retain the acoustical integrity of the system as well as provide the required fire stop between floors. One way to achieve this seal is to place a 4 pcf density mineral wool blanket between the floor slab and the exterior wall. Fig. 9.2.11 demonstrates the acoustical isolation ef-

fects of this treatment.

In exterior walls, the proper application of sealant and backup materials in the joints between units will not allow sound to flank the wall.

If the acoustical design is balanced, the maximum amount of acoustic energy reaching a space via flanking should not equal the energy transmitted through the primary barriers.

Although not easily quantified, an inverse relationship exists between the performance of an element as a primary barrier and its propensity to transmit flanking sound. In other words, the probability of existing flanking paths in a concrete structure is much less than in one of a steel or wood frame.

In addition to using basic structural materials, flanking paths can be minimized by:

1. Interrupting the continuous flow of energy with dissimilar materials, i.e., expansion or control joints or air gaps.

2. Increasing the resistance to energy flow with floating floor systems, full height and/or double partitions and suspended ceilings.

Fig. 9.2.10 Chart for calculating the effective transmission loss of a composite barrier. (For purposes of approximation STC values can be used in place of TL values.)

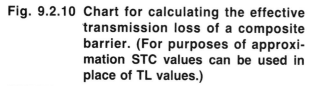

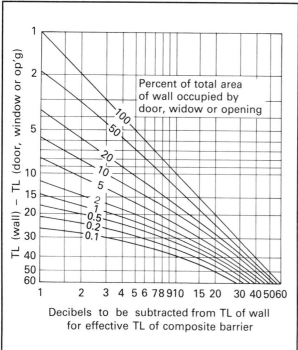

Fig. 9.2.11 Effect of safing insulation seals

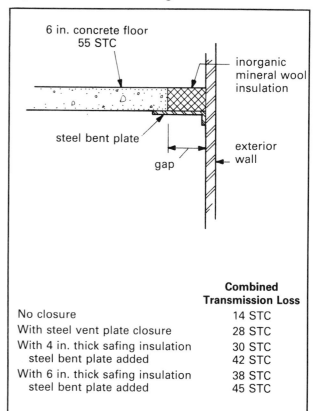

	Combined Transmission Loss
No closure	14 STC
With steel vent plate closure	28 STC
With 4 in. thick safing insulation steel bent plate added	30 STC 42 STC
With 6 in. thick safing insulation steel bent plate added	38 STC 45 STC

PCI Design Handbook

9.2.12 References

1. ASHRAE, *ASHRAE Systems Handbook for 1984*, American Society of Heating, Refrigerating & Air Conditioning Engineers, Inc., New York, 1984.

2. Blazier, W.E., "Revised Noise Criteria for Design and Rating of HVAC Systems", paper presented at ASHRAE Semiannual Meeting, Chicago, IL, Jan. 26, 1981.

3. Berendt, R.D., Winzer, G.E., Burroughs, C.B., "A Guide to Airborne, Impact and Structureborne Noise Control in Multifamily Dwellings", prepared for Federal Housing Administration, U.S. Government Printing Office, Washington, D.C., 1975.

4. Sabine, H.J., Lacher, M.B., Flynn, D.R., Quindry, T.L., "Acoustical and Thermal Performance of Exterior Residential Walls, Doors & Windows", National Bureau of Standards, U.S. Government Printing Office, Washington, D.C., 1975.

5. IITRI, "Compendium of Materials for Noise Control", U.S. Department of Health, Education & Welfare, U.S. Government Printing Office, Washington, D.C., 1980.

Additonal Bibliography

Beranek, L.L, *Noise Reduction*, McGraw-Hill Book Co., New York, 1960.

Ceramic Tile Institute of America and American Enka Company, unpublished floor/ceiling tests.

Harris, C.M., *Handbook of Noise Control*, McGraw-Hill Book Co., New York, 1967.

Harris, C.M., Crede, C.E., *Shock & Vibration Handbook - 2nd edition*, McGraw-Hill Book Co., New York, 1976.

Litvin, A., Belliston, H.W., "Sound Transmission Loss Through Concrete and Concrete Masonry Walls", *ACI Journal*, Dec. 1978.

FIRE RESISTANCE

9.3.1 Notation

Note: Subscript θ indicates the property as affected by elevated temperatures.

a = depth of equivalent rectangular compression stress block

A_{ps} = area of prestressing steel

A_s = area of non-prestressed reinforcement

A_s^- = area of reinforcement in negative moment region

b = width of member

d = distance from centroid of prestressing steel to the extreme compression fiber

f_c' = compressive strength of concrete

f_{ps} = stress in the prestressing steel at nominal strength

f_{pu} = ultimate tensile strength of prestressing steel

h = total depth of a member

ℓ = span length

M_n = nominal moment strength

$M_{n\theta}^+$ = positive and negative nominal moment
$M_{n\theta}^-$ strength at elevated temperatures, respectively

R = fire endurance of a composite assembly

$R_1,$ = fire endurance of individual courses
R_2, R_n

u = distance from prestressing steel to the fire exposed surface

w = uniform total load

w_d = uniform dead load

w_ℓ = uniform live load

$x, x_o,$ = horizontal distances as shown in Figs.
x_1, x_2 9.3.9, 9.3.10, and 9.3.11

θ_s = temperature of steel

ϕ = strength reduction factor

9.3.2 Glossary

Carbonate aggregate concrete — concrete made with aggregates consisting mainly of calcium or magnesium carbonate, e.g., limestone or dolomite.

Fire endurance — a measure of the elapsed time during which a material or assembly continues to exhibit fire resistance under specified conditions of test and performance. As applied to elements of buildings it shall be measured by the methods and to the criteria defined in ASTM E119. (Defined in ASTM E176).

Fire resistance — the property of a material or assembly to withstand fire or to give protection from it. As applied to elements of buildings, it is characterized by the ability to confine a fire or to continue to perform a given structural function, or both. (Defined in ASTM E176).

Fire resistance rating (sometimes called **fire rating, fire resistance classification,** or **hourly rating**) — a legal term defined in building codes, usually based on fire endurances. Fire resistance ratings are assigned by building codes for various types of construction and occupancies and are usually given in hourly increments.

Lightweight aggregate concrete — concrete made with aggregates of expanded clay, shale, slag, or slate or sintered fly ash, and weighing about 85 to 115 pcf.

Sand-lightweight concrete — concrete made with a combination of expanded clay, shale, slag, or slate or sintered fly ash and natural sand. Its unit weight is generally between 105 and 120 pcf.

Siliceous aggregate concrete — concrete made with normal weight aggregates consisting mainly of silica or compounds other than calcium or magnesium carbonate.

9.3.3 Introduction

Precast and prestressed concrete members can be provided with any degree of fire resistance that may be required by building codes, insurance companies, and other authorities. The fire resistance of building assemblies is determined from standard fire tests defined by the American Society for Testing and Materials.

To insure that fire resistance requirements are satisfied, the engineer can use tabulated information provided by various authoritative bodies, such as Underwriters Laboratories, Inc., the American Insurance Association and model building codes. This information is based on the results of standard fire tests of assemblies that may include ceilings and other building components. The 1983 edition of the UL *Fire Resistance Directory* alone provides information on more than 120 assemblies incorporating precast, prestressed concrete members.

In the absence of tabulated data, the fire resistance of precast and prestressed concrete members and assemblies can be determined in most cases by calculation. These calculations are based on engineering principles and take into account the conditions of a standard fire test. This is known as the Rational Design Method of determining fire resistance. It is based on extensive research sponsored in part by the Prestressed Concrete Institute and conducted by the Portland Cement Association and other laboratories.

After a discussion on fire tests in Sects. 9.3.4 and 9.3.5, calculations using the Rational Design Method in many common situations are presented in the following sections. Brief explanations of the underlying principles are also given. For additional examples, design charts and a complete explanation of the method, refer to the PCI manual, MNL 124-77, *PCI Design for Fire Resistance of Precast Prestressed Concrete,* and the references listed in that manual as well as the CRSI manual *Reinforced Concrete Fire Resistance.*

Research reports recognizing the use of PCI MNL 124-77 have been issued by the International Conference of Building Officials and the Building Officials and Code Administrators, International.

9.3.4 Standard Fire Tests

The fire resistance of building components is measured in standard fire tests defined by ASTM E119. During these tests the building assembly, such as a portion of floors, walls, roofs or columns, is subjected to increasing temperatures that vary with time as shown in Fig. 9.3.1. This time-temperature relation is used as a standard to represent the combustion of about 10 lb of wood (with a heat potential of 8,000 Btu per lb) per sq ft of exposed area per hour of test. Actually, the fuel consumption to maintain the standard time-temperature relation during a fire test depends on the design of the furnace and on the test specimen. When fire-tested, assemblies with exposed concrete members require considerably more fuel than other assemblies due to the favorable heat capacity. This fact is not recognized when evaluating fire resistance.

Fig. 9.3.1 Standard time-temperature curve

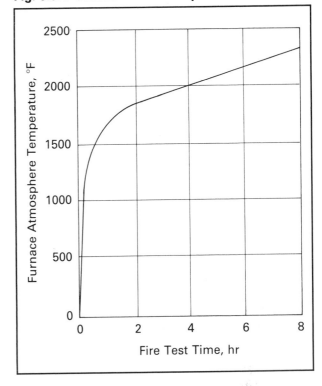

In addition to defining a standard time-temperature relationship, standard fire tests involve regulations concerning the size of the assemblies, the amount of applied load, the region of the assembly to be exposed to fire, and the end point criteria on which fire resistance (duration) is based.

The Standard, ASTM E119-83, specifies the minimum sizes of specimens to be exposed in fire tests.* For floors and roofs, at least 180 sq ft must be exposed to fire from beneath, and neither dimension can be less than 12 ft. For tests of walls, either load bearing or non-load bearing, the minimum specified area is 100 sq ft with neither dimension less than 9 ft. The minimum length for columns is specified to be 9 ft, while for beams it is 12 ft.

During the fire tests of floors, roofs, beams, load bearing walls, and columns, the maximum permissible superimposed load as required or permitted by nationally recognized standards is applied. A load other than the maximum load may be applied, but the test results then apply only to the restricted load condition.

Floor and roof specimens are exposed to fire from beneath, beams from the bottom and sides, walls from one side, and columns from all sides.

ASTM E119-83 distinguishes between ''restrained'' and ''unrestrained'' assemblies and defines them as follows:

---

*Much valuable data have been developed from tests on specimens smaller than the ASTM minimum sizes.

"Floor and roof assemblies and individual beams in buildings shall be considered restrained when the surrounding or supporting structure is capable of resisting substantial thermal expansion throughout the range of anticipated elevated temperatures. Constructions not complying with this definition are assumed to be free to rotate and expand and shall therefore be considered as unrestrained."

ASTM E119-83 includes a guide for classifying types of construction as restrained or unrestrained. The guide indicates that cast-in-place and most precast concrete constructions are considered to be restrained.

Fire endurance, end point criteria, and fire rating

The *fire resistance* of an assembly is measured by its fire endurance defined as the period of time elapsed before a prescribed condition of failure or end point is reached during a standard fire test. A *fire rating* or *classification* is a legal term for a fire endurance required by a building code authority.

End point criteria defined by ASTM E119 include:

1. Load bearing specimens must sustain the applied loading. Collapse is an obvious end point (structural end point).
2. Holes, cracks, or fissures through which flames or gases hot enough to ignite cotton waste must not form (flame passage end point).
3. The temperature increase of the unexposed surface of floors, roofs, or walls must not exceed an average of 250°F or a maximum of 325°F at any one point (heat transmission end point).

Unrestrained assembly classifications can be derived from tests of restrained floor, roof, or beam specimens provided that the average temperature of the tension steel at any section must not exceed 800°F for cold-drawn prestressing steel or 1100°F for reinforcing bars.

Additional end point criteria for restrained specimens are:

1. Beams more than 4 ft on centers: the above steel temperatures must not be exceeded for classifications of 1 hr or less; for classifications longer than 1 hr, the above temperatures must not be exceeded for the first half of the classification period or 1 hr, whichever is longer.
2. Beams 4 ft or less on centers or slabs are not subjected to steel temperature limitations.

Walls and partitions must meet the same structural, flame passage, and heat transmission end points described above. In addition, they must withstand a hose stream test (simulating, in a specified manner, a fire fighter's hose stream).

9.3.5 Fire Tests of Prestressed Concrete Assemblies

General

The first fire test of a prestressed concrete assembly in America was conducted in 1953 at the National Bureau of Standards. Since that time, more than 150 prestressed concrete assemblies have been subjected to standard fire tests in America. Although many of the tests were conducted for the purpose of deriving specific fire ratings, most of the tests were performed in conjunction with broad research studies whose objectives have been to understand the behavior of prestressed concrete subjected to fire. The knowledge gained from these tests has resulted in the development of (1) lists of fire resistive prestressed concrete building components, and (2) procedures for determining the fire endurance of prestressed concrete members by calculation.

Many different types of prestressed concrete elements have been fire tested. These elements include joists, double tees, mono-wing tees, single tees, solid slabs, hollow-core slabs, rectangular beams, ledger beams, and I-shaped beams. In addition, roofs with thermal insulation and load bearing wall panels have also been tested. Nearly all of these elements have been exposed directly to fire, but a few tests have been conducted on specimens that received additional protection from the fire by spray-applied coatings, ceilings, etc.

Fire tests of flexural elements

Tests have shown that the structural fire endurance of a flexural precast, prestressed concrete element depends on several factors, the most important of which is the method of support, i.e., restrained or unrestrained. Other factors include size and shape of the element, thickness of cover (or more precisely, the distance between the centers of the prestressing tendons and the nearest fire-exposed surface), aggregate type, and load intensity. The fire endurance as determined by the criteria for temperature rise of the unexposed surface (heat transmission) depends primarily on the concrete thickness and aggregate type.

Reports of a number of tests sponsored by the Prestressed Concrete Institute have been issued by Underwriters Laboratories, Inc. Most of the reports have been reprinted by the Prestressed Concrete Institute, and the results of the tests are the basis for UL's listings and specifications for nonproprietary products such as double tee and single tee floors and roofs, wet-cast hollow-core and solid slabs, and prestressed concrete beams.

The Portland Cement Association (PCA) conducted many fire tests of prestressed concrete assemblies. PCA's unique furnaces have made it possible to study in depth the effects of support conditions. Four series of tests dealt with simply supported slabs and beams; two series dealt with continuous slabs and beams; and one major series dealt with the effects of restrained thermal expansion on the behavior during fire of prestressed concrete floors and roofs. PCA has also conducted a number of miscellaneous fire tests of prestressed and reinforced concrete assemblies. Reports of these tests have been published and are available from the Portland Cement Association.

In addition to the tests sponsored by PCI and PCA, a number of fire tests of proprietary products, such as hollow-core slabs, have been sponsored by their manufacturers. Most of these tests have been performed by Underwriters Laboratories, Inc., but some have been conducted by Ohio State University, the Fire Prevention Research Institute, and the National Bureau of Standards. Reports of proprietary tests are generally available from test sponsors. The UL *Fire Resistance Directory* lists many proprietary assemblies.

Fig. 9.3.2 Fire endurance (heat transmission) of concrete slabs or panels. Interpolation for different concrete unit weights is reasonably accurate.

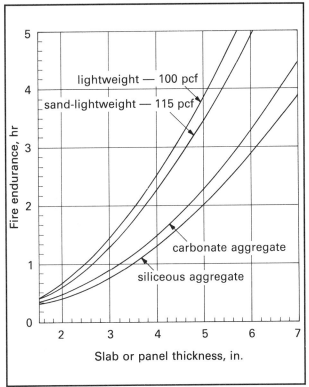

Fire tests of walls and columns

Not all of the tests conducted for Underwriters Laboratories, Inc., result in listings in UL's publications; some tests are conducted for research purposes. One such test was conducted on a double tee wall assembly. Fire was applied to the flat surface of the flange. The flange was only 1-1/2 in. thick. A load of about 10k per ft was applied at the top of the wall. The wall withstood a 2 hr fire and a subsequent hose stream test followed by a double load test without distress. Because the flange was only 1-1/2 in. thick, the heat transmission requirement was exceeded for most of the test. By providing adequate flange thickness or insulation, the heat transmission requirement would have been met in addition to the structural requirement.

Fire tests of loaded column assemblies (of any material) have not been conducted in the United States since the 1920's; therefore, prestressed concrete columns have not been fire tested. However, tests of reinforced and plain concrete columns indicate that the results are equally applicable to prestressed concrete columns.

9.3.6 Designing for Heat Transmission

As noted in Sect. 9.3.4, ASTM E119 imposes heat transmission criteria for floor, roof, and wall assemblies. Thus floors, roofs, or walls requiring a fire-resistance rating must satisfy the heat transmission requirements as well as the various structural criteria. The heat transmission fire endurance of a concrete assembly is essentially the same whether the assembly is tested as a floor (oriented horizontally) or as a wall (tested vertically). Because of this, and unless otherwise noted, the information which follows is applicable to floors, roofs, or walls.

Single course slabs or wall panels

For concrete slabs or wall panels, the temperature rise of the unexposed surface depends mainly on the thickness and aggregate type of the concrete. Other less important factors include unit weight, moisture condition, air content, and maximum aggregate size. Within the usual ranges, water-cement ratio, strength, and age have only insignificant effects.

Fig. 9.3.2 shows the fire endurance (heat transmission) of concrete slabs as influenced by aggregate type and thickness. For a hollow-core slab, this thickness may be obtained by dividing the net cross sectional area by its width. The curves represent air-entrained concrete made with air-dry aggregates having a nominal maximum size of 3/4 in. and fire tested when the concrete was at the standard moisture condition (75% R.H. at mid-depth). On the graph, concrete aggregates are designated as lightweight, sand-lightweight, carbonate, or siliceous. Lightweight aggregates include expanded clay, shale, and slate which produce concretes having unit weights of about 95 to 105 pcf without sand replacement. Lightweight concretes, in which sand is used as part or all of the fine aggregate and weigh no more than about 120 pcf, are designated as sand-lightweight. Carbonate aggregates include limestone and dolomite, i.e., those consisting mainly of calcium and/or magnesium carbonate. Siliceous aggregates include quartzite, granite, basalt, and most hard rocks other than limestone and dolomite.

Floors, roofs or walls faced with gypsum wallboard

Table 9.3.1 shows the fire endurance of concrete slabs with 5/8 in. gypsum wallboard (Type X) for two cases: (1) a 6-in. air space between the wallboard and the slab, and (2) no space between the wallboard and slab. Materials and techniques of attaching the wallboard should be similar to those used in the UL test[1] on which the data are based.

Table 9.3.1 Thickness of concrete slabs or wall panels (in inches) faced with 5/8-in. Type X gypsum wallboard to provide fire endurances of 2 and 3 hr

	Thickness (in.) of concrete panel for fire endurance of			
	With no air space		With 6-in. air space	
Aggregate	2 hr	3 hr	2 hr	3 hr
Sand-lightweight	2.0	3.0	1.2	2.4
Carbonate	2.3	3.7	1.3	2.7
Siliceous	2.5	3.9	1.3	2.8

Ribbed panels

Heat transmission through a ribbed panel is influenced by the thinnest portion of the panel and by the panel's "equivalent thickness." Here, equivalent thickness is defined as the net cross-sectional area of the panel divided by the width of the cross section. In calculating the net cross-sectional area of the panel, portions of ribs that

Fig. 9.3.3 Cross sections of ribbed wall panels

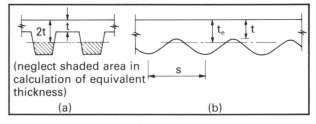

(neglect shaded area in calculation of equivalent thickness)
(a)   (b)

project beyond twice the minimum thickness should be neglected, as shown in Fig. 9.3.3(a).

The heat transmission fire endurance can be governed either by the thinnest section, by the average thickness, or by a combination of the two. The following rule-of-thumb expressions appear to give a reasonable guide as to when the minimum thickness governs and when the average thickness governs:

t = minimum thickness
t_e = equivalent thickness of panel
s = rib spacing

If $t \leq s/4$, fire endurance R is governed by t and is equal to R_t.

If $t \geq s/2$, fire endurance R is governed by t_e and is equal to R_{te}.

If $s/2 > t > s/4$:
$$R = R_t + (4t/s - 1)(R_{te} - R_t)$$

where R is the fire endurance of a concrete panel and subscripts t and t_e relate the corresponding R values to concrete slab thicknesses t and t_e, respectively.

These expressions apply to ribbed and corrugated panels, but for panels with widely spaced grooves or rustications they give excessively low results. Consequently, engineering judgment must be used when applying the above expressions.

Example 9.3.1 Fire endurance of a ribbed panel

Given:
The section of a wall panel shown.

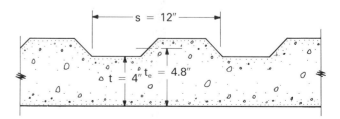

PCI Design Handbook

Problem:

Estimate the fire endurance if the minimum thickness is 4 in. and the equivalent thickness is 4.8 in. Assume that the panel is made of sand-lightweight concrete.

Solution:

t = 4 in., $s/2$ = 6 in., $s/4$ = 3 in.

Therefore, $s/2 > 4 > s/4$.

So $R = R_t + (4 t/s - 1) (R_{te} - R_t)$

R_t = fire endurance of 4 in. sand-lightweight panel = 135 min.

R_{te} = fire endurance of 4.8 in. sand-lightweight panel = 193 min.

R = 135 + [4(4)/12 − 1] [193 − 135]
= 154 min.

Multi-course assemblies

Floors and roofs often consist of concrete base slabs with overlays or undercoatings of other types of concrete or insulating materials. In addition, roofs generally have built-up roofing.

If the fire endurances of the individual courses are known, the fire endurance of the composite assembly can be estimated from the formula:

$$R = (R_1^{0.59} + R_2^{0.59} \ldots + R_n^{0.59})^{1.7} \qquad \text{(Eq. 9.3.1)}$$

where:

R = fire endurance of the composite assembly in minutes

R_1, R_2, R_n = fire endurances of the individual courses in minutes.

The following example illustrates the use of this equation:

Example 9.3.2 Fire endurance of an assembly

Problem:

Determine the fire endurance of a slab consisting of a 2 in. base slab of siliceous aggregate concrete with a 2-1/2 in. topping of sand-lightweight concrete (115 pcf).

Solution:

From Fig. 9.3.2, the fire endurance of a 2 in. thick slab of siliceous aggregate concrete and 2-1/2 in. of sand-lightweight aggregate concrete are 25 min and 54 min, respectively.

R = [(25)$^{0.59}$ + (54)$^{0.59}$]$^{1.7}$

R = (6.68 + 10.52)$^{1.7}$ = 126 min = 2 hr 6 min

Table 9.3.2 Values of R of various insulating materials for use in Eq. 9.3.1.

Roof Insulation Material[1]	Thickness (in.)	R (min)
Cellular plastic	≥ 1	5
Glass fiber board	¾	11
Glass fiber board	1½	35
Foam glass	2	55
Mineral board	1	19
Mineral board	2	62
Mineral board	3	123

1. Some of these materials may be used only where combustible construction is permitted.

Table 9.3.2 gives values which can be used in this equation for certain insulating materials. For heat transmission, three-ply built-up roofing contributes 10 minutes to the fire endurance.

Equation 9.3.1 has certain shortcomings in that it does not account for the location of the individual courses relative to the fired surface. Also, it is not possible to directly obtain the fire endurances of many insulating materials. Nevertheless, in a series of tests, the formula estimated the fire endurances within about 10% for most assemblies.

A paper on two-course floors and roofs[2] gives results of many fire tests. The paper also shows graphically the fire endurances of assemblies consisting of various thicknesses of two materials. Tables 9.3.3 through 9.3.5, which are based on test results, can be used to estimate the required thicknesses of two-course materials for various fire endurances.

Sandwich panels

Some wall panels are made by sandwiching an insulating material between two face slabs of concrete.

Several building codes require that where non-combustible construction is specified, combustible elements in walls shall be limited to thermal and sound insulation having a flame spread classification of not more than 75 when the insulation is sandwiched between two layers of non-combustible material such as concrete.

When insulation is not installed in this manner, it is required to have a flame spread of not more than 25. Data on flame spread classification are available from insulation manufacturers.

A fire test was conducted of one such panel that consisted of a 2 in. base slab of carbonate aggregate concrete, a 1 in. layer of cellular polystyrene insulation, and a 2 in. face slab of carbonate ag-

Table 9.3.3 Thickness of spray-applied insulation on fire-exposed surface of concrete[1] slabs or panels to resist transfer of heat through the assemblies

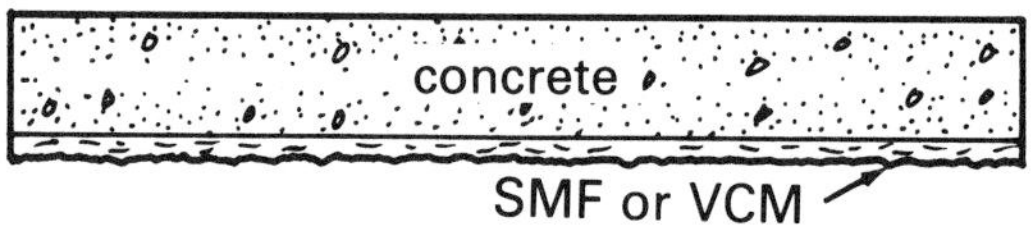

Slab equivalent thickness (in.)	Type of insulation[2]	Thickness (in.) for fire resistance rating of				
		1 hr	1½ hr	2 hr	3 hr	4 hr
1½	SMF	½	¾	1⅛	N.A.	N.A.
2	SMF	⅜	⅝	⅞	1⅜	N.A.
2½	SMF	¼	½	⅝	1⅛	1⅝
3	SMF	⅛	¼	½	⅞	1⅜
4	SMF	0	0	¼	⅝	1
1½	VCM	½	¾	1⅛	1¾	N.A.
2	VCM	⅜	⅝	⅞	1⅜	1¾
2½	VCM	¼	½	¾	1¼	1⅝
3	VCM	⅛	⅜	⅝	⅞	1⅜
4	VCM	0	0	¼	⅝	⅞

[1]Values shown are for siliceous aggregate concrete, and are conservative for other concretes.

[2]SMF = Sprayed mineral fiber consists of refined mineral fibers with inorganic binders and water added during the spraying operation. The density of the oven-dry material should be at least 13 pcf.

VCM = Vermiculite cementitious material consists of expanded vermiculite with inorganic binders and water. The density of the oven-dry material should be at least 14 pcf.

N.A. = Not applicable.

Table 9.3.4 Thickness of two-course roof assemblies consisting of concrete[1] slabs with insulating concrete overlays[2]

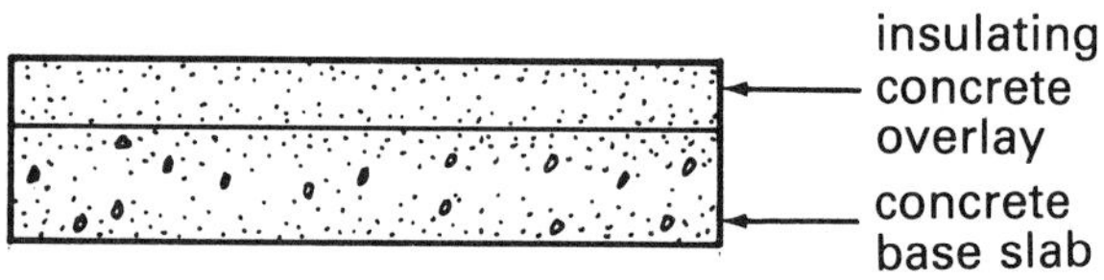

Base slab thickness (in.)	Thickness of overlay (in.) for fire-resistance rating of				
	1 hr	1½ hr	2 hr	3 hr	4 hr
1½	1⅛	1½	1⅞	2⅝	3⅛
2	1	1⅜	1¾	2½	3
3	⅜	¾	1¼	2	2⅝
4	0	0	⅝	1⅜	2

[1]Values shown are for siliceous aggregate concrete and are conservative for other concretes.

[2]Insulating concrete having a dry density less than 35 pcf.

Table 9.3.5 Thickness of roof assemblies consisting of concrete[1] slabs with insulation and built-up roofing

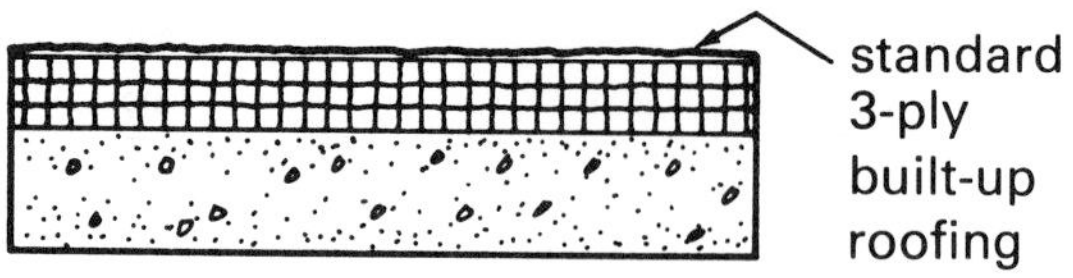

standard 3-ply built-up roofing

Base slab thickness (in.)	Insulation[2]	Thickness of insulation (in.) for fire-resistance rating of				
		1 hr	1½ hr	2 hr	3 hr	4 hr
1½	MB	¾	1¼	1⅞	2¾	N.A.
2	MB	½	1	1⅜	2¼	2⅞
3	MB	0	¾	¾	1⅜	1⅞
4	MB	0	0	¼	¾	1¼
1½	GFB	⅝	1⅜	2	N.A.	N.A.
2	GFB	¼	⅞	1½	2⅞	N.A.
3	GFB	0	⅜	¾	1½	2⅛
4	GFB	0	0	¼	¾	1¼

[1]Values shown are for siliceous aggregate concrete, and are conservative for other concretes.

[2]MB = Mineral board insulation composed of spherical cellular beads of expanded aggregate and fibers formed into rigid flat rectangular units with an integral waterproofing treatment.

GFB = Glass fiber board fibrous glass roof insulation consisting of inorganic glass fibers formed into rigid boards using a binder. The board has a top surface faced with glass fiber reinforced with asphalt and kraft.

N.A. = Not applicable.

gregate concrete. The resulting fire endurance was 2 hr 00 min. From Eq. 9.3.1, the contribution of the 1 in. layer of polystyrene was calculated to be 5 min.

It is likely that the comparable R value for a 1 in. layer of cellular polyurethane would be somewhat greater than that for a 1 in. layer of cellular polystyrene, but test values are not available. Until more definitive data are obtained, it is suggested that 5 min be used as the value for R for any thickness of cellular plastic greater than 1 in.

It should be noted that the cellular plastics melt and are consumed at about 400 to 600°F. Thus, additional thickness or changes in composition probably have only a minor effect on the fire endurance of sandwich panels. The danger of toxic fumes caused by the burning cellular plastics is practically eliminated when the plastics are completely encased within concrete sandwich panels.[8]

Table 9.3.6 lists fire endurances of sandwich panels with either cellular plastic, glass fiber board,

Table 9.3.6 Fire endurance of precast concrete sandwich walls [calculated, based on Eq. 9.3.1]

Outside and inside wythes	Insulation	Fire endurance, hr:min
1½ in. Sil	1 in. CP	1:23
1½ in. Carb	1 in. CP	1:23
1½ in. SLW	1 in. CP	1:45
2 in. Sil	1 in. CP	1:50
2 in. Carb	1 in. CP	2:00
2 in. SLW	1 in. CP	2:32
3 in. Sil	1 in. CP	3:07
1½ in. Sil	¾ in. GFB	1:39
a in. Sil	¾ in. GFB	2:07
2 in. SLW	¾ in. GFB	2:52
1½ in. Sil	1½ in. GFB	2:35
2 in. Sil	1½ in. GFB	3:08
2 in. SLW	1½ in. GFB	4:00
1½ in. Sil	1 in. IC	2:12
1½ in. SLW	1 in. IC	2:39
2 in. Carb	1 in. IC	2:56
2 in. SLW	1 in. IC	3:33
1½ in. Sil	1½ in. IC	2:54
1½ in. SLW	1½ in. IC	3:24
2 in. Sil	2 in. IC	4:25
1½ in. SLW	2 in. IC	4:19

Notes:
Carb = carbonate aggregate concrete
Sil = siliceous aggregate concrete
SLW = sand-lightweight concrete (115 pcf maximum)
CP = cellular plastic (polystyrene or polyurethane)
IC = lightweight insulating concrete (35 pcf maximum)
GFB = glass fiber board

or insulating concrete used as the insulating material. The values were obtained by use of Eq. 9.3.1.

Treatment of joints between wall panels

Joints between wall panels should be detailed so that passage of flame or hot gases is prevented, and transmission of heat does not exceed the limits specified in ASTM E119. Concrete wall panels expand when heated, so the joints tend to close during fire exposure. Non-combustible materials that are flexible, such as ceramic fiber blankets or asbestos rope, provide thermal, flame, and smoke barriers, and when used in conjunction with caulking materials they can provide the necessary weather-tightness while permitting normal volume change movements. Joints that do not move can be filled with mortar.

Joints between wall panels are similar to openings. Most building codes do not require openings to be protected against fire if the openings constitute only a small percentage of the wall area and if the spatial separation is greater than some minimum distance. In those cases, protection of joints would not be required.

In other cases, openings must be protected, but most codes permit a lesser degree of protection. For example, the Uniform Building Code requires that when openings are permitted and must be protected, the "openings shall be protected by a fire assembly having a 3/4-hour fire-protection rating." Where no openings are permitted, the fire resistance required for the wall should be provided at the joints.

Table 9.3.7 is based on results of fire tests of panels with butt joints.[3] The tabulated values apply to one-stage butt joints and are conservative for two-stage and ship-lap joints.

Table 9.3.7 Protection of joints between wall panels utilizing ceramic fiber felt

Panel equivalent thickness* (in.)	Thickness of ceramic fiber felt (in.) required for fire resistance ratings and joint widths shown							
	Joint width = ⅜ in.				Joint width = 1 in.			
	1 hr	2 hr	3 hr	4 hr	1 hr	2 hr	3 hr	4 hr
4	¼	N.A.	N.A.	N.A.	¾	N.A.	N.A.	N.A.
5	0	¾	N.A.	N.A.	½	2⅛	N.A.	N.A.
6	0	0	1⅛	N.A.	¼	1¼	3½	N.A.
7	0	0	0	1	¼	⅞	2	3¾

N.A. = Not applicable

Interpolation may be used for joint widths between ⅜ in. and 1 in.

*Panel equivalent thicknesses are for carbonate concrete. For siliceous aggregate concrete change "4, 5, 6, and 7" to "4.3, 5.3, 6.5, and 7.5". For sand-lightweight concrete change "4, 5, 6, and 7" to "3.3, 4.1, 4.9, and 5.7".

The tabulated values apply to one-stage butt joints and are conservative for two stage and ship-lap joints as shown below.

Joints between adjacent precast floor or roof elements may be ignored in calculating the slab thickness provided that a concrete topping at least 1-1/2 in. thick is used. Where no concrete topping is used, joints should be grouted to a depth of at least one-third the slab thickness at the joint, or the joints made fire-resistive in a manner acceptable to the authority having jurisdiction.

9.3.7 Designing for Structural Integrity

It was noted above that many fire tests and related research studies have been directed toward an understanding of the structural behavior of prestressed concrete subjected to fire. The information gained from that work has led to the development of calculation procedures which can be used in lieu of fire tests. The purpose of this section is to present an introduction to these calculation procedures. Because the method of support is the most important factor affecting structural behavior of flexural elements during a fire, the discussion that follows deals with three conditions of support: simply supported members, continuous slabs and beams, and members in which restraint to thermal expansion occurs.

Simply supported members

Assume that a simply supported prestressed concrete slab is exposed to fire from below, that the ends of the slab are free to rotate, and that expansion can occur without restriction. Also assume that the reinforcement consists of straight strands located near the bottom of the slab. With the underside of the slab exposed to fire, the bottom will expand more than the top causing the slab to deflect downward; also, the strength of the steel and concrete near the bottom will decrease as the temperature rises. When the strength of the steel diminishes to that required to support the slab, flexural collapse will occur. In essence, the applied moment remains practically constant during the fire exposure, but the resisting moment capacity is reduced as the steel weakens.

Fig. 9.3.4 illustrates the behavior of a simply supported slab exposed to fire from beneath, as described above. Because strands are parallel to the axis of the slab, the design moment strength is constant throughout the length:

$$\phi M_n = \phi A_{ps} f_{ps} (d - a/2) \qquad \text{(Eq. 9.3.2)}$$

f_{ps} can be determined from Fig. 4.10.3 or Eq.18-3 of ACI 318-83.

If the slab is uniformly loaded, the moment dia-

gram will be parabolic with a maximum value at midspan of:

$$M = \frac{w\ell^2}{8} \qquad \text{(Eq. 9.3.3)}$$

where:

w = dead plus live load per unit of length, k/in.

ℓ = span length, in.

As the material strengths diminish with elevated temperatures, the retained nominal strength becomes:

$$M_{n\theta} = A_{ps} f_{ps\theta} (d - a_\theta/2) \qquad \text{(Eq. 9.3.4)}$$

in which θ signifies the effects of high temperatures. Note that A_{ps} and d are not affected, but f_{ps} is reduced. Similarly a_θ is reduced, but the concrete strength at the top of the slab, f'_c, is generally not reduced significantly because of its lower temperature.

From Fig.4.10.3, it can be seen that when f_{pu} ($f_{pu\theta}$) is low, $C\omega_{pu}$ is low, so f_{ps}/f_{pu} can be assumed to be 0.98 for most cases of design for fire.

Flexural failure can be assumed to occur when $M_{n\theta}$ is reduced to M. Strength reduction factor, ϕ, is not applied because a safety factor is included in the required ratings.[4] From this expression, it

Fig. 9.3.4 Moment diagrams for simply supported beam or slab

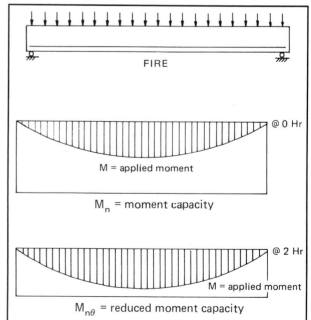

PCI Design Handbook

can be seen that the fire endurance depends on the applied loading and on the strength-temperature characteristics of the steel.

In turn, the duration of the fire before the "critical" steel temperature is reached depends on the protection afforded to the reinforcement.

To solve problems involving the above equations, it is necessary to utilize data on the strength-temperature relationships for steel and concrete, and information on temperature distributions within concrete members during fire exposures. Fig. 9.3.5 shows strengths of certain steels at elevated temperatures, and Fig. 9.3.6 shows similar data for various types of concrete.

Data on temperature distribution in concrete slabs during fire tests are shown in Figs. 9.3.7 and 9.3.8. These figures can also be used for beams wider than about 10 in. An "effective u", $\bar{u}$, is used, which is the average of the distances between the centers of the individual strands or bars and the nearest fire-exposed surface. The values for corner strands or bars are reduced one-half to account for the exposure from two sides (see example 9.3.4). The procedure does not apply to bundled bars or strands.

Example 9.3.3 Calculation of fire endurance

Determine the maximum safe superimposed load that can be supported by an 8 in. deep hollow-core slab with a simply supported unrestrained span of 25 ft and a fire endurance of 3 hr.

Given:

h = 8 in.
u = 1.75 in.
Eight 1/2 in. 250 ksi strands
$A_{ps} = 8(0.144) = 1.152$ sq in.
b = 48 in.
d = 8 − 1.75 = 6.25 in.
$w_d = 60$ psf
Carbonate aggregate concrete
$f'_c = 5000$ psi
$\ell = 25$ ft

Problem:

Determine the maximum safe superimposed load that can be supported.

Solution:

(a) Estimate strand temperature at 3 hr from Fig. 9.3.7: At 3 hr, carbonate aggregate,
   u = 1.75 in.
   θ_s = 925°F

(b) Determine $f_{pu\theta}$ from Fig. 9.3.5. For cold-drawn steel at 925°F, $f_{pu\theta} = 0.33 \, (f_{pu}) = 82.5$ ksi

(c) Determine $M_{n\theta}$ and w

$f_{ps\theta} = 0.98 \, f_{pu\theta} = 0.98 \, (82.5) = 80.9$ ksi

$$a_\theta = \frac{A_{ps} f_{ps}}{0.85 \, f'_c b} = \frac{1.152(80.9)}{0.85(5)(48)} = 0.46 \text{ in.}$$

$$\begin{aligned} M_{n\theta} &= A_{ps} f_{ps\theta} (d - a_\theta/2) = M \\ &= 1.152(80.9)(6.25 - 0.46/2)/12 \\ &= 46.8 \text{ ft-kips} \end{aligned}$$

$$w = \frac{8M}{\ell^2} = \frac{8(46.8)}{(25)^2} = 0.599 \text{ klf} = 150 \text{ psf}$$

$$w_\ell = w - w_d = 150 - 60 = 90 \text{ psf}$$

Example 9.3.4 Calculation of fire endurance

Given:

The 12RB24 shown.

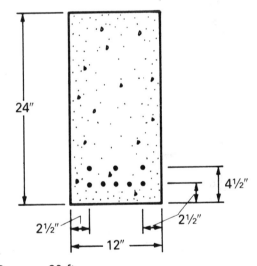

Span = 30 ft
Dead load (including bm. wt.) = 1100 plf
Live load = 1100 plf
Simple support, no restraint
Siliceous aggregate concrete, $f'_c = 5000$ psi
Stress-relieved strand, $f_{ps} = 270$ ksi

Problem:

Provide 4 hr fire endurance by adding strands and/or rebars to the beam.

Solution:

$A_{ps} = 8(0.153) = 1.224$ sq. in.

$y_s = [5(2.5) + 3(4.5)]/8 = 3.25$ in.

$d_p = 24 - 3.25 = 20.75$ in.

Fig. 9.3.5 Temperature-strength relationships for various steels

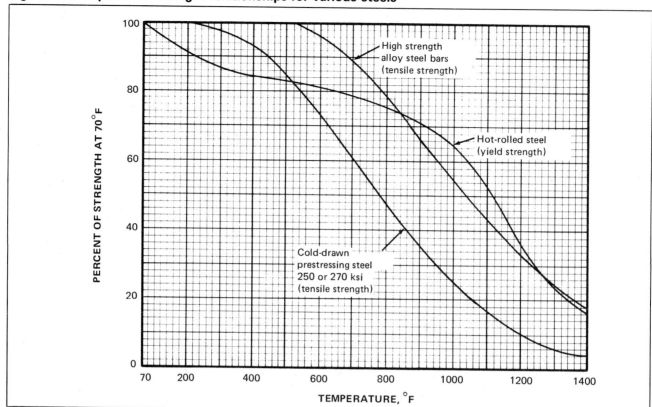

Fig. 9.3.6 Compressive strength of concrete at high temperatures

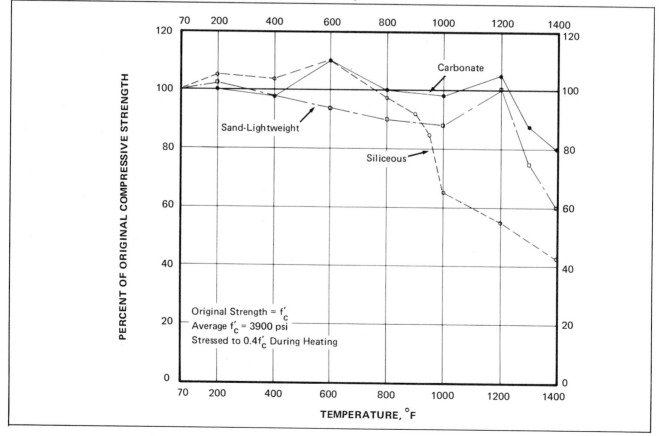

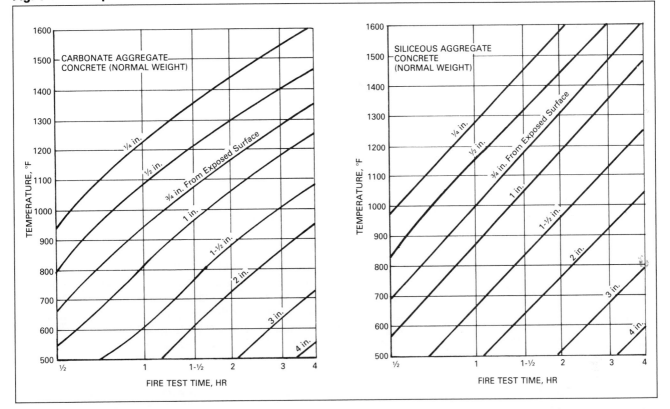

Fig. 9.3.8 Temperatures within concrete slabs or panels during fire tests - sand lightweight concrete

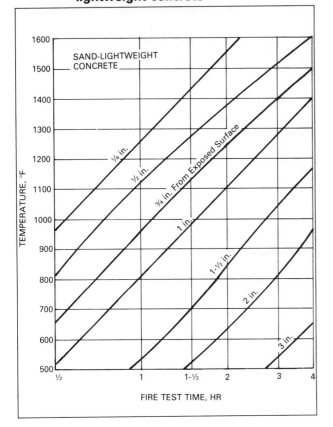

$\bar{u}$ = [5(2.5) + 1(4.5) + 2(2.5) (0.5)]/8 = 2.44 in.

From Fig. 9.3.7, siliceous aggregate:

at 4 hr, strand temp. = 920°F

From Fig. 9.3.5:

$f_{pu\theta}$ = 0.34(270) = 91.8 ksi

$f_{ps\theta}$ = 0.98(91.8) = 90.0 ksi

$a_\theta = \dfrac{A_{ps}f_{pso\theta}}{0.85f'_c b} = \dfrac{1.224(90.0)}{0.85(5)(12)}$

= 2.16 in.

$M_{n\theta}$ = $A_{ps}f_{ps\theta}$ ($d_p - a_\theta/2$)

= 1.224(90.0) (20.75 − 2.16/2)

= 2167 in.-kips = 181 ft-kips

Service load moment:

w = 2.20 kips/ft

M = $w\ell^2/8$ = 2.20(30)²/8

= 247.5 ft-kips > 181 ft-kips

Therefore, beam will not satisfy criteria for a 4-hr fire endurance.

Solution No. 1:

Try adding 2 additional strands at 4½ in. from bottom:

A_{ps} = 10(0.153) = 1.53 sq in.

y_s = [5(2.5) + 5(4.5)]/10 = 3.5 in.
d_p = 24 − 3.5 = 20.5 in.
$\bar{u}$ = [5(2.5) + 1(4.5) + 2(4.25) +
2(2.5)(0.5)]/10
= 2.80 in.

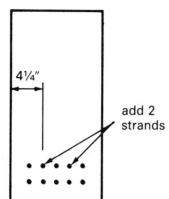

add 2
strands

From Fig. 9.3.7:
Strand temp. = 840°F

From Fig. 9.3.5:

$f_{pu\theta}$ = 0.42(270) = 113.4 ksi
$f_{ps\theta}$ = 0.98(113.4) = 111.1 ksi
a_θ = $\dfrac{1.53(111.1)}{0.85(5)(12)}$ = 3.33 in.
$M_{n\theta}$ = 1.53(111.1)(20.5 − 3.33/2)
= 3201 in.-kips = 266 ft-kips
>247.5 ft-kips OK

Solution No. 2:

Try adding 2 - #7 at 4.5 in. from bottom:

A_s = 1.20 sq in.

d = 24 − 4.5 = 19.5 in.

$\bar{u}$(rebar) = 4.25 in.

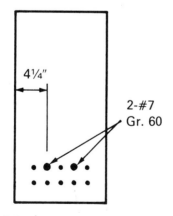

2-#7
Gr. 60

From Fig. 9.3.7:
Bar temp. at 4 hr = 500°F
From Fig. 9.3.5:
$f_{y\theta}$ = f_y = 60 ksi

a_θ = $\dfrac{1.224(90.0) + 1.20(60)}{0.85(5)(12)}$ = 3.57 in.

$M_{n\theta}$(strand) = 1.224(90)(20.75 − 3.57/2)
= 2089 in.-kips = 174 ft-kips

$M_{n\theta}$ (bars) = 1.20(60)(19.5 − 3.57/2)
= 1275 in.-kips = 106 ft-kips

174 + 106 = 280 ft-kips > 247.5 OK

Continuous members

Continuous members undergo changes in stresses when subjected to fire. These stresses result from temperature gradients within the structural members, or changes in strength of the materials at high temperatures, or both.

Fig. 9.3.9 shows a two-span continuous beam whose underside is exposed to fire. The bottom of the beam becomes hotter than the top and tends to expand more than the top. This differential temperature effect causes the ends of the beam to tend to lift from their supports thereby increasing the reaction at the interior support. This action results in a redistribution of moments, i.e., the negative moment at the interior support increases while the positive moments decrease.

During a fire, the negative moment reinforcement (Fig. 9.3.9) remains cooler than the positive moment reinforcement because it is better protected from the fire. In addition, the redistribution that occurs is sufficient to cause yielding of the negative moment reinforcement. Thus, a relatively large increase in negative moment can be accommodated throughout the test. The resulting decrease in positive moment means that the positive moment reinforcement can be heated to a higher temperature before failure will occur. Therefore, the fire endurance of a continuous concrete beam is generally significantly longer than that of a simply supported beam having the same cover and the same applied loads.

It is possible to design the reinforcement in a continuous beam or slab for a particular fire endurance period. From Fig. 9.3.9 the beam can be expected to collapse when the positive moment capacity, $M_{n\theta}^+$, is reduced to the value of the maximum redistributed positive moment at a distance x_1 from the outer support.

Fig. 9.3.10 shows a uniformly loaded beam or slab continuous (or fixed) at one support and simply supported at the other. Also shown is the redistributed applied moment diagram at failure.

It can be shown that at the point of positive moment, x_1,

$$x_1 = \frac{\ell}{2} - \frac{M_{n\theta}^-}{w\ell}$$

(Eq. 9.3.5)

Fig. 9.3.9 Moment diagram for two-span continuous beam

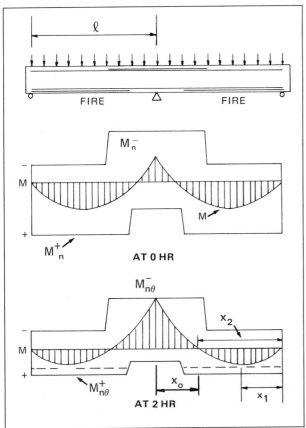

at $x = x_2$, $M_x = 0$ and $x_2 = 2 x_1$

$$x_o = \frac{2 M_{n\theta}^-}{w\ell} \qquad \text{(Eq. 9.3.6)}$$

$$M_{n\theta}^- = \frac{w\ell^2}{2} \pm w\ell^2 \sqrt{\frac{2 M_{n\theta}^+}{w\ell^2}} \qquad \text{(Eq. 9.3.7)}$$

In most cases, redistribution of moments occur early during the course of a fire and the negative moment reinforcement can be expected to yield before the negative moment capacity has been reduced by the effects of fire. In such cases, the length of x_o is increased, i.e, the inflection point moves toward the simple support. If the inflection point moves beyond the point where the bar stress cannot be developed in the negative moment reinforcement, sudden failure may result.

Fig. 9.3.11 shows a symmetrical beam or slab in which the end moments are equal.

$$M_{n\theta}^- = w\ell^2 / 8 - M_{n\theta}^+ \qquad \text{(Eq. 9.3.8)}$$

$$\frac{w x_2^2}{8} = M_{n\theta}^+$$

Fig. 9.3.10 Uniformly loaded member continuous at one support

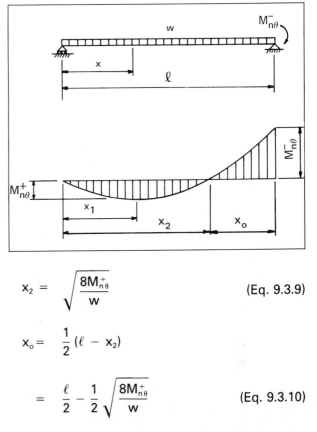

$$x_2 = \sqrt{\frac{8 M_{n\theta}^+}{w}} \qquad \text{(Eq. 9.3.9)}$$

$$x_o = \frac{1}{2} (\ell - x_2)$$

$$= \frac{\ell}{2} - \frac{1}{2} \sqrt{\frac{8 M_{n\theta}^+}{w}} \qquad \text{(Eq. 9.3.10)}$$

To determine the maximum value of x_o, the value of w should be the minimum service load anticipated, and $(w\ell^2/8 - M_n^-)$ should be substituted for $M_{n\theta}^+$ in Eq. 9.3.10.

For any given fire endurance period, the value of $M_{n\theta}^+$ can be calculated by the procedures given in Sect. 9.3.7. Then the value of $M_{n\theta}^-$ can be calculated by the use of Eqs. 9.3.7 or 9.3.8 and the necessary lengths of the negative moment reinforcement can be determined from Eqs. 9.3.6 or 9.3.10. Use of these equations is illustrated in Example 9.3.5.

Fig. 9.3.11 Symmetrical uniformly loaded member continuous at both supports

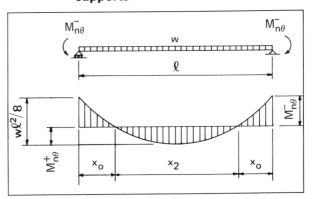

It should be noted that the amount of moment redistribution that can occur is dependent on the amount of negative moment reinforcement. Tests have clearly demonstrated that in most cases the negative moment reinforcement will yield, so the negative moment capacity is reached early during a fire test, regardless of the applied loading. The designer must exercise care to ensure that a secondary type of failure will not occur. To avoid a compression failure in the negative moment region, the amount of negative moment reinforcement should be small enough so that $\omega_\theta = A_s f_{y\theta} / b_\theta d_\theta f'_{c\theta}$ is less than 0.30, before and after reductions in f_y, b, d and f'_c are taken into account. Furthermore, the negative moment bars or mesh must be long enough to accommodate the complete redistributed moment and change in the inflection points. It should be noted that the worst condition occurs when the applied loading is smallest, such as the dead load plus partial or no live load. It is recommended that at least 20% of the maximum negative moment reinforcement extend throughout the span.

Example 9.3.5 Calculation of fire endurance

Problem:

Design a floor using hollow-core slabs and topping for 22 ft span for 4 hr fire endurance. Service loads = 175 psf dead (including structure) and 150 psf live. Use 4 ft wide, 10 in. deep slabs with 2 in. topping, carbonate aggregate concrete. Continuity can be achieved at both ends. Use f'_c (precast) = 5000 psi, f_{pu} = 250 ksi, and f'_c (topping) = 3000 psi, sixteen 3/8 in., 250 ksi strands at u = 1.75 in. Provide negative moment reinforcement needed for fire resistance.

Solution:

A_{ps} = 16(0.080) = 1.28 sq in.

u = 1.75 in.

d = 12 − 1.75 = 10.25 in.

From Fig. 9.3.7, θ_s = 1010°F

From Fig. 9.3.5, f_{pu_θ} = 0.24f_{pu} = 60 ksi

Using Table 4.10.2:*

$$\omega_{p\theta} = \frac{16(0.080)(60)}{48(10.25)(3)} = 0.052$$

$$K'_u = 133/0.9 = 148$$

---
*The values for K'_u in Table 4.10.2 include ϕ = 0.9. Since in the design for fire, ϕ = 1.0, the value of K'_u must be divided by 0.9.

$$M^+_{n\theta} = K'_u bd^2 / 12,000$$

$$= 148(48)(10.25)^2 / 12,000$$

$$= 62.2 \text{ ft-kips/unit}$$

$$= 15.5 \text{ ft-kips/ft}$$

For simply supported members:

$$M = 0.325(22)^2 / 8 = 19.7 \text{ ft-kips/ft}$$

$$\text{Req'd } M^-_{n\theta} = 19.7 − 15.5 = 4.2 \text{ ft-kips/ft}$$

Assume d − a_θ / 2 = 10.25 in., and f_y = 60 ksi

$$A^-_s = \frac{4.2(12)}{60(10.25)} = 0.082 \text{ in.}^2/\text{ft}$$

Use 20% of required A^-_s throughout span:

Try 6 × 6-W1.4 × W1.4 continuous plus 6 × 6-W2.9 × W2.9 over supports

$$A^-_s = 0.028 + 0.058 = 0.086 \text{ in.}^2 /\text{ft}$$

Neglect concrete above 1400°F in negative moment region, i.e., from Fig. 9.3.7, neglect bottom 5/8 in. Also, concrete within compressive zone will be about 1350 to 1400°F, so use $f'_{c\theta}$ = 0.81 f'_c (see Fig. 9.3.6) = 4.05 ksi.

Check $M^-_{n\theta}$, assuming that the temperature of the negative steel does not rise above 200°F. If greater than 200°F, steel strength should be reduced according to Fig. 9.3.5.

$$a_\theta = \frac{0.086(60)}{0.85(4.05)(12)} = 0.125 \text{ in.}$$

$$M^-_{n\theta} = 086(60)(10.37 − 0.063) / 12$$

$$= 4.43 \text{ ft-kips/ft}$$

With dead load + 1/2 live load, w = 0.25 ksf, M = 15.12 ft-kips/ft, and M^-_n = 4.71 ft-kips/ft (calculated for room temperature)

$$M^+_{min} = 15.12 − 4.71 = 10.41 \text{ ft-kips/ft}$$

From Eq. 9.3.10

$$\max x_o = \frac{22}{2} − \frac{1}{2} \sqrt{\frac{8(10.41)}{0.25}} = 1.87 \text{ ft}$$

Use 6 × 6-W1.4 × W1.4 continuous throughout plus 6 × 6-W2.9 × W2.9 for a distance of 3 ft from the support. Mesh must extend into walls which must be designed for the moment induced at the top.

PCI Design Handbook

Members restrained against thermal expansion

If a fire occurs beneath an interior portion of a large reinforced concrete slab, the heated portion will tend to expand and push against the surrounding part of the slab. In turn, the unheated part of the slab exerts compressive forces on the heated portion. The compressive force, or thrust, acts near the bottom of the slab when the fire test occurs but, as the fire progresses, the line of action of the thrust rises as the mechanical properties of the heated concrete changes. This thrust is generally great enough to increase the fire endurance significantly.

The effects of restraint to thermal expansion can be characterized as shown in Fig. 9.3.12. The thermal thrust acts in a manner similar to an external prestressing force, which, in effect, increases the positive moment capacity.

The increase in bending moment capacity is similar to the effect of added reinforcement located along the line of action of the thrust. It can be assumed that the added reinforcement has a yield strength (force) equal to the thrust. By this approach, it is possible to determine the magnitude and location of the required thrust to provide a given fire endurance.

The above explanation is greatly simplified because in reality restraint is quite complex, and can be likened to the behavior of a flexural member subjected to an axial force. Interaction diagrams similar to those for columns can be constructed for a given cross-section at a particular stage of a fire, e.g., 2 hr of a standard fire exposure.[5]

The guidelines in ASTM E119-83 given for determining conditions of restraint are useful for preliminary design purposes. Most interior and many exterior bays of multi-bay floors or roofs can be considered to be restrained and the magnitude and location of the thrust are generally of academic interest only. In such cases, the fire endurance is governed by heat transmission rather than by structural considerations.

Shear resistance

Many fire tests have been conducted in America on simply supported reinforced or prestressed concrete elements as well as on elements in which restraint to thermal expansion occurred. Shear failures did not occur in any of those tests.

It should be noted that when beams which are continuous over one support (e.g., such as that shown in Fig. 9.3.9) are exposed to fire, both the moment and the shear at the interior support increase. Such a redistribution of moment and shear

Fig. 9.3.12 Longitudinally restrained beam during fire exposure

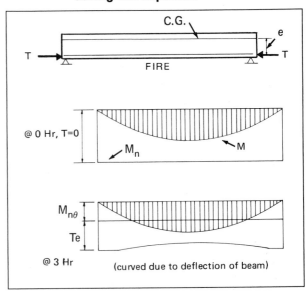

results in a severe stress condition. However, of the several fire tests of reinforced concrete beams in which that condition was simulated, failure occured only in one beam.[6] In that test, the shear reinforcement was inadequate, even for service load conditions without fire, as judged by the shear requirements of ACI 318-83. Thus, it appears from available test data that members which are designed for shear strength in accordance with ACI 318-83 will perform satisfactorily in fire situations, i.e., failure will not occur prematurely due to a shear failure.

9.3.8 Protection of Connections

Many types of connections in precast concrete construction are not vulnerable to the effects of fire and, consequently, require no special treatment. For example, gravity-type connections, such as the bearings between precast concrete panels and concrete footings or beams which support them, do not generally require special fire protection.

If the panels rest on elastomeric pads or other combustible materials, protection of the pads is not generally needed because deterioration of the pads will not cause collapse.

Connections that can be weakened by fire and thereby jeopardize the structure's load carrying capacity should be protected to the same degree as that required for the supported member. For example, an exposed steel bracket supporting a panel or spandrel beam will be weakened by fire and might fail causing the panel or beam to collapse. Such a bracket should be protected. The amount of protection depends on (a) the stress-strength ratio in the steel at the time of the fire and (b) the intensity and duration of the fire. The thickness of protection material required is greater as the stress level and fire severity increase.

Fig. 9.3.13 shows the thickness of various commonly used fire protection materials required for fire endurances up to 4 hr. The values shown are based on a critical steel temperature of 1000°F, i.e., a stress-strength ratio (f_s/f_y) of about 65 percent. Values in Fig. 9.3.13 (b) are applicable to concrete or dry-pack mortar encasement of structural steel shapes used as brackets or lintels.

9.3.9 Precast Concrete Column Covers

Steel columns are often clad with precast concrete panels or covers for architectural reasons. Such covers also provide fire protection for the columns.

Fig. 9.3.14 shows the relationship between the thickness of concrete column covers and fire endurance for various steel column sections. The fire endurances shown are based on an empirical relationship developed by Lie and Harmathy.[7]

The above authors also found that the air space between the steel core and the column covers has only a minor effect on the fire endurance. An air space will probably increase the fire endurance but only by an insignificant amount.

Most precast concrete column covers are 3 in. or more in thickness, but some are as thin as

2-1/2 in. From Fig. 9.3.14, it can be seen that precast concrete column covers can qualify for fire endurances of at least 2-1/2 hr, and usually more than 3 hr. For steel column sections other than those shown, including shapes other than wide flange beams, interpolation between the curves on the basis of weight per foot will generally give reasonable results.

For example, the fire endurance afforded by a 3 in thick column cover of normal weight concrete for a 8 × 8 × 1/2 in. steel tube column will be about 3 hr 20 min (the weight of the section is 47.35 lb per ft).

Precast concrete column covers (Fig. 9.3.15) are made in various shapes such as (a) four flat panels with butt or mitered joints that fit together to enclose the steel column, (b) four L-shaped units, (c) two L-shaped units, (d) two U-shaped units, and (e) and (f) U-shaped units and flat closure panels. There are, of course, many combinations to accommodate isolated columns, corner columns, and columns in walls.

To be fully effective the column covers must remain in place without severe distortion. Many types of connections are used to hold the column covers in place. Some connections consist of bolted or welded clip angles attached to the tops and bottoms of the covers. Others consist of steel plates embedded in the covers that are welded to angles, plates, or other shapes which are, in turn, welded or bolted to the steel column. In any case, the connections are used primarily to position the column covers and as such are not highly stressed. As a result, temperature limits need not be ap-

Fig. 9.3.13 Thickness of protection materials applied to connections consisting of structural steel shapes (IM = intumescent mastic, SMF = sprayed mineral fiber, VCM = vermiculite cementitious material)

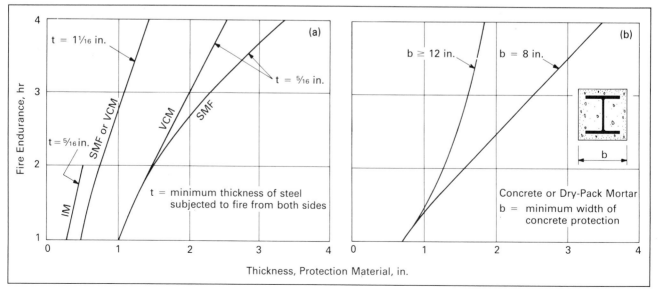

Fig. 9.3.14 Fire endurance of steel columns afforded protection by concrete column covers

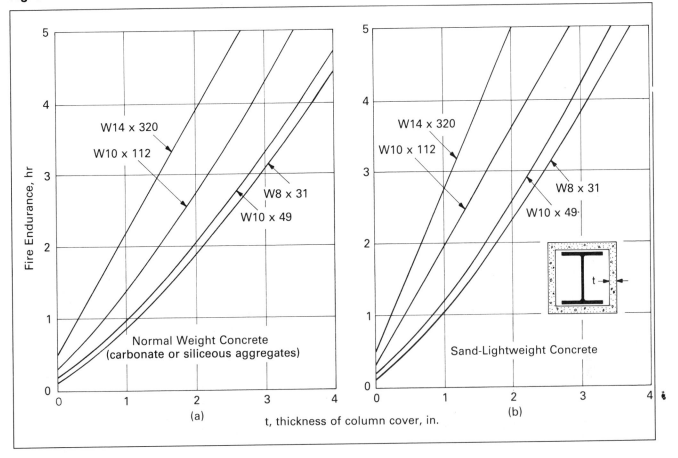

(a)

t, thickness of column cover, in.

(b)

plied to the steel in most column cover connections.

If restrained, either partially or fully, concrete panels tend to deflect or bow when exposed to fire. For example, for a steel column that is clad with four flat panels attached top and bottom, the column covers will tend to bulge at midheight thus tending to open gaps along the sides. The gap size drecreases as the panel thickness increases.

With L, C, or U-shaped panels, the gap size is further reduced. The gap size can be further minimized by connections at midheight. In some cases, ship-lap joints can be used to minimize the effects of joint openings.

Joints should be sealed in such a way to prevent passage of flame to the steel column. A noncombustible material such as sand-cement mortar, ceramic fiber blanket, or asbestos rope can be used to seal the joint.

Precast concrete column covers should be installed in such a manner that if they are exposed to fire, they will not be restrained vertically. As the covers are heated they tend to expand. Connections should accommodate such expansion without subjecting the cover to additional loads.

Fire resistive compressible materials, such as mineral fiber safing, can be used to seal the tops

Fig. 9.3.15 Types of precast concrete column covers; (a) would probably be most vulnerable to bowing during fire exposure while (f) would probably be the least vulnerable

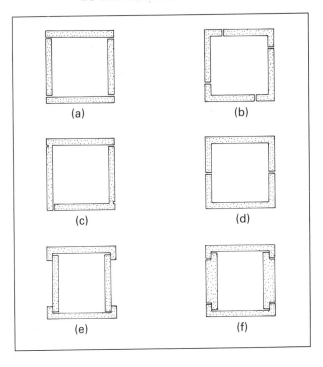

or bases of the column covers, thus permitting the column covers to expand.

9.3.10 Code and Economic Considerations

An important aspect of dealing with fire resistance is to understand what the benefits are to the owner of a building in the proper selection of materials incorporated in his structure. These benefits fall into two areas, code and economics.

Building codes are laws that must be satisfied regardless of any other considerations and the manner in which acceptance of code requirements is achieved is explained in the preceding pages. The designer, representing the owner, has no option in the code regulations, only in the materials and assemblies that meet these regulations.

Economic benefits are often overlooked by the designer/owner team at the time decisions are made on the structural system. Proper consideration of fire resistive construction through life-cycle cost analysis will provide the owner economic benefits over other types of construction in many areas, e.g., lower insurance costs, larger allowable gross areas under certain types of building construction, fewer stairwells and exits, increased value for loan purposes, longer mortgage terms, and better resale value. To ensure an owner of the best return on his investment, a life-cycle cost analysis using fire resistive construction must be prepared.

Beyond the theoretical considerations is the history of excellent performance of prestressed concrete in actual fires. Structural integrity has been provided, fires are contained in the area of origin, and, in many instances, repairs consist of "cosmetic" treatment only leading to early re-occupancy of structure.

9.3.11 References

1. "UL Report on Floor and Ceiling Assembly Consisting of Prestressed, Precast Concrete Double Tee Units with a Wallboard Ceiling," File R1319-131, Underwriters Laboratories, Inc., Feb. 21, 1973.
2. Abrams, M. S. and Gustaferro, A. H., "Fire Endurance of Two-Course Floors and Roofs," *Journal of the American Concrete Institute*, Feb., 1969.
3. Gustaferro, A. H. and Abrams, M. S., "Fire Tests of Joints Between Precast Wall Panels: Effect of Various Joint Treatments," *PCI Journal*, Sept.-Oct., 1975.
4. "Design for Fire Resistance of Precast Prestressed Concrete," PCI MNL 124–77, Prestressed Concrete Institute.
5. Abrams, M. S., et al, "Fire Tests of Concrete Joist Floors and Roofs," Portland Cement Association Publication RD 006.
6. Lin, T. D., et al, "Fire Endurance of Continuous Reinforced Concrete Beams," PCA Publication RD 072.
7. Lie, T. T. and Harmathy, T. Z., "Fire Endurance of Concrete-Protected Steel Columns," *ACI Journal*, Proceedings Vol. 71, No. 1, 1974.
8. Lie, T.T., "Contribution of Insulation in Cavity Walls to Propagation of Fire," *Fire Study No. 29*, Division of Building Research, National Research Council of Canada.

SANDWICH PANELS

9.4.1 General

Sandwich panels are composed of two concrete wythes separated by a layer of insulation. One of the concrete wythes may be a standard shape, as shown in Fig. 9.4.1, or any architectural concrete section produced for a single project. In place, sandwich panels provide the dual function of transferring load and insulating the structure. They may be used only for cladding, or they may act as beams, bearing walls or shear walls.

Fig. 9.4.1 Typical precast concrete load-bearing insulated wall panels

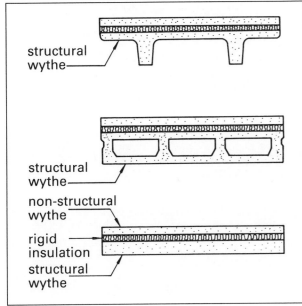

9.4.2 Structural Design

The structural design of sandwich panels is the same as design of other wall panels once the section properties of the panel have been determined. Three different assumptions may be used for the properties, depending on the construction:

1. The two concrete layers act independently (non-composite, Fig. 9.4.2a). The wythes are connected by ties and/or hangers which are flexible enough that they offer insignificant resistance to shrinkage and temperature movement. Positive steps are taken to assure that one wythe does not bond to the insulation, usually by placing a sheet of polyethylene or reinforced paper over the insulation before the final concrete wythe is placed, by applying a retarder or form release agent to one side of the insulation, or by placing insulation in two layers (Fig. 9.4.9a). When polyethelene is placed

on the warm side of the insulation, it also serves as a vapor barrier. One wythe is usually assumed to be "structural" and all loads are carried by that wythe, both during handling and in service, although wind loads and slenderness may in some cases be designed as in (3) below.

2. The two concrete layers act as a fully composite unit for the full life of the structure (Fig. 9.4.2b). The wythes are connected by rigid ties or regions of solid concrete, plus what bond there is between concrete and insulation.

Fig. 9.4.2 Non-composite and composite panels

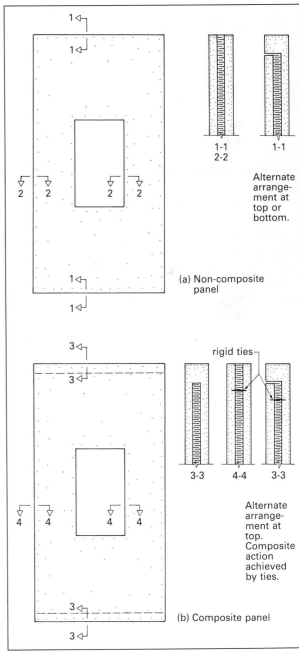

3. The panel acts as a fully or partially composite unit during handling and erection, but non-compositely for loads in service. The composite action is largely because of bond to the insulation combined with relatively flexible ties, but this bond is considered unreliable for the long term. Wind loads are distributed to each wythe in proportion to stiffness. For slenderness effects the sum of the moments of inertia may be used.

Composite panels larger than about 40 sq ft often exhibit cracking due to differential shrinkage and temperature exposure, unless both wythes are prestressed, or positive crack control measures, such as the use of dummy joints, are taken. In general, it is recommended that, unless positive steps are taken to de-bond the insulation from one wythe (usually the outer), both wythes be prestressed with a minimum of 150 psi after losses. Prestress force should be as nearly concentric as possible, or the camber caused by eccentricity must be considered.

Example 9.4.1 Section properties of sandwich panels.

Given:

The sandwich panel shown below.

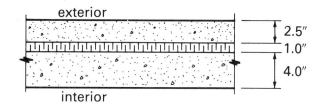

Problem:

Calculate the moment of inertia and section modulus if the section is (a) non-composite and (b) composite. Also find the load distribution for the non-composite case.

Solution:

(a) The non-composite properties are as follows:

Interior (structural) wythe:
$I_i = bd^3/12 = 12(4)^3/12 = 64$ in.^{4}/ft width
$S_i = I/c = 64/2 = 32$ in.^{3}/ft width

Exterior (non-structural) wythe:
$I_e = 12(2.5)^3/12 = 15.6$ in.^{4}/ft
$S_e = 15.6/1.25 = 12.5$ in.^{4}/ft

Distribution:
$I_i + I_e = 64 + 15.6 = 79.6$ in.
Lateral load resisted by:
Interior: $64(100)/79.6 = 80\%$
Exterior: $15.6(100)/79.6 = 20\%$

(b) The composite properties are as follows:

	A	y	Ay	$\bar{y}$	A$\bar{y}^2$	I
Interior	48	2.00	96.0	1.63	127.5	15.6
Exterior	30	6.25	187.5	2.62	205.9	64.0
	78		283.5		333.4	79.6

$$y_b = 283.5/78 = 3.63 \text{ in.}$$
$$I_c = 333.4 + 79.6 = 413.0 \text{ in.}^4/\text{ft width}$$

9.4.3 Connections

Panel connections

Analysis and design of connections are described in Parts 3 and 6. Panels will usually have a lateral tie near the top and a mid-height connection to adjacent panels to prevent differential bowing. Hung panels should be avoided if possible.

Wythe connectors

When one wythe is non-structural, its weight must be transferred to the structural wythe. This may be accomplished by using shear connectors or solid concrete ribs at the top and/or bottom of the panel. If the panel is non-composite, the shear connector should be a single element or a closely spaced pair of elements placed as near the center of gravity of the panel as possible. This permits the non-structural wythe to contract and expand with the least restraint.

When shear connectors are used for weight transfer between wythes, the assumptions shown in Fig. 9.4.3 may be used, where:

t = thickness of the structural wythe
a = vertical distance between panel support points
b = horizontal distance between panel support points
f = distance between wythe connectors

Shear connectors may be bent reinforcing bars, sleeve anchors, expanded metal, or welded wire trusses, as illustrated in Fig. 9.4.4.

For ribbed panels, the shear connector is placed in the rib to assure proper embedment depth. In non-composite panels, it is preferable to have only one anchoring center. In a panel with two ribs the shear connector can be positioned in either one of the ribs, and in the other, a flat anchor with the same vertical shear capacity is used (Fig. 9.4.5). Since the flat anchor has little or no horizontal shear capacity, restraint of the exterior wythe is minimized. In a multi-ribbed panel, the shear connector is placed as near the center as possible,

Fig. 9.4.3 Effective beam widths for panel design

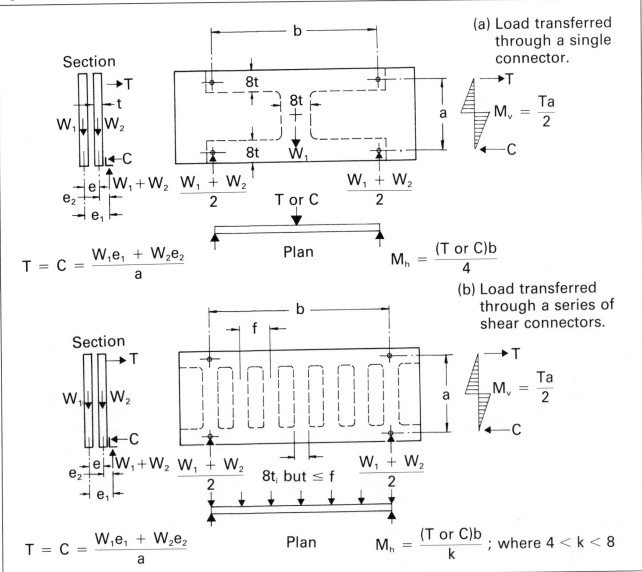

Section

T

t

W_1 W_2

C

e_2 e $W_1 + W_2$

e_1

(a) Load transferred through a single connector.

b

$8t$

$8t$

a

$8t$

W_1

$\dfrac{W_1 + W_2}{2}$ $\dfrac{W_1 + W_2}{2}$

T or C

$M_v = \dfrac{Ta}{2}$

$T = C = \dfrac{W_1e_1 + W_2e_2}{a}$ Plan $M_h = \dfrac{(T \text{ or } C)b}{4}$

(b) Load transferred through a series of shear connectors.

Section

T

W_1 W_2

C

e_2 e $W_1 + W_2$

e_1

b

f

a

$\dfrac{W_1 + W_2}{2}$ $8t_i$ but $\leq f$ $\dfrac{W_1 + W_2}{2}$

$M_v = \dfrac{Ta}{2}$

$T = C = \dfrac{W_1e_1 + W_2e_2}{a}$ Plan $M_h = \dfrac{(T \text{ or } C)b}{k}$; where $4 < k < 8$

Fig. 9.4.4 Typical shear connectors

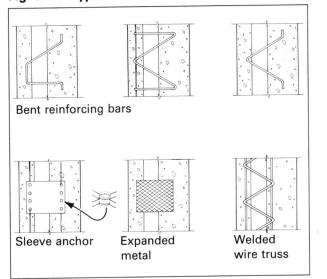

Bent reinforcing bars

Sleeve anchor   Expanded metal   Welded wire truss

Fig. 9.4.5 Anchorage for ribbed panels

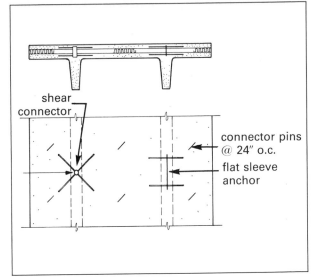

shear connector

connector pins @ 24" o.c.

flat sleeve anchor

and flat anchors are used in the other ribs.

To complete the connection, metal tension/compression ties passing through the insulation are spaced at regular intervals to prevent the wythes from separating. Functions of wythe connectors are shown in Fig. 9.4.6. Typical tie details are shown in Fig. 9.4.7, and arrangements and spacing in Fig. 9.4.8. Wire tie connectors are usually 12 to 14 gauge, and preferably of stainless steel. Galvanized metal or plastic ties may also be acceptable. Ties of welded wire fabric and reinforcing bars are sometimes used. Shaped, crimped, or bent ties should be cold bent.

Tension/compression ties should be flexible enough so as not to resist temperature and shrinkage parallel to the panel surface, yet strong enough to resist a lifetime of stress reversals caused by temperature strains. Ties should be ar-

Fig. 9.4.6 Functional behavior of connectors

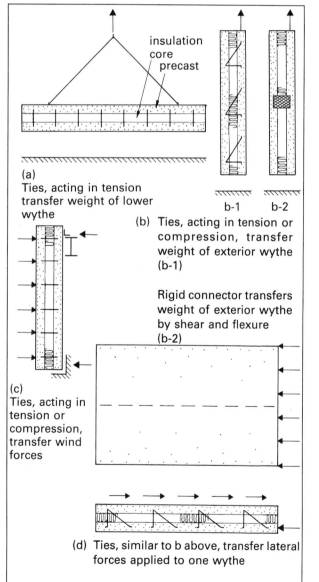

(a) Ties, acting in tension transfer weight of lower wythe

(b) Ties, acting in tension or compression, transfer weight of exterior wythe (b-1)

Rigid connector transfers weight of exterior wythe by shear and flexure (b-2)

(c) Ties, acting in tension or compression, transfer wind forces

(d) Ties, similar to b above, transfer lateral forces applied to one wythe

Fig. 9.4.7 Tension/compression ties

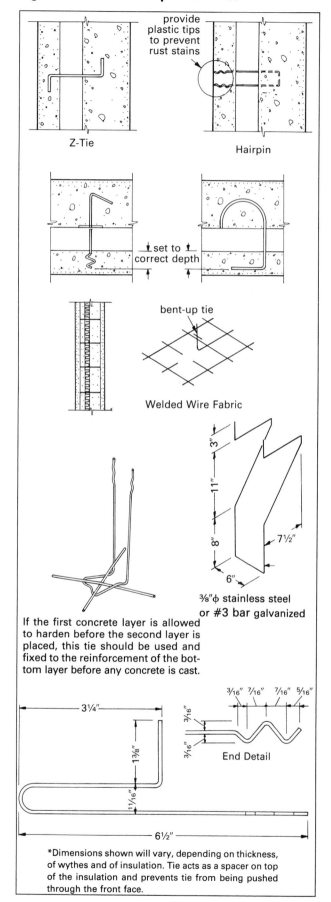

provide plastic tips to prevent rust stains

Z-Tie

Hairpin

set to correct depth

bent-up tie

Welded Wire Fabric

3⁄8"φ stainless steel or #3 bar galvanized

If the first concrete layer is allowed to harden before the second layer is placed, this tie should be used and fixed to the reinforcement of the bottom layer before any concrete is cast.

End Detail

*Dimensions shown will vary, depending on thickness, of wythes and of insulation. Tie acts as a spacer on top of the insulation and prevents tie from being pushed through the front face.

Fig. 9.4.8 Arrangement of connectors between wythes

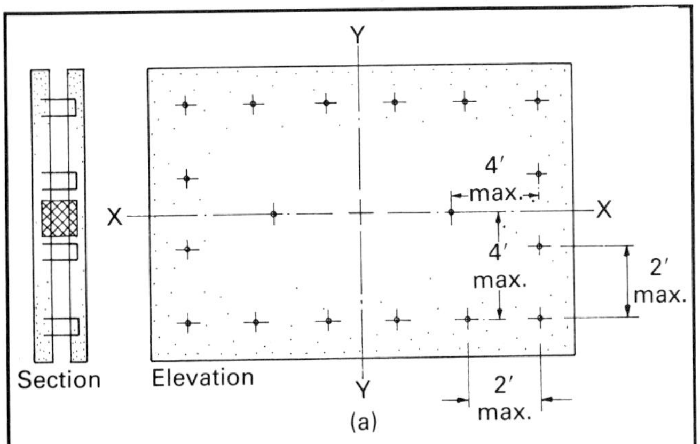

Section   Elevation

4' max.

4' max.

2' max.

2' max.

(a)

One device transfers weight of non-structural wythe to structural wythe, as well as any racking shear.

Ties spaced in the field of the panel transfer direct wind forces and stripping loads from the non-structural wythe to the structural wythe.

very flexible in vert. dir.

very flexible in horiz. dir.

Section (vert.)   Elevation

Section (horiz.)

(b)

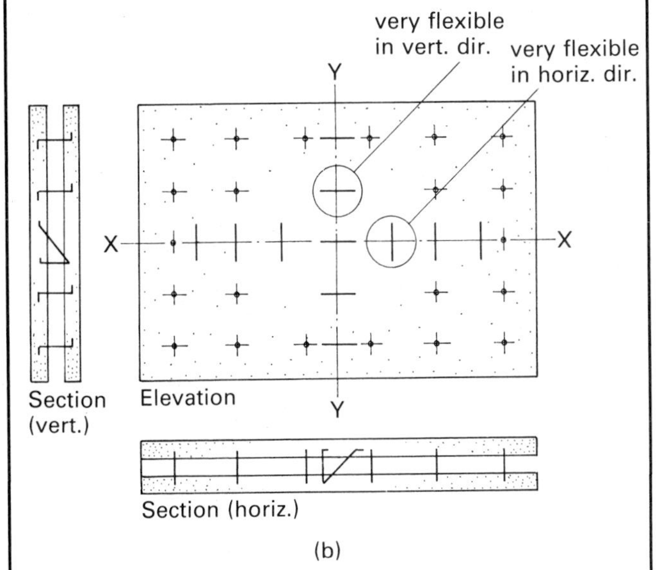

Devices spaced along the x-x axis transfer weight of non-structural wythe to the structural wythe. Similar devices along the Y-Y axis transfer racking shear from non-structural wythe to structural wythe.

Ties spaced in the field of the panel transfer direct wind loads and stripping forces from the non-structural wythe to the structural wythe.

ranged, or coated, so that galvanic reaction between the tie and reinforcement will not occur (see Sect. 9.5).

The spacing of ties should be approximately 2 ft on centers, but not more than 4 ft, or at least 2 ties per 10 sq ft of panel area.

9.4.4 Suggested Minimum Criteria

Wythes should be no thinner than 3 times the maximum aggerate size or 2 in., with 2½ in. preferred for non-structural and prestressed structural wythes, and 3 in. for structural wythes with nonprestressed reinforcement. Wythes should be thick enough to meet structural criteria, and to provide adequate cover (for reinforcement and other embedded steel) and anchorage (for connection hardware).

Except as otherwise required by analysis or experience, the minimum reinforcement in each wythe should be 0.001 of the cross-sectional area of the wythe in each direction. Panels exposed to the environment which are less than 2 ft wide in one direction need not be reinforced in that direction — 4 ft for panels not exposed to the environment. The recommended maximum spacing of reinforcement in wythes exposed to the environment is 6 in. for welded wire fabric, or 5 times the wythe thickness (18 in. maximum) for bars.

9.4.5 Insulation

Physical properties of insulation materials are listed in Table 9.4.1. Thermal properties are discussed in Sect. 9.1. The insulation should have low absorption or a water-repellent coating to minimize absorption of water from fresh concrete.

The thickness of insulation is determined as described in Sect. 9.1. A minimum of 1 in. is recommended. While there is no upper limit on the thickness, the deflection characteristics of the wythe connectors should be considered.

Openings in the insulation around connectors should be packed with insulation to avoid forming thermal bridges between wythes.

Fig. 9.4.9 Preferred installation of insulation sheets

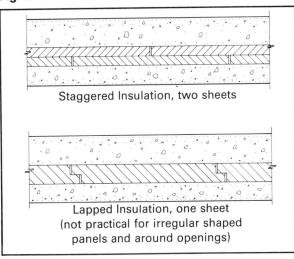

Staggered Insulation, two sheets

Lapped Insulation, one sheet
(not practical for irregular shaped
panels and around openings)

Table 9.4.1 Properties of insulation[1]

Properties of insulation	Glass, cellular rigid block	Perlite, expanded bonded rigid board	Polystyrene, expanded beads board cut from molded block	Polystyrene, cellular foam board cut from molded board	Polystyrene, cellular foam board cut from slab stock	Polyurethane, cellular foam rigid boards	Vinyl chloride, cellular foam
Density (pcf)	7.0 to 9.5	9.5 to 11.5	1.0 to 2.0	1.1 to 3.3	1.5 to 3.5	1.5 to 4.0	1.5 to 3.0
Capillarity	None	Negl.	None	None	None	None	None
Water absorption (% Vol., 24 hr. submersion)	Surface only	3.5 to 4.0	1.0 to 2.5	0.10 to 1.25	0.10 to 1.25	1.5 to 3.0	4.5
Comp. strength (psi @ 5% deform.)	100	80	6 to 17	8 to 16	20 to 100	18 to 50	28 to 50
Tensile strength (psi)	50	32	24		70	20 to 90	60
Linear coefficient thermal expansion ($\times 10^{-6}/°F$)	4.6		35	18	35	50	
Modulus of elasticity ($\times 10^{5}$ psi)	1.5			0.001			
Shear strength (psi)	50		17	127	27 to 36	16 to 55	
Flexural strength[2] (psi)	80		28 to 90	31	42 to 61	30 to 60	
Bond strength to concrete (psi)			6	10			

1. Physical properties listed are typical. Any actual design should be based on properties as furnished or verified by the material supplier.
2. ASTM C204, C240.

Using the maximum size of insulation sheets, consistent with the panel shape, is recommended. This will minimize joints and the resulting cold sinks. Lapped abutting ends of single layer insulation, or staggered joints with double layer insulation, will effectively remove cold sinks at joints (Fig. 9.4.9).

9.4.6 Thermal Bowing

Thermal bowing of exterior composite panels must be considered. Calculation of bow and the forces required to restrain it are discussed in Sect.

3.3.2. Differential movement between panels is seldom a problem, except at corners or where abrupt changes in the building occur. Crazing or cracking is sometimes caused by bowing, but will be minimized if the recommendations of Sect. 9.4.2 are followed.

9.4.7 Typical Details

As with all precast concrete construction, satisfactory performance usually depends on proper details. Some typical sandwich panel details are shown in Fig. 9.4.10.

Fig. 9.4.10 Architectural and structural details

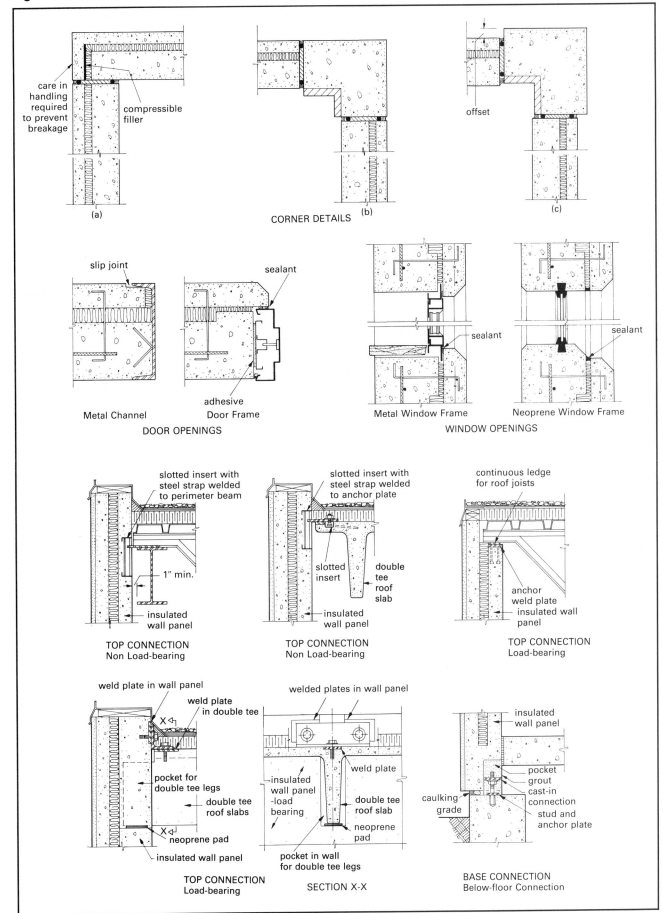

(a)

care in handling required to prevent breakage

compressible filler

(b)

CORNER DETAILS

(c)

offset

slip joint

Metal Channel

sealant

adhesive

Door Frame

DOOR OPENINGS

Metal Window Frame

sealant

Neoprene Window Frame

sealant

WINDOW OPENINGS

slotted insert with steel strap welded to perimeter beam

1″ min.

insulated wall panel

TOP CONNECTION
Non Load-bearing

slotted insert with steel strap welded to anchor plate

slotted insert

double tee roof slab

insulated wall panel

TOP CONNECTION
Non Load-bearing

continuous ledge for roof joists

anchor weld plate

insulated wall panel

TOP CONNECTION
Load-bearing

weld plate in wall panel

weld plate in double tee

pocket for double tee legs

double tee roof slabs

neoprene pad

insulated wall panel

TOP CONNECTION
Load-bearing

welded plates in wall panel

weld plate

insulated wall panel -load bearing

double tee roof slab

neoprene pad

pocket in wall for double tee legs

SECTION X-X

insulated wall panel

pocket

grout

cast-in connection

stud and anchor plate

caulking grade

BASE CONNECTION
Below-floor Connection

CORROSIVE ENVIRONMENTS

9.5.1 Corrosion of Embedded Steel

A potential problem which must be addressed is the deterioration of concrete structures due to corrosion of embedded steel materials. The by-products of corrosion expand and increase the volume once occupied by the original material, and cracking and spalling of the concrete results. An example of this problem is the deterioration of parking structure floor slabs where water and road salts from automobiles penetrate to the reinforcement and cause corrosion.

Portland cement concrete provides protection from steel corrosion under normal conditions, due to its high alkalinity and relatively high electrical resistivity. These properties degrade quickly in the presence of moisture, oxygen and chloride. Chlorides are often found in concrete aggregates, water and admixtures, and hence may be present in the concrete when it is cast. Moisture and oxygen alone, however, can cause corrosion.

Cracking is not necessary for corrosion damage to occur, but cracks may allow moisture, air and contaminants to reach the steel more quickly.

Several applications that require special consideration for corrosion protection are:

1. Parking structures
2. Food processing facilities
3. Storage tanks
4. Cooling towers
5. Paper mills
6. Marine structures
7. Enclosed swimming pools
8. Open stadia
9. Bridge decks

Generally, the higher cement content, plant controlled procedures, and improved density due to concrete compaction in precast and prestressed products provide adequate protection. However, in more severe environments, greater cover, use of entrained air, epoxy coated reinforcement, special cements and protective coatings should be considered either singly or in combination.

ACI 318-83, Sect. 4.5.4, has the following maximum water soluble chloride ion limits in hardened concrete at 28 days, expressed as a percent by weight of cement:

Prestressed concrete 0.06%

Reinforced concrete exposed
to chloride in service 0.15%

Reinforced concrete that will
be dry or protected from moisture
in service 1.00%

Other reinforced concrete
construction 0.30%

The low limit for prestressed concrete emphasizes the need for specifications to ban calcium chloride, or chemicals containing chloride ions, in concrete mixes.

Several practices will improve the resistance of reinforced or prestressed concrete to potential corrosion damage:

1. Use of high quality, low permeability, low water-cement ratio concrete
2. Good workmanship
3. Good curing
4. Adequate steel cover
5. Limiting chlorides in the concrete mix
6. Limiting intrusion of corrosive materials along with wetting and drying
7. Surface protection through sealers, coatings or special wearing surfaces
8. Assessment of potential problems, and correponding precautions in design and detailing.

Further discussion of prevention and repair of damage from corrosion of embedded steel may be found in the references listed in Sect. 9.5.4

9.5.2 Corrosion of Other Metals

Several other embedded materials may corrode in concrete. This is discussed in detail in Refs. 1 and 2. Briefly:

Aluminum. Conditions conducive to corrosion are created if steel is in contact with the aluminum, appreciable concentrations of chlorides are present, or the cement is high in alkali. Since it is virtually impossible to avoid contact with reinforcement, embedment of aluminum items such as electrical conduit is discouraged.

Lead. Lead is attacked by fresh concrete and mortar, but this corrosion tends to stop as the concrete cures and dries. Protection of embedded lead with organic coatings is suggested; however, damage to the concrete seldom occurs because the softness of the lead absorbs the expansive pressures caused by the formation of corrosion products.

Copper and copper alloys. Copper is not normally corroded by concrete, except in the presence of soluble chlorides. Copper in contact with or in close proximity to steel can accelerate the corrosion of the steel.

Zinc. Zinc is susceptible to attack by fresh concrete and mortar. The most frequent use of zinc in concrete is galvanizing of steel. This is discussed in Part 1.

Most other metals have good resistance to corrosion in concrete, but the presence of soluble chlorides may reduce the corrosion resistance.

9.5.3 Sulfate Attack

Another chemical attack of concrete materials occurs from sulfates found in soils, ground water, processing liquids or sewage. Sulfate resistant cements are available, and ACI 201.2R-77 provides detailed discussion and recommendations on this problem.

9.5.4 References

1. "Guide to Durable Concrete," ACI 201.2R-77, *ACI Manual of Concrete Practice*, 1984, Part 1.

2. "Corrosion of Nonferrous Metals in Contact with Concrete," IS136.04T, Portland Cement Association, 1969.

3. "Catalog of NACE Technical Committee Report and Standards," National Association of Corrosion Engineers, Technical Practices Committee, Katy, TX.

4. "Corrosion of Reinforcement and Prestressing Tendons – A State-of-the-Art Report," RILEM Technical Committee, Corrosion of Reinforcing Steel in Concrete.

5. "Durability of Concrete Bridge Decks," NCHRP Synthesis 57, Transportation Research Board, Washington, DC.

6. "Corrosion of Reinforcement in Concrete Construction," edited by Alan P. Crane, Society of Chemical Industry, London, 437 pp., John Wiley & Sons, Inc., New York, NY.

QUALITY CONTROL

9.6.1 Introduction

The use and application of precast and prestressed concrete demands a high level of quality in its design and production.

Each producer must have a total quality assurance program that ensures structural integrity and desired esthetic appearance in the final structure. However, the owner or the architect/engineer must be satisfied that materials, methods, products and quality control meet all the requirements of the project specifications. He can do this by conducting his own inspections of the plant, by engaging an independent agency to conduct inspections to prescribed criteria, or by specifying manufacturing facilities that are certified by the Prestressed Concrete Institute Plant Certification Program (see Part 10, Guide Specifications).

9.6.2 Plant Certification

Plants certified under the PCI Plant Certification Program have a confirmed capability to produce quality precast or prestressed concrete products. This is based on three inspections per year by a structural engineering firm engaged by PCI; two of the inspections are unannounced, and one is a scheduled in-depth inspection requiring two or more days. Inspection criteria and grading are based on the industry's standards for quality control. These standards are presented in PCI's two quality control manuals, one for structural precast, prestressed concrete products and the other for architectural precast concrete products. Inspections cover all phases of production including materials, production methods, product handling and storage, shipping, product appearance, testing, record keeping, personnel and safety practices. Failure to maintain a production facility at or above required minimum standards results in mandatory decertification.

9.6.3 PCI Quality Control Manuals

The two PCI manuals listed below contain industry-approved requirements for quality control for both structural and architectural precast concrete products. Any in-house quality assurance program should be based on these manuals. They should also be used by any outside inspection of production facilities. They form the basis for the Plant Certification Program and the detailed grading system on which a plant can become certified.

9.6.4 References

1. *Manual for Quality Control for Plants and Production of Precast and Prestressed Concrete Products,* MNL-116-85, Prestressed Concrete Institute, Chicago, IL.

2. *Manual for Quality Control for Plants and Production of Architectural Precast Concrete Products,* MNL-117-77, Prestressed Concrete Institute, Chicago, IL.

CONCRETE COATINGS AND JOINT SEALANTS

9.7.1 Coatings for Horizontal Deck Surfaces

Horizontal surfaces of concrete are frequently coated to protect them from ingress of salt-laden water or other harmful liquids, from abrasion, or to provide a more pleasing appearance. Adverse environments can be better tolerated, dusting is lessened, and coatings of many colors are available. Chemical, industrial, food and drug processing plants, parking structures, stadia and many other types of facilities can benefit from application of coatings. The references in Sect. 9.7.4 contain guides to various coatings and their uses.

The products most commonly used include linseed oil, epoxies, polyurethanes, a number of other synthetic resins, and chemically modified concrete or bituminous surfacing, which may be used with a rubberized membrane. Combinations of materials are often used, for example, a polyurethane base membrane with an epoxy wearing surface. Most coatings are expensive, and application requires special expertise. Some materials are hazardous to workers during application.

The quality of coatings in terms of long-term effectiveness, durability, change of appearance, and ease of application varies, as do costs. The owner often must decide between a costly, more durable coating and a less expensive one that will require frequent reapplication or patching. Choice of material is best based on previous *successful* use under similar conditions. Rarely are coatings a permanent "cure-all", and for some exposures and uses, a regular program of maintenance is recommended.

9.7.2 Clear Surface Sealers for Architectural Precast Panels

Clear surface coatings or sealers are sometimes used on precast concrete wall panels to improve weathering qualities or to reduce attack of the concrete surface by airborne pollutants. Because of the quality of concrete normally achieved in plant-cast precast concrete, even with very thin sections, sealers are not required for waterproofing. Because the results are uncertain, use of sealers in locations having little or no air pollution is not recommended.

A careful evaluation should be made before deciding on the type of sealer. This includes consultation with the local precasters. In the absence of near-identical experience, it is desirable to test sealers on reasonably sized samples of varying age to verify performance over a suitable period of exposure or usage, based on prior experience under similar exposure conditions.

Sealers are usually applied to wall panels after erection to avoid problems with adhesion of sealants.

Any coating used should be guaranteed by the supplier or applicator not to stain, soil, or discolor the precast finish. Also, some clear coatings may cause joint sealants to stain concrete. Consult manufacturers of both sealants and coatings or pretest before applying the coating.

Sealers should be applied in accordance with manufacturer's recommendations. Generally, good airless spray equipment is used for uniformity and to prevent surface rundown. Two coats are usually required to provide a uniform coating, because the first coat is absorbed into the concrete. The second coat does not penetrate as much and provides a more uniform surface color. Care must be taken to keep sealer off glass surfaces.

9.7.3 Joint Sealants

Joint sealants include viscous liquids, mastics or pastes, and tapes, gaskets and foams. Generally, a sealant is any material placed in a joint for the purpose of preventing the passage of moisture, air, heat or dirt into or through the joint.

Successful performance of a wall system is often dependent on good joint details and sealant selection. Too often, because the sealant is a relatively minor part of the total structure, the responsibility for selection is left to the contractor or erector who does not fully understand the task it is to perform.

The selection of the proper sealant is confused by the method of supply. Most of the raw materials are manufactured by a few major chemical companies. These basic materials are then compounded with other ingredients by literally hundreds of sealant manufacturers. The quality of finished product varies widely because the expertise of the formulators varies widely.

Viscous liquid sealants are often used in pourable form and normally are used in horizontal joints. Mastics are applied with a gun and are compounded to prevent sagging or flowing when used in vertical joints. Tapes are most often used around glass. Elastomeric gaskets are used in joints which experience considerable movement. Foams are used as air seals or backup for more durable surface seals.

Specification of sealants is difficult because many of the numerous formulations are not covered by standard specifications, either for materials or installation. Many of the "standard" specifications are also out of date, and often two standard-writing agencies have specifications that do not agree.

Proven performance is still the procedure that is often relied upon. References 5 through 9, Sect. 9.7.4., will aid in the design and selection of joint sealants. Reference 6 is also a very valuable aid for the design and detailing of architectural precast panel joints.

9.7.4 References

1. "A Guide to the Use of Waterproofing, Dampproofing, Protective, and Decorative Barrier Systems for Concrete," ACI 515.1R-79, *ACI Manual of Concrete Practice,* 1984, Part 5.

2. "Coatings, Penetrants and Specialty Concrete Overlays for Concrete Surfaces," D. W. Pfeifer and W. F. Perenchio, National Association of Corrosion Engineers Seminar, September 1982, Chicago, IL.

3. "Concrete Sealers for Protection of Bridge Structures," NCHRP Report 244, Transportation Research Board, Washington, DC.

4. Litvin, Albert, "Clear Coatings for Architectural Concrete Surfaces," PCA Development Department Bulletin D 137, Portland Cement Association, Skokie, IL, 1968.

5. "Guide Specifications, Section 07900 (Sealants)", Sealant and Waterproofers Institute, Glenview, IL, 1982.

6. "Sealants: The Professionals' Guide", Sealant and Waterproofers Institute, Glenview, IL, 1984.

7. "Architectural Precast Concrete Joint Details," JR-155, Prestressed Concrete Institute, Chicago, IL, *PCI Journal*, V. 18, No. 2, March-April, 1973.

8. "Guide to Joint Sealants for Concrete Structures," ACI 504R-77, *ACI Manual of Concrete Practice,* 1984, Part 5.

9. *Construction Sealants and Adhesives,* J. P. Cook, Wiley-Interscience, New York, NY.

VIBRATION IN CONCRETE STRUCTURES

9.8.1 Human Response to Building Vibrations

Modern buildings often use components with low weight-to-strength ratios, which allow longer spans with less mass. This trend increasingly results in transient vibrations which are annoying to the occupants. Unlike equipment vibration, a person often causes the vibration and also senses it. These vibrations usually have very small amplitudes (less than 0.05 in.) and were not noticed in older structures with heavier framing and more numerous and heavier partitions, which provided greater damping and other beneficial dynamic characteristics.

This problem is not well understood. Predicting human response to floor motion and the dynamic response of a floor system to moving loads are developing technologies. A number of discomfort criteria have been published,[4-7] but they often give contradictory results.[3]

The vibration problem is most effectively treated by modifying the structural system. The natural period (or its inverse, frequency), stiffness, mass, and damping are the structural parameters related to vibration control. Stiffness is increased by providing greater section properties than may be required for supporting loads. An increase in mass improves the natural frequency, but increases defletions and stresses, so by itself is only partially effective in controlling vibrations. For example, increasing the depth of a flexural member will aid greatly in vibration control, but increasing the width will not.

Recent research has emphasized the effect that damping plays in the human perception of vibration. In a study of 91 floor systems it was concluded that with damping greater than 5.5 to 6 percent of critical, structural systems were acceptable; systems with less were not.[3] Damping is usually attributed to the existence of partitions, supported mechanical work, ceilings and similar items, but is really not well understood. Guides for quantifying damping effect are scarce, and those that are available are very approximate.[6-8]

9.8.2 Vibration Isolation for Mechanical Equipment

Vibration produced by equipment with unbalanced operating or starting forces can usually be isolated from the structure by mounting on a heavy concrete slab placed on resilient supports. This type of slab, called an inertia block, provides a low center of gravity to compensate for thrusts such as those generated by large fans.

For equipment with less unbalanced weight, a "housekeeping" slab is sometimes used below the resilient mounts to provide a rigid support for the mounts and to keep them above the floor so they are easier to clean and inspect. This slab may also be mounted on pads of precompressed glass fiber or neoprene.

The natural frequency of the total load on resilient mounts must be well below the frequency generated by the equipment. The required weight of an inertia block depends on the total weight of the machine and the unbalanced force. For a long-stroke compressor, five to seven times its weight might be needed. For high pressure fans, one to five times the fan weight is usually sufficient.

A floor supporting resiliently mounted equipment must be much stiffer than the isolation system. If the static deflection of the floor approaches the static deflection of the mounts, the floor becomes a part of the vibrating system, and little vibration isolation is achieved. In general, the floor deflection should be limited to about 15 percent of the deflection of the mounts.

Simplified theory shows that for 90% vibration isolation, a single resilient supported mass (isolator) should have a natural frequency of about ⅓ the driving frequency of the equipment. The natural frequency of this mass can be calculated by:[9]

$$f_n = 188 \sqrt{1/\Delta_i} \qquad \text{(Eq. 9.8.1)}$$

where:

f_n = natural frequency of the isolator, CPM
Δ_i = static deflection of the isolator, in.

From the above, the required static deflection of an isolator can be determined as follows:

$$f_n = f_d/3 = 188\sqrt{1/\Delta_i}, \text{ or}$$

$$\Delta_i = (564/f_d)^2 \qquad \text{(Eq. 9.8.2)}$$

and:

$$\Delta_f \leq 0.15\Delta_i \qquad \text{(Eq. 9.8.3)}$$

where:

f_d = driving frequency of the equipment
Δ_f = static deflection of the floor system caused by the weight of the equipment, including inertia block, at the location of the equipment

Example 9.8.1 Vibration isolation

Given:
A piece of mechanical equipment has a driving frequency of 800 CPM.

Problem:
Determine the approximate minimum deflection of the isolator and the maximum deflection of the floor system that should be allowed.

Solution:
From Eq. 9.8.2:
Isolator, $\Delta_i = (564/800)^2 = 0.50$ in.
Floor, $\Delta_f = 0.15(0.50) = 0.07$ in.

9.8.3 References

1. "Vibrations of Concrete Structures", Publication SP-60, American Concrete Institute, Detroit, MI.

2. Galambos, T.V., Gould, P. C., Ravindra, M. R., Surgoutomo, H., and Crist, R. A., "Structural Deflections — A Literature and State-of-the-Art Survey", *Building Science Series,* Oct., 1973, National Bureau of Standards, Washington, DC.

3. Murray, T. M., "Acceptability Criterion for Occupant-Induced Floor Vibration", *Sound and Vibration,* Nov. 1979.

4. "Design and Evaluation of Operation Breakthrough Housing Systems", NBS Report 10200, Amendment 4, Sept., 1970, U. S. Department of Housing and Urban Development, Washington, DC.

5. Wiss, J. F. and Parmelee, R. H., "Human Perception of Transient Vibrations", Journal of the Structural Division, ASCE, Vol. 100, No. ST4, April, 1974.

6. "Guide to Floor Vibrations", *Steel Structures for Buildings-Limit States Design,* CSA S16.1-1974, Appendix G, Canadian Standards Association, Rexdale, Ontario.

7. "Guide for the Evaluation of Human Exposure to Whole-Body Vibration", International Standard 2631, International Organization for Standardization, 1974.

8. Murray, T. M., "Design to Prevent Floor Vibration", Engineering Journal, AISC, Third Quarter, 1975.

9. Harris, C.M. and Crede, C. E., *Shock and Vibration Handbook,* 2nd Edition, McGraw-Hill, New York, NY, 1976.

CRACKING, REPAIR AND MAINTENANCE

9.9.1 Cracking

Minor cracking may occur in precast concrete without being detrimental, and it is impractical to impose specifications that prohibit all cracking. (Note: In other sections, the term "crack-free" design is used. This refers to a design in which concrete tension is kept within limits, and should not be construed as being a guarantee that there will be no cracking.) However, in addition to being unsightly, cracks are potential locations of concrete deterioration, and should be avoided if possible. Prestressing and proper handling procedures are two of the best methods of keeping cracks to a minimum. To evaluate the acceptability of a crack, the cause and service conditions of the precast unit should be determined.

Crazing, i.e., fine, random (commonly called "hairline") cracks, may occur in the cement film on the surface of concrete. The primary cause is the shrinkage of the surface with respect to the mass of the unit. Crazing has no structural significance and does not significantly affect durability, but may be visually accentuated if dirt settles in these minute cracks. Crazing should not be cause for rejection.

Tension cracks are sometimes caused by temporary loads during production, transportation, or erection of the products (see Part 5). These cracks may extend through to the reinforcement. If the crack width is narrow, the structural adequacy of the casting will remain unimpaired, as long as corrosion of the reinforcement is prevented. See Sect. 4.2.2.1 for recommended limits on crack width. The acceptability of cracks wider than recommended maximums should be governed by the function of the unit. Most can be effectively repaired and sealed.

Long-term volume changes can also cause cracking after the member is in place in the building, if the connections provide enough restraint to the member (see Sect. 3.3). Internal causes, such as corrosion of reinforcement or cement-aggregate reactivity, can also lead to long-term cracking and should be considered when materials are selected.

9.9.2 Repair and Maintenance

Products which are damaged prior to or during placement in the structure can usually be repaired, but major repairs should not be made until it is determined that the unit will be structurally sound. Patching of architectural panels is an art requiring expert craftsmanship and careful selection and mixing of materials.

Concrete surfaces normally require little maintenance, except for those that are subjected to harsh environments, such as parking structures in northern climates. Routine inspection and maintenance can do much to extend the life of such buildings.

Small cracks (under 0.005 to 0.010 in.) may not need repair, unless failure to do so can cause corrosion of reinforcement. If repair is required for the restoration of structural integrity or esthetics, they may be pressure injected with a low viscosity epoxy. Many firms specialize in this process, with varying experience with precast and prestressed concrete.

Spalls exposed in the completed structure should be patched. Again, a number of proprietary products and procedure specialists are available. Experience and care are required to place a patch that is permanent and esthetically acceptable.

9.9.3 References

1. "Control of Cracking in Concrete Structures," ACI 224R-80, 1980, by ACI Committee 224, American Concrete Institute, Detroit, MI.

2. "Development and Distribution of Cracks in Rectangular Prestressed Beams During Static and Fatigue Loading," Abeles, P. W., Brown, E. L., et al, *PCI Journal*, Vol. 13, No. 5, October 1968, p. 36.

3. "Cracking Induced by Environmental Effects," Mather, Bryant, *Causes of Cracking in Concrete*, SP-20, American Concrete Institute, Detroit, MI.

4. "Guide to Durable Concrete," ACI Committee 201, *ACI Journal, Proceedings*, Vol. 14, No. 12, December 1977, p. 573.

5. "Manual for Quality Control for Plants and Production of Architectural Precast Concrete," MNL-117–77, Prestressed Concrete Institute, Chicago, IL.

6. "Fabrication and Shipment Cracks in Prestressed Hollow-Core Slabs and Double Tees," PCI Committee on Quality Control Performance Criteria, *PCI Journal*, Vol. 28, No. 1, January-February 1983.

PRECAST SEGMENTAL CONSTRUCTION

9.10.1 General

Segmental construction is a method of construction in which primary load carrying members are composed of individual segments post-tensioned together. This allows precast concrete to be used for long horizontal or vertical spans within size limitations imposed by manufacturing, transportation, and handling equipment. With proper planning and element selection, a large re-use of forms is possible, with the resulting economy.

The method is best known for a number of landmark bridges,[1,10] but has also been used in the construction of airport control towers, storage tanks for various solids and liquids, including liquid natural gas, a long-span exterior Vierendeel truss on a high-rise building, an Olympic stadium and other non-bridge structures.[2-9]

Segmental construction requires that the designer give special consideration to:

1. Size and weight of the precast elements.
2. Configuration and behavior of the joints between elements.
3. Construction sequence, and the loads and deflections imposed at various stages.
4. The effect of normal tolerances and deviations upon the joints.

9.10.2 Joints

Joints may be either "open", to permit completion by a field placed grout, or "closed", where the joint is either dry or bonded by a thin layer of adhesive (Fig. 9.10.1).

Open joints

The individual segments are separated by an amount sufficient to place (usually by pressure) a grout mix, but not more than about 2 in. Prior to placing the segments, the joint surface is thoroughly cleaned and wire brushed or sand-blasted.

The perimeter of the joint is sealed with a gasket which is compressed by use of "come-alongs" or by a small amount of prestress. Gaskets are also provided around the post-tensioning elements to prevent leakage into the ducts, blocking passage of the tendons. Vents are provided at the top to permit escape of entrapped air during grouting. Prior to filling the joint, the surfaces should be thoroughly wetted, or coated with a bonding agent. Grout strength should be at least equal to that of the precast segments, but not less than 4000 psi.

After grouting, vents are closed and pressure increased to assure full grout intrusion. After a few days, the vent is reopened, and filled with grout as required.

Closed joints

If a closed joint is used, the segment is usually "match-cast", i.e., each segment is cast against its previously cast neighbor. A bond breaker is applied to the joint during casting. Thus, the connecting surfaces fit each other accurately, so that little or no filling material is needed at the joint. The sharpness of line of the assembled construction depends mainly on the accuracy of the manufacture of the segments.

Match cast segments are usually joined by coating the abutting surfaces with a thin layer of epoxy adhesive, and then using the post-tensioning to draw the elements together and hold them in position. In some structures, dry joints have been used successfully, in which case, the compression provided by the post-tensioning is relied upon for

Fig. 9.10.1 Types of joints

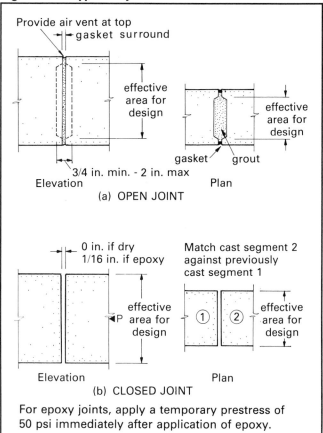

(a) OPEN JOINT

(b) CLOSED JOINT

For epoxy joints, apply a temporary prestress of 50 psi immediately after application of epoxy.

weather-proofing, as well as for transfer of forces.

Surface preparation of closed joints is extremely important. They should be sound and clean, free from all traces of form release agents, curing compounds, laitance, oil, dirt and loose concrete.

A small piece of foreign material in a joint, or slightly imperfect alignment will frequently cause the concrete to spall around the edges of the contact area of a closed joint. Thus, care in joint preparation and segment alignment cannot be overemphasized.

9.10.3 Design

Analysis of precast segmental structures usually assumes monolithic behavior of the members under service loads, except that if dry joints are used, tension is not permitted between segments. Some designers also prefer to not allow tension in grouted joints, unless tests have indicated otherwise.

The behavior during construction is of particular importance in segmental structures. Stresses which may be caused by settlement and shortening of scaffolding, temperature changes, elastic shortening from post-tensioning and other construction related stresses and movements may need to be considered in the design.

Shear stresses across joints are resisted by epoxy adhesives or grout, sometimes in combination with shear keys. In dry joints, shear stresses are resisted by friction (and shear keys if present), with the post-tensioning providing the normal force. In the absence of test data, the coefficient of friction may be assumed to be 0.8 (Sect. 6.6).

Individual segments are designed for handling and erection stresses (Part 5). They may be pretensioned or reinforced with mild steel, depending on size and manufacturing procedure.

9.10.4 Post-Tensioning

Information on various post-tensioning systems and their applications is given in the *PTI Post-Tensioning Manual.*[11] Nearly any type of bonded system can be used.

Tendon and duct placement

In addition to the design parameters for in-service conditions, the tendon layout must consider the sequence of construction and the changes in load conditions during the various construction stages.

Ducts for the tendons are placed in the seg-ments prior to concrete placement. In some cases, tendons are installed in the ducts before the segment is erected, or even before concrete is placed, and subsequently coupled at each joint. In others, the tendons are placed after the segments are erected, either full length of the member or in shorter lengths, again coupled at the joint. Special attention must be given to the alignment of the ducts, especially at the joints. They must be large enough in diameter to adequately place the tendon, allowing some tolerance in alignment, and receive the grout placed subsequent to stressing. Special attention must be given to the corrosion protection of the post-tensioning steel, if it is to be unbonded at any stage of construction. Also, drains should be installed in the ducts at low points of the tendon profile.

Post-tensioning tendons crossing joints should be approximately perpendicular to the joint surface in the smaller dimension of the segment, e.g., flange or web thickness, but may be inclined to the direction of the larger dimension (e.g., depth). This is to minimize any unbalanced shearing force which could lead to dislocation of edges at the joints.

Couplers

Couplers are designed to develop the full strength of the tendons they connect. Adjacent to the coupler, the tendons should be straight, or have very minor curvature for a minimum length of 12 times the diameter of the coupler. Adequate provisions must be made to assure that the couplers can move during prestressing.

Bearing areas

Prestress force is transferred to the joined segments through bearing plates or other anchorage devices. This causes high load concentrations at the bearing areas, usually requiring vertical and horizontal reinforcement in several locations:

1. Under end surfaces, not more than 3/4 in. deep, to control possible surface cracking around anchorages.
2. Internally, to prevent splitting between individual anchors. Size and location of this area and the magnitude of splitting (bursting) force depends on the type of anchorage and the force in the post-tensioning tendon.
3. Internally, to prevent splitting between groups of anchors.
4. Adjacent to joint surfaces, to decrease the possibility of damage to segments during post-tensioning or handling.

Grouting of the ducts is provided for corrosion protection and to develop bond between the prestressing steel and the surrounding concrete. Grouting procedures should follow the *Recommended Practice for Grouting of Post-Tensioned Prestressed Concrete*, contained in Ref. 11.

9.10.5 References

1. "Recommended Practice for Precast Post-Tensioned Segmental Construction" Joint PCI-PTI Committee on Segmental Construction, *PCI Journal*, Vol. 27, No. 1, January-February, 1982, p. 15.

2. Anderson, Arthur R., "World's Largest Prestressed LPG Floating Vessel," *PCI Journal*, Vol. 22, No. 1, January-February, 1977.

3. Pery, William E., "Precast Prestressed Clinker Storage Silo Saves Time and Money," *PCI Journal*, Vol. 21, No. 1, January-February, 1976, pp. 50–67.

4. Martynowicz, A., McMillan, C. B., "Large Precast Prestressed Vierendeel Trusses Highlight Multistory Building," *PCI Journal*, Vol. 20, No. 6, November-December, 1975, pp. 50–65.

5. Arafat, Mahmoud Z., "Giant Precast Prestressed LNG Storage Tanks at Staten Island," *PCI Journal*, Vol. 20, No. 3, May-June, 1975, pp. 22–33.

6. Muller, Jean, "Ten Years Experience in Precast Segmental Construction," *PCI Journal*, Vol. 20, No. 1, January-February, 1975, p. 28.

7. Lamberson, Eugene A., "Post-Tensioned Structural Systems - Dallas-Ft. Worth Airport," *PCI Journal*, Vol. 18, No. 6, November-December, 1973, pp. 72–91.

8. Zielinski, Z. A., "Prestressed Slabs and Shells Made of Prefabricated Components," *PCI Journal*, Vol. 9, No. 4, August, 1964, pp. 69–80.

9. Zielinski, Z. A., "Prefabricated Building Made of Triangular Prestressed Components," *ACI Journal*, Proceedings, Vol. 61, No. 4, April, 1964.

10. *Precast Segmental Box Girder Bridge Manual*, published jointly by the Prestressed Concrete Institute (Chicago, IL) and the Post-Tensioning Institute (Phoenix, AZ), 1978, 116 pp.

11. *PTI Post-Tensioning Manual*, Post-Tensioning Institute, Phoenix, Arizona, 1981.

COORDINATION WITH MECHANICAL, ELECTRICAL AND OTHER SUB-SYSTEMS

9.11.1 Introduction

Prestressed concrete is used in a wide variety of buildings, and its integration with lighting, mechanical, plumbing, and other services is of importance to the designer. Because of increased environmental demands, the ratio of costs for mechanical and electrical installations to total building cost has increased substantially in recent years. This section is intended to provide the designer with the necessary perspective to economically satisfy mechanical and electrical requirements, and to describe some standard methods of providing for the installation of other sub-systems.

9.11.2 Lighting and Power Distribution

For many applications, the designer can take advantage of the fire resistance, reflective qualities and appearance of prestressed concrete by leaving the columns, beams, and ceiling structure exposed. To achieve uniform lighting free from distracting shadows, the lighting system should parallel the stems of tee members.

By using a reflective paint and properly spaced high-output fluorescent lamps installed in a continuous strip, the designer can achieve a high level of illumination at a minimum cost. In special areas, lighting troffers can be enclosed with diffuser panels fastened to the bottom of the tee stems providing a flush ceiling (see Fig. 9.11.1). By using reflective paints, these precast concrete lighting channels can be made as efficient as conventional fluorescent fixtures.

9.11.3 Electrified Floors

The increasing use of business machines, telephones, and other communication systems stresses the need for adequate and flexible means of supplying electricity and communication service. Since a cast-in-place topping is usually placed on prestressed floor members, conduit runs and floor outlets can be readily buried within this topping. With shallow height electrical systems, a comprehensive system can be provided in a reasonably thin topping. The total height of conduits for these comprehensive electrical systems is as little as 1-3/8 in. Most systems can easily be included in a 2 to 4 inch thick slab. Voids in hollow-core slabs can also be used as electrical raceways (see Fig. 9.11.2).

When the system is placed in a structural composite slab, the effect of ducts and conduits must be carefully examined and their location coordinated with reinforcing steel. Tests on slabs with buried ductwork have shown that structural strength is not normally impaired by the voids.

Because of the high load-carrying capacity of prestressed concrete members, it is possible to locate high-voltage substations, with heavy transformers, near the areas of consumption with little or no additional expense. For extra safety, distri-

bution feeds can also be run within those channels created by stemmed members. Such measures also aid the economy of the structure by reducing the overall story height and minimizing maintenance expenses.

9.11.4 Ductwork

The designer may also utilize the space within stemmed members or the holes inside hollow-core slabs for distribution ducts for heating, air-conditioning, or exhaust systems. In stemmed members three sides of the duct are provided by the bottom of the flange and the sides of the stems. The bottom of the duct is completed by attaching a metal panel to the tee stems in the same fashion as the lighting diffusers (see Fig. 9.11.1).

Fig. 9.11.1 Metal panels attached at the bottoms of the stems create ducts, and diffuser panels provide a flush ceiling

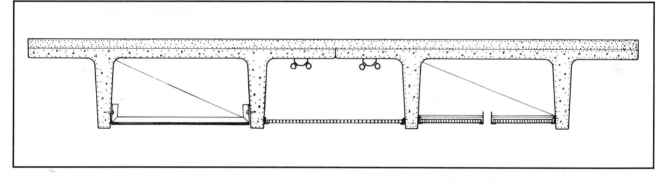

Fig. 9.11.2 Underfloor electrical ducts can be embedded within a concrete topping

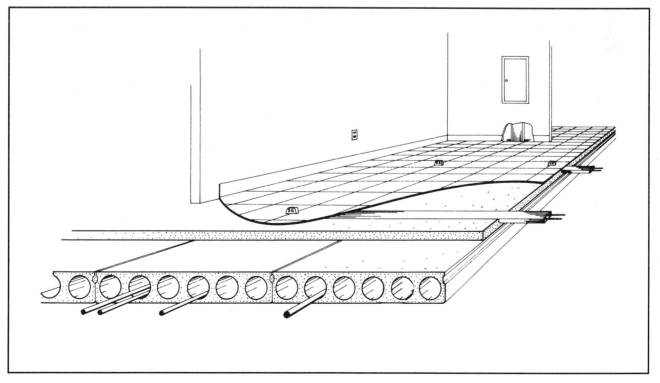

Connections can be made by several means, among them powder-activated fasteners, cast-in inserts or reglets. Field installed devices generally offer the best economy and ensure placement in the exact location where the connecting devices will be required. Inserts should only be cast-in when they can be located in the design stage of the job, well in advance of casting the precast members.

With hollow-core slabs, additional duct work can be eliminated. These members have oval, round, or rectangular voids of varying size which can provide ducts or raceways for the various systems. Openings core drilled in the field can provide access and distribution. The voids in the slabs are aligned and connected to provide continuity of the system. Openings can also be provided in intermediate supporting beams such as inverted tees, to allow duct continuity.

If high velocity air movement is utilized, the enclosed space becomes a long plenum chamber with uniform pressure throughout its length. Diffusers are installed in the ceiling to distribute the air. Branch runs, when required, can be standard ducts installed along the column lines.

When ceilings are required, proper selection of precast components can result in shallow ceiling spaces as shown in Fig. 9.11.3. This figure also illustrates the flexibility of space arrangements possible with long span prestressed concrete members.

Branch ducts of moderate size can also be accommodated by providing block-outs in the stems of tees or beams. To achieve best economy and performance in prestressed concrete members, particularly stemmed members, such block-outs should be repeated in size and location to handle all conditions demanded by mechanical, electrical, or plumbing runs. While this may lead to slightly larger openings in some cases, the end result will probably be more economical. It should also be noted that sufficient tolerance should also be allowed in sizing the openings to provide for necessary field assembly considerations (see Part 8).

Prestressed concrete box girders have been used to serve a triple function as air conditioning distribution ducts, conduit for utility lines and structural supporting members for the roof deck units. Conditioned air can be distributed within the void area of the girders and then introduced into the building work areas through holes cast into the sides and bottoms of the box girders. The system is balanced by plugging selected holes.

Vertical supply and return air trunks can be carried in the exterior walls, with only small ducts needed to branch out into the ceiling space. In some cases the exterior wall cavities are replaced with three or four sided precast boxes stacked to provide vertical runs for the mechanical and electrical systems. These stacked boxes can also be used as columns or lateral bracing elements for

Fig. 9.11.3 Where ceilings are required, ducts, piping, and lighting fixtures can be accommodated within a shallow depth

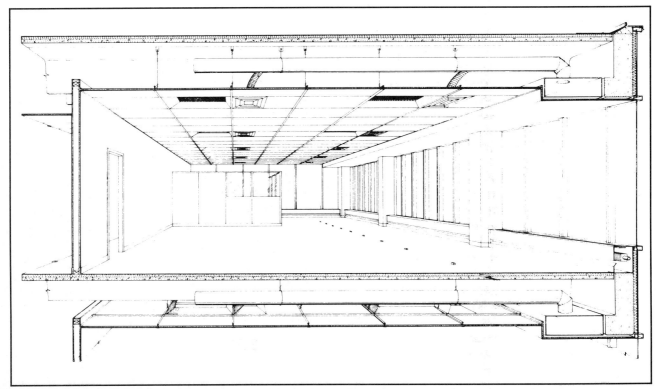

Fig. 9.11.4 Large openings in floors and roofs are made during manufacture of the units; small openings are field drilled. Some common types of openings are shown here

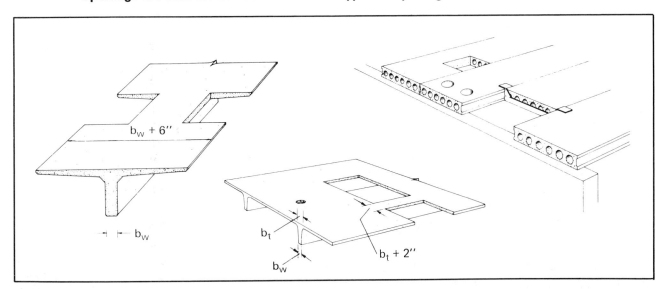

the structure.

In some cases it may be required to provide openings through floor and roof units. Large openings are usually made by block-outs in the forms during the manufacture; smaller ones (up to about 8 in.) are usually field drilled. Openings in flanges of stemmed members should be limited to the "flat" portion of the flange, that is, beyond 1 in. of the edge of the stem on double tees and 3 in. of the edge of the stem on single tees. Angle headers are often used for framing large openings in hollow-core floor or roof systems (see Fig. 9.11.4).

9.11.5 Other Sub-Systems

Suspended ceilings, crane rails, and other sub-systems can be easily accommodated with standard manufactured hardware items and embedded plates as shown in Fig. 9.11.5.

Architectural precast concrete wall panels can be adapted to combine with preassembled window or door units. Door or window frames properly braced to prevent bowing during concrete placement, can be cast in the panels and then the glazing or doors can be installed prior to or after delivery to the job site. If the glazing or doors are properly protected, they can also be cast into the panel at the plant. When casting-in aluminum window frames, particular attention should be made to properly coat the aluminum so that it will not react with the concrete. It should be noted that repetition is one of the real keys to economy in a precast concrete wall assembly. Windows and doors should be located in identical places for all panels wherever possible.

Insulated wall panels can be produced by embedding an insulating material such as expanded polyurethane between layers of concrete. These "sandwich panels" are described in detail in Sect. 9.4. Sandwich panels are normally cast on flat beds or tilting tables. The inside surface of the concrete panel can be given a factory troweled finish followed by minor touch up work. The interior face is completed by painting, or by wall papering to achieve a finished wall. The formed surface of the panel can also be treated in a similar manner when used as the interior face.

9.11.6 Systems Building

As more and more complete systems buildings are built with precast and prestressed concrete, and as interest in this method of construction increases, we can expect that more of the building sub-systems will be prefabricated and pre-coordinated with the structure.

This leads to the conclusion that those parts of the structure that require the most labor skills should logically be prefabricated prior to installation in the field. The prefabricated components can be preassemblies of basic plumbing systems or electrical/mechanical systems plus lighting.

For housing systems, electrical conduits and boxes can be cast in the precast wall panels. This process requires coordination with the electrical contractor; and savings on job-site labor and time are possible. The metal or plastic conduit is usually pre-bent to the desired shape and delivered to the casting bed already connected to the electrical boxes. It is essential that all joints and connections be thoroughly sealed and the boxes

Fig. 9.11.5 Methods of attaching suspended ceilings, crane rails, and other sub-systems

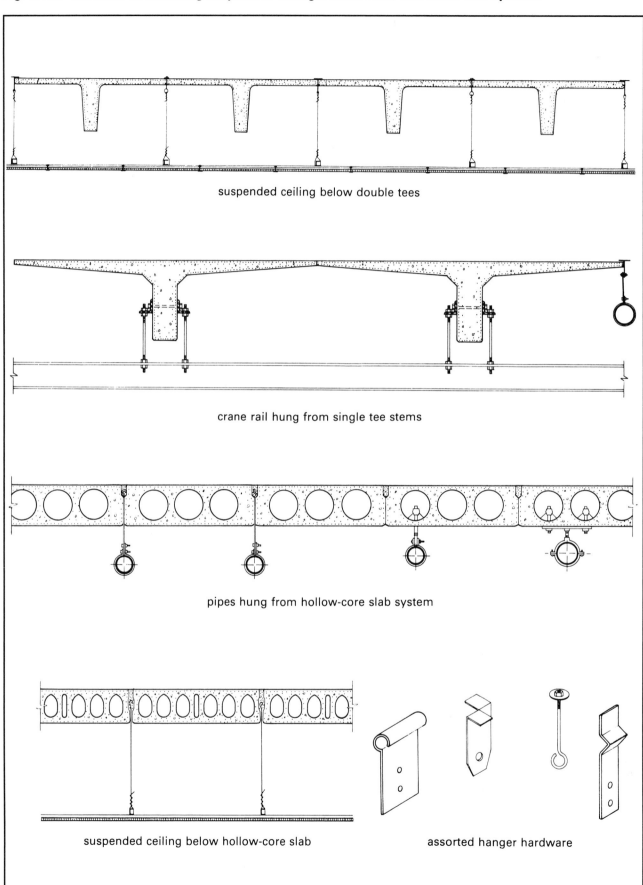

suspended ceiling below double tees

crane rail hung from single tee stems

pipes hung from hollow-core slab system

suspended ceiling below hollow-core slab

assorted hanger hardware

enclosed prior to casting in order to prevent the system from becoming clogged. The wires are usually pulled through at the job site. Television antennae and telephone conduits have also been cast-in using the same procedure.

To reduce on-site labor, prefabricated bathroom units or combination bathroom/kitchen modules have been developed (see Fig. 9.11.6). Such units include bathroom fixtures, kitchen cabinets and sinks, as well as wall, ceiling, and floor surfaces. Plumbing units are often connected and assembled prior to delivery to the job sites. These bathroom/kitchen modules can be molded plastic units or fabricated from drywall components. To eliminate a double floor, the module can be plant built on the structural member or the walls of the unit can be designed strong enough for all fixtures to be wall hung. In the latter case, the units

Fig. 9.11.6 Kitchen/bathroom modules can be pre-assembled on precast prestressed slabs ready for installation in systems buildings

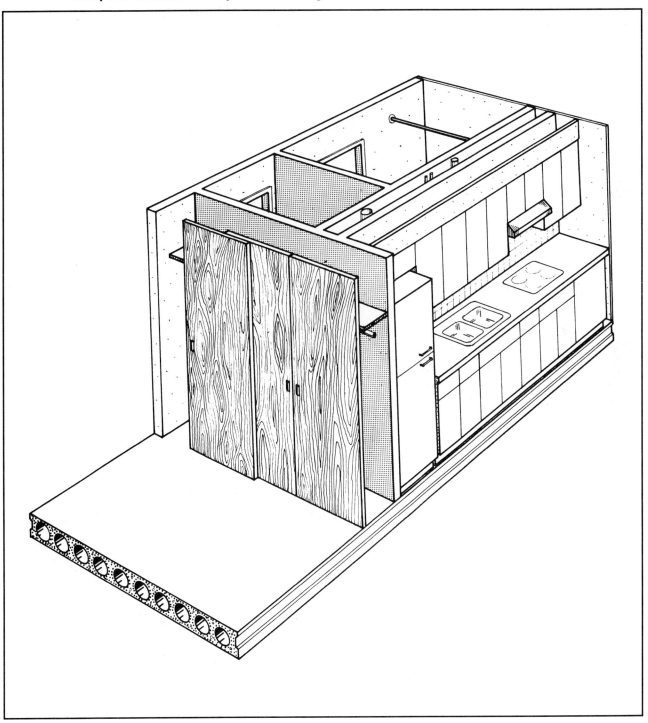

are placed directly on a precast floor and in multi-story construction are located in a stack fashion with one bathroom directly over the one below. A block-out for a chase is provided in the precast floor and connections are made from each unit to the next to provide a vertical plumbing stack. Prefabricated wet-wall plumbing systems, as shown in Fig. 9.11.7, incorporate preassembled piping systems using snap-on or no-hub connections made up of a variety of materials. These units only require a block-out in the prestressed flooring units and are also arranged in a stack fashion. Best economy results when bathrooms are backed up to each other, since a common vertical run can service two bathrooms.

Some core modules not only feature bath and kitchen components, but also HVAC components all packaged in one unit. These modules can also be easily accommodated in prestressed structural systems by placing them directly on the prestressed members with shimming and grouting as required.

Fig. 9.11.7 Prefabricated wet-wall plumbing systems incorporate pre-assembled piping

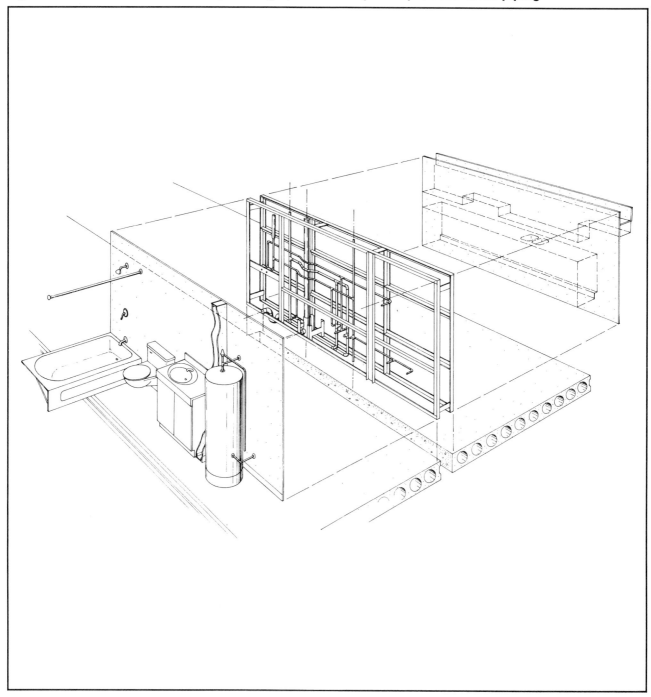

PCI Design Handbook

PART 10
SPECIFICATIONS AND STANDARD PRACTICES

	Page No.
10.1 Guide Specification for Precast, Prestressed Concrete	10–2
10.2 Guide Specification for Architectural Precast Concrete	10–11
10.3 Code of Standard Practice for Precast Concrete	10–26
10.4 Inspection Checklist for Architectural Precast Concrete	10–36

GUIDE SPECIFICATION FOR PRECAST, PRESTRESSED CONCRETE

This Guide Specification is intended to be used as a basis for the development of an office master specification or in the preparation of specifications for a particular project. In either case, this Guide Specification must be edited to fit the conditions of use.

Particular attention should be given to the deletion of inapplicable provisions. Necessary items related to a particular project should be included. Also, appropriate requirements should be added where blank spaces have been provided.

The Guide Specifications are on the left. *Notes to Specifiers are on the right.*

GUIDE SPECIFICATIONS

NOTES TO SPECIFIERS

1. GENERAL

1.01 Description

A. Work included:

1. These specifications cover precast and precast, prestressed structural concrete construction, including product design not shown on contract drawings, manufacture, transportation, erection, and other related items such as anchorage, bearing pads, storage and protection of precast concrete.

B. Related work specified elsewhere:

1. Cast-in-place concrete: Section _____ .
2. Precast architectural concrete: Section _____ .
3. Post-tensioning: Section _____ .
4. Masonry bearing walls: Section _____ .
5. Miscellaneous steel: Section _____ .
6. Waterproofing: Section _____ .
7. Flashing and sheet metal: Section _____ .
8. Sealants and caulking: Section _____ .
9. Painting: Section _____ .
10. Holes for other trades: Sections _____ .

C. Work installed but furnished by others:

1. Receivers or reglets for flashing: Section _____ .

2. Elevator guides: Section _____ .

1.01.A. This Section is to be in Division 3 of Construction Specifications Institute format.
Verify that plans clearly differentiate between this work and architectural precast concrete if both are on the same job. One may need to list items such as beams, purlins, girders, lintels, columns, slab or deck members, etc. Some items such as prestressed wall panels could be included in either this or the Specifications for Architectural Precast Concrete, depending on the desired finish.

1.02 Quality Assurance

A. Manufacturer qualifications:
The precast concrete manufacturing plant shall be certified by the Prestressed Concrete Institute, Plant Certification Program, prior to the start of production.

* * OR * *

The manufacturer shall, at his expense, meet the following requirements:

1. Retain an independent testing or consulting firm approved by the architect/engineer and/or owner.

2. The basis of inspection shall be the Prestressed Concrete Institute *Manual for Quality Control for Plants and Production of Precast and Prestressed Concrete Products*, MNL-116 and *Manual for Quality Control for Plants and Production of Architectural Precast Concrete Products*, MNL-117.

3. This firm shall inspect the precast plant at two-week intervals during production and issue a report, certified by a registered engineer verifying that materials, methods, products and quality control meet all the requirements of the specifications, drawings, and MNL-116 and/or MNL-117. If the report indicates to the contrary, the architect/engineer, at the precaster's expense, will inspect and, at architect's option, may reject any or all products produced during the period of non-compliance with the above requirements.

* * * * * *

B. Erector qualifications:
Regularly engaged for at least _____ years in the erection of precast structural concrete similar to the requirements of this project.

C. Welder qualifications:
In accordance with AWS D1.1.

D. Testing:
In general compliance with testing provisions in MNL-116, *Manual for Quality Control for Plants and Production of Precast and Prestressed Concrete Products.*

E. Requirements of regulatory agencies:
All local codes plus the following specifications, standards and codes are a part of these specifications:

1.02.B. Usually 2 to 5 years.

1.02.C Qualified within the past year.

1.02.E. Always include the specific year or edition of the specifications, codes and standards used in the design of the project and made part of the specifications.

1. ACI 318 – Building Code Requirements for Reinforced Concrete.
2. AWS D1.1 – Structural Welding Code – Steel.
3. AWS D1.4 – Structural Welding Code – Reinforcing Steel.
4. ASTM Specifications – As referred to in Part 2 – Products, of this Specification.
5. AASHTO Standard Specifications for Highway Bridges.

1.03 Submittals

A. Shop drawings:

  1. Erection drawings:

    a. Plans and/or elevations locating and defining all material furnished by manufacturer.

    b. Sections and details showing connections, cast-in items and their relation to the structure.

    c. Description of all loose, cast-in and field hardware.

    d. Field installed anchor location drawings.

    e. Erection sequences and handling requirements.

    f. All dead, live and other applicable loads used in the design.

  2. Production drawings:

    a. Elevation view of each member.

    b. Sections and details to indicate quantities and position of reinforcing steel, anchors, inserts, etc.

    c. Lifting and erection inserts.

    d. Dimensions and finishes.

    e. Prestress for strand and concrete strengths.

    f. Estimated cambers.

    g. Method of transportation.

B. Product design criteria:

  1. Loadings for design:

    a. Initial handling and erection stresses.

    b. All dead and live loads as specified on the contract drawings.

For projects in Canada, standards from the National Building Code of Canada should be listed in addition to, or in place of the U.S. Standards. Fire ratings are generally a code requirement. When required, fire-rated products shall be clearly identified on the design drawings.

1.03.A.2 Production drawings are normally submitted only upon request.

1.03.B and C. The contract drawings normally will be prepared using a local precast, prestressed concrete manufacturer's design data and load tables. Dimensional changes that would not materially affect architectural and structural properties or details usually are permissible.

GUIDE SPECIFICATIONS	NOTES TO SPECIFIERS

GUIDE SPECIFICATIONS

c. All other loads specified for member, where applicable.

2. Design calculations of products not completed on the contract drawings shall be performed by a registered engineer experienced in precast, prestressed concrete design and submitted for approval upon request.

3. Design shall be in accordance with applicable codes, ACI 318, or AASHTO Standard Specifications for Highway Bridges.

C. Permissible design deviations:

1. Design deviations will be permitted only after the architect/engineer's written approval of the manufacturer's proposed design supported by complete design calculations and drawings.

2. Design deviations shall provide an installation equivalent to the basic intent without incurring additional cost to the owner.

NOTES TO SPECIFIERS

Most precast, prestressed concrete is cast in continuous steel forms, therefore connection devices on the formed surfaces must be contained within the member since penetration of the form is impractical.

Camber will generally occur in prestressed concrete members having eccentricity of the stressing force. If camber considerations are important, check with local prestressed concrete manufacturer to secure estimates of the amount of camber and of camber movement with time and temperature change.

Architectural details must recognize the existence of camber and camber movement in connection with:

1. *Closures to interior non-load bearing partitions.*

2. *Closures parallel to prestressed concrete members (whether masonry, windows, curtain walls or others) must be properly detailed for appearance.*

3. *Floor slabs receiving cast-in-place topping. The elevation of top of floor and amount of concrete topping must allow for camber of prestressed concrete members.*

Design cambers less than obtained under normal design practices are possible but this usually requires the addition of tendons or non-prestressed steel reinforcement and price should be checked with the local manufacturer.

As the exact cross section of precast, prestressed members might vary somewhat from producer to producer, permissible deviations in member shape from that shown on the contract drawings might enable more manufacturers to quote on the project. Manufacturing procedures also vary between plants and permissible modifications to connection details, inserts, etc., will allow the manufacturer to use devices he can best adapt to his manufacturing procedure.

Be sure that loads shown on the contract drawings are easily interpreted. For instance, on members which are to receive concrete topping, be sure to state whether all superimposed dead and live loads on precast, prestressed members do or do not include the weight of the concrete topping.

It is best to list the live load, superimposed dead load, topping weight, and weight of the member, all as separate loads. Where there are two different live loads (e.g., roof level of a parking structure) indicate how they are to be combined.

D. Test reports:
Reports of tests on concrete and other materials upon request.

2. PRODUCTS

2.01 Materials

A. Portland cement:

    1. ASTM C150 – Type I or III.

B. Admixtures:

    1. Air-entraining admixtures: ASTM C260.

    2. Water reducing, retarding, accelerating, high range water reducing admixtures: ASTM C494.

C. Aggregates:

    1. ASTM C33 or C330.

D. Water:

Potable or free from foreign materials in amounts harmful to concrete and embedded steel.

E. Reinforcing steel:

    1. Bars:

    Deformed billet-steel: ASTM A615.

    Deformed rail-steel: ASTM A616.

    Deformed axle-steel: ASTM A617.

    Deformed low-alloy steel: ASTM A706.

    2. Wire:

    Cold-drawn steel: ASTM A82.

    3. Wire fabric:

    Welded steel: ASTM A185.

    Welded deformed steel: ASTM A497.

F. Strand:

    1. Uncoated, 7-wire, stress-relieved strand: ASTM A416 (including supplement) – Grade 250K or 270K.

G. Anchors and inserts:

    1. Materials:

        a. Structural steel: ASTM A36.

        b. Malleable iron.

        c. Stainless steel: ASTM A666.

    2. Finish:

        a. Shop primer: Manufacturer's standards.

2.01. Delete or add materials that may be required for the particular job.

2.01.E.1. When welding of bars is required, weldability must be established to conform to AWS D1.4.

2.01.G.1.b. Usually specified by type and manufacturer.

b. Hot dipped galvanized: ASTM A153.

c. Zinc-rich coating: MIL-P-2135, self curing, one component, sacrificial.

d. Cadmium coating.

H. Grout:

   1. Cement grout: Portland cement, sand, and water sufficient for placement and hydration.

   2. Non-shrink grout: Premixed, packaged ferrous and non-ferrous aggregate shrink-resistant grout.

   3. Epoxy-resin grout: Two-component mineral-filled epoxy-polysulfide, FS MMM-G-560 _____, Type _____, Grade C.

I. Bearing pads:

   1. Chloroprene (Neoprene): Conform to Division II, Section 25 of AASHTO Standard Specifications for Highway Bridges.

   2. Random oriented fiber reinforced: Shall support a compressive stress of 3000 psi with no cracking, splitting or delaminating in the internal portions of the pad. One specimen shall be tested for each 200 pads used in the project.

   3. Duck layer reinforced pad: Conform to Division II, Section 10.3.12 of AASHTO Standard Specifications for Highway Bridges or Military Specification MIL-C-882C.

   4. Plastic: Multi-monomer plastic strips shall be non-leaching and support construction loads with no visible overall expansion.

   5. Tetraflouroethylene (TFE) reinforced with glass fibers and applied to stainless or structural steel plates.

J. Welded studs:

   1. AWS D1.1.

2.02 Concrete Mixes

A. 28-day compressive strength: Minimum of _____psi.

2.01.H. *Indicate required strengths on contract drawings.*

2.01.H.2. *Non-ferrous grouts with a gypsum base should not be exposed to moisture. Ferrous grouts should not be used where possible staining would be undesirable.*

2.01.H.3. *Check with local suppliers to determine availability and types of epoxy-resin grouts.*

2.01.I.1 *AASHTO grade pads utilize 100 percent chloroprene as the elastomer. Less expensive commercial grade pads are available, but are not recommended.*

2.01.I.2 *Standard guide specifications are not available for random-oriented, fiber-reinforced pads. Proof testing of a sample from each group of 200 pads is suggested. Normal design working stresses are 1500 psi, so the 3000 psi test load provides a factor of 2 over design stress. The shape factor for the test specimens should not be less than 2.*

2.01.I.4 *Plastic pads are widely used with concrete plank. Compression stress in use is not normally over a few hundred psi and proof testing is not considered necessary. No standard guide specifications are available.*

2.01.I.5 *ASTM D2116 applies only to basic TFE resin molding and extrusion material in powder or pellet form. Physical and mechanical properties must be specified by naming manufacturer or other methods.*

2.02.A. and B. *Verify with local manufacturer. 5000 psi for prestressed products is normal prac-*

B. Release strength: Minimum of _____psi.

C. Use of calcium chloride, chloride ions or other salts is not permitted.

2.03 Manufacture

A. Manufacturing procedures shall be in general compliance with PCI MNL-116.

B. Manufacturing tolerances shall comply with PCI MNL-116.

C. Finishes:

1. Standard underside: Resulting from casting against approved forms using good industry practice in cleaning of forms, design of concrete mix, placing and curing. Small surface holes caused by air bubbles, normal color variations, normal form joint marks, and minor chips and spalls shall be tolerated, but no major or unsightly imperfections, honeycomb, or other defects shall be permitted.

2. Standard top: Result of vibrating screed and additional hand finishing at projections. Normal color variations, minor indentations, minor chips and spalls shall be permitted. No major imperfections, honeycomb, or defects shall be permitted.

3. Exposed vertical ends: Strands shall be recessed and the ends of the member shall receive sacked finish.

4. Special finish: If required, listed as follows:

D. Openings:
Primarily on thin sections, the manufacturer shall provide for those openings 10 in. round or square or larger as shown on the structural drawings. Other openings shall be located and field drilled or cut by the trade requiring them after the precast, prestressed products have been erected. Openings shall be approved by architect/engineer before drilling or cutting.

tice, with release strength of 3500 psi.

2.03.C. Other formed finishes that may be specified are:
Commercial Finish. Concrete may be produced in forms that impart a texture to the concrete (e.g., plywood or lumber). Fins and large protrusions shall be removed and large holes shall be filled. All faces shall have true, well-defined surfaces. Any exposed ragged edges shall be corrected by rubbing or grinding.
Architectural Grade B Finish. All air pockets and holes over 1/4 in. in diameter shall be filled with a sand-cement paste. All form offsets or fins over 1/8 in. shall be ground smooth.
Architectural Grade A Finish. In addition to the requirements for Architectural Grade B Finish, all exposed surfaces shall be coated with a neat cement paste using an acceptable float. After thin pastecoat has dried, the surface shall be rubbed vigorously with burlap to remove loose particles. These requirements are not applicable to extruded products using zero-slump concrete in their process.

2.03.C.4. Special finishes, if required, should be described in this section of the specifications and noted on the contract drawings, pointing out which members require special finish. Such finishes will involve additional cost and consultation with the manufacturer is recommended. A sample of such finishes should be made available for review prior to bidding.

2.03.D This paragraph requires other trades to field drill holes needed for their work, and such trades should be alerted to this requirement through proper notation in their sections of the specifications. Some manufacturers prefer to install openings smaller than 10 in. which is acceptable if their locations are properly identified on the contract drawings.

E. Patching:
Shall be acceptable providing the structural adequacy of the product and the appearance are not impaired.

F. Fasteners:
Manufacturer shall cast in structural inserts, bolts and plates as detailed or required by the contract drawings.

2.03.F. Exclude this requirement from extruded sections.

3. EXECUTION

3.01 Product Delivery, Storage, and Handling

A. Delivery and handling:

1. Precast concrete members shall be lifted and supported during manufacturing, stockpiling, transporting and erection operations only at the lifting or supporting point, or both, as shown on the shop drawings, and with approved lifting devices. Lifting inserts shall have a minimum safety factor of 4. Exterior lifting hardware shall have a minimum safety factor of 5.

2. Transportation, site handling, and erection shall be performed with acceptable equipment and methods, and by qualified personnel.

B. Storage:

1. Store all units off ground.
2. Place stored units so that identification marks are discernible.
3. Separate stacked members by battens across full width of each bearing point.
4. Stack so that lifting devices are accessible and undamaged.
5. Do not use upper member of stacked tier as storage area for shorter member or heavy equipment.

3.02 Erection

A. Site access:
General contractor shall be responsible for providing suitable access to the building, proper drainage and firm, level bearing for the hauling and erection equipment to operate under their own power.

B. Preparation:
General contractor shall be responsible for:

1. Providing true, level bearing surfaces on all field placed bearing walls and other field placed supporting members.

2. Placement and accurate alignment of anchor bolts, plates or dowels in column footings, grade beams and other field placed supporting members.

3. All shoring required for composite beams and slabs. Shoring shall have a minimum load factor of 1.5 × (dead load plus construction loads).

3.02.B. Construction tolerances for cast-in-place concrete, masonry, etc., should be specified in those sections of the specifications.

C. Installation:
Installation of precast, prestressed concrete shall be performed by the manufacturer or a competent erector. Members shall be lifted by means of suitable lifting devices at points provided by the manufacturer. Temporary shoring and bracing, if necessary, shall comply with manufacturer's recommendations.

D. Alignment:
Members shall be properly aligned and leveled as required by the approved shop drawings. Variations between adjacent members shall be reasonably leveled out by jacking, loading, or any other feasible method as recommended by the manufacturer and acceptable to the architect/engineer.

3.02.D. The following erection tolerance may be specified if other requirements do not control: Individual pieces are considered plumb, level and aligned if the error does not exceed 1:500 excluding structural deformations caused by loads.

3.03 Field Welding

A. Field welding is to be done by qualified welders using equipment and materials compatible to the base material.

3.04 Attachments

A. Subject to approval of the architect/engineer, precast, prestressed products may be drilled or "shot" provided no contact is made with the prestressing steel. Should spalling occur, it shall be repaired by the trade doing the drilling or the shooting.

3.05 Inspection and Acceptance

A. Final inspection and acceptance of erected precast, prestressed concrete shall be made by the architect/engineer.

GUIDE SPECIFICATION FOR
ARCHITECTURAL PRECAST CONCRETE

This Guide Specification is intended to be used as a basis for the development of an office master specification or in the preparation of specifications for a particular project. In either case, this Guide Specification must be edited to fit the conditions of use.

Particular attention should be given to the deletion of inapplicable provisions. Necessary items related to a particular project should be included. Also, appropriate requirements should be added where blank spaces have been provided.

The Guide Specifications are on the left. *Notes to Specifiers are on the right.*

GUIDE SPECIFICATIONS

NOTES TO SPECIFIERS

1. GENERAL

1.01 Description

A. Related work specified elsewhere:

1. Concrete reinforcement: Section _____ .

    1.01.A.1 Architectural precast concrete steel requirements are different from cast-in-place reinforcement and should be specified in this section.

2. Cast-in-place concrete: Section _____.

    1.01.A.2 Placement of anchorage devices in cast-in-place concrete for precast panels.

3. Precast, prestressed concrete: Section _____ .

    1.01.A.3 Precast floor, roof slabs, beams, columns and other structural elements. Some items, such as prestressed wall panels on industrial buildings, could be included in either specification, depending on the desired finish.

4. Structural steel framing: Section _____ .

    1.01.A.4 Steel supporting structure and sometimes loose anchors.

5. Dampproofing: Section _____ .

    1.01.A.5 For exposed face of panels.

6. Insulation: Section _____ .

    1.01.A.6 Insulation job-applied to precast panels.

7. Flashing and sheet metal: Section _____ .

    1.01.A.7 Counterflashing inserts and receivers, unless included in this section.

8. Sealants and caulking: Section _____ .

    1.01.A.8 Panel joint caulking and sealing.

9. Painting: Section _____ .

    1.01.A.9 Field touch-up painting. Delete when specified in this section.

B. Work installed but furnished by others:

    1. Counterflashing receivers or reglets: Section _____ .

    2. Inserts or attachments for _____ : Section _____ .

C. Testing agency provided by owner.

1.02 Quality Assurance

A. Manufacturer qualifications:
The precast concrete manufacturing plant shall be certified by the Prestressed Concrete Institute, Plant Certification Program, prior to the start of production.

* * OR * *

The manufacturer shall, at his expense, meet the following requirements:

    1. Retain an independent testing or consulting firm approved by the architect/engineer and/or owner.

    2. The basis of inspection shall be the Prestressed Concrete Institute *Manual for Quality Control for Plants and Production of Precast and Prestressed Concrete Products*, MNL-116 and *Manual for Quality Control for Plants and Production of Architectural Precast Concrete Products*, MNL-117.

    3. This firm shall inspect the precast plant at two-week intervals during production and issue a report, certified by a registered engineer verifying that materials, methods, products and quality control meet all the requirements of the specifications, drawings, and MNL-116 and/or MNL-117. If the report indicates to the contrary, the architect/engineer, at the precaster's expense, will inspect and, at architect's option, may reject any or all products produced during the period of non-compliance with the above requirements.

* * OR * *

Acceptable manufacturers:

    1. _____ .

    2. _____ .

    3. _____ .

* * * * * *

1.01.B Delete when furnished by precast manufacturer. Add additional items as may be required for the particular project.

1.01.B.2 May include inserts/attachments for window/door frames, window washing equipment, etc.

1.01.C Delete when testing agency is provided by precast manufacturer or contractor. Coordinate with Division 1, General Conditions.

B. Erector qualifications:
Regularly engaged for at least _____years in erection of architectural precast concrete units similar to those required on this project.

1.02.B   *Usually 2 to 5 years.*

C. Welder qualifications:
In accordance with AWS D1.1.

1.02.C   *Qualified within the past year. Delete when welding is not required.*

D. Testing:
In general compliance with testing provisions in MNL-117, *Manual for Quality Control for Plants and Production of Architectural Precast Concrete Products.*

E. Testing agency:

1. Not less than _____ years experience in performing concrete tests of type specified in this section.

2. Capable of performing testing in accordance with ASTM E 329.

3. Inspected by Cement and Concrete Reference Laboratory of the National Bureau of Standards.

1.02.E   *Delete when provided by Owner.*

1.02.E.1   *Usually 2 to 5 years.*

F. Requirements of regulatory agencies:
Manufacture and installation of architectural precast concrete to meet requirements of ___

_____

1.02.F   *Local building code or other governing code relating to precast concrete. For projects in Canada, standards from the National Building Code of Canada should be listed in addition to or in place of the U.S. Standards.*

G. Allowable tolerances:

1. Manufacture and install wall panels so that each panel after erection complies with the dimensional tolerances listed in MNL 117.

H. Job mock-up:

1. After standard samples are accepted for color and texture, submit full-scale unit meeting design requirements.

1.02.G   *For manufacturing, and after manufacturing. The tolerances listed in PCI MNL 117 are also listed in Part 8 of this manual. Most manufacturers can meet closer tolerances, if required, but closer tolerances normally increase costs. The normal tolerances of the support system should also be recognized.*

1.02.H.1   *Full-scale samples or inspection of the first production unit are sometimes required, especially when a new design concept or new manufacturing process or other unusual circumstance indicates that proper evaluation cannot otherwise be made. It is difficult to assess appearance from small samples.*

2. Mock-up to be standard of quality for architectural precast concrete work, when accepted by architect/engineer.

3. Incorporate mock-up into work in location reviewed by architect/engineer after keeping unit in plant _____ for checking purpose.

1.02.H.2   *Use to determine range of acceptability with respect to color and texture variations, surface defects and overall appearance.*

1.02.H.3   *Delete when mock-up is not to be included in work. State how long unit should be kept.*

I. Source quality control:

1. Quality control and inspection procedures to comply with applicable sections of MNL 117.

2. Water absorption test on unit shall be conducted in accordance with MNL 117.

1.02.I.2 Water absorption test is an early indication of weather staining (rather than durability). Verify the water absorption of the proposed face mix, which for average exposures and based upon normal weight concrete (150 lbs per cubic foot) should not exceed 5% to 6% by weight. As an improved weathering (staining) precaution, lower absorption between 3% to 4% (by weight) is feasible with some concrete mixes and consolidation methods. In order to establish comparable absorption figures for all materials the current trend is to specify absorption percentages by volumes. The stated limits for absorption would in volumetric terms correspond to 12% to 14% for average exposures and 8% to 10% for special conditions.

1.03 Submittals

A. Samples:

1.03.A Number of samples and submittal procedures should be specified in Division 1.

1. Submit samples representative of finished exposed face showing typical range of color and texture prior to commencement of manufacture.

1.03.A.1 If the back face of a precast unit is to be exposed, samples of the workmanship, color, and texture of the backing should be shown as well as the facing.

2. Sample size: approximately 12 in. × 12 in. and of appropriate thickness, representative of the proposed finished product.

B. Shop drawings:

1. Content:

a. Unit shapes (elevations and sections), and dimensions.

b. Finishes.

c. Reinforcing, joint, and connection details.

d. Lifting and erection inserts.

e. Location and details of hardware attached to structure.

f. Other items cast into panels.

g. Handling procedures and sequence of erection for special conditions.

h. Relationship to adjacent material.

2. Show location of unit by same identification mark placed on panel.

1.03.B When erection drawings contain all information sufficient for design approval, production drawings, except for shape drawings, need not be submitted for approval, except in special cases. However, record copies are frequently requested. Guidelines for the preparation of drawings are given in the "PCI Architectural Precast Concrete Drafting Handbook".

1.03.B.1.g If sequence of erection is critical to the structural stability of the structure, or for access to connections at certain locations, it should be noted on the contract plans and specified.

C. Test reports:
Submit, on request, reports on materials, compressive strength tests on concrete and water absorption tests on units.

D. Design calculations:
Submit, on request, structural design calculations.

1.04 Product Delivery, Storage, and Handling

A. Delivery and handling:

1. Handle and transport units in a position consistent with their shape and design in order to avoid stresses which would cause cracking or damage.

2. Lift or support units only at the points shown on the shop drawings.

3. Place nonstaining resilient spacers of even thickness between each unit.

4. Support units during shipment on nonstaining shock-absorbing material.

5. Do not place units directly on ground.

B. Storage at jobsite:

1. Store and protect units to prevent contact with soil, staining, and physical damage.

2. Store units, unless otherwise specified, with nonstaining, resilient supports located in same positions as when transported.

3. Store units on firm, level, and smooth surfaces.

4. Place stored units so that identification marks are discernible, and so that product can be inspected.

2. PRODUCTS

2.01 Materials

A. Concrete:

1. Portland cement:

    a. ASTM C 150, type _____ , _____ color.

    b. For exposed surfaces use same brand, type, and source of supply throughout.

2. Air entraining agent: ASTM C 260.

2.01.A.1.a Type: [I(General use)], [III(High early strength)]. Color: (gray), (white), (buff).

2.01.A.1.b To minimize color variation. Specify source of supply when color shade is important.

2.01.A.2 Delete if air entrainment is not required.

3. Water reducing, retarding, accelerating, high range water reducing admixtures: ASTM C 494.

2.01.A.3 Delete if water reducing, retarding, or accelerating admixtures are not required. Calcium chloride, or admixtures containing significant amounts of calcium chloride, should not be allowed.

4. Coloring agent:

   a. Synthetic mineral oxide.

   b. Harmless to concrete set and strength.

   c. Stable at high temperature.

   d. Sunlight and alkali-fast.

2.01.A.4 Investigate use of naturally colored fine aggregate in lieu of coloring agent.

2.01.A.4.b Consider effects upon concrete prior to final selection.

5. Aggregates:

   a. Provide fine and coarse aggregates for each type of exposed finish from a single source (pit or quarry) for entire job. They shall be clean, hard, strong, durable, and inert, free of staining or deleterious material.

   b. ASTM C 33 or C 330.

   c. Material and color: _____.

   d. Maximum size and gradation: _____ .

2.01.A.5.a Base choice on visual inspection of concrete sample and on assessment of certified test reports. Use same type and source of supply to minimize color variation. Fine aggregate is not always from same source as coarse aggregate.

2.01.A.5.b Grading requirements are generally waived or modified.

2.01.A.5.c Specify type of stone desired such as crushed marble, quartz, limestone, granite, or locally available gravel. Some lightweight aggregates, limestones, and marbles may not be acceptable as facing aggregates. Omit where sample is to be matched.

2.01.A.5.d State required sieve analysis. Omit where sample is to be matched.

6. Water: Free from deleterious matter that may interfere with the color, setting or strength of the concrete.

2.01.A.6 Potable water is ordinarily acceptable.

B. Reinforcing steel:

   1. Material:

2.01.B.1 Grades of reinforcing are determined by the structural design of the precast units. Panels are designed as crack free sections so benefit of higher grade steel is not utilized.

   a. Bars:

2.01.B.1.a State plain or galvanized. Use galvanizing only where corrosive environment or severe exposure conditions justify extra cost. Availability of galvanized bars should be verified.

   (1) Deformed steel: ASTM A 615, grade _____ .

2.01.B.1.a.(1) Grades 40 or 60. Grade 60 is preferred.

(2) Weldable deformed steel: ASTM A 706.

2.01.B.1.a(2)   Availability should be checked. When not available establish weldability in accordance with AWS D1.4.

b. Wire fabric:

(1) Welded steel: ASTM A 185.

(2) Welded deformed steel: ASTM A 497.

2.01.B.1.b   Should be sheets, not rolls. State plain or galvanized. Use galvanizing where corrosive environment or exposure conditions justify extra cost.

c. Fabricated steel bar or rod mats: ASTM A 184.

d. Prestressing strand: ASTM A 416, grade _____ .

2.01.B.1.d   Occasionally used in long and/or thin panels. Grades 250 or 270.

C. Cast-in anchors:

1. Materials:

2.01.C   Loose attachment hardware usually specified under Miscellaneous Metals.

a. Carbon steel bars: ASTM A 306, grade 65.

2.01.C.1.a   For completely encased anchors.

b. Structural steel: ASTM A 36.

2.01.C.1.b   For carbon steel clip anchors.

c. Stainless steel: ASTM A 666, type 304, grade _____ .

2.01.C.1.c   Stainless steel anchors for use when resistance to staining merits extra cost. (A), (B).

d. Carbon steel plate: ASTM A 283, grade ____ .

2.01.C.1.d   (A), (B), (C), (D).

e. Malleable iron castings:

2.01.C.1.e   Usually specified by type and manufacturer.

f. Carbon steel castings: ASTM A 27, grade 60-30.

2.01.C.1.f   For cast steel casting clamps.

g. Bolts: ASTM A 307 or A 325.

2.01.C.1.g   For steel bolts, nuts and washers.

h. Welded headed studs: AWS D1.1, Chap. 4, Part F.

2. Finish:

a. Shop primer: FS TT-P-86, oil base paint, type I, or SSPC-Paint 14, or manufacturer's standard.

2.01.C.2.a   For exposed carbon steel anchors.

| GUIDE SPECIFICATIONS | NOTES TO SPECIFIERS |

GUIDE SPECIFICATIONS

    b. Hot-dip galvanized:
       ASTM A 153.

    c. Cadmium coating:
       ASTM A 165.

    d. Zinc rich coating: MIL-P-21035, self curing, one component, sacrificial organic coating.

D. Receivers for flashing: 28 ga. formed _____ _____ , or polyvinyl chloride extrusions.

E. Sandwich panel insulation: _____ .

F. Grout:

    1. Cement grout: Portland cement, sand, and water sufficient for placement and hydration.

    2. Nonshrink grout: Premixed, packaged ferrous and non-ferrous aggregate shrink-resistant grout.

    3. Epoxy-resin grout: Two-component mineral-filled epoxy-polysufide, FS MMM-G-560 _____, type _____, grade C.

G. Bearing Pads:

    1. Chloroprene (Neoprene): Conform to Division II, Section 25 of AASHTO Standard Specifications for Highway Bridges.

    2. Random oriented fiber reinforced: Shall support a compressive stress of 3000 psi with no cracking, splitting or delaminating in the internal portions of the pad. One specimen shall be tested for each 200 pads used in the project.

    3. Duck layer reinforced pad: Conform to Division II, Section 10.3.12 of AASHTO Standard Specifications for Highway Bridges or Military Specification MIL-C-882C.

NOTES TO SPECIFIERS

2.01.C.2.b For exposed carbon steel anchors where corrosive environment justifies the additional cost. Field welding should generally not be permitted on galvanized element.

2.01.C.2.c Particularly appropriate for threaded fasteners.

2.01.C.2.d For field spot painting.

2.01.D (stainless steel), (copper), (zinc). Coordinate with flashing specification to avoid dissimilar metals. Delete when included in flashing and sheet metal section. Specify whether precaster or others furnish.

2.01.E Specify type of insulation such as foamed plastic (polystyrene and polyurethane), glasses (foamed glass and fiberglass), foamed or cellular lightweight concretes, or lightweight mineral agregate concretes. Thickness of sandwich panel insulation governed by wall U-value requirements.

2.01.F Indicate required strengths on contract drawings.

2.01.F.3 Check with local suppliers to determine availability and types of epoxy-resin grouts.

2.01.G.1 AASHTO grade pads utilize 100 percent chloroprene as the elastomer. Less expensive commercial grade pads are available, but are not recommended.

2.01.G.2 Standard guide specifications are not available for random-oriented, fiber-reinforced pads. Proof testing of a sample from each group of 200 pads is suggested. Normal design working stresses are 1500 psi, so the 3000 psi test load provides a factor of 2 over design stress. The shape factor for the test specimens should not be less than 2.

4. Plastic: Multi-monomer plastic strips shall be non-leaching and support construction loads with no visible overall expansion.

2.0.1.G.4 Plastic pads are widely used with concrete plank. Compression stress in use is not normally over a few hundred psi and proof testing is not considered necessary. No standard guide specifications are available.

5. Tetraflouroethylene (TFE) reinforced with glass fibers and applied to stainless or structural steel plates.

2.01.G.5 ASTM D2116 applies only to basic TFE resin molding and extrusion material in powder or pellet form. Physical and mechanical properties must be specified by naming manufacturer or other methods.

2.02 Mixes

A. Concrete properties:

2.02.A The back-up concrete and the surface finish concrete can be of one mix design, depending upon resultant finish, or the surface finish (facing mix) concrete can be separate from the back-up concrete. Clearly indicate specific requirements or allow manufacturer's option.

1. Water-cement ratio: Maximum 40 lbs. of water to 100 lbs. of cement.

2.02.A.1 Keep to a minimum consistent with strength and durability requirements and placement needs.

2. Air entrainment: Amount produced by adding dosage of air entraining agent that will provide 19% ± 3% of entrained air in standard 1:4 sand mortar as tested according to ASTM C 185; or minimum 3%, maximum 6%.

2.02.A.2 Gradation characteristics of most facing mix concrete will not allow use of a given percentage of air. PCI recommends a range of air entraining be stated in preference to specified percentage.

3. Coloring agent: Not more than 10% of cement weight.

2.02.A.3 Amount used should not have any detrimental effects on concrete qualities.

4. 28 day compressive strength: Minimum of 5000 psi when tested by 6 × 12 or 4 × 8 in. cylinders; or minimum 6250 psi when tested on 4 in. cubes.

2.02.A.4 Vary strength to match requirements. Strength requirements for facing mixes and backup mixes may differ. Also the strength at time of removal from the forms should be stated if critical to the engineering design of the units. The strength level of the concrete should be considered satisfactory if the average of each set of any three consecutive cylinder strength tests equals or exceeds the specified strength and no individual test falls below the specified value by more than 500 psi.

B. Facing mix:

2.02.B Delete if separate face mix is not used.

1. Minimum thickness of face mix after consolidation shall be at least one inch or a

2.02.B.1 Minimum thickness should be sufficient to prevent bleeding through of the backup mix.

minimum of 1 1/2 times the maximum size of aggregates used; whichever is larger.

2. Water-cement and cement-aggregate ratios of face and back-up mixes shall be similar.

2.02.B.2 Similar behavior with respect to shrinkage is necessary in order to avoid undue bowing and warping.

2.03 Fabrication

A. Manufacturing procedures shall be in general compliance with PCI MNL-117.

B. Finishes:

2.03.B Finishing techniques used in individual plants may vary considerably from one part of the continent to another, and between individual plants. Many plants have developed specific techniques supported by skilled operators or special facilities.

1. Exposed face to match approved sample or mockup panel.

* * OR * *

2.03.B.1 Preferable to match sample rather than specify method of exposure.

1. Smooth finish:

a. As cast using flat smooth non-porous molds.

* * OR * *

2.03.B.1.a Difficult to obtain satisfactory finish.

1. Smooth finish:

a. As cast using fluted, sculptured, board finish or textured form liners.

* * OR * *

2.03.B.1.a Many standard shapes of plastic form liners are readily available.

1. Textured finish:

a. Achieve finish on face surface of precast concrete units by form liners applied to inside of forms.

b. Distress finish by breaking off portion of face of each flute.

2.03.B.1.b Delete if distressed finish is not desired.

c. Achieve uniformity of cleavage by alternately striking opposite sides of flute.

* * OR * *

1. Exposed aggregate finish:

a. Apply even coat of retardant to face of mold.

b. Remove units from forms after concrete hardens.

c. Expose coarse aggregate by washing and brushing or lightly sandblasting away surface mortar.

d. Expose aggregate to depth of _____ .

2.03.B.1.d Finishes obtained vary from light etch

* * OR * *

1. Exposed aggregate finish:

    a. Immerse unit in tank of acid solution.

* * OR * *

to a depth of reveal of 1/2 in., but must relate to the size of aggregates. Matrix can be removed to a maximum depth of one-third the average diameter of coarse aggregate but not more than one-half the diameter of smallest sized coarse aggregate.

2.03.B.1.a Use reasonably acid resistant aggregates such as quartz or granite.

1. Exposed aggregate finish:

    a. Treat surface of unit with brushes which have been immersed in acid solution.

    b. Protect hardware, connections and insulation from acid attack.

* * OR * *

2.03.B.1.a Use with softer aggregates such as dolomite and marble.

1. Exposed aggregate finish:

    a. Use power or hand tools to remove mortar and fracture aggregates at the surface of units (bushhammer).

* * OR * *

1. Exposed aggregate finish:

    a. Hand place large facing aggregate or brick, or cobblestones over form bottom.

    b. Produce mortar joints by keeping cast concrete 1/2 in. to 1 in. from face of unit.

* * OR * *

1. Sandblasted finish:

    a. Sandblast away _____ of cement-sand matrix to expose aggregate face.

* * OR * *

2.03.B.1 Exposure of aggregate by sandblasting can vary from 1/16 in. or less to over 3/8 in. Remove matrix to a maximum depth of one-third the average diameter of coarse aggregate but not more than one-half the diameter of smallest sized coarse aggregate. Depth of sandblasting should be adjusted to suit the aggregate hardness and size.

1. Honed or polished finish:

2.03.B.1 Honing and polishing of concrete are techniques which require highly skilled personnel. Use with aggregates such as marble, onyx, and granite.

    a. Polish surface by continued mechanical abrasion with fine grit, followed by special treatment which includes filling of all surface holes and rubbing.

2.03.B.1.a Delete if polished surface not desired.

* * OR * *

1. Veneer faced finish:

   a. Cast concrete over ceramic tile, brick or cut stone placed in the bottom of the mold.

2.03.B.1.a Full scale mock-up units with cut stone in actual production sizes, along with casting and curing of the units under realistic production conditions are essential for each new or major application or configuration of the cut stones. Bowing should be carefully measured over several weeks in the normal storage area and the final details of stone sizes and fastening determined to suit the observed behavior.

   b. Connection of cut stone face material to concrete shall be by mechanical means.

2.03.B.1.b Provide a complete bond-breaker between the cut stone face material and the concrete. Ceramic tile and brick are bonded to the concrete.

* * * * * *

2. _____back surfaces of precast concrete units after striking surfaces flush to form finish lines.

2.03.B.2 (Smooth float finish), (Smooth steel trowel), (Light broom), (Stippled finish). Use for exposed back surfaces of units.

C. Cover:

   1. Provide at least 3/4 in. cover for reinforcing steel.

2.03.C.1 Increase cover requirements when units are exposed to corrosive environment or severe exposure conditions.

   2. Do not use metal chairs, with or without coating, in the finished face.

2.03.C.2 For smooth cast facing, stainless steel chairs may be permitted.

   3. Provide embedded anchors, inserts, plates, angles and other cast-in items with sufficient anchorage and embedment for design requirements.

D. Curing:

2.03.D A wide variation exists in acceptable curing methods, ranging from no curing in some warm humid areas, to carefully controlled moisture-pressure-temperature-curing. Consult with local panel manufacturers to avoid unrealistic curing requirements.

   1. Cure precast units until 2000 psi minimum compressive strength has developed before removing the units from the form, unless greater strength is required for stripping.

2.03.D.1 Stripping strength should be set by the plant based on the characteristics of the product and plant facilities.

E. Panel identification:

   1. Mark each precast panel to correspond to identification mark on shop drawings for panel location.

   2. Mark each precast panel with date cast.

F. Acceptance: Architectural precast units which do not meet the color and texture range or the dimensional tolerances may be rejected at the option of the architect, if they cannot be satisfactorily corrected.

2.04 Concrete Testing

A. Make one compression test at 28 days for each day's production of each type of concrete.

2.04.A This test should be only a part of an in-plant quality control program.

B. Specimens:

1. Provide two test specimens for each compression test.

2.04.B.1 One test specimen may be used to check the stripping strength.

2. Obtain concrete for specimens from actual production batch.

3. 6 in. x 12 in. or 4 in. x 8 in. concrete test cylinder, ASTM C 31 – _____ .

* * OR * *

3. _____ sized concrete cube, _____ _____ .

* * * * * *

2.04.B.3 Specify size. Cube specimens are usually 4 in. units, but 2 in. or 6 in. units are sometimes required. Larger specimens give more accurate test results than smaller ones. Source: (molded individually), (sawed from slab).

4. Cure specimens using the same methods used for the precast concrete units until the units are stripped, then moist cure specimens until test.

C. Keep quality control records available for the architect upon request for two years after final acceptance.

2.04.C These records should include mix designs, test reports, inspection reports, member identification numbers along with date cast, shipping records and erection reports.

3. EXECUTION

3.01 Inspection

A. Before erecting architectural precast concrete the general contractor shall verify that structure and anchorage inserts not within tolerances required to erect panels have been corrected.

B. Determine field conditions by actual measurements.

3.02 Erection

A. Clear, well-drained unloading areas and road access around and in the structure (where appropriate) shall be provided and maintained by the general contractor to a degree that the hauling and erection equipment for the architectural precast concrete products are able to operate under their own power.

B. General contractor shall erect adequate barricades, warning lights or signs to safeguard traffic in the immediate area of hoisting and handling operations.

C. Set precast units level, plumb, square and true within the allowable tolerances. General contractor shall be responsible for providing lines, center and grades in sufficient detail to allow installation.

D. Provide temporary supports and bracing as required to maintain position, stability and alignment as units are being permanently connected.

E. Non-cumulative tolerances for location of precast units shall be in accordance with MNL 117.

F. Set non-load bearing units dry without mortar, attaining specified joint dimension with lead, plastic or asbestos cement spacing shims.

3.02.F Shims should be near the back of the unit to prevent their causing spall on face of unit if shim is loaded. The selection of the width and depth of field-molded sealants, for the computed movement in a joint, should be based on the maximum allowable strain in the sealant.

G. Fasten precast units in place by bolting or welding, or both.

3.02.G The erector shall protect units from damage caused by field welding or cutting operations and provide non-combustible shields as necessary during these operations. Precast units shall be fastened in place as indicated on the approved erection drawings.

3.03 Patching

A. Mix and place patching mixture to match color and texture of surrounding concrete and to minimize shrinkage.

3.03.A Patching is normally accomplished prior to final cleaning and caulking. It is recommended that the precaster execute all repairs or approve the methods proposed for such repairs by other qualified personnel. The precaster should be compensated for repairs of any damage for which he is not responsible. Patching should be acceptable providing the structural adequacy of the product and the appearance is not impaired.

B. Adhere large patch to hardened concrete with bonding agent.

3.03.B Bonding agent should not be used with small patches.

3.04 Cleaning

A. After installation: _____shall clean soiled precast concrete surfaces with detergent and water, using fiber brush and sponge, and rinse thoroughly with clean water.

3.04.A State whether erector or general contractor responsible for cleaning.

** OR **

A. Clean precast concrete panels with _____ .

* * * * *

B. Use acid solution only to clean particularly stubborn stains after more conservative methods have been tried unsuccessfully.

C. Use extreme care to prevent damage to precast concrete surfaces and to adjacent materials.

D. Rinse thoroughly with clean water immediately after using cleaner.

3.05 Protection

A. The erector shall be responsible for any chipping, spalling, cracking or other damage to the units after delivery to the job site unless damage is caused in site storage by others. After installation is completed, any further damage shall be the responsibility of the general contractor.

The following should be placed in the General Conditions section of the project specification and renumbered for that section:

A.1 Payments

A.1.1 Monthly progress payments equal to 90% of the in plant value will be made to the manufacturer for all products fabricated and stocked in the manufacturer's plant prior to delivery. Progress payments shall not relieve the manufacturer from compliance with terms of his contract with the buyer.

A.1.2 Full payment for all products delivered and/ or installed will be made within 35 days of completion of all work under the contract and acceptance by the architect.

3.04.A (acid-free commercial cleaners), (steam cleaning), (water blasting), (sand blasting). Use sand blasting only for units with original sand blasted finish. Ensure that materials of other trades are protected when cleaning panels.

Industry practice is to fabricate, in advance, products for each individual job in accordance with design requirements and dimensions for that job. Such products cannot normally be used on any other project. As monthly payment would be made for such products if they were fabricated on the site, monthly progress payments for such material fabricated off site and stored for delivery and erection as scheduled is justified.

CODE OF STANDARD PRACTICE
FOR PRECAST CONCRETE

The precast concrete industry has grown rapidly and certain practices relating to the design, manufacture and erection of precast concrete have become standard in many areas of North America. This "Code of Standard Practice" is a compilation of these practices, and others deemed worthy of consideration, in the form of recommendations for the guidance of those involved with the use of structural and architectural precast concrete.

The goal of this Code is to build better understanding by suggesting standards which more clearly define procedures and responsibilities, thus resulting in fewer problems for everyone involved in the planning, preparation and completion of any project.

As the precast concrete industry continues to evolve, and it becomes apparent that additional practices have become standard in the industry or that current standards require modification, it is the intent of the Prestressed Concrete Institute to enlarge and revise this Code.

1. DEFINITIONS OF PRECAST CONCRETE

1.1 Structural Precast Concrete

Structural precast concrete usually includes beams, tees, joists, purlins, girders, lintels, columns, posts, piers, piles, slab or deck members, and wall panels. In order to avoid misunderstandings, it is important that the contract documents for each project list all the elements that are considered to be structural precast concrete.

1.2 Architectural Precast Concrete

Architectural precast concrete usually includes precast elements that require architectural finishes and/or exhibit decorative exposed surfaces. Typical architectural precast concrete elements include wall panels, window wall panels, mullions and column covers. In order to avoid misunderstandings, it is important that the contract documents for each project list all the elements that are considered to be architectural precast concrete.

1.3 Prestressed Concrete

Both types of precast concrete may be prestressed or non-prestressed. All structural precast concrete products referred to herein which are prestressed, are specifically referred to as prestressed concrete.

2. SAMPLES, MOCKUPS, AND QUALIFICATION OF MANUFACTURERS

2.1 Samples and Mockups

Samples, mockups, etc., are rarely required for structural prestressed concrete. If samples are required, they should be described in the contract documents and the samples should be manufactured in accordance with Section 5.1.4, *PCI Architectural Precast Concrete.**

2.2 Qualification of Manufacturer

Manufacture, transportation, erection and testing should be accomplished by a company, firm, corporation, or similar organization specializing in providing precast products and services normally associated with structural or architectural precast concrete construction.

The manufacturer may be requested to list similar and comparable work successfully completed, and demonstrate the adequacy of plant capability and facilities for performance of contract requirements.

Standards of performance are given in the PCI quality control manuals MNL 116-85 and MNL 117-77. Current certification under the PCI Plant Certification Program is normally accepted as fulfilling experience and plant capability requirements.

3. CONTRACT DOCUMENTS AND DESIGN RESPONSIBILITY

3.1 Contract Documents

Prior to initiation of the engineering-drafting function, the manufacturer should have the following contract documents at his disposal:

1. Architectural drawings.

2. Structural drawings.

3. Electrical, mechanical and plumbing drawings (if pertinent).

4. Specifications (complete with addenda).

---

*Available from Prestressed Concrete Institute

Other pertinent drawings may also be desirable, such as approved shop drawings from other trades, roofing requirements, alternates, etc.

3.2 Design Responsibilities

It is the responsibility of the owner* to keep the manufacturer supplied with up-to-date documents and written information. The manufacturer should not be held responsible for problems arising from the use of outdated or obsolete contract documents. If updated documents are furnished, it may also be necessary to modify the contract.

The contract documents should clearly define the following:

1. Items furnished by manufacturer.

2. Size, location and function of all openings, blockouts, and cast-in items.

3. Production and erection schedule requirements and restrictions.

4. Design intent including connections and reinforcement.†

5. Allowable tolerances. Normal field tolerances should be recommended by the manufacturer.

6. Dimension, material and quantity requirements.

7. General and supplemental general conditions.

8. Any other special requirements and conditions.

9. Site plan showing storage areas to be used, parking areas for trucks and equipment, etc.

Other design responsibility relationships are described in Table 3.10.1, *PCI Architectural Precast Concrete.*

4. SHOP DRAWINGS

Shop drawings consist of production and erection drawings. Different areas of the country may use different terminology.

---

*The owner of the proposed structure or his designated representatives, who may be the architect, engineer, general contractor, public authority or others contracting with the precast manufacturer.

†When the manufacturer accepts design responsibility, the area or amount of responsibility must be clearly defined in the contract documents. The engineer or architect of record must be identified and it is understood that all designs are submitted through him for his approval and acceptance. The manufacturer's responsibility can be limited to member design only or it may include the entire structure. When the manufacturer is responsible for product design only, all loads which are to be applied to precast, including forces developed by restraint, should be provided to the manufacturer by the owner unless otherwise agreed.

4.1 Erection Drawings

The information provided in the contract documents is used by the manufacturer to prepare erection drawings for approval and field use. They contain:

1. Plans and/or elevations locating and dimensioning all members furnished by manufacturer.

2. Sections and details showing connections, finishes, openings, blockouts and cast-in items and their relationship to the structure.

3. Description of all loose and cast-in hardware including designation of who furnishes it.

4. Drawings showing location of anchors installed in the field.

5. Erection sequences and handling requirements.

4.2 Production Drawings

The contract documents are also used to prepare production drawings for manufacturing showing all dimensions together with locations and quantities for all cast-in materials (reinforcement, inserts, etc.) and completely defining all finish requirements.

Normal practices for the preparation of drawings for architectural precast concrete are described in the *PCI Architectural Precast Concrete Drafting Handbook.*

4.3 Discrepancies

When discrepancies or omissions are discovered on the contract documents, the manufacturer has the responsibility to check with the owner to resolve the problem. If this is not possible, the following procedures are normally followed:

1. Contract terms govern over specifications and drawings.

2. Specifications govern over drawings.

3. Structural drawings govern over architectural drawings.

4. Written dimensions govern over scale dimensions.

5. Sections govern over plans or elevations.

6. Details govern over sections.

Graphic verification should be requested for any unclear condition.

4.4 Approvals

Completed erection drawings, usually in reproducible form, should be submitted for approval. The exact sequence is dictated by construction schedules and erection sequences, and is determined when the contract is awarded.

Production drawings should not be started prior to receipt of approved or approved-as-noted erection drawings. Production drawings should be submitted for approval only when so requested.*

Corrections should be noted on the reproducible erection drawings and copies made for distribution.

The following approval interpretation is normal practice:†

1. **Approved** – The approvers‡ have completely checked and verified the drawings for conformance with contract documents and all expected loading conditions. Such approval should not relieve the manufacturer from responsibility for his design when that responsibility is placed upon him by the contract. The manufacturer may then proceed with production drawings and production without resubmittal. Erection drawings may then be released for field use and plant use.

2. **Approved as Noted** – Same as above except that noted changes should be made and corrected erection drawings issued. Production drawings and production may be started after noted changes have been made.

3. **Not Approved** – Drawings must be corrected and resubmitted. Production drawings should not be started until "approved" or "approved as noted" erection drawings are returned.

5. MATERIALS

The relevant ASTM Standards that apply to materials for a project should be listed in the contract documents together with any special requirements that are not included in the ASTM Standards.

Note: Additional informaton regarding material specifications can be found in "Guide Specification for Precast, Prestressed Concrete" and "Guide Specification for Architectural Precast Concrete" in this Part.

---

*When production drawings are the only drawings showing reinforcment, representative drawings should be submitted for approval.

†When production drawings are submitted, the same applies.

‡The contract should state who has approval authority.

6. TESTS AND INSPECTIONS

6.1 Tests of Materials

Manufacturers generally keep the test records required by the PCI manuals for quality control (MNL-116 and MNL-117). The contract documents may require the precast concrete manufacturer to make these records available for inspection by the owner's representative upon his request.

When the manufacturer is required to submit copies of test records to the owner and/or required to perform or have performed tests not required by MNL-116 and MNL-117, these special testing requirements should be clearly described in the contract documents along with the responsibility for payment.

6.2 Inspections

On certain projects the owner may require inspection of precast concrete products in the manufacturer's yard by persons other than the manufacturer's own quality control personnel. Such inspections are normally made at the owner's expense. The contract documents should describe how, when and by whom the inspections are to be made, the responsibility of the inspection agency, and who is to pay for them. Alternatively, the owner may accept plant certification in lieu of outside inspection, such as provided in the PCI Plant Certification Program.

6.3 Fire Rated Products

If the manufacturer is expected to provide a fire rated product and/or labels, these requirements should be clearly stated in the contract documents.

7. FINISHES

Finishes of precast concrete products, both structural and architectural, are probably the cause of more misunderstandings between the various members of the building team than any other question concerning product quality.

It is therefore extremely important that the contract documents describe clearly and completely the required finishes for all surfaces of all members, and that the erection drawings also include this information. When finish is not specified, the standard finish described in "Guide Specification for Precast, Prestressed Concrete" should normally be furnished.

For descriptions of the usual finishes for structural precast concrete, see "Guide Specification for Precast, Prestressed Concrete," and for architectural precast concrete, see Sect. 4.3, *PCI Architectural Precast Concrete*. Where special or critical requirements exist or where large expanses of exposed precast will occur on a project, samples are essential and, if required, should be so stated and described in the contract documents.

8. DELIVERY OF MATERIALS

8.1 Manner of Delivery

The manufacturer should deliver the precast concrete to the erector* in a manner to facilitate the speed of erection of the building or as mutually agreed upon between the owner, manufacturer and erector. Special requirements of the owner for the delivery of materials or the mode of transport, should be stated in the contract documents.

8.2 Marking and Shipping of Materials

The precast concrete members should be separately marked in accordance with approved erection drawings in such a manner as to distinguish varying pieces and to facilitate erection of the structure. Any members which require a sequential erection should be properly marked.

The owner should give the manufacturer sufficient time to fabricate and ship any special plates, bolts, anchorage devices, etc., contractually agreed to be furnished by the manufacturer.

8.3 Precautions During Delivery

Special protection or precautions beyond that required in MNL-116 and MNL-117 should not be expected unless stated in the bid invitation or specifications. The manufacturer is not responsible for the product, including loose material, after delivery to the site unless required by the contract documents.

8.4 Access to Jobsite

Free and easy access to the delivery site should be provided to the manufacturer, including backfilling and compacting, drainage and snow re-

moval, so that delivery trucks can operate under their own power.

8.5 Unloading Time Allowance

Delivery of product includes a reasonable unloading time allowance. Any delay beyond a reasonable time is normally paid for by the party which is responsible for the delay.

9. ERECTION

9.1 Special Erection Requirements

When the owner requires a particular method or sequence of erection, this information should be stated in the contract documents. In the absence of such stated restrictions, the erector will proceed using the most efficient and economical method and sequence available to him, consistent with the contract documents.

9.2 Tolerances

Some variation is to be expected in the overall dimensions of any building or other structure. It is common practice for the manufacturer and erector to work within the tolerances recommended by the American Concrete Institute and Prestressed Concrete Institute.

The owner, by whatever agencies he may elect, immediately upon completion of the erection, should determine if the work is plumb, level, aligned and properly fastened. Discrepancies should immediately be brought to the attention of the erector so that proper corrective action can be taken.

The work of the manufacturer and erector is complete once the precast product has been properly plumbed, leveled and aligned within the established tolerances. Acceptance for this work should be secured from the authorized representative of the general contractor (see Section 11.2).

9.3 Foundations, Piers, Abutments and Other Bearing Surfaces

The invitation to bid should state the anticipated time when all foundations, piers, abutments and other bearing surfaces will be ready and accessible to the erector.

9.4 Building Lines and Bench Marks

The precast manufacturer should be furnished

---

*The erector may be either the manufacturer or a subcontractor engaged by the manufacturer, or the general contractor.

all building lines and bench marks at the site of the structure.

9.5 Anchor Bolts and Bearing Plates

9.5.1 The precast manufacturer does not normally furnish or install anchor bolts, plates, etc. that are to be installed in cast-in-place concrete or masonry for connection with precast members. However, if this is to be the responsibility of the precast manufacturer, it should be so defined in the specifications. It is important that such items be installed true to line and grade, and that installation be completed in time to avoid delays or interference with the precast erection.

9.5.2 Anchor bolts and foundation bolts are set by the owner in accordance with an approved drawing. They must not vary from the dimensions shown on the erection drawings by more than the following:

(a) 1/8-inch center to center of any two bolts within an anchor bolt group, where an anchor bolt group is defined as the set of anchor bolts which receive a single fabricated steel shipping piece.

(b) 1/4-inch center to center of adjacent anchor bolt groups.

(c) Maximum accumulation of 1/4-inch per hundred feet along the established column line of multiple anchor bolt groups, but not to exceed a total of 1 inch, where the established column line is the actual field line most representative of the centers of the as-built anchor bolt groups along a line of columns.

(d) 1/4-inch from the center of any anchor bolt group to the established column line through that group.

(e) The tolerances of paragraphs b, c and d apply to offset dimensions shown on the plans, measured parallel and perpendicular to the nearest established column line for individual columns shown on the plans to be offset from established column lines.

9.5.3 Erectors should check both line and grade in sufficient time before erection is scheduled to permit any necessary corrections. Corrections, if any, should be made by the general contractor before erection begins. Proposed corrections should be submitted to the owner for approval.

9.6 Utilities

Water and electricity should be furnished for erection and grouting operations by the owner.

9.7 Working Space

The owner should furnish adequate, properly drained, graded, and convenient working space for the erector and access for his equipment necessary to assemble the structure. The owner should provide adequate storage space for the precast products to enable the erector to operate at the speed required to meet the established schedule. Unusual hazards such as high voltage lines, buried utilities, or areas of restricted access should be declared in the invitation to bid.

9.8 Materials of Other Trades

Other building materials or work of other trades should not be built up above the bearing of the precast concrete until after erection of the precast.

9.9 Correction of Errors

Corrections of minor misfits are considered a part of erection even if the precast concrete is not erected by the manufacturer. Any error in manufacturing which prevents proper connection or fitting should be immediately reported to the manufacturer and the owner so that corrective action can be taken.

9.10 Field Assembly

The size of precast concrete pieces may be limited by transportation requirements for weight and clearance dimensions. Unless agreed upon between the manufacturer and owner, the manufacturer should provide for such field connections that will meet required loads and forces without altering the function or appearance of the structure.

All loose materials for temporary and permanent connection of structural precast members are normally furnished by the erector. The manufacturer furnishes only those items embedded in the precast. Temporary guys, braces, falsework, shims, and cribbing are the property of the erector and are removed only by the erector or with the erector's approval upon completion of the erection of the structure, unless otherwise agreed.

9.11 Blockouts, Cuts and Alterations

Neither the manufacturer nor the erector is responsible for the blockouts, cuts or alterations by or for other trades unless so specified in the contract documents. Whenever such additional work is required, all information regarding size, location and number of alterations is furnished by the owner prior to preparation of the precast production and erection drawings.

The general contractor is responsible for warning other trades against cutting of precast concrete members without prior approval of the owner.

9.12 Temporary Floors and Access

The manufacturer or erector is not required to furnish temporary flooring for access unless so specified in the contract documents.

9.13 Painting, Grouting, Caulking and Closure Panels

Painting, grouting, caulking and placing of closure panels between stems of flanged concrete members are services not ordinarily supplied by the manufacturer or erector. If any of these services are required of the manufacturer, it should be stated in the contract documents.

9.14 Patching

A certain amount of patching of product is to be expected to repair minor spalls and chips. Patching should meet the finish requirements of the project and color should be reasonably matched. Responsibility for accomplishing this work should be resolved between the manufacturer and erector.

9.15 Safety

Safety procedures for the erection of the precast concrete members is the responsibility of the erector and must be in accordance with all local, state or Federal rules and regulations which have jurisdiction in the area where the work is to be performed, but not less than required in ANSI Standard A 10.9, *American National Standard Safety Requirements for Concrete Construction and Masonry Work.**

9.16 Security Measures

Security protection at the job site should be the responsibility of the general contractor.

---

*American National Standards Institute, New York, New York.

10. INTERFACE WITH OTHER TRADES

Coordination of the requirements for other trades to be included in the precast should be the responsibility of the owner unless clearly defined otherwise in the contract documents.

The PCI manuals for quality control (MNL-116 and MNL-117) specify manufacturing tolerances for precast concrete members. Interfaces with other materials and trades must take these tolerances into account. Unusual requirements or allowances for interfacing should be stated in the contract documents.

11. WARRANTY AND ACCEPTANCE

11.1 Warranties

Warranties of product and workmanship have become a widely accepted practice in this industry, as in most others. Warranties given by the precast concrete manufacturer and erector should indicate that their product and work meet the specifications for the project.

In no case should the warranty of the manufacturer and erector be in excess of the warranty required by the specifications. Warranties should in all instances include a time limit and it is recommended that this should not exceed one year.

In order to protect the interests of all parties concerned, warranties should also state that any deviations in the designed use of the product, modifications of the product by the owner and/or contractor or changes in other products used in conjunction with the precast will cause said warranty to become null and void.

Warranty may be included as a part of the conditions of the contract agreement, or it may be presented in letter form, as requested by the owner. A sample warranty follows:

> Manufacturer warrants that all materials furnished have been manufactured in accordance with the specifications for this project. Manufacturer further warrants that if erection of said material is to be performed by those subject to his control and direction, work will be completed in accordance with the same specifications.
>
> In no event shall manufacturer be held responsible for any damages, liability or costs of any kind or nature occasioned by or arising out of the actions or omissions of others, or for work, including design, done by others; or for material manufactured, supplied or installed by others; or for inadequate construction of foundations, bearing walls, or other units to which materials furnished by the precast manufacturer are attached or affixed.
>
> This warranty ceases to be in effect beyond the date of _____. Should any defect develop during the contract warranty period, which can be directly attrib-

uted to defect in quality of product or workmanship, precast manufacturer shall, upon written notice, correct defects or replace products without expense to owner and/or contractor.

COMPANY NAME

_____
Signature   Title

11.2 Acceptance

Manufacturer should request approval and acceptance for all materials furnished and all work completed by him periodically as deemed necessary in order to adequately protect the interests of everyone involved in the project. The size and nature of the project will dictate the proper intervals for securing approval and acceptance. Periodic approval in writing should be considered when it appears that such action will minimize possible problems which would seriously affect the progress of the project. A sample acceptance form follows:

FIELD INSPECTION REPORT

Project # _____

On this _____ day of _____, 19 _____
_____ of _____
Company Field Superintendent   Precast Manufacturer

and _____ of _____
      General Contractor Superintendent   General Contractor

_____ have inspected _____

_____
portion of building being inspected

All of the work performed by the above indicated company in the above described portion of the project has been performed to the satisfaction of the above named General Contractor's Superintendent with the exception of the following: _____

_____

The General Contractor's Superintendent hereby releases the Precast Manufacturer of its responsibility to perform any other work in the above described portion of the project except as detailed herein.

The Precast Manufacturer in turn hereby releases the above described portion of the project to the General Contractor.

_____   _____
Precaster's Superintendent   Gen'l. Contr. Superintendent

12. CONTRACT ADMINISTRATION

12.1 General Statement

Information relative to invoicing, payment, bonding and other data pertinent to a project or

material sale should be specifically provided for in the major provisions of the contract documents or in the special terms and conditions applicable to all contractual agreements between manufacturer and owner.

Contract agreements may vary widely from area to area, but the objective should be the same in all instances. The contract agreement should be written to protect the interests of all parties concerned and at the same time, be specific enough in content to avoid misunderstandings once the project begins.

The intent of this section is to recommend those matters which ought to be considered, but not necessarily the form in which they should be expressed. The final statement of policies should be the result of careful consideration of all pertinent factors as well as of the normal practices in the area.

12.2 Retentions

Although retentions have been used for many years as a means of assuring a satisfactory job performance, it is apparent that they directly contribute to the cost of construction, frequently lead to disputes, and often result in job delays. In view of the unfavorable consequences of retentions and possible abuse, it is recommended that the following procedure be followed:

1. Wherever possible, retentions should be eliminated and bonding should be used as the single, best source of protection. This should apply to prime contractors and subcontractors equally.

2. Where there are no bonding requirements, the retention percentage should be as low as possible. It is recommended that this be not more than 5 percent of the work invoiced.

3. The percentage level of any retention should be the same for subcontractors as for prime contractors on a job.

4. Release of retained funds and final payment, as well as computing the point of reduction of the retention, should be done on a line item basis, that is, each contractor or subcontractor's work considered as a separate item and the retention reduced by 50 percent upon substantial completion and the balance released within 30 days after final completion of that work.

5. Retained funds should be held in an escrow account with interest accruing to the benefit of the party to whom the funds are due.

6. When materials are furnished FOB plant or job-

site, it is recommended that there be no retentions.

12.3 Contract Agreement

1. Contract agreement should fully describe the project involved, including job location, project name, name of owner/developer, architect or other design professionals and all reference numbers identifying job relation information such as plans, specifications, addenda, bid number, etc.

2. Contract agreement should fully describe the materials to be furnished and/or all work to be completed by the seller.

3. All exclusions should be stated to avoid the possibility of any misunderstanding.

4. Price quoted should be stated to eliminate any possibility of misunderstanding.

5. Reference should be made to the terms and conditions governing the proposed contract agreement. The terms and conditions may best be stated on the reverse side of the contract form. Special terms or conditions should be stated in sufficient detail to avoid the possibility of misunderstanding.

6. The terms of payment should be specifically detailed so there is no doubt as to intent. Special care should be exercised where the terms of payment will differ from those normally in effect or where they deviate from the general terms and conditions appearing on the reverse side of the contract form.

7. A statement of policy should be made with reference to the inclusion or exclusion of taxes in the stated price.

8. The proposal form stating the full intent and conditions under which the project will be performed may contain an acceptance clause to be signed by the purchaser. At such time as said acceptance clause is signed, the proposal form then becomes the contract agreement.

9. Seller should clearly state the limits of time within which an accepted proposal will be recognized as a binding contract.

10. A statement indicating the classification of labor to perform the work in the field is advisable to eliminate later dispute over jurisdiction of work performed.

12.4 Terms and Conditions

The terms and conditions stated on the proposal contract agreement should include, but are not necessarily limited to, the following:

1. **Lien Laws** – Where the lien laws of a state specifically require advance notice of intent, it is advisable to include the required statement in the general terms and conditions.

2. **Specifications** – Seller should make a specific declaration of material and/or work specifications, but normally this should not be in excess of the specifications required by the contract agreement.

3. **Contract Control** – A statement should be made indicating that the agreement when duly signed by both parties supersedes and invalidates any verbal agreement and can only be modified in writing with the approval of those signing the original agreement.

4. **Terms of Payment** – Terms of payment should be specifically stated either on the face of the contract or in the general terms and conditions. Mode and frequency of invoicing should be so stated, indicating time within which payment is expected.

5. **Late Payment Charges** – The contract may provide for legal interest charges for late payments not made in accordance with contract terms, and if this is desired, it should be stated in the general terms and conditions. A statement indicating seller is entitled to reasonable attorney's fees and related costs should collection proceedings be necessary may also be included.

6. **Overtime Work** – Prices quoted in the proposal should be based on an 8-hour day and a 5-day week under prevailing labor regulations. Provisions should be included in the contract agreement to provide for recovery of overtime costs plus a reasonable markup when the seller is requested to provide such service.

7. **Financial Responsibility** – General terms and conditions may indicate the right of the seller to suspend or terminate material delivery and/or work on a project if there is a reasonable doubt of the ability of the purchaser to fulfill his financial responsibility.

8. **Payment for Inventory**

   (a) It has become common practice to include in the contract terms and conditions provisions for the invoicing and payment of all materials stored at the plant or job-site when deliveries or placement of said materials are delayed for more than a stipulated time beyond the originally scheduled date because of purchaser's inability either to accept delivery of materials or to provide proper job access.

(b) Under certain conditions, it may be necessary to purchase special materials or to produce components well in advance of job requirements to insure timely deliveries. When job requirements are of such a nature, it is advisable to include provisions for payment of such raw and finished inventories stored in seller's plant or on jobsites on a current basis.

9. **Payment for Suspended or Discontinued Projects** – The terms and conditions should provide that in the event of a discontinued or suspended project, seller shall be entitled to payment for all material purchased and/or manufactured including costs, overhead and profit, and not previously billed, as well as reasonable engineering and other costs incurred.

10. **Job Extras** – Requests for job extras should be confirmed in writing. Invoicing should be presented immediately following completion of the extra work with payment subject to the terms and conditions of the contract agreement, or as otherwise stated in the change order.

11. **Claims for Shortages, Damages or Delays** – Seller should, upon immediate notification in writing on the face of the delivery ticket of rejected material or shortage, acknowledge and furnish replacement material at no cost to purchaser. It is normal practice that the seller should not be responsible for any loss, damage, detention or delay caused by fire, accident, labor dispute, civil or military authority, insurrection, riot, flood or by occurrences beyond his control.

12. **Back Charges** – Back charges should not be binding on the seller, unless the condition is promptly reported in writing, and opportunity is given seller to inspect and correct the problem.

13. **Permits, Fees and Licenses** – Costs of permits, fees, licenses and other similar expenses are normally assumed by the purchaser.

14. **Bonds** – Cost of bonds is normally assumed by the purchaser.

15. **Taxes** – Federal, State, County or Municipal, occupation or similar taxes which may be imposed are normally paid by the purchaser.

16. **Insurance** – Seller shall carry Workmen's Compensation, Public Liability, Property Damage and Auto Insurance and certificates of insurance will be furnished to purchaser upon request. Additional coverage required over and above that provided by the seller is normally paid by the purchaser.

17. **Services** – Heat, water, light, electricity, toilet, telephone, watchmen and general services of a similar nature are normally the responsibility of the purchaser unless specifically stated otherwise in the contract agreement.

18. **Safety Equipment** – The purchaser is normally responsible for necessary barricades, guard rails and warning lights for the protection of vehicular and pedestrian traffic and seller's equipment. Purchaser is also normally responsible for furnishing, installing and maintaining all safety appliances and devices required on the project under U.S. Department of Labor, *Safety and Health Regulations for Construction,* as well as all other safety regulations imposed by other agencies having jurisdiction over the project.

19. **Warranty** – Seller should provide specific information relative to warranties given, including limitations, exclusions and methods of settlement. Warranties should not be in excess of warranty required by the specific project.

20. **Title** – Contract should provide for proper identification of title to material furnished. It is normal practice for title and risk of loss or damage to the product furnished to pass to the purchaser at the point of delivery, except in cases of FOB factory, in which event title to and risk of loss of damage to the product normally should pass to purchaser at factory pickup.

21. **Shop Drawings Approval** – Seller should prepare and submit to purchaser for approval all shop drawings* necessary to describe the work to be completed. Shop drawings approval should constitute final agreement to quantity and general description of material to be supplied. No work should be done upon material to be furnished by seller until approved shop drawings are in his possession.

_____

*See Section 4 for definition of shop drawings.

22. **Delivery** – Delivery times or schedules set forth in contract agreements should be computed from the date of delivery to the seller of approved shop drawings. Where materials are specified to be delivered FOB to jobsite, the purchaser should provide labor, cranes or other equipment to remove the materials from the trucks and should pay seller for truck expense for time at the jobsite in excess of a specified time for each truck. On shipments to be delivered by trucks, delivery should be made as near to the construction site as the truck can travel under its own power. In the event delivery is required beyond the curb line, the purchaser should assume full liability for damages to sidewalks, driveways or other properties and should secure in advance all necessary permits or licenses to effect such deliveries.

23. **Builder's Risk Insurance** – Purchaser should provide Builder's Risk Insurance without cost to seller, protecting seller's work, materials and equipment at the site from loss or damage caused by fire or the standard perils of extended coverage, including vandalism and malicious acts.

24. **Erection** – Purchaser should assure that the proposed project will be accessible to all necessary equipment including cranes and trucks, and that the operation of this equipment will not be impeded by construction materials, water, presence of wires, pipes, poles, fences or framings. Purchaser should further indemnify and save harmless the seller and his respective representatives, including subcontractors, vendors, assigns and successors from any and all liability, fine, penalty or other charge, cost or expense and defend any action or claim brought against seller for any failures by purchaser to provide suitable access for work to be performed. Seller also reserves the right to discontinue the work for failure of purchaser to provide suitable access and the purchaser should be responsible for all expenses and costs incurred.

25. **Exclusions of Work to be Performed** – Unless otherwise stated in the contract, all shoring, forming, framing, cutting holes, openings for mechanical trades and other modifications of seller's products should not be performed by the seller nor are they included in the contract price. Seller should not be held responsible for modifications made by others to his product unless said modifications are previously approved by him.

26. **Sequence of Erection** – Sequence of erection should be as agreed upon between seller and purchaser and expressly stated in the contract agreement. Purchaser should have ready all foundations, bearing walls or other units to which seller's material is to be affixed, connected or placed, prior to start of erection. Purchaser should be responsible for the accuracy of all job dimensions, bench marks, and true and level bearing surfaces. Claims or expenses arising from the purchaser's neglect to fulfill this responsibility should be assumed by the purchaser.

27. **Arbitration** – In view of the difficulties and misunderstandings which may occur due to misinterpretation of contractual documents, it is recommended that the seller stipulate that all claims, disputes and other matters in question, arising out of or related to the contract, be decided by arbitration in accordance with the Construction Industry Arbitration Rules of the American Arbitration Association then obtaining, or some other rules acceptable to both parties. The location for such arbitration should be stipulated.

28. **Contract Form** – Contract documents should stipulate policy governing acceptance of proposal on other than the seller's form. In the event purchaser does not accept the seller's proposal and/or contract agreement, but requires the execution of a contract on his own form, it is advisable that seller stipulate in writing on the contract agreement that the contract will be fulfilled according to his proposal originally submitted. All identifying information such as proposal number, dates, etc. should be included so there can be no question of the document referred to.

Inspection Checklist for Architectural Precast Concrete

During Manufacture:

A. Check qualifications of testing agencies.

B. Where in-plant testing is to be performed, see that manufacturer's testing equipment is calibrated by testing agency personnel, and that manufacturer's testing personnel are qualified to perform work.

C. Check casting and after-casting tolerances. Inspectors should be provided with quality instruments for measuring tolerances.

D. Verify that production panels match texture and color of accepted job mock-up or panel samples.

E. Examine concrete test results.

Before Erection:

A. Check precast concrete panel erector's experience and qualifications.

B. Check welder's qualifications.

C. Assure that precast concrete panels are stored on site with resilient and stain resistant spacers.

D. Verify that panel identification marks are easily discernible.

E. Check field dimensions affecting erection.

During Erection:

A. Check erection tolerances.

B. See that walls are clean and exposed metal spot painted.

After Erection:

A. Inspect repairs for accurate color and texture match of surrounding concrete.

B. Inspect panels after cleaning to see that they are properly prepared to receive caulking and dampproofing.

References:

Manual for Quality Control for Plants and Production of Architectural Precast Concrete Products, PCI MNL-117, Prestressed Concrete Institute, 1977.

Architectural Precast Concrete, PCI MNL-122, Prestressed Concrete Institute, 1973.

Architectural Precast Concrete Drafting Handbook, PCI MNL-119, Prestressed Concrete Institute, 1975.

PCI Design Handbook-Precast and Prestressed Concrete, 3rd Edition, Prestressed Concrete Institute, 1985.

PART 11
GENERAL DESIGN INFORMATION

			Page No.
11.1	Design Information		11–2
	11.1.1	Dead weights of floors, ceilings, roofs, and walls	11–2
	11.1.2	Recommended minimum floor live loads	11–3
	11.1.3	Beam design equations and diagrams	11–4
	11.1.4	Camber (deflection) and rotation coefficients for prestress force and loads	11–12
	11.1.5	Moments in beams with fixed ends	11–13
	11.1.6	Moving load placement for maximum moment and shear	11–14
11.2	Material Properties		11–15
	11.2.1	Table of concrete stresses	11–15
	11.2.2	Concrete modulus of elasticity as affected by unit weight and strength	11–15
	11.2.3	Properties and design strengths of prestressing strand and wire	11–16
	11.2.4	Properties and design strengths of prestressing bars	11–17
	11.2.5	Typical stress-strain curve, 7-wire stress-relieved and low-relaxation prestressing strand	11–18
	11.2.6	Reinforcing bar data	11–19
	11.2.7	Required development and lap lengths for Grade 60 bars	11–20
	11.2.8	Required embedment lengths for standard end hooks on Grade 60 bars	11–21
	11.2.9	Common stock styles of welded wire fabric	11–22
	11.2.10	Special welded wire fabric for double tee flanges	11–23
	11.2.11	Wires used in welded wire fabric	11–23
11.3	Section Properties		11–24
	11.3.1	Properties of geometric sections	11–24
	11.3.2	Coefficients for determining section properties of T-beams	11–27
11.4	Metric Conversion		11–29

DESIGN INFORMATION

11.1.1 Dead weights of floors, ceilings, roofs, and walls

Floorings	Weight (psf)
Normal weight concrete topping, per inch of thickness	12
Sand-lightweight (120 pcf) concrete topping, per inch	10
Lightweight (90-100 pcf) concrete topping, per inch	8
7/8″ hardwood floor on sleepers clipped to concrete without fill	5
1 1/2″ terrazzo floor finish directly on slab	19
1 1/2″ terrazzo floor finish on 1″ mortar bed	30
1″ terrazzo finish on 2″ concrete bed	38
3/4″ ceramic or quarry tile on 1/2″ mortar bed	16
3/4″ ceramic or quarry tile on 1″ mortar bed	22
1/4″ linoleum or asphalt tile directly on concrete	1
1/4″ linoleum or asphalt tile on 1″ mortar bed	12
3/4″ mastic floor	9
Hardwood flooring, 7/3″ thick	4
Subflooring (soft wood), 3/4″ thick	2 1/2
Asphaltic concrete, 1 1/2″ thick	18

Ceilings	
1/2″ gypsum board	2
5/8″ gypsum board	2 1/2
3/4″ plaster directly on concrete	5
3/4″ plaster on metal lath furring	8
Suspended ceilings	2
Acoustical tile	1
Acoustical tile on wood furring strips	3

Roofs	
Ballasted inverted membrane	16
Five-ply felt and gravel (or slag)	6 1/2
Three-ply felt and gravel (or slag)	5 1/2
Five-ply felt composition roof, no gravel	4
Three-ply felt composition roof, no gravel	3
Asphalt strip shingles	3
Rigid insulation, per inch	1/2
Gypsum, per inch of thickness	4
Insulating concrete, per inch	3

Walls	Un-Plastered	One side Plastered	Both sides Plastered
4″ brick wall	40	45	50
8″ brick wall	80	85	90
12″ brick wall	120	125	130
4″ hollow normal weight concrete block	28	33	38
6″ hollow normal weight concrete block	36	41	46
8″ hollow normal weight concrete block	51	56	61
12″ hollow normal weight concrete block	59	64	69
4″ hollow lightweight block or tile	19	24	29
6″ hollow lightweight block or tile	22	27	32
8″ hollow lightweight block or tile	33	38	43
12″ hollow lightweight block or tile	44	49	54
4″ brick 4″ hollow normal weight block backing	68	73	78
4″ brick 8″ hollow normal weight block backing	91	96	101
4″ brick 12″ hollow normal weight block backing	119	124	129
4″ brick 4″ hollow lightweight block or tile backing	59	64	69
4″ brick 8″ hollow lightweight block or tile backing	73	78	83
4″ brick 12″ hollow lightweight block or tile backing	84	89	94
4″ brick, steel or wood studs, 5/8″ gypsum board	43		
Windows, glass, frame and sash	8		
4″ stone	55		
Steel or wood studs, lath, 3/4″ plaster	18		
Steel or wood studs, 5/8″ gypsum board each side	6		
Steel or wood studs, 2 layers 1/2″ gypsum board each side	9		

DESIGN INFORMATION

11.1.2 Recommended minimum floor live loads*

Uniformly Distributed Loads

Occupancy or Use	Live Load (psf)
Apartments (see Residential)	
Armories and drill rooms	150
Assembly halls and other places of assembly:	
Fixed seats	60
Movable seats	100
Platforms (assembly)	100
Balcony (exterior)	100
On one and two family residences only and not exceeding 100 sq ft	60
Bowling alleys, poolrooms, and similar recreational areas	75
Corridors:	
First floor	100
Other floors, same as occupancy served except as indicated	
Dance halls and ballrooms	100
Dining rooms and restaurants	100
Dwellings (see Residential)	
Fire escapes	100
On multi- or single-family residential buildings only	40
Garages (passenger cars only)	50
For trucks and buses use AASHTO lane loads (1)	
Grandstands (see Stadium and arena bleachers)	
Gymnasiums, main floors and balconies	100
Hospitals:	
Operating rooms, laboratories	60
Private rooms	40
Wards	40
Corridors, above first floor	80
Hotels (see Residential)	
Libraries:	
Reading rooms	60
Stack rooms (books & shelving at 65 pcf) but not less than	150
Corridors, above first floor	80
Manufacturing:	
Light	125
Heavy	250
Marquees and canopies	75
Office buildings:	
Offices	50
Lobbies	100
File and computer rooms require heavier loads based upon anticipated occupancy	
Penal institutions:	
Cell blocks	40
Corridors	100
Residential:	
Dwellings (one-and two family)	
Uninhabitable attics without storage	10
Uninhabitable attics with storage	20
Habitable attics and sleeping areas	30
All other areas	40

Occupancy or Use	Live Load (psf)
Residential (cont.)	
Hotels and multifamily houses:	
Private rooms and corridors serving them	40
Public rooms and corridors serving them	100
Schools:	
Classrooms	40
Corridors above first floor	80
Sidewalks, vehicular driveways, and yards, subject to trucking (2)	250
Stadiums and arena bleachers (3)	100
Stairs and exitways	100
Storage warehouse:	
Light	125
Heavy	250
Stores:	
Retail:	
First floor	100
Upper floors	75
Wholesale, all floors	125
Walkways and elevated platforms (other than exitways)	60

Concentrated Loads

Location	Load (lb)
Elevator machine room grating (on area of 4 sq in)	300
Finish light floor plate construction (on area of 1 sq in)	200
Garages	(4)
Office floors	2000
Scuttles, skylight ribs, and accessible ceilings	200
Sidewalks	8000
Stair treads (on area of 4 sq in at center of tread)	300

(1) American Association of State Highway and Transportation Officials.
(2) AASHTO lane loads should also be considered where appropriate.
(3) For detailed recommendations, see Assembly Seating, Tents and Air Supported Structures, ANSI/NFPA 102-1978 [Z20.3].
(4) Floors in garages or portions of buildings used for storage of motor vehicles shall be designed for the uniformly distributed live loads shown or the following concentrated loads: (1) for passenger cars accommodating not more than nine passengers, 2000 pounds acting on an area of 20 sq in; (2) mechanical parking structures without slab or deck, passenger cars only, 1500 pounds per wheel; (3) for trucks or buses, maximum axle load on an area of 20 sq in.

*Source: American National Standard ANSI A58.1-1982
Local building codes take precedence.

DESIGN INFORMATION

11.1.3 Beam design equations and diagrams

(1) Simple Beam — uniformly distributed load

$R = V$ .... $= \dfrac{wl}{2}$

V_x .... $= w\left(\dfrac{l}{2} - x\right)$

M max. (at center) .... $= \dfrac{wl^2}{8}$

M_x .... $= \dfrac{wx}{2}(l - x)$

$\triangle$ max. (at center) .... $= \dfrac{5\,wl^4}{384\,EI}$

$\triangle_x$ .... $= \dfrac{wx}{24\,EI}(l^3 - 2lx^2 + x^3)$

(2)

Simple Beam — concentrated load at center

$R = V$ .... $= \dfrac{P}{2}$

M max. (at point of load) .... $= \dfrac{Pl}{4}$

$M_x \left(\text{when } x < \dfrac{l}{2}\right)$ .... $= \dfrac{Px}{2}$

$\triangle$ max. (at point of load) .... $= \dfrac{Pl^3}{48\,EI}$

$\triangle_x \left(\text{when } x < \dfrac{l}{2}\right)$ .... $= \dfrac{Px}{48\,EI}(3l^2 - 4x^2)$

(3) Simple Beam — concentrated load at any point

$R_1 = V_1$ (max. when $a < b$) .... $= \dfrac{Pb}{l}$

$R_2 = V_2$ (max. when $a > b$) .... $= \dfrac{Pa}{l}$

M max. (at point of load) .... $= \dfrac{Pab}{l}$

M_x (when $x < a$) .... $= \dfrac{Pbx}{l}$

$\triangle$ max. $\left(\text{at } x = \sqrt{\dfrac{a(a+2b)}{3}} \text{ when } a > b\right)$ .... $= \dfrac{Pab(a+2b)\sqrt{3a(a+2b)}}{27\,EI\,l}$

$\triangle a$ (at point of load) .... $= \dfrac{Pa^2b^2}{3\,EI\,l}$

$\triangle_x$ (when $x < a$) .... $= \dfrac{Pbx}{6\,EI\,l}(l^2 - b^2 - x^2)$

(4) Simple Beam — two equal concentrated loads symmetrically placed

$R = V$ .... $= P$

M max. (between loads) .... $= Pa$

M_x (when $x < a$) .... $= Px$

$\triangle$ max. (at center) .... $= \dfrac{Pa}{24\,EI}(3l^2 - 4a^2)$

$\triangle_x$ (when $x < a$) .... $= \dfrac{Px}{6\,EI}(3la - 3a^2 - x^2)$

$\triangle_x$ (when $x > a$ and $< (l - a)$) .... $= \dfrac{Pa}{6\,EI}(3lx - 3x^2 - a^2)$

PCI Design Handbook

DESIGN INFORMATION

11.1.3 (Cont.) Beam design equations and diagrams

(5) Simple Beam — two unequal concentrated loads unsymmetrically placed

$$R_1 = V_1 \ldots = \frac{P_1(l-a) + P_2 b}{l}$$

$$R_2 = V_2 \ldots = \frac{P_1 a + P_2(l-b)}{l}$$

$$V_x \quad (\text{when } x > a \text{ and } < (l-b)) \ldots = R_1 - P_1$$

$$M_1 \quad (\text{max. when } R_1 < P_1) \ldots = R_1 a$$

$$M_2 \quad (\text{max. when } R_2 < P_2) \ldots = R_2 b$$

$$M_x \quad (\text{when } x < a) \ldots = R_1 x$$

$$M_x \quad (\text{when } x > a \text{ and } < (l-b)) \ldots = R_1 x - P_1(x-a)$$

---

(6) Simple Beam — uniform load partially distributed

$$R_1 = V_1 \ (\text{max. when } a < c) \ldots = \frac{wb}{2l}(2c + b)$$

$$R_2 = V_2 \ (\text{max. when } a > c) \ldots = \frac{wb}{2l}(2a + b)$$

$$V_x \quad (\text{when } x > a \text{ and } < (a+b)) \ldots = R_1 - w(x-a)$$

$$M \ \text{max.} \quad \left(\text{at } x = a + \frac{R_1}{w}\right) \ldots = R_1\left(a + \frac{R_1}{2w}\right)$$

$$M_x \quad (\text{when } x < a) \ldots = R_1 x$$

$$M_x \quad (\text{when } x > a \text{ and } < (a+b)) \ldots = R_1 x - \frac{w}{2}(x-a)^2$$

$$M_x \quad (\text{when } x > (a+b)) \ldots = R_2(l-x)$$

---

(7) Simple Beam — load increasing uniformly to one end

$$R_1 = V_1 \ldots = \frac{W}{3}$$

$$R_2 = V_2 \ \text{max.} \ldots = \frac{2W}{3}$$

$$V_x \ldots = \frac{W}{3} - \frac{Wx^2}{l^2}$$

$$M \ \text{max.} \quad \left(\text{at } x = \frac{l}{\sqrt{3}} = .5774l\right) \ldots = \frac{2Wl}{9\sqrt{3}} = .1283\,Wl$$

$$M_x \ldots = \frac{Wx}{3l^2}(l^2 - x^2)$$

$$\Delta \ \text{max.} \quad \left(\text{at } x = l\sqrt{1 - \sqrt{\tfrac{8}{15}}} = .5193l\right) \ldots = .01304\frac{Wl^3}{EI}$$

$$\Delta_x \ldots = \frac{Wx}{180\,EI\,l^2}(3x^4 - 10l^2 x^2 + 7l^4)$$

---

(8) Simple Beam — load increasing uniformly to center

$$R = V \ldots = \frac{W}{2}$$

$$V_x \quad \left(\text{when } x < \frac{l}{2}\right) \ldots = \frac{W}{2l^2}(l^2 - 4x^2)$$

$$M \ \text{max.} \quad (\text{at center}) \ldots = \frac{Wl}{6}$$

$$M_x \quad \left(\text{when } x < \frac{l}{2}\right) \ldots = Wx\left(\frac{1}{2} - \frac{2x^2}{3l^2}\right)$$

$$\Delta \ \text{max.} \quad (\text{at center}) \ldots = \frac{Wl^3}{60\,EI}$$

$$\Delta_x \ldots = \frac{Wx}{480\,EI\,l^2}(5l^2 - 4x^2)^2$$

11.1.3 (Cont.) Beam design equations and diagrams

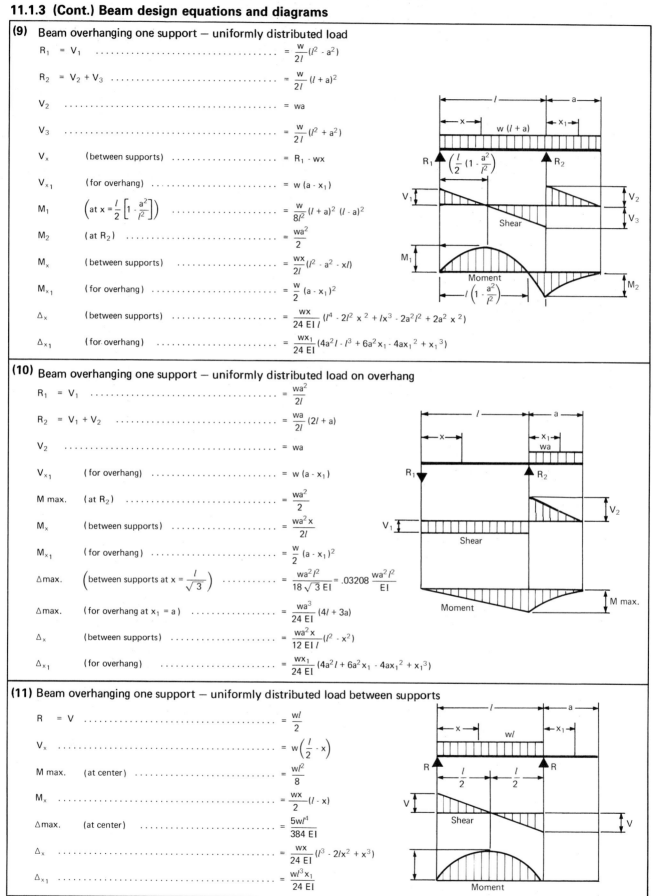

(9) Beam overhanging one support — uniformly distributed load

$R_1 = V_1$ .... $= \dfrac{w}{2l}(l^2 - a^2)$

$R_2 = V_2 + V_3$ .... $= \dfrac{w}{2l}(l + a)^2$

V_2 .... $= wa$

V_3 .... $= \dfrac{w}{2l}(l^2 + a^2)$

V_x (between supports) .... $= R_1 - wx$

V_{x_1} (for overhang) .... $= w(a - x_1)$

M_1 $\left(\text{at } x = \dfrac{l}{2}\left[1 - \dfrac{a^2}{l^2}\right]\right)$ .... $= \dfrac{w}{8l^2}(l + a)^2(l - a)^2$

M_2 (at R_2) .... $= \dfrac{wa^2}{2}$

M_x (between supports) .... $= \dfrac{wx}{2l}(l^2 - a^2 - xl)$

M_{x_1} (for overhang) .... $= \dfrac{w}{2}(a - x_1)^2$

Δ_x (between supports) .... $= \dfrac{wx}{24\,EI\,l}(l^4 - 2l^2 x^2 + lx^3 - 2a^2l^2 + 2a^2 x^2)$

Δ_{x_1} (for overhang) .... $= \dfrac{wx_1}{24\,EI}(4a^2l - l^3 + 6a^2 x_1 - 4ax_1^2 + x_1^3)$

(10) Beam overhanging one support — uniformly distributed load on overhang

$R_1 = V_1$ .... $= \dfrac{wa^2}{2l}$

$R_2 = V_1 + V_2$ .... $= \dfrac{wa}{2l}(2l + a)$

V_2 .... $= wa$

V_{x_1} (for overhang) .... $= w(a - x_1)$

M max. (at R_2) .... $= \dfrac{wa^2}{2}$

M_x (between supports) .... $= \dfrac{wa^2 x}{2l}$

M_{x_1} (for overhang) .... $= \dfrac{w}{2}(a - x_1)^2$

Δ max. $\left(\text{between supports at } x = \dfrac{l}{\sqrt{3}}\right)$ .... $= \dfrac{wa^2 l^2}{18\sqrt{3}\,EI} = .03208\,\dfrac{wa^2 l^2}{EI}$

Δ max. (for overhang at $x_1 = a$) ...... $= \dfrac{wa^3}{24\,EI}(4l + 3a)$

Δ_x (between supports) .... $= \dfrac{wa^2 x}{12\,EI\,l}(l^2 - x^2)$

Δ_{x_1} (for overhang) .... $= \dfrac{wx_1}{24\,EI}(4a^2l + 6a^2 x_1 - 4ax_1^2 + x_1^3)$

(11) Beam overhanging one support — uniformly distributed load between supports

$R = V$ .... $= \dfrac{wl}{2}$

V_x .... $= w\left(\dfrac{l}{2} - x\right)$

M max. (at center) .... $= \dfrac{wl^2}{8}$

M_x .... $= \dfrac{wx}{2}(l - x)$

Δ max. (at center) .... $= \dfrac{5wl^4}{384\,EI}$

Δ_x .... $= \dfrac{wx}{24\,EI}(l^3 - 2lx^2 + x^3)$

Δ_{x_1} .... $= \dfrac{wl^3 x_1}{24\,EI}$

11.1.3 (Cont.) Beam design equations and diagrams

(12)

Beam overhanging one support — concentrated load at any point between supports

$R_1 = V_1$ (max. when $a < b$) $\dots\dots\dots\dots\dots\dots = \dfrac{Pb}{l}$

$R_2 = V_2$ (max. when $a > b$) $\dots\dots\dots\dots\dots\dots = \dfrac{Pa}{l}$

M max. (at point of load) $\dots\dots\dots\dots\dots\dots = \dfrac{Pab}{l}$

M_x (when $x < a$) $\dots\dots\dots\dots\dots\dots = \dfrac{Pbx}{l}$

Δ max. $\left(\text{at } x = \sqrt{\dfrac{a\,(a+2b)}{3}} \text{ when } a > b\right) \dots\dots = \dfrac{Pab\,(a+2b)\sqrt{3a\,(a+2b)}}{27\,EI\,l}$

Δa (at point of load) $\dots\dots\dots\dots\dots\dots = \dfrac{Pa^2 b^2}{3\,EI\,l}$

Δ_x (when $x < a$) $\dots\dots\dots\dots\dots\dots = \dfrac{Pbx}{6\,EI\,l}\,(l^2 - b^2 - x^2)$

Δ_x (when $x > a$) $\dots\dots\dots\dots\dots\dots = \dfrac{Pa\,(l-x)}{6\,EI\,l}\,(2lx - x^2 - a^2)$

Δ_{x_1} $\dots\dots\dots\dots\dots\dots\dots\dots\dots\dots\dots = \dfrac{Pabx_1}{6\,EI\,l}\,(l+a)$

(13)

Beam overhanging one support — concentrated load at end of overhang

$R_1 = V_1$ $\dots\dots\dots\dots\dots\dots\dots\dots\dots\dots = \dfrac{Pa}{l}$

$R_2 = V_1 + V_2$ $\dots\dots\dots\dots\dots\dots\dots\dots\dots = \dfrac{P}{l}\,(l+a)$

V_2 $\dots\dots\dots\dots\dots\dots\dots\dots\dots\dots\dots\dots = P$

M max. (at R_2) $\dots\dots\dots\dots\dots\dots\dots\dots = Pa$

M_x (between supports) $\dots\dots\dots\dots\dots\dots = \dfrac{Pax}{l}$

M_{x_1} (for overhang) $\dots\dots\dots\dots\dots\dots\dots = P\,(a - x_1)$

Δ max. $\left(\text{between supports at } x = \dfrac{l}{\sqrt{3}}\right) \dots\dots = \dfrac{Pal^2}{9\sqrt{3}\,EI} = .06415\,\dfrac{Pal^2}{EI}$

Δ max. (for overhang at $x_1 = a$) $\dots\dots\dots\dots = \dfrac{Pa^2}{3\,EI}\,(l+a)$

Δ_x (between supports) $\dots\dots\dots\dots\dots = \dfrac{Pax}{6\,EI\,l}\,(l^2 - x^2)$

Δ_{x_1} (for overhang) $\dots\dots\dots\dots\dots\dots = \dfrac{Px_1}{6\,EI}\,(2al + 3ax_1 - x_1^2)$

DESIGN INFORMATION

11.1.3 (Cont.) Beam design equations and diagrams

(14)

Cantilever Beam — uniformly distributed load

$R = V$ .... $= wl$

V_x .... $= wx$

M max. (at fixed end) .... $= \dfrac{wl^2}{2}$

M_x .... $= \dfrac{wx^2}{2}$

Δ max. (at free end) .... $= \dfrac{wl^4}{8\,EI}$

Δ_x .... $= \dfrac{w}{24\,EI}(x^4 - 4l^3 x + 3l^4)$

(15)

Cantilever Beam — concentrated load at free end

$R = V$ .... $= P$

M max. (at fixed end) .... $= Pl$

M_x .... $= Px$

Δ max. (at free end) .... $= \dfrac{Pl^3}{3\,EI}$

Δ_x .... $= \dfrac{P}{6\,EI}(2l^3 - 3l^2 x + x^3)$

(16)

Cantilever Beam — concentrated load at any point

$R = V$ .... $= P$

M max. (at fixed end) .... $= Pb$

M_x (when $x > a$) .... $= P(x - a)$

Δ max. (at free end) .... $= \dfrac{Pb^2}{6\,EI}(3l - b)$

Δa (at point of load) .... $= \dfrac{Pb^3}{3\,EI}$

Δ_x (when $x < a$) .... $= \dfrac{Pb^2}{6\,EI}(3l - 3x - b)$

Δ_x (when $x > a$) .... $= \dfrac{P(l-x)^2}{6\,EI}(3b - l + x)$

(17)

Cantilever Beam — load increasing uniformly to fixed end

$R = V$ .... $= W$

V_x .... $= W\dfrac{x^2}{l^2}$

M max. (at fixed end) .... $= \dfrac{Wl}{3}$

M_x .... $= \dfrac{Wx^3}{3l^2}$

Δ max. (at free end) .... $= \dfrac{Wl^3}{15\,EI}$

Δ_x .... $= \dfrac{W}{60\,EI\,l^2}(x^5 - 5l^4 x + 4l^5)$

11.1.3 (Cont.) Beam design equations and diagrams

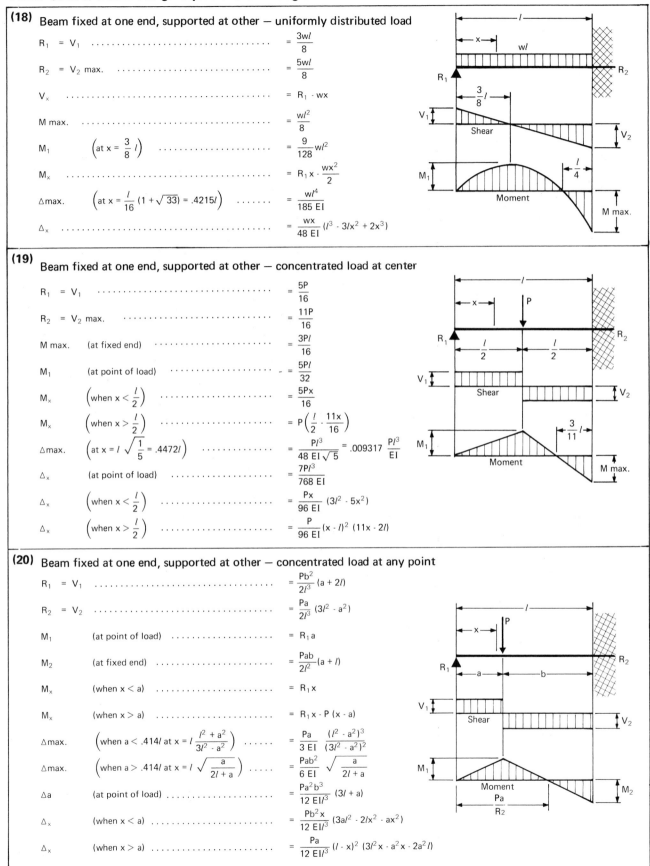

(18) Beam fixed at one end, supported at other — uniformly distributed load

$R_1 = V_1$ .... $= \dfrac{3wl}{8}$

$R_2 = V_2$ max. .... $= \dfrac{5wl}{8}$

V_x .... $= R_1 - wx$

M max. .... $= \dfrac{wl^2}{8}$

M_1 $\left(\text{at } x = \dfrac{3}{8}\,l\right)$ .... $= \dfrac{9}{128}wl^2$

M_x .... $= R_1 x - \dfrac{wx^2}{2}$

Δmax. $\left(\text{at } x = \dfrac{l}{16}(1+\sqrt{33}) = .4215l\right)$ ...... $= \dfrac{wl^4}{185\,EI}$

Δ_x .... $= \dfrac{wx}{48\,EI}(l^3 - 3lx^2 + 2x^3)$

(19) Beam fixed at one end, supported at other — concentrated load at center

$R_1 = V_1$ .... $= \dfrac{5P}{16}$

$R_2 = V_2$ max. .... $= \dfrac{11P}{16}$

M max. (at fixed end) .... $= \dfrac{3Pl}{16}$

M_1 (at point of load) .... $= \dfrac{5Pl}{32}$

M_x $\left(\text{when } x < \dfrac{l}{2}\right)$ .... $= \dfrac{5Px}{16}$

M_x $\left(\text{when } x > \dfrac{l}{2}\right)$ .... $= P\left(\dfrac{l}{2} - \dfrac{11x}{16}\right)$

Δmax. $\left(\text{at } x = l\sqrt{\dfrac{1}{5}} = .4472l\right)$ ...... $= \dfrac{Pl^3}{48\,EI\sqrt{5}} = .009317\,\dfrac{Pl^3}{EI}$

Δ_x (at point of load) .... $= \dfrac{7Pl^3}{768\,EI}$

Δ_x $\left(\text{when } x < \dfrac{l}{2}\right)$ .... $= \dfrac{Px}{96\,EI}(3l^2 - 5x^2)$

Δ_x $\left(\text{when } x > \dfrac{l}{2}\right)$ .... $= \dfrac{P}{96\,EI}(x-l)^2(11x - 2l)$

(20) Beam fixed at one end, supported at other — concentrated load at any point

$R_1 = V_1$ .... $= \dfrac{Pb^2}{2l^3}(a + 2l)$

$R_2 = V_2$ .... $= \dfrac{Pa}{2l^3}(3l^2 - a^2)$

M_1 (at point of load) .... $= R_1 a$

M_2 (at fixed end) .... $= \dfrac{Pab}{2l^2}(a + l)$

M_x (when $x < a$) .... $= R_1 x$

M_x (when $x > a$) .... $= R_1 x - P(x - a)$

Δmax. $\left(\text{when } a < .414l \text{ at } x = l\,\dfrac{l^2+a^2}{3l^2-a^2}\right)$ ...... $= \dfrac{Pa}{3\,EI}\dfrac{(l^2 - a^2)^3}{(3l^2 - a^2)^2}$

Δmax. $\left(\text{when } a > .414l \text{ at } x = l\sqrt{\dfrac{a}{2l+a}}\right)$ ..... $= \dfrac{Pab^2}{6\,EI}\sqrt{\dfrac{a}{2l+a}}$

Δa (at point of load) .... $= \dfrac{Pa^2 b^3}{12\,EIl^3}(3l + a)$

Δ_x (when $x < a$) .... $= \dfrac{Pb^2 x}{12\,EIl^3}(3al^2 - 2lx^2 - ax^2)$

Δ_x (when $x > a$) .... $= \dfrac{Pa}{12\,EIl^3}(l - x)^2(3l^2 x - a^2 x - 2a^2 l)$

DESIGN INFORMATION

11.1.3 (Cont.) Beam design equations and diagrams

(21)

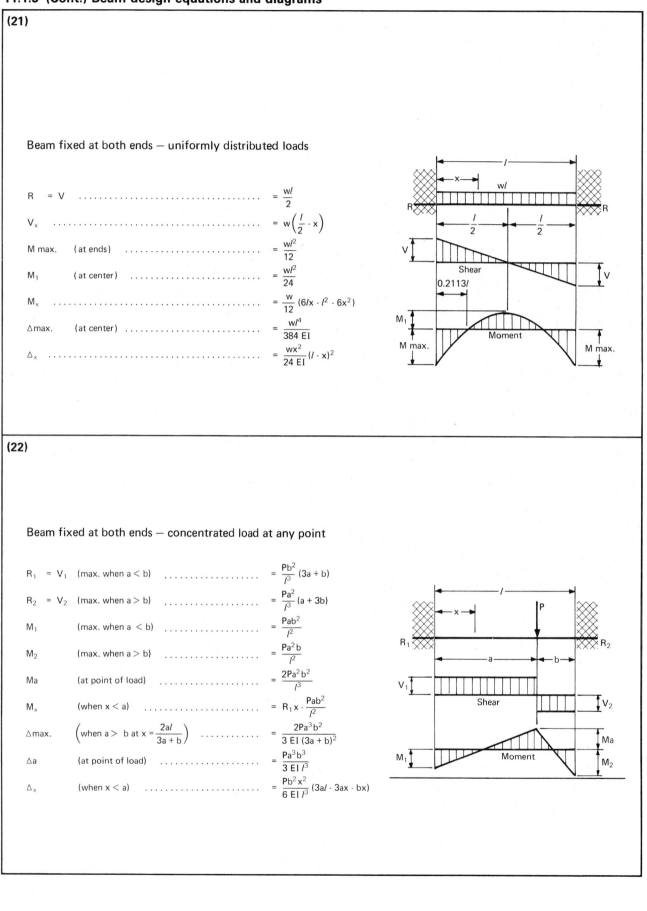

Beam fixed at both ends — uniformly distributed loads

$R = V$ .... $= \dfrac{wl}{2}$

V_x .... $= w\left(\dfrac{l}{2} - x\right)$

M max.　(at ends) .... $= \dfrac{wl^2}{12}$

M_1　(at center) .... $= \dfrac{wl^2}{24}$

M_x .... $= \dfrac{w}{12}(6lx - l^2 - 6x^2)$

$\triangle$ max.　(at center) .... $= \dfrac{wl^4}{384\,EI}$

$\triangle_x$ .... $= \dfrac{wx^2}{24\,EI}(l - x)^2$

(22)

Beam fixed at both ends — concentrated load at any point

$R_1 = V_1$　(max. when $a < b$) .... $= \dfrac{Pb^2}{l^3}(3a + b)$

$R_2 = V_2$　(max. when $a > b$) .... $= \dfrac{Pa^2}{l^3}(a + 3b)$

M_1　(max. when $a < b$) .... $= \dfrac{Pab^2}{l^2}$

M_2　(max. when $a > b$) .... $= \dfrac{Pa^2 b}{l^2}$

M_a　(at point of load) .... $= \dfrac{2Pa^2 b^2}{l^3}$

M_x　(when $x < a$) .... $= R_1 x - \dfrac{Pab^2}{l^2}$

$\triangle$ max.　$\left(\text{when } a > b \text{ at } x = \dfrac{2al}{3a + b}\right)$ .... $= \dfrac{2Pa^3 b^2}{3\,EI\,(3a + b)^2}$

$\triangle a$　(at point of load) .... $= \dfrac{Pa^3 b^3}{3\,EI\,l^3}$

$\triangle_x$　(when $x < a$) .... $= \dfrac{Pb^2 x^2}{6\,EI\,l^3}(3al - 3ax - bx)$

11.1.3 (Cont.) Beam design equations and diagrams

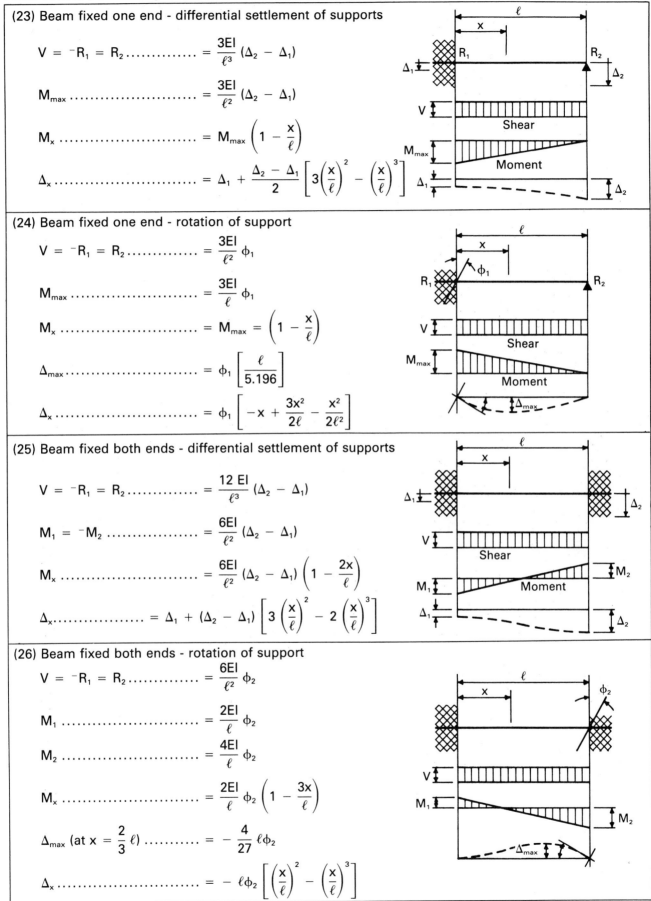

(23) Beam fixed one end - differential settlement of supports

$$V = {}^-R_1 = R_2 \ldots\ldots\ldots\ldots = \frac{3EI}{\ell^3}(\Delta_2 - \Delta_1)$$

$$M_{max} \ldots\ldots\ldots\ldots\ldots\ldots = \frac{3EI}{\ell^2}(\Delta_2 - \Delta_1)$$

$$M_x \ldots\ldots\ldots\ldots\ldots\ldots = M_{max}\left(1 - \frac{x}{\ell}\right)$$

$$\Delta_x \ldots\ldots\ldots\ldots\ldots\ldots = \Delta_1 + \frac{\Delta_2 - \Delta_1}{2}\left[3\left(\frac{x}{\ell}\right)^2 - \left(\frac{x}{\ell}\right)^3\right]$$

(24) Beam fixed one end - rotation of support

$$V = {}^-R_1 = R_2 \ldots\ldots\ldots\ldots = \frac{3EI}{\ell^2}\phi_1$$

$$M_{max} \ldots\ldots\ldots\ldots\ldots\ldots = \frac{3EI}{\ell}\phi_1$$

$$M_x \ldots\ldots\ldots\ldots\ldots\ldots = M_{max} = \left(1 - \frac{x}{\ell}\right)$$

$$\Delta_{max} \ldots\ldots\ldots\ldots\ldots\ldots = \phi_1\left[\frac{\ell}{5.196}\right]$$

$$\Delta_x \ldots\ldots\ldots\ldots\ldots\ldots = \phi_1\left[-x + \frac{3x^2}{2\ell} - \frac{x^2}{2\ell^2}\right]$$

(25) Beam fixed both ends - differential settlement of supports

$$V = {}^-R_1 = R_2 \ldots\ldots\ldots\ldots = \frac{12\,EI}{\ell^3}(\Delta_2 - \Delta_1)$$

$$M_1 = {}^-M_2 \ldots\ldots\ldots\ldots\ldots = \frac{6EI}{\ell^2}(\Delta_2 - \Delta_1)$$

$$M_x \ldots\ldots\ldots\ldots\ldots\ldots = \frac{6EI}{\ell^2}(\Delta_2 - \Delta_1)\left(1 - \frac{2x}{\ell}\right)$$

$$\Delta_x \ldots\ldots\ldots\ldots\ldots\ldots = \Delta_1 + (\Delta_2 - \Delta_1)\left[3\left(\frac{x}{\ell}\right)^2 - 2\left(\frac{x}{\ell}\right)^3\right]$$

(26) Beam fixed both ends - rotation of support

$$V = {}^-R_1 = R_2 \ldots\ldots\ldots\ldots = \frac{6EI}{\ell^2}\phi_2$$

$$M_1 \ldots\ldots\ldots\ldots\ldots\ldots = \frac{2EI}{\ell}\phi_2$$

$$M_2 \ldots\ldots\ldots\ldots\ldots\ldots = \frac{4EI}{\ell}\phi_2$$

$$M_x \ldots\ldots\ldots\ldots\ldots\ldots = \frac{2EI}{\ell}\phi_2\left(1 - \frac{3x}{\ell}\right)$$

$$\Delta_{max}\ \left(\text{at } x = \frac{2}{3}\ell\right) \ldots\ldots\ldots = -\frac{4}{27}\ell\phi_2$$

$$\Delta_x \ldots\ldots\ldots\ldots\ldots\ldots = -\ell\phi_2\left[\left(\frac{x}{\ell}\right)^2 - \left(\frac{x}{\ell}\right)^3\right]$$

11.1.4 Camber (deflection) and rotation coefficients for prestress force and loads*

Prestress Pattern	Equivalent Moment or Load	Equivalent Loading	Camber $+\ \uparrow$	End Rotation (+)	End Rotation (+)
(1)	$M = Pe$		$+\dfrac{Ml^2}{16\,EI}$	$+\dfrac{Ml}{3\,EI}$	$-\dfrac{Ml}{6\,EI}$
(2)	$M = Pe$		$+\dfrac{Ml^2}{16\,EI}$	$+\dfrac{Ml}{6\,EI}$	$-\dfrac{Ml}{3\,EI}$
(3)	$M = Pe$		$+\dfrac{Ml^2}{8\,EI}$	$+\dfrac{Ml}{2\,EI}$	$-\dfrac{Ml}{2\,EI}$
(4)	$N = \dfrac{4Pe'}{l}$		$+\dfrac{Nl^3}{48\,EI}$	$+\dfrac{Nl^2}{16\,EI}$	$-\dfrac{Nl^2}{16\,EI}$
(5)	$N = \dfrac{Pe'}{bl}$		$+\dfrac{b\,(3-4b^2)\,Nl^3}{24\,EI}$	$+\dfrac{b\,(1-b)\,Nl^2}{2\,EI}$	$-\dfrac{b\,(1-b)\,Nl^2}{2\,EI}$
(6)	$w = \dfrac{8Pe'}{l^2}$		$+\dfrac{5wl^4}{384\,EI}$	$+\dfrac{wl^3}{24\,EI}$	$-\dfrac{wl^3}{24\,EI}$
(7)	$w = \dfrac{8Pe'}{l^2}$		$+\dfrac{5wl^4}{768\,EI}$	$+\dfrac{9wl^3}{384\,EI}$	$-\dfrac{7wl^3}{384\,EI}$
(8)	$w = \dfrac{8Pe'}{l^2}$		$+\dfrac{5wl^4}{768\,EI}$	$+\dfrac{7wl^3}{384\,EI}$	$-\dfrac{9wl^3}{384\,EI}$
(9)	$w = \dfrac{4Pe'}{(0.5-b)\,l^2}$; $\ w_1 = \dfrac{w}{b}\,(0.5-b)$		$\left[\dfrac{5}{8}-\dfrac{b}{2}\,(3-2b^2)\right]\dfrac{wl^4}{48\,EI}$	$+\dfrac{(1-b)\,(1-2b)\,wl^3}{24\,EI}$	$-\dfrac{(1-b)\,(1-2b)\,wl^3}{24\,EI}$
(10)	$w = \dfrac{4Pe'}{(0.5-b)\,l^2}$; $\ w_1 = \dfrac{w}{b}\,(0.5-b)$		$\left[\dfrac{5}{16}-\dfrac{b}{4}\,(3-2b^2)\right]\dfrac{wl^4}{48\,EI}$	$\left[\dfrac{9}{8}-b\,(2-b)^2\right]\dfrac{wl^3}{48\,EI}$	$-\left[\dfrac{7}{8}+b\,(2-b^2)\right]\dfrac{wl^3}{48\,EI}$
(11)	$w = \dfrac{4Pe'}{(0.5-b)\,l^2}$; $\ w_1 = \dfrac{w}{b}\,(0.5-b)$		$\left[\dfrac{5}{16}-\dfrac{b}{4}\,(3-2b^2)\right]\dfrac{wl^4}{48\,EI}$	$\left[\dfrac{7}{8}-b\,(2-b^2)\right]\dfrac{wl^3}{48\,EI}$	$-\left[\dfrac{9}{8}+b\,(2-b)^2\right]\dfrac{wl^3}{48\,EI}$

*The tabulated values apply to the effects of prestressing. By adjusting the directional notation, they may also be used for the effects of loads.

For patterns 4-11, superimpose on 1, 2, or 3 for other C.G. locations.

DESIGN INFORMATION

11.1.5 Moments in beams with fixed ends

Loading	Moment at A	Moment at center	Moment at B
(1) A ⊢ l/2 ⊢ P ⊢ l/2 ⊢ B, span l	$-\dfrac{Pl}{8}$	$+\dfrac{Pl}{8}$	$-\dfrac{Pl}{8}$
(2) al, P	$-Pla(1-a)^2$		$-Pla^2(1-a)$
(3) $l/3$, P, $l/3$, P, $l/3$	$-\dfrac{2Pl}{9}$	$+\dfrac{Pl}{9}$	$-\dfrac{2Pl}{9}$
(4) $l/4$, P, $l/4$, P, $l/4$, P, $l/4$	$-\dfrac{5Pl}{16}$	$+\dfrac{3Pl}{16}$	$-\dfrac{5Pl}{16}$
(5) W (uniform full span)	$-\dfrac{Wl}{12}$	$+\dfrac{Wl}{24}$	$-\dfrac{Wl}{12}$
(6) al, W, al	$-\dfrac{Wl(1+2a-2a^2)}{12}$	$+\dfrac{Wl(1+2a+4a^2)}{24}$	$-\dfrac{Wl(1+2a-2a^2)}{12}$
(7) al, $W/2$, $W/2$, al	$-\dfrac{Wl(3a-2a^2)}{12}$	$+\dfrac{Wla^2}{6}$	$-\dfrac{Wl(3a-2a^2)}{12}$
(8) al, W	$-\dfrac{Wla(6-8a+3a^2)}{12}$		$-\dfrac{Wla^2(4-3a)}{12}$
(9) $l/2$, $l/2$, W (triangular, peak at center)	$-\dfrac{5Wl}{48}$	$+\dfrac{3Wl}{48}$	$-\dfrac{5Wl}{48}$
(10) W (triangular)	$-\dfrac{Wl}{10}$		$-\dfrac{Wl}{15}$

W = Total load on beam

DESIGN INFORMATION

11.1.6 Moving load placement for maximum moment and shear

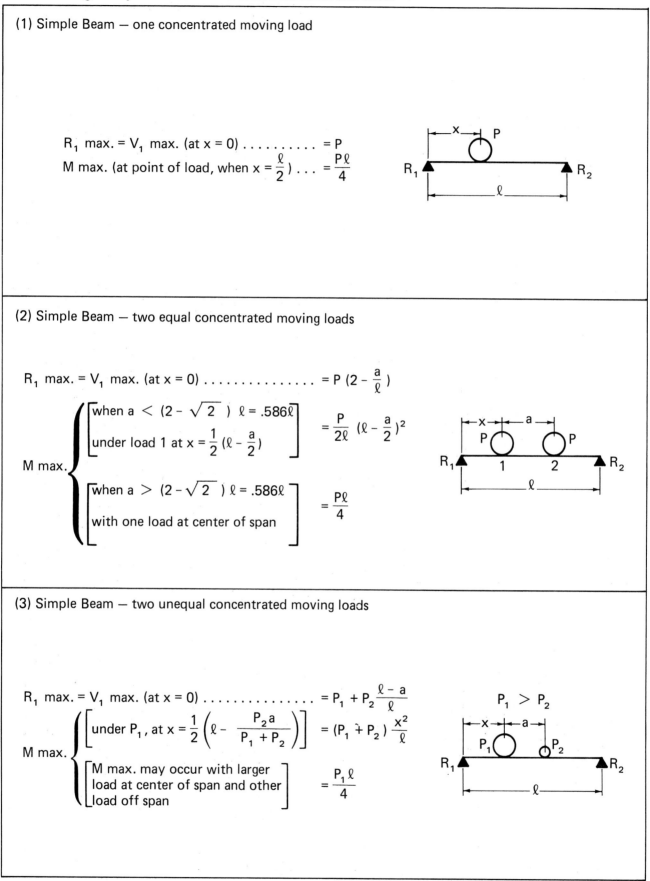

(1) Simple Beam — one concentrated moving load

R_1 max. $= V_1$ max. (at $x = 0$) . . . . $= P$

M max. (at point of load, when $x = \frac{\ell}{2}$) . . . $= \frac{P\ell}{4}$

(2) Simple Beam — two equal concentrated moving loads

R_1 max. $= V_1$ max. (at $x = 0$) . . . . $= P\left(2 - \frac{a}{\ell}\right)$

M max. $\begin{cases} \begin{bmatrix} \text{when } a < (2 - \sqrt{2})\,\ell = .586\ell \\ \text{under load 1 at } x = \frac{1}{2}\left(\ell - \frac{a}{2}\right) \end{bmatrix} = \frac{P}{2\ell}\left(\ell - \frac{a}{2}\right)^2 \\ \\ \begin{bmatrix} \text{when } a > (2 - \sqrt{2})\,\ell = .586\ell \\ \text{with one load at center of span} \end{bmatrix} = \frac{P\ell}{4} \end{cases}$

(3) Simple Beam — two unequal concentrated moving loads

R_1 max. $= V_1$ max. (at $x = 0$) . . . . $= P_1 + P_2\dfrac{\ell - a}{\ell}$

M max. $\begin{cases} \left[\text{under } P_1, \text{ at } x = \frac{1}{2}\left(\ell - \frac{P_2 a}{P_1 + P_2}\right)\right] = (P_1 + P_2)\,\dfrac{x^2}{\ell} \\ \\ \begin{bmatrix} \text{M max. may occur with larger} \\ \text{load at center of span and other} \\ \text{load off span} \end{bmatrix} = \dfrac{P_1\ell}{4} \end{cases}$

CONCRETE MATERIAL PROPERTIES

11.2.1 Table of concrete stresses

f'_c	$0.45 f'_c$	$0.6 f'_c$	$\sqrt{f'_c}$	$0.6\sqrt{f'_c}$	$2\sqrt{f'_c}$	$3.5\sqrt{f'_c}$	$4\sqrt{f'_c}$	$5\sqrt{f'_c}$	$6\sqrt{f'_c}$	$7.5\sqrt{f'_c}$	$12\sqrt{f'_c}$
3000	1350	1800	55	33	110	192	219	274	329	411	657
3500	1575	2100	59	35	118	207	237	296	355	444	710
4000	1800	2400	63	38	126	221	253	316	379	474	759
4500	2025	2700	67	40	134	235	268	335	402	503	805
5000	2250	3000	71	42	141	247	283	354	424	530	849
5500	2475	3300	74	44	148	260	297	371	445	556	890
6000	2700	3600	77	46	155	271	310	387	465	581	930
6500	2925	3900	81	48	161	281	322	403	484	605	967
7000	3150	4200	84	50	167	293	335	418	502	627	1004
7500	3375	4500	87	52	173	303	346	433	519	650	1039
8000	3600	4800	89	54	179	313	358	447	537	671	1073

11.2.2 Concrete modulus of elasticity as affected by unit weight and strength

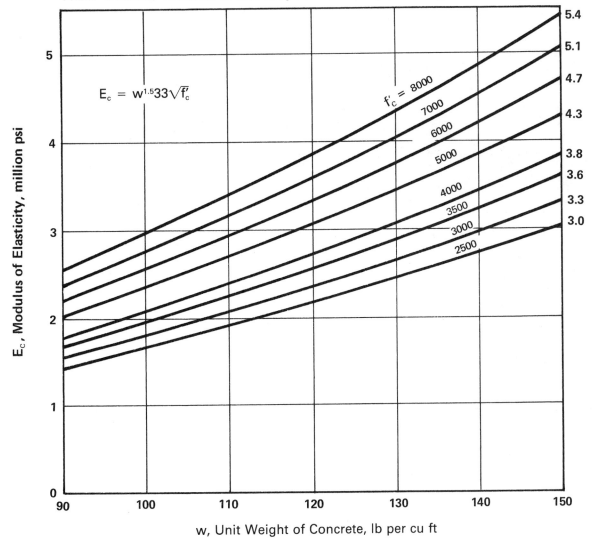

$$E_c = w^{1.5}33\sqrt{f'_c}$$

$f'_c = 8000$
7000
6000
5000
4000
3500
3000
2500

E_c, Modulus of Elasticity, million psi

w, Unit Weight of Concrete, lb per cu ft

11.2.3 Properties and design strengths of prestressing strand and wire

Seven-Wire Strand, f_{pu} = 270 ksi					
Nominal Diameter, in.	3/8	7/16	1/2	9/16	0.600
Area, sq in.	0.085	0.115	0.153	0.192	0.215
Weight, plf	0.29	0.40	0.53	0.65	0.74
0.7 f_{pu} A_{ps}, kips	16.1	21.7	28.9	36.3	40.7
0.75 f_{pu} A_{ps}, kips	17.2	23.3	31.0	38.9	43.5
0.8 f_{pu} A_{ps}, kips	18.4	24.8	33.0	41.4	46.5
f_{pu} A_{ps}, kips	23.0	31.0	41.3	51.8	58.1

Seven-Wire Strand, f_{pu} = 250 ksi					
Nominal Diameter, in.	1/4	5/16	3/8	7/16	1/2	0.600
Area, sq in.	0.036	0.058	0.080	0.108	0.144	0.215
Weight, plf	0.12	0.20	0.27	0.37	0.49	0.74
0.7 f_{pu} A_{ps}, kips	6.3	10.2	14.0	18.9	25.2	37.6
0.8 f_{pu} A_{ps}, kips	7.2	11.6	16.0	21.6	28.8	43.0
f_{pu} A_{ps}, kips	9.0	14.5	20.0	27.0	36.0	53.8

Three- and Four-Wire Strand, f_{pu} = 250 ksi				
Nominal Diameter, in.	1/4	5/16	3/8	7/16
No. of wires	3	3	3	4
Area, sq in.	0.036	0.058	0.075	0.106
Weight, plf	0.13	0.20	0.26	0.36
0.7 f_{pu} A_{ps}, kips	6.3	10.2	13.2	18.6
0.8 f_{pu} A_{ps}, kips	7.2	11.6	15.0	21.2
f_{pu} A_{ps}, kips	9.0	14.5	18.8	26.5

Prestressing Wire										
Diameter	0.105	0.120	0.135	0.148	0.162	0.177	0.192	0.196	0.250	0.276
Area, sq in.	0.0087	0.0114	0.0143	0.0173	0.0206	0.0246	0.0289	0.0302	0.0491	0.0598
Weight, plf	0.030	0.039	0.049	0.059	0.070	0.083	0.098	0.10	0.17	0.20
Ult. strength, f_{pu}, ksi	279	273	268	263	259	255	250	250	240	235
0.7 f_{pu} A_{ps}, kips	1.70	2.18	2.68	3.18	3.73	4.39	5.05	5.28	8.25	9.84
0.8 f_{pu} A_{ps}, kips	1.94	2.49	3.06	3.64	4.26	5.02	5.78	6.04	9.42	11.24
f_{pu} A_{ps}, kips	2.43	3.11	3.83	4.55	5.33	6.27	7.22	7.55	11.78	14.05

MATERIAL PROPERTIES
PRESTRESSING STEEL

11.2.4 Properties and design strengths of prestressing bars

Smooth Prestressing Bars, f_{pu} = 145 ksi*

Nominal Diameter, in.	3/4	7/8	1	1 1/8	1 1/4	1 3/8
Area, sq in.	0.442	0.601	0.785	0.994	1.227	1.485
Weight, plf	1.50	2.04	2.67	3.38	4.17	5.05
0.7 f_{pu} A_{ps}, kips	44.9	61.0	79.7	100.9	124.5	150.7
0.8 f_{pu} A_{ps}, kips	51.3	69.7	91.0	115.3	142.3	172.2
f_{pu} A_{ps}, kips	64.1	87.1	113.8	144.1	177.9	215.3

Smooth Prestressing Bars, f_{pu} = 160 ksi*

Nominal Diameter, in.	3/4	7/8	1	1 1/8	1 1/4	1 3/8
Area, sq in.	0.442	0.601	0.785	0.994	1.227	1.485
Weight, plf	1.50	2.04	2.67	3.38	4.17	5.05
0.7 f_{pu} A_{ps}, kips	49.5	67.3	87.9	111.3	137.4	166.3
0.8 f_{pu} A_{ps}, kips	56.6	77.0	100.5	127.2	157.0	190.1
f_{pu} A_{ps}, kips	70.7	96.2	125.6	159.0	196.3	237.6

Deformed Prestressing Bars

Nominal Diameter, in.	5/8	1	1	1 1/4	1 1/4	1 3/8
Area, sq. in.	0.28	0.85	0.85	1.25	1.25	1.58
Weight, plf	0.98	3.01	3.01	4.39	4.39	5.56
Ult. strength, f_{pu}, ksi	157	150	160*	150	160*	150
0.7 $f_{pu}A_{ps}$, kips	30.5	89.3	95.2	131.3	140.0	165.9
0.8 $f_{pu}A_{ps}$, kips	34.8	102.0	108.8	150.0	160.0	189.6
$f_{pu}A_{ps}$, kips	43.5	127.5	136.0	187.5	200.0	237.0

Stress-strain characteristics (all prestressing bars):

For design purposes, following assumptions are satisfactory:

E_s = 29,000 ksi

f_y = 0.95 f_{pu}

*Verify availability before specifying

11.2.5 Typical stress-strain curve, 7-wire stress-relieved and low-relaxation prestressing strand

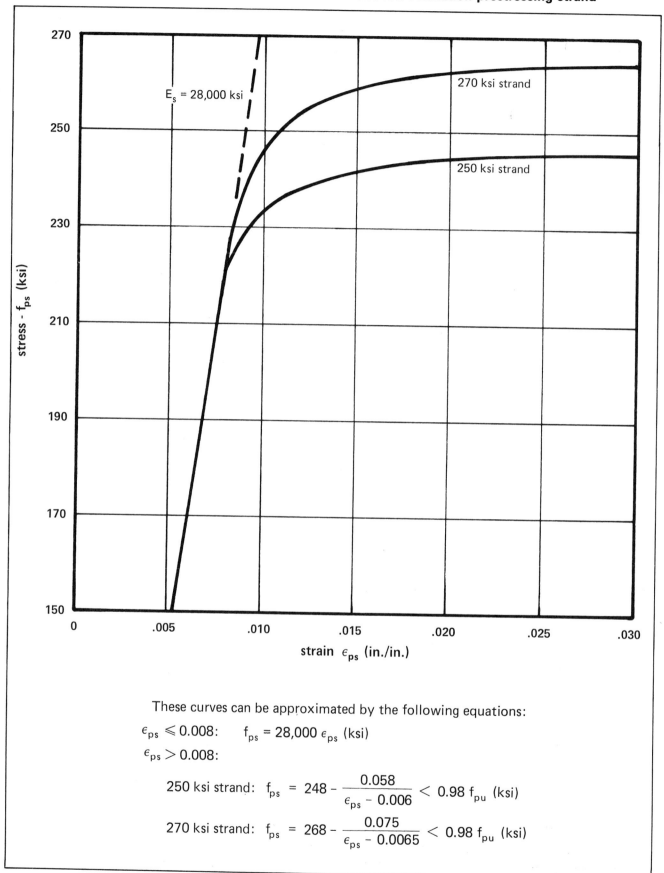

These curves can be approximated by the following equations:

$\epsilon_{ps} \leqslant 0.008$: $f_{ps} = 28,000\,\epsilon_{ps}$ (ksi)

$\epsilon_{ps} > 0.008$:

250 ksi strand: $f_{ps} = 248 - \dfrac{0.058}{\epsilon_{ps} - 0.006} < 0.98\,f_{pu}$ (ksi)

270 ksi strand: $f_{ps} = 268 - \dfrac{0.075}{\epsilon_{ps} - 0.0065} < 0.98\,f_{pu}$ (ksi)

MATERIAL PROPERTIES
REINFORCING BARS

11.2.6 Reinforcing bar data

ASTM STANDARD REINFORCING BARS

Bar Size Designation	Weight (lb per foot)	NOMINAL DIMENSIONS		
		Diameter (in.)	Area (sq. in.)	Perimeter (in.)
#3	0.376	0.375	0.11	1.178
#4	0.668	0.500	0.20	1.571
#5	1.043	0.625	0.31	1.963
#6	1.502	0.750	0.44	2.356
#7	2.044	0.875	0.60	2.749
#8	2.670	1.000	0.79	3.142
#9	3.400	1.128	1.00	3.544
#10	4.303	1.270	1.27	3.990
#11	5.313	1.410	1.56	4.430
#14	7.65	1.693	2.25	5.32
#18	13.60	2.257	4.00	7.09

STANDARD HOOKS / STIRRUP AND TIE-HOOKS

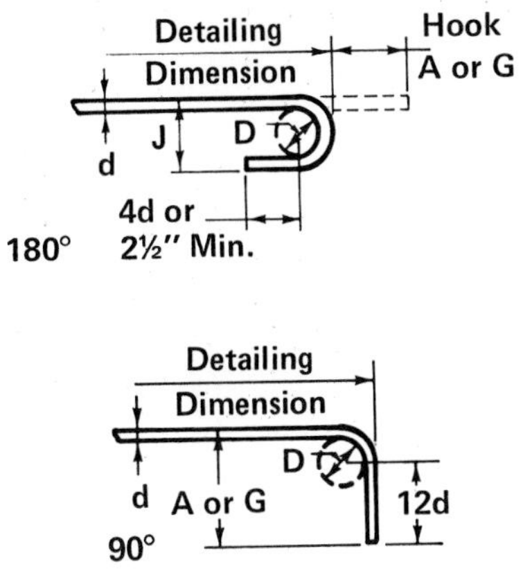

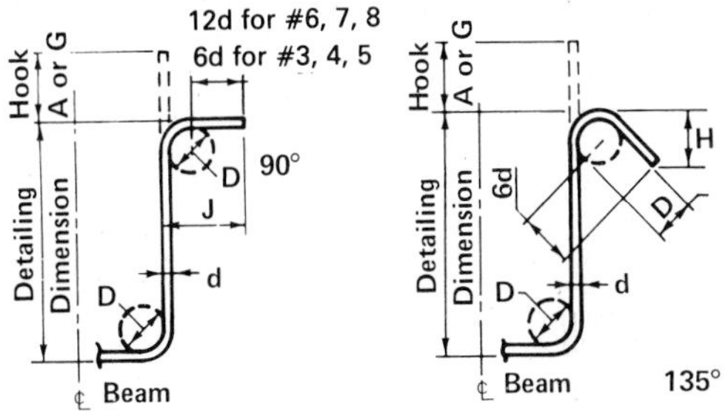

Bar Size	D	180 degree		90 deg	D	90 deg	135 degree	
		A or G	J	A or G		A or G	A or G	H
#3	2¼	5	3	6	1½	4	4	2½
#4	3	6	4	8	2	4½	4½	3
#5	3¾	7	5	10	2½	6	5½	3¾
#6	4½	8	6	1-0	4½	1-0	7¾	4½
#7	5¼	10	7	1-2	5¼	1-2	9	5¼
#8	6	11	8	1-4	6	1-4	10¼	6
#9	9½	1-3	11¾	1-7				
#10	10¾	1-5	1-1¼	1-10				
#11	12	1-7	1-2¾	2-0				
#14	18¼	2-3	1-9¾	2-7				
#18	24	3-0	2-4½	3-5				

MATERIAL PROPERTIES
REINFORCING BARS

11.2.7 Required development and lap lengths for Grade 60 bars*

Tension:

$$\ell_d = 2400\, A_b/\sqrt{f'_c};\ \text{min. } 24d_b \text{ or } 12 \text{ in.}$$

Compression development length:
$$\ell_d = 1200\, d_b/\sqrt{f'_c};\ \text{min. } 18d_b \text{ or } 8 \text{ in.}$$

Compression splice length:
compression ℓ_d; min. $30d_b$ or 12 in.

where:
A_b = area of individual bar, in.²
d_b = diameter of bar, in.

For limitations, see ACI 318-83, Chapter 12.

Multiply table values by:

1.4 for top reinforcement
1.33 for "all lightweight" concrete
1.18 for "sand-lightweight" concrete
0.8 for bar spacing 6" or more
     (3" from member face)

$\dfrac{A_{s\ req'd}}{A_{s\ prov'd}}$ for excess reinforcement

*for Grade 40 bars, required lengths are
two-thirds of the table values, but not
less than the required minimum lengths

Development and lap lengths in inches

Bar size	f'_c = 3000 psi Tension ℓ_d	$1.3\,\ell_d$	$1.7\,\ell_d$	Compression ℓ_d	f'_c = 4000 psi Tension ℓ_d	$1.3\,\ell_d$	$1.7\,\ell_d$	Compression ℓ_d	f'_c = 5000 psi Tension ℓ_d	$1.3\,\ell_d$	$1.7\,\ell_d$	Compression ℓ_d	Min. Comp. Splice
3	12	12	15	8	12	12	15	8	12	12	15	8	12
4	12	16	20	11	12	16	20	9	12	16	20	9	15
5	15	20	26	14	15	20	26	12	15	20	26	11	19
6	19	25	33	16	18	23	31	14	18	23	31	14	23
7	26	34	45	19	23	30	39	17	21	27	36	16	26
8	35	45	59	22	30	39	51	19	27	35	46	18	30
9	44	57	74	25	38	49	65	21	34	44	58	20	34
10	56	72	95	28	48	63	82	24	43	56	73	23	38
11	68	89	116	31	59	77	101	27	53	69	90	25	42

Bar size	f'_c = 6000 psi Tension ℓ_d	$1.3\,\ell_d$	$1.7\,\ell_d$	Compression ℓ_d	f'_c = 7000 psi Tension ℓ_d	$1.3\,\ell_d$	$1.7\,\ell_d$	Compression ℓ_d	f'_c = 8000 psi Tension ℓ_d	$1.3\,\ell_d$	$1.7\,\ell_d$	Compression ℓ_d	Min. Comp. Splice
3	12	12	15	8	12	12	15	8	12	12	15	8	12
4	12	16	20	9	12	16	20	9	12	16	20	9	15
5	15	20	26	11	15	20	26	11	15	20	26	11	19
6	18	23	31	14	18	23	31	14	18	23	31	14	23
7	21	27	36	16	21	27	36	16	21	27	36	16	26
8	24	32	42	18	24	31	41	18	24	31	41	18	30
9	31	40	53	20	29	37	49	20	27	35	46	20	34
10	39	51	67	23	36	47	62	23	34	44	58	23	38
11	48	63	82	26	45	58	76	26	42	54	71	26	42

11.2.8 Required embedment length for standard end hooks on Grade 60 bars*

$\ell_{dh} = 1200\, d_b/\sqrt{f'_c}$; min. $8d_b$ or 6 in.

where:
d_b = diameter of bar, in.

For limitations, see ACI 318-83, Sect. 12.5

Multiply table values by:

0.7 for side cover $\geq$ 2.5 in.
    and end cover (90° hook only) $\geq$ 2 in.
0.8 for ties or stirrups spaced $\geq 3d_b$
1.3 for lightweight concrete

$\dfrac{A_{s\ req'd}}{A_{s\ prov'd}}$ for excess reinforcement

*for Grade 40 bars, embedment
lengths are two-thirds of the
table values, but not less than
the minimum ℓ_{dh}

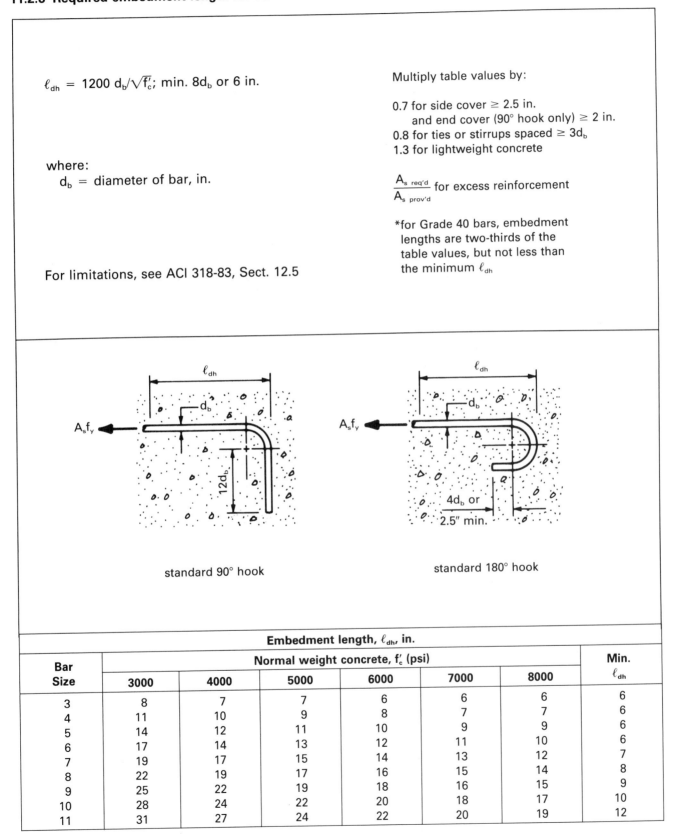

standard 90° hook   standard 180° hook

Embedment length, ℓ_{dh}, in.

Bar Size	Normal weight concrete, f'_c (psi)						Min. ℓ_{dh}
	3000	4000	5000	6000	7000	8000	
3	8	7	7	6	6	6	6
4	11	10	9	8	7	7	6
5	14	12	11	10	9	9	6
6	17	14	13	12	11	10	6
7	19	17	15	14	13	12	7
8	22	19	17	16	15	14	8
9	25	22	19	18	16	15	9
10	28	24	22	20	18	17	10
11	31	27	24	22	20	19	12

MATERIAL PROPERTIES
WELDED WIRE FABRIC

11.2.9 Common stock styles of welded wire fabric

Style Designation		Steel Area		Approx. Weight
		sq in. per ft		
Old Designation (By Steel Wire Gage)	New Designation (By W-Number)	Longit.	Trans.	lb per 100 sq ft
6x6-10x10	6x6-W1.4xW1.4	.029	.029	21
4x12-8x12**	4x12-W2.1xW0.9	.062	.009	25
6x6-8x8	6x6-W2.1xW2.1	.041	.041	30
4x4-10x10	4x4-W1.4xW1.4	.043	.043	31
4x12-7x11**	4x12-W2.5xW1.1	.074	.011	31
6x6-6x6*	6x6-W2.9xW2.9	.058	.058	42
4x4-8x8	4x4-W2.1xW2.1	.062	.062	44
6x6-4x4*	6x6-W4.0xW4.0	.080	.080	58
4x4-6x6	4x4-W2.9xW2.9	.087	.087	62
6x6-2x2*	6x6-W5.5xW5.5***	.110	.110	80
4x4-4x4*	4x4-W4.0xW4.0	.120	.120	85
4x4-3x3*	4x4-W4.7xW4.7	.141	.141	102
4x4-2x2*	4x4-W5.5xW5.5***	.165	.165	119

* Commonly available in 8 ft x 12 ft or 8 ft x 15 ft sheets.

** These items may be carried in sheets by various manufacturers in certain parts of the U.S. and Canada.

*** Exact W-Number size for 2 gage is 5.4.

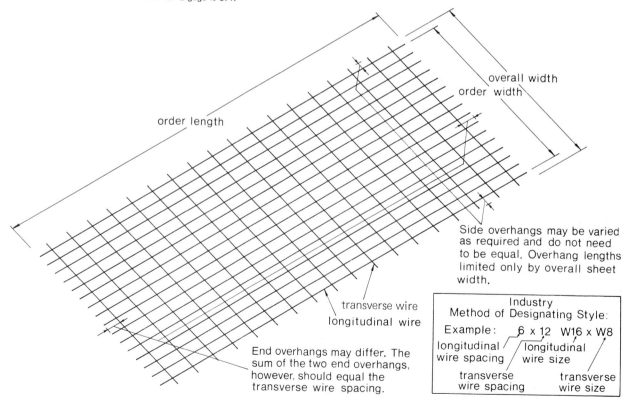

overall width
order width

order length

Side overhangs may be varied as required and do not need to be equal. Overhang lengths limited only by overall sheet width.

transverse wire
longitudinal wire

End overhangs may differ. The sum of the two end overhangs, however, should equal the transverse wire spacing.

Industry Method of Designating Style:

Example: 6 x 12 W16 x W8

longitudinal wire spacing / longitudinal wire size

transverse wire spacing / transverse wire size

MATERIAL PROPERTIES
WELDED WIRE FABRIC

11.2.10 Special welded wire fabric for double tee flanges*

Application	Style Designation	Steel Area sq. in. per ft.		Approx. Weight lb per 100 sq. ft.
		Longit.	Trans.	
8-ft wide DT, 2-in. flange	12 X 6-W1.4 X W2.5	.014	.050	23
10-ft wide DT, 2-in. flange	12 X 6-W2.0 X W4.0	.020	.080	35
10-ft wide DT, 2½-in. flange	12 X 6-W1.4 X W2.9	.014	.058	27

*See "Standardization of Welded Wire Fabric," *PCI Journal*, July/Aug., 1976

11.2.11 Wires used in welded wire fabric

Wire Size Number		Nominal Diameter in.	Nominal Weight plf	Area — sq in. per ft of width Center to Center Spacing, in.						
Smooth	Deformed			2	3	4	6	8	10	12
W31	D31	0.628	1.054	1.86	1.24	.93	.62	.465	.372	.31
W30	D30	0.618	1.020	1.80	1.20	.90	.60	.45	.36	.30
W28	D28	0.597	.952	1.68	1.12	.84	.56	.42	.336	.28
W26	D26	0.575	.934	1.56	1.04	.78	.52	.39	.312	.26
W24	D24	0.553	.816	1.44	.96	.72	.48	.36	.288	.24
W22	D22	0.529	.748	1.32	.88	.66	.44	.33	.264	.22
W20	D20	0.504	.680	1.20	.80	.60	.40	.30	.24	.20
W18	D18	0.478	.612	1.08	.72	.54	.36	.27	.216	.18
W16	D16	0.451	.544	.96	.64	.48	.32	.24	.192	.16
W14	D14	0.422	.476	.84	.56	.42	.28	.21	.168	.14
W12	D12	0.390	.408	.72	.48	.36	.24	.18	.144	.12
W11	D11	0.374	.374	.66	.44	.33	.22	.165	.132	.11
W10.5		0.366	.357	.63	.42	.315	.21	.157	.126	.105
W10	D10	0.356	.340	.60	.40	.30	.20	.15	.12	.10
W9.5		0.348	.323	.57	.38	.285	.19	.142	.114	.095
W9	D9	0.338	.306	.54	.36	.27	.18	.135	.108	.09
W8.5		0.329	.289	.51	.34	.255	.17	.127	.102	.085
W8	D8	0.319	.272	.48	.32	.24	.16	.12	.096	.08
W7.5		0.309	.255	.45	.30	.225	.15	.112	.09	.075
W7	D7	0.298	.238	.42	.28	.21	.14	.105	.084	.07
W6.5		0.288	.221	.39	.26	.195	.13	.097	.078	.065
W6	D6	0.276	.204	.36	.24	.18	.12	.09	.072	.06
W5.5		0.264	.187	.33	.22	.165	.11	.082	.066	.055
W5	D5	0.252	.170	.30	.20	.15	.10	.075	.06	.05
W4.5		0.240	.153	.27	.18	.135	.09	.067	.054	.045
W4	D4	0.225	.136	.24	.16	.12	.08	.06	.048	.04
W3.5		0.211	.119	.21	.14	.105	.07	.052	.042	.035
W3		0.195	.102	.18	.12	.09	.06	.045	.036	.03
W2.9		0.192	.098	.174	.116	.087	.058	.043	.035	.029
W2.5		0.178	.085	.15	.10	.075	.05	.037	.03	.025
W2.1		0.162	.070	.126	.084	.063	.042	.031	.025	.021
W2		0.159	.068	.12	.08	.06	.04	.03	.024	.02
W1.5		0.138	.051	.09	.06	.045	.03	.022	.018	.015
W1.4		0.135	.049	.084	.056	.042	.028	.021	.017	.014

SECTION PROPERTIES

11.3.1 Properties of geometric sections

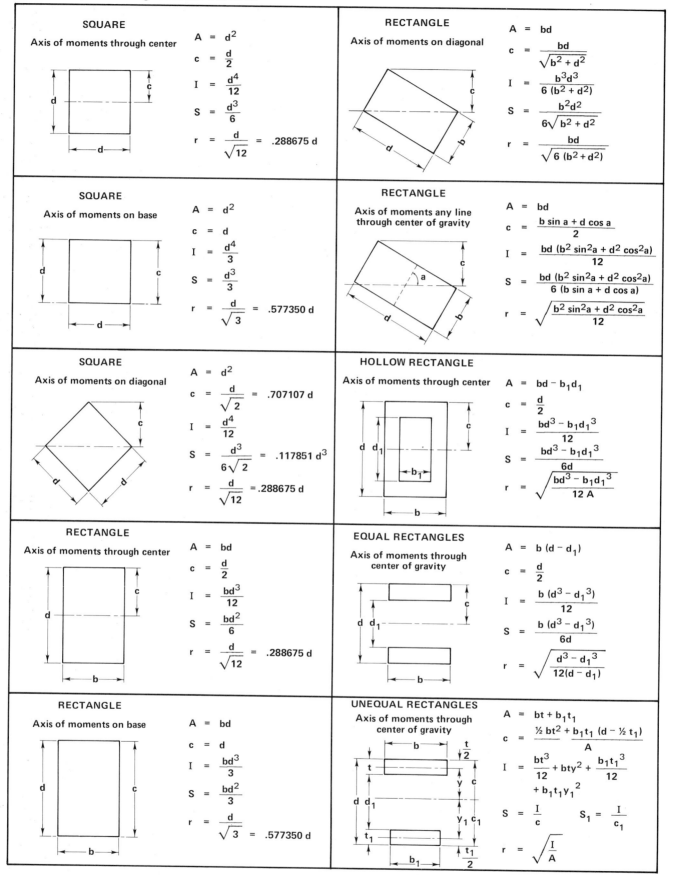

SQUARE
Axis of moments through center

$A = d^2$
$c = \dfrac{d}{2}$
$I = \dfrac{d^4}{12}$
$S = \dfrac{d^3}{6}$
$r = \dfrac{d}{\sqrt{12}} = .288675\,d$

RECTANGLE
Axis of moments on diagonal

$A = bd$
$c = \dfrac{bd}{\sqrt{b^2 + d^2}}$
$I = \dfrac{b^3 d^3}{6\,(b^2 + d^2)}$
$S = \dfrac{b^2 d^2}{6\sqrt{b^2 + d^2}}$
$r = \dfrac{bd}{\sqrt{6\,(b^2 + d^2)}}$

SQUARE
Axis of moments on base

$A = d^2$
$c = d$
$I = \dfrac{d^4}{3}$
$S = \dfrac{d^3}{3}$
$r = \dfrac{d}{\sqrt{3}} = .577350\,d$

RECTANGLE
Axis of moments any line through center of gravity

$A = bd$
$c = \dfrac{b \sin a + d \cos a}{2}$
$I = \dfrac{bd\,(b^2 \sin^2 a + d^2 \cos^2 a)}{12}$
$S = \dfrac{bd\,(b^2 \sin^2 a + d^2 \cos^2 a)}{6\,(b \sin a + d \cos a)}$
$r = \sqrt{\dfrac{b^2 \sin^2 a + d^2 \cos^2 a}{12}}$

SQUARE
Axis of moments on diagonal

$A = d^2$
$c = \dfrac{d}{\sqrt{2}} = .707107\,d$
$I = \dfrac{d^4}{12}$
$S = \dfrac{d^3}{6\sqrt{2}} = .117851\,d^3$
$r = \dfrac{d}{\sqrt{12}} = .288675\,d$

HOLLOW RECTANGLE
Axis of moments through center

$A = bd - b_1 d_1$
$c = \dfrac{d}{2}$
$I = \dfrac{bd^3 - b_1 d_1^3}{12}$
$S = \dfrac{bd^3 - b_1 d_1^3}{6d}$
$r = \sqrt{\dfrac{bd^3 - b_1 d_1^3}{12\,A}}$

RECTANGLE
Axis of moments through center

$A = bd$
$c = \dfrac{d}{2}$
$I = \dfrac{bd^3}{12}$
$S = \dfrac{bd^2}{6}$
$r = \dfrac{d}{\sqrt{12}} = .288675\,d$

EQUAL RECTANGLES
Axis of moments through center of gravity

$A = b\,(d - d_1)$
$c = \dfrac{d}{2}$
$I = \dfrac{b\,(d^3 - d_1^3)}{12}$
$S = \dfrac{b\,(d^3 - d_1^3)}{6d}$
$r = \sqrt{\dfrac{d^3 - d_1^3}{12(d - d_1)}}$

RECTANGLE
Axis of moments on base

$A = bd$
$c = d$
$I = \dfrac{bd^3}{3}$
$S = \dfrac{bd^2}{3}$
$r = \dfrac{d}{\sqrt{3}} = .577350\,d$

UNEQUAL RECTANGLES
Axis of moments through center of gravity

$A = bt + b_1 t_1$
$c = \dfrac{\frac{1}{2} bt^2 + b_1 t_1\,(d - \frac{1}{2} t_1)}{A}$
$I = \dfrac{bt^3}{12} + bty^2 + \dfrac{b_1 t_1^3}{12} + b_1 t_1 y_1^2$
$S = \dfrac{I}{c} \qquad S_1 = \dfrac{I}{c_1}$
$r = \sqrt{\dfrac{I}{A}}$

SECTION PROPERTIES

11.3.1 (Cont.) Properties of geometric sections

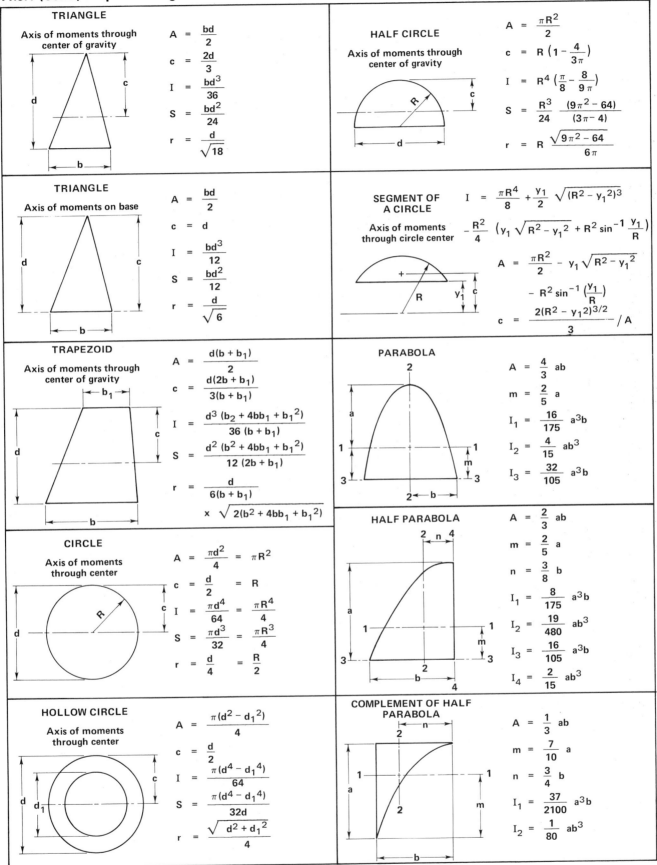

TRIANGLE

Axis of moments through center of gravity

$A = \dfrac{bd}{2}$

$c = \dfrac{2d}{3}$

$I = \dfrac{bd^3}{36}$

$S = \dfrac{bd^2}{24}$

$r = \dfrac{d}{\sqrt{18}}$

HALF CIRCLE

Axis of moments through center of gravity

$A = \dfrac{\pi R^2}{2}$

$c = R\left(1 - \dfrac{4}{3\pi}\right)$

$I = R^4\left(\dfrac{\pi}{8} - \dfrac{8}{9\pi}\right)$

$S = \dfrac{R^3}{24}\dfrac{(9\pi^2 - 64)}{(3\pi - 4)}$

$r = R\dfrac{\sqrt{9\pi^2 - 64}}{6\pi}$

TRIANGLE

Axis of moments on base

$A = \dfrac{bd}{2}$

$c = d$

$I = \dfrac{bd^3}{12}$

$S = \dfrac{bd^2}{12}$

$r = \dfrac{d}{\sqrt{6}}$

SEGMENT OF A CIRCLE

Axis of moments through circle center

$I = \dfrac{\pi R^4}{8} + \dfrac{y_1}{2}\sqrt{(R^2 - y_1^2)^3}$
$\quad - \dfrac{R^2}{4}\left(y_1\sqrt{R^2 - y_1^2} + R^2\sin^{-1}\dfrac{y_1}{R}\right)$

$A = \dfrac{\pi R^2}{2} - y_1\sqrt{R^2 - y_1^2}$
$\quad - R^2\sin^{-1}\left(\dfrac{y_1}{R}\right)$

$c = \dfrac{2(R^2 - y_1^2)^{3/2}}{3}/A$

TRAPEZOID

Axis of moments through center of gravity

$A = \dfrac{d(b + b_1)}{2}$

$c = \dfrac{d(2b + b_1)}{3(b + b_1)}$

$I = \dfrac{d^3(b^2 + 4bb_1 + b_1^2)}{36(b + b_1)}$

$S = \dfrac{d^2(b^2 + 4bb_1 + b_1^2)}{12(2b + b_1)}$

$r = \dfrac{d}{6(b + b_1)}$
$\quad \times \sqrt{2(b^2 + 4bb_1 + b_1^2)}$

PARABOLA

$A = \dfrac{4}{3}ab$

$m = \dfrac{2}{5}a$

$I_1 = \dfrac{16}{175}a^3 b$

$I_2 = \dfrac{4}{15}ab^3$

$I_3 = \dfrac{32}{105}a^3 b$

CIRCLE

Axis of moments through center

$A = \dfrac{\pi d^2}{4} = \pi R^2$

$c = \dfrac{d}{2} = R$

$I = \dfrac{\pi d^4}{64} = \dfrac{\pi R^4}{4}$

$S = \dfrac{\pi d^3}{32} = \dfrac{\pi R^3}{4}$

$r = \dfrac{d}{4} = \dfrac{R}{2}$

HALF PARABOLA

$A = \dfrac{2}{3}ab$

$m = \dfrac{2}{5}a$

$n = \dfrac{3}{8}b$

$I_1 = \dfrac{8}{175}a^3 b$

$I_2 = \dfrac{19}{480}ab^3$

$I_3 = \dfrac{16}{105}a^3 b$

$I_4 = \dfrac{2}{15}ab^3$

HOLLOW CIRCLE

Axis of moments through center

$A = \dfrac{\pi(d^2 - d_1^2)}{4}$

$c = \dfrac{d}{2}$

$I = \dfrac{\pi(d^4 - d_1^4)}{64}$

$S = \dfrac{\pi(d^4 - d_1^4)}{32d}$

$r = \dfrac{\sqrt{d^2 + d_1^2}}{4}$

COMPLEMENT OF HALF PARABOLA

$A = \dfrac{1}{3}ab$

$m = \dfrac{7}{10}a$

$n = \dfrac{3}{4}b$

$I_1 = \dfrac{37}{2100}a^3 b$

$I_2 = \dfrac{1}{80}ab^3$

SECTION PROPERTIES

11.3.1 (Cont.) Properties of geometric sections

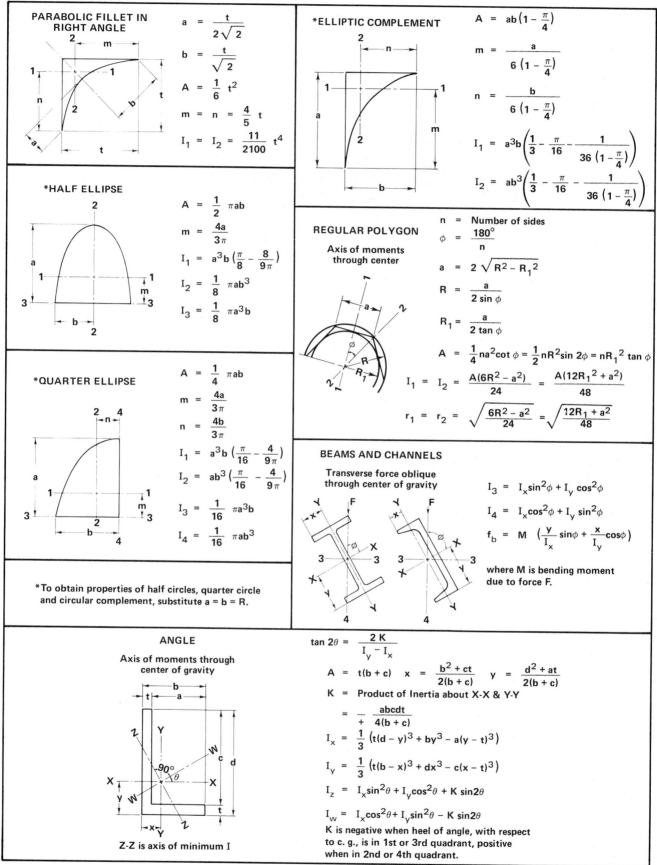

PARABOLIC FILLET IN RIGHT ANGLE

$$a = \frac{t}{2\sqrt{2}}$$

$$b = \frac{t}{\sqrt{2}}$$

$$A = \frac{1}{6} t^2$$

$$m = n = \frac{4}{5} t$$

$$I_1 = I_2 = \frac{11}{2100} t^4$$

*HALF ELLIPSE

$$A = \frac{1}{2} \pi ab$$

$$m = \frac{4a}{3\pi}$$

$$I_1 = a^3 b \left(\frac{\pi}{8} - \frac{8}{9\pi} \right)$$

$$I_2 = \frac{1}{8} \pi ab^3$$

$$I_3 = \frac{1}{8} \pi a^3 b$$

*QUARTER ELLIPSE

$$A = \frac{1}{4} \pi ab$$

$$m = \frac{4a}{3\pi}$$

$$n = \frac{4b}{3\pi}$$

$$I_1 = a^3 b \left(\frac{\pi}{16} - \frac{4}{9\pi} \right)$$

$$I_2 = ab^3 \left(\frac{\pi}{16} - \frac{4}{9\pi} \right)$$

$$I_3 = \frac{1}{16} \pi a^3 b$$

$$I_4 = \frac{1}{16} \pi ab^3$$

*To obtain properties of half circles, quarter circle and circular complement, substitute a = b = R.

*ELLIPTIC COMPLEMENT

$$A = ab \left(1 - \frac{\pi}{4} \right)$$

$$m = \frac{a}{6 \left(1 - \frac{\pi}{4} \right)}$$

$$n = \frac{b}{6 \left(1 - \frac{\pi}{4} \right)}$$

$$I_1 = a^3 b \left(\frac{1}{3} - \frac{\pi}{16} - \frac{1}{36 \left(1 - \frac{\pi}{4} \right)} \right)$$

$$I_2 = ab^3 \left(\frac{1}{3} - \frac{\pi}{16} - \frac{1}{36 \left(1 - \frac{\pi}{4} \right)} \right)$$

REGULAR POLYGON

Axis of moments through center

$$n = \text{Number of sides}$$

$$\phi = \frac{180°}{n}$$

$$a = 2 \sqrt{R^2 - R_1^2}$$

$$R = \frac{a}{2 \sin \phi}$$

$$R_1 = \frac{a}{2 \tan \phi}$$

$$A = \frac{1}{4} na^2 \cot \phi = \frac{1}{2} nR^2 \sin 2\phi = nR_1^2 \tan \phi$$

$$I_1 = I_2 = \frac{A(6R^2 - a^2)}{24} = \frac{A(12R_1^2 + a^2)}{48}$$

$$r_1 = r_2 = \sqrt{\frac{6R^2 - a^2}{24}} = \sqrt{\frac{12R_1^2 + a^2}{48}}$$

BEAMS AND CHANNELS

Transverse force oblique through center of gravity

$$I_3 = I_x \sin^2\phi + I_y \cos^2\phi$$

$$I_4 = I_x \cos^2\phi + I_y \sin^2\phi$$

$$f_b = M \left(\frac{y}{I_x} \sin\phi + \frac{x}{I_y} \cos\phi \right)$$

where M is bending moment due to force F.

ANGLE

Axis of moments through center of gravity

Z-Z is axis of minimum I

$$\tan 2\theta = \frac{2K}{I_y - I_x}$$

$$A = t(b + c) \qquad x = \frac{b^2 + ct}{2(b + c)} \qquad y = \frac{d^2 + at}{2(b + c)}$$

$$K = \text{Product of Inertia about X-X \& Y-Y}$$

$$= \underset{+}{-} \frac{abcdt}{4(b + c)}$$

$$I_x = \frac{1}{3} \left(t(d - y)^3 + by^3 - a(y - t)^3 \right)$$

$$I_y = \frac{1}{3} \left(t(b - x)^3 + dx^3 - c(x - t)^3 \right)$$

$$I_z = I_x \sin^2\theta + I_y \cos^2\theta + K \sin 2\theta$$

$$I_w = I_x \cos^2\theta + I_y \sin^2\theta - K \sin 2\theta$$

K is negative when heel of angle, with respect to c. g., is in 1st or 3rd quadrant, positive when in 2nd or 4th quadrant.

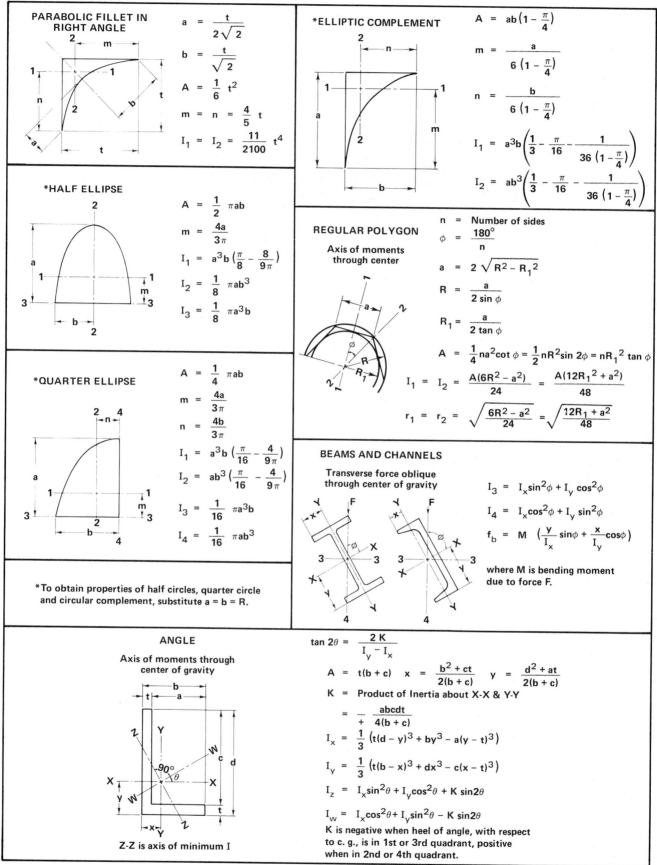

SECTION PROPERTIES

11.3.2 Coefficients for determining section properties of T-Beams

Moment of Inertia $\quad I_g = C_I \dfrac{b_w h^3}{12}$

Section Modulus $\quad Z_b = C_Z \dfrac{b_w h^2}{6}$

Distance to Bottom Fiber $\quad y_b = C_y h$

Center of Gravity of Section

h/t	b/b_w									
	1	2	3	4	5	6	7	8	9	10
Moment Of Inertia — Values Of C_I										
2	1.000	1.375	1.625	1.825	2.000	2.160	2.312	2.458	2.600	2.738
3	1.000	1.370	1.607	1.777	1.910	2.018	2.111	2.192	2.265	2.333
4	1.000	1.353	1.593	1.770	1.906	2.015	2.106	2.183	2.250	2.308
5	1.000	1.328	1.564	1.744	1.885	2.000	2.095	2.176	2.245	2.306
6	1.000	1.302	1.530	1.708	1.851	1.970	2.069	2.154	2.227	2.291
7	1.000	1.278	1.495	1.669	1.813	1.932	2.034	2.122	2.198	2.266
8	1.000	1.257	1.463	1.632	1.773	1.893	1.996	2.085	2.164	2.233
9	1.000	1.238	1.433	1.596	1.734	1.853	1.956	2.046	2.126	2.197
10	1.000	1.221	1.407	1.563	1.698	1.815	1.917	2.007	2.088	2.160
11	1.000	1.207	1.382	1.533	1.664	1.778	1.879	1.969	2.049	2.122
12	1.000	1.194	1.361	1.505	1.632	1.744	1.843	1.932	2.012	2.085
13	1.000	1.183	1.341	1.480	1.603	1.712	1.809	1.897	1.977	2.049
14	1.000	1.172	1.324	1.457	1.576	1.682	1.778	1.864	1.943	2.015
15	1.000	1.163	1.308	1.436	1.551	1.654	1.748	1.833	1.911	1.982
Section Modulus — Values Of C_Z										
2	1.000	1.178	1.300	1.403	1.500	1.592	1.681	1.770	1.857	1.943
3	1.000	1.174	1.269	1.333	1.383	1.424	1.461	1.494	1.526	1.555
4	1.000	1.176	1.275	1.339	1.386	1.422	1.452	1.477	1.500	1.519
5	1.000	1.171	1.273	1.341	1.390	1.428	1.458	1.483	1.504	1.523
6	1.000	1.163	1.266	1.336	1.388	1.428	1.460	1.486	1.508	1.527
7	1.000	1.154	1.256	1.328	1.382	1.424	1.457	1.485	1.509	1.528
8	1.000	1.145	1.245	1.317	1.372	1.416	1.451	1.480	1.505	1.526
9	1.000	1.137	1.234	1.306	1.362	1.406	1.443	1.473	1.499	1.521
10	1.000	1.129	1.223	1.294	1.350	1.396	1.433	1.464	1.491	1.514
11	1.000	1.122	1.213	1.283	1.339	1.385	1.422	1.455	1.482	1.506
12	1.000	1.115	1.203	1.272	1.328	1.373	1.412	1.444	1.472	1.497
13	1.000	1.109	1.194	1.262	1.317	1.362	1.401	1.434	1.462	1.487
14	1.000	1.104	1.186	1.252	1.306	1.352	1.390	1.424	1.452	1.478
15	1.000	1.099	1.178	1.243	1.296	1.341	1.380	1.413	1.422	1.468
Bottom Fiber Distance — Values Of C_y										
2	0.500	0.583	0.625	0.650	0.666	0.678	0.687	0.694	0.700	0.704
3	0.500	0.583	0.633	0.666	0.690	0.708	0.722	0.733	0.742	0.750
4	0.500	0.575	0.625	0.660	0.687	0.708	0.725	0.738	0.750	0.759
5	0.500	0.566	0.614	0.650	0.677	0.700	0.718	0.733	0.746	0.757
6	0.500	0.559	0.604	0.638	0.666	0.689	0.708	0.724	0.738	0.750
7	0.500	0.553	0.595	0.628	0.655	0.678	0.697	0.714	0.728	0.741
8	0.500	0.548	0.587	0.619	0.645	0.668	0.687	0.704	0.718	0.731
9	0.500	0.544	0.580	0.611	0.636	0.658	0.677	0.694	0.709	0.722
10	0.500	0.540	0.575	0.603	0.628	0.650	0.668	0.685	0.700	0.713
11	0.500	0.537	0.569	0.597	0.621	0.642	0.660	0.676	0.691	0.704
12	0.500	0.535	0.565	0.591	0.614	0.634	0.652	0.668	0.683	0.696
13	0.500	0.532	0.561	0.586	0.608	0.628	0.645	0.661	0.675	0.688
14	0.500	0.530	0.558	0.581	0.603	0.622	0.639	0.654	0.668	0.681
15	0.500	0.529	0.554	0.577	0.598	0.616	0.633	0.648	0.662	0.675

SECTION PROPERTIES

11.3.2 (Cont.) Coefficients for determining section properties of T-Beams b/b_w = 11 to 20

Moment of Inertia $I_g = C_I \dfrac{b_w h^3}{12}$

Section Modulus $Z_b = C_Z \dfrac{b_w h^2}{6}$

Distance to Bottom Fiber $Y_b = C_y h$

h/t	b/b_w									
	11	12	13	14	15	16	17	18	19	20
Moment Of Inertia — Values Of C_I										
2	2.875	3.009	3.142	3.275	3.406	3.536	3.666	3.796	3.925	4.053
3	2.396	2.455	2.511	2.564	2.616	2.666	2.715	2.762	2.809	2.855
4	2.361	2.409	2.453	2.493	2.531	2.566	2.600	2.631	2.661	2.690
5	2.360	2.408	2.451	2.490	2.526	2.560	2.590	2.619	2.646	2.672
6	2.348	2.398	2.444	2.485	2.523	2.557	2.589	2.618	2.645	2.671
7	2.325	2.379	2.427	2.470	2.510	2.546	2.579	2.610	2.639	2.666
8	2.295	2.351	2.401	2.447	2.488	2.527	2.562	2.595	2.625	2.653
9	2.261	2.318	2.370	2.418	2.462	2.502	2.538	2.573	2.604	2.634
10	2.225	2.283	2.337	2.386	2.431	2.473	2.511	2.547	2.580	2.611
11	2.188	2.247	2.302	2.352	2.398	2.441	2.481	2.518	2.552	2.584
12	2.151	2.211	2.267	2.318	2.365	2.409	2.449	2.487	2.522	2.556
13	2.115	2.176	2.232	2.284	2.331	2.376	2.417	2.456	2.492	2.526
14	2.081	2.142	2.198	2.250	2.298	2.343	2.385	2.424	2.461	2.496
15	2.048	2.108	2.165	2.217	2.265	2.311	2.353	2.393	2.430	2.466
Section Modulus — Values Of C_Z										
2	2.029	2.114	2.200	2.284	2.369	2.454	2.538	2.622	2.706	2.790
3	1.583	1.611	1.637	1.663	1.689	1.714	1.739	1.763	1.787	1.811
4	1.537	1.554	1.570	1.584	1.598	1.612	1.625	1.637	1.649	1.661
5	1.539	1.553	1.566	1.578	1.589	1.600	1.609	1.618	1.627	1.635
6	1.544	1.558	1.571	1.583	1.593	1.603	1.612	1.620	1.628	1.635
7	1.546	1.561	1.574	1.586	1.597	1.607	1.616	1.624	1.632	1.639
8	1.544	1.560	1.574	1.587	1.598	1.609	1.618	1.627	1.634	1.642
9	1.540	1.557	1.572	1.585	1.597	1.608	1.618	1.627	1.635	1.643
10	1.534	1.552	1.567	1.581	1.594	1.605	1.616	1.625	1.634	1.642
11	1.527	1.545	1.561	1.576	1.589	1.601	1.612	1.622	1.631	1.640
12	1.518	1.537	1.554	1.569	1.583	1.596	1.607	1.618	1.627	1.636
13	1.509	1.529	1.547	1.562	1.577	1.589	1.601	1.612	1.622	1.631
14	1.500	1.520	1.538	1.554	1.569	1.583	1.595	1.606	1.617	1.626
15	1.491	1.511	1.530	1.546	1.561	1.575	1.588	1.600	1.610	1.620
Bottom Fiber Distance — Values Of C_y										
2	0.708	0.711	0.714	0.716	0.718	0.720	0.722	0.723	0.725	0.726
3	0.756	0.761	0.766	0.770	0.774	0.777	0.780	0.783	0.785	0.787
4	0.767	0.775	0.781	0.786	0.791	0.796	0.800	0.803	0.806	0.809
5	0.766	0.775	0.782	0.788	0.794	0.800	0.804	0.809	0.813	0.816
6	0.760	0.769	0.777	0.785	0.791	0.797	0.803	0.807	0.812	0.816
7	0.752	0.761	0.770	0.778	0.785	0.792	0.798	0.803	0.808	0.813
8	0.743	0.753	0.762	0.770	0.778	0.785	0.791	0.797	0.802	0.807
9	0.733	0.744	0.753	0.762	0.770	0.777	0.784	0.790	0.796	0.801
10	0.725	0.735	0.745	0.754	0.762	0.770	0.776	0.783	0.789	0.794
11	0.716	0.727	0.737	0.746	0.754	0.762	0.769	0.775	0.782	0.787
12	0.708	0.719	0.729	0.738	0.746	0.754	0.761	0.768	0.775	0.780
13	0.700	0.711	0.721	0.730	0.739	0.747	0.754	0.761	0.767	0.774
14	0.693	0.704	0.714	0.723	0.732	0.740	0.747	0.754	0.761	0.767
15	0.686	0.697	0.707	0.716	0.725	0.733	0.740	0.747	0.754	0.760

PCI Design Handbook

METRIC CONVERSION

11.4 Conversion to International System of Units (SI)

To convert from	to	multiply by
Length		
inch (in.)	millimeter (mm)	25.4
inch (in.)	meter (m)	0.0254
foot (ft)	meter (m)	0.3048
yard (yd)	meter (m)	0.9144
Area		
square foot (sq ft)	square meter (sq m)	0.09290
square inch (sq in.)	square millimeter (sq mm)	645.2
square inch (sq in.)	square meter (sq m)	0.0006452
square yard (sq yd)	square meter (sq m)	0.8361
Volume		
cubic inch (cu in.)	cubic meter (cu m)	0.00001639
cubic foot (cu ft)	cubic meter (cu m)	0.02832
cubic yard (cu yd)	cubic meter (cu m)	0.7646
gallon (gal) Can. liquid*	liter	4.546
gallon (gal) Can. liquid*	cubic meter (cu m)	0.004546
gallon (gal) U.S. liquid*	liter	3.785
gallon (gal) U.S. liquid*	cubic meter (cu m)	0.003785
Force		
kip	kilogram (kgf)	453.6
kip	newton (N)	4448.0
pound (lb)	kilogram (kgf)	0.4536
pound (lb)	newton (N)	4.448
Pressure or Stress		
kips/square inch (ksi)	megapascal (MPa)**	6.895
pound/square foot (psf)	kilopascal (kPa)**	0.04788
pound/square inch (psi)	kilopascal (kPa)**	6.895
pound/square inch (psi)	megapascal (MPa)**	0.006895
pound/square foot (psf)	kilogram/square meter (kgf/sq m)	4.882
Mass		
pound (avdp)	kilogram (kg)	0.4536
ton (short, 2000 lb)	kilogram (kg)	907.2
ton (short, 2000 lb)	tonne (t)	0.9072
grain	kilogram (kg)	0.00006480
tonne (t)	kilogram (kg)	1000
Mass (weight) per Length		
kip/linear foot (klf)	kilogram/meter (kg/m)	0.001488
pound/linear foot (plf)	kilogram/meter kg/m	1.488
pound/linear foot (plf)	newton/meter (N/m)	14.593

*One U.S. gallon equals 0.8321 Canadian gallon
**A pascal equals one newton/square meter

METRIC CONVERSION

To convert from	to	multiply by
Mass per volume (density)		
pound/cubic foot (pcf)	kilogram/cubic meter (kg/cu m)	16.02
pound/cubic yard (pcy)	kilogram/cubic meter (kg/cu m)	0.5933
Bending Moment or Torque		
inch-pound (in.-lb)	newton-meter	0.1130
foot-pound (ft-lb)	newton-meter	1.356
foot-kip (ft-k)	newton-meter	1356
Temperature		
degree Fahrenheit (deg F)	degree Celsius (C)	$t_C = (t_F - 32)/1.8$
degree Fahrenheit (deg F)	degree Kelvin (K)	$t_K = (t_F + 459.7)/1.8$
Energy		
British thermal unit (Btu)	joule (j)	1056
kilowatt-hour (kwh)	joule (j)	3,600,000
Power		
horsepower (hp) (550 ft lb/sec)	watt (W)	745.7
Velocity		
mile/hour (mph)	kilometer/hour	1.609
mile/hour (mph)	meter/second (m/s)	0.4470
Other		
Section modulus (in.^{3})	mm^3	16.387
Moment of inertia (in.^{4})	mm^4	416.231
Coefficient of heat transfer (Btu/ft^2/h/°F)	W/m^2/°C	5.678
Modulus of elasticity (psi)	MPa	0.006895
Thermal conductivity (Btu-in./ft^2/h/°F)	Wm/m^2/°C	0.1442
Thermal expansion in./in./°F	mm/mm/°C	1.800
Area/length (in.^{2}/ft)	mm^2/m	2116.80

INDEX

Page No.

Absorption...7-22
Absorption, sound..................................9-26
Acoustical properties.............................9-23
Aggregates 1-2
Aircraft cable 5-11
Air entrainment.......................................1-3
Alignment connections...........................7-13
Anchorage...................................4-39, 6-23
Angles.............................6-17, 6-61, 7-9
Architectural precast concrete 7-2

Balance point...4-47
Base plates 6-21, 6-62
Beam design 5-15, 11-4, 11-13
Beam ledges..6-32
Beam load tables....................................2-44
Bearing 6-19, 6-41, 7-7
Bearing pads 6-14
Bilinear behavior 4-44
Bolts 6-10, 6-59, 7-11
Bracing 5-25
Brackets, concrete.....................6-23, 6-63, 7-8
Brackets, steel 6-12, 6-25, 6-68, 7-8
Brick veneers 7-21
Buckling 3-22, 5-14

Cable 5-11, 5-27
Camber......................2-4, 4-42, 4-73, 11-12
Camber multipliers..................................4-46
Cantilever design....................................4-58
Carbon equivalent...................................6-4
Caulking 9-64
Cazaly hanger 6-34
Cement 1-2
Certification, plant...................................9-63
Chemical resistance 1-5
Chloride ions 9-62
Chloroprene pads 6-15
Clay product veneers.............................7-21
Clearances...8-7
Coatings 9-64
Coefficient of expansion 1-4, 3-9
Coefficient of friction.....................6-15, 6-18
Column bases 3-43, 6-21, 6-62
Column covers 7-18
Column design 3-20, 3-22, 4-47
Columns, end reinforcement....................4-50
Columns, fixed base.................................3-43
Columns, precast....................................2-49
Columns, prestressed 2-47
Composite members 4-16, 4-27, 4-59, 9-55
Compression members, design...................4-47
Compressive strength............................. 1-2
Computer modeling.........................3-48, 3-55
Concentrated loads.................................4-57

Page No.

Concrete materials 1-2, 10-6, 10-15, 10-28
Concrete strengths, load tables.....................2-4
Concrete stresses....................................11-15
Condensation..9-20
Connections 3-28, 3-59, 6-3, 7-4, 9-56
Construction loads 5-27, 7-14
Continuity 4-58
Contracts 10-32
Corbels 6-23, 6-63, 7-8
Corrosion 1-5, 9-62
Coupled shear walls................................3-32
Couplers 6-6, 9-69
Cover 1-5, 9-44, 9-62
Crack control 4-14, 5-3, 9-67
Creep 1-4, 3-6, 4-39
Curing 1-2

Dapped-ends 6-28
Dead loads 11-2
Deflection.........................4-42, 4-73, 11-12
Deformed bar anchors 6-10
Degree days 9-3, 9-12
Delivery...........................10-9, 10-15, 10-29
Design equations...........................11-4, 11-13
Development length.................4-20, 11-20, 11-21
Diaphragm design..................................3-27
Double tee load tables 2-6
Double tee wall panels............................2-51
Draft 5-3
Drill-in anchors 6-11
Drypack..1-7
Dy-Core..2-35
Dynamic stresses....................................5-17
Dynaspan..2-35

Earthquake analysis................................3-56
Edge distance 6-6, 6-18
Effective length..3-25
Effective moment of inertia 4-45
Effective shear-fiction coefficient...............6-18
Effective width..4-52
Elastic shortening 4-39
Elastomeric pads 6-14
Electrical coordination.............................9-70
End point criteria.....................................9-38
Equivalent volume change.......................3-48
Erection........................5-19, 10-9, 10-23, 10-29
Erection bracing......................................5-25
Erection tolerances 8-7
Expansion inserts 6-11
Expansion joints......................................3-14

Fabric bearing pads 6-14
Facing aggregates..................................1-2
Factors of safety, handling and erection......5-5, 5-27

Fire resistance 9-36
Fixed-end moments 11-13
Flat slabs, load tables 2-38
Flat wall panels 2-53
Flexicore 2-36
Flexural member design . . 4-6, 4-61, 11-4, 11-13, 11-14
Floor loads 11-2, 11-3
Footing-soil interaction 3-44
Force transfer 7-5
Forms, precast concrete 7-14
Frame distortion 3-16
Freeze-thaw resistance 1-4
Friction 6-15, 6-18

Galvanized reinforcement 1-7
Geometric sections 11-24
Gergely-Lutz equation 4-15
Granite veneers 7-24
Grout 1-7

Handling of products 5-2, 10-9
Hanger connections 6-34
Hauling 5-17
Haunches, concrete 6-23, 6-63, 7-8
Haunches, steel 6-25, 6-68, 7-8
Headed concrete anchors 6-6, 6-52
Hollow-core load tables 2-29
Hollow-core section properties 2-35
Hollow-core wall panels 2-52
Hooks 11-19, 11-21
Horizontal shear 4-27
Humidity, annual average 3-7

Impact insulation class 9-23, 9-24
Inserts 6-10
Insulated wall panels 9-4, 9-41, 9-55
Insulation properties 9-6, 9-60
Interaction curves 2-47, 4-47
Interfacing tolerances 8-17
Inverted tee load table 2-46

Joints 6-39, 6-41, 9-64, 9-68
Joint sealants 9-64

Keyed joints 6-40

L-shaped beam, load table 2-45
Lap lengths 11-20
Lateral loads 3-5, 3-16, 3-56
Lateral stability 5-14
Ledger beams 2-45, 6-32
Ledges 6-32
Lifting loops 5-10
Limestone veneers 7-25
Line loads 4-57
Live loads 11-3, 11-14
Load distribution 4-57
Load factors 5-5, 6-3
Loop inserts 6-10
Loov hanger 6-36
Loss, prestress 4-39

Low-relaxation strand .1-5, 2-4

Maintenance 9-67
Manufacturing 10-8, 10-20
Marble veneers 7-23
Mechanical coordination 9-70
Mesh 1-5, 11-22, 11-23
Metric units 11-29
Modulus of elasticity 1-3, 11-15, 11-18
Modulus of rupture 1-3, 4-8
Moment connections 6-38
Moment magnifier 3-23
Moment-resisting frames 3-43, 3-72
Mortar 1-7
Mullions 7-18

Natural stone veneers 7-22
Neoprene pads 6-14
Noise control 9-23

Openings 4-58, 9-73

P-delta analysis 3-22
Paint 9-64
Partial fixity 3-48
Partial prestressing 4-6
Permeability 9-3, 9-20
Piles 2-55, 4-54
Pipe braces 5-27
Plain concrete bearing 6-19
Plant certification 9-63
Plastic design, steel 6-27, 6-60
Plastic section modulus 6-60
Poisson's ratio 1-3
Post-tensioning 5-6, 9-69
Precast, prestressed columns 2-47
Precast, reinforced columns 2-49
Preliminary analysis 3-4
Prestress loss 4-39
Prestress transfer 4-20
Prestressing steel properties 1-5, 11-16
Prestressing strands, load tables 2-4
Product information and capability 2-2
Product tolerances 8-3

Quality control 9-63, 10-3, 10-12, 10-28

Rational design for fire 9-36
Recessed ends 6-28
Rectangular beam load table 2-44
Reinforced concrete bearing 6-20
Reinforcement 1-5, 11-20
Reinforcing bar welding 6-4
Relative humidity 3-7
Relaxation 4-39
Release reinforcement 4-19
Repair 9-67

Safety factors, handling and erection 5-5, 5-27
Sandwich panels 9-4, 9-41, 9-55
Sealers 9-64

Second-order analysis.............................3-22
Section properties..............................11-24
Segmental construction..........................9-68
Seismic analysis.................................3-56
Service load design 4-14
Shear cone 6-6
Shear design 4-20, 4-65
Shear-friction...............................6-18, 6-20
Shear strength of concrete....................... 1-3
Shear transfer between members 3-27
Shear walls..............................3-29, 3-55
Sheet piles.............................2-56, 4-56
Shop drawings 10-4, 10-14, 10-27
Shrinkage........................1-4, 3-6, 4-39
Single tee load tables 2-25
SI units 11-29
Slenderness effects.........................3-22
Solid slabs, load tables....................2-38
Sound control 9-23
Sound transmission class...............9-23, 9-24
Spancrete.................................2-36
Span-Deck................................2-37
Span-depth ratios 3-5
Spandrels..............................3-18, 3-19
Specifications.........................10-2, 10-11
Spiroll, Corefloor.........................2-37
Stability.................................5-26
Standard practice........................10-36
Static friction 6-18
Steel haunches....................6-25, 6-68, 7-8
Stone veneers 7-22
Storage..............................5-16, 10-9
Strain compatibility 4-9
Strand lifting loops........................5-10
Strength design 4-6
Stress-relieved strand 1-5, 2-4
Stress-strain relationship...................11-18
Stresses 4-16
Stripping.................................. 5-3
Structural integrity 6-43

Structural steel 6-11, 6-25
Studs 6-6, 6-52
Systems building..........................9-73

Teflon pads..............................6-15
Temperature change map......................3-7
Temperature effects.................1-4, 3-6, 9-36
Temporary bracing..........................5-25
Temporary loads 5-27
Tensile strength of concrete 1-3
Testing..................10-3, 10-13, 10-28
TFE pads...............................6-15
Thermal bowing..........................3-11
Thermal properties..........................9-2
Thermal storage effects 9-14
Threaded rods........................6-59, 7-12
Tie-back connections 7-10
Tolerances................................8-2
Torsion design 4-29, 4-31
Transportation..................5-17, 10-9, 10-29

U-values 9-4
Ultimate strength design.................... 4-6
Unbraced frames.......................3-26, 3-44

Varying section properties 4-52
Veneered panels..........................7-19
Vibration................................9-65
Volume changes.............1-3, 3-5, 3-6, 3-48
Volume-surface ratio 1-4

Wall panels.......2-51, 3-16, 3-22, 3-58, 4-47, 6-39, 7-2
Wall weights..............................11-2
Warpage..............................5-16, 8-5
Welded headed studs 6-6, 6-52
Welded wire fabric 1-5, 11-22
Weld groups.............................6-12
Welding.................6-4, 6-11, 6-48, 7-10
Wire 1-5, 11-23
Wire rope..............................5-27